MW00386804

THE HEAT TRANSFER PROBLEM SOLVER®

REGISTERED TRADEMARK

A Complete Solution Guide to Any Textbook

Staff of Research and Education Association
Dr. M. Fogiel, Director

special chapter reviews by
Harlan H. Bengtson, Ph.D.
Assistant Dean, School of Engineering
Southern Illinois University at Edwardsville
Edwardsville, Illinois

REA

Research and Education Association
61 Ethel Road West
Piscataway, New Jersey 08854

THE HEAT TRANSFER PROBLEM SOLVER ®

Printed in the United States of America

Library of Congress Catalog Card Number 91-67028

International Standard Book Number 0-87891-557-5

REVISED PRINTING, 1992

WHAT THIS BOOK IS FOR

Students have generally found heat transfer a difficult subject to understand and learn. Despite the publication of hundreds of textbooks in this field, each one intended to provide an improvement over previous textbooks, students continue to remain perplexed as a result of the numerous conditions that must often be remembered and correlated in solving a problem. Various possible interpretations of terms used in heat transfer have also contributed to much of the difficulties experienced by students.

In a study of the problem, REA found the following basic reasons underlying students' difficulties with heat transfer taught in schools:

(a) No systematic rules of analysis have been developed which students may follow in a step-by-step manner to solve the usual problems encountered. This results from the fact that the numerous different conditions and principles which may be involved in a problem, lead to many possible different methods of solution. To prescribe a set of rules to be followed for each of the possible variations, would involve an enormous number of rules and steps to be searched through by students, and this task would perhaps be more burdensome than solving the problem directly with some accompanying trial and error to find the correct solution route.

(b) Textbooks currently available will usually explain a given principle in a few pages written by a professional who has an insight in the subject matter that is not shared by students. The explanations are often written in an abstract manner which leaves the students confused as to the application of the principle. The explanations given are not sufficiently detailed and extensive to make the student aware of the wide range of applications and different aspects of the principle being studied. The numerous possible variations of principles and their applications are usually not discussed, and it is left for the students to discover these for themselves while doing exercises. Accordingly, the average student is expected to rediscover that which has been long known and practiced, but not published or explained extensively.

(c) The examples usually following the explanation of a topic are too few in number and too simple to enable the student to obtain a thorough grasp of the principles involved. The explanations do not provide sufficient basis to enable a student to solve problems that may be subsequently assigned for homework or given on examinations.

The examples are presented in abbreviated form which leaves out much material between steps, and requires that students derive the omitted material themselves. As a result, students find the examples difficult to understand--contrary to the purpose of the examples.

Examples are, furthermore, often worded in a confusing manner. They do not state the problem and then present the solution. Instead, they pass through a general discussion, never revealing what is to be solved for.

Examples, also, do not always include diagrams/graphs, wherever appropriate, and students do not obtain the training to draw diagrams or graphs to simplify and organize their thinking.

(d) Students can learn the subject only by doing the exercises themselves and reviewing them in class, to obtain experience in applying the principles with their different ramifications.

In doing the exercises by themselves, students find that they are required to devote considerably more time to heat transfer than to other subjects of comparable credits, because they are uncertain with regard to the selection and application of the theorems and principles involved. It is also often necessary for students to discover those "tricks" not revealed in their texts (or review books), that make it possible to solve problems easily. Students must usually resort to methods of trial-and-error to discover these "tricks", and as a result they find that they may sometimes spend several hours to solve a single problem.

(e) When reviewing the exercises in classrooms, instructors usually request students to take turns in writing solutions on the boards and explaining them to the class. Students often find it difficult to explain in a manner that holds the interest of the class, and enables the remaining students to follow the material written on the boards. The remaining students seated in the class are, furthermore, too occupied with copying the material from the boards, to listen to the oral explanations and concentrate on the methods of solution.

This book is intended to aid students in heat transfer to overcome the difficulties described, by supplying detailed illustrations of the solution methods which are usually not apparent to students. The solution methods are illustrated by problems selected from those that are most often assigned for class work and given on examinations. The problems are arranged in order of complexity to enable students to learn and understand a particular topic by reviewing the problems in sequence. The problems are illustrated with detailed step-by-step explanations, to save the students the large amount of time that is often needed to fill in the gaps that are usually found between steps of illustrations in textbooks or review/outline books.

The staff of REA considers heat transfer a subject that is best learned by allowing students to view the methods of analysis and solution techniques themselves. This approach to learning the subject matter is similar to that practiced in various scientific laboratories, particularly in the medical fields.

In using this book, students may review and study the illustrated problems at their own pace; they are not limited to the time allowed for explaining problems on the board in class.

When students want to look up a particular type of problem and solution, they can readily locate it in the book by referring to the index which has been extensively prepared. It is also possible to locate a particular type of problem by glancing at just the material within the boxed portions. To facilitate rapid scanning of the problems, each problem has a heavy border around it. Furthermore, each problem is identified with a number immediately above the problem at the right-hand margin.

To obtain maximum benefit from the book, students should familiarize themselves with the section, "How To Use This Book," located in the front pages.

To meet the objectives of this book, staff members of REA have selected problems usually encountered in assignments and examinations, and have solved each problem meticulously to illustrate the steps which are difficult for students to comprehend. Special gratitude is expressed to them for thier efforts in this area, as well as to the numerous contributors who devoted brief periods of time to this work.

Gratitude is also expressed to the many persons involved in the difficult task of typing the manuscript with its endless changes, and to the REA art staff who prepared the numerous detailed illustrations together with the layout and physical features of the book.

The difficult task of coordinating the efforts of all persons was carried out by Carl Fuchs. His conscientious work deserves much appreciation. He also trained and supervised art and production personnel in the preparation of the book for printing.

Finally, special thanks are due to Helen Kaufmann for her unique talents to render those difficult border-line decisions and constructive suggestions related to the design and organization of the book.

Max Fogiel, Ph.D.
Program Director

HOW TO USE THIS BOOK

This book can be an invaluable aid to students in heat transfer as a supplement to their textbooks. The book is subdivided into 11 chapters, each dealing with a separate topic. The subject matter is developed beginning with steady and unsteady state heat conduction, radiation, forced and free convection and extending through thermal boundary layer theory, radiation and combined heat transfer mechanisms. Also included are problems in laminar and turbulent flow on flat plates and through pipes, numerical analysis and three dimensional problems. An extensive number of applications have been included, since these appear to be more troublesome to students.

TO LEARN AND UNDERSTAND A TOPIC THOROUGHLY

1. Refer to your class text and read the section pertaining to the topic. You should become acquainted with the principles discussed there. These principles, however, may not be clear to you at that time.

2. Then locate the topic you are looking for by referring to the "Table of Contents" in front of this book, "The Heat Transfer Problem Solver."

3. Turn to the page where the topic begins and review the problems under each topic, in the order given. For each topic, the problems are arranged in order of complexity, from the simplest to the more difficult. Some problems may appear similar to others, but each problem has been selected to illustrate a different point or solution method.

To learn and understand a topic thoroughly and retain its contents, it will be generally necessary for students to review the problems several times. Repeated review is essential in order to gain experience in recognizing the principles that should be applied, and in selecting the best solution technique.

TO FIND A PARTICULAR PROBLEM

To locate one or more problems related to a particular subject matter, refer to the index. In using the index, be certain to note that the numbers given there refer to problem numbers, not page numbers. This arrangement of the index is intended to facilitate finding a problem more rapidily, since two or more problems may appear on a page.

If a particular type of problem cannot be found readily, it is recommended that the student refer to the "Table of Contents" in the front pages, and then turn to the chapter which is applicable to the problem being sought. By scanning or glancing at the material that is boxed, it will generally be possible to find problems related to the one being sought, without consuming considerable time. After the problems have been located, the solutions can be reviewed and studied in detail. For this purpose of locating problems rapidly, students should acquaint themselves with the organization of the book as found in the "Table of Contents".

In preparing for an exam, it is useful to find the topics to be covered in the exam from the "Table of Contents," and then review the problems under those topics several times. This should equip the student with what might be needed for the exam.

CONTENTS

6 FREE CONVECTION HEAT TRANSFER

7 FORCED CONVECTION HEAT TRANSFER

8 RADIATION HEAT TRANSFER

UNITS CONVERSION FACTORS

This section includes a particularly useful and comprehensive table to aid students and teachers in converting between systems of units.

The problems and their solutions in this book use SI (International System) as well as English units. Both of these units are in extensive use throughout the world, and therefore students should develop a good facility to work with both sets of units until a single standard of units has been found acceptable internationally.

In working out or solving a problem in one system of units or the other, essentially only the numbers change. Also, the conversion from one unit system to another is easily achieved through the use of conversion factors that are given in the subsequent table. Accordingly, the units are one of the least important aspects of a problem. For these reasons, a student should not be concerned mainly with which units are used in any particular problem. Instead, a student should obtain from that problem and its solution an understanding of the underlying principles and solution techniques that are illustrated there.

To convert	To	Multiply by	For the reverse, multiply by
acres	square feet	4.356×10^{4}	2.296×10^{-5}
acres	square meters	4047	2.471×10^{-4}
ampere-hours	coulombs	3600	2.778×10^{-4}
ampere-turns	gilberts	1.257	0.7958
ampere-turns per cm.	ampere-turns per inch	2.54	0.3937
angstrom units	inches	3.937×10^{-9}	2.54×10^{8}
angstrom units	meters	10^{-10}	10^{10}
atmospheres	feet of water	33.90	0.02950
atmospheres	inch of mercury at 0°C	29.92	3.342×10^{-2}
atmospheres	kilogram per square meter	1.033×10^{4}	9.678×10^{-5}
atmospheres	millimeter of mercury at 0°C	760	1.316×10^{-3}
atmospheres	pascals	1.0133×10^{5}	0.9869×10^{-5}
atmospheres	pounds per square inch	14.70	0.06804
bars	atmospheres	9.870×10^{-7}	1.0133
bars	dynes per square cm.	10^{6}	10^{-6}
bars	pascals	10^{5}	10^{-5}
bars	pounds per square inch	14.504	6.8947×10^{-2}
Btu	ergs	1.0548×10^{10}	9.486×10^{-11}
Btu	foot-pounds	778.3	1.285×10^{-3}
Btu	joules	1054.8	9.480×10^{-4}
Btu	kilogram-calories	0.252	3.969
calories, gram	Btu	3.968×10^{-3}	252
calories, gram	foot-pounds	3.087	0.324
calories, gram	joules	4.185	0.2389
Celsius	Fahrenheit	$(°C \times 9/5) + 32 = °F$	$(°F - 32) \times 5/9 = °C$

To convert	To	Multiply	For the reverse, multiply by
Celsius	kelvin	°C + 273.1 = K	K – 273.1 = °C
centimeters	angstrom units	1×10^{8}	1×10^{-8}
centimeters	feet	0.03281	30.479
centistokes	square meters per second	1×10^{-6}	1×10^{6}
circular mils	square centimeters	5.067×10^{-6}	1.973×10^{5}
circular mils	square mils	0.7854	1.273
cubic feet	gallons (liquid U.S.)	7.481	0.1337
cubic feet	liters	28.32	3.531×10^{-2}
cubic inches	cubic centimeters	16.39	6.102×10^{-2}
cubic inches	cubic feet	5.787×10^{-4}	1728
cubic inches	cubic meters	1.639×10^{-5}	6.102×10^{4}
cubic inches	gallons (liquid U.S.)	4.329×10^{-3}	231
cubic meters	cubic feet	35.31	2.832×10^{-2}
cubic meters	cubic yards	1.308	0.7646
curies	coulombs per minute	1.1×10^{12}	0.91×10^{-12}
cycles per second	hertz	1	1
degrees (angle)	mils	17.45	5.73×10^{-2}
degrees (angle)	radians	1.745×10^{-2}	57.3
dynes	pounds	2.248×10^{-6}	4.448×10^{5}
electron volts	joules	1.602×10^{-19}	0.624×10^{18}
ergs	foot-pounds	7.376×10^{-8}	1.356×10^{7}
ergs	joules	10^{-7}	10^{7}
ergs per second	watts	10^{-7}	10^{7}
ergs per square cm.	watts per square cm.	10^{-3}	10^{3}
Fahrenheit	kelvin	(°F + 459.67)/1.8	1.8K – 459.67
Fahrenheit	Rankine	°F + 459.67 = °R	°R – 459.67 = °F
faradays	ampere-hours	26.8	3.731×10^{-2}
feet	centimeters	30.48	3.281×10^{-2}
feet	meters	0.3048	3.281
feet	mils	1.2×10^{4}	8.333×10^{-5}
fermis	meters	10^{-15}	10^{15}
foot candles	lux	10.764	0.0929
foot lamberts	candelas per square meter	3.4263	0.2918
foot-pounds	gram-centimeters	1.383×10^{4}	1.235×10^{-5}
foot-pounds	horsepower-hours	5.05×10^{-7}	1.98×10^{6}
foot-pounds	kilogram-meters	0.1383	7.233
foot-pounds	kilowatt-hours	3.766×10^{-7}	2.655×10^{6}
foot-pounds	ounce-inches	192	5.208×10^{-3}
gallons (liquid U.S.)	cubic meters	3.785×10^{-3}	264.2
gallons (liquid U.S.)	gallons (liquid British Imperial)	0.8327	1.201
gammas	teslas	10^{-9}	10^{9}
gausses	lines per square cm.	1.0	1.0
gausses	lines per square inch	6.452	0.155
gausses	teslas	10^{-4}	10^{4}
gausses	webers per square inch	6.452×10^{-8}	1.55×10^{7}
gilberts	amperes	0.7958	1.257
grads	radians	1.571×10^{-2}	63.65
grains	grams	0.06480	15.432
grains	pounds	$^{1}/_{7000}$	7000
grams	dynes	980.7	1.02×10^{-3}
grams	grains	15.43	6.481×10^{-2}

To convert	To	Multiply	For the reverse, multiply by
grams	ounces (avdp)	3.527×10^{-2}	28.35
grams	poundals	7.093×10^{-2}	14.1
hectares	acres	2.471	0.4047
horsepower	Btu per minute	42.418	2.357×10^{-2}
horsepower	foot-pounds per minute	3.3×10^{4}	3.03×10^{-5}
horsepower	foot-pounds per second	550	1.182×10^{-3}
horsepower	horsepower (metric)	1.014	0.9863
horsepower	kilowatts	0.746	1.341
inches	centimeters	2.54	0.3937
inches	feet	8.333×10^{-2}	12
inches	meters	2.54×10^{-2}	39.37
inches	miles	1.578×10^{-5}	6.336×10^{4}
inches	mils	10^{3}	10^{-3}
inches	yards	2.778×10^{-2}	36
joules	foot-pounds	0.7376	1.356
joules	watt-hours	2.778×10^{-4}	3600
kilograms	tons (long)	9.842×10^{-4}	1016
kilograms	tons (short)	1.102×10^{-3}	907.2
kilograms	pounds (avdp)	2.205	0.4536
kilometers	feet	3281	3.408×10^{-4}
kilometers	inches	3.937×10^{4}	2.54×10^{-5}
kilometers per hour	feet per minute	54.68	1.829×10^{-2}
kilowatt-hours	Btu	3413	2.93×10^{-4}
kilowatt-hours	foot-pounds	2.655×10^{6}	3.766×10^{-7}
kilowatt-hours	horsepower-hours	1.341	0.7457
kilowatt-hours	joules	3.6×10^{6}	2.778×10^{-7}
knots	feet per second	1.688	0.5925
knots	miles per hour	1.1508	0.869
lamberts	candles per square cm.	0.3183	3.142
lamberts	candles per square inch	2.054	0.4869
liters	cubic centimeters	10^{3}	10^{-3}
liters	cubic inches	61.02	1.639×10^{-2}
liters	gallons (liquid U.S.)	0.2642	3.785
liters	pints (liquid U.S.)	2.113	0.4732
lumens per square foot	foot-candles	1	1
lumens per square meter	foot-candles	0.0929	10.764
lux	foot-candles	0.0929	10.764
maxwells	kilolines	10^{-3}	10^{3}
maxwells	webers	10^{-8}	10^{8}
meters	feet	3.28	30.48×10^{-2}
meters	inches	39.37	2.54×10^{-2}
meters	miles	6.214×10^{-4}	1609.35
meters	yards	1.094	0.9144
miles (nautical)	feet	6076.1	1.646×10^{-4}
miles (nautical)	meters	1852	5.4×10^{-4}
miles (statute)	feet	5280	1.894×10^{-4}
miles (statute)	kilometers	1.609	0.6214
miles (statute)	miles (nautical)	0.869	1.1508
miles per hour	feet per second	1.467	0.6818
miles per hour	knots	0.8684	1.152
millimeters	microns	10^{3}	10^{-3}

To convert	To	Multiply	For the reverse, multiply by
mils	meters	2.54×10^{-5}	3.94×10^{4}
mils	minutes	3.438	0.2909
minutes (angle)	degrees	1.666×10^{-2}	60
minutes (angle)	radians	2.909×10^{-4}	3484
newtons	dynes	10^{5}	10^{-5}
newtons	kilograms	0.1020	9.807
newtons per sq. meter	pascals	1	1
newtons	pounds (avdp)	0.2248	4.448
oersteds	amperes per meter	7.9577×10	1.257×10^{-2}
ounces (fluid)	quarts	3.125×10^{-2}	32
ounces (avdp)	pounds	6.25×10^{-2}	16
pints	quarts (liquid U.S.)	0.50	2
poundals	dynes	1.383×10^{4}	7.233×10^{-5}
poundals	pounds (avdp)	3.108×10^{-2}	32.17
pounds	grams	453.6	2.205×10^{-3}
pounds (force)	newtons	4.4482	0.2288
pounds per square inch	dynes per square cm.	6.8946×10^{4}	1.450×10^{-5}
pounds per square inch	pascals	6.895×10^{3}	1.45×10^{-4}
quarts (U.S. liquid)	cubic centimeters	946.4	1.057×10^{-3}
radians	mils	10^{3}	10^{-3}
radians	minutes of arc	3.438×10^{3}	2.909×10^{-4}
radians	seconds of arc	2.06265×10^{5}	4.848×10^{-6}
revolutions per minute	radians per second	0.1047	9.549
roentgens	coulombs per kilogram	2.58×10^{-4}	3.876×10^{3}
slugs	kilograms	1.459	0.6854
slugs	pounds (avdp)	32.174	3.108×10^{-2}
square feet	square centimeters	929.034	1.076×10^{-3}
square feet	square inches	144	6.944×10^{-3}
square feet	square miles	3.587×10^{-8}	27.88×10^{6}
square inches	square centimeters	6.452	0.155
square kilometers	square miles	0.3861	2.59
stokes	square meter per second	10^{-4}	10^{-4}
tons (metric)	kilograms	10^{3}	10^{-3}
tons (short)	pounds	2000	5×10^{-4}
torrs	newtons per square meter	133.32	7.5×10^{-3}
watts	Btu per hour	3.413	0.293
watts	foot-pounds per minute	44.26	2.26×10^{-2}
watts	horsepower	1.341×10^{-3}	746
watt-seconds	joules	1	1
webers	maxwells	10^{8}	10^{-8}
webers per square meter	gausses	10^{4}	10^{-4}

CHAPTER 1

INTRODUCTION TO HEAT TRANSFER

Basic Attacks and Strategies for Solving Problems in this Chapter. See pages 1 to 42 for step-by-step solutions to problems.

Heat transfer occurs whenever there is a difference in temperature. Heat always flows from a higher temperature region to a lower temperature region. Heat may be transferred by conduction, convection, or radiation. Also, heat transfer may occur by some combination of two of the mechanisms or by all three. In this introductory section, conduction, convection, and radiation are each considered briefly. More details about each are given in later chapters.

Heat transfer by conduction is governed by Fourier's Law, which for one-dimensional cases is

$$q = -\,kA\,dT\,/\,dx$$

One class of problems involves integration of this equation for conduction through a solid of some particular shape. The approach to that type problem is to express the area, A, in terms of the variable, x, then separate variables, and integrate the equation. This may be done for rectangular, cylindrical or spherical shapes. Conduction problems may also require substitution of values into an integrated form of Fourier's Law to calculate one of the variables, such as a temperature or heat transfer rate.

When conduction takes place through a solid which is made up of more than one type of material, an electrical analog to heat transfer can often be used. The solid materials cause a resistance to heat transfer which is analogous to electrical resistance. The different materials may be arranged to cause resistances in series and/or in parallel. The overall resistance to heat transfer can then be calculated from the resistances of individual materials in the same way that overall electrical resistance is calculated for series and parallel circuits. The expression for heat transfer resistance for individual materials depends upon the shape of the solid and can be found by integrating Fourier's Law for the particular shape.

Convection heat transfer occurs due to the motion of a fluid. The basic equation for convection heat transfer is Newton's Law of cooling:

$$q = hA\,(T_2 - T_1)$$

The heat transfer coefficient, h, *depends upon properties of the fluid and upon the shape and size of any solid surfaces involved. A major part of convection heat transfer calculations is often determination of a value for a heat transfer coefficient from an empirical correlation. The empirical equations typically are expressed in terms of dimensionless numbers, such as the Nusselt Number, Reynold's Number, Prandtl Number and Grashof Number.*

For cases where both conduction and convection are important, an electrical analogy can again be used. The thermal resistance for convection is given by:

convective resistance = 1 / (hA) .

For a combination of conductive and convective resistances in series, an overall heat transfer coefficient is often used. The overall heat transfer coefficient, U, *can be calculated from the general relationship that* 1/UA *equals the sum of all the individual thermal resistances.*

Radiation heat transfer is due to electromagnetic waves traveling between two objects which are at different temperatures. There will always be a net transfer of heat from the higher temperature object to the lower temperature object. The rate of heat transfer depends upon the two temperatures, the geometry of the two objects, and the nature of the two surfaces. Introductory problems involving radiation heat transfer usually can be solved using an equation which applies to the particular geometry given in the problem, such as two parallel plates or two concentric cylinders. The equation will typically relate the heat transfer rate to the temperatures of the two surfaces, the emissivity of the two surfaces, and perhaps areas and some other size dimensions. Any of the above variables can thus be the unknown which you are asked to determine for the problem.

Step-by-Step Solutions to Problems in this Chapter, "Introduction to Heat Transfer"

MECHANISM OF HEAT TRANSFER

• PROBLEM 1-1

What is steady state heat transfer?

Solution: The conservation of energy law states 'the rate of energy in - the rate of energy out = rate of energy accumulation'. If there is no accumulation of energy, then 'rate of energy in = rate of energy out'. When this condition is met and heat transfer takes place, it is known as steady state heat transfer.

• PROBLEM 1-2

What are the three basic mechanisms of heat transfer? Explain the mode of heat transfer in each of the three cases and give examples for each.

Solution: Heat transfer may occur by any one or more of the following mechanisms: conduction, convection, or radiation.

1. Conduction: When heat is transferred by the energy of motion between adjacent molecules, heat transfer by conduction is said to take place. In gases, molecules, which have greater energy and motion ('hotter molecules') impart energy to the adjacent molecules which are at lower energy levels. This type of heat transfer takes place to a certain extent in all solids, liquids and gases. In metallic solids, energy or heat transfer may also take place by 'free electrons'. Examples of conduction are: heat transfer through walls, pipes, freezing of the ground during winter, etc.

2. Convection: The transfer of heat by convection takes place by bulk transport and the mixing of macroscopic parts of hot and cold elements of a fluid. This also includes the transfer of heat between a solid surface and a fluid. Convection heat transfer is of two types; forced convection heat transfer, where a fluid is forced to flow past a surface by any mechanical means, and natural or free convection, where a warmer or cooler fluid next to the surface causes circulation because of density difference due to a temperature difference.

Examples of heat transfer by convection are: cooling or heating of fluids in heat exchangers, cooling a cup of coffee by blowing over the surface, baking a cake in a gas oven, etc.

3. Radiation: Radiation heat transfer differs from conduction and convection heat transfer because no physical medium is required, and can take place in a vacuum. It is the mode of heat transfer through electromagnetic waves similar to the way electromagnetic waves transfer light and hence radiation heat transfer is governed by the same laws that govern the transfer of light. Solids and liquids have a tendency to absorb the radiation that is being transferred through them, therefore, radiation is of primary importance in the transfer of heat through gases and space.

Examples of radiation heat transfer are: the transfer of heat from the sun to the earth, cooking of food over electric coil radiators, heating of tubes in a furnace, etc.

• PROBLEM 1-3

State Fourier's Law of heat conduction.

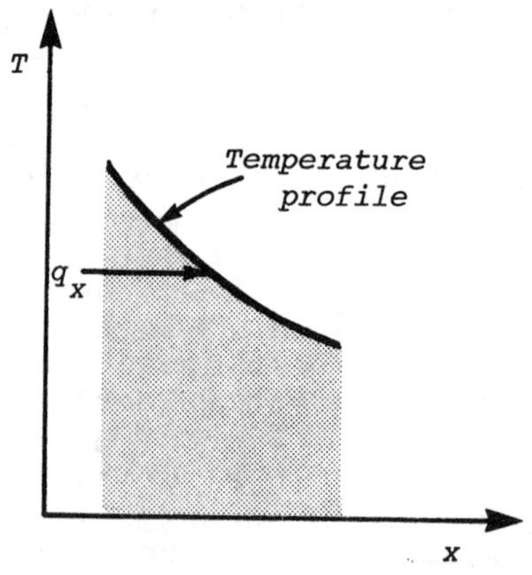

Sketch showing direction of heat flow.

Solution: Fourier's Law states that for conduction to take place, a temperature gradient must exist in the body. Also the heat transfer rate is proportional to the normal temperature gradient

$$\frac{q}{A} \alpha \frac{\partial T}{\partial x}$$

where q is the heat transfer rate and $\frac{\partial T}{\partial x}$ is the temperature gradient. When the proportionality constant (k) is inserted,

$$q = -kA \frac{\partial T}{\partial x}$$

This equation is called Fourier's Law of heat conduction. The minus sign is inserted in accordance with the second law of thermodynamics, i.e., if the heat flows in a positive direction, the temperature decreases in that direction (becomes negative).

• PROBLEM 1-4

One face of a copper slab is maintained at 1000^0F, and the other face is at 200^0F. How much heat is conducted through the slab per unit area if the slab is 3 inches thick? The thermal conductivity of copper may be taken as k = 215Btu/hr-ft^{0}F.

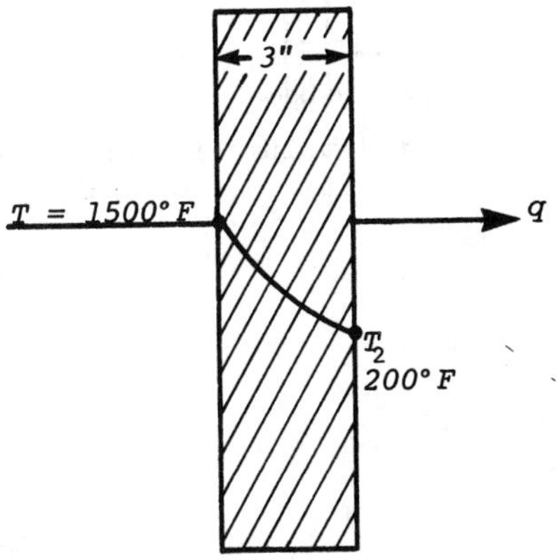

Solution: From Fourier's Law of heat conduction

$$\frac{q}{A} = -k \frac{dT}{dx}$$

Integrating

$$\frac{q}{A} = -k \frac{\Delta T}{\Delta x}$$

$$= -215 \frac{(T_2 - T_1)}{\Delta x}$$

$$= \frac{-215(200-1000)}{\frac{3}{12}}$$

$$= 6.88 \times 10^5 \text{ Btu/hr-ft}^2 \quad (3.9 \times 10^6 \text{w/m}^2\text{-}^0\text{C})$$

• PROBLEM 1-5

Derive the three-dimensional heat conduction equation, starting with Fourier's Law of heat transfer. Also write the special cases of the conduction equation and its transformation into cylindrical and spherical coordinates.

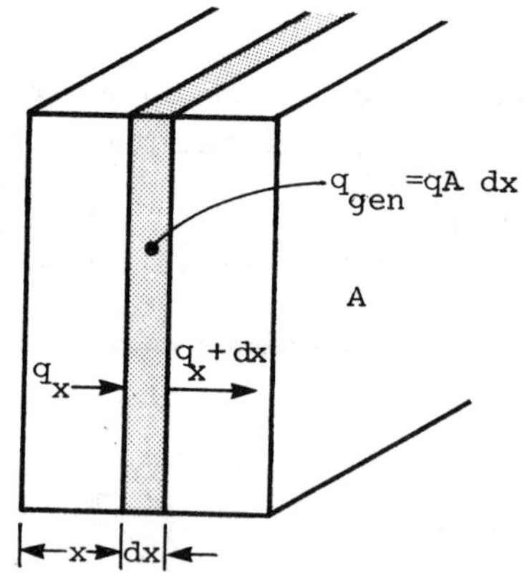

Fig. 1 Elemental volume for one-dimensional heat-conduction.

Solution: Fourier's Law is

$$q = -kA \frac{\partial T}{\partial x} \tag{1}$$

Consider the one-dimensional system shown in Fig. 1. If the system is in a steady state, i.e., if the temperature does not change with time, then the problem is a simple one and need only integrate eq.(1) and substitute the appropriate values to solve for the desired quantity. However, if the temperature of the solid is changing with time, or if there are heat sources or sinks within the solid, the situation is more complex. We consider the general case where the temperature may be changing with time and heat sources may be present within the body. For the element of thickness dx the following energy balance may be made:

Energy conducted in left face + heat generated within element
= change in internal energy + energy conducted out right face

These energy quantities are given as follows:

$$\text{Energy in left face} = q_x = -kA\frac{\partial T}{\partial x}$$

$$\text{Energy generated within element} = \dot{q}A\,dx$$

$$\text{Change in internal energy} = \rho cA\frac{\partial T}{\partial \tau}\,dx$$

$$\text{Energy out right face} = q_{x+dx} = -kA\left.\frac{\partial T}{\partial x}\right|_{x+dx}$$

$$= -A\left[k\frac{\partial T}{\partial x} + \frac{\partial}{\partial x}\left(k\frac{\partial T}{\partial x}\right)dx\right]$$

where $\dot{q}$ = energy generated per unit volume
c = specific heat of material
ρ = density

Combining the relations above,

$$-kA\frac{\partial T}{\partial x} + \dot{q}A\,dx = \rho cA\frac{\partial T}{\partial \tau}\,dx - A\left[k\frac{\partial T}{\partial x} + \frac{\partial}{\partial x}\left(k\frac{\partial T}{\partial x}\right)dx\right]$$

or

$$\frac{\partial}{\partial x}\left(k\frac{\partial T}{\partial x}\right) + \dot{q} = \rho c\frac{\partial T}{\partial \tau} \qquad (2)$$

This is the one-dimensional heat-conduction equation. To treat more than one-dimensional heat flow, only consider the heat conducted in and out of a unit volume in all three coordinate directions, as shown in Fig. 2a. The energy balance yields

$$q_x + q_y + q_z + q_{gen} = q_{x+dx} + q_{y+dy} + q_{z+dz} + \frac{dE}{d\tau}$$

and the energy quantities are given by

$$q_x = -k\,dy\,dz\frac{\partial T}{\partial x}$$

$$q_{x+dx} = -\left[k\frac{\partial T}{\partial x} + \frac{\partial}{\partial x}k\left(\frac{\partial T}{\partial x}\right)dx\right]dy\,dz$$

$$q_y = -k\,dx\,dz\frac{\partial T}{\partial y}$$

$$q_{y+dy} = -\left[k\frac{\partial T}{\partial y} + \frac{\partial}{\partial y}\left(k\frac{\partial T}{\partial y}\right)dy\right]dx\,dz$$

$$q_z = -k\,dx\,dy\frac{\partial T}{\partial z}$$

$$q_{z+dz} = -\left[k\frac{\partial T}{\partial z} + \frac{\partial}{\partial z}\left(k\frac{\partial T}{\partial z}\right)dz\right]dx\,dy$$

$$q_{gen} = \dot{q}\ dx\ dy\ dz$$

$$\frac{dE}{d\tau} = \rho c\ dx\ dy\ dz\ \frac{\partial T}{\partial \tau}$$

so that the general three-dimensional heat-conduction equation is

$$\frac{\partial}{\partial x}\left(k\ \frac{\partial T}{\partial x}\right) + \frac{\partial}{\partial y}\left(k\ \frac{\partial T}{\partial y}\right) + \frac{\partial}{\partial z}\left(k\ \frac{\partial T}{\partial z}\right) + \dot{q} = \rho c\ \frac{\partial T}{\partial \tau} \qquad (3)$$

For constant thermal conductivity eq. (3) is written

$$\frac{\partial^2 T}{\partial x^2} + \frac{\partial^2 T}{\partial y^2} + \frac{\partial^2 T}{\partial z^2} + \frac{\dot{q}}{k} = \frac{1}{\alpha}\ \frac{\partial T}{\partial \tau} \qquad (4)$$

where the quantity $\alpha = k/\rho c$ is called the thermal diffusivity of the material. The larger the value of α, the faster will heat diffuse through the material. This may be seen by examining the quantities which make up α.

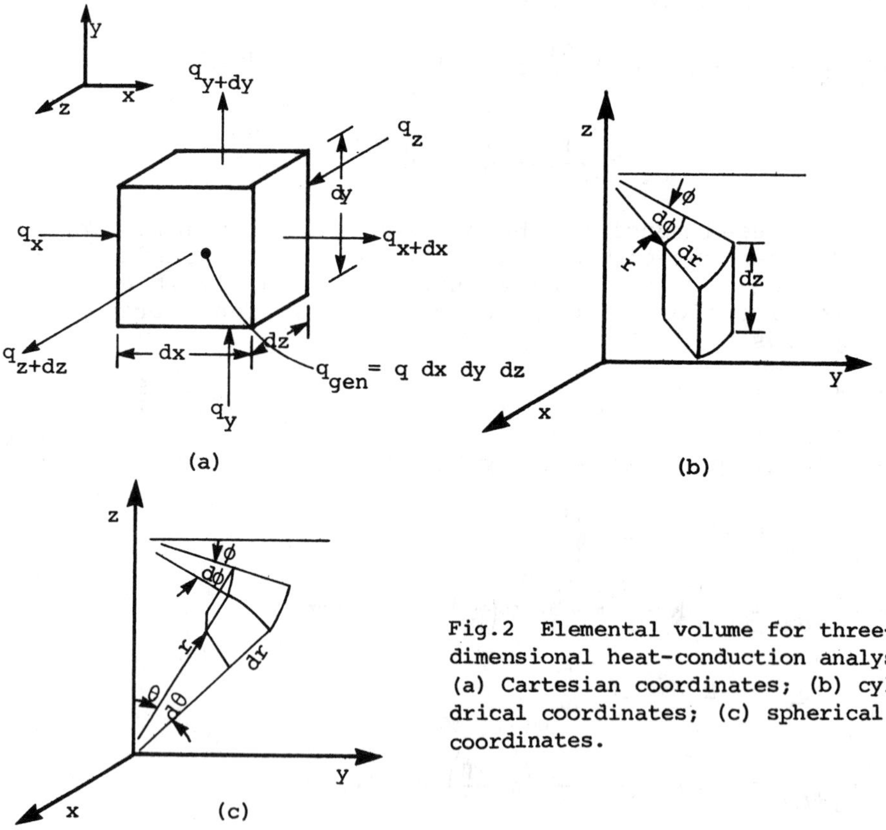

Fig.2 Elemental volume for three-dimensional heat-conduction analysis. (a) Cartesian coordinates; (b) cylindrical coordinates; (c) spherical coordinates.

Special Cases of the Conduction Equation

1. Fourier equation (no internal energy conversion)

$$\frac{\partial^2 T}{\partial x^2} + \frac{\partial^2 T}{\partial y^2} + \frac{\partial^2 T}{\partial z^2} = \frac{1}{\alpha}\frac{\partial T}{\partial t}$$

2. Poisson equation (steady state with internal energy conversion)

$$\frac{\partial^2 T}{\partial x^2} + \frac{\partial^2 T}{\partial y^2} + \frac{\partial^2 T}{\partial z^2} + \frac{q'''}{k} = 0$$

3. Laplace equation (steady state and no internal energy conversion)

$$\frac{\partial^2 T}{\partial x^2} + \frac{\partial^2 T}{\partial y^2} + \frac{\partial^2 T}{\partial z^2} = 0$$

Equation (4), the general conduction equation in cylindrical and spherical coordinates is given by:

Cylindrical coordinates (Fig. 2b)

$$\frac{\partial^2 T}{\partial r^2} + \frac{1}{r}\frac{\partial T}{\partial r} + \frac{1}{r^2}\frac{\partial^2 T}{\partial \phi^2} + \frac{\partial^2 T}{\partial z^2} + \frac{\dot{q}}{k} = \frac{1}{\alpha}\frac{\partial T}{\partial \tau}$$

Spherical coordinates (Fig. 2c)

$$\frac{1}{r}\frac{\partial^2}{\partial r^2}(rT) + \frac{1}{r^2 \sin\theta}\frac{\partial}{\partial \theta}\left(\sin\theta \frac{\partial T}{\partial \theta}\right) + \frac{1}{r^2 \sin^2\theta}\frac{\partial^2 T}{\partial \phi^2} + \frac{\dot{q}}{k} = \frac{1}{\alpha}\frac{\partial T}{\partial \tau}$$

ANALOGY BETWEEN HEAT CONDUCTION AND SYSTEM

• PROBLEM 1-6

Derive the equation for heat transfer by conduction through a composite wall (Fig. 1a, 2a) and draw an analogy between heat transfer through walls and an electric resistance circuit.

Solution: Fourier's Law of heat conduction is

$$q = -kA\frac{dT}{dx}$$

Integrating the equation gives the heat transferred through a single wall

$$q = -kA\frac{\Delta T}{\Delta x}.$$

$$q = \frac{-kA}{\Delta x}(T_2 - T_1) \qquad (1)$$

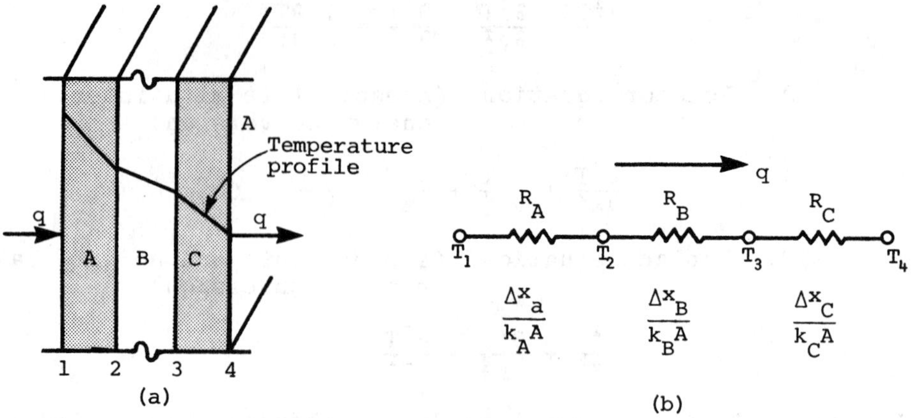

Fig. 1 One-dimensional heat transfer through a composite wall and electrical analog.

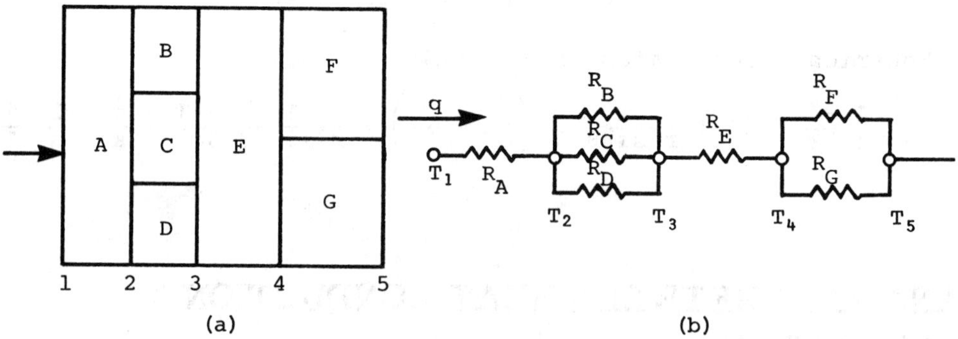

Fig. 2 Series and parallel one-dimensional heat transfer through a composite wall and electric analog.

In the case of a composite wall, where more than one material is present (Fig. 1a), the analysis is as follows: The temperature gradient is shown in Fig. 1a and the heat flow, which is the same through all sections, is written as

$$q = -k_A \; A \; \frac{T_2 - T_1}{\Delta x_A} = -k_B \; A \; \frac{T_3 - T_2}{\Delta x_B} = -k_C \; A \; \frac{T_4 - T_3}{\Delta x_C}$$

Solving these equations simultaneously gives the heat transfer rate through a composite wall

$$q = \frac{T_1 - T_4}{\Delta x_A / k_A A + \Delta x_B / k_B A + \Delta x_C / k_C A} \tag{2}$$

Now, heat transfer rate (q) may be considered to be a flow, the expression $\Delta x/kA$ (which consists of the thickness, thermal conductivity and area) treated as a resistance and the temperature difference (ΔT) as the potential. The Fourier's Law of heat conduction may now be written as

$$\text{Heat Flow } (q) = \frac{\text{Thermal Potential } (\Delta T)}{\text{Thermal Resistance } \Delta x/kA}$$

This relation is analogous to Ohm's Law in electric circuit theory.

In equation (1), the resistance is $\Delta x/kA$, but in equation (2), the resistance is the sum of three terms,

$$\Delta x_A/k_A A + \Delta x_B/k_B A + \Delta x_C/k_C A$$

and can be taken as three resistances in series. An equivalent electric circuit for the composite wall in Fig. 1a is shown in Fig. 1b.

This electrical analogy is very useful in solving complex problems involving thermal resistances in series and parallel as shown in Fig. 2a. The equivalent electric circuit is shown (Fig. 2b).

Therefore a one-dimensional heat flow equation for cases of this type, may be written as

$$\text{Heat Flow } q = \frac{\Delta T}{\sum R_{\text{Thermal}}}$$

• PROBLEM 1-7

> Derive the expression for heat transfer through a cylinder. Then go on to find the expression for a multilayered cylinder.

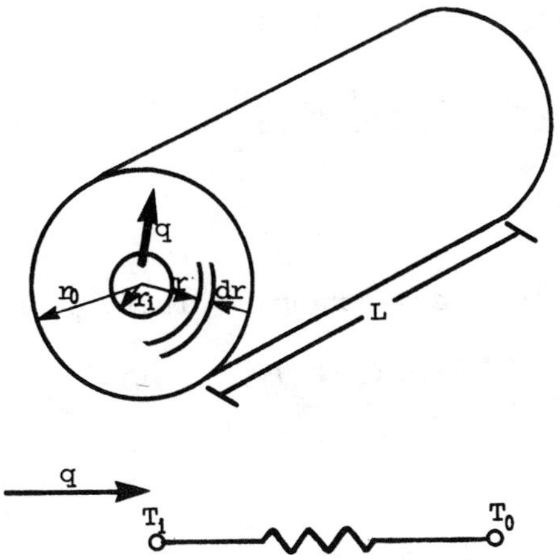

$$R_{th} = \frac{\ln (r_0 / r_i)}{2\pi k L}$$

Fig. 1 One-dimensional heat flow through a hollow cylinder and electrical analog.

Solution: Consider a long cylinder having an inside radius r_i, outer radius r_o, and length L, as shown in Fig. 1. The temperature on the inside is T_i and T_o on the outside. It may be assumed that the heat flow is radial and therefore only one space coordinate r is required to specify the system.

The area through which heat flows is

$$A_r = 2\pi rL \ .$$

Fourier's Law is now written as

$$q_r = -kA_r \frac{dT}{dr}$$

$$= -2\pi rLk \frac{dT}{dr}$$

The boundary conditions are

$$r = r_i \qquad T = T_i$$

$$r = r_o \qquad T = T_o$$

Integrating with q constant gives

$$T_o - T_i = \frac{-q}{2\pi kL} \ln \frac{r_o}{r_i}$$

The expression for heat transfer through a cylinder is

$$q = \frac{2\pi kL(T_i - T_o)}{\ln(r_o/r_i)}$$

The thermal resistance in the case of a cylinder is

$$R_{th} = \frac{\ln(r_o/r_i)}{2\pi kL}$$

$$q$$

$$T_1 \quad R_A \quad T_2 \quad R_B \quad T_3 \quad R_C \quad T_4$$

$$\frac{\ln(r_2 l r_1)}{2\pi k_A L} \qquad \frac{\ln(r_3 l r_2)}{2\pi k_B L} \qquad \frac{\ln(r_4 l r_3)}{2\pi k_C L}$$

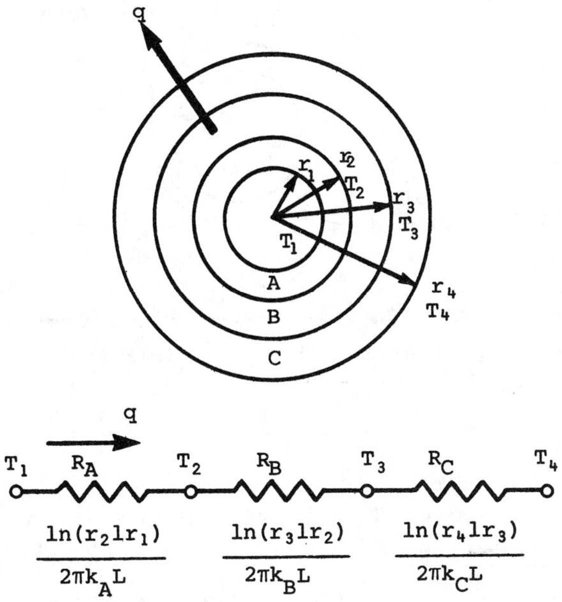

Fig. 2 One-dimensional heat flow through multiple
cylindrical sections and electrical analog.

For any system, heat flow is given by

$$q = \frac{\Delta T}{\sum R_{th}}$$

where ΔT is the temperature difference. Hence for heat flow
through a multilayered cylinder as shown in Fig. 2.

$$q = \frac{(T_1 - T_4)2\pi L}{\ln(r_2/r_1)/k_A + \ln(r_2/r_3)/k_B + \ln(r_3/r_4)/k_C}$$

● **PROBLEM** 1-8

A thick walled tube of chrome steel, 1 inch inner diameter and 2 inch outer diameter, having a thermal conductivity k=13 Btu/hr ft ^{0}F is covered with a 2" layer
of asbestos (k=0.048). If the temperature on the inside of the tube is 500 ^{0}F and on the outside of the
tube is 100 ^{0}F, find the heat transfer (q) per unit
length.

Solution: Heat flow per unit length (q/L) is given by

$$\frac{q}{L} = \frac{2\pi(T_i - T_o)}{\ln(r_2/r_1)/k_s + \ln(r_3/r_2)/k_a}$$

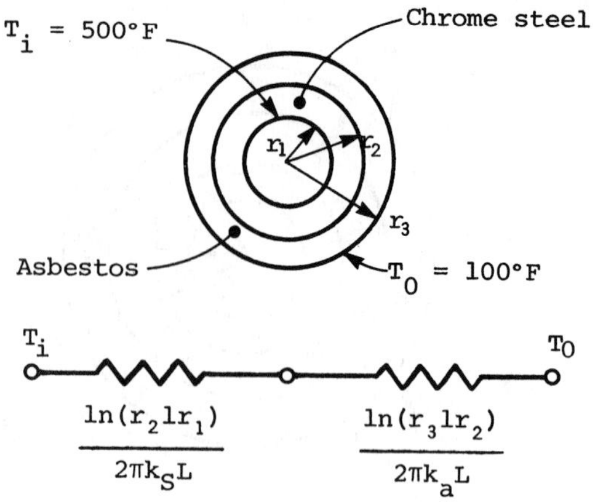

where, T_i and T_o are the inner & outer surface temperatures

r_1 and r_2 are the inner & outer radii of the steel tube

r_2 and r_3 are the inner & outer radii of the insulation

k_s and k_a are the thermal conductivities of steel & asbestos

Substituting the values

$$\frac{q}{L} = \frac{2\pi(500-100)}{\ln(2/1)/13+\ln(4/2)/0.048} = 173.4 \text{ Btu/hr ft} \quad (167 \text{ W/m})$$

• PROBLEM 1-9

> Find the expression for heat transfer by conduction through a sphere of inner radius r_1 and outer radius r_2. Also find the expression for heat transfer through a multilayered sphere. The temperatures on the inside and outside surfaces of the sphere are T_1 and T_2.

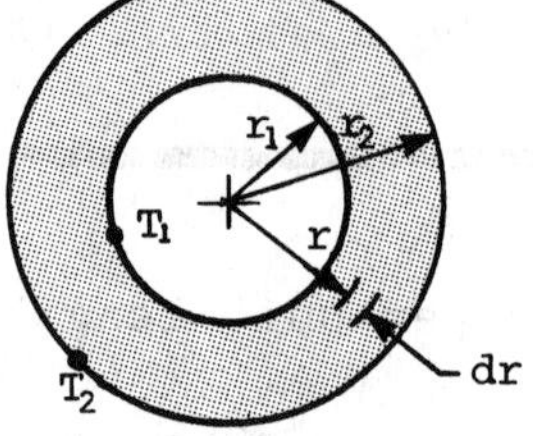

Fig. 1

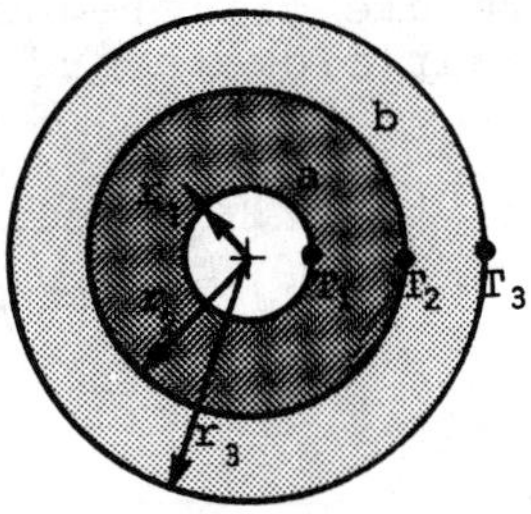

Fig. 2

Solution: For a sphere (Fig. 1), the area (A_r) through which heat transfer by conduction occurs is

$$A_r = 4\pi r^2.$$

Fourier's Law is written as

$$q_r = -k\ A_r \frac{dT}{dr}$$

$$= -4\pi r^2\ k \frac{dT}{dr}$$

The boundary conditions are

$$r = r_1 \qquad T = T_1$$

$$r = r_2 \qquad T = T_2$$

Integrating with q constant and between the boundary conditions given

$$T_2 - T_1 = \frac{-q}{4\pi k}\left[\frac{1}{r_2} - \frac{1}{r_1}\right]$$

the expression for conduction heat transfer through a sphere is

$$q = \frac{4\pi k(T_1 - T_2)}{\left[\frac{1}{r_1} - \frac{1}{r_2}\right]}$$

The thermal resistance in this case is

$$R_{th} = \frac{\left[\frac{1}{r_1} - \frac{1}{r_2}\right]}{4\pi k}$$

For any system, heat flow is given by

$$q = \frac{\Delta T}{\sum R_{th}}$$

where ΔT is the temperature difference. Hence using the above expression for heat flow through a multilayered sphere, as shown in Fig. 2,

$$q = \frac{(T_1 - T_3)4\pi}{\left(\frac{1}{r_1} - \frac{1}{r_2}\right)\Big/k_a + \left(\frac{1}{r_2} - \frac{1}{r_3}\right)\Big/k_b}$$

COMPOSITE SYSTEM

• PROBLEM 1-10

Compute the rate of heat transfer through the walls of a room having inside and outside temperatures 255.4 k and 297.1 k respectively. The walls of the room are made up of 3 in. concrete, 4 in. cork board, and ½ in. wood. The thermal conductivities of the wall materials are $k_{concrete}$ = 0.762 w/m k, $k_{cork\text{-}board}$ = 0.0433 w/m k and k_{wood} = 0.151 w/m k. Also, find the temperature at the interface between the wood and cork board.

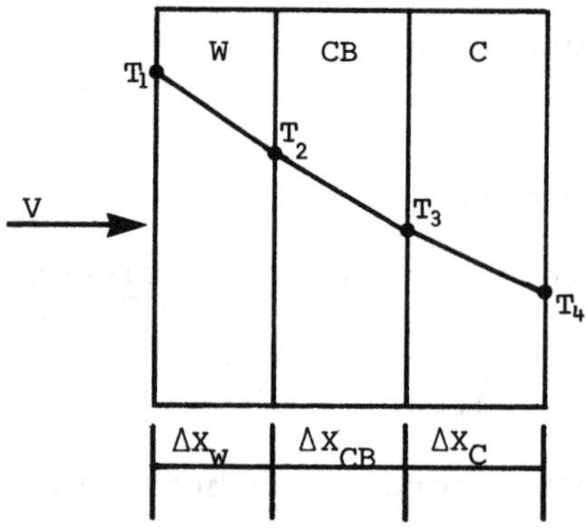

Fig. 1 Heat flow through a composite wall.

Solution: Let T_1 - room inner wall temperature = 255.4^0k

T_2 - room outer wall temperature = 297.1^0k

The first letter of the materials are used as suffixes in order to indentify each material

$$k_W = 0.151, \quad k_{CB} = 0.0433, \quad k_C = 0.762$$

$\Delta x_W = \frac{1}{2}$ in. or 0.0127 m

$\Delta x_{CB} = 4$ in. or 0.1016 m

$\Delta x_C = 3$ in. or 0.0762 m

The resistances to heat flow for each material per unit surface area ($1m^2$) are

$$R_W = \frac{\Delta x_W}{k_W} = \frac{0.0127}{(0.151)} = 0.0841 \text{ k/W}$$

$$R_{CB} = \frac{\Delta x_{CB}}{k_{CB}} = \frac{0.1016}{0.0433} = 2.346 \text{ k/W}$$

$$R_C = \frac{\Delta x_C}{k_C} = \frac{0.0762}{0.762} = 0.100 \text{ k/W}$$

From Fourier Equation of heat transfer

$$q = \frac{T_1 - T_4}{R_W + R_{CB} + R_C} = \frac{255.4 - 297.1}{0.0841 + 2.346 + 0.100}$$

$$= -16.48 \text{ W}$$

The negative sign indicates that the heat flows in from outside.

The interface temperature between wood and cork board can be calculated from the equation

$$q = \frac{T_1 - T_2}{R_W}$$

or $$-16.48 = \frac{255.4 - T_2}{0.0841}$$

or $$T_2 = 256.79 \text{ k}$$

• PROBLEM 1-11

Calculate the thickness of magnesia insulation necessary to restrict the heat loss to 5 Btu/hr ft^2 through the walls of a furnace having inside and ambient temperatures 1500°F and 150°F, respectively. The furnace wall is $\frac{1}{4}$ in. steel plate with 3 in. refractory lining.

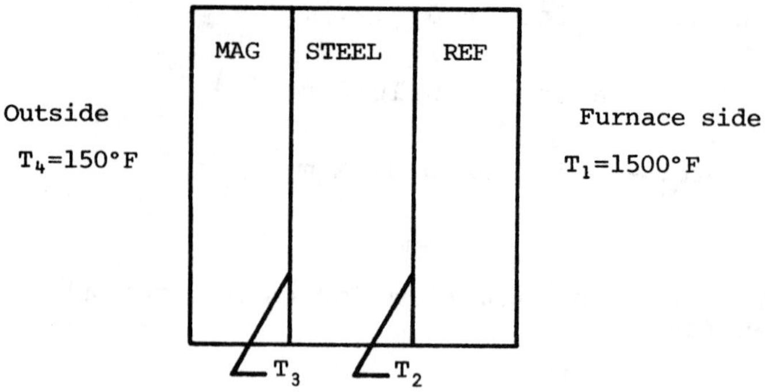

Fig. 1

Solution: The rate of heat transfer per unit surface area (q/A) is the same for each of the three materials

$$q_x = \left[-\frac{k}{\Delta x}(T_2 - T_1)\right]_{\text{Ref.}} \tag{1}$$

$$q_x = \left[-\frac{k}{\Delta x}(T_3 - T_2)\right]_{\text{Steel}} \tag{2}$$

$$q_x = \left[-\frac{k}{\Delta x}(T_4 - T_3)\right]_{\text{Mag.}} \tag{3}$$

T_1 - Furnace inner wall temperature ^{0}F
T_2 - Interface wall temperature ^{0}F (refractory and steel)
T_3 - Interface wall temperature ^{0}F (steel and magnesium)
T_4 - Furnace outer wall temperature ^{0}F
Δx - Thickness of magnesium coating

The three unknowns T_2, T_3 and Δx can be calculated by solving the three simultaneous equations (1), (2) and (3).

Considering the maximum limit of heat loss 5 Btu/hr ft^2, from equation (1)

$$5 = -0.6 \times \frac{4}{1}(T_2 - 1500)$$

$$\text{or } T_2 = -2.08 + 1500 = 1498^0\text{F}$$

Substituting $T_2 = 1498^0$F in equation (2)

$$\frac{5}{(26)(4)(12)} = 1498 - T_3$$

$$\text{or } T_3 = 1498 + 0.004 = 1498^0\text{F}$$

Again, substituting $T_3 = 1498^0$F in equation (3)

$$x = \frac{-0.03}{5}(150 - 1498) = 8.08 \text{ Ft.}$$

This thickness of insulation is not practical.

• PROBLEM 1-12

Steam at 350°F flows through a $\frac{1}{32}$ in. glass lined steel pipe. The outer surface of the pipe is coated with 1 in. thick refractory material. Compute the rate of heat loss per unit length of the pipe surface if the outside ambient temperature is 120°F. The thermal conductivities are:

$$k_{glass} = 0.5 \text{ Btu/hr ft}^0\text{F}, \quad k_{steel} = 26 \text{ Btu/hr ft}^0\text{F}$$

$$k_{ref.} = 0.03 \text{ Btu/hr ft }^0\text{ F}$$

The outside and inside diameters of pipe are 3.5 in. and 3 in., respectively.

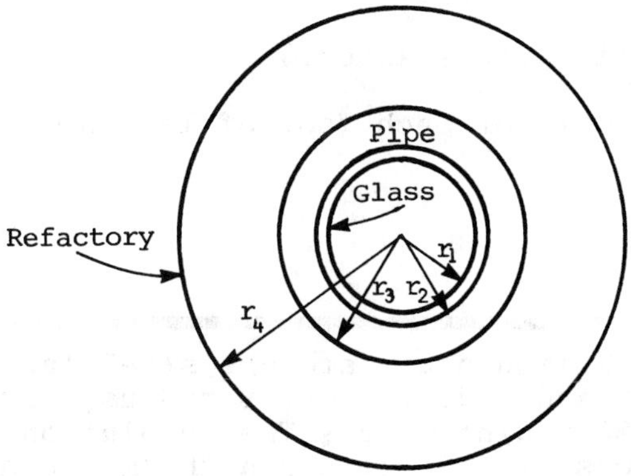

Fig. 1

Solution: The rate of heat conduction through the concentric cylindrical surfaces is

$$-Q = \frac{(\Delta T)\text{overall}}{\sum_{i=1}^{n} \frac{\Delta r}{kA}}$$

$(\Delta T)_{overall}$ - Overall temperature difference in °F

Δr - thickness of the conducting material in inches

A - Log mean area in Ft²

Assuming r = 1 ft.

$$A_{ref.} = \frac{2\pi r_4(1) - 2\pi r_3(1)}{\ln(2\pi r_4/2\pi r_3)}$$

$$= \frac{[2\pi(4.5)(1)-2\pi(3.5)(1)]\ ^1/_{12}}{\ln(4.5/3.5)}$$

$$= 2.1\ ft^2$$

$$A_{pipe} = \frac{[2\pi(3.5)(1)-2\pi(3.0)(1)]\ ^1/_{12}}{\ln[(3.5)(1)/(3.0)(1)]} = 1.71\ ft^2$$

$$A_{glass} = \frac{2\pi(3.0)}{12}\ 1 = 1.57\ Ft^2$$

Substituting in equation (a):

$$Q_r = \frac{350 - 120}{1/[(12)(0.03)(2.1)]+0.5/[(12)(26)(1.71)]+1/[(32)(12)(0.5)(1.57)]}$$

$$= \frac{230}{1.32+0.000938+0.332}$$

(Notice the relative resistances.)

Q_r = 140 Btu/hr for each foot of length

• PROBLEM 1-13

The figure shows a liquid air storage steel tank of 3 ft. inside radius and 3.25 ft. outer radius, having a conductivity k_a=25 Btu/hr ft^{0}F. The insulation layer (b) has a thickness of 1.5 ft. and a thermal conductivity k_b=0.048 Btu/hr ft^{0}F. The temperature on the inside is -200^{0}F and the outside is at 70^{0}F. Determine the steady state heat transfer rate through the tank to the liquid air inside.

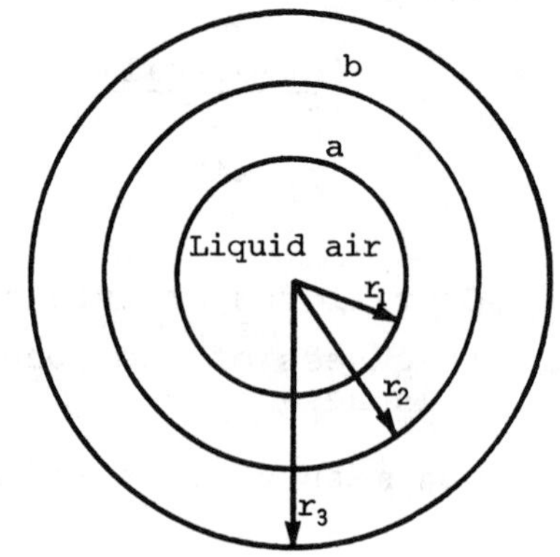

Fig. 1 The spherical liquid air storage tank.

Solution: Heat flow through an insulated sphere is given by

$$q = \frac{(T_1 - T_3)4\pi}{\left(\frac{1}{r_1} - \frac{1}{r_2}\right)\Big/k_a + \left(\frac{1}{r_2} - \frac{1}{r_3}\right)\Big/k_B}$$

where T_1 and T_3 are the inner and outer surface temperatures, r_1, r_2 and r_2, r_3 are the inner and outer diameters of the sphere and the insulation, respectively. k_a and k_b are the thermal conductivities of the sphere and the insulation, respectively.

Substituting the values in the given equation

$$q = \frac{(-200 - 70)4\pi}{\left(\frac{1}{3} - \frac{1}{3.25}\right)\Big/25 + \left(\frac{1}{3.25} - \frac{1}{4.75}\right)\Big/0.048}$$

$$= \frac{(-270)4\pi}{0.0256 + 2.024} = -1655 \text{ Btu/hr}$$

The negative sign indicates that the heat flow is in the negative radial direction i.e., towards the center of the sphere.

• PROBLEM 1-14

Calculate the safe thickness of the shield coating necessary on a radioactive waste spherical ball of 1 ft. radius to be dumped in the ocean with ambient water temperature of 50°F. The waste ball gives off heat at the rate of 440 Btu/hr. The thermal conductivity of the shielding material is k=0.02 Btu/hr ft°F.

Solution: Let r = outer radius of the spherical ball. The rate of heat transfer is

$$Q_r = -kA_{gm}\frac{\Delta T}{\Delta r}$$

or $$440 \text{ Btu/hr} = -0.02(\text{Btu/hr ft}^0\text{F})4\pi r(1)\text{ft}^2 \ \frac{50 - 400}{r - 1}$$

$$352r = 440$$

$$r = 1.25 \text{ ft}$$

The coating is 1.25 - 1 = 0.25 ft or 3 in.

CONVECTION

• PROBLEM 1-15

What is Newton's Law of cooling as applied to convective heat transfer?

Solution: Newton's Law of cooling is used to determine the convective heat transfer rates and is given by

$$q = h\ A\ \Delta T$$

where A is the boundary or interface area between a fluid and solid or between two fluids, ΔT is the temperature difference between the ambient or upstream temperature (T_∞) and the interface temperature (T_w), and h is the surface coefficient of heat transfer (sometimes referred to as unit surface conductance or film coefficient or convective coefficient). The units of h are Btu/hr ft^2 - ^0F or W/m^2 - ^0k.

In the case of fluid motion being induced by an external source e.g., pump, blower, fan, the process of heat transfer is known as forced convection. If the fluid motion is due to density gradients caused by the temperature field itself, then this process of heat transfer is called free or natural convection. Newton's Law of cooling is used to calculate heat transfer rates in both forced and free convection cases, although the expression for surface coefficient will be different.

• PROBLEM 1-16

Compute the convective heat transfer coefficient between the surface of the pipe insulation at 56^0c and the surrounding air at 20^0c. Steam flows through the pipe having outer diameter of insulation 10 cm.

Use: $k = 0.0266$ W/m k, $\rho = 1.14$ kg/m^3

$\beta = 0.00322\ k^{-1}$, $\mu = 1.92 \times 10^{-5}$ Pa.s

$C_p = 1000$ J/kg·k

Solution: The convective heat transfer coefficient can be obtained from the correlation between Grashof and Prandtl numbers i.e. Rayleigh number

$$Ra = Pr \times Gr$$

or

$$\frac{hD}{k} = 0.53\left(\frac{D^3\rho^2 g\beta\Delta t}{\mu^2}\right)^{1/4}\left(\frac{C_p\mu}{k}\right)^{1/4} \tag{1}$$

$$Gr = \frac{(0.1)^3(1.14)^2(9.8)(0.00322)(36)}{(1.92\times 10^{-5})^2}$$

$$= 4.00 \times 10^6$$

$$Pr = \frac{(1000)(1.92\times 10^{-5})}{0.0266}$$

$$= 0.722$$

From equation (1)

$$h = \frac{(0.53)(0.0266)}{0.1}(4\times 10^6)^{1/4}(0.722)^{1/4}$$

$$= 5.81 \text{ W/m}^2\cdot\text{K}$$

The product Pr Gr is evidently within 10^3 to 10^9, for which range, the equation is applicable.

• PROBLEM 1-17

Steam flows through a 0.375 ft. diameter pipe with a surface temperature of 200 °F.
Compute the rate of heat convection per unit pipe surface area to the surrounding air at 70 °F.

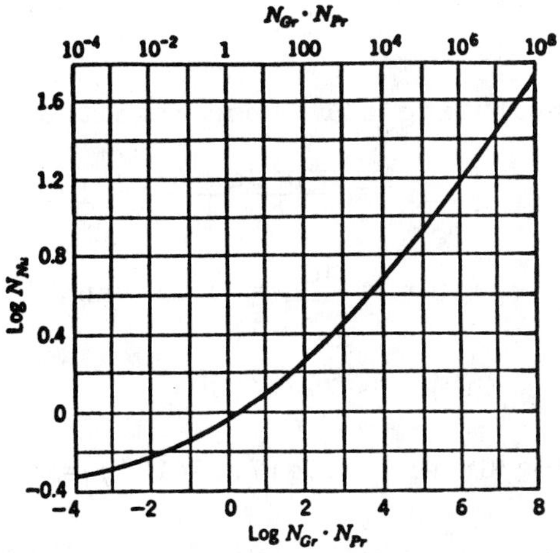

Correlated data for determining coefficients of free convection between horizontal cylinders and gases or liquids.

Solution: The physical properties of air at mean temperature $t = \frac{200 + 70}{2} = 135\,^0F$ are

$$\rho = \frac{0.0667}{32.2} = 0.00208 \text{ slug ft}^{-3}$$

$$\mu = 0.0482/32.2 = 0.00150 \text{ slug ft}^{-1}\text{hr}^{-1}$$

$$C_p = 0.241(32.2) = 7.76 \text{ Btu slug}^{-1}\text{F}^{-1}$$

$$k = 0.0164 \text{ Btu slug}^{-1}\text{ F}^{-1}$$

Further, the relation for thermal expansion β is

$$\beta = \frac{1}{T}$$

T is the absolute temperature. This holds for ideal gases. For air at room temperature, it is reasonable to assume that air behaves as an ideal gas.

Therefore, $\beta = \frac{1}{T_1} = \frac{1}{460+70} = 0.00189 \text{ F}^{-1}$

Wherever pounds occur, they are converted to the mass unit slug, by dividing by 32.2. This is necessary because in the derivation of the Grashof number, mass units are used. The usual value of the specific heat is related to 1 lb mass; in order to relate it to 1 slug it must be multiplied by 32.2.

In addition to the properties of the air, the gravity constant g occurs in Gr. Because the unit of time used above is the hour, it is necessary to take $g = 32.2(3600)^2 \text{ft hr}^{-2}$. Otherwise the consistency of the units in the dimensionless group would be disturbed.

Now, substituting all these magnitudes,

$$Gr = \frac{D^3\rho^2\beta g \cdot \Delta t}{\mu^2} = 10.33(10^6)$$

$$Pr = \frac{\mu C_p}{k} = 0.71, \text{ and}$$

$$Gr \cdot Pr = 7.34(10^6)$$

$$\log\left(Gr \cdot Pr\right) = \log\left(7.34(10^6)\right) = 6.866$$

Using these values and referring to the accompanying figure it is found that

$$Nu = 25.2$$

But $Nu = hD/k$, therefore,

$$h = \frac{25.2(0.0164)}{0.375}$$

$$= 1.10 \text{ Btu/hr ft}^{2\,\circ}\text{F}$$

and heat transfer per square foot is

$$q = 1.10(200 - 70)$$

$$= 143 \text{ Btu/hr ft}^2$$

• PROBLEM 1-18

> Water enters a single-pass heat exchanger at 61.4°F and leaves at 69.9°F. The average water velocity through the 0.902 in. diameter tube is 7 ft/sec. Compute the average film heat transfer coefficient on the water side.

Solution: The physical properties of water at mean temperature

$$t = \frac{61.4 + 69.9}{2} = 65.7^\circ\text{F}$$

are

$$\rho = \frac{62.3}{32.2} = 1.94 \text{ slug ft}^{-3}$$

$$\mu = 2.51 \text{ lb}_m\text{ft}^{-1}\text{hr}^{-1} = 2.51/32.2 \text{ slug ft}^{-1}\text{hr}^{-1}$$

$$= 0.0780 \text{ slug ft}^{-1}\text{hr}^{-1}$$

$$C_p = 1 \text{ Btu lb}_m^{-1}\text{F}^{-1} = 1(32.2) \text{ Btu slug}^{-1}\text{F}^{-1}$$

$$k = 0.340 \text{ Btu hr}^{-1}\text{ft}^{-1}\text{F}^{-1}$$

Now,

$$Re = \frac{vD\rho}{\mu} = \frac{25,200(0.0752)1.94}{0.0780} = 47,100$$

$$Pr = \frac{\mu C_p}{k} = \frac{0.0780(32.2)}{0.340} = 7.39$$

The film heat transfer coefficient can be found from equation

$$\frac{hD}{k} = 0.023\left(\frac{vD\rho}{\mu}\right)^{0.8}\left(\frac{\mu C_p}{k}\right)^n$$

where for heating n = 0.4 and for cooling n = 0.3 may be used.

Substituting the values

$$= \frac{hD}{k} = 0.023(47{,}100)^{0.8}\,(7.39)^{0.4} = 280.3, \text{ and}$$

$$h = \frac{280.3(0.340)}{0.0752} = 1267 \text{ Btu hr}^{-1}\text{ft}^{-2}\text{F}^{-1}$$

• PROBLEM 1-19

> Compute the rate of heat transfer per unit length of a 2.54 in. diameter tube to air at 2 atm and 200°C flowing at a velocity of 10 meter/sec. The tube wall is maintained at a temperature 20°C higher than the air temperature all along the length of the tube. Also, compute the bulk temperature increase over a tube length of 3 meters.

Solution: The physical properties of air at a bulk temperature 200°C

$$\rho = \frac{p}{RT} = \frac{(2)(1.0132\times 10^5)}{(287)(473)} = 1.493 \text{ kg/m}^3 (0.0932 \text{ lb}_m/\text{ft}^3)$$

$$\mu = 2.57\times 10^{-5} \text{kg/m·s } (0.0622 \text{ lb}_m/\text{h·ft})$$

$$k = 0.0386 \text{ W/m·}^0\text{C} (0.0223 \text{ Btu/h·ft·}^0\text{F})$$

$$C_p = 1.025 \text{ kJ/kg·}^0\text{C}$$

The flow characteristics can be known from the Reynold's number

$$\text{Re} = \frac{\rho u_m d}{\mu} = \frac{(1.493)(10)(0.0254)}{2.57\times 10^{-5}} = 14{,}756$$

thus, the flow is turbulent and, the relevant empirical relation for Nussett number is

$$\text{Nu} = 0.023(\text{Re})^{0.8}\,(\text{Pr})^{0.4}$$

$$\text{Pr} = \frac{\mu C_p}{k} = 0.681$$

$$\text{Nu} = \frac{hd}{k} = (0.023)(14{,}756)^{0.8}\,(0.681)^{0.4} = 42.67$$

$$h = \frac{k}{d}\text{Nu} = \frac{(0.0386)(42.67)}{0.0254} = 64.85 \text{W/m}^2\ {}^0\text{C} (11.42 \text{ Btu/h·ft}^2\ {}^0\text{F})$$

The heat flow per unit length is then

$$\frac{q}{L} = h\pi d(T_w - T_b) = (64.85)\pi(0.0254)(20)$$

$$= 103.5 \text{ W/m} (107.7 \text{ Btu/ft})$$

Now make an energy balance to calculate the increase in bulk temperature in a 3.0m length of tube

$$q = \dot{m}c_p\Delta T_b = L\left(\frac{q}{L}\right)$$

Also

$$\dot{m} = \rho u_m \frac{\pi d^2}{4} = (1.493)(10)\pi\frac{(0.0254)^2}{4}$$

$$= 7.565 \times 10^{-3}\text{kg/s } (0.0167 \text{ lb}_m/\text{S})$$

Substituting the values in energy balance equation

$$(7.565 \times 10^{-3})(1025)\Delta T_b = (3.0)(103.5)$$

or

$$\Delta T_b = 40.04^0\text{C}(72.07^0\text{F})$$

OVERALL HEAT TRANSFER COEFFICIENT

• PROBLEM 1-20

Explain and derive the expression for overall heat transfer coefficient, in the following cases

1) Plane wall
2) Radial systems.

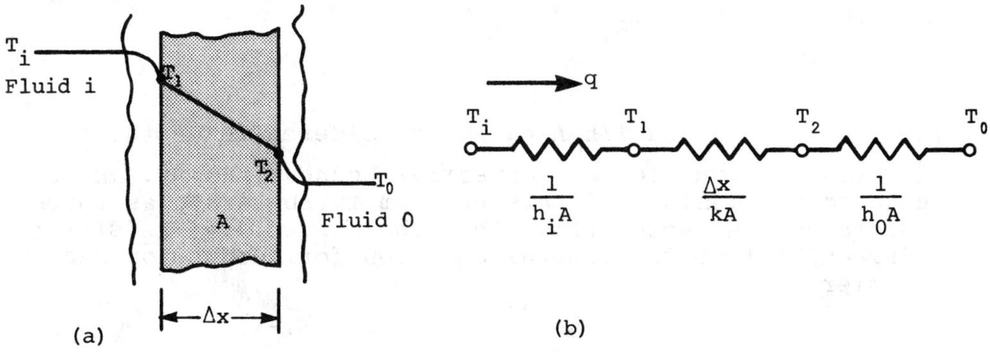

Fig. 1 Overall heat transfer through a plane wall.

Solution: When the overall heat transfer rate is to be determined for a system where heat transfer is taking place by both conduction and convection, h the surface coefficient is replaced by U the overall heat transfer coefficient in Newton's equation for cooling,

$$q = h\ A\ \Delta T$$

which becomes

$$q = UA(\Delta T)_{overall}$$

where $(\Delta T)_{overall}$ is the overall temperature difference, A the area through which heat flows. The units of the overall heat transfer coefficient (U) are the same as the units of h, i.e. Btu/hr ft^{2} ^{0}F(English) or W/m^{2} ^{0}C.

1) PLANE WALL

Consider a plane wall, as shown in Fig. (1a), made of a homogenous material having constant thermal conductivity. One side is exposed to fluid i at temperature T_i, and the other side to fluid o at temperature T_o. It has a thickness Δx.

In most cases, the temperatures known are away from the wall surface and are not affected by the heat transfer process and the wall surface temperatures T_1 and T_2 are not known.

Applying Newton's law of cooling, at the two surfaces gives the heat transferred per unit area

$$\frac{q}{A} = h_i(T_i - T_1) = h_o(T_2 - T_o)$$

This is because the heat transferred from one side of the wall is equal to the heat transferred to the other side. On rearranging the equation

$$q = \frac{T_i - T_1}{1/h_i A} = \frac{T_2 - T_o}{1/h_o A}$$

The terms $1/h_i A$ and $1/h_o A$ can be considered to be the thermal resistance due to the convective boundary layer. Hence the electric analogy of this problem may be taken as three resistances in series (Fig. 1b), the conductive resistance being $\Delta x/kA$ from the general equation for conduction heat transfer

$$q = kA\frac{\Delta T}{\Delta x} \qquad \Delta T = T_1 - T_2$$

The heat transfer rate through each layer of resistance of a plain wall is

$$q = \frac{T_i - T_1}{1/h_i A} = \frac{T_1 - T_2}{\frac{\Delta x}{kA}} = \frac{T_2 - T_o}{1/h_o A}$$

The conductive heat flow is exactly equal to convective heat flow on either side of the plane wall.

Combining these equations to find the heat transfer per unit area

$$\frac{q}{A} = \frac{T_i - T_o}{1/h_i + \Delta x/k + 1/h_o} = \frac{(\Delta T)_{overall}}{\sum R_{thermal}}$$

U for the plane wall Fig. (1a) is given by

$$U = \frac{1}{1/h_i + \Delta x/k + 1/h_o} \quad \frac{Btu}{hr \cdot ft^2 \cdot {}^0F} \quad \text{or} \quad \frac{W}{m^2 \cdot {}^0k}$$

and for a multilayered wall of layers a, b,..., of thickness x, y,..., is

$$U = \frac{1}{1/h_i + \Delta x/k_a + \Delta y/k_b + \ldots + 1/h_o}$$

2) RADIAL SYSTEMS

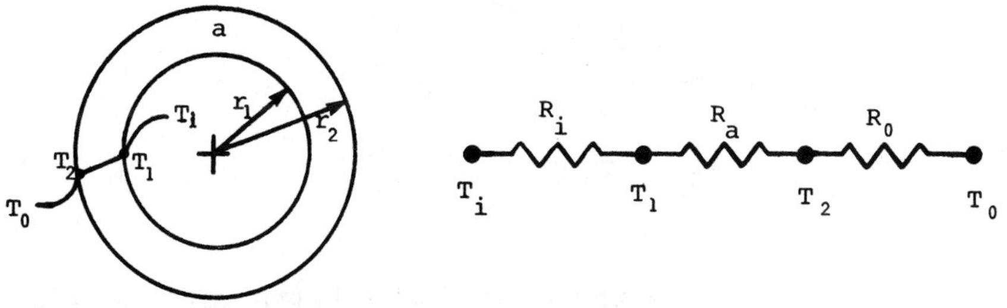

(a) (b)

Fig. 2

Consider the cylindrical system as shown in Fig. 2(a). T_i and T_o are the temperatures of the fluids flowing on the inner and outer sides of the system and r_1 and r_2 are the inner and outer radii, respectively. T_1 and T_2 are the surface temperatures as shown in Fig. 2(a).

Heat flow through any system is

$$q = \frac{(\Delta T)_{overall}}{\sum R_{thermal}} = \frac{T_i - T_o}{R_i + R_a + R_o}$$

where

R_i = inner convective thermal resistance = $1/h_i A_i = 1/2\pi r_i L h_i$

R_a = conductive thermal resistance $= \dfrac{\ln(r_2/r_1)}{2\pi k_a L}$

R_o = outer convective thermal resistance = $1/h_o A_o = 1/2\pi r_o L h_o$

L is the length of the cylindrical system.

Thus the heat transfer rate through each layer of resistance of the cylindrical system shown is

$$q = \frac{T_i - T_1}{1/h_i A_i} = \frac{T_1 - T_2}{\ln(r_2/r_1)/2\pi k_a L} = \frac{T_2 - T_o}{1/h_o A_o}$$

This is because the heat flow through each resistive layer is exactly the same. Hence the heat transfer through the system is

$$q = \frac{T_i - T_o}{1/2\pi r_i L h_i + \ln(r_2/r_1)/2\pi k_a L + 1/2\pi r_o h_o L}$$

The overall heat transfer coefficient for a cylindrical system is based on either the inner surface area (A_i) or the outer surface area (A_o) of the cylinder. Hence

$$q = U_i A_i (T_i - T_o) = U_o A_o (T_i - T_o) = \frac{T_i - T_o}{\sum R}$$

$$U_o = \frac{1}{\dfrac{r_2}{r_1 h_i} + \dfrac{r_2 \ln(r_2/r_1)}{k_a} + \dfrac{1}{h_o}}$$

$$U_i = \frac{1}{\dfrac{1}{h_i} + \dfrac{r_1 \ln(r_2/r_1)}{k_a} + \dfrac{r_1}{r_2 h_o}}$$

where U_o and U_i are the overall heat transfer coefficients based on outer and inner surface area, respectively.

• PROBLEM 1-21

The inside temperature of a furnace is 2400°F and outside wall temperature is 100°F. The furance wall is made up of 9 in. chrome brick, $4\frac{1}{2}$ in. fire brick and $4\frac{1}{2}$ in. of masonry brick. The convective heat transfer coefficient at the inside and outside furance walls are 10 Btu/hr ft^2°F and 1 Btu/hr ft^2°F respectively. Compute the following:

1. Overall heat transfer coefficient
2. Rate of heat transfer per unit wall area
3. The surface temperature at each layer of brick
4. Ascertain, if, the temperature of the fire brick is within the allowable limit of 2000°F.

Use $k_{chrome} = 0.96$ Btu/hr ft°F

$k_{fire} = 0.183$ Btu/hr ft°F

$k_{mas} = 0.40$ Btu/hr ft°F

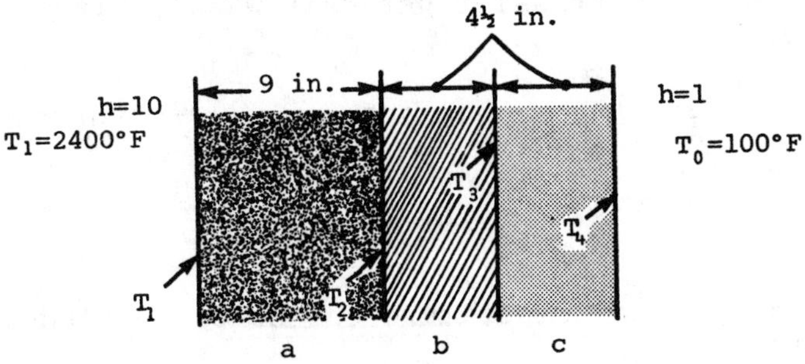

a - firebrick $k_a = 0.96$
b - insulating brick $k_b = 0.183$
c - building brick $k_c = 0.4$

Solution: 1. The overall heat transfer coefficient is expressed as U_o

$$U_o = \frac{1}{1/h_i + \frac{\Delta x_{ch}}{k_{ch}} + \frac{\Delta x_f}{k_f} + \frac{\Delta x_m}{k_m} + 1/h_o}$$

where

h_i and h_o - are inside and outside convective heat transfer coefficients

Δx = thickness of the material

$$\frac{1}{h_i} = (1 \text{ hr ft}^2{}^\circ\text{F})/10 \text{ Btu} = 0.1 \text{ hr ft}^2{}^\circ\text{F/Btu}$$

$$\frac{1}{h_o} = (1 \text{ hr ft}^2{}^\circ\text{F})/(1 \text{ Btu}) = 1 \text{ hr ft}^2{}^\circ\text{F/Btu}$$

$$\frac{\Delta x_{ch}}{k_{ch}} = \frac{9/12 \text{ ft}}{(0.96 \text{ Btu/hr ft}^\circ\text{F})} = 0.78 \text{ hr ft}^2{}^\circ\text{F/Btu}$$

$$\frac{\Delta x_f}{k_f} = \frac{(4.5/12)\text{ft}}{(0.183 \text{ Btu/hr ft}^\circ\text{F})} = 2.06 \text{ hr ft}^2{}^\circ\text{F/Btu}$$

$$\frac{\Delta x_m}{k_m} = \frac{(4.5/12)\text{ft}}{(0.4 \text{ Btu/hr ft}^\circ\text{F})} = 0.94 \text{ hr ft}^2{}^\circ\text{F/Btu}$$

Substituting the values

$$U_o = \frac{1}{0.1 + 0.78 + 2.06 + 0.94 + 1} = 0.205 \text{ Btu/hr ft}^2{}^\circ\text{F}$$

2. Rate of heat transfer per unit wall surface area is given by

$$\frac{Q}{A_o} = U_o\ \Delta t = \left(0.205\ \text{Btu/hr ft}^{2\,0}\text{F}\right)\left((2400-100)\,^0\text{F}\right)$$

$$= 472\ \text{Btu/hr ft}^2$$

3. The interface temperature can be calculated from the equation

$$Q = A\ h\ \Delta t$$

$$\Delta t = 2400 - t_1 = \frac{Q}{Ah_i} = \frac{472\ \text{Btu/hr ft}^{2\,0}\text{F}}{\text{hr ft}^2 (10)\ \text{Btu}}$$

$$= 47.2\ \text{F}$$

or

$$t_1 = 2353\,^0\text{F}$$

$$2353 - t_2 = \frac{Q(\Delta x_{ch}/k_{ch})}{A}$$

$$= (472)(0.78)$$

$$= 368\,^0\text{F}$$

or $t_2 = 1985\,^0\text{F}$

$$1985 - t_3 = (472)(2.06)$$

$$= 972\,^0\text{F}$$

$$t_3 = 1063\,^0\text{F}$$

$$1063 - t_4 = (472)(0.94)$$

$$= 443\,^0\text{F}$$

$$t_4 = 570\,^0\text{F}$$

The temperature profile is sketched in Figure 2.

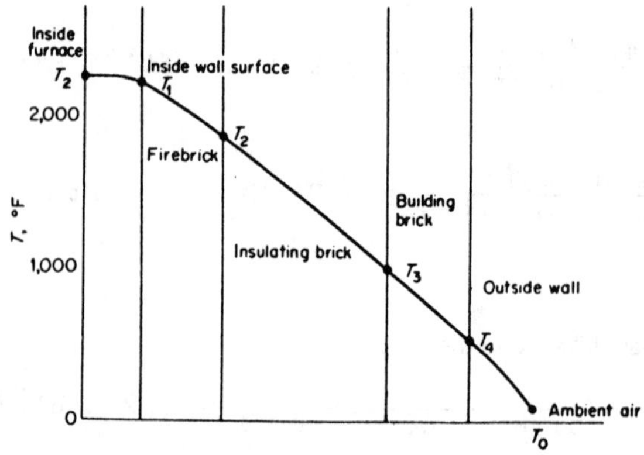

Fig. 2

4. The temperature of the fire brick remains below the maximum allowable temperature 2000°F as can be seen in the Figure 2.

• PROBLEM 1-22

Develop an expression for the "overall heat transfer coefficient", based on the inner surface, for a composite cylindrical wall system.

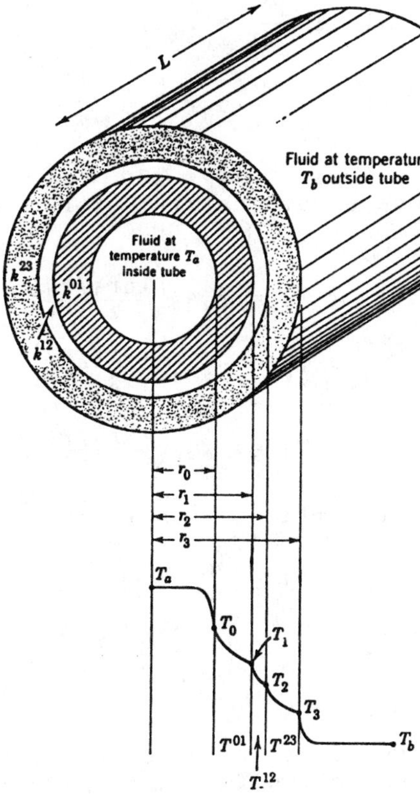

Fig. 1 Heat conduction through a laminated tube with fluid at temperature T_a inside and fluid at temperature T_b outside.

Solution: Energy balance on a cylindrical shell of volume $2\pi rL\Delta r$ for region "01" is

$$q_r^{01}\big|_r \cdot 2\pi rL - q_r^{01}\big|_{r+\Delta r} \cdot 2\pi(r+\Delta r)L = 0 \tag{1}$$

Dividing by $2\pi L\Delta r$ and taking the limit as $\Delta r \to 0$ gives

$$\frac{d}{dr}(rq_r^{01}) = 0 \qquad (2)$$

Integration of this equation gives for constant k^{01}, k^{12}, and k^{23}

$$rq_r^{01} = r_0 q_0 \quad \text{(a constant)} \qquad (3)$$

in which r_0 is the inner radius of region "01" and q_0 is the heat flux there. Hence for steady state

$$-k^{01} r \frac{dT^{01}}{dr} = r_0 q_0 \qquad (4)$$

$$-k^{12} r \frac{dT^{12}}{dr} = r_0 q_0 \qquad (5)$$

$$-k^{23} r \frac{dT^{23}}{dr} = r_0 q_0 \qquad (6)$$

Integration of these equations gives for constant k^{01}, k^{12}, and k^{23}

$$T_0 - T_1 = r_0 q_0 \left(\frac{\ln r_1/r_0}{k^{01}}\right) \qquad (7)$$

$$T_1 - T_2 = r_0 q_0 \left(\frac{\ln r_2/r_1}{k^{12}}\right) \qquad (8)$$

$$T_2 - T_3 = r_0 q_0 \left(\frac{\ln r_3/r_2}{k^{23}}\right) \qquad (9)$$

At the fluid-solid interface it is required that

$$T_a - T_0 = \frac{q_0}{h_0} \qquad (10)$$

and

$$T_3 - T_b = \frac{q_3}{h_3} = \frac{q_0}{h_3}\frac{r_0}{r_3} \qquad (11)$$

Addition of all five equations gives an expression for $T_a - T_b$, which may be solved for q_0 to give

$$Q_0 = 2\pi L r_0 q_0 = \left(\frac{2\pi L(T_a - T_b)}{\frac{1}{r_0 h_0} + \frac{\ln r_1/r_0}{k^{01}} + \frac{\ln r_2/r_1}{k^{12}} + \frac{\ln\ _3/\ _2}{k^{23}} + \frac{1}{r_3 h_3}}\right) \qquad (12)$$

Now U_0 is defined as "over-all heat transfer coefficient based on the inner surface"

$$Q_0 = U_0(2\pi r_0 L)(T_a - T_b) \qquad (13)$$

Combination of eqs. 12 and 13 yields

$$U_0 = r_0^{-1}\left(\frac{1}{r_0 h_0} + \sum_{i=1}^{3} \frac{\ln r_i/r_{i-1}}{k^{i-1,i}} + \frac{1}{r_3 h_3}\right)^{-1} \qquad (14)$$

The subscript "0" on U_0 indicates that the over-all heat transfer coefficient is referred to the radius r_0. This result may be generalized to include a cylindrical tube made of n laminae by replacing "3" by "n" in three places in eq. 14.

• PROBLEM 1-23

Compute the rate of heat loss per unit length through a 3/4 in. thick steel pipe carrying saturated steam at 267°F at an ambient temperature of 80°F, using resistance method. The pipe outside diameter is 1.05 in. and is covered with 1.5 in. thick insulation. Also, compute the overall heat transfer coefficient.

The mean thermal conductivity of the steel is 26 Btu/hr ft°F and of the insulation is 0.037 Btu/hr ft°F. Use inside and outside convective heat transfer coefficient as

$$h_i = 1000 \text{ Btu/hr ft}^2\,{}^0\text{F} \quad \text{and,}$$

$$h_o = 2 \text{ Btu/hr ft}^2\,{}^0\text{F}.$$

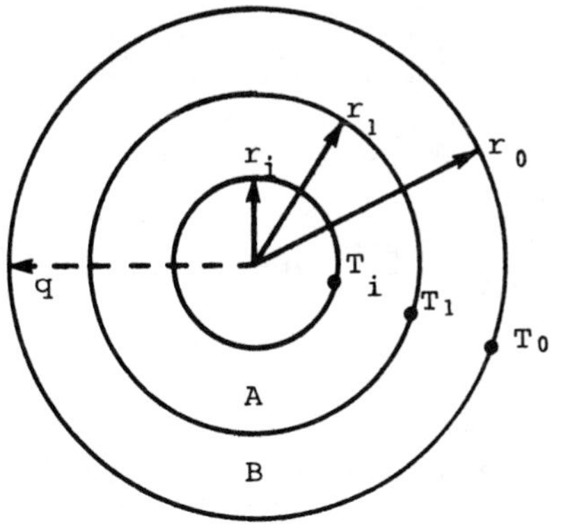

Fig. 1

Solution: The pipe surface areas at different radii as shown in the figure are

$$A_i = 2\pi L r_i = 2\pi(1)\left(\frac{0.412}{12}\right) = 0.2157 \text{ ft}^2$$

$$A_1 = 2\pi L r_1 = 2\pi(1)\left(\frac{0.525}{12}\right) = 0.2750 \text{ ft}^2$$

$$A_0 = 2\pi L r_0 = 2\pi(1)\left(\frac{2.025}{12}\right) = 1.060 \text{ ft}^2$$

The log mean areas for the steel (P) pipe and insulation (I) are

$$A_{Plm} = \frac{A_1 - A_i}{\ln(A_1/A_i)} = \frac{0.2750 - 0.2157}{\ln(0.2750/0.2157)} = 0.245$$

$$A_{Ilm} = \frac{A_0 - A_1}{\ln(A_0/A_1)} = \frac{1.060 - 0.2750}{\ln(1.060/0.2750)} = 0.583 \quad (1)$$

The resistances are

$$R_i = \frac{1}{h_i A_i} = \frac{1}{1000(0.2157)} = 0.00464$$

$$R_P = \frac{r_1 - r_i}{k_P A_{Plm}} = \frac{(0.525 - 0.412)/12}{26(0.245)} = 0.00148 \quad (2)$$

$$R_I = \frac{r_0 - r_1}{k_I A_{Ilm}} = \frac{(2.025 - 0.525)/12}{0.037(0.583)} = 5.80$$

$$R_o = \frac{1}{h_0 A_0} = \frac{1}{2(1.060)} = 0.472$$

Using equation

$$q = \frac{T_i - T_o}{1/h_i A_i + (r_o - r_i)/k_P A_{Plm} + 1/h_o A_o} \quad (3)$$

or substituting the values in equation (3)

$$q = \frac{T_i - T_o}{R_i + R_P + R_I + R_o} = \frac{267 - 80}{0.00464 + 0.00148 + 5.80 + 0.472}$$

$$= \frac{267 - 80}{6.278} = 29.8 \text{ Btu/hr}$$

The overall heat transfer coefficient U can be calculated from the equation

$$q = U_i A_i (T_i - T_o) = U_o A_o (T_i - T_o) = \frac{T_i - T_o}{\Sigma R} \quad (4)$$

or

$$U_i = \frac{1}{A_i \ \Sigma R}$$

substituting the values

$$U_i = \frac{1}{0.2157(6.278)} = 0.738 \ \frac{\text{Btu}}{\text{hr}\cdot\text{ft}^2\cdot{}^0\text{F}}$$

Then,

$$q = U_i A_i (T_i - T_o) = 0.738(0.2157)(267-80) = 29.8\text{Btu/h}(8.73\text{W})$$

CRITICAL RADIUS

• **PROBLEM 1-24**

What is critical thickness of insulation?

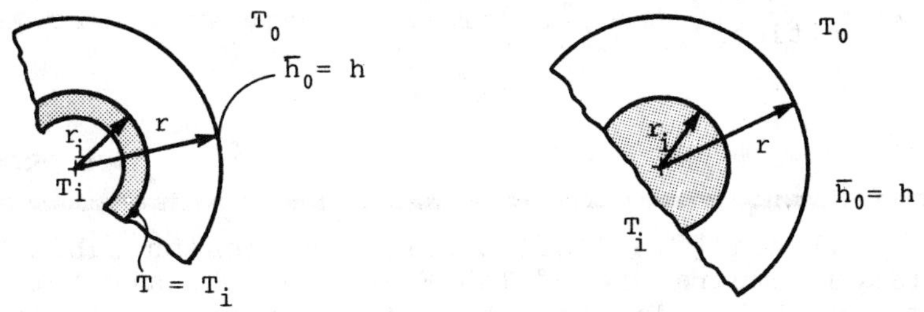

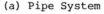

(a) Pipe System (b) Rod or Wire System

Solution: Metal pipes, duct walls, etc. have a very small thermal resistance, compared to the insulation layer around them. In most cases the temperature of the pipe wall is nearly the same as that of the fluid flowing inside the pipe. Heat transfer rate per unit length is given by

$$\frac{q}{L} = U_0 \frac{A}{L} \Delta T = \frac{2\pi(T_i - T_0)}{\ln(r/r_i)/k + \frac{1}{hr}}$$

To maximize the heat transfer rate 'q' with respect to r, differentiate the equation for heat transfer rate and equate it to 0.

$$q = \frac{2\pi L(T_i - T_0)}{\ln(r/r_i)/k + 1/hr}$$

$$q = \frac{2\pi L(T_i - T_0)}{\frac{\log r}{k} - \frac{\log r_i}{k} + \frac{1}{hr}}$$

Differentiating

$$\frac{dq}{dr} = \frac{-2\pi L(T_i - T_0)[1/r\,k - 1/hr^2]}{(\log r/k - \log r_i/k + 1/hr)^2} = 0$$

$$(1/r\,k - 1/hr^2) = 0$$

$$r = \frac{k}{h}$$

Therefore the heat transfer rate (q), taken as a function of the radius (r) is maximum at

$$r = k/h = r_{critical}$$

and this radius is called the critical radius or critical thickness of insulation.

If the radius $r_i < r_{critical}$, then the heat transfer rate (q) will increase with the addition of insulation, till q reaches a maximum at

$$r_i = r_{critical}$$

then for any further addition of insulation, i.e., $r_i > r_{critical}$, the heat transfer rate will decrease.

• PROBLEM 1-25

An insulated ½ in. (OD) power transmission cable has a surface temperature of 150°F and ambient air temperature is 70°F. Analyze the influence of insulation on heat transfer, if the convective heat transfer coefficient is h_0 = 1.5 Btu/hr ft² °F and thermal conductivity of insulation is k = 0.09 Btu/hr ft°F.

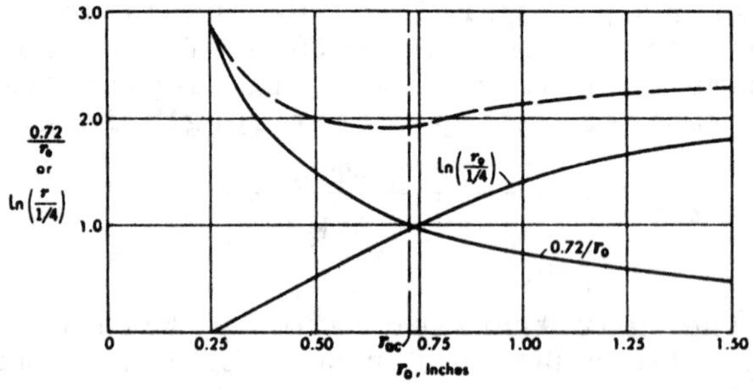

Fig. 1 Variation of thermal resistance with outside radius for an insulated type wire.

Solution: The rate of heat transfer through the cylindrical surface per unit length is given by equation

$$q = \frac{2\pi\ k(t_i - t_0)}{\ln(r_0/r_i) + k/h_0 r_0} \quad \text{Btu/hr ft}$$

$$\text{or} \quad q = \frac{2\pi(0.09)(150-70)}{\ln(r_0/\tfrac{1}{4}) + (0.09)(12/1.5r_0)} \text{ Btu/hr ft}$$

$$= \frac{45.2}{\ln 4r_0 + (0.72/r_0)} \text{ Btu/hr ft}$$

The first term of the denominator is proportional to the thermal resistance of the insulation, the second term to the surface resistance. In the figure, each of these terms is plotted against the outer radius r_0. The dotted line, representing the sum of both terms, has a minimum at $r_{oc} = 0.72$ in. This is the critical radius at which the rate of heat dissipation reaches a maximum value of

$$q = \frac{45.2}{2.08} = 22.8 \text{ Btu/hr ft length}$$

In absence of insulation the rate of heat transfer would have reduced to 15.7 Btu/hr ft.

• PROBLEM 1-26

Find the critical radius of insulation for a pipe surrounded by asbestos [$k = 0.181$ W/m. ^{0}C] and exposed to air at a temperature of 10^{0}C with $L = 3.5$ W/m^2. ^{0}C. Also, find the heat loss from the pipe at 275^{0}C for the following cases:

a) Pipe with critical radius of insulation
b) Pipe without insulation
c) Pipe with critical radius of insulation + 1.5 cm thick insulation.
d) Pipe with critical radius of insulation -1.5 cm insulation

Also plot the values of q/L versus radius for each case.

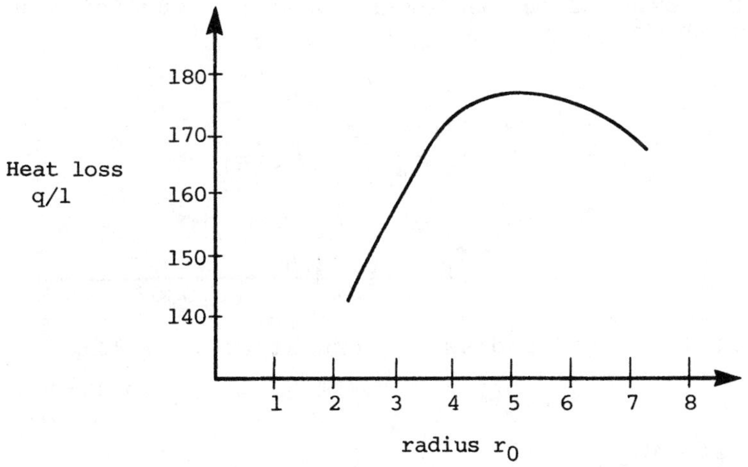

Solution: The critical radius of insulation is given by the equation

$$r_c = \frac{k}{h}$$

$$r_c = \frac{0.181}{3.5} = 0.05171\text{m} = 5.17 \text{ cm}$$

a. The outer radius of the pipe is $\frac{5.0}{2}$ = 2.5 cms. The heat transfer is found by calculating the heat conducted through the insulation using the equation

$$q = \frac{2\pi L(T_S - T_\infty)}{\left[\frac{\ln(r_c/r_i)}{k} + \frac{1}{r_c h}\right]}$$

where T_s = Surface temperature of pipe

T_∞ = Temperature of surrounding air

r_i = Inner radius of insulation

$$\frac{q}{L} = \frac{2\pi(275 - 10)}{\left[\frac{\ln(5.17/2.5)}{0.181} + \frac{1}{(0.517)(3.5)}\right]} = 174.52\text{W/m}.$$

b. Since there is no insulation on the pipe, the heat loss is calculated by finding the heat transfer by convection from the outer surface of the pipe

$$q = (2\pi rL)h(T_s - T_\infty)$$

$$\frac{q}{L} = (2\pi)(0.025)(3.5)(275 - 10) = 145.69\text{W/m}$$

c. In this case the outer radius $r_o = 5.17 - 1.5 = 3.67$ cm = 0.0367m. Using the equation for heat lost per unit length for a cylinder

$$\frac{q}{L} = \frac{2\pi(T_S - T_\alpha)}{\left[\frac{\ln(r_o/r_i)}{k} + \frac{1}{r_o h}\right]}$$

$$= \frac{2\pi(275 - 10)}{\left[\frac{\ln(3.67/2.5)}{0.181} + \frac{1}{(0.0367)(3.5)}\right]} = 168.08\text{W/m}$$

d. Here the outer radius of insulation $r_o = r_c + 1.5$

$= 5.17 + 1.5 = 6{,}67$cm

$= 0.0667$m

Hence heat loss is

$$\frac{q}{L} = \frac{2\pi(275 - 10)}{\left[\frac{\ln(6.67/2.5)}{0.181} + \frac{1}{(0.0667)(3.5)}\right]}$$

$$= 171.56 \text{ W/m}.$$

The values of the heat lost per unit length (q/L) are plotted versus radius in the figure shown.

RADIATION

• **PROBLEM 1-27**

Explain the nature of radiation heat transfer.

Solution: When two objects at different temperatures are placed at finite distance apart in a vacuum, a net energy transfer occurs from the body at the higher temperature to the object at the lower temperature, even though there is no medium between them. The net energy transfer mode is radiation heat transfer.

Radiation heat transfer is the transfer of heat by electromagnetic radiation. It is a type of radiation similar to x-rays, light waves, microwaves, radio waves (AM, FM, TV, etc.), and is different from them only in wavelength. It obeys all the laws of electromagnetic waves; travels in a straight line, can be transmitted through a vacuum and if there is a medium through which the heat is transferred, the medium is usually not heated.

This is a very important mode of heat transfer, especially when there are large temperature differences between the source and the receptor. A furnace with boiler tubes, oven baking, solar furnaces, etc., are examples of radiation heat transfer.

Radiation heat transfer often occurs in conjunction with conduction and convection. Heat transfer by radiation takes place as follows:

1. The thermal energy from a hot source such as a heating element, is converted into electromagnetic energy in the form of radiation waves.

2. These waves, travelling in a straight line through the intervening space, strike the cold body such as a vessel containing water to be heated.

3. The electromagnetic waves that strike the body are absorbed and converted back to thermal energy or heat.

• PROBLEM 1-28

Explain the following terms:

a) Absorptivity
b) Black body
c) Emissivity
d) Grey body.

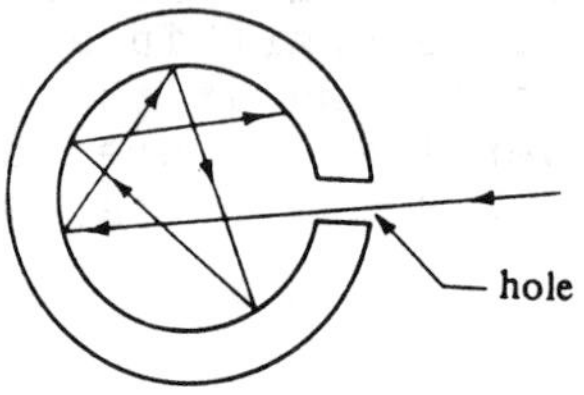

Fig. 1 Concept of a perfect black body.

Solution: a) Absorptivity: When thermal radiation (i.e. light waves) falls on a body, part is absorbed by the body in the form of heat, part is reflected back and part may be transmitted through the body. For most cases of heat transfer, the bodies are opaque and hence the transmission of energy through the body is neglected. Hence, in cases of objects that are opaque to transmission

$$\alpha + \rho = 1.0$$

where α is the absorptivity or fraction of energy absorbed and ρ is the reflectivity or fraction of energy reflected.

b) Black body: A body that absorbs all radiation that falls on it and reflects none, is known as a black body. Hence, the reflectivity $\rho=0$, and the absorptivity $\alpha=1.0$. Similarly, a black surface is a surface which has the above mentioned properties. No perfectly black surfaces or bodies exist in nature, although there are some real surfaces that come very close to being black.

A close approximation to a black body, is a small hole in a spherical hollow body, as shown in the figure. The inside surface is blackened with a coating of a substance with a very low reflectivity. The radiation is allowed to enter through the hole and impinges on the rear wall. Part of the radiation is absorbed and part is reflected in all directions. These reflected radiations impinge again, part is absorbed and part reflected. This process carries on until, essentially all the entering radiations have been absorbed. Hence, the area of the hole acts as a perfect black body.

c) Emissivity: When a body emits radiation, the ratio of the emissive power of the body to that of a black body is known as its emissivity.

A black body absorbs all radiant energy that falls on it and reflects none. A black body also emits radiation, depending on its temperature, while it does not reflect any. The emissivity of a black body $\varepsilon = 1.0$. Hence, at the same temperature, the absorptivity and the emissivity are the same for a black body, or

$$\alpha = \varepsilon$$

This is also true for non-black bodies at the same temperature.

for a perfect black body with an emissivity $\varepsilon = 1.0$

$$q = A\sigma \; T^4$$

where q = heat flow in Watts (Btu/hr)
σ = constant = 5.676×10^{-8} W/m^2.^{0}k^4
= 0.1714×10^{-8} Btu/hr ft^2 ^{0}R^4
T = temperature of the black body ^{0}k(^{0}R)

d) Grey body: Any body that has an emissivity $\varepsilon < 1.0$ is called a grey body. For a non-black body or a grey body that has an emissivity $\varepsilon < 1.0$, the emissive power is reduced ε and the equation for heat transfer by radiation becomes

$$q = A\varepsilon\sigma T^4$$

• PROBLEM 1-29

The inner sphere of a Dewar flask containing liquified oxygen is 1.2 ft diameter and the outer sphere is 1.8 ft diameter. Both spheres are plated for which $\varepsilon = 0.04$. Determine the rate at which liquid oxygen would evaporate at its boiling point -297^0F when the outer sphere temperature is 25^0F. Assume other modes of heat transfer are absent except radiation.

Solution: Evaporation of oxygen takes place at constant temperature. Heat transfer between two concentric cylinders is given by

$$q = \frac{\sigma A_1 [T_1^4 - T_2^4]}{1/\varepsilon_1 + \left(\frac{1}{\varepsilon_2} - 1\right)\frac{A_1}{A_2}}$$

$$= \frac{\sigma 4\Pi r_1^2 \; [T_1^4 - T_2^4]}{\frac{1}{\varepsilon_1} + \left(\frac{1}{\varepsilon_2} - 1\right)\left(\frac{r_1}{r_2}\right)^2},$$ where r_1, r_2 are

the radii of the inner and outer spheres and T_1, T_2 are their absolute temperatures.

$$= \frac{0.1713 \times 10^{-8} \times 4 \times \Pi \times 0.6^2 \; [(460+(-297))^4-(460+30)^4]}{\frac{1}{0.04} + \left(\frac{1}{0.04} - 1\right)\left(\frac{0.6}{0.9}\right)^2}$$

$$= -12.37 \text{ Btu/hr}$$

The negative sign implies that heat flows inside.

• PROBLEM 1-30

Consider a double layered wall. The emissivities of the outer and inner layer are 0.85 and 0.46, respectively. The ambient temperature is 110°F and the temperature inside the room is 60°F. Calculate the radiant heat flux per unit area. Also, calculate the radiant heat flux when an aluminium radiation shield having $\varepsilon_s = 0.05$ is used.

Solution: Let T_a, T_b and T_s be the temperatures of the outer wall, inner wall and the shield, respectively, when the shield is not used

$$\frac{q}{A} = \frac{\sigma(T_a^4 - T_b^4)}{1/\varepsilon_a + 1/\varepsilon_b - 1} \qquad (1)$$

$$= \frac{0.1713 \times 10^{-8}[(570)^4-(520)^4]}{1/0.85 + 1/0.46 - 1}$$

$$= 23.64 \text{ Btu/hr ft}^2$$

When the use of shield is made we have the following relation

$$\frac{q}{A} = \frac{\sigma(T_a^4 - T_s^4)}{1/\varepsilon_a + 1/\varepsilon_s - 1} = \frac{\sigma(T_s^4 - T_b^4)}{1/\varepsilon_s + 1/\varepsilon_b - 1}$$

$$\frac{(570)^4 - T_s^4}{1/0.85 + 1/0.05 - 1} = \frac{T_s^4 - (520)^4}{1/0.05 + 1/0.46 - 1}$$

$$T_s = 547$$

Hence,

$$\frac{q}{A} = \frac{0.1713 \times 10^{-8}[(570)^4-(547)^4]}{1/0.85 + 1/0.05 - 1}$$

$$= 1.36 \text{ Btu/hr ft}^2$$

CHAPTER 2

STEADY STATE HEAT CONDUCTION

Basic Attacks and Strategies for Solving Problems in this Chapter. See pages 43 to 131 for step-by-step solutions to problems.

Steady state heat conduction problems involve some type of temperature variation in a solid which causes a flow of heat, or in other words, heat transfer. The governing equation is Fourier's Law, which was introduced in Chapter 1. Most problems require calculation of either a heat transfer rate or the temperature at one or more points in the solid. Problems may also be set up to require determination of an unknown thermal conductivity or some size parameter for the solid, such as thickness of a slab or diameter of a cylinder.

The simplest type of problem involves only one-dimensional heat transfer, such as through a flat slab, through the wall of a cylinder, or through the wall of a sphere. Such problems may require integration of Fourier's Law and/or substitution of values into the integrated form to calculate some unknown. They may also involve thermal resistance in series or in parallel, as considered in Chapter 1. Some complicating factors which may be present are variable thermal conductivity or internal generation of heat within the solid. Variable thermal conductivity is usually expressed in terms of the temperature. In this case the variables in Fourier's Law can still be separated and the equation can usually still be integrated.

If internal heat generation is present, the differential heat conduction equation must be the starting point rather than Fourier's Law. The steady state conduction equation in one dimension with constant thermal conductivity is:

$$\frac{d^2T}{dx^2} + \frac{g}{k} = 0 .$$

The internal heat generation, g, may be constant, or it may be a function of some parameter such as electrical current. In any case, integration of the above equation for the particular geometry of the problem is usually required.

Two-dimensional heat transfer occurs whenever a coordinate system cannot be set up so that the temperature only varies along one axis. In this case, the two-dimensional heat conduction equation is required as the starting point. The steady state form of that equation for constant thermal conductivity is:

$$\frac{d^2T}{dx^2} + \frac{d^2T}{dy^2} + \frac{g}{k} = 0 .$$

If there is no internal heat generation, the parameter, g*, is, of course, equal to zero. For some two-dimensional solid shapes, the differential equation above can be solved to find the temperature distribution as a function of* x *and* y*. If the solid is cylindrical instead of rectangular, then the conduction equation in cylindrical coordinates must be used instead.*

For some two-dimensional configurations the solutions to the differential conduction equation have been expressed in terms of a conduction shape factor, S*. The shape factor,* S*, is usually calculated from a generalized equation for the particular configuration, and it can then be used in the following equation:*

$$q = kS(T_2 - T_1) .$$

Problems involving fins extending from a heat transfer surface may sometimes be solved by using a fin efficiency, which is the ratio of the heat transfer flux from the fin surface to the heat transfer flux from the main heat transfer surface. In other cases, the solution may require integration over the surface of the fin.

Numerical solution of the two-dimensional conduction equation is sometimes needed, because that differential equation can only be solved for fairly simple geometries. Numerical solution involves setting up a grid over the solid cross-section and solving simultaneous algebraic equations to find the temperature at each node of the grid.

Step-by-Step Solutions to Problems in this Chapter, "Steady State Heat Conduction"

HEAT CONDUCTION THROUGH A PLANE WALL

• PROBLEM 2-1

The inside temperature of a furnace is 1500°F and its walls are lined with a 4 in. thick refractory brick. The ambient temperature is 100°F. Compute the rate of heat flow through the walls and the temperature profile across the wall, if the variation of thermal conductivity of the refractory is given by

$$k = 0.10 + 5 \times 10^{-5}T$$

Solution: The rate of heat transfer is expressed by the Fourier equation

$$q = -k\frac{dT}{dx}$$

For steady state
$$q\int_0^L dx = -\int_{T_1}^{T_2} k\,dT$$

or
$$q\int_0^{1/3\ ft} dx = -\int_{1,500}^{100} (0.10 + 5 \times 10^{-5}T)dT$$

$$q = 3[(0.1)(1,500 - 100) + (5 \times 10^{-5})(1.12 \times 10^6)]$$

$$q = 588\ \text{Btu/hr ft}^2$$

For obtaining the profile

$$q\int_0^x dx = -\int_{1,500}^{T} (0.10 + 5 \times 10^{-5}T)dT$$

$$588x = 0.1(1,500 - T) + \frac{5 \times 10^{-5}}{2}(2.25 \times 10^6 - T^2)$$

$$x = 0.352 - 1.7\times 10^{-4}T - 4.28\times 10^{-8}T^2$$

The above equation is a quadratic one and shows the non-linear characteristic of temperature variation across the furnace wall.

• PROBLEM 2-2

A furnace wall consists of two layers (see fig.). The inner layer is 4.5" thick and is made of silica bricks (k=0.08), and the outer layer is 9" of common brick (k=0.8). The inner and outer surface temperatures are 1400°F and 170°F, respectively. Calculate the heat loss per sq. ft. through this wall.

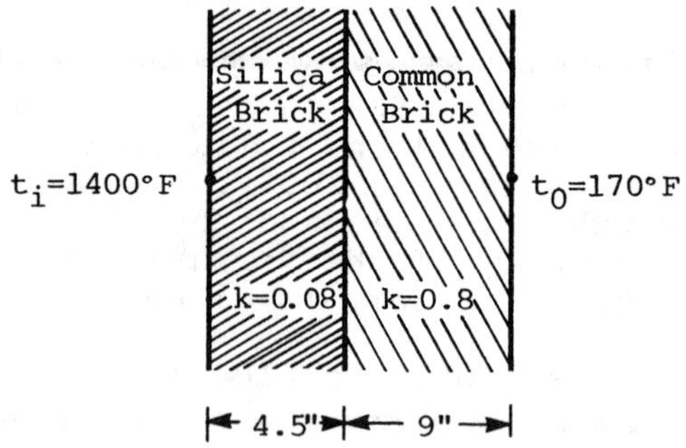

Solution: The thermal resistance of a plane wall is given by $R_{th} = L/kA$.

Therefore the thermal resistances are for the silica bricks

$$R_{th_1} = \frac{4.5/12}{(0.08)(1)}$$

$$= 4.687$$

for the common bricks

$$R_{th_2} = \frac{9.0/12}{(0.8)(1)}$$

$$= 0.938$$

The total resistance is equal to the sum of the individual resistances.

$$R = 4.687 + 0.938 = 5.625$$

Therefore, Rate of heat flow = temperature drop/resistance

Therefore, $\frac{q}{A} = \frac{1400 - 170}{5.625} = 219$ Btu/(hr)(sq ft)

• PROBLEM 2-3

Calculate the contact surface temperatures of a furnace wall made up of 3 layers of brick. The innermost layer consists of 9" thick fire brick (k=0.72 Btu $hr^{-1}ft^{-1}F^{-1}$), the second layer is 5" of insulating brick (k=0.08 Btu $hr^{-1}ft^{-1}F^{-1}$), and the outer layer is 7.5" of red brick (k=0.5 Btu $hr^{-1}ft^{-1}F^{-1}$). The inner and outer temperatures of the composite wall are 1500°F and 150°F, respectively.

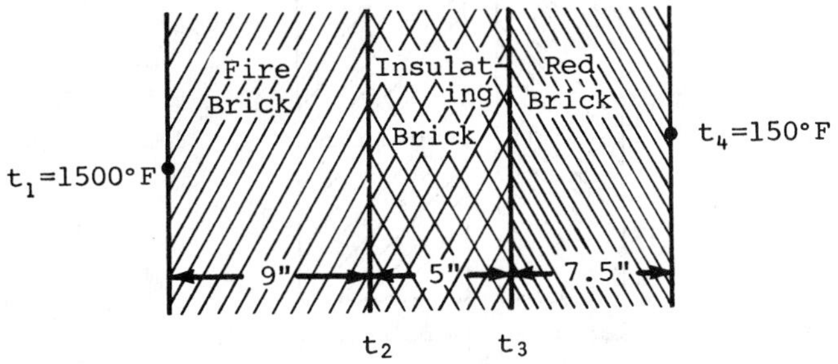

Solution: The rate of heat flow (q), per unit area (sq. ft.) is determined using the following equation:

$$q = \frac{A(t_1 - t_4)}{\frac{(\Delta x)_{12}}{k_{12}} + \frac{(\Delta x)_{23}}{k_{23}} + \frac{(\Delta x)_{34}}{k_{34}}}$$

$$q = \frac{1(1500 - 150)}{\frac{9/12}{0.72} + \frac{5/12}{0.08} + \frac{7.5/12}{0.5}} = 180 \text{ Btu/hr}$$

This value of q is the heat flow through any section of the wall, and is used to determine the contact surface temperatures. Using this value in the equation to determine t_2,

$$q = kA\,\frac{t_1 - t_2}{x_2 - x_1}$$

$$180 \text{ Btu/hr} = \frac{(0.72)\,1\,(1500 - t_2)}{9/12}$$

$$t_2 = 1312^0\text{F}$$

The above procedure is repeated to obtain $t_3 = 375^0$F.

Consider a double layered wall. The inner layer is made of a 0.5 inch thick iron wall (k = 30) and the outer layer consists of a 1.0 inch thick aluminium wall (k=118). A clearance of 0.0005 inch exists between the two layers. Find the thermal resistance of the wall with and without the clearance. The thermal conductivity of air may be taken as 0.015 Btu/hr ft^{0}F.

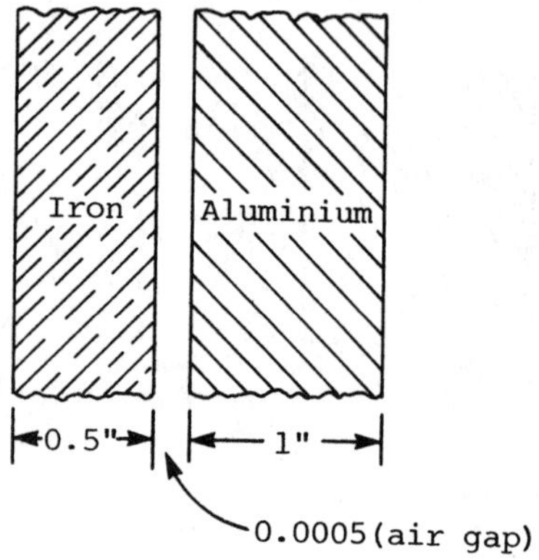

Solution: The thermal resistance of a plane wall is given by

$$R_{th} = \frac{1}{A}\left(\frac{b}{k}\right)$$

where A is the area, sq. ft.
b is the thickness, ft.
and k is the thermal conductivity

In the first case, with the clearance, the thermal resistance is

$$R_{th} = \frac{1}{A}\left(\frac{b_1}{k_1} + \frac{b_2}{k_2} + \frac{b_3}{k_3}\right)$$

$$= \frac{1}{A}\left(\frac{0.5}{12\times 30} + \frac{0.0005}{12\times 0.015} + \frac{1}{12\times 118}\right)$$

$$= \frac{1}{A}\left(\frac{1}{720} + \frac{1}{360} + \frac{1}{1,416}\right)$$

$$= \frac{1}{A}\left(\frac{1}{205}\right) \text{ hr}^0\text{F/Btu}$$

In the second case, without the clearance, the thermal resistance is

$$R_{th} = \frac{1}{A}\left(\frac{b_1}{k_1} + \frac{b_2}{k_2}\right)$$

$$= \frac{1}{A}\left(\frac{0.5}{12\times30} + \frac{1}{12\times118}\right)$$

$$= \frac{1}{A}\left(\frac{1}{720} + \frac{1}{1416}\right)$$

$$= \frac{1}{A}\left(\frac{1}{477}\right) \text{ hr}^0\text{F/Btu.}$$

On comparing the results for both cases, it is found that in the absence of the small (0.005 in.) air space, the thermal resistance decreases by more than one-half the value of the case with the air space.

• PROBLEM 2-5

A flat piece of plastic, area $A=1$ ft^2 and thickness $b=0.252$ inch, was tested and found to conduct heat at a rate of 3.0 watts in steady state with temperatures of $T_0=24^0$C and $T_1=26^0$C on the two main sides of the panel. Find the thermal conductivity of the plastic in cal sec^{-1}cm^{-1}(^{0}K)$^{-1}$at 25^{0}C.

Solution: The first step is to convert the units for the properties given, to S.I

$$\text{Area } A = 144 \text{ in.}^2\times(2.54)^2 = 929 \text{ cm}^2$$

$$\text{Thickness } b = 0.252 \text{ in.} \times 2.54 = 0.640 \text{ cm}^2$$

Heat conduction

$$\text{rate} = 3.0 \text{ watts} = 3.0 \text{ joules sec}^{-1}$$

$$= 3.0 \text{ joules sec}^{-1} \times 0.23901 = 0.717 \text{ cal sec}^{-1}$$

Temp. difference

$$\Delta T = 26 - 24 = 2^0\text{C}$$

The equation for heat transfer by conduction under steady state conditions is given by

$$\frac{Q}{A} = \frac{k\ \Delta T}{\Delta x}$$

where k is the thermal conductivity
ΔT is the temperature difference
$\Delta x=b$ the thickness

Rearranging this equation to find the value of k,

$$k = \frac{Q\ \Delta x}{A\ \Delta T} = \frac{Q\ b}{A\ \Delta T} = \frac{0.717\times0.640}{929\times2.00}\ \frac{(\text{cal sec}^{-1})(\text{cm})}{(\text{cm}^2)(^0\text{K})^{-1}}$$

$$= 2.47\times10^{-4} \text{ cal sec}^{-1} \text{ cm}^{-1} (^0\text{K})^{-1}$$

For ΔT as low as in this case ($\Delta T = 2.0^\circ C$), it is reasonable to assume that the value of k found is applicable at the average temperature $\left(\frac{t_1 + t_2}{2} = \frac{26+24}{2} = 25^\circ C\right)$. This is the temperature at which the thermal conductivity was to be determined.

• PROBLEM 2-6

Determine the rate of heat flow per unit area through a composite wall, as shown in the figure. A 2 in. layer of fire brick ($k_1 = 1.0$ Btu/hr ft $^\circ$F) is placed between two 0.25 in. thick steel plates ($k_3 = 30$ Btu/hr ft $^\circ$F). The surface of the brick face adjoining the steel plate is rough and has only 30 percent of the brick area in contact with the steel plates. The average height of the asperities is 1/32 inch. If the temperatures on the two outer steel plates are 200 $^\circ$F and 800 $^\circ$F respectively, find the rate of heat flow per unit area.

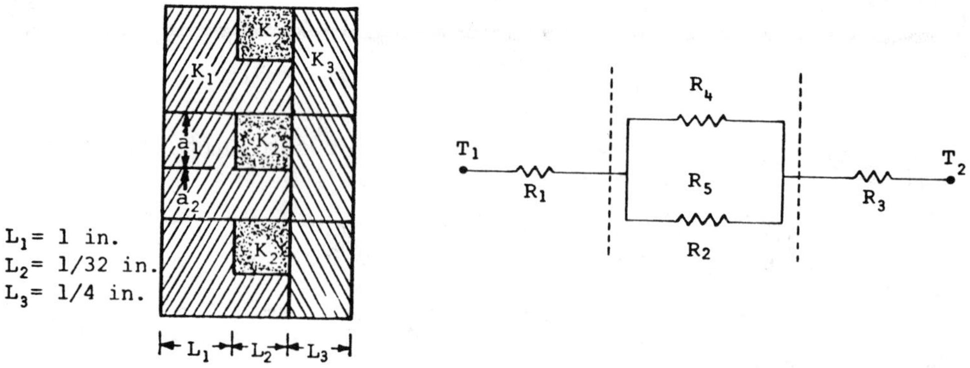

Fig. 1 Thermal circuit for a parallel-series composite wall. (T_1 is at the center.)

Solution: For the wall section shown it is assumed that the asperities are evenly distributed. Then the overall heat transfer coefficient for the composite wall is

$$u = \frac{1/2}{R_1 + \frac{R_4 R_5}{R_4 + R_5} + R_3} \tag{1}$$

The thermal resistance of the steel plate (R_3), on the basis of unit area is

$$R_3 = \frac{L_3}{k_3} = \frac{0.25}{(12)(30)}$$

$$= 0.694 \times 10^{-3} \text{hr sq ft}^\circ\text{F/Btu}$$

The thermal resistance of the brick asperities, per unit area is

$$R_4 = \frac{L_2}{0.3k_1} = \frac{1/32}{12\times0.03\times1}$$

$$= 8.7\times10^{-3} \text{ hr sq ft}^0\text{F/Btu}$$

The air trapped in between the steel plate and the fire brick is very less and the effects of convection are negligible. Hence, the heat transfer through the air is assumed to take place by conduction alone. The thermal conductivity of air is taken at 300^0F and is equal to k_a=0.02 Btu/hr ft^{0}F. Then the thermal resistance of air (R_5) is

$$R_5 = \frac{L_2}{0.7\ k_a} = \frac{1/32}{12\times0.7\times0.02}$$

$$= 187\times10^{-3} \text{ hr sq ft}^0\text{F/Btu}$$

The factors 0.3 and 0.7 used in calculating R_4 and R_5 are the percent areas of the total areas for the two separate heat flow paths.

The total thermal resistance for the two paths, R_4 and R_5, is

$$R_2 = \frac{R_4R_5}{R_4+R_5}$$

$$= \frac{8.7\times187\times10^{-6}}{(8.7+187)\times10^{-3}} = 8.3\times10^{-3}\text{hrsqft}^0\text{F/Btu}$$

Therefore, the thermal resistance of one-half the fire brick R_1 is

$$R_1 = \frac{1}{2}\frac{L_1}{k_1} = \frac{1}{2}\frac{2}{12\times1.0} = 83.5\times10^{-3}\text{hr sq ft}^0\text{F/Btu}$$

Then, the value of u, the overall heat transfer coefficient is determined by substituting the values of R_1, $R_2(R_4+R_5)$ and R_3 in equation 1,

$$u = \frac{1/2}{R_1+R_2+R_3}$$

$$= \frac{1/2}{83.5+8.3+0.69} = 5.4 \text{ Btu/hr sq ft}^0\text{F}$$

Heat transfer rate (q/A) is given by

$$q/A = u\ \Delta T$$

$$= 5.4(800-200) = 3250 \text{ Btu/hr sq ft}$$

The thermal resistance due to the roughness of the adjoining surfaces is called contact resistance.

• **PROBLEM 2-7**

Determine the temperature distribution in a semi-infinite two-dimensional flat plate shown in the figure, if the base temperature is F(x) and the ambient temperature is T_∞.

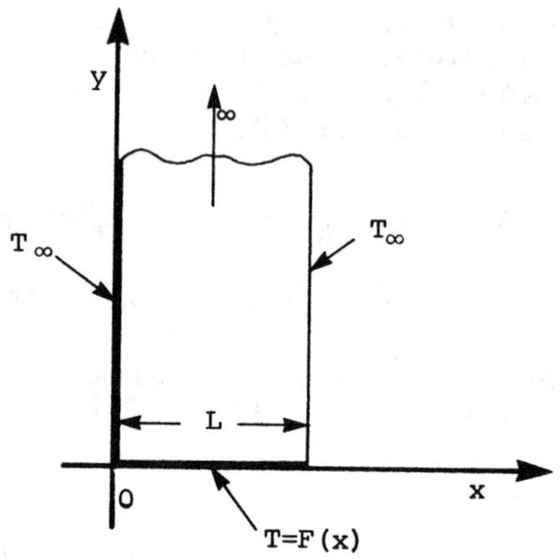

Solution: The controlling differential equation is

$$\frac{\partial^2 T}{\partial x^2} + \frac{\partial^2 T}{\partial y^2} = 0$$

and the given boundary conditions are

$$T(x,0) = F(x); \quad T(L,y) = T_\infty$$

$$T(0,y) = T_\infty \quad ; \quad T(x,\infty) = T_\infty$$

Since the boundary conditions are not homogeneous, the differential equation cannot be solved by the method of separation of variables. It requires the differential equation and three of the boundary conditions to be homogeneous. This can be achieved by a simple transformation

$$\theta = T - T_\infty$$

Writing the differential equation and boundary conditions in terms of θ,

$$\frac{\partial^2 \theta}{\partial x^2} + \frac{\partial^2 \theta}{\partial y^2} = 0 \tag{1}$$

$$\theta(x,0) = F(x) - T_\infty = f(x) \tag{2}$$

$$\theta(0,y) = 0 \tag{3}$$

$$\theta(L,y) = 0 \tag{4}$$

$$\theta(x,\infty) = 0 \tag{5}$$

The solution to the equation (1) by the method of separation of variables is

$$\theta(x,y) = (A\cos\lambda x + B\sin\lambda x)(Ce^{\lambda y} + De^{-\lambda y})$$

where

A, B, C, D and λ are constants.

Using the boundary condition (3)

$$\theta(0,y) = (A+0)(Ce^{\lambda y} + De^{-\lambda y}) = 0$$

Therefore A = 0

so, $$\theta(x,y) = B\sin\lambda x(Ce^{\lambda y} + De^{-\lambda y})$$

The constant B can be absorbed into the constants C and D.

Therefore,

$$\theta(x,y) = \sin\lambda x(Ce^{\lambda y} + De^{-\lambda y})$$

Using the boundary condition (5)

$$\theta(x,\infty) = \sin\lambda x(Ce^{\infty} + De^{-\infty}) = 0$$

From the above equation C = 0. Therefore the differential equation becomes

$$\theta(x,y) = \sin\lambda x(De^{-\lambda y})$$

Now using the boundary condition (4)

$$\theta(L,y) = De^{-\lambda y}\sin\lambda L = 0$$

Since $D \neq 0$, $\sin\lambda L = 0$

Therefore $\lambda L = n\pi \qquad n = 1,\ldots\infty$

or

$$\lambda = \frac{n\pi}{L} \tag{6}$$

The solution $\theta(x,y) = De^{-\lambda y}\sin\lambda x$, is a particular solution of the equation (1).

The general solution is obtained by summing all the particular solutions.

Therefore $\theta(x,y) = \sum_{n=L}^{\infty} D_n e^{-\lambda_n y}\sin\lambda_n x$ is the general solution.

Using the non-homogeneous boundary condition (2)

$$f(x) = \sum_{n=L}^{\infty} D_n \sin\lambda_n x$$

This is a Fourier series. The value of D_n is

$$D_n = \frac{\int_o^L f(x)\sin\lambda_n x \, dx}{\int_o^L \sin^2\lambda_n x \, dx}$$

Simplifying, this reduces to

$$D_n = \frac{2}{L}\int_o^L f(x)\sin\lambda_n x \, dx$$

Substituting the value of D_n into the general solution,

$$\theta(x,y) = \frac{2}{L} \sum_{n=1}^{\infty} \left[\int_o^L f(\eta)\sin\lambda_n \eta \, d\eta\right] e^{-\lambda_n y} \sin\lambda_n x$$

where η is a dummy variable.

Equation (7) is the required temperature distribution.

• PROBLEM 2-8

A carbon heating element in the form of a bar 3 in. × 1/2 in. × 3 ft., has a uniform surface temperature of 1400 °F when a potential of 12V is applied to the ends of the bar. Find the temperature at the center of the bar. The electrical resistivity of the bar is 1.3×10^{-4} Ωft and the thermal conductivity is 2.9 Btu/hr ft °F.

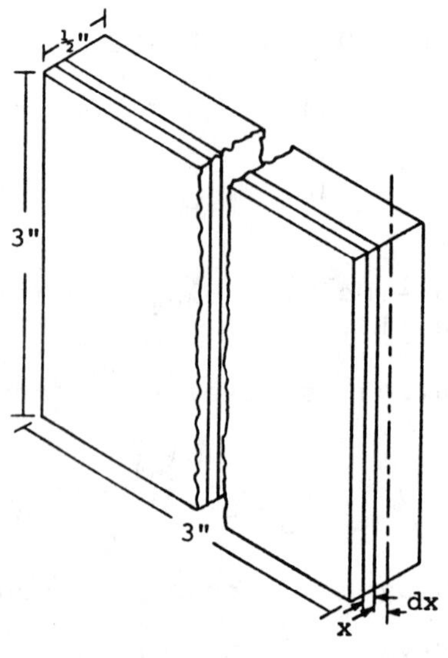

Fig. 1. Heat conduction in a bar with generation.

Solution: A differential equation is obtained for heat conduction normal to the largest faces of the bar, by doing an energy balance on an infinitesimal segment dx, as shown in the figure.

Thus, heat flow into the element

$$= -kA \frac{dt}{dx} \text{ Btu/hr} \qquad (1)$$

where $A = \text{Area} = (3/12)(3) = 0.75 \text{ ft}^2$.

Heat flow out of element

$$= -kA\left(\frac{dt}{dx} + \frac{d}{dx}\frac{dt}{dx}\right)$$

$$= -kA\left(\frac{dt}{dx} + \frac{d^2t}{dx^2}\right) \text{ Btu/hr} \qquad (2)$$

In order to obtain the heat generation in the element, the electrical resistance of the bar is required. The electrical resistance is

$$= \frac{(1.3\times10^{-4})(3)}{(1/24)(3/12)}$$

$$= 0.0375\ \Omega$$

The rate of heat generation in the entire bar is

$$= \frac{(12)^2}{0.0357}$$

$$= 3830\text{W}$$

$$[419{,}000 \text{ Btu/hr ft}^3]$$

Therefore, rate of heat generation in the element

$$= 419{,}000 \text{ A dx Btu/hr} \qquad (3)$$

Considering a heat balance on the element gives the equation,

heat flow in + rate of heat generation = heat flow out

The terms in this expression are the same as equations (1), (2) and (3), which on substitution give

$$-kA\frac{dt}{dx} + 419000A\,dx = -kA\left(\frac{dt}{dx} + \frac{d^2t}{dx^2}\,dx\right)$$

Simplifying the equation, we have

$$\frac{d^2t}{dx^2} = -\frac{419000}{k} \qquad (4)$$

or

$$\frac{d^2t}{dx^2} = -\frac{419000}{2.9}$$

$$\frac{d^2t}{dx^2} = -145000^0 F/ft^2 \tag{5}$$

Equation (5) on integration becomes

$$\frac{dt}{dx} = -145{,}000\,x + C_1$$

The bar is symmetrical, therefore $\frac{dt}{dx} = 0$ at x=0, which gives c=0.

Integrating equation (5) twice, the relation for temperature is obtained as follows.

$$t = \frac{-145000x^2}{2} + C_2 \tag{6}$$

Using another boundary condition

$$\text{at} \quad x = 1/48 \text{ ft.} \qquad t = 1400^0F$$

and applying it to equation (6) gives

$$C_2 = 1400 + (145000/2)(1/48)^2$$
$$= 1431$$

Hence the expression for the temperature distribution in the bar is

$$t = -72500\, x^2 + 1431$$

At x=0, i.e. the center of the bar, the temperature is 1431^0F.

• PROBLEM 2-9

> The walls of a furnace 3 ft.× 4 ft.× 2.5 ft., are made from a refractory material 4½ in. thick and having an average thermal conductivity of 0.8 Btu/hr ft^{0}F.
>
> The inner and outer surface temperatures are 400^0F and 100^0F, respectively. Determine the heat loss by conduction over a period of 24 hours.

Solution: For an arbitrary number of plane sections having a total area of $\sum A$ and M edges, the shape factor is

$$S = \frac{\sum A}{\Delta y} + 0.54\, aM \tag{1}$$

where the shape factor for an edge section, s=0.54a has been experimentally determined. The quantities Δy and a are shown in the figure.

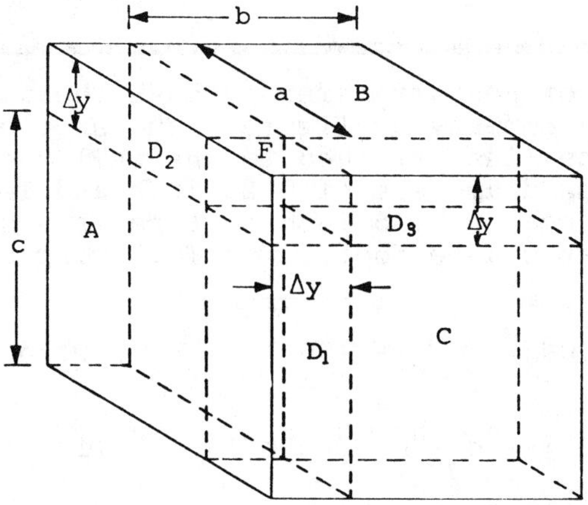

Fig. 1 Corner section of a furnace wall.

Similarly, the shape factor for corner sections has been experimentally determined. For N corners,

$$S = 0.15(\Delta y)N \tag{2}$$

Since the inside dimensions are greater than 1/5 the thickness of the wall, the combined shape factor for plane sections and corners is determined by combining equations (1) and (2), which gives

$$S = \left(\frac{2(3\times4) + 2(4\times2.5) + 2(3\times2.5)}{4.5/12}\right) + (4(0.54)[4+3+2.5]) + \left(8(0.15)\ \frac{4.5}{12}\right)$$

$$= 157.5 + 20.5 + 0.45$$

The first, second and third terms represent the individual shape factor for the plane sections, the edge sections and the corners, respectively.

The heat flow per hour is given by

$$Q = Sk(t_1 - t_2)$$

$$q = 178.5(0.8)(400 - 100)$$

$$= 42{,}800 \text{ Btu/hr}$$

Therefore, after 24 hours, the heat loss is

$$Q = 42{,}800 \times 24$$

$$= 1{,}028{,}000 \text{ Btu.}$$

Find the rate of heat transfer through the wall section shown in the accompanying figure. The section measures 1 ft.×1ft. The fins attached to the wall section are pin fins having diameters of 0.25 inch and lengths of 2 inches, arranged on 0.75 inch centers in a square pitch or 256 pins per square foot. The following data may be used.

$T_h = 200\,^0F \qquad T_c = 100\,^0F \qquad x_w = 0.25$ in.

$k_w = 24 \qquad h_c = h_h = 2 = h_{eff\,c} = h_{eff\,h}$.

Neglect the scale resistance.

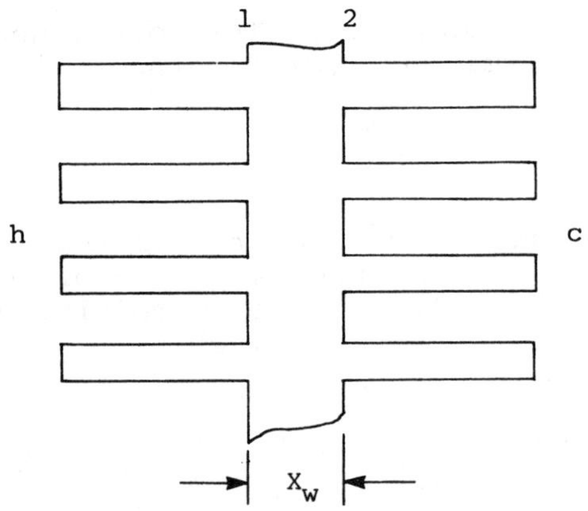

Finned wall.

Solution: The total finned area is

$$A_f = \frac{\pi(0.25)(2.0625)(256)}{144}$$

$$= 2.885 \text{ ft}^2$$

The total area of the wall is $A_w = 1 \text{ ft}^2$.

The remaining unfinned area is

$$A_u = 1 - \frac{256(\pi)(0.25)^2}{4(144)}$$

$$= 0.912 \text{ ft}^2$$

The total area is $A = A_u + A_f$

$$= 0.912 \text{ ft}^2 + 2.885 \text{ ft}^2$$

$$= 3.797 \text{ ft}^2.$$

For a fin

$$\eta = \frac{\tan h\,(BL)}{BL}$$

$$\text{where} \quad B = \sqrt{\frac{hP}{kS}},$$

P is the perimeter, S is the cross sectional area

$$\eta = \frac{\tan h\, \eta[(4)(2.0625/12)]}{(4)(2.0625/12)}$$

$$= 0.867$$

The overall surface efficiency is given by

$$\varepsilon \equiv \frac{q}{Ah\ \Delta T} = \eta\,\frac{A_f}{A} + \frac{A_u}{A}$$

where $A \equiv A_u + A_f$ (u⇒ unfinned, f ⇒ finned, w ⇒ wall)

Therefore,

$$\varepsilon = \frac{0.867(2.885)+0.912}{3.797}$$

$$= 0.90$$

Then from the equation

$$\frac{T_h - T_c}{dq/dA_r} = \frac{1}{u_r} = \frac{1}{\varepsilon_c h_{effc}}\frac{dA_r}{dA_c} + R_w\frac{dA_r}{dA_w} + \frac{1}{\varepsilon_h h_{effh}}\frac{dA_r}{dA_h}$$

$$\left(\frac{1}{h_{eff}} \equiv \frac{1}{h_{fluid}} + \frac{1}{h_{scale}}\right),$$

$$\text{where} \quad \frac{T_h - T_c}{q} = \frac{1}{A_r U_r}$$

$$= 2\left(\frac{1}{(0.90)(3.797)(2)}\right) + \frac{0.25}{24(12)(1)} = 0.294$$

or

$$q = \frac{1}{0.294}\,(200 - 100)$$

$$= 340 \text{ Btu/hr}.$$

• PROBLEM 2-11

Find the solution for a region, with steady state heat conduction, having boundaries at x=0, x=a, y=0 and y=f(x) and boundary conditions as shown in the figure. Use the Galerkin method of partial integration, with respect to the y variable. Take the heat generation rate to be constant.

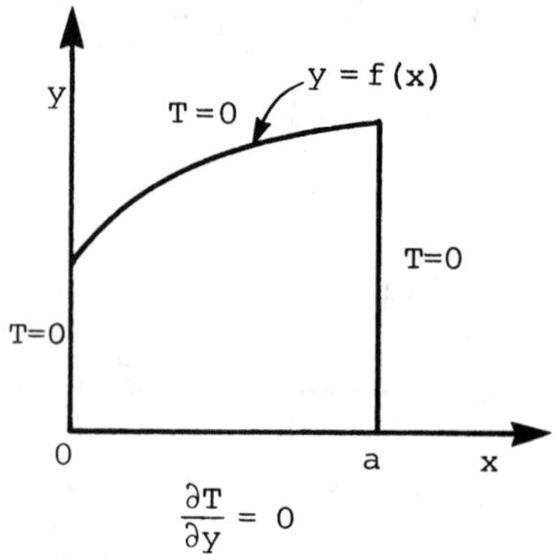

Solution: The appropriate equations and boundary conditions for this problem are

$$\frac{\partial^2 T}{\partial x^2} + \frac{\partial^2 T}{\partial y^2} + \frac{g}{k} = 0 \quad \text{in } 0 < x < a,\ 0 < y < f(x) \tag{1a}$$

$$T = 0 \quad \text{at } x=0, x=a, \text{ and } y=f(x) \tag{1b}$$

$$\frac{\partial T}{\partial y} = 0 \quad \text{at } y=0 \tag{1c}$$

Using the Galerkin method, partial integration of equation (1a) with respect to y, yields

$$\int_{y=0}^{f(x)} \left[\frac{\partial^2 T}{\partial x^2} + \frac{\partial^2 T}{\partial y^2} + \frac{g}{k}\right] \phi_i(y)dy = 0 \tag{2}$$

Now, taking a one-term trial solution

$$T_1(x,y) = X(x) \cdot \phi_1(y) \tag{3a}$$

where

$$\phi_1(y) = [y^2 - f^2(x)] \tag{3b}$$

The boundary conditions y=0 and y=f(x) are satisfied by this solution, but the function X(x) is not known. Substituting the trial solution into equation (2) and doing a partial integration with respect to the y variable, gives an ordinary differential equation which can be used to determine the function X(x),

$$2/5\ f^2X'' + 2ff'X' + (ff''+f'^2 - 1)X = -\frac{g}{2k} \text{ in } 0<x<a \quad (4a)$$

$$\text{for } X = 0 \quad \text{at} \quad x = 0 \quad \text{and} \quad x = a \quad (4b)$$

The function X(x) can be determined, once the function f(x), defining the form of the boundary, is specified. There are two special cases:

1. $y = f(x) = b$: The region is rectangular and the equation (4a) becomes

$$X'' - \frac{5}{2b^2}X = \frac{-5g}{4b^2k} \quad \text{in } 0 < x < a \quad (5a)$$

$$X = 0 \quad \text{at } x = 0 \text{ and } x = a, \quad (5b)$$

and the one term approximate solution becomes

$$T_1(x,y) = (y^2 - b^2)X(x) \quad (6)$$

where $$X(x) = \frac{g}{2k}\left[1 - \frac{\cos h\left(\sqrt{2.5}\ \frac{x}{b}\right)}{\cos h\left(\sqrt{2.5}\ \frac{a}{b}\right)}\right]$$

2. $y = f(x) = \beta x$: For this case eq.(4a) becomes

$$x^2X'' + 5xX' + \frac{5(\beta^2-1)}{2\beta^2}X = \frac{5g}{4\beta^2k} \quad \text{in } 0 < x < a \quad (7a)$$

$$x = 0 \quad \text{at} \quad x = 0 \quad \text{and} \quad x = a \quad (7b)$$

Equation (7a) is an Euler type equation that can be solved by finding a solution for X(x) in the form x^n. Substitution of $X = x^n$ into the homogenous part of eq.(7a), gives the expression

$$n^2 + 4n + \frac{5(\beta^2-1)}{2\beta^2} = 0 \quad (8a)$$

Hence

$$n_1,\ n_2 = -2 \pm \sqrt{4 - B} \quad (8b)$$

$$\text{where} \quad B \equiv \frac{5(\beta^2-1)}{2\beta^2} \quad (8c)$$

Hence the complete solution for X(x) can be written as

$$X(x) = c_1x^{n_1} + c_2x^{n_2} - P(x) \quad (8d)$$

where P(x) is the particular solution of equation (7a)

$$P(x) = g/2k(\beta^2-1)$$

and the coefficients c_1 and c_2 are determined by the application of the boundary conditions given in equation (7b). Since the solution of X(x) is now known, the one term approximate solution is

$$T_1(x,y) = (y^2-\beta^2x^2)\ X(x)$$

RADIAL HEAT CONDUCTION THROUGH A HOLLOW SPHERE AND A HOLLOW CYLINDER

• PROBLEM 2-12

A 30 ft. long and 2 inch nominal diameter pipe is covered by a 1 inch thick insulation (k=0.0375 Btu/hr.ft.°F). Determine the heat loss if the inner and outer temperatures of the insulation are 380°F and 80°F, respectively.

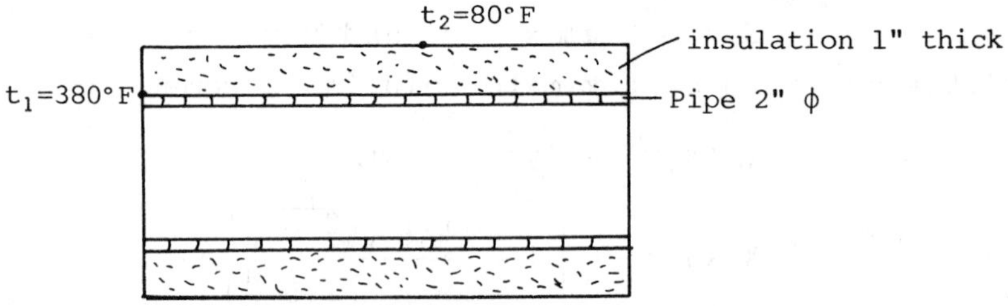

Solution: A pipe having a 2 inch nominal diameter has an outside diameter of 2.375 in. This is the inside diameter of the insulation. Adding 2" (1" all around) to the inside diameter of the insulation, gives the outer diameter of the insulation.

Hence, outer diameter of the insulation

$$= 2 + 2.375 = 4.375 \text{ in.}$$

To determine the heat transfer per unit length of a cylinder, the equation is

$$q = \frac{2\pi k(t_1 - t_2)}{\ln r_2/r_1}$$

$$= \frac{2\pi k(t_1 - t_2)}{\ln D_1/D_2}$$

where D_1 and D_2 are the inner and outer diameters.

Substituting the values in the equation for q

$$q = \frac{2\pi\ 0.0375(380 - 80)}{\ln \frac{4.375/12}{2.375/12}} = 116 \text{ Btu/hr ft}$$

This is the heat loss for 1 foot length of the pipe. Therefore, for 30 feet of pipe, the heat loss is

$$q = 116 \times 30 = 3480 \text{ Btu/hr}$$

• PROBLEM 2-13

A 3 inch schedule 40 pipe is covered with two layers of insulations. The inner layer (k_1=0.050) is 2 inches thick and the outer layer (k_2=0.037) is $1\frac{1}{4}$ inches thick. Calculate the heat loss, in Btu/hr per unit length, if the outer surface temperature of the pipe is 670^{0}F and the outer surface temperature of the outer layer of insulation is 100^{0}F.

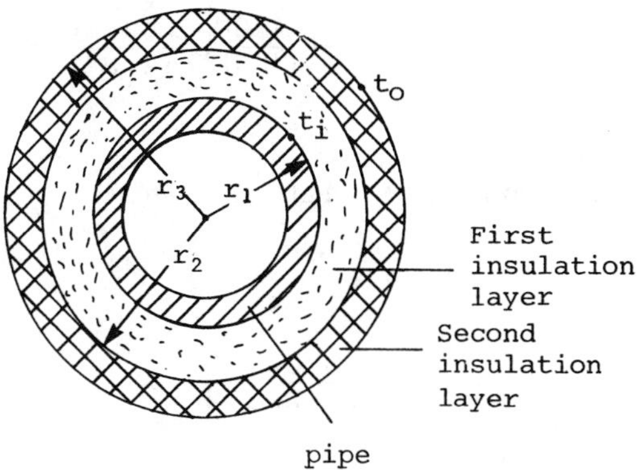

Solution: The outer diameter of the pipe is the inner diameter of the first layer of insulation. Therefore,

Outer diameter of pipe = 3.50 in., r_1=1.75 in.

Outer diameter of first layer = 3.50 + 4.00 = 7.50 in.

r_2=3.75 in.

Outer diameter of second layer = 7.50 + 2.50 = 10.0 in.,

r_3=5.0 in.

For the first layer of insulation, the mean radius is

$$r_{m_1} = \frac{r_2-r_1}{\ln(r_2/r_1)}$$

$$= \frac{3.75-1.75}{\ln(3.75/1.75)} = 2.62 \text{ in.}$$

For the second layer of insulation, the mean radius is

$$r_{m_2} = \frac{r_3-r_2}{\ln(r_3/r_2)}$$

$$= \frac{5.00-3.75}{\ln(5.00/3.75)} = 4.345 \text{ in.}$$

The thermal resistance of the first layer of insulation is

$$R_1 = \frac{r_2 - r_1}{2\pi k_1 r_{m_1}}$$

$$= \frac{3.75-1.75}{2\pi(0.05)(2.62)} = 2.43$$

The thermal resistance of the second layer of insulation

$$R_2 = \frac{r_3 - r_2}{2\pi k_2 r_{m_2}}$$

$$= \frac{5.00-3.75}{2\pi(0.037)4.345} = 1.24$$

The heat loss (q), per unit length, is given by

$$q = \frac{\Delta T}{R_1 + R_2}$$

$$= \frac{670-100}{2.43+1.24}$$

$$q = 155.3 \text{ Btu/hr} \quad \text{per foot length of pipe}$$

• PROBLEM 2-14

> Consider a 1 inch steel pipe insulated with a 2 inch thick layer of magnesia, carrying saturated steam at 150 psig. Determine the heat loss per 100 ft. if the pipe is exposed to a room at 80^0F.

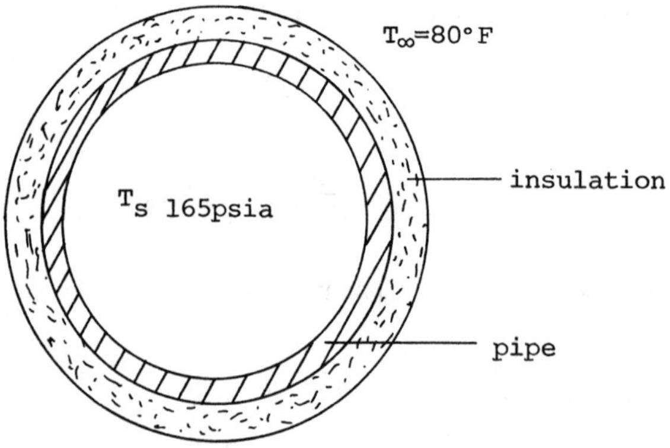

Fig. 1

Solution: The absolute pressure of the steam, in the pipe, is equal to the sum of the gauge pressure and the atmospheric pressure. Hence the absolute pressure (P_{abs}) is

$$P_{abs} = P_{gauge} + P_{atm}$$

$$= 150 + 15$$

$$P_{abs} = 165 \text{ psi.}$$

The saturation temperature of the steam passing through the pipe, as obtained from the steam tables, corresponding to the absolute pressure of 165 psi is

$$T_{s_{165\ psi}} = 366°F$$

Hence the overall temperature difference (ΔT) is

$$\Delta T = T_s - T_\infty = 366 - 80 = 286°F$$

For a 1 inch nomial diameter pipe,

the inner diameter $D_i = 1.049$ in.
the outer diameter $D_o = 1.315$ in.
the wall thickness $x_w = 0.133$ in.

Since the insulation thickness (x) is 2 inches, the outside diameter of the insulation over the pipe is

$$D_s = 1.315 + 4.00 = 5.315 \text{ in.}$$

Due to the fact that resistance on the steam side will be small relative to that of the other resistance, the value of h need not be very accurate; hence, the combined film coefficient (h_i) for the condensate film and dirt deposit is taken as 1000.

Then the thermal resistance (R_i) is given by

$$R_i = \frac{1}{h_i(\pi D_i)} = \frac{1}{(1000)(3.14 \times 1.049/12)}$$

$$= 0.0036$$

The thermal conductivity of steel is k_w=26 and the resistance of the steel wall

$$R_w = \frac{x_w}{k_w \pi D_w}$$

where D_w is the mean diameter in ft.

$$= \frac{(D_i + D_o)}{2 \times 12}$$

$$= \frac{1.049 + 1.315}{2 \times 12} = 1.182/12$$

Therefore $$R_w = \frac{0.133/12}{26 \times \pi \times 1.182/12} = 0.0014$$

If the temperature drop in the insulation is 90 percent of the overall drop, then the mean temperature of the insulation is 238°F. The thermal conductivity of the insulation (k_m), at 238°F, is found to be 0.0418, from the property tables.

The thermal resistance of the insulation is given by

$$R_x = \frac{x}{k_m(\pi\ D_m)}$$

where D_m is the logarithmic mean of the outer diameter (D_o=1.315) and the outer diameter of the insulation (D_s=5.315). Thus the value of D_m = 2.64 in.

$$R_x = \frac{2/12}{0.0418 \times \pi \times 2.64/12} = 5.76$$

Table 1 VALUES OF $h_c + h_r$

(For horizontal bare or insulated standard steel pipe of various sizes in a room at 80°F)

Nominal pipe diam, in.	$(\Delta t)_s$, temperature difference, def F, from surface to room														
	50	100	150	200	250	300	400	500	600	700	800	900	1000	1100	1200
½	2.12	2.48	2.76	3.10	3.41	3.75	4.47	5.30	6.21	7.25	8.40	9.73	11.20	12.81	14.65
1	2.03	2.38	2.65	2.98	3.29	3.62	4.33	5.16	6.07	7.11	8.25	9.57	11.04	12.65	14.48
2	1.93	2.27	2.52	2.85	3.14	3.47	4.18	4.99	5.89	6.92	8.07	9.38	10.85	12.46	14.28
4	1.84	2.16	2.41	2.72	3.01	3.33	4.02	4.83	5.72	6.75	7.89	9.21	10.66	12.27	14.09
8	1.76	2.06	2.29	2.60	2.89	3.20	3.88	4.68	5.57	6.60	7.73	9.05	10.50	12.10	13.93
12	1.71	2.01	2.24	2.54	2.82	3.13	3.83	4.61	5.50	6.52	7.65	8.96	10.42	12.03	13.84
24	1.64	1.93	2.15	2.45	2.72	3.03	3.70	4.48	5.37	6.39	7.52	8.83	10.28	11.90	13.70

With 10 percent (29°) of the total drop through the surface resistance, referring to table 1 at Δt_s=50°F (since there are no values for Δt_s less than 50° and in this problem Δt_s is 29°), and diameter D_m = 2.65 in., gives h_c+h_r=1.90

$$R_s = \frac{1}{(h_c+h_r)(\pi\ D_s)}$$

$$= \frac{1}{1.9 \times \pi \times 5.315/12} = 0.38$$

The total resistance $\sum R$ is

$$= R_i + R_w + R_x + R_s$$

$$= 0.0036 + 0.0014 + 5.76 + 0.38$$

$$= 6.145$$

Hence the resistances R_i and R_w are negligible compared with the total resistance. The estimated resistance of R_s is $(0.38/6.145)(100) = 6.2$ percent of the total resistance, instead of the 10 percent assumed above, but the discrepancy is too small and a second calculation is not necessary.

Thus the predicted heat flow per foot is

$$q = \Delta T / \sum R$$

$$= \frac{286}{6.145} = 46.5 \text{ Btu/hr}$$

Hence, for 100 ft. length of pipe, heat loss from the pipe is

$$Q = 46.5 \times 100 = 4650 \text{ Btu/hr}$$

• PROBLEM 2-15

Determine the heat flow rate from a tube having an inside diameter of 0.742 in. and a wall thickness of 0.154 in. The inside and outside surface temperatures are $T_i = 200\,^0F$ and $T_o = 160\,^0F$, respectively (see fig.). Also, find the heat flow rate based on the inner surface area and that based on the outer surface area.

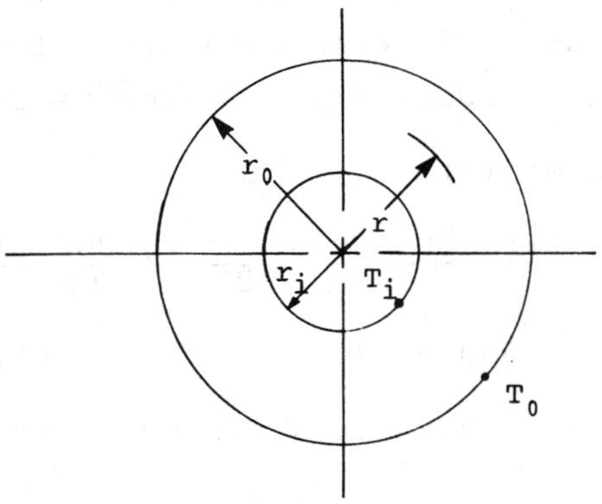

Heat conduction in a radial direction with uniform surface temperature.

Solution: The first law of thermodynamics applies to this problem, which reduces it to the form

$$\frac{\delta Q}{dt} = 0$$

which shows that the rate of heat transfer into the control volume, is equal to the rate leaving the control volume.

The heat flow is in the radial direction, with the singular variable being the radius, r. The form of Fourier's conduction equation is

$$q_r = -kA \frac{dT}{dr}$$

Putting the value of A = 2π r L in the equation gives

$$q_r = -k(2\pi rL) \frac{dT}{dr}$$

where q_r is constant and may be separated and solved as follows:

$$q_r \int_{r_i}^{r_o} \frac{dr}{r} = -2\pi kL \int_{T_i}^{T_o} dT = 2\pi kL \int_{T_o}^{T_i} dT$$

$$q_r \ln \frac{r_o}{r_i} = 2\pi kL(T_i - T_o)$$

$$q_r = \frac{2\pi kL}{\ln r_o/r_i} (T_i - T_o)$$

The values to be substituted in the above equation are

r_i = The inner radius = 0.742/2 = 0.371 in.
r_o = The outer radius = ½(0.742+(0.154×2)) = 0.525 in.
T_i = The inside surface temperature = 200°F
T_o = The outside surface temperature = 160°F
k = The thermal conductivity of steel = 24.8 Btu/hr ft°F

Therefore, heat transfer rate

$$q_r = \frac{2\pi(24.8 \text{ Btu/hr ft}^\circ\text{F})(200 - 160)^\circ\text{F}}{\ln(0.525/0.371)}$$

$$= 17950 \text{ Btu/hr} \qquad (17250 \text{ W/m})$$

The inside and outside surface areas per unit length of tube are

$$A_i = \pi \frac{0.742}{12} (1) = 0.194 \text{ ft}^2 \qquad (0.018\text{m}^2)$$

$$A_o = \frac{1.050}{12} (1) = 0.275 \text{ ft}^2 \qquad (0.0255\text{m}^2)$$

which gives the heat transfer rate based on the inner surface area

$$q_r/A_i = \frac{17950}{0.194} = 92500 \text{ Btu/hr ft}^2$$

$$(291000 \text{W/m}^2)$$

The heat transfer rate based on the outer surfaces is

$$q_r/A_o = \frac{17950}{0.275} = 65300 \text{ Btu/hr ft}^2$$

$$(206000 \text{W/m}^2)$$

The results obtained above show the importance of specifying the area upon which a heat flux value is based. Note: for the same amount of heat flow, the flux based on the inside and outside surface areas differ by approximately 42%.

• PROBLEM 2-16

A right circular cylinder, having a constant thermal conductivity k and no heat generation, has all its surfaces at 0 except for the surface at z=L which has the temperature $T = T_0 J_0(2.405r/R)$. The temperature T is a function of the radial coordinate r, T_0 is a known constant and $J_0(2.405\ r/R)$ is the ordinary Bessel function of the first kind and zero order. Derive an exact analytical expression for the steady state temperature distribution within the cylinder.

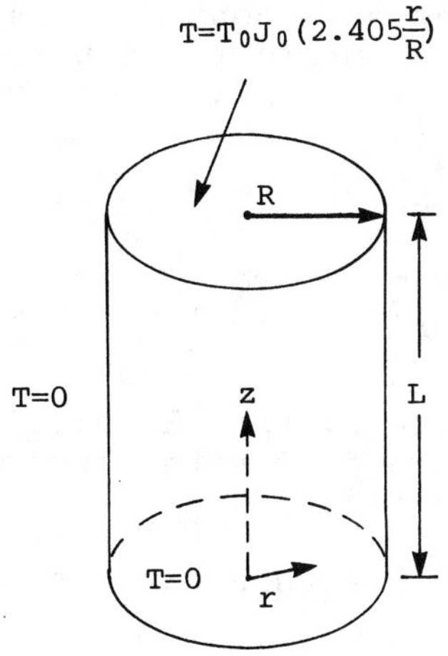

The circular cylinder

Solution: The most appropriate equation for this problem is the general conduction equation in the circular cylindrical coordinate system

$$\frac{\partial^2 T}{\partial r^2} + \frac{1}{r}\frac{\partial T}{\partial r} + \frac{1}{r^2}\frac{\partial^2 T}{\partial \theta^2} + \frac{\partial^2 T}{\partial z^2} + \frac{q'''}{k} = \frac{1}{\alpha}\frac{\partial T}{\partial \tau}$$

Dropping the generation term, the transient term and the circumferential conduction term, the equation reduces to

$$\frac{\partial^2 T}{\partial r^2} + \frac{1}{r}\frac{\partial T}{\partial r} + \frac{\partial^2 T}{\partial z^2} = 0 \qquad (1)$$

The boundary conditions are

(a) At $r = 0$ and $0 < z < L$ $\quad \partial T/\partial r = 0$

(b) At $r = R$ and $0 < z < L$ $\quad T = 0$

(c) At $z = 0$ and $0 < r < R$ $\quad T = 0$

(d) At $z = L$ and $0 < r < R$ $\quad T = T_0 J_0(2.405\ r/R)$

Equation (1) is solved by the separation of variables technique. Let the solution be of the form $T=Q(r)\ P(z)$. Substitute in eq.(1) and separate the functions

$$\frac{d^2Q/dr^2 + (1/r)dQ/dr}{Q} = \frac{-d^2P/dz^2}{P}$$

Since Q depends only on r and P depends only on z, the left-hand side depends only on r and the right-hand side depends only on z. The left-hand side of the equation must equal the right-hand side for all values of r and z in the conduction region, even though r and z can be varied independently of each other.

This equality can hold only if both sides of the equation are equal to the same constant, say $-\lambda^2$. This is called the separation constant. Equate both sides of equation (1) to $-\lambda^2$, which gives

$$\frac{d^2P}{dz^2} - \lambda^2 P = 0 \qquad (2)$$

$$\frac{d^2Q}{dr^2} + \frac{1}{r}\frac{dQ}{dr} + \lambda^2 Q = 0 \qquad (3)$$

Equation (2) is solved by the classical operator technique, and since the z domain is finite, the results are converted to the hyperbolic functions

$$P = A_1 \sinh \lambda z + A_2 \cosh \lambda z$$

Equation (3) is a Bessel equation whose solution is given as

$$Q = B_1 J_0(\lambda r) + B_2 y_0(\lambda r)$$

Hence,

$$T = [B_1 J_0(\lambda r) + B_2 y_0(\lambda r)](A_1 \sinh\lambda z + A_2 \cosh\lambda z) \quad (4)$$

Applying boundary condition (a) to eq.(4), the derivative with respect to r gives a term $J_1(\lambda r)$ and $y_1(\lambda r)$. Since $J_1(0)=0$ and $y_1(0) \to -\infty$, $B_2=0$ is taken to satisfy boundary condition (a).

Let $B_1 A_2 = C_2$
and
$B_1 A_1 = C_1$

Equation (4) becomes

$$T = J_0(\lambda r)(C_1 \sinh \lambda z + C_2 \cosh \lambda z) \quad (5)$$

For boundary condition (b) to exist, it is required that at r=R, T=0 for all z. Hence,

$$0 = J_0(\lambda R)(C_1 \sinh \lambda z + C_2 \cosh \lambda z)$$

The functions of z cannot satisfy the condition, which must hold for all z in the region

$$J_0(\lambda R) = 0 \quad (6)$$

Equation (6) is a transcendental equation for the positive roots of the equation, that is, for the values of $\lambda_n R$ that make J_0 equal to zero. To determine these roots, plot $J_0(\lambda R)$ versus λR. The points at which the curve cuts the λR axis are those values of λR which satisfy equation (6). Now equation (5) becomes

$$T = J_0(\lambda_n r)(C_1 \sinh \lambda_n z + C_2 \cosh \lambda_n z) \quad (7)$$

Applying the boundary condition (c) to equation (7) gives $C_2=0$; hence, equation (7) becomes

$$T = C_1 \sinh \lambda_n z J_0(\lambda_n r) \quad (8)$$

There are an infinite number of solutions, because there are an infinite number of λ_n which satisfy all the conditions, except condition (d) which has not been considered as yet. Boundary condition (d) is: at z=L and $0< r < R$

$$T = T_0 J_0(2.405\ r/R) \quad (9)$$

Setting z=L and multiplying the arguments in equation (8) of both $\sinh \lambda_n z$ and $J_0(\lambda_n r)$ above and below by R and equating it to equation (9), yields

$$C_1 \sinh(\lambda_n RL/R) J_0(\lambda_n R\ r/R) = T_0 J_0(2.405\ r/R) \quad (10)$$

This equation can only be solved if 2.405 is a value of $\lambda_n R$ that satisfies equation (6), else the solution of equation (8), in the form of equation (11) has to be used

$$T = \sum_{n=1}^{\infty} C_n \sinh \lambda_n z J_0(\lambda_n r) \tag{11}$$

Now taking $\lambda_n R=2.405$, it is seen from equation (10) that

$$C_1 = \frac{T_0}{\sinh(2.405\ L/R)}$$

Hence, the steady state temperature distribution is

$$T(r,z) = T_0 \frac{\sinh(2.405\ z/R)}{\sinh(2.405\ L/R)} J_0(2.405\ r/R)$$

• PROBLEM 2-17

Superheated steam flows through a pipe of diameter 10 cm, at a temperature and pressure of 315°C and 1.00kg/cm^2; respectively. An iron thermometer well having a 1.50cm diameter is placed in the tube. If the flow velocity is 20 m/sec, determine the length of the well which gives an error in temperature measurement of less than 0.5 percent of the difference between the steam temperature and the tube wall temperature. The film heat transfer coefficient between the steam and the tube wall is h=105 W/m^{2}K. The wall thickness of the well is s=0.1cm. Take the thermal conductivity of iron to be k=55 W/m.

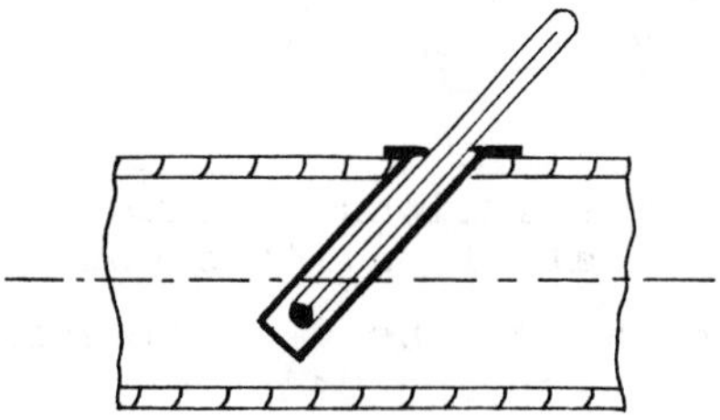

Fig. 1 Temperature measurement in flow in a tube.

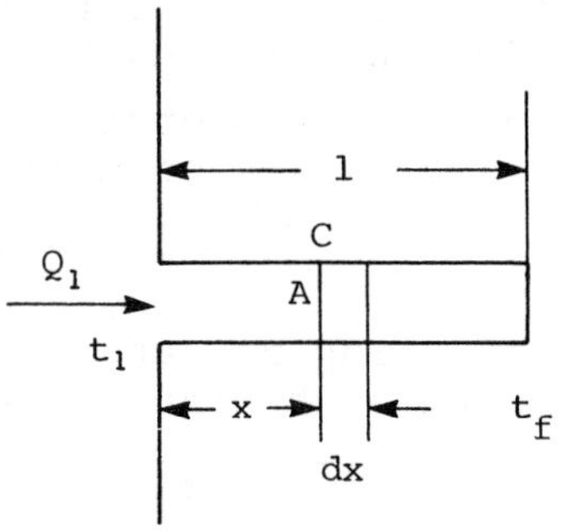

Fig. 2 Steady heat conduction in a thin rod.

Solution: The fluid temperature is much higher than the ambient temperature and hence the tube wall will have a lower temperature than the fluid as heat will flow by conduction from the well to the tube wall. This will also lead to a lower indicated temperature at the end of the well where the thermometer bulb is placed.

The equation for temperature distribution in the rod (see fig. 2) is derived from the steady heat conduction differential equation for a thin rod, and is given as

$$\frac{\theta}{\theta_1} = \frac{e^{m(\ell-x)}+e^{-m(\ell-x)}}{e^{m\ell}+e^{-m\ell}} = \frac{\cosh m(\ell-x)}{\cosh m\ell} \qquad (1)$$

Referring to figure 2, θ_1 is defined as t_1-t_f, and C is the circumference of the cross-section of the rod. The excess temperature at the end of the rod is determined by taking $x=\ell$, which yields

$$\theta_2 = \frac{\theta_1}{\cosh m\ell} \qquad (2)$$

θ_2 being $t-t_f$ at $x=\ell$.

The cross-sectional area for heat flow in the well is

$$A = \pi ds$$

and the circumference is

$$C = \pi d$$

Then,

$$m = \sqrt{\frac{hC}{kA}} = \sqrt{\frac{h}{ks}}$$

$$= \sqrt{\frac{105(100)}{55(0.1)}} = 45.7\ m^{-1}$$

Since θ_2/θ_1 must be 0.005(0.5%) or less, find the appropriate value of $\cosh m\ell$.

For $m\ell = 6$, $\cosh m\ell = 201.7$

and

$$\frac{\theta_2}{\theta_1} = \frac{1}{\cosh mL} = 0.005$$

Therefore, the length of the well is

$$\ell = 6/m$$

$$= 6/0.457 = 13.1\text{cm}.$$

As this length is greater than the diameter of the tube, it is necessary to place the thermometer well obliquely in the tube as shown in fig. (1).

• PROBLEM 2-18

> A pipe having a diameter of 15cm is buried in the ground at a depth of 20cm in a horizontal position. The pipe wall temperature is 75^0C and that of the ground surface is 5^0C. The length of the tube is 4m. Taking the thermal conductivity of the earth to be $k=0.8\ W/m^0C$, determine the heat lost by the pipe.

Conduction Shape Factors,

Physical system	Schematic	Shape factor	Restrictions
Isothermal cylinder of radius r buried in semi-infinite medium having isothermal surface	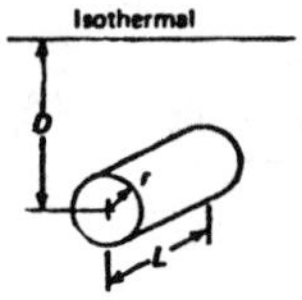	$\frac{2\pi L}{\cosh^{-1}(D/r)}$ $\frac{2\pi L}{\ln(2D/r)}$ $\frac{2\pi L}{\ln\frac{L}{r}\left[1-\frac{\ln(L/2D)}{\ln(L/r)}\right]}$	$L \gg r$ $L \gg r$, $D > 3r$ $D \gg r$, $L \gg D$
Isothermal sphere of radius r buried in infinite medium		$4\pi r$	
Isothermal sphere of radius r buried in semi-infinite medium having isothermal surface	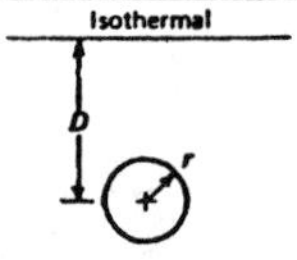	$\frac{4\pi r}{1-r/2D}$	
Conduction between two isothermal cylinders buried in infinite medium		$\frac{2\pi L}{\cosh^{-1}\left(\frac{D^2-r_1^2-r_2^2}{2r_1r_2}\right)}$	$L \gg r$ $L \gg D$

Solution: (a) In this problem, the physical system is that of a cylinder buried in a semi-infinite medium having an isothermal surface. Also, the restriction that applies is

$$D < 3r$$

and

$$L >> r.$$

The shape factor for this system is given by

$$S = \frac{2\pi\ L}{\cosh^{-1}(D/r)}$$

where D is the depth

Substituting the given values,

$$S = \frac{2\pi(4)}{\cosh^{-1}(20/7.5)}$$

$$= 15.35 \text{ cm}$$

The equation for heat loss (q) is

$$q = k\, S\, \Delta T$$

where ΔT is the temperature difference

$$= (0.8)(15.35)(75-5)$$

$$= 859.6 \text{ W } (2933 \text{ Btu/hr}).$$

• PROBLEM 2-19

(a) A steam pipe, having an outside temperature of 180^0F, is buried in the earth (k=0.60 Btu/hr ft^{0}F) at a depth of 2 feet. The diameter of the pipe is 6 inches. If the soil surface temperature is at 50^0F, calculate the heat transfer rate per foot from the pipe.

(b) If the heat transfer rate, from the pipe in part (a), is to be reduced to 100 Btu/hr ft by adding insulation (k_a=0.03 Btu/hr ft^{0}F), while the physical system remains the same, determine the required outside diameter of the insulation.

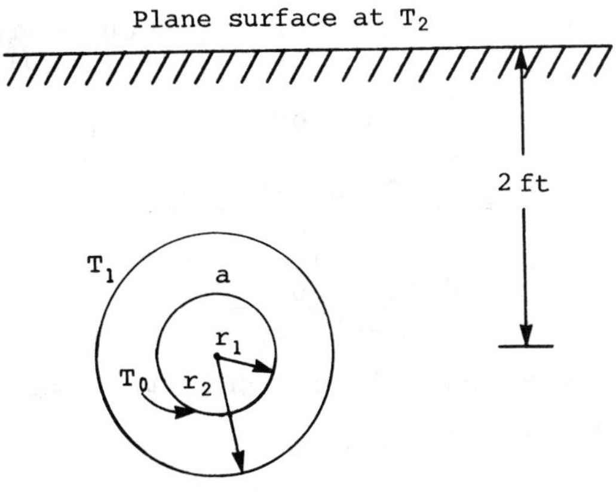

The buried steampipe.

Solution: The physical system is that of a horizontal pipe buried in a semi-infinite medium having an isothermal surface. The conduction shape factor for this system, having the restrictions $L \gg D$ and $D > 3r$ is

$$S = \frac{2\pi}{\cosh^{-1}(2z/D)}$$

The shape factor for this system is

$$S = \frac{2\pi}{\ln(2D/r)}$$

where D = Depth = 2 ft.
r = Radius = 3in.= 0.25ft.

Therefore

$$S = \frac{2\pi}{\ln(4/0.25)}$$

Shape factor $S = 2.26$

Heat loss $q = kS(T_1 - T_2)$

where T_1 = Outer surface temp=180 ^{0}F
T_2 = Ground surface temp=50 ^{0}F

Therefore

$$q = 0.6 \times 2.26(180 - 50)$$

$$q = 176.28 \text{ Btu/hr ft}$$

(b) In this case, the outside surface temperature of the insulation is unknown. An expression for heat loss (q) in terms of the shape factor for the system is

$$q = kS(T_1 - T_2) \tag{1}$$

and the expression for the conduction heat transfer rate across 'a', in the absence of transients, heat generation, or heat flow in any other direction other than radial is given as

$$q = \frac{2\pi k_a(T_0 - T_1)}{\ln(r_2/r_1)} \tag{2}$$

where r_2 = outer radius of insulation

Since $r_2/r_1 = d_2/d_1 = d_2/0.5 = 2d_2$, equation (2) becomes

$$q = \frac{2\pi(0.03)(180 - T_1)}{\ln(2d_2)}$$

$$= \frac{0.1884(180 - T_1)}{\ln(2d_2)} \tag{3}$$

Rearranging equation (1) and substituting in equation (3) we get,

$$T_1 = T_2 + q/kS \tag{1}$$

$$q = \frac{0.1884(180 - T_2 - q/kS)}{\ln(2d_2)} \tag{3}$$

$$0 = 180 - T_2 - q/kS - q \ln 2d_2/0.1884$$

$$q = \frac{180 - T_2}{5.3\ln(2d_2) + 1/kS} \qquad (4)$$

Since the depth D = 2ft., the shape factor is

$$S = \frac{2\pi}{\cosh^{-1}(2D/r_2)}$$

because in this case the restriction is only

$$L \gg r$$

The second restriction $D > 3r$, which was a valid restriction in part(a), is not applicable as the outer radius of the insulation is not known.

$$S = \frac{2\pi}{\cosh^{-1}(2/\tfrac{1}{2}d_2)}$$

$$= \frac{2\pi}{\cosh^{-1}(4/d_2)}$$

$$kS = \frac{0.6 \times 2\pi}{\cosh^{-1}(4/d_2)}$$

$$\text{and} \quad 1/kS = \frac{\cosh^{-1}(4/d_2)}{1.2\pi}$$

$$= 0.266 \cosh^{-1}(4/d_2)$$

Substituting this value of 1/kS into equation (4) gives

$$q = \frac{180 - T_2}{5.3\ln(2d_2) + 0.266\cosh^{-1}(4/d_2)}$$

substituting the values

$$T_2 = 50^0\text{F}$$

$$\text{and } q = 100 \text{ Btu/hr ft}$$

$$100 = \frac{180 - 50}{5.3\ln(2d_2) + 0.266\cosh^{-1}(4/d_2)} \qquad (5)$$

Equation (5) must be solved by the method of trial and error.

$$5.3\ln(2d_2) + 0.266 \cosh^{-1}(4/d_2) = 1.3$$

1st trial

let d_2 = 1 ft.

$$5.3\ln(2) + 0.266 \cosh^{-1}(4) = 4.22.$$

This value of d_2 is too high.

2nd trial

let d_2 = 0.60 ft.

$$5.3\ln(1.2) + 0.266 \cosh^{-1}(6.67) - 1.65$$

The value of d_2 is still high.

3rd trial

$$\text{let } d_2 = 0.55$$

$$5.3\ln(1.1) + 0.266 \cosh^{-1}(7.27) = 1.22$$

This value is very close, but lower than the required value of d_2.

4th trial

$$\text{let } d_2 = 0.56$$

$$5.3\ln(1.12) + 0.266 \cosh^{-1}(7.14) = 1.3$$

The outside diameter of the insulation should therefore be about $d_2 = 0.56$ ft. = 6.72 in.

• PROBLEM 2-20

An electro-magnetic coil of circular cross-section and constant thermal conductivity dissipates electrical energy at the rate per unit volume of $b_0(1+\beta T)$, where b_0 and β are known constants. The dissipation rate is temperature-dependent since the resistivity of the coil depends upon temperature. The outside surface of the coil is maintained at a constant temperature T_0 by a coolant. If steady state conditions prevail and axial temperature gradients are neglected, find the temperature distribution as a function of the general coordinate r.

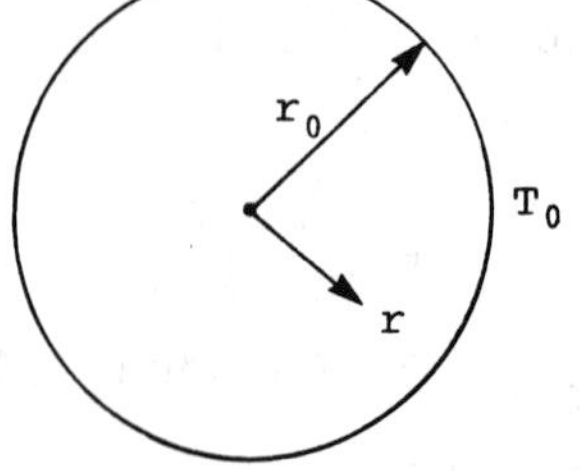

Fig. 1 Cross section of the electrical coil.

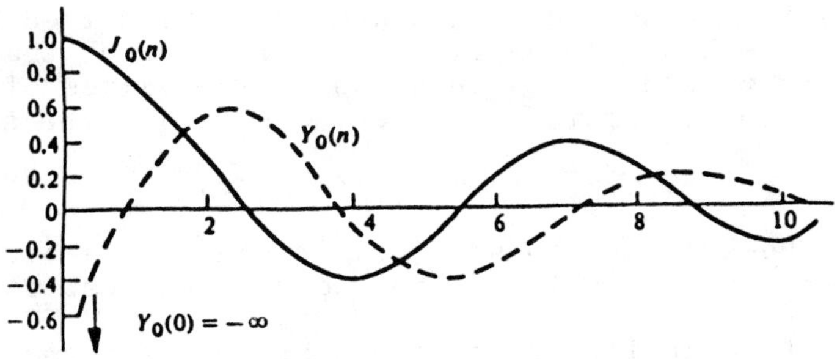

Fig. 2 Graph of the ordinary Bessel functions of the first and second kind and of zero order.

<u>Solution</u>: The governing partial differential equation for heat conduction in the circular cylindrical coordinate system is

$$\frac{\partial^2 T}{\partial r^2} + \frac{1}{r}\frac{\partial T}{\partial r} + \frac{1}{r^2}\frac{\partial^2 T}{\partial \theta^2} + \frac{\partial^2 T}{\partial z^2} + \frac{q'''}{k} = \frac{1}{\alpha}\frac{\partial T}{\partial \tau} \qquad (1)$$

Since steady state conditions prevail and there are no axial (z direction) and circumferential (θ direction) temperature gradients, eq.(1) reduces to

$$\frac{d^2 T}{dr^2} + \frac{1}{r}\frac{dT}{dr} + \frac{q'''}{k} = 0 \qquad (2)$$

The heat generation rate $q''' = b_0(1+\beta T)$, and defining $a_0 = b_0/k$, eq.(2) becomes

$$\frac{d^2 T}{dr^2} + \frac{1}{r}\frac{dT}{dr} + a_0(1+\beta T) = 0 \qquad (3)$$

Eq.(3) is simplified by making it homogenous, using the transformation

$$\phi = 1 + \beta T,$$

then

$$\frac{dT}{dr} = \frac{1}{\beta}\frac{d\phi}{dr}$$

and

$$\frac{d^2 T}{dr^2} = \frac{1}{\beta}\frac{d^2\phi}{dr^2}$$

Substituting the above values in eq.(3) yields

$$\frac{d^2\phi}{dr^2} + \frac{1}{r}\frac{d\phi}{dr} + a_0\beta\phi = 0.$$

This is a second order, linear, homogenous ordinary differential equation with variable coefficients. Multiplying every term by r^2 gives

$$r^2\frac{d^2\phi}{dr^2} + r\frac{d\phi}{dr} + a_0\beta r^2\phi = 0 \qquad (4)$$

Equation (4) is a Bessel equation and is to be solved by infinite series method. The various infinite series that satisfy it are called Bessel functions. The values of Bessel functions and solutions to Bessel equations give the solution

$$\phi = AJ_0(\sqrt{a_0\beta}\ r) + B\ y_0(\sqrt{a_0\beta}\ r) \qquad (5)$$

The ordinary Bessel functions $J_0(\sqrt{a_0\beta}\ r)$ and $y_0(\sqrt{a_0\beta}\ r)$ are of zero order and of the first and second kind respectively.

A qualitative indication of the behavior of these functions at various values of the argument n are plotted in figure 2 (where the argument $n = \sqrt{a_0\beta}\ r$).

Replacing the value of $\phi = 1+\beta T$ in equation (5), and rearranging the terms gives

$$T = AJ_0(\sqrt{a_0\beta}\ r) + \beta y_0(\sqrt{a_0\beta}\ r) - 1/\beta \qquad (6)$$

where the constants A and B absorb the $1/\beta$ term.

The boundary condition at $r = r_0 \quad T = T_0$ applied to equation (6) gives

$$T_0 = AJ_0(\sqrt{a_0\beta}\ r_0) + By_0(\sqrt{a_0\beta}\ r_0) - 1/\beta \qquad (7)$$

The boundary conditions determine the integration constants A and B, and can be found from the radial symmetry of the temperature distribution about the origin, i.e.

$$\text{at } r = 0 \qquad \frac{dT}{dr} = 0$$

Also, in figure 2, it is seen that $y_0(0) \to -\infty$, so to keep the temperature bounded, at $r = 0$, $B = 0$.

Substituting the value $B = 0$ in eq.(7)

$$A = \frac{T_0 + 1/\beta}{J_0(\sqrt{a_0\beta}\ r)} \qquad (8)$$

Combining equations (6) and (8) gives the temperature distribution in the coil as a function of r, i.e. in the radial direction

$$T = (T_0+1/\beta)\ \frac{J_0(\sqrt{a_0\beta}\ r)}{J_0(\sqrt{a_0\beta}\ r_0)} - 1/\beta \qquad (9)$$

Suppose the value of $\sqrt{a_0\beta}\ r_0$ is 2.405, then from the tabulated values of Bessel function $(J_0(n))$, the corresponding value is $J_0(2.405) = 0$. Thus the temperature in eq.(9) would be infinite, resulting in the coil material having melted or being destroyed. For a fixed value of r_0, the value of the term $\sqrt{a_0\beta}$ determines whether $\sqrt{a_0\beta}\ r_0=2.405$, and since a_0 is dependent on the electric current carried by the coil. The value of $\sqrt{a_0\beta}\ r_0$ is used to estimate the maximum or 'blow out' current for the coil under consideration.

A 1/8" steel tube well is placed in a in. diameter steam-carrying pipe as shown in the figure. The temperature of the steam is measured by inserting a mercury thermometer in the well which has been filled with oil. The indicated temperature of the steam is 355°F and the actual line pressure is 153 psia. From steam tables, at 153 psia, the temperature of saturated steam is 360°F. Thus, the indicated temperature would seem to be in error. For a steel wall temperature of 200°F and a heat transfer coefficient of 50 Btu/hr ft²°F, show that the temperature reading is not inconsistent with the pressure reading.

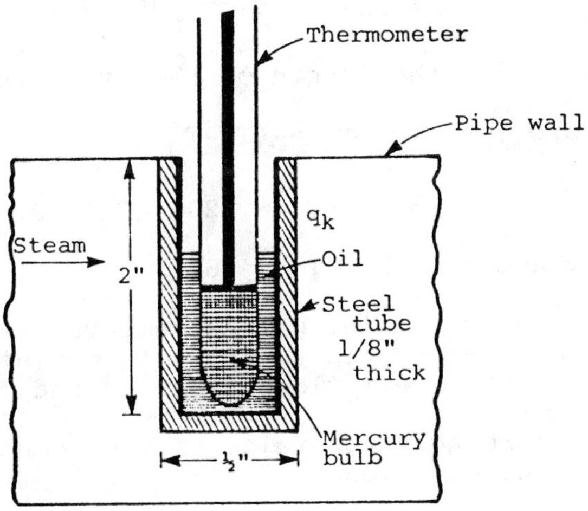

Fig. 1 Thermometer Well

Solution: Since the thermometer well is essentially a hollow tube with one end closed, heat flows from the steam at temperature T_∞ towards the cooler pipe walls at 200°F and therefore does not give the temperature of the steam, but rather the temperature at the bottom of the well.

The thermometer well may be assumed to be a hollow rod or fin, and neglect the conduction along the glass rod, since it is small as compared to the heat flowing along the steel well. The cross-sectional area of the fin is

$$A = \frac{\pi}{4}\left[d_0^2 - d_i^2\right]$$

$$= (\pi/4)\left[(1/2)^2 - (1/4)^2\right]$$

$$= 0.1475 \text{ in}^2$$

The perimeter of the tube is $= \pi D = \pi(0.5)$

$$= 1.57$$

The thermal conductivity of steel is 25 Btu/hr ft^{0}F. Hence

$$m = \sqrt{\frac{hP}{kA}}$$

$$= \sqrt{\frac{(50)(1.57)}{(25)(0.1475)}(12)}$$

$$= 16 \text{ ft.}$$

An approximate value of temperature can be obtained using the following equation,

$$T - T_\infty = (T_s - T_\infty)e^{-mx}$$

where x, the length of the steel tube, is 1/6 ft.

Therefore, $T - T_\infty = (T_s - T_\infty)e^{-16/6}$

gives $355 - T_\infty = (200 - T_\infty)0.07$

on solving, $T_\infty = 367^0F$

This is a reasonable value of the temperature, but boundary conditions that were employed are not consistent with the physical system, as heat flows from the end of the fin.

Hence a more accurate value of the temperature is obtained using the equation

$$\frac{T - T_\infty}{T_s - T_\infty} = \frac{\cosh m(L-x) + (\bar{h}L/mk)\sin h\ m(L-x)}{\cosh mL + (\bar{h}L/mk)\sinh mL}$$

For the condition $x = L$, the equation reduces to

$$T - T_\infty = (T_s - T_\infty)\frac{1}{\cosh mL + (\bar{h}L/mk)\sinh mL}$$

Substituting the numerical values,

$$T - T_\infty = (T_s - T_\infty)\frac{1}{7.18 + [50/(25)(16)]7.04}$$

$$355 - T_\infty = (200 - T_\infty)(0.124)$$

$$T_\infty = 377^0F$$

Thus the steam in the pipe is superheated and there is an error of 22^{0}F(5%). This error could be reduced by increasing the length of the well by 3 inches and placing it at an angle of 45^0 to prevent it from touching the tube wall.

• PROBLEM 2-22

> A bare steel wire (k=13 Btu/hr ft°F) is carrying a current of 1000 amp. The diameter of the wire is 0.75 in. The air temperature is 70°F and the surface heat transfer coefficient is 5.0 Btu/hr ft²°F. The electrical resistance of the wire per foot length is 0.0001 ohm. Find the centerline temperature of the wire.

Solution: The heat transfer rate from the wire is equal to the heat produced by the current (I) which is given by

$$I^2R = (1000)^2(0.0001)$$

$$= 100 \text{ watt/ft} = 341.5 \text{ Btu/hr ft}$$

This is the heat produced per foot length of wire. In terms of heat transferred per unit (ft^2) surface area, it becomes

$$q/A = \frac{341.5}{\pi} \times \frac{12}{0.75}$$

$$= 1740 \text{ Btu/hr ft}^2$$

From the equation for the heat transfer coefficient

$$h = \frac{q/A}{T_s - T_f} = \frac{-k(\partial T/\partial y)_s}{T_s - T_f}$$

the surface temperature is

$$T_s = 70 + \frac{1740}{5.0}$$

$$= 418^\circ F$$

The energy equation in cylindrical coordinates, including the heat generation per unit volume, W_i, is

$$\frac{d}{dr}\left(r \frac{dT}{dr}\right) + \frac{W_i r}{k} = 0$$

Integrating this equation twice and applying the appropriate boundary conditions gives

$$T - T_s = \frac{W_i}{4k}(r_0^2 - r^2) \qquad (1)$$

Also, it is clear that the heat generated is equal to the heat lost through the surface of the wire

$$W_i \pi r_0^2 L = \left(\frac{q}{A}\right) 2\pi r_0 L$$

or $$W_i = \frac{2}{r_0}\frac{q}{A} \tag{2}$$

Combining equations (1) and (2) gives

$$T = T_s + \frac{q}{A}\frac{r_0}{2k}$$

$$= 418 + 1740\ \frac{(0.375/12)}{2(13)}$$

$$= 420^0\text{F}$$

• PROBLEM 2-23

The lateral surface of a copper rod is insulated and its two ends are maintained at a constant temperature of 70°F. The rod is 0.2 in. in diameter and 1 ft. long and its thermal conductivity is 220 Btu/hr ft°F. Conduction in the rod is one-dimensional (along the length). Calculate the maximum electrical current that may be carried by the rod, such that the temperature at any point should not exceed 250°F. Do the calculations for two values of electrical resistivity

a) Constant at 1.73×10^{-6} ohm-cm

b) equal to $1.73[1+0.002(t-70)]\times 10^{-6}$ ohm-cm

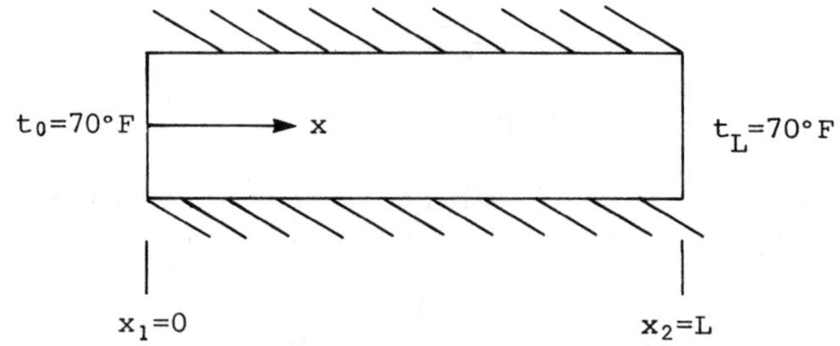

Solution: The resistance of the wire R is given as

$$R = \frac{\text{resistivity} \times \text{length}}{\text{cross-sectional area}}$$

$$= 1.73 \times 10^{-6} \times \frac{1}{2.54} \times \frac{12}{\pi\frac{(0.2)^2}{4}}$$

$$= 2.60\ \ 10^{-4}\ \text{ohm.}$$

The differential equation applicable to this case is

$$\frac{d^2t}{dx^2} + \frac{q^*}{k} = 0$$

The boundary conditions are

$$\text{at } x = x_1 \qquad t = t_1$$

$$x = x_2 \qquad t = t_2$$

Using these conditions, the differential equation can be solved. Integrating twice gives

$$t + \frac{q^*}{k}\frac{x^2}{2} = Bx + C$$

Solving for the constants by inserting the boundary conditions,

$$B = \frac{t_2-t_1}{x_2-x_1} + \frac{q^*}{2k}(x_2+x_1)$$

$$C = \frac{x_2t_1-x_1t_2}{x_2-x_1} - \frac{q^*}{2k}x_1x_2$$

Inserting these constants in the integrated expression,

$$t = \left[t_i + \frac{t_2-t_1}{x_2-x_1}(x-x_1)\right] + \left[\frac{q^*(x_2-x_1)^2}{2k}\right]\left[\frac{x-x_1}{x_2-x_1} - \left(\frac{x-x_1}{x_2-x_1}\right)^2\right]$$

Substituting the numerical values into this equation,

$$250 = \left[70 + \frac{70-70}{1}(0.5)\right] + q^*\frac{(1)^2}{2\times 220}\left[\frac{0.5}{1} - \frac{0.5}{1}^2\right]$$

from which

$$q^* = 3.168 \times 10^5 \text{ Btu/hr ft}^3$$

Also,

$$q^* = \frac{RI^2}{\text{Volume}}$$

Equating these two expressions for q^*,

$$3.168 \times 10^5 = \frac{2.60\text{x}10^{-4}\text{x}I^2\text{x}3.413}{[12\text{x}\pi\text{x}\frac{(0.2)^2}{4}]/1728}$$

from which the maximum current I is found to be 279 Amp.

(b) The given linear resistivity law yields a linear resistance

$$R = 2.60 \times 10^{-4}[1 + 0.002(t-70)]$$

The heat generation will also depend linearly on temperature, with the rate constant $b=0.002\ 1/^\circ F$. This is because linearly dependent heat generation follows the relation

$$q^* = q^*_L[1+b(t-t_L)]$$

The differential equation in this case is

$$\frac{d^2t}{dx^2} + \frac{q_L^*}{k}[1+b(t-t_L)] = 0$$

The boundary conditions are

At x=0: $\frac{dt}{dx} = 0$

At x=L: $t = t_L$

This differential equation when solved gives the expression for q* and also the expression

$$t_0 = t_L + \frac{1}{b}\left(\frac{1}{\cos mL} - 1\right)$$

where $m = \sqrt{b\frac{q_L^*}{k}}$

Substituting the numerical values in this equation gives

$$250 = 70 + \frac{1}{0.002}\left[\frac{1}{\cos(0.5m)} - 1\right]$$

and

$$m = 1.489 \text{ 1/ft.}$$

Therefore,

$$q_L^* = (1.489)^2 \frac{220}{0.002}$$

$$= 2.44 \times 10^5 \text{ Btu/hr ft}^3$$

The quantity q_L^* is the dissipated energy corresponding to the coefficient in the above resistance law. Again, equating both the expressions for q_L^*

$$q_L^* = 2.44 \times 10^5 = \frac{2.60\text{x}10^{-4}\text{x}I^2\text{x}3.413}{[12\text{x}\pi\text{x}\frac{(0.2)^2}{4}]/1728}$$

or

$$I = 245 \text{ Amp.}$$

• PROBLEM 2-24

A heat exchanger is to be designed for the following specifications:

Hot gas temperature = 1900°F
Cold gas temperature = 100°F
Unit surface conductance on the hot side $\bar{h}_1$
= 40 Btu/hr sq ft°F
Unit surface conductance on the cold side $\bar{h}_3$
= 50 Btu/hr sq ft°F

Find the maximum thermal resistance permissible per square foot of metal wall between the hot gas and the cold gas, so that the maximum temperature of the wall does not exceed 1000°F.

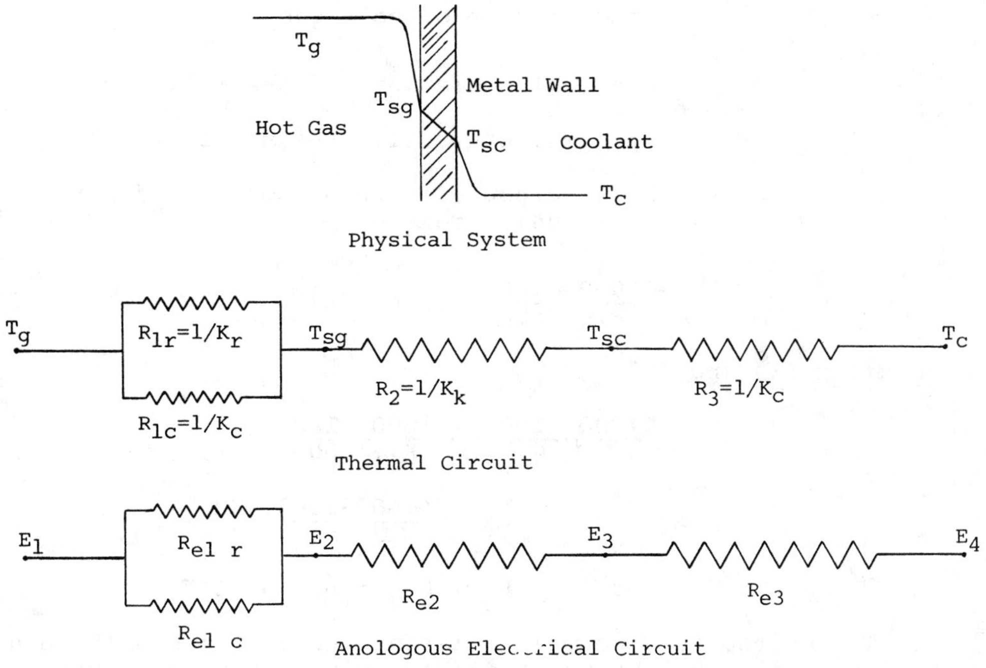

Thermal circuit and analogous electric circuit for heat flow from hot gas through a metal wall to a coolant.

Solution: Under steady state conditions the heat transfer rate from the hot gas to the hot side of the wall is equal to the heat transfer rate through the wall to the cold gas.

Heat transfer from the hot gas to the hot side of the wall is given by

$$\frac{q}{A} = \frac{t_g - t_{sq}}{AR_1}$$

where R_1 = thermal convective resistance = $1/h_1$

T_g = temperature of the hot gas

T_{sq} = the maximum surface temperature on the hot gas side

$$= \frac{1900 - 1000}{1/40} \text{ Btu/hr ft}^2 \qquad (1)$$

Heat dissipation from the wall (cold side) to the cold gas is given by

$$\frac{q}{A} = \frac{T_{sq} - T_c}{A(R_2 + R_3)}$$

T_c = temperature of cold gas

R_2 = thermal resistance of the wall

R_3 = thermal convective resistance = $1/h_3$

A =1 ft^2 since q is to be determined per unit area

$$= \frac{1000 - 100}{R_2 + 1/50} \text{ Btu/hr ft}^2 \qquad (2)$$

Equating (1) and (2)

$$\frac{1900 - 1000}{1/40} = \frac{1000 - 100}{R_2 + 1/50}$$

$$R_2 = \frac{(1000-100)1/40}{(1900-1000)} - \frac{1}{50}$$

or $R_2 = 0.005$ hr sq ft^{0}F/Btu

Therefore, the thermal resistance (R_2) of the wall should be less than 0.005 hr sq ft ^{0}F/Btu, else the temperature of the wall would increase above the permissible limit of 1000^{0}F.

• PROBLEM 2-25

A sphere having equispaced nodes at a distance Δr is in a steady state condition with a heat generation rate q'''. Derive the finite difference equation for node 0 if the heat flow is in the radial direction. Also find the finite difference equation for node 3, if the sphere is immersed in a fluid at temperature T_f. The average surface heat transfer coefficient is h and the thermal conductivity (k) is constant.

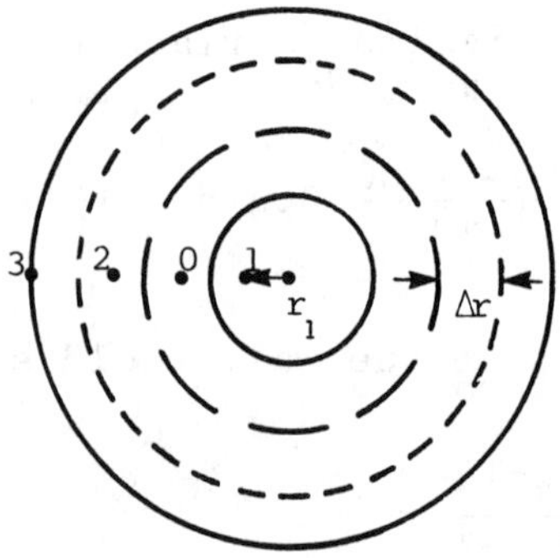

Solution: The volume (V_0) of material enclosed by the dashed surface such that the node 0 is the center point of the shell having radii ($r_0+\frac{1}{2}\Delta r$) and ($r_0-\frac{1}{2}\Delta r$), is given by

$$V_0 = \frac{4}{3}\pi\left[(r_0+\tfrac{1}{2}\Delta r)^3-(r_0-\tfrac{1}{2}\Delta r)^3\right]$$

Now an energy balance is made on volume V_0. The conservation of energy equation is

$$Q_{in} + Q_{gen} = Q_{out} + Q_{stor} \tag{1}$$

$$Q_{gen} = q_0''' V_0 = q_0'''\left(\frac{4}{3}\right)\pi\left[(r_0+\tfrac{1}{2}\Delta r)^3-(r_0-\tfrac{1}{2}\Delta r)^3\right] \tag{2}$$

The heat (Q_{in}) enters the shell associated with node 0, by conduction through the material of the surrounding nodes 1 and 2. The heat transfer rate by conduction from node 1 to 0 is given by the equation

$$q_{10} = \frac{4\pi k(T_1-T_0)}{1/r_1-1/r_0} \tag{3}$$

Similarly, the heat conducted into the shell of node 1 through the material surrounding node 2, is given by

$$q_{20} = \frac{4\pi k(T_2-T_0)}{1/r_0-1/r_2} \tag{4}$$

Since $Q_{stor} = 0$, substituting equations (2), (3) and (4) into equation (1) yields

$$\frac{4\pi k(T_1-T_0)}{1/r_1-1/r_0} + \frac{4\pi k(T_2-T_0)}{1/r_0-1/r_2} + \frac{4q_0'''}{3k}\pi\left[(r_0+\tfrac{1}{2}\Delta r)^3-(r_0-\tfrac{1}{2}\Delta r)^3\right] = 0$$

On rearranging the terms in the equation, the governing finite difference equation for the node 0 becomes

$$\frac{T_1-T_0}{1/r_1-1/r_0} + \frac{T_2-T_0}{1/r_0-1/r_2} + \frac{q_0'''}{3k}\left[(r_0+\tfrac{1}{2}\Delta r)^3-(r_0-\tfrac{1}{2}\Delta r)^3\right] = 0$$

For node 3, the volume (V_{03}) of the material associated with it is given by

$$V_{03} = \frac{4}{3}\pi\left[r_3^3-(r_3-\tfrac{1}{2}\Delta r)^3\right]$$

Application of equation (1) to the control volume V_{03} gives

$$Q_{gen} = q_0''' \ \frac{4}{3}\pi\left[r_3^3-(r_3-\tfrac{1}{2}\Delta r)^3\right] \tag{5}$$

Heat enters the control volume by conduction from node 2 and by convection from the fluid at temperature T_f. Hence, the heat conducted into the control volume is

$$q_{23} = \frac{4\pi k(T_2-T_3)}{1/r_2-1/r_3} \quad (6)$$

and the heat transferred by convection from the fluid is determined by the Newton's law of cooling, giving

$$q_c = h(4\pi r_3^2)(T_f-T_3) \quad (7)$$

Substituting equations (5 to 7) into equation (1) gives the governing finite difference equation for node 3,

$$\frac{T_2-T_3}{1/r_2-1/r_3} + \frac{hr_3^2}{k}(T_f-T_3) + \frac{q_3'''}{3k}\left[r_3^3 - (r_3-\tfrac{1}{2}\Delta r)^3\right] = 0$$

HEAT TRANSFER THROUGH FINS

• PROBLEM 2-26

From economy considerations, suggest the use of either aluminium alloy or copper alloy fins. The following data is given:

Density of the aluminium alloy = 2400 kg/m^3

Density of the copper alloy = 8400 kg/m^3

Conductivity of the aluminium alloy = 78 $kCal/m\text{-}hr^0C$

Conductivity of the copper alloy = 190 $kCal/m\text{-}hr^0C$

The cost of aluminium alloy is twice the cost of copper alloy.

Solution: The ratio of the weights of copper and aluminum alloys for fins is proportional to the ratio of their densities. Also, from basic heat transfer considerations, this ratio is inversely proportional to the ratio of their thermal conductivities. Hence, we have

$$\frac{w_1}{w_2} = \frac{\rho_1}{\rho_2}\,\frac{k_2}{k_1} \quad (1)$$

where suffix 1 and 2 represent copper and aluminium alloys, respectively.

From the relation (1)

$$w_1 = w_2 \times \frac{8400}{2400} \times \frac{78}{190}$$

$$= 1.43\ w_2$$

It is clear that the weight of copper alloy would be 1.43 times the weight of aluminium alloy for the same amount of heat transfer. If the cost of copper alloy is P per unit weight, then the cost of aluminium alloy will be 2P per unit weight.

Therefore, the cost of aluminium alloy $= 2P\ w_2$

cost of copper alloy $= Pw_1$

$w_1 = 1.43\ w_2$, hence the cost of

copper alloy $= 1.43\ w_2P$

Finally, we have

cost of aluminium alloy $= 2Pw_2$

cost of copper alloy $= 1.43\ Pw_2$

It is clear that the choice of copper alloy would be more economical.

• PROBLEM 2-27

The maximum heat dissipation from a rectangular profile fin, designed for minimum weight, is 500 k cal per hour per meter width of the fin. The fin base temperature and the ambient air temperature are 110°C and 10°C, respectively. The heat transfer coefficient on the fin surface is 90 k cal/m^2-hr-°C and the thermal conductivity of the fin material is 35 k cal/m-hr-°C. Determine the height and thickness of the fin required. Assume the thermal gradient at the fin base is zero.

Solution: The weight of a fin, for a given maximum heat flow, is minimum when the fin profile area is minimum. The expression relating heat transfer to area for a fin of minimum profile area is given as follows.

$$A_f = \frac{2.109}{4w^3h^2k}\left(\frac{Q}{\theta_0}\right)^3 \qquad (1)$$

where

w = width of the fin

θ_0 = temperature difference between fin base and the surroundings

The condition for maximum heat flow for a given profile area is expressed as follows.

$$\frac{L}{\delta} = \frac{1.419}{\sqrt{B_n}} \qquad (2)$$

where

$$B_n = \frac{h\delta}{k} \text{ is the "Biot Number"}$$

$$2\delta = \text{thickness of the fin in meters}$$

From equation (2)

$$L = 1.419\sqrt{\frac{ks}{h}}$$

$$= 1.419\sqrt{\frac{35\times1000\delta}{90}}$$

$$= 27.98\ \sqrt{\delta}$$

$$A_f = L \times 2\delta$$

$$= \frac{2.109}{4\times1^3\times90\times90\times35}\left(\frac{500}{100}\right)^3 \ m^2$$

$$= 2.32 \times 10^{-4} m^2$$

$$= 232\ mm^2$$

Using $L = 27.98\sqrt{\delta}$ and $A_f = L \times 2\delta = 232\ mm^2$ we have

$$27.98\sqrt{\delta} \times 2\delta = 232$$

or

$$\delta^{3/2} = 4.14$$

or

$$\delta = 2.57$$

Hence, the thickness of the fin = 2δ = 2×2.57 = 5.14

Length of the fin $L = 27.98\sqrt{\delta}$

$$= 27.98\sqrt{2.57}$$

$$= 44.85\ mm$$

• **PROBLEM 2-28**

A fin of length L having an insulated tip, has a uniform initial temperature T_∞, equal to that of the ambient. The base temperature is suddenly changed to T_0 and is maintained at that temperature. Find a first order solution that gives the approximate temperature variation in the fin.

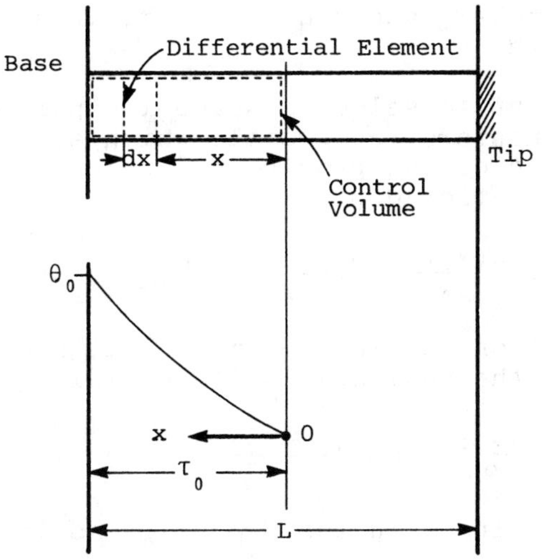

Fig. 1

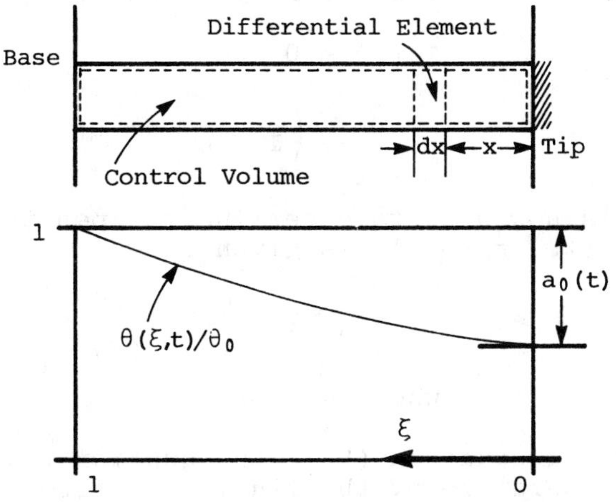

Fig. 2

<u>Solution</u>: The solution involves the penetration depth , and its formulation may be given in two successive time domains. In the first domain, the penetration depth is less than or equal to the length of the fin,and in the second domain, the temperature of the tip rises from zero to a steady value.

Consider the lumped control volume and the differential system shown in Fig. 1, the integral formulation of the problem for the first time domain is

$$\frac{1}{a}\frac{d}{dt}\int_0^{\tau_0}\theta dx = \left(\frac{\partial\theta}{\partial x}\right)_{x=\tau_0} - m_2\int_0^{\tau_0}\theta dx, \tag{1}$$

where $\theta = T-T_\infty$ and x has its origin at the penetration depth.

It is convenient to select a spacewise parabolic, timewise unspecified Kantorovich profile which satisfies the boundary conditions of the problem and which is expressed in terms of the penetration depth

$$\frac{\theta(x,t)}{\theta_0} = \left(\frac{x}{\tau_0}\right)^2. \tag{2}$$

Insertion of equation (2) into equation (1) and subsequent integration gives the nonlinear differential equation

$$\frac{d\tau_0}{dt} + am^2\tau_0 = \frac{6a}{\tau_0},$$

which can be made linear in terms of $\tau_0^2/2$ as follows:

$$\frac{d(\tau_0^2/2)}{dt} + 2am^2(\tau_0^2/2) = 6a. \tag{3}$$

Equation (3) is to be satisfied by the condition

$$\tau_0(0) = 0 \tag{4}$$

and the solution is

$$\tau_0^2 = \frac{6}{m^2}\left(1 - e^{-2am^2t}\right), \tag{5}$$

This equation is now used to determine the penetration time, t_0, at which $\tau_0(t_0) = L$. It is given as

$$t_0 = \frac{1}{2am^2}\ln\left(\frac{1}{1-\mu^2/6}\right), \tag{6}$$

where $\mu = mL$.

Now, inserting equation (5) into equation (2) and rearranging, the temperature of the fin in the first domain is obtained,

$$\frac{\theta(x,t)}{\theta_0} = \frac{m^2x^2}{6[1 - \exp(-2am^2t)]} \tag{7}$$

The differential and lumped systems for the second time domain are shown in Fig. 2.

Since the tip of the fin is insulated, rather than referring to figure (2), the integral formulation of the problem for the second time domain is obtained by replacing τ_0 by L in equation (3).

$$\frac{1}{a}\frac{d}{dt}\int_0^L \theta dx = \left(\frac{\partial\theta}{\partial x}\right)_{x=L} - m^2\int_0^L \theta dx.$$

Now, introducing the dimensionless variable $\xi = x/L$ and the parameter $\mu = mL$ in equation (8), which on rearranging gives

$$\frac{L^2}{a}\frac{d}{dt}\int_0^1 \theta d\xi = \left(\frac{\partial\theta}{\partial\xi}\right)_{\xi=1} - \mu^2 \int_0^1 \theta d\xi.$$

The first order Ritz profile is employed, a_0 is a time dependent parameter, and the function $a_0(t)$ is to be determined, therefore

$$\theta(\xi,t)/\theta_0 = 1 - (1 - \xi^2)a_0(t). \tag{10}$$

This is a first order Kantorovich profile. Inserting equation (10) into equation (9) and integrating, we get

$$\frac{da_0}{dt} + \frac{3a}{L^2}\left(1 + \frac{\mu^2}{3}\right)a_0 = \frac{3a}{L^2}\left(\frac{\mu^2}{2}\right). \tag{11}$$

The initial condition to be employed on equation (11) is

$$a_0(t_0) = 1 \tag{12}$$

Then the solution of equation (11) which satisfies eq. (12) is

$$a_0(t) = \frac{\mu^2/2}{1+\mu^2/3} + \frac{1 - \mu^2/6}{1 + \mu^2/3} \exp\left[-3\left(1 + \frac{\mu^2}{3}\right)\frac{a(t-t_0)}{L^2}\right]. \tag{13}$$

Introducing equation (13) into equation (10) gives the temperature of the fin in the second time domain,

$$\frac{\theta(\xi,t)}{\theta_0} = 1 - (1-\xi^2)\left\{\frac{\mu^2/2}{1+\mu^2/3} + \frac{1 - \mu^2/6}{1 + \mu^2/3}\exp\left[-3\left(1 + \frac{\mu^2}{3}\right)\frac{a(t-t_0)}{L^2}\right]\right\}. \tag{14}$$

As $t \to t_0$, equation (14) approaches the upper limit of the first time domain solution, equation (2) for τ_0=L, and as $t \to \infty$, it tends to the steady solution given by the equation

$$\frac{\theta(\xi)}{\theta_0} = 1 - \frac{\mu^2/2}{1 + \mu^2/3}(1 - \xi^2) \tag{15}$$

The error of equation (14) is of the order of that obtained in eq.(15).

• PROBLEM 2-29

Determine whether it is advantageous to have thin iron fins (say 0.30 cm), on a heating surface.

Solution: The heat flow through the base of a fin is given by the equation

$$Q_1 = m\,KA\,\theta_1\left[\frac{h_2/mk+\tanh m\ell}{1+(h_2/mk)\tanh m\ell}\right] \tag{1}$$

where h_2 is the film heat transfer coefficient at the end of the fin. When the heat loss through the end is insignificant compared to the other surfaces, this equation reduces to a simpler and more familiar form,

$$Q_1 = m\theta_1 \tan h\, m\ell$$

The condition for which fins are most advantageous is given by

$$\frac{dQ_1}{d\ell} = 0$$

Taking the factors k, A, m and θ_1 as constants, and differentiating equation (1), the resulting differentiation becomes zero when either the numerator is zero or the denominator is infinite. The second condition results in a trivial statement only. Hence considering the numerator

$$\left(1 + \frac{h_2}{km} \tanh m\ell\right) \frac{m}{\cosh^2 m\ell} - \left(\frac{h_2}{km} + \tanh m\ell\right) \frac{h_2/k}{\cosh^2 m\ell} = 0$$

which on simplification gives

$$m - \frac{h_2^2}{k^2 m} = 0$$

where

$$m = \sqrt{\frac{hc}{kA}} = \sqrt{\frac{2h}{kb}}$$

and when

$$h = h_2$$

$$\frac{2k}{hb} = 1$$

or

$$\frac{1}{h} = \frac{b/2}{k}$$

The term 1/h is the film heat transfer resistance, and $\frac{b/2}{k}$ is the thermal resistance of conduction for a plane wall of thickness equal to one-half the fin thickness. Beyond the point at which both resistances are equal, there is no advantage in using fins. The condition for advantageous use of fins is given by

$$\frac{2k}{hb} > 1$$

A reasonable cutoff point which is generally used is

$$\frac{2k}{hb} > 5$$

The average value of the thermal conductivity of iron is taken to be $k=55\ W/m^0K$. When heat is transferred to air, the range of the heat transfer coefficient is

$$10 < h < 100\ W/m^{2\,0}K.$$

Using the higher value

$$h = 100\ W/m^{2\,0}K.$$

The value of the characteristic factor is

$$\frac{2k}{hb} = \frac{2 \times 55}{100 \times 0.0030}$$

$$= 367$$

Thus for heat transfer to air, the usage of fins is advantageous. For heat transfer to water, the range of the heat transfer coefficient is

$$550 < h < 5500\ W/m^2\ {}^0K$$

Taking the higher value of $h=5500\ W/m^{2\,0}K$, the value of the characteristic factor is given by

$$\frac{2k}{hb} = \frac{2 \times 55}{5500 \times 0.0030}$$

$$= 6.67$$

The addition of fins to the heating surface,having water as the fluid will not substantially increase the rate of heat transfer. For heat transfer to liquids, thin fins of metals having high thermal conductivity should be used.

• PROBLEM 2-30

A radial fin is attached to the outside of a steel tube which has an outside surface temperature of $T_0=200^0F$. The fin is of rectangular section having a thickness of 5/16 inch, and the inside and outside radii are 2 inches and 4 inches, respectively. The heat transfer coefficient is given by

$$h = 0.29(T - T_\infty)^{0.25}$$

where h is in $Btu/hr\ ft^{2\,0}F$ and T and T_∞ are in ^0F. Find the temperature distribution in the fin by finite differences using a 19 node model. The temperature of the surroundings is 100^0F and the fin material has a thermal conductivity $k=1.0\ Btu/hr\ ft^0F$.

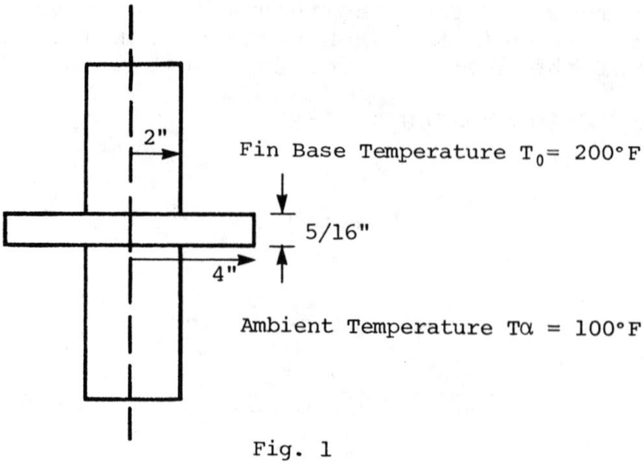

Fig. 1

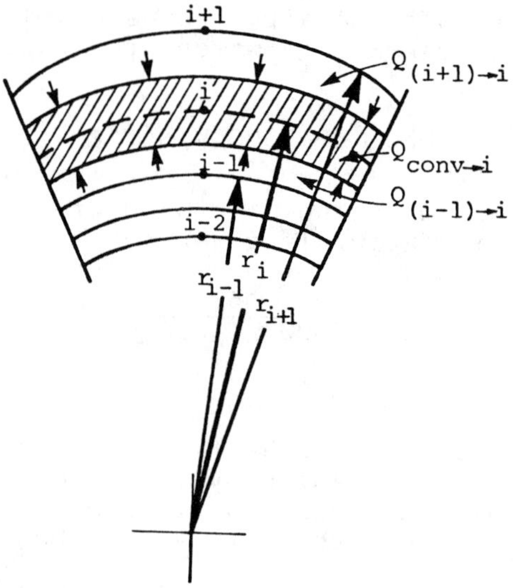

Fig. 2

Solution: The fin is divided into a grid and the layout is shown in Fig. 2.

Assuming the fin to be a ring of width Δr and thickness t, and doing an energy balance at the ith node gives

$$Q_{(i-1)\to i} = kt2\pi\left(\frac{r_{i-1}+r_i}{2}\right)\left(\frac{T_{i-1}-T_i}{\Delta r}\right)$$

$$Q_{(i+1)\to i} = kt2\pi\left(\frac{r_i+r_{i+1}}{2}\right)\left(\frac{T_{i+1}-T_i}{\Delta r}\right)$$

$$Q_{conv \to i} = 2h_i 2\pi r_i \Delta r(T_\infty - T_i)$$

The law of conservation of energy as applied to steady state conditions may be written as

$$Q_{(i-1)\to i} + Q_{(i+1)\to i} + Q_{conv\to i} = 0$$

Substituting and simplifying, we get

$$kt\left(\frac{r_{i-1}+r_i}{\Delta r}\right)T_{i-1} - \left[\frac{kt}{\Delta r}(r_{i-1} + 2r_i + r_{i+1}) + 4h_i r_i \Delta r\right] T_i$$

$$+ kt\left(\frac{r_{i+1}+r_i}{\Delta r}\right)T_{i+1} = -4h_i r_i \Delta r T_\infty, \text{ for } i=2,3,\ldots,18$$

At the radius r_1=2 inches, the temperature T_1=200°F. Performing an energy balance on the outermost node, i.e. at r_{19}=4 inches, yields

$$kt2\pi\left(\frac{r_{18} + r_{19}}{2}\right)\left(\frac{T_{18} - T_{19}}{\Delta r}\right) + h_{19}\,2\pi r_{19}\,t(T_\infty - T_{19})$$

$$+ 2h_{19}\,2\pi\underbrace{\left[\frac{1}{2}\left(\frac{r_{18} + r_{19}}{2} + r_{19}\right)\right]}_{\text{This is the mean radius of the outermost element.}}\frac{\Delta r}{2}(T_\infty - T_{19}) = 0$$

which on rearranging and simplification gives the equation

$$\frac{kt}{\Delta r}(r_{19} + r_{18})T_{18} - \left[\frac{kt}{\Delta r}(r_{19} + r_{18}) + 2h_{19}r_{19}t + 2h_{19}\left(\frac{1}{4}r_{18} + \frac{3}{4}r_{19}\right)\cdot\Delta r\right]T_{19}$$

$$= -\left[2h_{19}r_{19}t + 2h_{19}\left(\frac{1}{4}r_{18} + \frac{3}{4}r_{19}\right)\Delta r\right]T_\infty$$

The temperature distribution in the fin can be obtained by writing a program for the derived equations and running the program on a computer. The convective heat transfer coefficient is not constant and varies, nonlinearly, with the surface temperature. The solution is obtained by an iterative method where the error or tolerance between the successive iterations are obtained and checked with the given tolerance level.

The program starts by reading the temperatures T_0 and T_∞ and solving for h_0, by assuming $h=0.29(T_0-T_\infty)^{0.25} = h_0$ and then calculating all the temperatures at nodes 2 through 19. At this juncture, the difference in the value of T_is of two successive iterations would not be within the prescribed tolerance.

Next, h for nodes 1 through 19 are recalculated using the new temperatures and the two sets of values of h_is are compared to check if they are within the tolerance limit. In the first iteration they may not be within the limit, then the new set of h_is is used to calculate the temperatures. This procedure is repeated until the values of h_is and T_is are within the tolerance level.

Variables in the Fortran Program:

Given below is a list of variables to be used in the computer program.

T(I): Temperatures T_i, i = 1 to 19, ^0F

TN(I): New set of iterated values of temperatures, ^0F

TZERO: Temperature T_0, at the base of the fin, ^0F

TINF: Temperature of ambient air, T_∞, ^0F

H(I): Convective heat transfer coefficients, h_i, for various nodes, Btu/hr-ft^2 ^0F

HH(I): New set of iterated values of heat transfer coefficients, Btu/hr-ft^2 ^0F

R(I): Radius of the ith node, r_i, ft(Input in inches)

DELR: Radial distance between two successive nodes, Δr,ft

TH: Thickness of the radial fin, s, ft(Input in inches)

TOL: Tolerence on the values of T_is

HTOL: Tolerence on the values of h_is

QFIN: Rate of heat dissipation from the radial fin, Btu/hr

The following points are to be noted in the given program.

Line 16: ERR, which is the criterion for the convergence of T_is is set at 0.01. ERR equals TOL in line 46, when the convergence of h_is is sufficient.

Line 17: ITEST is assigned a value of 1 until convergence on h_is is obtained.

Line 18: ICOUNT is a counter for the number of iterations.

Lines 39 & 40: Convergence test is applied to iterated values of T_is.

Lines 44 & 45: Convergence test is applied to all iterated values of h_is.

Given below is a computer program for solving the problem using the iterative method.

```
C  RADIAL FIN ANALYSIS - VARIABLE HEAT TRANSFER COEFFICIENT
      DIMENSION R(19),T(19),TN(19),H(19),HH(19)
      REAL K
      PI=3.141592654
      IR=105
      IW=108
      READ (IR,100) TH,R(1),R(19),TZERO,TINF,K,TOL,HTOL
  100 FORMAT(F10.4)
      WRITE (IW,101) TH,R(1),R(19),TZERO,TINF,K
  101 FORMAT (1H1, 'THICKNESS OF THE FIN       =',F6.4,'    INCH',//
     1,'INNER RADIUS OF THE FIN    =',F6.2,' INCHES',//
     2,'OUTER RADIUS OF THE FIN    =',F6.2,' INCHES',//
     3,'FIN-BASE TEMPERATURE       =',F6.2,' DEG. F',//
     4,'AMBIENT TEMPERATURE        =',F6.2,' DEG. F',//
     5,'THERMAL CONDUCTIVITY       =',F7.2,' BTU/DEG.F,HR,FT',//)
      ERR=0.01
      ITEST=1
      ICOUNT=0
      T(1)=TZERO
      TN(1)=TZERO
      R(1)=R(1)/12.                      }
      R(19)=R(19)/12.                    }  Conversion of units.
      TH=TH/12.                          }
      DELR=(R(19)-R(1))/18.  ----------->   Defines Δr.
      DO 200 I=1,19                         Calculates initial guess for T_i's.
      T(I)=T(1)+(TINF-TZERO)*(I-1)/18.      Calculates r_i's.
      R(I)=R(1)+(I-1)*DELR                  Calculates initial guess for h_i's.
  200 H(I)=0.29*((T(I)-TINF)**0.25)
  600 DO 201 I=2,18                      }
      C1=K*TH*(R(I-1)+R(I))/DELR         }  Coefficients in the equations for
      C2=K*TH*(R(I+1)+R(I))/DELR         }  internal nodes are calculated.
      C3=4.*H(I)*R(I)*DELR               }
      C4=C1+C2+C3                        }
  201 TN(I)=(C1*T(I-1)+C2*T(I+1)+C3*TINF)/C4
      C1=K*TH*(R(19)+R(18))/DELR
      C2=2.*H(19)*R(19)*TH+H(19)*(.25*R(18)+.75*R(19))*DELR*2.
      C3=C1+C2
      TN(19)=(C1*T(18)+C2*TINF)/C3
      DO 202 J=2,19
  202 IF (ABS((TN(J)-T(J))/TN(J)) .GT. ERR) GO TO 310
      IF (ITEST.EQ.0) GO TO 700
      DO 203 J=1,19                      }
  203 HH(J)=.29*((TN(J)-TINF)**.25)      }  Calculate new h_i's.
      DO 204 KK=1,19
  204 IF (ABS((HH(KK)-H(KK))/HH(KK)) .GT. HTOL) GO TO 300
      ITEST=0
      ERR=TOL
  300 DO 250 J=1,19                      }
  250 H(J)=HH(J)                         }  Sets h_old equal to h_new
  310 DO 205 I=2,19                      }  and T_old equal to T_new.
  205 T(I)=TN(I)                         }
      ICOUNT=ICOUNT+1                    }
      GO TO 600
  700 WRITE(IW,102)
  102 FORMAT(///,'    I        T(I)',//)
      WRITE (IW,103)((I,TN(I)),I=1,19)
  103 FORMAT(I4,4X,F8.2,/)
      TZ=.75*T(1)+.25*T(2)
      QFIN=((T(1)-T(2))/DELR)*K*TH*2*PI*(R(1)+R(2))/2.
     1+.29*((TZ-TINF)**.25)*(TZ-TINF)*4.*PI*(.75*R(1)
     2+.25*R(2))*DELR
      TOL=TOL*100.
      WRITE(IW,104) ICOUNT,TOL
  104 FORMAT(//,'SOLUTION IS OBTAINED AFTER  ',I6,'  ITERATIONS
     1',//,'SUCCESSIVE TEMPERATURE DISTRIBUTIONS ARE WITHIN  ',
     2F6.3,'  PERCENT',//)
      WRITE(IW,105) QFIN
  105 FORMAT(//,'RATE OF HEAT TRANSFER IS  ',F8.2,'  BTUS/HR')
      STOP
      END
```

Results obtained from a run of the above program are as follows:

Thickness of the Fin	= 0.3125 Inch
Inner Radius of the Fin	= 2.00 Inches
Outer Radius of the Fin	= 4.00 Inches
Fin-Base Temperature	= 200.00 Deg. F
Ambient Temperature	= 100.00 Deg. F
Thermal Conductivity	= 1.00 Btu/Deg.F, Hr, Ft

1	T(I)	10	154.03
1	200.00	11	151.51
2	192.00	12	149.34
3	184.93	13	147.48
4	178.69	14	145.92
5	173.18	15	144.64
6	168.32	16	143.61
7	164.02	17	142.82
8	160.24	18	142.26
9	156.03	19	141.92

Solution is obtained after 1441 iterations

Successive temperature distributions are within .0001%

Rate of heat transfer is 25.99 Btu/hr

The temperature variation along the radius of the fin is shown graphically in the figure below.

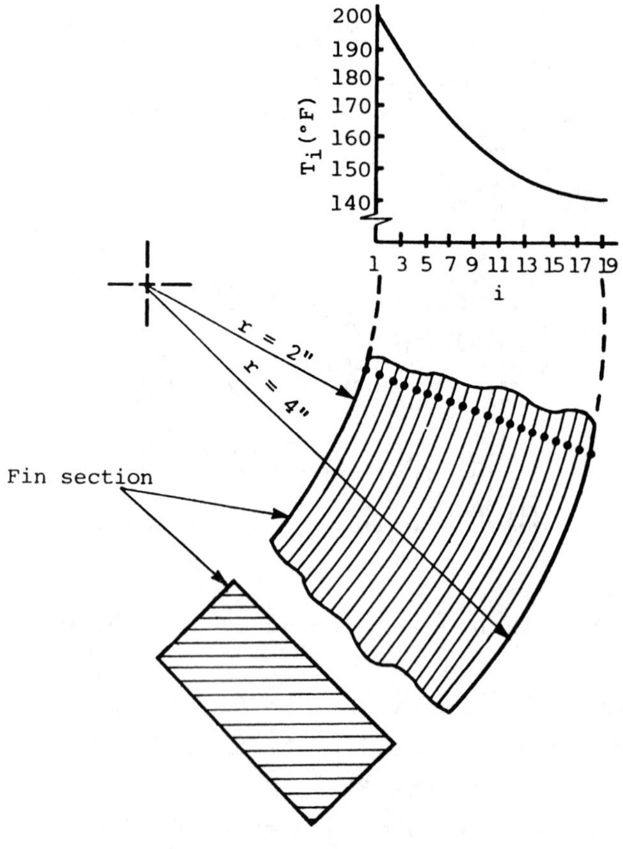

Fig. 3 Variation in fin temperature for a circumferential fin.

A two-dimensional fin of thickness ℓ is moving with a constant velocity V in the x-direction. The temperature at the base of the fin is F(y) and the surrounding temperature is T_∞. Assuming that the fin is infinitely long and the heat transfer coefficient is large, determine with respect to the stationary coordinate system shown in Fig. 1, the steady temperature. The effects of conduction along the x-direction should also be considered.

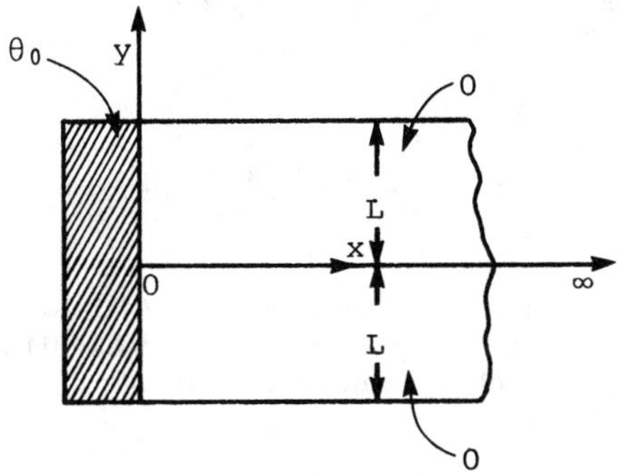

Fig. 1

Solution: The differential equation for the conditions given in the problem is

$$2s\frac{\partial\theta}{\partial x} = \frac{\partial^2\theta}{\partial y^2} \qquad \text{where} \qquad 2s = \frac{\rho c V}{k}$$

and the boundary conditions are

$$\theta(0,y) = \theta_0, \qquad \frac{\partial\theta(x,0)}{\partial y} = 0, \qquad \theta(x,\ell) = 0$$

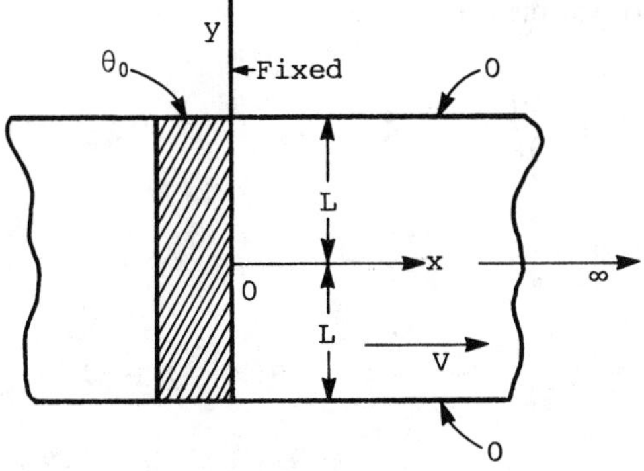

Fig. 2

This results in the product solution

$$\frac{d^2Y}{dy^2} + \lambda^2 Y = 0; \qquad \frac{dY(o)}{dy} = 0, \quad Y(\ell) = 0$$

$$2s\frac{dX}{dx} + \lambda^2 X = 0$$

Thus, when the boundary conditions and the other variables remain unchanged, the separation of variables gives the same equation for this direction, irrespective of the boundary conditions of the other direction.

The solution in the x-direction is

$$X_n(x) = B_n\, e^{-\lambda^2_n x/2s}$$

and the product solution is

$$\theta(x,y) = \sum_{n=0}^{\infty} a_n e^{-\lambda_n x/2s} \cos\lambda_n y \qquad (1)$$

Application of the non-homogeneous boundary condition to equation (1) gives the value of a_n. Substituting this value back into equation (1), the temperature of the fin is obtained in the following form:

$$\frac{\theta(x,y)}{\theta_0} = \frac{2}{\ell}\sum_{n=0}^{\infty} \frac{(-1)^n}{\lambda_n} e^{-\lambda_n^2 x/2s} \cos\lambda_n y \qquad (2)$$

The first of the two effects of conduction in the axial direction, often taken as the only effect, is the addition of $\partial^2\theta/\partial x^2$ to the differential equation; the second effect is the penetration of conduction outside the region of integration, and the problem becomes one of two regions or domains. The location x=0 is now the interface between the two domains and hence the temperature at that location can no longer be specified. The natural boundary condition, equality of heat fluxes and temperature, is also required to be specified. At large distances from $x/_{x=0}$, in the negative x-direction and also at the fin surface for $x < 0$, the temperature may be taken to be θ_0. The temperature distribution, in the region $x < 0$, is given by $\theta(x,y)$ and in the region $x > 0$ by $\Psi(x,y)$. Using this, the problem becomes

$$\frac{\partial^2\psi}{\partial x^2} - 2s\frac{\partial\psi}{\partial x} + \frac{\partial^2\psi}{\partial y^2} = 0, \qquad x < 0$$

$$\frac{\partial^2\theta}{\partial x^2} - 2s\frac{\partial\theta}{\partial x} + \frac{\partial^2\theta}{\partial y^2} = 0, \qquad x > 0$$

$$\psi(-\infty,y) = \theta_0, \qquad \theta(\infty,y) = 0,$$

$$\psi(0,y) = \theta(0,y), \qquad \frac{\partial\psi(0,y)}{\partial x} = \frac{\partial\theta(0,y)}{\partial x},$$

$$\frac{\partial\psi(x,0)}{\partial y} = 0, \qquad \frac{\partial\theta(x,0)}{\partial y} = 0,$$

$$\psi(x,\ell) = \theta_0, \qquad \theta(x,\ell) = 0.$$

There are now two homogeneous differential equations, and four homogeneous and four non-homogeneous boundary conditions. One of the non-homogeneous conditions $\psi(x,\ell) = \theta_0$, introduces a non-homogeneity into the only possible finite orthogonal direction, and thus precludes the use of separation of variables for the region $x < 0$. This difficulty is eliminated for $x < 0$, by the transformation

$$\psi(x,y) = \theta_0 + \phi(x,y)$$

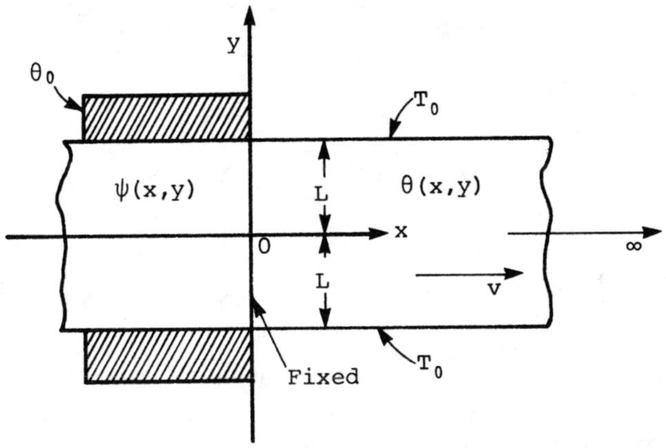

Fig. 3

Thus the region $x < 0$, in terms of $\phi(x,y)$, is satisfied by the equations

$$\frac{\partial^2\phi}{\partial x^2} - 2s\frac{\partial\phi}{\partial x} + \frac{\partial^2\phi}{\partial y^2} = 0, \qquad x < 0$$

$$\phi(-\infty,y) = 0,$$

$$\theta_0 + \phi(0,y) = \theta(0,y), \qquad \frac{\partial\phi(0,y)}{\partial x} = \frac{\partial\theta(0,y)}{\partial x},$$

$$\frac{\partial\phi(x,0)}{\partial y} = 0, \qquad \phi(x,\ell) = 0,$$

and the equations for the region $x > 0$ are the same.

The general solution in the x-direction (3)

$$X_n(x) = A_n e^{[s+(s^2+\lambda_n^2)^{1/2}]x} + B_n e^{[s-(s^2+\lambda_n^2)^{1/2}]x}$$

is valid in both domains. The first term satisfies the region $x < 0$ and the second is for the region $x > 0$. The solution of $\psi(x,y)$ and that of $\theta(x,y)$ are

$$\psi(x,y) = \theta_0 + \sum_{n=0}^{\infty} a_n e^{[s+(s^2+\lambda_n^2)^{1/2}]x} \cos\lambda_n y, \tag{4}$$

$$\theta(x,y) = \sum_{n=0}^{\infty} b_n e^{[s-(s^2+\lambda_n^2)^{1/2}]x} \cos\lambda_n y. \tag{5}$$

The values of a_n and b_n are determined from the conditions at the interface x=0; the equality of interface temperatures leads to

$$\theta_0 + \sum_{n=0}^{\infty} a_n \cos\lambda_n y = \sum_{n=0}^{\infty} b_n \cos\lambda_n y \tag{6}$$

Series expansion of θ_0 gives

$$\theta_0 = \frac{2\theta_0}{\ell} \sum_{n=0}^{\infty} \frac{(-1)^n}{\lambda_n} \cos \lambda_n y \tag{7}$$

Substituting equation (7) in equation (6) yields

$$(-1)^n \frac{2\theta_0}{\lambda_n \ell} + a_n = b_n \tag{8}$$

and the equality of fluxes gives

$$\left[s + (s^2+\lambda_n^2)^{\frac{1}{2}}\right] a_n = \left[s - (s^2+\lambda_n^2)^{\frac{1}{2}}\right] b_n \tag{9}$$

The values of a_n and b_n are determined by solving equations (8) and (9), and substituting their values into equations (4) and (5). The dimensionless parameters used are given by

$$\mathrm{Pe(or\ P)} = V\ell/a \qquad Gz = 1/\xi = Pe/(x/\ell)$$

$$\eta = y/\ell \qquad \mu = (2n+1)\pi/2 \quad n=0,1,2,\ldots,$$

The dimensionless temperatures on either side of the interface are

$$\frac{\psi(\xi,\eta,P)}{\theta_0} = 1 - \sum_{n=0}^{\infty} \frac{(-1)^n}{\mu_n} \left[1 - \frac{1}{[1+(2\mu_n/P)^2]^{1/2}}\right]$$

$$\times \cos\mu_n \eta \exp\left(\frac{1}{2}\left\{1 + \left[1 + \left(\frac{2\mu_n}{P}\right)^2\right]^{1/2}\right\} P^2 \xi\right), \xi < 0, \tag{10}$$

$$\frac{\theta(\xi,\eta;P)}{\theta_0} = \sum_{n=0}^{\infty} \frac{(-1)^n}{\mu_n} \left\{1 + \frac{1}{[1+(2\mu_n/P)^2]^{1/2}}\right\}$$

$$\times \cos\mu_n\eta \exp\left(\frac{1}{2}\left\{1 - \left[1 + \left(\frac{2\mu_n}{P}\right)^2\right]^{1/2}\right\}P^2\xi\right), \quad \xi > 0. \tag{11}$$

The Peclet number is given as

$$\frac{\text{Axial enthalpy flow}}{\text{Axial conduction}} = Pe,$$

As $Pe \to \infty$, the effect of axial conduction decreases and equations (10) and (11) reduce to

$$\frac{\psi(\xi,\eta)}{\theta_0} = 0, \qquad \xi < 0, \tag{12}$$

$$\frac{\theta(\xi,\eta)}{\theta_0} = 2 \sum_{n=0}^{\infty} \frac{(-1)^n}{\mu_n} e^{-\mu_n^2\xi} \cos\mu_n\eta, \qquad \xi > 0. \tag{13}$$

Equation (13) is the dimensionless form of equation (2). The interface temperature (at $\xi=0$, $Gz=\infty$) is plotted against η, for different values (0.05, 0.3, 1,2, 5 and 50) of the Peclet number (Pe), in figure 4. The plots show the progressive effect of axial conduction as the Peclet number decreases.

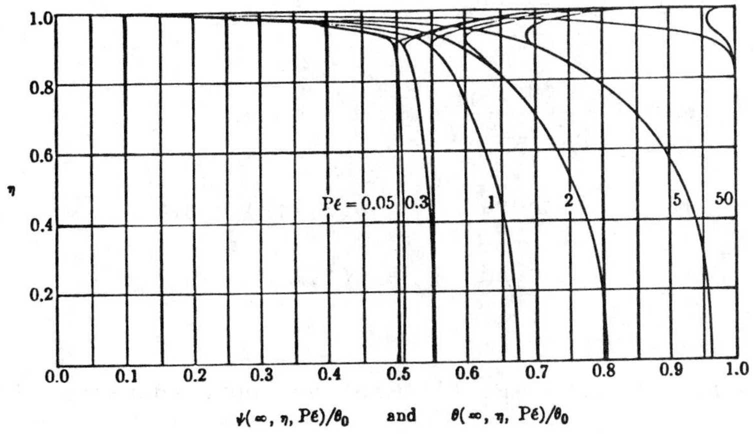

Fig. 4

The plot of the middle-plane temperature versus 1/Gz for different values (0.05, 0.1, 0.2, 0.3, 1 and 2) of Pe (fig. 5), shows that the penetration of the temperature distribution into the region $x < 0$ increases as the value of the Peclet number decreases.

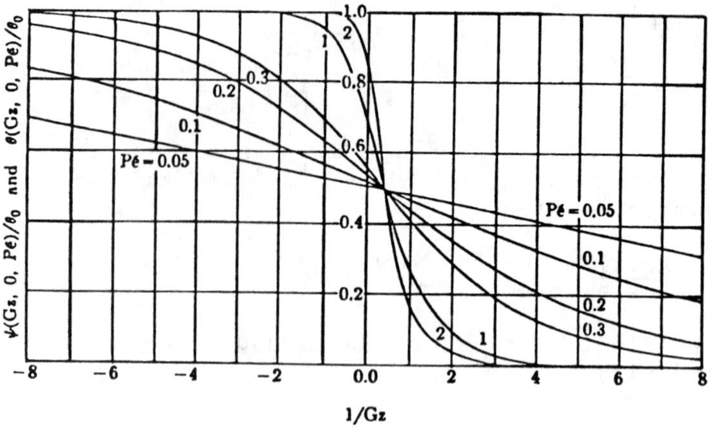

Fig. 5

Therefore, the heat transfer coefficients based on the length (ℓ) and in terms of

$$\theta_b = \int_0^1 \theta \, d\eta \qquad \text{and} \qquad \psi_b = \int_0^1 \psi d\eta$$

are given by the Nusselt number (Nu):

$$Nu = \frac{(\partial\psi/\partial\eta)_{\eta=1}}{\psi_w - \psi_b}, \qquad \xi < 0,$$

$$Nu = \frac{-(\partial\theta/\partial\eta)_{\eta=1}}{\theta_b - \theta_w} \qquad \xi > 0,$$

where $\psi_w = \psi(\xi,1;P) = \theta_0$

and $\theta_w = \theta(\xi,1;P) = 0$

Inserting equations (10) and (11) in the above equation gives

$$\frac{\sum_{n=0}^{\infty}\left\{1 - \frac{1}{[1+(2\mu_n/P)^2]^{1/2}}\right\} \exp\left(\frac{1}{2}\left\{1+\left[1+\left(\frac{2\mu_n}{P}\right)^2\right]^{1/2}\right\}P^2\xi\right)}{\sum_{n=0}^{\infty}\frac{1}{\mu_n^2}\left\{1+\frac{1}{[1+(2\mu_n/P)^2]^{1/2}}\right\}\exp\left(\frac{1}{2}\left\{1-\left[1+\left(\frac{2\mu_n}{P}\right)^2\right]^{1/2}\right\}P^2\xi\right)},$$

$$\xi < 0. \qquad (15)$$

These equations are plotted in Fig. (6) versus 1/Gz for the values 0.05, 0.1, 0.2, 0.3, 1 and 2 of Pe.

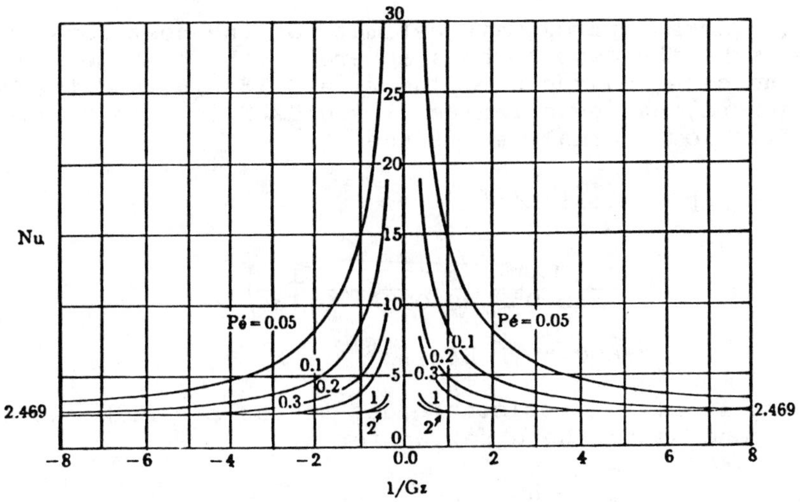

Fig. 6

It can be seen from Fig. (6) that the development of the heat transfer coefficient decreases as the value of the axial conduction increases.

• PROBLEM 2-32

A semiconductor chip, in the shape of a cube of size 0.04 in., is encased in an electric insulator, from which electrical connectors protrude. The dimensions of the case are 0.25 in × 0.25 in. × 0.1 in (see Fig. 1). The power used by the device is 120mW and the maximum operating temperature is 200°F. Describe the heat transfer process. Take $k_{insulation}$ = 0.058 Btu/ft hr°F, and the ambient temperature T_{amb} = 100°F.

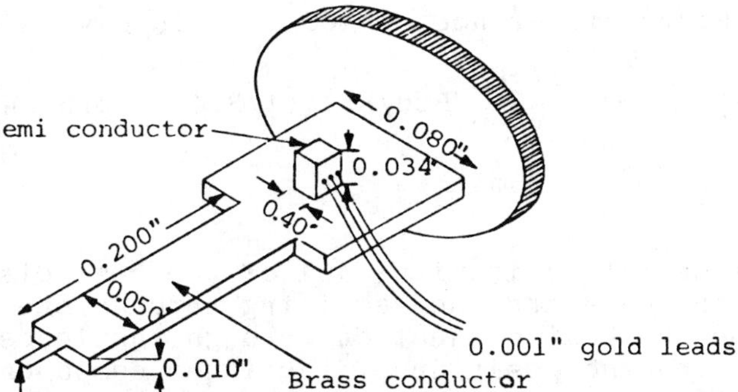

Fig. 1. Copper conductor connected to the circuit board.

An integrated circuit incorporating a thermal sink lead.

Solution: For an approximate estimate of the heat loss from the case, take the case to be a sphere, with the same surface area as the case, having a radius r_2 = 0.134 in. and the chip to be a sphere, having a radius r_1 = 0.0233 in. Then using the equation for thermal resistance

$$R_c = \frac{(1/r_1 - 1/r_2)}{4\pi k}$$

$$= \frac{12}{4\pi(0.058)}\left(\frac{1}{0.023} - \frac{1}{0.134}\right)$$

$$\simeq 600 \text{ hr}^0\text{F/Btu}$$

Since there is also a resistance due to the air film, an estimation is done using the equation for h,

$$Nu = \frac{hD}{k} = 2$$

$$h_{min} = 2\frac{k_{air}}{2r_2}$$

$$= \frac{0.016}{0.134/12} = 1.43 \text{ hr ft}^{2\,0}\text{F/Btu.}$$

This value is doubled, to take into account free convection and radiation. Therefore the film has a resistance

$$R_f = \frac{1}{hA_2}$$

$$= 144/(2.86)(4\pi)(0.134)^2 = 220 \text{ hr}^0\text{F/Btu}$$

The total resistance (R) is

$$R = R_c + R_f$$

$$= 600 + 220$$

$$= 820 \text{ hr}^0\text{F/Btu}$$

A temperature difference of 100^0F would have a dissipation of

$$Q = \left(\frac{100}{820} \text{ Btu/hr}\right) (1/3.41 \text{ Btu/hr W})$$

$$= 36\text{mW}$$

In integrated circuits, 40% of the heat dissipation is through the case and the remaining through the leads. To determine the heat conduction through the leads, the case is cut open and the position of the inner components are as shown in figure 1.

The chip is made of a beryllia oxide compound, which has a high thermal conductivity, approximately equal to that of aluminium. Thus the resistance of the chip would be

$$R_1 = \frac{L}{kA}$$

$$= \frac{0.034/12}{(125)(0.040/12)^2} = 2.04 \text{ hr}^0\text{F/Btu}$$

and the resistance of the electrical leads is

$$R_2 = \frac{L}{kA} = \frac{(0.200/12)}{(64)(0.010/12)(0.050/12)} = 75.0 \text{ hr}^0\text{F/Btu}$$

Now the thermal resistance of the circuit board is required since the lead is soldered to it. The circuit board is made up of copper foil (0.003 in. thick) laminated on a phenolic filled cloth board (Fig. 2). The copper strip on the board acts like a fin and transfers heat to the phenolic and from there to the ambient air.

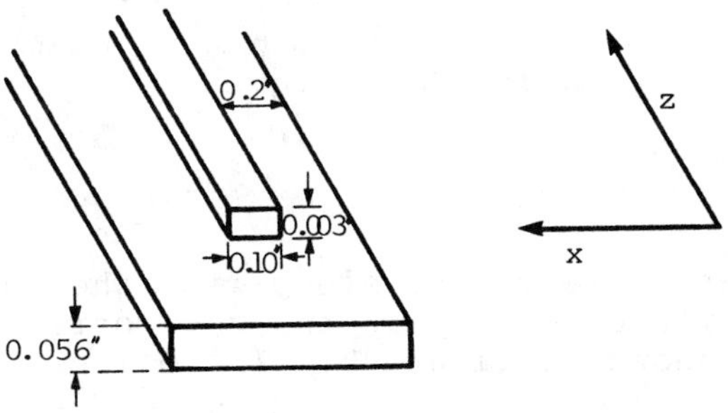

Fig. 2 Copper conductor soldered on a circuit board.

Assume that the copper strip to which the lead has been soldered is 0.10 in. wide. The spacing between the copper foil strips is 0.5 in., as shown in figure 2. Consider a small length dz, along the width of the fin. Then the effectiveness is determined by combining the equations,

$$\beta^2 = \frac{h\ P}{kA_c} \quad (1)$$

and

$$\eta = \frac{1}{\beta L} \tan h\ \beta L \quad (2)$$

to get the equation

$$\eta = \left(\frac{h\ PL^2}{kA_c}\right)^{-\frac{1}{2}} \tan h \left(\frac{h\ PL^2}{kA_c}\right)^{\frac{1}{2}} \quad (3)$$

The numerical values for the equation are

$L = 1$ Btu/hr ft$^{2\,0}$F $\qquad$ $P = 2$ dz

$L = 0.20/12$ ft $\qquad$ $k = 0.10$ Btu/hr ft°F

$A_c = \delta_w \, dz$ $\qquad$ $\delta_w = 0.056/12$ ft.

Substitution of these values in the equation for effectiveness gives

$$\eta = 0.73$$

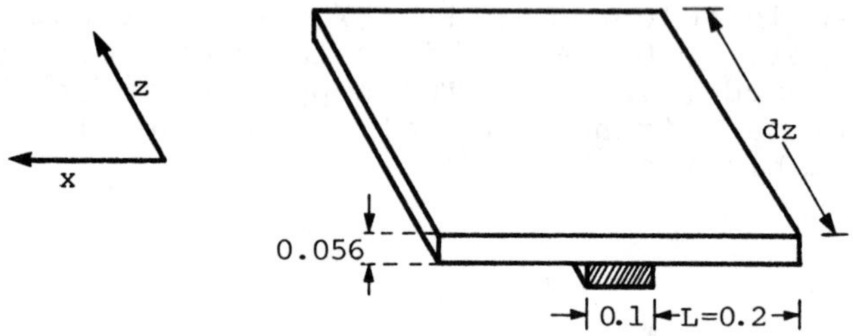

Fig. 3 An element of circuit board acting as a fin for the conductor.

The element of length dz along with the conductor is taken to be a fin extending in the z-direction, having a length L_c, as shown in figure 3. The effective perimeter is

$$P = 2(0.10) + 2(0.40)(0.73)$$

$$= 0.784 \text{ in.}$$

Conduction in the z-direction is mainly given by the product of the area and the conductivity,

$$kA_c = 223(0.003/12)(0.10/12)$$

$$= 4.6 \times 10^{-4} \text{ Btu ft/hr °F.}$$

The thermal resistance of the circuit board is given by

$$R_3 = \frac{1}{(hPkA_c)^{\frac{1}{2}} \tanh (hPL_c^2/kA_c)^{\frac{1}{2}}} \qquad (4)$$

The resistance of the fin is determined by combining equations (1) and (3),

$$R_{fin} = \frac{1}{k\,A_c\,\beta \tanh \beta L} = \frac{1}{h\,PL\,\eta} \qquad (5)$$

If, in equation (4), L_c is long enough so that the tanh term is nearly one, then

$$R_3 = 1/(h\ Pk\ A_c)^{\frac{1}{2}}$$

$$= 1/[(1)(0.784/12)(4.6 \times 10^{-4})]^{\frac{1}{2}}$$

$$= 182[\text{Btu/hr }^0\text{F}]^{-1}$$

Thus, for the tanh term to be almost one, the term $(h\ PL_c^2/kA_c)^{\frac{1}{2}}$ should be four. Therefore

$$hPL_c^2/kA = 16$$

or

$$L_c = 4[(4.6 \times 10^{-4})(12)/(0.784)]^{1/3}$$

$$= 0.34 \text{ ft} \simeq 4.0 \text{ inches}$$

This is a reasonable length for the conductor.

The total resistance (R) of the leads and the circuit board is given by

$$R = R_1 + R_2 + R_3$$

$$= 2.0 + 75.0 + 182$$

$$= 259[\text{Btu/hr.}^0\text{F}]^{-1}$$

For a temperature difference of 100^0F, the heat loss is

$$\dot{Q} = \frac{100(1000)}{(259)(3.41)} = 113 \text{ mW}$$

Thus, including the heat loss from the casing, the total heat loss is

$$Q = 113 + 36 = 149 \text{ W}$$

This is higher than the given rating of 120 Watts.

NUMERICAL METHODS IN HEAT CONDUCTION

• PROBLEM 2-33

A wall, as shown in the figure, has the following properties:

$$k = 1 \text{ Btu/hr ft}^0\text{F}$$

$$h = 1 \text{ Btu/hr ft}^0\text{F}$$

One end (4) is maintained at a temperature of 200^0F (t_4) and the points 1, 2, 3 and 4 are spaced 1 foot apart. The temperature of the air surrounding the point 1 is 600^0F (t_0). Determine the temperature distribution at points marked 1, 2 and 3, in the figure.

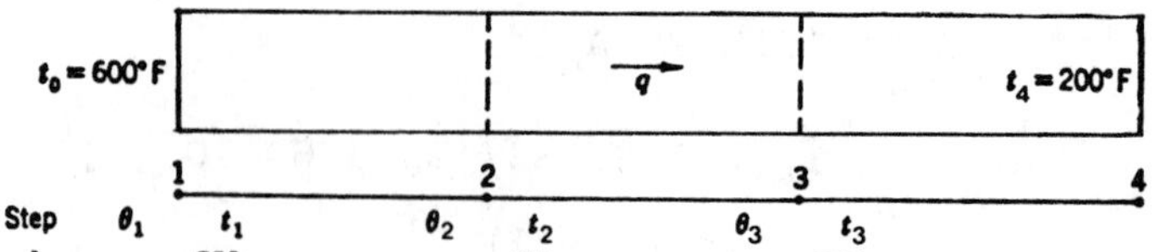

Step	θ_1	t_1	θ_2	t_2	θ_3	t_3
1		550		450		350
2	-50	550	0	450	-50	350
3	-50	550	-25	450	0	325
4	0	525	-50	450	0	325
5	-25	525	0	425	-25	325
6	0	512.5	-12.5	425	-25	325
7	0	512.5	-25	425	0	312.5
8	-12.5	512.5	0	412.5	-12.5	312.5
9	0	506.2	-6.3	412.5	-12.5	312.5
10	0.1	506.2	-12.5	412.5	0	306.3
11	-6.1	506.2	0	406.3	-6.3	306.3
12	-6.1	506.2	-3.3	406.3	0	303.1
13	0	503.1	-6.4	406.3	-0.1	303.1
14	-3.1	503.1	0	403.1	-3.1	303.1

Continue until all sink strengths are zero (that is, all θ values are zero)

Determination of the steady-state temperature distribution across a wall by means of the relaxation method.

Solution: This problem is solved by using the relaxation method. The equation for determining the residuals is

$$\theta_1 = \frac{Nt_0+t_2-(N+1)t_1}{N}$$

where $N = \frac{h\ \Delta x}{k}$

$h = 1$ Btu/hr ft^2 ^{0}F

$k = 1$ Btu/hr ft ^{0}F

$\Delta x = 1$ ft

therefore $N = \frac{1\cdot(1)}{1} = 1$

Also, $\theta_1 = t_0 + t_2 - 2t_1$

This equation is set equal to zero, so that when all the residuals are equal to zero,

$$t_1 = \frac{t_0+t_2}{2}$$

This is because N is equal to 1.

The solution is shown in tabulated form in the figure. The original assumptions of temperatures 1, 2 and 3 are taken as 550, 450 and 350, respectively.

The first few sample calculations are shown

STEP 1. $\theta_1'' = 600 + 450 - 2(550) = -50$

$\theta_2'' = 550 + 350 - 2(450) = 0$

$\theta_3'' = 450 + 200 - 2(350) = -50$

STEP 2. In the next set of calculations, t_3 is changed from 350^0F to 325^0F and the three calculations are repeated.

$\theta_1''' = 600 + 450 - 2(550) = -50$

$\theta_2''' = 325 + 550 - 2(450) = -25$

$\theta_3''' = 450 + 200 - 2(325) = 0$

• PROBLEM 2-34

For the chimney section shown in Fig. 1, find the temperature distribution if the inside and outside surface temperatures are 100^0F and 500^0F, respectively. Also determine the heat flow rate per foot of chimney height. Take the thermal conductivity of the wall to be 1.0 Btu/hr ft^{0}F.

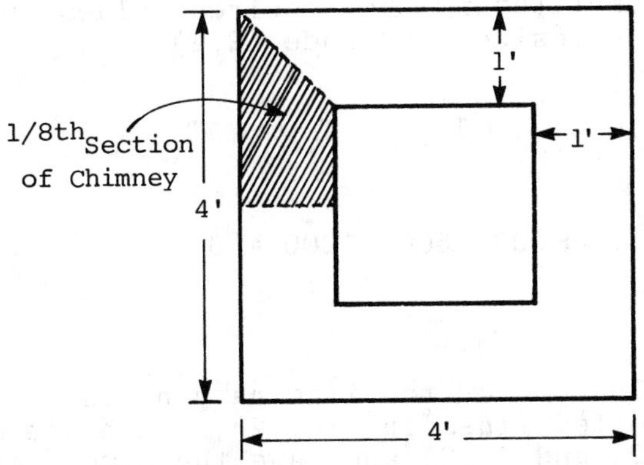

Fig. 1 Plan view of chimney.

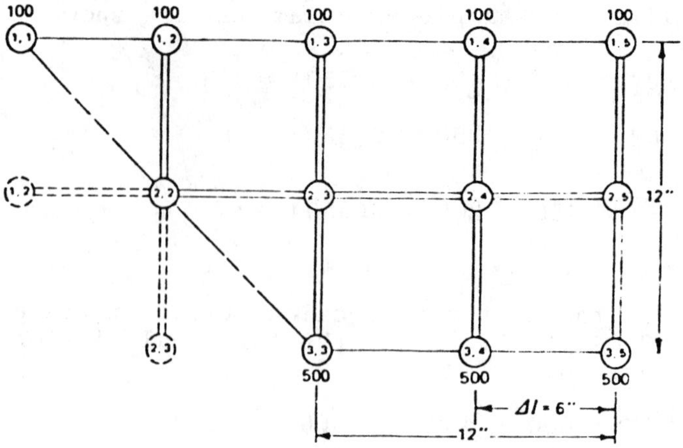

Fig. 2

<u>Solution</u>: Due to the symmetry of the cross-section being considered, only one-eighth section (shown enlarged in Fig. 2) is analyzed. The coarse mesh is being employed so as to reduce the number of steps in the relaxation method. For a higher accuracy, the mesh should be further subdivided after the original mesh has been relaxed.

The boundary temperatures are fixed, therefore only the nodes at (2.2), (2.3), (2.4) and (2.5) need to be relaxed. Since nodes (2.3) and (2.4) are simple interior points, the following equation is employed for determining the residuals.

$$Q'_{m,n} = \frac{Q_{m,n}}{kb} = T_{m-1,n} + T_{m,n-1} + T_{m+1,n} + T_{m,n+1} - 4T_{m,n}$$

The first step is to assume the temperatures at the nodes to be relaxed. Taking the initial temperatures at these points to be 300°F and substituting the required values in the above equation, gives the residual at node (2,4)

$$Q'_{(2,4)} = T_{(2,3)} + T_{(1,4)} + T_{(2,5)} + T_{3(4)} - 4T_{(2,4)}$$

$$= 300 + 100 + 300 + 500 - 1200 = 0$$

Node (2,2) is influenced by the imaginary nodes (1,2) and (2,3) (shown by dotted lines in Fig. 2), which are mirror imaged of nodes (1,2) and (2,3) and have the same temperatures. The same applies for node (2.5). The residual at node (2,2) is given by the equation

$$Q'_{(2,2)} = (T_{(2,3)} + T_{(1,2)}) - 4T_{(2,2)}$$

$$= 2(300 + 100) - 1200$$

$$= -400$$

The relaxation is carried out for all the nodes to be relaxed and the process is shown in the table below.

Table 1: Relaxation solution

	$T_{2,2}$	$Q'_{2,2}$	$T_{2,3}$	$Q'_{2,3}$	$T_{2,4}$	$Q'_{2,4}$	$T_{2,5}$	$Q'_{2,5}$
Assumed Initial Temperature Distribution	300		300		300		300	
Initial Residuals		−400		0		0		0
Decrease $T_{2,2}$ by 100	200	0		−100				
Decrease $T_{2,3}$ by 25		−50	275	0		−25		
Decrease $T_{2,2}$ by 12	188	−2		−12				
Decrease $T_{2,4}$ by 6				−18	294	−1		−12
Decrease $T_{2,3}$ by 5		−12	270	+2		−6		
Decrease $T_{2,2}$ by 3	185	0		−1				
Decrease $T_{2,5}$ by 3						−9	297	0
Decrease $T_{2,4}$ by 2				−3	292	−1		−2
Decrease $T_{2,3}$ by 1		−2	269	+1		−2		−2
Final Temperatures	185		269		292		297	
Final Residuals		−2		+1		−2		−2

The heat flow rate is determined by adding the individual flow rates through the rods joining the nodes and is found to be

$$q = 8k\left[(500 - 269) + (500 - 292) + \frac{1}{2}(500 - 297)\right]$$

$$= 4324 \text{ Btu/hr.}$$

Approximate the steady state temperature distribution for the L-shaped corner shown in the figure, using the relaxation method. The constant boundary temperatures are marked in the figure, in a larger-sized type.

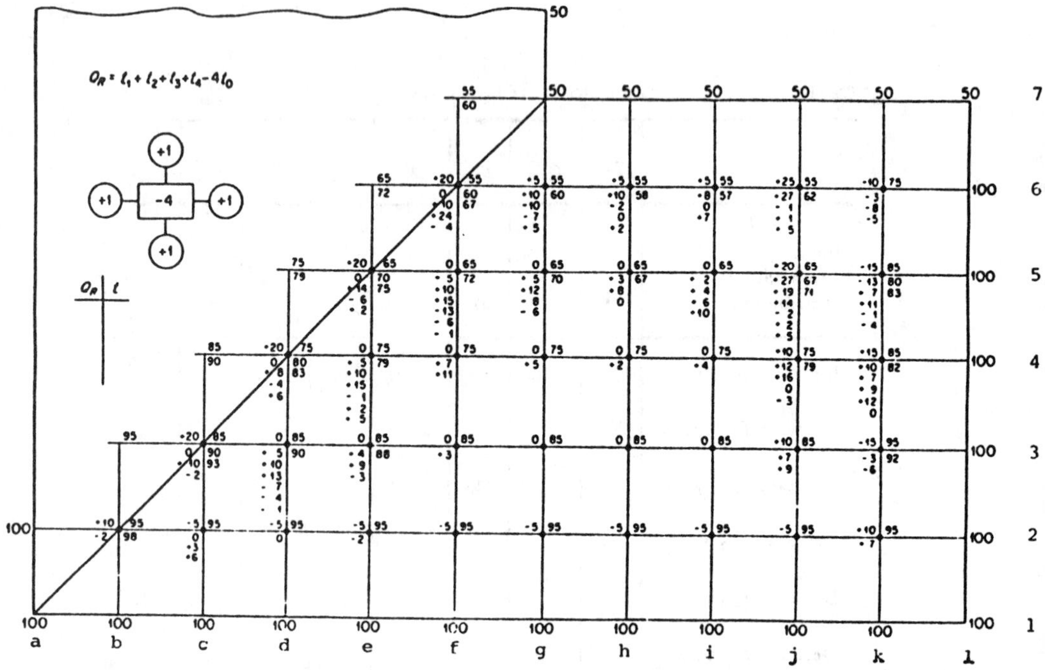

Relaxation network for the L-shaped corner.

Solution: The partial differential equation for the L-shaped corner under steady state conditions, with no internal energy conversion, is

$$\frac{\partial^2 t}{\partial x^2} + \frac{\partial^2 t}{\partial y^2} = 0 \qquad \text{(Laplace equation)}$$

The second derivative, $\frac{\partial^2 T}{\partial x^2}$, can be approximated as

$$\frac{\partial^2 t}{\partial x^2} \simeq \frac{t_1 + t_3 - 2t_0}{a^2}$$

where a is the distance between the adjoining nodes. Similarly,

$$\frac{\partial^2 t}{\partial y^2} \simeq \frac{t_2 + t_4 - 2t_0}{a^2}$$

These two equations add up to give an approximate form of the Laplace equation,

$$0 = \frac{\partial^2 t}{\partial x^2} + \frac{\partial^2 t}{\partial y^2} \simeq \frac{t_1+t_2+t_3+t_4-4t_0}{a^2}$$

For the term on the right to be equal to zero, the numerator must be zero. In the relaxation method, Q_0 is defined as the residual at node 0. In this method, the objective is to make the residual at all the nodes equal to zero. Initially, an approximation of the temperatures at each node is required. These assumed temperatures are shown to the right of the nodal points. Now the residuals are calculated for each of these nodes by using the approximate form of the Laplace equation. A sample set of calculations for the nodes at 6 is given below:

$$Q_{f6} = 65 + 55 + 55 + 65 - 4(55) = 20$$

$$Q_{g6} = 55 + 50 + 55 + 65 - 4(55) = 5$$

$$Q_{h6} = 55 + 50 + 55 + 65 - 4(55) = 5$$

$$Q_{i6} = 55 + 50 + 55 + 65 - 4(55) = 5$$

$$Q_{j6} = 55 + 50 + 75 + 65 - 4(55) = 25$$

The residuals appear to the left of the nodal points.

Next, attempt to reduce the residual at each node using the method described earlier, starting at the node with the largest residue. At times, it is better to get a higher value, so that it compensates the adjacent nodes. To illustrate this point, consider the node c3. The initial residual was +20. The temperature was adjusted to 90^0, which gave a residual of

$$Q_{c3} = 95 + 85 + 85 + 95 - 4(90) = 0$$

Since the temperature at node c3 was adjusted, the residual at nodes c2 and d3 was found to be

$$Q_{c2} = 95 + 90 + 95 + 100 - 4(95) = 0$$

$$Q_{d3} = 90 + 75 + 85 + 95 - 4(85) = 5$$

Now all the nodes that were affected by node c3 have been adjusted, therefore, find a node with a large residue, and repeat the procedure. It is merely coincidental that the residuals at nodes c3 and c2 turned out to be zero. The residuals do not have to remain zero throughout the process. The order in which one proceeds is not fixed, but the nodes with the larger residuals should be adjusted first. Only one half of the corner was used in the calculations because of the symmetry of the object.

• PROBLEM 2-36

Formulate the nodal equations necessary to analyze the heat dissipated by a square plate (see Fig. 1). Use numerical techniques to solve the problem. The following boundary conditions exist:

1) at x=0, q=0

2) at x=a and y=0, heat is convected to surrounding at T_∞

3) at y=a, there are four given nodal temperatures t_k, k=1 to 4

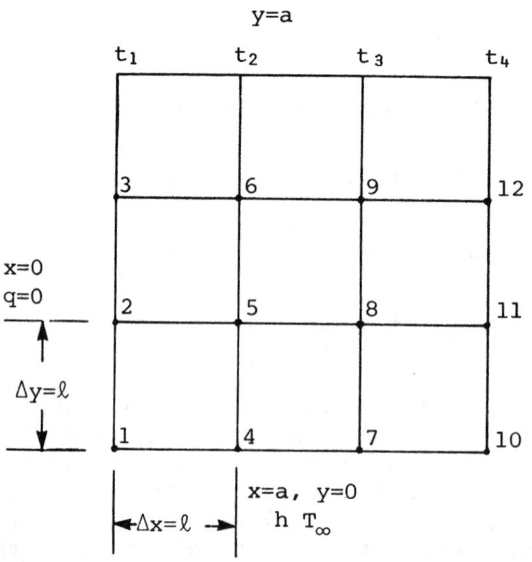

Fig. 1

Solution: For two-dimensional heat flow analysis, the Laplace equation applies. For steady state conditions:

$$\frac{\partial^2 T}{\partial x^2} + \frac{\partial^2 T}{\partial y^2} + \frac{\ell}{k} g = 0 \tag{1}$$

where g is fixed value.

Since the temperatures at y=a are given, there are only 12 nodal temperatures to be determined.

The following equations are derived from eq.(1) for the given boundary conditions:

For the interior nodes 5, 6, 8, 9 and the insulated boundary nodes 2 and 3, the governing equation can be written as

$$T_{i-1,j} + T_{-(i-1),j} + T_{i,j-1} + T_{i,j+1} - 4T_{i,j}) + \frac{\ell^2}{k} g_{i,j} = 0 \tag{2}$$

For the convection boundary nodes, 4, 7, 11, 12:

$$\left[2T_{i-1,j} + T_{i,j-1} + T_{i,j+1} - \left(4 + \frac{2\ell h}{k}\right) T_{i,j}\right] + \left(\frac{2\ell h}{k} T_\infty + \frac{\ell^2}{k} g_{i,j}\right) = 0 \tag{3}$$

For the corner nodes 1 and 10:

$$\left[2T_{i-1,j} + 2T_{i,j+1} - \left(4 + \frac{2\ell h_1}{k} + \frac{2\ell h_2}{k}\right) T_{i,j}\right] + \left(\frac{2\ell h_1}{k} T_{\infty 1} + \frac{2\ell h_2}{k} T_{\infty 2} + \frac{\ell^2}{k} g_{i,j}\right) = 0 \tag{4}$$

Substituting the numerics for i and j in eq.(2), (3) and (4), we obtain

node 5: $T_2 + T_8 + T_4 + T_6 - 4T_5 + \frac{g\ell^2}{k} = 0$

node 6: $T_3 + T_9 + T_5 + f_2 - 4T_6 + \frac{g\ell^2}{k} = 0$

node 8: $T_5 + T_{11} + T_7 + T_9 - 4T_8 + \frac{g\ell^2}{k} = 0$

node 9: $T_6 + T_{12} + T_8 + f_3 - 4T_9 + \frac{g\ell^2}{k} = 0$

node 2: $T_1 + T_3 + 2T_5 - 4T_2 + \frac{g\ell^2}{k} = 0$

node 3: $T_2 + f_1 + 2T_6 - 4T_3 + \frac{g\ell^2}{k} = 0$

node 4: $T_1 + T_7 + 2T_5 - \left(4 + \frac{2\ell h}{k}\right) T_4 + \left(\frac{2\ell h}{k} T_\infty + \frac{g\ell^2}{k}\right) = 0$

node 7: $T_4 + T_{10} + 2T_8 - \left(4 + \frac{2\ell h}{k}\right) T_7 + \left(\frac{2\ell h}{k} T_\infty + \frac{g\ell^2}{k}\right) = 0$

node 11: $2T_8 + T_{10} + T_{12} - \left(4 + \frac{2\ell h}{k}\right) T_{11} + \left(\frac{2\ell h}{k} T_\infty + \frac{g\ell^2}{k}\right) = 0$

node 12: $2T_9 + T_{11} + f_4 - \left(4 + \frac{2\ell h}{k}\right) T_{12} + \left(\frac{2\ell h}{k} T_\infty + \frac{g\ell^2}{k}\right) = 0$

node 1: $2T_2 + 2T_4 - 4T_1 + \frac{g\ell^2}{k} = 0$

node 10: $2T_7 + 2T_{11} - \left(4 + \frac{4\ell h}{k}\right) T_{10} + \left(\frac{4\ell h}{k} T_\infty + \frac{g\ell^2}{k}\right) = 0$

These sets of equations for T_k(k=1 to 12), may be solved using an iteration method. The Gauss-Seidel technique is recommended since a large number of nodes are considered.

• PROBLEM 2-37

A steel cantilever beam of circular cross-section is heated at the wall with a constant temperature T_i=300°F (see Fig. 1). The diameter, length, and conductivity of the beam are d=1.5", L=18", and k=25 Btu/hr ft°F, respectively. The beam is cooled by air at 70°F with a convection coefficient h_{air}=2.3 Btu/hr ft²°F. Using (a) numerical techniques and (b) matrix inversion, compute the temperature distribution of the beam and the total energy transferred.

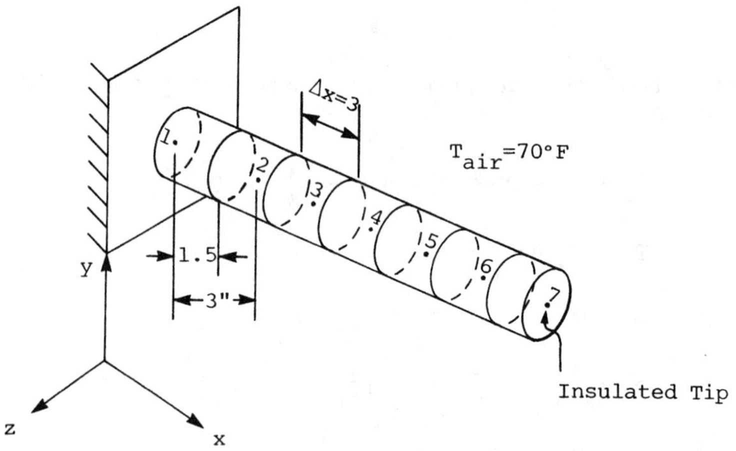

Fig. 1

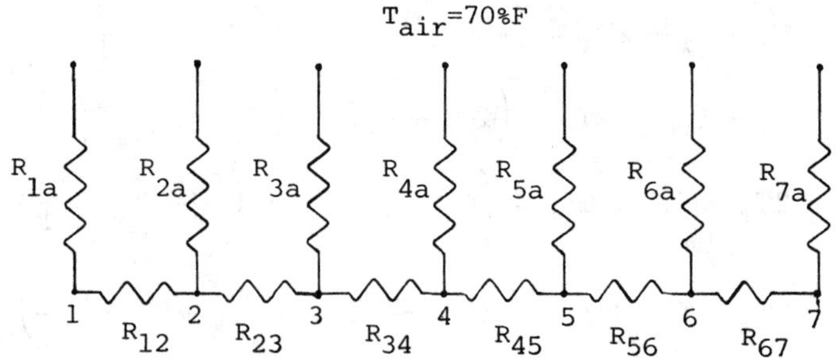

Fig. 2 Analog of 1-D system.

Solution: (a) In this problem, we will only consider the conduction in the x-direction or the length of the beam. From Fig. 1, we can see that the beam is divided into seven nodes spaced 3" from each other.

The cross-sectional area for conduction is:

$$A_k = \frac{\pi d^2}{4} = \frac{(3.14)(1.5/12)^2}{4} = 0.01226 \text{ ft}^2$$

Since the beam is of uniform cross-section, the resistance of each element is the same:

$$R_{12} = R_{23} = R_{34} = R_{45} = R_{56} = R_{67} = \frac{\Delta x}{kA_k}$$

$$= \frac{3.0/12}{25(0.01226)} = 0.8157 \text{ hr.}^0\text{F/Btu}$$

The surface area of each element exposed to air is:

$$A_c = \pi d\Delta x = 3.14(1.5/12)(3/12) = 0.0981 \text{ ft}^2$$

The resistance of the air for the elements is also identical:

$$R_{2a} = R_{3a} = R_{4a} = R_{5a} = R_{6a} = \frac{1}{h_{air}A_c} = \frac{1}{2.3(0.0981)} = 4.432 \text{ hr}^0\text{F/Btu}$$

The surface area of nodes 1 and 7 for convection is half that of the interior nodes, therefore the resistance is twice the amount of nodes 2 to 6:

$$R_{1a} = R_{7a} = 8.864 \text{ hr.}^0\text{F/Btu}$$

A numerical method frequently used is the Gauss-Seidel iteration which makes use of the following equations:

$$q_i = \sum_j \frac{-T_j + T_i}{R_{ij}} \qquad (1)$$

$$T_i = \frac{q_i + \sum_j (T_j/R_{ij})}{\sum_j (1/R_{ij})} \qquad (2)$$

Since only node 1 has an energy source, we obtain from eq.(1)

$$q_1 = \frac{-T_2+T_1}{R_{12}} = \frac{-T_2+300^0\text{F}}{0.8157} = -1.226\ T_2 + 367.8$$

Applying equation (2) we obtain the equations for the temperatures at nodes 2 to 7:

$$T_2 = \frac{T_1/R_{12} + T_{air}/R_{2a} + T_3/R_{23}}{\frac{1}{R_{12}} + \frac{1}{R_{2a}} + \frac{1}{R_{23}}} = \frac{300/0.8157+70/4.432+T_3/0.8157}{\frac{1}{0.8157} + \frac{1}{4.432} + \frac{1}{0.8157}}$$

$T_2 = 0.4579\ T_3 + 143.27$

$T_3 = 0.4579(T_2+T_4) + 5.90$

$T_4 = 0.4579(T_3+T_5) + 5.90$

$T_5 = 0.4579(T_4+T_6) + 5.90$

$T_6 = 0.4579(T_5+T_7) + 5.90$

$T_7 = 0.9158\ T_6 + 5.90$

Begin the iteration by assuming $T_2 = 270^0F$, $T_3 = 240^0F$, $T_4 = 210^0F$, $T_5 = 180^0F$, $T_6 = 150^0F$, and $T_7 = 120^0F$. The results are tabulated in table 1 with a convergence test of $\varepsilon = 1.0^0F$. Fur further accuracy, the iteration may be taken to a convergence of $\varepsilon = 0.1^0F$.

$T_1 = 300^0F$, $T_2 = 222.7^0F$, $T_3 = 173.4^0F$, $T_4 = 142.0^0F$

$T_5 = 123.8^0F$, $T_6 = 113.3^0F$, $T_7 = 110.8^0F$

number of iteration	constant T_1	T_2	T_3	T_4	T_5	T_6	T_7
0	300	270	240	210	180	150	120
1	300	266.9	225.7	198.2	170.7	143.3	143.3
2	300	246.6	218.9	187.4	162.3	149.7	137.1
3	300	243.5	204.6	180.5	160.3	143.0	143.0
4	300	237.0	200.0	173.0	154.0	144.8	136.8
5	300	243.9	193.6	168.0	151.4	139.1	138.5
6	300	232.2	190.4	163.9	140.6	138.6	133.3
7	300	230.5	187.3	157.5	144.4	131.3	132.8
8	300	229.0	183.6	157.8	138.1	132.8	126.1
9	300	227.3	183.0	153.2	138.9	126.9	127.5
10	300	227.1	180.1	153.3	134.2	127.9	122.1
11	300	225.7	180.1	149.8	134.7	123.3	123.0
12	300	225.7	177.8	150.0	131.0	123.9	118.8
13	300	224.7	177.9	147.3	131.3	120.3	119.4
14	300	224.7	176.2	147.5	128.4	120.7	116.1
15	300	223.9	176.3	145.4	128.7	117.9	116.4
16	300	224.0	175.0	145.5	126.5	118.1	113.9
17	300	223.4	175.1	144.0	126.6	116.0	114.0
18	300	223.4	174.1	144.0	125.0	116.1	112.1
19	300	223.0	174.1	142.9	125.0	114.5	112.2
20	300	223.0	173.0	142.9	123.8	114.5	110.8
21	300	222.7	173.4	142.0	123.8	113.3	110.8

(b) The matrix solution to the problem is in the form

$$\begin{bmatrix} T_2 \\ T_3 \\ T_4 \\ T_5 \\ T_6 \\ T_7 \end{bmatrix} = [A]^{-1}[C]$$

The heat flux will not be considered for this solution, therefore, only 6 unknowns are present. Referring to the nodal equations in part (a), we obtained:

$$[A] = \begin{bmatrix} 1 & -0.4579 & 0 & 0 & 0 & 0 \\ -0.4579 & 1 & -0.4579 & 0 & 0 & 0 \\ 0 & -0.4579 & 1 & -0.4579 & 0 & 0 \\ 0 & 0 & -0.4579 & 1 & -0.4579 & 0 \\ 0 & 0 & 0 & -0.4579 & 1 & -0.4579 \\ 0 & 0 & 0 & 0 & -0.4579 & 1 \end{bmatrix}$$

and

$$[C] = \begin{bmatrix} 143.27 \\ 5.90 \\ 5.90 \\ 5.90 \\ 5.90 \\ 5.90 \end{bmatrix}$$

Next, the inverse of matric [A]is found to be:

$$A^{-1} = \begin{bmatrix} 1.487 & 1.020 & 0.734 & 0.5814 & 0.491 & 0.2375 \\ 1.020 & 2.102 & 1.68 & 1.29 & 1.085 & 0.491 \\ 0.734 & 1.68 & 2.78 & 2.14 & 1.795 & 0.843 \\ 0.5814 & 1.29 & 2.14 & 3.32 & 2.71 & 1.295 \\ 0.491 & 1.085 & 1.795 & 2.71 & 4.130 & 1.92 \\ 0.457 & 1.015 & 1.70 & 2.58 & 3.84 & 2.79 \end{bmatrix}$$

Multiplying $[A]^{-1}$ by $[C]$, we obtain the nodal temperatures.

$T_2 = 223.1^0F$, $T_3 = 173.5^0F$, $T_4 = 142.6^0F$

$T_5 = 123.5^0F$, $T_6 = 113.9^0F$, $T_7 = 110.2^0F$

• PROBLEM 2-38

Find the temperature distribution in the interior of a rectangular plate, shown in figure 1, which is insulated on the top and bottom surfaces. The temperature along the edges AD, DC and CB is zero. The temperature along AB is

$$t = t_E \sin \frac{\pi x}{AB} \quad \text{on} \quad y = \overline{AD}$$

if the origin is at D. Also, the temperature is maximum at E and zero at A and B. Take the section ADFE to be a square.

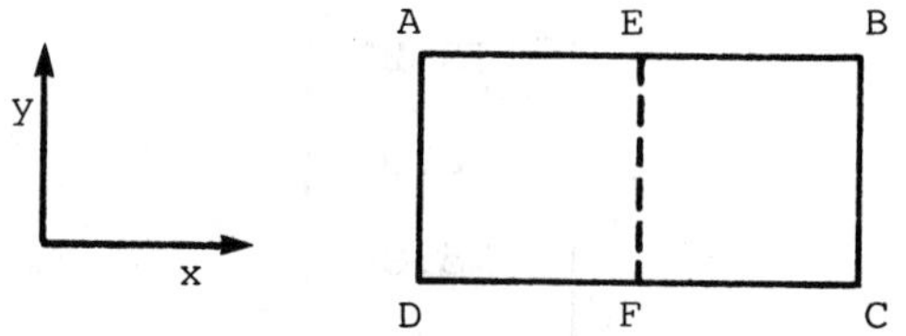

Fig. 1 A rectangular plate.

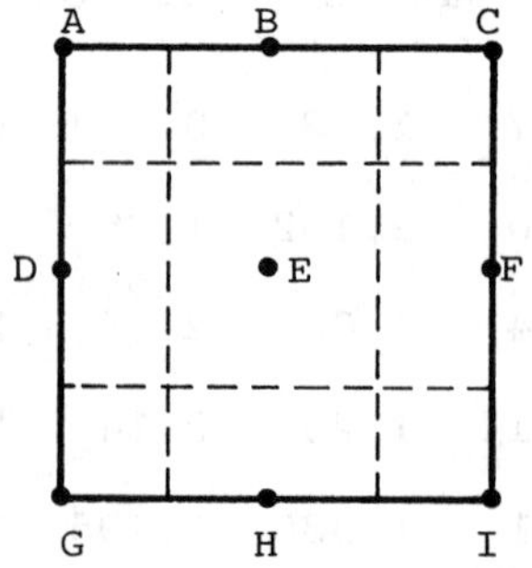

Fig. 2 Subdivision of the plate.

Solution: Since the plate is symmetric, only the part ADFE is considered. Also the section at EF is taken to be adiabatic. For an approximate solution, divide the section ADFE as shown in figure 2.

Take the value of t_c as 10, then to the nearest whole number

$$t_B = 10 \sin 45 = 7$$

and all the temperatures except at E and F are zero. If the thickness of the plate is taken to be unity, a heat balance at E gives

$$k(t_B-t_E) + k(t_D-t_E) + k(t_H-t_E) + k(t_F-t_E) = 0$$

or $$t_B + t_D + t_H + t_F - 4t_E = R_E \qquad (1)$$

where R_E is the residual and is to be minimized.

Similarly, a heat balance at F yields

$$\tfrac{1}{2} k (t_C-t_F) + k(t_E-t_F) + \tfrac{1}{2}k(t_I-t_F) = 0$$

or $$t_C + 2t_E + t_I - 4t_F = R_F \qquad (2)$$

Equations (1) and (2) can be understood very easily by examining figure 3, where (a) applies to a central nodal point (e.g. point E) and (b) to a point on the adiabatic boundary (e.g. point F).

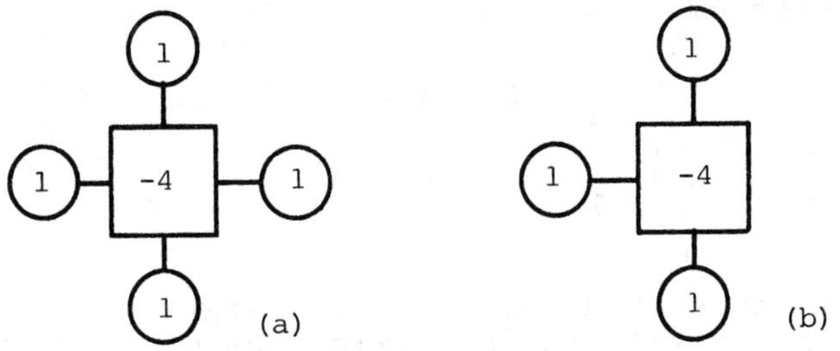

Fig. 3 Two-dimensional operation diagrams.

Both show the effect of a unit increase of temperature on the residual at the indicated point or adjacent points. These scaled diagrams of a two-dimensional region are very useful while carrying out the relaxation procedure, by avoiding the need of a large and complicated computation guide. The step-by-step procedure for the above described method is as follows (see figure 4).

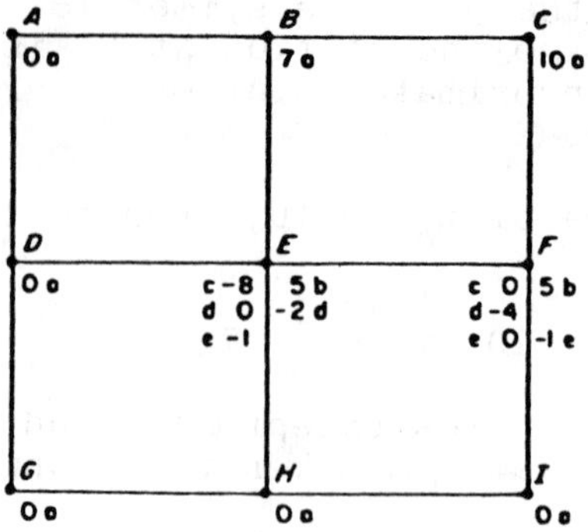

Fig. 4

(1) Write in the known temperatures

(2) Estimate the unknown temperatures and note them down (In this case, both E and F were taken to be 5)

(3) Compute the corresponding residuals and index each step

(4) Try to reduce the residual. By a change of (-2) at point E, and in accordance with figure 3a, this gives a change of (-2)(-4) = +8 in the residual. Thus the residual at E becomes (+8)+(-8) = 0. At F (Fig. 3b) the residual becomes (-2)(+2) = -4.

(5) A change of -1 at point F gives

$$t_F = 5 - 1 = 4$$

$$t_E = 5 - 2 = 3$$

The results were obtained without assigning any value of the thermal conductivity (k) or fixing the dimensions of the plate. The heat flow will be proportional to the temperature difference and the geometrical factors for the flow path. The results can be tabulated as follows:

HEAT FLOW FROM SOURCE

Path	AB	BE	CF	Total
Factor....................	1/2	1	1/2	
Temperature difference.....	7	4	6	
Product...................	3.5	4	3	10.5

HEAT FLOW TO SINK

Path	AB	ED	EH	FI	Total
Factor....................	1/2	1	1	1/2	
Temperature difference.....	7	3	3	4	
Product....................	3.5	3	3	2	11.5

A more precise method of finding the temperature distribution is to subdivide the plate into sixteen subsections, and use a maximum temperature (t_E) = 100. The approach is similar to the earlier method and one of the many possible procedures is as follows:

1) Set the temperature $t_D = t_E \sin 67.5$

2) The values of t_M and t_O are obtained from the earlier method and the temperatures at the other points are approximated.

3) The residuals are calculated using the operation diagrams of Fig. 3. Figure 3b applies to the points J, O and T, and figure 3a applies to all the other interior points.

The steps of the above procedure are shown in figure 5.

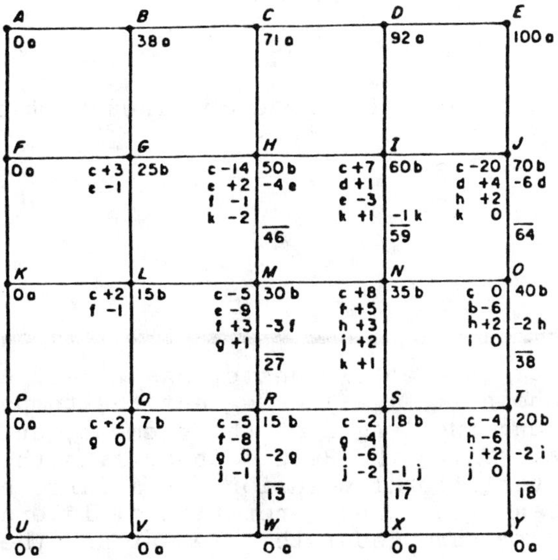

Fig. 5

The temperature distribution obtained is

0	38	71	92	100
0	25	46	59	64
0	15	27	35	38
0	7	13	17	18
0	0	0	0	0

There is also an analytical solution to the problem of the flat rectangular plate (Fig. 1). This is given by the equation

$$t = \frac{t_E \sinh(\pi y/\overline{DA})\sin(\pi x/\overline{DC})}{\sinh(\pi\ \overline{DA}/\overline{DC})}$$

where D is the origin of the system coordinates.

On solving the equation, the temperature distribution obtained is

0	28	71	92	100
0	25	47	61	66
0	15	28	36	39
0	7	14	18	19
0	0	0	0	0

The results are close to those obtained by the earlier method.

• PROBLEM 2-39

The element of an electric heater has a rectangular cross-section as shown in Fig. 1. Neglect the temperature variation along the length of the element, and the surround-of curvature. Using the Ritz method, find the approximate steady variational solution for a large heat transfer coefficient (h). The generation of internal energy (u''') is uniform throughout the element and the surrounding air temperature is T_∞.

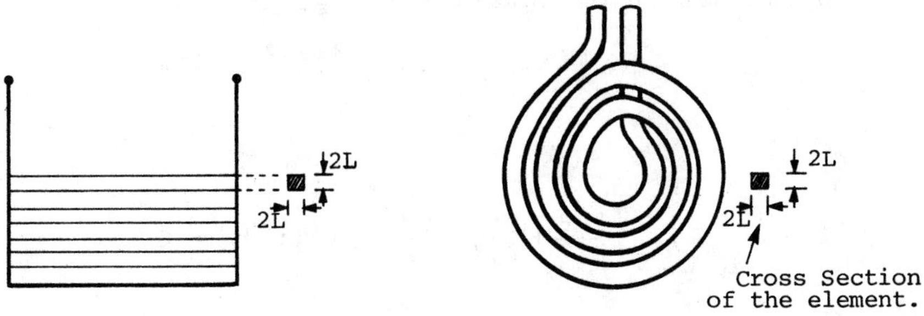

Fig. 1

Solution: The polynomials which satisfy the boundary conditions are represented by the series

$$\theta(x,y) = (L^2-x^2)(\ell^2-y^2)(a_0+a_1x^2+a_2y^2+....) \tag{1}$$

$$\Theta(x,y) = a_0(L^2-x^2)(\ell^2-y^2) + a_1(L^2-x^2)^2(\ell^2-y^2)^2 + ... \tag{2}$$

The section being analyzed is shown in figure 2, along with the reference frame. Due to symmetry (both thermal and geometric) of the section with respect to the x-y axis, the odd powers of x and y are zero and are not included in equations (1) and (2).

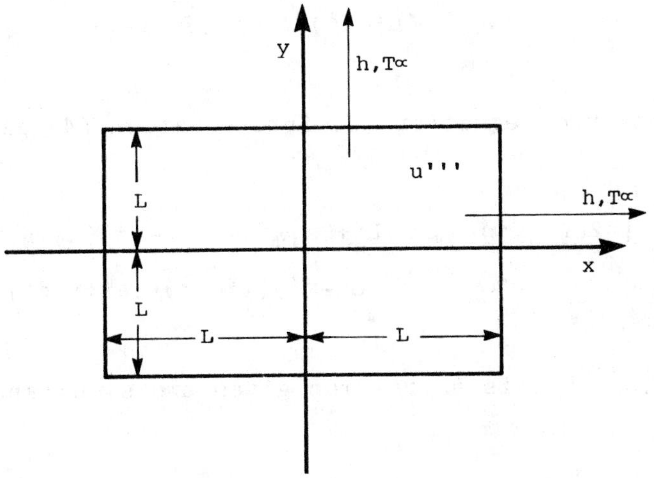

Fig. 2

Approximation of equation (1) gives

$$\theta(x,y) = (L^2-x^2)(\ell^2-y^2)a_0$$

Also,

$$\delta I = \int_{-L}^{L} \int_{-\ell}^{\ell} \left(\frac{\partial^2\theta}{\partial x^2} + \frac{\partial^2\theta}{\partial y^2} + \frac{u'''}{k}\right) \delta\theta \; dx \; dy = 0 \quad (4)$$

Inserting equation (3) into equation (4) yields the expression

$$4\int_0^L \int_0^\ell [-2(\ell^2-y^2)a_0 - 2(L^2-x^2)a_0 + u'''/k] \times (L^2-x^2)(\ell^2-y^2)\delta a_0 dx \; dy = 0.$$

Integration of this expression gives

$$a_0 = \frac{5}{8}\frac{u'''/k}{(\ell^2+L^2)}$$

Hence the first order solution, based on the variational formulation by Ritz method is

$$\frac{\theta(x,y)}{u''' \ell^2/k} = \frac{5}{8}\frac{[1-(x/L)^2][1-(y/\ell)^2]}{1+(\ell/L)^2} .$$

The second approximation of equation (1) is

$$\theta(x,y) = (L^2-x^2)(\ell^2-y^2)(a_0+a_1x^2) \quad (5)$$

Substitution of equation (5) into equation (4) gives

$$4\int_0^L \int_0^\ell [-2(\ell^2-y^2)(a_0-a_1L^2+6a_1x^2)-2(L^2-x^2)(a_0+a_1x^2)+u'''/k] \times (L^2-x^2)(\ell^2-y^2)(\delta a_0+x^2\delta a_1)dx \; dy = 0.$$

Integration of this expression gives two simultaneous equations.

$$(\ell^2+L^2)a_0 + \left(\frac{\ell^2}{5}+\frac{L^2}{7}\right)L^2a_1 = \frac{5}{8}\frac{u'''}{k} ,$$

$$\left(\frac{\ell^2}{5} + \frac{L^2}{7}\right)a_0 + \frac{1}{7}\left(\frac{11}{5}\,\ell^2 + \frac{L^2}{3}\right)L^2 a_1 = \frac{1}{8}\,\frac{u'''}{k}\,.$$

Solving these equations gives

$$a_0 = \frac{(u'''/16k)(L^2/3 + 24\ell^2/5)}{(L^4/21 + 8L^2\ell^2/15 + 12\ell^4/25)},$$

$$a_1 = \frac{(u'''/16k)}{(L^4/21 + 8L^2\ell^2/15 + 12\ell^4/25)}.$$

Thus the second order steady variational formulation, based on the Ritz method, is

$$\frac{\theta(x,y)}{u'''\ell^2/k} = \frac{1}{16}\left[1 - \left(\frac{x}{L}\right)^2\right]\left[1 - \left(\frac{y}{\ell}\right)^2\left[\frac{1/3 + (24/5)(\ell/L)^2 + (x/L)^2}{1/21 + (8/15)(\ell/L)^2 + (12/25)(\ell/L)^4}\right]\right.$$

CHAPTER 3

UNSTEADY STATE HEAT CONDUCTION: LUMPED PARAMETER ANALYSIS

Basic Attacks and Strategies for Solving Problems in this Chapter. **See pages 132 to 209 for step-by-step solutions to problems.**

Unsteady state conduction problems involve variation of temperature with time. There may or may not be variation of temperature with position in the solid also. Although, the general case of unsteady state conduction is quite complicated, there are several types of problems for which analytical solutions of the heat conduction equation are possible.

One such case, referred to as lumped parameter analysis, occurs when the temperature of a solid object is changing with time, but its temperature is nearly the same throughout the object at any given time. In other words, the temperature of the entire object changes together as a "lump". This situation would occur, in general, for small objects with high thermal conductivity and low surface convection coefficient. When using lumped parameter analysis, only temperature variation with time is needed, because a single temperature represents the entire object. The lumped parameter approach can be used when the Biot number is less than 0.1, where the biot number, Bi, *is given by:*

$$Bi = hV / kA .$$

The general equation which is used for lumped parameter analysis and the definition of the variables is given in the solution to problem 3.1 in this chapter.

If temperature is varying with position within the object as well as with the time, then the unsteady state conduction equation must be solved to find the temperature distribution as a function of both position and time. In order to set up the partial differential equation to be solved, it must be determined whether it is a one-dimensional or two-dimensional situation and whether rectangular, cylindrical, or spherical coordinates are needed. Then initial conditions and boundary conditions must be determined for the described physical situation. Finally a method of solving the differential equation subject to the initial and boundary conditions must be found. Example solutions of one-dimensional and two-dimensional forms of the unsteady conduction equation are given in this chapter.

A particular type of one-dimensional unsteady state case is that in which the solid object can be treated as a semi-infinite body. The semi-infinite body assumption is appropriate when the interior of the solid remains unaffected by the temperature variations taking place at its surface. The distinctive feature in the setup of the equations for a semi-infinite body problem is the use of a boundary condition stating that the temperature at $x = \infty$ *remains at the initial temperature. A variety of different surface boundary conditions may be present.*

Step-by-Step Solutions to Problems in this Chapter, "Unsteady State Heat Conduction: Lumped Parameter Analysis"

NON-PERIODIC CONDUCTION

• PROBLEM 3-1

Determine the average surface conductance per unit length of a long rod, 0.33 in. diameter, exposed to hot air blast at 110°F. The temperature of the rod has increased from 50°F to 70°F in 60 seconds. Assume the thermal conductivity for rod material as k=300 Btu/hr. ft. °F, ρ=600 lbm/ft^3, and Cp=0.18 Btu/lbm.°F.

Solution: The solution for the temperature variation with time is given by the equation

$$\frac{T - T_\infty}{T_0 - T_\infty} = e^{-hAt/\rho CpV} \qquad (1)$$

where,
- D = Diameter of the rod = 0.33 in. = 0.0275 ft.
- A = Surface area of the rod
- V = Volume of the rod
- T_∞ = Ambient temperature = 110°F
- T_0 = Initial temperature of the rod = 50°F
- T = Temperature at any instant of time
  = 70°F after 1.0 minutes of being exposed to hot air at 110°F.

The validity of the above equation depends on the magnitude of the Biot number. In other words, the temperature of the rod is uniform throughout.

The Biot number is defined as

$$Bi = \frac{hV}{kA} = \frac{h}{k}\frac{\pi D^2 L}{4\pi DL} \leq 0.1$$

$$= \frac{hD}{k4} = \frac{h \times 0.0275}{300 \times 4}$$

$$= 2.29 \times 10^{-5}h \leq 0.1$$

Taking the limiting value of the Biot number

i.e. Bi = 0.1

$$h = \frac{0.1}{2.29 \times 10^{-5}} = 4367 \text{ Btu/hr. ft}^2. \, ^0\text{F}.$$

Hence lumped-parameter analysis is applicable as h is less than 42000 Btu/hr. $ft^2.^0F$

Substituting the values in

$$h = \frac{\rho C p V}{tA} \ln \frac{T_0 - T_\infty}{T - T_\infty}$$

where $\rho = 600 \text{ lbm/ft}^3.$ $Cp = 0.18 \text{ Btu/lbm. } ^0F$

$$\frac{V}{A} = \frac{\text{Volume}}{\text{Area}} = \frac{\frac{\pi}{4} D^2 L}{\pi DL} = \frac{D}{4} = \frac{0.0275}{4} = 0.006875$$

$$t = 60 \text{ sec.} = \frac{60}{3600} \text{ hr.} = \frac{1}{60} \text{ hr.}$$

$$T_0 = 50\,^0F \quad T = 70\,^0F \quad \text{and} \quad T_\infty = 110\,^0F$$

Therefore

$$h = \frac{600 \times 0.18 \times \frac{0.0275}{4}}{\frac{1}{60}} \ln\left(\frac{50-110}{70-110}\right)$$

$$= 18.06 \text{ Btu/hr. ft}^2. \, ^0\text{F}.$$

• PROBLEM 3-2

A sudden discharge from a capacitor of an electrical device causes a time varying energy dissipation rate given by the equation $\dot{Q} = \dot{Q}_0^{-\beta\tau}$ where $\dot{Q}_0 = 30$ Btu/hr. and $\beta = 35(\text{hr})^{-1}$. The initial temperature of the device is the same as the ambient temperature and is equal to $80\,^0F$. The device has the following physical and thermal characteristics -

$h = 1.08 \text{ Btu/hr.ft}^2.^0F$, $Cp = 0.23 \text{ Btu/lbm. } ^0F$,

$k = 12\text{Btu/hr.ft. } ^0F$, mass = 0.08 lbm.,

surface area = 0.06 ft^2., volume = 0.03 ft^3.

Find (1) the relation between time and temperature in terms of the given data

and (2) the temperature of the device after 6 minutes have elapsed.

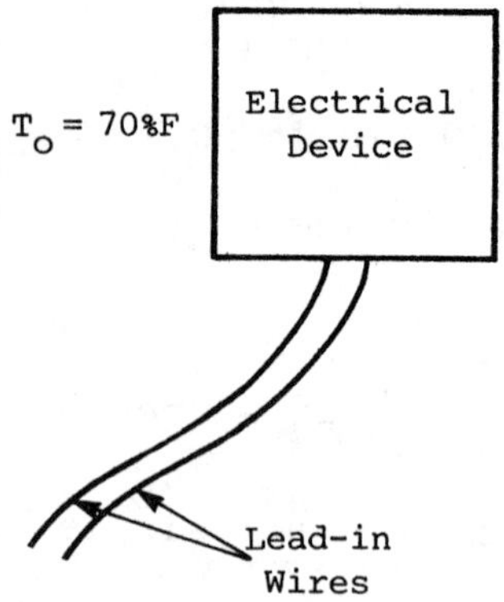

Solution: The criteria for applying lumped parameter method is

$$\frac{hs'}{k} \leq 0.10$$

where $s' = V_o/A_s$

V_o - Total volume of conduction region

A_s - Surface area for heat convection to the surroundings

Substituting the values in the equation

$$\frac{1 \times (0.03)}{(0.06)10} = 0.05 < 0.1$$

Therefore, the lumped parameter analysis can be applied.

By the energy balance principle

$$\dot{Q}_o\, e^{-\beta\tau} = hA_s(T - T_f) + mc_p\,\frac{dT}{d\tau}\,,$$

or with $\phi = hA_s/mc_p$ and $\theta = T - T_f$, Eq. (1) becomes

$$\frac{d\theta}{d\tau} + \phi\theta = \frac{\dot{Q}_0}{mc_p}\, e^{-\beta\tau}. \qquad (1)$$

The complementary function from the homogeneous equation

$$\frac{d\theta_c}{d\tau} + \phi\,\theta_c = 0 \quad \text{is}$$

$$\theta_c = B\, e^{-\phi\tau}$$

And, the particular solution can be constructed as

$$\theta_p = C_1 e^{-\beta\tau} + C_2 \frac{d}{d\tau}(e^{-\beta\tau}) + C_3 \frac{d^2}{d\tau^2}(e^{-\beta\tau}) + \ldots$$

As the expression $e^{-\beta\tau}$ is present in the first term, repeated differentiation results in no new functions of time, therefore

$$\theta_p = C_1 e^{-\beta\tau}$$

For the left side of equation (1) both the conditions,

$$d\theta_p/d\tau = -\beta C_1 e^{-\beta\tau}, \quad \text{and} \quad \phi\theta_p = \phi C_1 e^{-\beta\tau}$$

must be true. Thus equation (2) can be expressed as

$$-\beta C_1 e^{-\beta\tau} + \phi C_1 e^{-\beta\tau} = \frac{\dot{Q}_0}{mc_p} e^{-\beta\tau},$$

or

$$C_1(\phi - \beta)e^{-\beta\tau} = \frac{\dot{Q}_0}{mc_p} e^{-\beta\tau}. \tag{2}$$

This equation must hold for all values of time, therefore, like coefficients on either side of the equation can be equated

$$C_1(\phi - \beta) = \frac{\dot{Q}_0}{mc_p} \quad \text{or} \quad C_1 = \frac{\dot{Q}_0}{mc_p(\phi - \beta)}.$$

Thus,

$$\theta_p = \frac{\dot{Q}_0 e^{-\beta\tau}}{mc_p(\phi - \beta)}.$$

The complete solution of equation (1) is given by

$$\theta = \theta_c + \theta_p$$

or

$$\theta = Be^{-\phi\tau} + \frac{\dot{Q}_0 e^{-\beta\tau}}{mc_p(\phi - \beta)} \tag{3}$$

The constant B can be evaluated by applying the initial condition at $\tau = 0$, $T = T_0 = T_f$, hence $\theta = 0$.

Substituting in equation (3) yields

$$0 = B + \frac{\dot{Q}_0}{mc_p(\phi - \beta)}$$

or

$$B = -\frac{\dot{Q}_0}{mc_p(\phi - \beta)}$$

Substituting B into equation (3) yields

$$\theta = T - T_f = \frac{\dot{Q}_0}{mc_p(\phi - \beta)}(e^{-\beta\tau} - e^{-\phi\tau}).$$

(2) Temperature of the device after 5 min.

$$\phi = \frac{hA_s}{mC_p} = \frac{(1.08)\times(0.06)}{(0.08)\times(0.23)} = 3.52 \quad \text{and } \tau = 6 \text{ min} = \frac{6}{60} \text{ hr, after 6 min,}$$

Substituting the values, therefore

$$T - 80 = \frac{30}{(0.08)(0.23)(3.522 - 35)} \times \left[e^{-30\times\frac{6}{60}} - e^{-3.522\times\frac{6}{60}} \right]$$

$$= 33.84 \simeq 34$$

Therefore $T = 80 + 34 = 114\,^0F.$

• PROBLEM 3-3

A wire, 5/8 in. diameter, at a temperature of $500\,^0F$ is suddenly exposed to a cooling medium which is at $110\,^0F$. Evaluate the temperature response when the cooling medium is

a) water ($h = 15$ Btu/hr.ft^2. ^0F.) and

b) air ($h = 2$ Btu/hr.ft^2.^0F.)

Given the properties of the wire material

$$\rho = 560 \text{ lb/ft}^3, \quad k = 216 \text{ Btu/hr.ft.}^0F$$

and $C_p = 0.09$ Btu/lb.^0F.

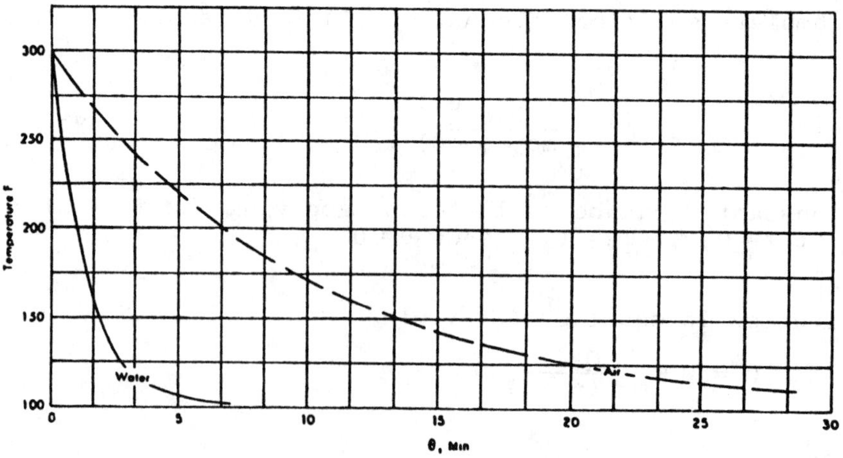

Cooling of wire in air and water.

Solution: The Biot number is defined as

$$Bi = \frac{hV}{kA}$$

where h - convective heat transfer coefficient

k - thermal conductivity of the wire material

V - volume of the wire

A - surface area of the wire

Therefore for water

$$Bi = \frac{(15) \times \pi/4\ D^2L}{216 \times \pi\ DL}$$

$$= \frac{15}{216} \times \frac{D}{4} = \frac{15}{216} \times \frac{(5/8 \times 12)}{4}$$

$$= 9.04 \times 10^{-4}$$

which is less than 0.1, hence the lumped parameter analysis is valid for the temperature history of the wire. The Biot number will be less for air as the cooling medium.

The temperature history can be found from the lumped parameter analysis equation as

$$\frac{T - T_\infty}{T_0 - T_\infty} = e^{-\frac{h\ A\ t}{\rho C_p\ V}} = e^{-\frac{h \times t}{\rho \times C_p} \times \frac{A}{V}}$$

where T is the temperature of the cooling wire at any instant of time t

T_0 is the temperature of the wire at time $t = 0$

T_∞ is the temperature of the cooling fluid and remains constant during the process of cooling.

Substituting the known values in the equation, yields

$$\frac{T - 110.0}{500 - 110.0} = \frac{T - 110.0}{390} = e^{-\frac{ht}{0.09 \times 560} \times \frac{A}{V}}$$

$$T - 110 = 390\ e^{-\frac{ht \times 4 \times 12}{0.09 \times 560 \times 5/8}}\ {}^0F$$

$$= 390\ e^{-1.5238 \times ht}\ {}^0F$$

For air $$= 390\ e^{-1.5238 \times 2 \times t}\ {}^0F$$

$$= 390\ e^{-3.04t}\ {}^0F$$

and for water $$= 390\ e^{-22.8t}\ {}^0F$$

Thus, the temperature history for cooling wire

in air $T_a = 110 + 390\ e^{-3.04t}\ {}^0F$ and

in water $T_w = 110 + 390\ e^{-22.8t}\ {}^0F$

Cooling curves are shown in the figure for both air and water.

Find the temperature distribution in the slab having initial uniform temperature $T = 0$. For time $t > 0$, one end of the slab ($x = L$) is maintained at the same initial temperature while the other end ($x = 0$) is exposed to a time varying temperature $T = f(t)$.

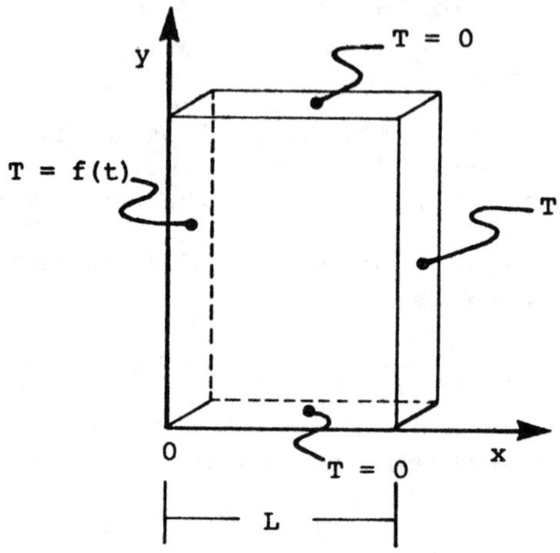

Fig. 1

Solution: The differential equation for uni-directional heat conduction is expressed as

$$\frac{\partial^2 T(x,t)}{\partial x^2} = \frac{1}{\alpha}\frac{\partial T(x,t)}{\partial t} \quad \text{at } t > 0$$

The given boundary conditions are

Boundary conditions
$$\begin{cases} T(x,t) = f(t) & \text{at} \quad x = 0, \quad t > 0 \\ T(x,t) = 0 & \text{at} \quad x = L, \quad t > 0 \end{cases}$$

Initial condition
$$T(x,0) = 0 \qquad 0 \le x \le L \text{ at } t = 0$$

Considering the Laplace transform of the heat conduction equation

$$\frac{d^2 T(x,s)}{dx^2} - \frac{s}{\alpha}\,\overline{T}(x,s) = 0 \quad \text{in } 0 < x < L$$

$$\overline{T}(x,s) = \overline{f}_1(s) \quad \text{at } x = 0$$

$$\overline{T}(x,s) = 0 \qquad \text{at } x = L$$

The solution of the above transformed equation can be expressed as

$$\overline{T}(x,s) = \overline{f}(s) \cdot \overline{g}(x,s)$$

where $$\overline{g}(x,s) \quad \frac{\sinh[(L-x)\sqrt{s/\alpha}]}{\sinh(L\sqrt{s/\alpha})} \tag{1}$$

As the form of the function $\overline{f}(s)$ is not defined, the Laplace transform of a convolution is used to invert this transform.

$$\overline{T}(x,s) = \overline{f}(s) \cdot \overline{g}(x,s) = L[f(t) * g(x,t)]$$

and the inversion

$$T(x,t) = f(t) * g(x,t)$$

Further, by the definition of convolutions

$$T(x,t) = \int_o^t f(\tau)\ g(x,t-\tau)d\tau \tag{2}$$

The Laplace transform of the function $\overline{g}(x,t)$ is given by equation (1). The inversion of $\overline{g}(x,s)$ is obtained from the inversion formula

$$F(t) = \frac{1}{2\pi i} \int_{(s=\gamma-i\infty)}^{(\gamma+i\infty)} e^{st}\ \overline{F}(s)ds$$

which yields

$$g(x,t) = \frac{1}{2\pi i} \int_{s=\gamma-i\infty}^{\gamma+i\infty} e^{st}\ \frac{\sinh[(L-x)\sqrt{s/\alpha}]}{\sinh(L\sqrt{s/\alpha})} ds \tag{3}$$

The term $\sinh(L\sqrt{s/\alpha})$ is zero when $\sqrt{s/\alpha}$ is purely imaginary, therefore let $\sqrt{s/\alpha} = i\beta$.

$$\sinh(L\sqrt{s/\alpha}) = \sinh(i\beta L) = i\ \sin\beta L = 0$$

The function $\sin\beta L$ is zero for

$$\beta_n L = n\pi \quad \text{or} \quad \beta_n = \frac{n\pi}{L}$$

where $n = 1,2,3,\ldots.$ then the poles are at the locations

$$\sqrt{s_n/\alpha} = i\beta_n \quad \text{or} \quad s_n = -\alpha\beta_n^2 .$$

The expression inside the integral sign in equation (3) has simple poles at $s_n = -\alpha\beta_n^2$. Everywhere else on the s-plane, the expression is analytic. The integral around the curve ABC in Fig. 2 vanishes, leaving

$$g(x,t) = \sum_{n=1}^{\infty} \text{res}(s_n)$$

The $res(s_n)$ correspond to the poles at $s_n = -\alpha\beta_n^2$. The expression $res(s_n)$ can be evaluated by rewriting the expression in integral sign in equation (3) in the form

$$(\text{Integrand}) = \frac{N(s)}{M(s)} \tag{4}$$

where

$$N(s) = e^{st} \sinh\left[(L-x)\sqrt{\frac{s}{\alpha}}\right]$$

$$M(s) = \sinh\left(L\sqrt{\frac{s}{\alpha}}\right)$$

$M(s)$ has simple roots at $s_n = -\alpha\beta_n^2$, $\beta_n = n\pi/L$, $n = 1,2,3\ldots$ Then, the residues are determined according to the equation

$$\sum_{k=1}^{m} res(s = a_k) = \sum_{k=1}^{m} \frac{N(a_k)}{a_k^r [dM(s)/ds]_{s=a_k}}$$

results in

$$g(x,t) = \sum_{n=1}^{\infty} res(s_n) = \sum_{n=1}^{\infty} \frac{N(s_n)}{[dM(s)/ds]_{s=s_n}} \tag{5}$$

where

$$s_n = -\alpha\beta_n^2$$

$$N(s_n) = e^{-\alpha\beta_n^2 t} \sinh[i\beta_n(L-x)] = -i(-1)^n e^{-\alpha\beta_n^2 t} \sin\beta_n x$$

$$\left[\frac{dM(s)}{ds}\right]_{s=s_n} = \left[\frac{d}{ds}\sinh\left(L\sqrt{\frac{s}{\alpha}}\right)\right]_{s=s_n} = (-1)^n \frac{L}{2i\alpha\beta_n}$$

Thus, the function $g(x,t)$ becomes

$$g(x,t) = \frac{2\alpha}{L}\sum_{n=1}^{\infty} e^{-\alpha\beta_n^2 t}\beta_n \sin\beta_n x \tag{6}$$

Substituting $(t-\tau)$ for t in equation (6) and putting the result into equation (2) yields the solution for temperature distribution

$$T(x,t) = \frac{2\alpha}{L}\sum_{n=1}^{\infty}\beta_n \sin\beta_n x \int_{\tau=0}^{t} f(\tau)e^{-\alpha\beta_n^2(t-\tau)}d\tau \tag{7}$$

where

$$\beta_n = \frac{n\pi}{L}$$

Further, considering a condition $f(t) = T_1$ = constant as a case the equation (7) reduces to

$$\frac{T(x,t)}{T_1} = \frac{2}{L} \sum_{n=1}^{\infty} \frac{\sin\beta_n x}{\beta_n} (1 - e^{-\alpha\beta_n^2 t}) \tag{8}$$

Now, as $t \to \infty$, the equation represents the steady state condition

$$T(x,\infty) = T_1(1 - x/L)$$

or

$$1 - \frac{x}{L} = \frac{2}{L} \sum_{n=1}^{\infty} \frac{\sin\beta_n x}{\beta_n} \tag{9}$$

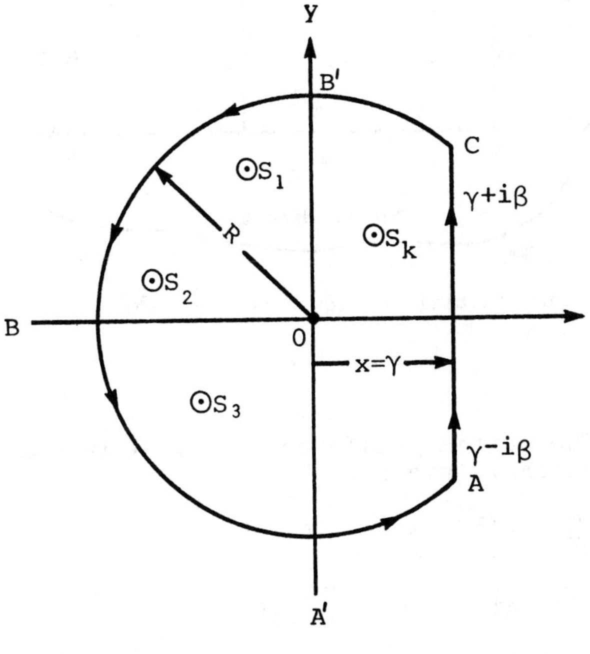

Fig. 2 Contour for the inversion problem.

Substitution of equation (9) into equation (8), gives the solution for $f(\tau) = T_1$ as

$$\frac{T(x,t)}{T_1} = \left(1 - \frac{x}{L}\right) - \frac{2}{L} \sum_{n=1}^{\infty} e^{-\alpha\beta_n^2 t} \frac{\sin\beta_n x}{\beta_n}$$

• PROBLEM 3-5

Find the temperature distribution T(x,y,t) for a rectangular plate with the following boundary conditions. Initially the plate is maintained at temperature F(x,y). Then for time $t > 0$ the boundary at $x = 0$ is insulated and that at $y = 0$ is kept at zero temperature. At the other two ends, $x = a$ and $y = b$, heat is dissipated by convection to the surroundings at temperature equal to zero.

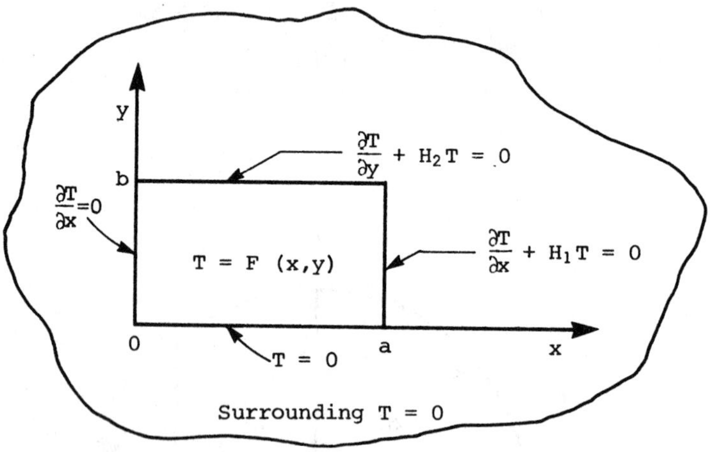

Fig. Initial configuration at t = 0.

Solution: The partial differential two-dimensional heat conduction equation is given by

$$\frac{\partial^2 T}{\partial x^2} + \frac{\partial^2 T}{\partial y^2} = \frac{1}{\alpha}\frac{\partial T}{\partial t} \tag{A}$$

B.C's at $x = 0$, $\frac{\partial T}{\partial x} = 0$

at $y = 0$, $T = 0$

at $x = a$, $t > 0$, $\frac{\partial T}{\partial x} + H_1 T = 0$

at $y = b$, $t > 0$, $\frac{\partial T}{\partial y} + H_2 T = 0$

and I.C. at $t = 0$, $T = F(x,y)$

Let the solution be of the form

$$T(x,y,t) = \Gamma(t) * X(x) * Y(y)$$

Separating the variables of equation (A) yields

$$\frac{d^2X(x)}{dx^2} + \beta^2 X(x) = 0 \quad \text{for} \quad 0 < x < a \tag{1}$$

$$\frac{dX}{dx} = 0 \quad \text{at } x = 0; \quad \frac{dX}{dx} + H_1 X = 0 \quad \text{at } x = a \tag{2}$$

and

$$\frac{d^2Y(y)}{dy^2} + \gamma^2 Y(y) = 0 \quad \text{in } 0 < y < b \tag{3}$$

$$Y = 0 \quad \text{at } y = 0; \quad \frac{dY}{dy} + H_2 Y = 0 \quad \text{at } y = b \tag{4}$$

Solution for $\Gamma(t)$ can be written as

$$\Gamma(t) = e^{-\alpha(\beta^2+\gamma^2)t} \tag{5}$$

The complete solution can be written as

$$T(x,y,t) = \sum_{m=1}^{\infty} \sum_{n=1}^{\infty} C_{mn} e^{-\alpha(\beta_m^2+\gamma_n^2)t} X(\beta_m,x)Y(\gamma_n,y) \tag{6}$$

Applying the initial condition at $t = 0$, equation (6) reduces to

$$F(x,y) = \sum_{m=1}^{\infty} \sum_{n=1}^{\infty} C_{mn} X(\beta_m,x)Y(\gamma_n,y) \quad \text{in } 0<x<a, 0<y<b \tag{7}$$

C_{mn} can be determined by successively multiplying both sides of equation (7) by

$$\int_0^a X(\beta_m,x)dx \quad \text{and} \int_0^b Y(\gamma_n,y)dx$$

Then applying the condition of orthogonality of the eigen functions results in

$$C_{mn} = \frac{1}{N(\beta_m)N(\gamma_n)} \int_{x'=0}^{a} \int_{y'=0}^{b} X(\beta_m,x')Y(\gamma_n,y')F(x',y')dx'dy' \tag{8}$$

where $\quad N(\beta_m) \equiv \int_0^a X^2(\beta_m,x)dx \quad$ and $\quad N(\gamma_n) \equiv \int_0^b Y^2(\gamma_n,y)dy$

Inserting equation (8) in (6) yields the complete solution for temperature distribution in a rectangular plate.

$$T(x,y,t) = \sum_{m=1}^{\infty} \sum_{n=1}^{\infty} e^{-\alpha(\beta_m^2+\gamma_n^2)t} \frac{1}{N(\beta_m)N(\gamma_n)} X(\beta_m x)Y(\gamma_n,y)$$

$$\times \int_0^a \int_0^b X(\beta_m,x')\cdot Y(\gamma_n,y')F(x',y')dx'dy' \tag{9}$$

The values for eigenfunctions, eigenvalues, and the norms can be found from the table.

Table 1. The Solution X (β_m, x), the Norm N (β_m) and the Eigenvalues β_m of the Differential Equation

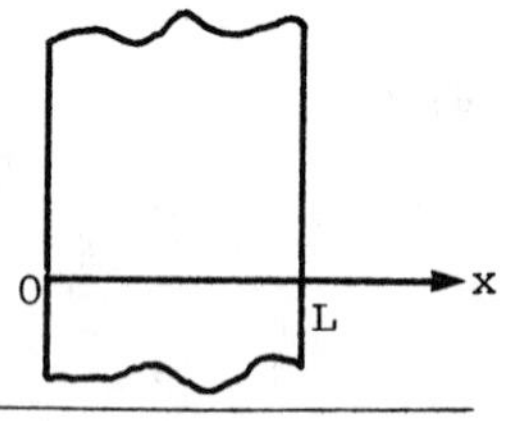

$$\frac{d^2X(x)}{dx^2} + \beta^2 X(x) = 0 \qquad 0 < x < L$$

Subject to Boundary Conditions given in the Table

No	Boundary Condition at $x = 0$	Boundary Condition at $x = L$	$X(\beta_m, x)$	$1/N(\beta_m)$	Eigenvalues β_m's are Positive Roots of
1	$-\frac{dX}{dx} + H_3 X = 0$	$\frac{dX}{dx} + H_1 X = 0$	$\beta_m \cos\beta_m x + H_3 \sin\beta_m x$	$2\left[(\beta_m^2 + H_3^2)\left(L + \frac{H_1}{\beta_m^2 + H_1^2}\right) + H_3\right]^{-1}$	$\tan\beta_m L = \frac{\beta_m(H_3 + H_1)}{\beta_m^2 - H_3 H_1}$
2	$-\frac{dX}{dx} + H_3 X = 0$	$\frac{dX}{dx} = 0$	$\cos\beta_m(L - x)$	$2\frac{\beta_m^2 + H_3^2}{L(\beta_m^2 + H_3^2) + H_3}$	$\beta_m \tan\beta_m L = H_3$
3	$-\frac{dX}{dx} + H_3 X = 0$	$X = 0$	$\sin\beta_m(L - x)$	$2\frac{\beta_m^2 + H_3^2}{L(\beta_m^2 + H_3^2) + H_3}$	$\beta_m \cot\beta_m L = -H_3$
4	$\frac{dX}{dx} = 0$	$\frac{dX}{dx} + H_1 X = 0$	$\cos\beta_m x$	$2\frac{\beta_m^2 + H_1^2}{L(\beta_m^2 + H^2) + H_1}$	$\beta_m \tan\beta_m L = H_1$
5	$\frac{dX}{dx} = 0$	$\frac{dX}{dx} = 0$	*$\cos\beta_m x$	$\frac{2}{L}$ for $\beta_m \neq 0$; $\frac{1}{L}$ for $\beta_0 = 0$*	$\sin\beta_m L = 0$*
6	$\frac{dX}{dx} = 0$	$X = 0$	$\cos\beta_m x$	$\frac{2}{L}$	$\cos\beta_m L = 0$
7	$X = 0$	$\frac{dX}{dx} + H_1 X = 0$	$\sin\beta_m x$	$2\frac{\beta_m^2 + H_1^2}{L(\beta_m^2 + H_1^2) + H_1}$	$\beta_m \cot\beta_m L = -H_1$
8	$X = 0$	$\frac{dX}{dx} = 0$	$\sin\beta_m x$	$\frac{2}{L}$	$\cos\beta_m L = 0$
9	$X = 0$	$X = 0$	$\sin\beta_m x$	$\frac{2}{L}$	$\sin\beta_m L = 0$

For example, $X(\beta_m, x) = \cos(\beta_m x)$ corresponding to boundary conditions at $x = 0$ and at $x = L$ i.e.

$$\frac{dX}{dx} = 0 \quad \text{and} \quad \frac{dX}{dx} + H_1 x = 0$$

also,

$$\frac{1}{N(\beta_m)} = 2\,\frac{\beta_m^2 + H_1^2}{a(\beta_m^2 + H_1^2) + H_1} \tag{10}$$

and β_m's are the positive roots of

$$\beta_m \tan\beta_m a = H_1$$

Similarly for boundary conditions at $x = 0$, $\frac{dX}{dx} + H_1 x = 0$, the function is

$$Y(\gamma_n, y) = \sin\gamma_n y$$

$$\frac{1}{N(\gamma_n)} = 2\ \frac{\gamma_n^2+H_2^2}{b(\gamma_n^2+H_2^2)+H_2} \tag{11}$$

and γ_n's are the positive roots of

$$\gamma_n \cot\gamma_n b = -H_2$$

Substituting equations (10) and (11) into equation (9), the expression for temperature distribution transforms to

$$T(x,y,t) = 4 \sum_{m=1}^{\infty} \sum_{n=1}^{\infty} e^{-\alpha(\beta_m^2+\gamma_n^2)t} \cdot \frac{\beta_m^2+H_1^2}{a(\beta_m^2+H_1^2)+H_1}\ \frac{\gamma_n^2+H_2^2}{b(\gamma_n^2+H_2^2)+H_2}$$

$$\cdot \cos\beta_m x \sin\gamma_n y \int_{x'=0}^{a} \int_{y'=0}^{b} \cos\beta_m x' \sin\gamma_n y' F(x',y')dx'dy'$$

where β_m and γ_n are the positive roots of the equations (10) and (11) respectively.

• PROBLEM 3-6

Derive an equation for the temperature distribution $T(r,\phi,t)$ in a segment of a solid cylinder initially, at time $t = 0$, at a temperature $T = F(r,\phi)$. For time $t > 0$ the boundaries of the segment at $r = b$, $\phi = 0$ and $\phi = \phi_0$ are maintained at constant temperature $T = 0$.

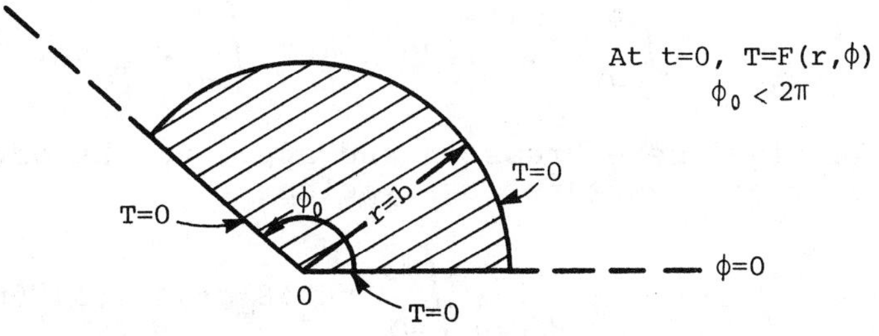

Fig. 1 Segment of a cylinder.

Solution: The Fourier equation for heat conduction is expressed as

$$\frac{\partial^2 T}{\partial r^2} + \frac{1}{r}\frac{\partial T}{\partial r} + \frac{1}{r^2}\frac{\partial^2 T}{\partial \phi^2} = \frac{1}{\alpha}\frac{\partial T}{\partial t} \quad \text{for } 0 \le r < b,\ 0 < \phi < \phi_0 \text{ at } t > 0$$

$$T = 0 \qquad \text{at } r=b,\ \phi=0 \text{ and } \phi=\phi_0 \text{ for } t>0$$

$$T = F(r,\phi) \qquad \text{for the segment at } t=0$$

Assuming the separated solutions as

$$e^{-\alpha\beta_m^2 t}, \quad \Phi(\nu,\phi), \quad \text{and} \quad R_\nu(\beta_m,r)$$

where $\Phi(\nu,\phi)$ is the solution of equation

$$\frac{d^2\Phi}{d\phi^2} + \nu^2\,\Phi = 0 \quad \text{for } 0<\phi<\phi_0 (\phi_0 < 2\pi)$$

with boundary conditions $\Phi(\nu,\phi)=0$ at $\phi=0$ and $\phi=\phi_0$; the expression $R_\nu(\beta_m,r)$ is the solution of the Bessel's differential equation subject to the boundary conditions $R_\nu(\beta_m,r)=0$ at $r=b$ and $R_\nu(\beta_m,r)$ remain finite at $r=0$.

The complete solution for temperature distribution in the segment $T(r,\phi,t)$ can be obtained by summing up the separated solutions and can be expressed as

$$T(r,\phi,t) = \sum_{m=1}^{\infty}\sum_{\nu} C_{m\nu} R_\nu(\beta_m,r)\Phi(\nu,\phi)e^{-\alpha\beta_m^2 t} \tag{1}$$

At $t=0$ the temperature is $T=F(r,\phi)$, then

$$F(r,\phi) = \sum_{m=1}^{\infty}\sum_{\nu} C_{m\nu} R_\nu(\beta_m,r)\Phi(\nu,\phi) \text{ in } 0\leq r<b,\ 0<\phi<\phi_0$$

In order to obtain the coefficient $C_{m\nu}$, both sides of the equation are successively operated by

$$\int_{\phi=0}^{\phi_0} \Phi(\nu',\phi)d\phi \quad \text{and} \quad \int_{r=0}^{b} rR_\nu(\beta_{m'},r)dr$$

Simplifying the expression and applying the orthogonality property of eigenfunctions yields

$$C_{m\nu} = \frac{1}{N(\beta_m)N(\nu)}\int_{r=0}^{b}\int_{\phi=0}^{\phi_0} rR_\nu(\beta_m,r)\Phi(\nu,\phi)F(r,\phi)d\phi dr$$

Substituting the expression for $C_{m\nu}$ into equation (1) gives

$$T(r,\phi,t) = \sum_{m=1}^{\infty}\sum_{\nu} \frac{e^{-\alpha\beta_m^2 t}}{N(\beta_m)N(\nu)} R_\nu(\beta_m,r)\Phi(\nu,\phi)$$

$$\cdot \int_{r'=0}^{b}\int_{\phi'=0}^{\phi_0} r'R_\nu(\beta_m,r')\Phi(\nu,\phi')F(r',\phi')d\phi' dr' \tag{5}$$

Table 1 The Eigenfunctions $R_\nu(\beta_m, r)$, the Norm $N(\beta_m)$, and the Eigenvalues β_m of the Differential Equation

$$\frac{d^2R_\nu}{dr^2}+\frac{1}{r}\frac{dR_\nu}{dr}+\left(\beta^2-\frac{\nu^2}{r^2}\right)R_\nu=0 \quad \text{in} \quad 0\le r<b$$

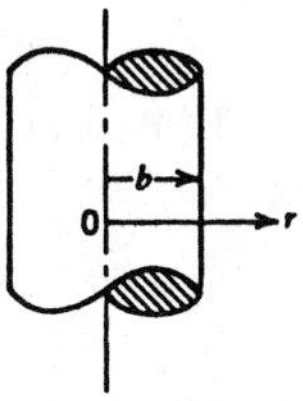

Subject to the Shown Boundary Conditions

No	Boundary Condition at $r=b$	$R_\nu(\beta_m, r)$	$\frac{1}{N(\beta_m)}$	Eigenvalues β_m's are the Positive Roots of
1	$\frac{dR_\nu}{dr}+HR_\nu=0$	$J_\nu(\beta_m r)$	$\frac{2}{J_\nu^2(\beta_m b)}\cdot\frac{\beta_m^2}{b^2(H^2+\beta_m^2)-\nu^2}$	$\beta_m J'_\nu(\beta_m b)+HJ_\nu(\beta_m b)=0$
2	$\frac{dR_\nu}{dr}=0$	$J_\nu(\beta_m r)$*	$\frac{2}{J_\nu^2(\beta_m b)}\cdot\frac{\beta_m^2}{b^2\beta_m^2-\nu^2}$ *	$J'_\nu(\beta_m b)=0$*
3	$R_\nu=0$	$J_\nu(\beta_m r)$	$\frac{2}{b^2J'^2_\nu(\beta_m b)}$	$J_\nu(\beta_m b)=0$

where $R_\nu(\beta_m,r)$, $N(\beta_m)$, and β_m's are obtained from Table (1), Case 3, as

$$R_\nu(\beta_m,r) = J_\nu(\beta_m r), \quad \frac{1}{N(\beta_m)} = \frac{2}{b^2J'^2_\nu(\beta_m b)} \tag{6}$$

and β_m's are the positive roots of

$$J_\nu(\beta_m b) = 0$$

From table (2) the solution, the norm and eigenvalues are obtained and expressed with appropriate change of notations as

$$\Phi(\nu,\phi) = \sin \nu\phi, \quad \frac{1}{N(\nu)} = \frac{2}{\phi_0} \tag{7}$$

and ν's are the roots of

$$\sin \nu\phi_0 = 0$$

The complete solution for temperature distribution can be obtained by substituting equations (6) and (7) into equation (5) giving

$$T(r,\phi,t) = \frac{4}{b^2\phi_0}\sum_{m=1}^{\infty}\sum_{\nu} e^{-\alpha\beta_m^2 t}\,\frac{J_\nu(\beta_m r)}{J_\nu'^2(\beta_m b)}\sin\nu\phi \tag{8}$$

$$\cdot \int_{r'=0}^{b}\int_{\phi'=0}^{\phi_0} r' J_\nu(\beta_m r')\sin\nu\phi' F(r',\phi')\,d\phi'\,dr'$$

where β_m's are roots of $J_\nu(\beta_m b) = 0$, and ν's are given by

$$\nu = \frac{n\pi}{\phi_0}, \quad n = 1,2,3\ldots$$

Now, if the initial temperature $F(r,\phi)$ is constant, i.e. $F(r,\phi) = T_0$ then the solution equation (8) changes to

$$T(r,\phi,t) = \frac{8T_0}{b^2\phi_0}\sum_{m=1}^{\infty}\sum_{\nu} e^{-\alpha\beta_m^2 t}\,\frac{J_\nu(\beta_m r)}{J_\nu'^2(\beta_m b)}\,\frac{\sin\nu\phi}{\nu}\int_{r'=0}^{b} r' J_\nu(\beta_m r')\,dr'$$

where β_m's are the roots of $J_\nu(\beta b) = 0$, and ν's are given by

$$\nu = \frac{(2n-1)\pi}{\phi_0}, \quad n = 1,2,3\ldots$$

Table 2 The Solution $X(\beta_m, x)$, the Norm $N(\beta_m)$ and the Eigenvalues β_m of the Differential Equation

$$\frac{d^2X(x)}{dx^2} + \beta^2 X(x) = 0 \quad \text{in} \quad 0 < x < L$$

Subject to the Boundary Conditions Shown in the Table

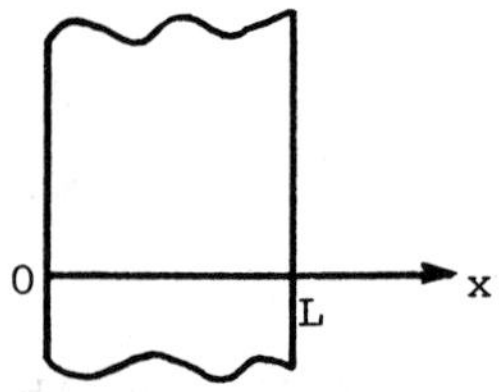

No	Boundary Condition at $x = 0$	Boundary Condition at $x = L$	$X(\beta_m, x)$	$1/N(\beta_m)$	Eigenvalues β_m's are Positive Roots of
1	$-\frac{dX}{dx} + H_1X = 0$	$\frac{dX}{dx} + H_2X = 0$	$\beta_m\cos\beta_m x + H_1\sin\beta_m x$	$2\left[(\beta_m^2 + H_1^2)\left(L + \frac{H_2}{\beta_m^2 + H_2^2}\right) + H_1\right]^{-1}$	$\tan\beta_m L = \frac{\beta_m(H_1 + H_2)}{\beta_m^2 - H_1H_2}$
2	$-\frac{dX}{dx} + H_1X = 0$	$\frac{dX}{dx} = 0$	$\cos\beta_m(L - x)$	$2\frac{\beta_m^2 + H_1^2}{L(\beta_m^2 + H_1^2) + H_1}$	$\beta_m\tan\beta_m L = H_1$
3	$-\frac{dX}{dx} + H_1X = 0$	$X = 0$	$\sin\beta_m(L - x)$	$2\frac{\beta_m^2 + H_1^2}{L(\beta_m^2 + H_1^2) + H_1}$	$\beta_m\cot\beta_m L = -H_1$
4	$\frac{dX}{dx} = 0$	$\frac{dX}{dx} + H_2X = 0$	$\cos\beta_m x$	$2\frac{\beta_m^2 + H_2^2}{L(\beta_m^2 + H_2^2) + H_2}$	$\beta_m\tan\beta_m L = H_2$
5	$\frac{dX}{dx} = 0$	$\frac{dX}{dx} = 0$	*$\cos\beta_m x$	$\frac{2}{L}$ for $\beta_m \neq 0$; $\frac{1}{L}$ for $\beta_0 = 0$*	$\sin\beta_m L = 0$*
6	$\frac{dX}{dx} = 0$	$X = 0$	$\cos\beta_m x$	$\frac{2}{L}$	$\cos\beta_m L = 0$
7	$X = 0$	$\frac{dX}{dx} + H_2X = 0$	$\sin\beta_m x$	$2\frac{\beta_m^2 + H_2^2}{L(\beta_m^2 + H_2^2) + H_2}$	$\beta_m\cot\beta_m L = -H_2$
8	$X = 0$	$\frac{dX}{dx} = 0$	$\sin\beta_m x$	$\frac{2}{L}$	$\cos\beta_m L = 0$
9	$X = 0$	$X = 0$	$\sin\beta_m x$	$\frac{2}{L}$	$\sin\beta_m L = 0$

*For this particular case $\beta_m = 0$ is also an eigenvalue corresponding to $X = 1$.

• PROBLEM 3-7

Find the temperature distribution in a composite block made up of two dissimilar materials (Fig.) initially at temperatures $T_1=F_1(x)$ and $T_2=F_2(x)$, where subscript 1 and 2 refer to first and second block. Through the outer surface of the block 2 heat is convected to the surrounding, while the outer surface of block 1 at x=0 is maintained at the surrounding temperature T=0. Assume the interface between the two blocks to be in perfect thermal contact.

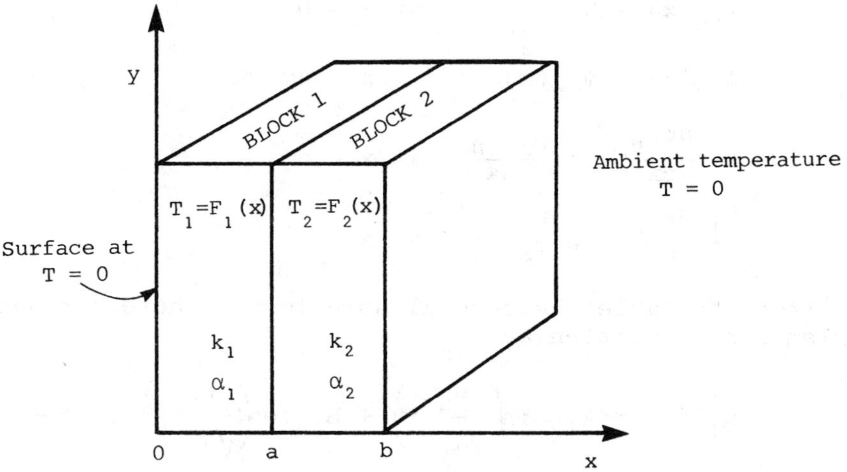

Solution: The heat transfer differential equation for the two slabs is expressed as

$$\alpha_1 \frac{\partial^2 T_1}{\partial x^2} = \frac{\partial T_1(x,t)}{\partial t} \quad \text{in } 0 < x < a, \quad t > 0$$

$$\alpha_2 \frac{\partial^2 T_2}{\partial x^2} = \frac{\partial T_2(x,t)}{\partial t} \quad \text{in } a < x < b, \quad t > 0 \tag{1}$$

The boundary conditions are

$$T_1(x,t) = 0 \qquad \text{at } x = 0, \ t > 0$$

$$T_1(x,t) = T_2(x,t) \qquad \text{at } x = a, \ t > 0$$

$$k_1 \frac{\partial T_1}{\partial x} = k_2 \frac{\partial T_2}{\partial x} \qquad \text{at } x = a, \ t > 0$$

$$k_2^* \frac{\partial T_2}{\partial x} + h_3 T_2 = 0 \qquad \text{at } x = b, \ t > 0$$

and the initial conditions

$$T_1(x,t) = F_1(x) \qquad \text{for } t = 0,\ 0 < x < a$$

$$T_2(x,t) = F_2(x) \qquad \text{for } t = 0,\ a < x < b$$

Expressing the equation (1) in terms of eigenvalues

$$\frac{d^2\psi_{1n}}{dx^2} + \frac{\beta_n^2}{\alpha_1}\psi_{1n}(x) = 0 \qquad \text{in } 0 < x < a$$

$$\frac{d^2\psi_{2n}}{dx^2} + \frac{\beta_n^2}{\alpha_2}\psi_{2n}(x) = 0 \qquad \text{in } a < x < b$$

and the corresponding boundary conditions are

$$\psi_{1n}(x) = 0 \qquad \text{at } x = 0$$

$$\psi_{1n}(x) = \psi_{2n}(x) \qquad \text{at } x = a$$

$$k_1 \frac{d\psi_{1n}}{dx} = k_2 \frac{d\psi_{2n}}{dx} \qquad \text{at } x = a$$

$$k_2^* \frac{d\psi_{2n}}{dx} + h_3\psi_{2n} = 0 \text{ at } x = b$$

From the table the general solution of the eigenvalue problem can be written as

$$\psi_{in}(x) = A_{in}\sin\left(\frac{\beta_n}{\sqrt{\alpha_i}}x\right) + B_{in}\cos\left(\frac{\beta_n}{\sqrt{\alpha_i}}x\right), \qquad i = 1,2$$

Linearly independant solutions to eigenvalue problems - ϕ in (x) and θ in (x) for different geometrical shapes - slabs, cylinders and spheres.

Geometry	$\phi_{in}(x)$	$\theta_{in}(x)$
Slab	$\sin\left(\frac{\beta_n}{\sqrt{\alpha_i}}x\right)$	$\cos\left(\frac{\beta_n}{\sqrt{\alpha_i}}x\right)$
Cylinder	$J_0\left(\frac{\beta_n}{\sqrt{\alpha_i}}x\right)$	$Y_0\left(\frac{\beta_n}{\sqrt{\alpha_i}}x\right)$
Sphere	$\frac{1}{x}\sin\left(\frac{\beta_n}{\sqrt{\alpha_i}}x\right)$	$\frac{1}{x}\cos\left(\frac{\beta_n}{\sqrt{\alpha_i}}x\right)$

Applying the boundary conditions the coefficient $B_{1n} = 0$ and $A_{1n} = 1$, the solution reduces to

$$\psi_{1n}(x) = \sin\left(\frac{\beta_n}{\sqrt{\alpha_1}} x\right) \qquad \text{in } 0 < x < a$$

$$\psi_{2n}(x) = A_{2n}\sin\left(\frac{\beta_n}{\sqrt{\alpha_2}} x\right) + B_{2n}\cos\left(\frac{\beta_n}{\sqrt{\alpha_2}} x\right) \text{in } a < x < b$$

Applying the other boundary conditions, the solution equation can be rewritten as

$$\sin\left(\frac{\beta_n a}{\sqrt{\alpha_1}}\right) = A_{2n}\sin\left(\frac{\beta_n a}{\sqrt{\alpha_2}}\right) + B_{2n}\cos\left(\frac{\beta_n a}{\sqrt{\alpha_2}}\right)$$

$$\frac{k_1}{k_2}\sqrt{\frac{\alpha_2}{\alpha_1}}\cos\left(\frac{\beta_n a}{\sqrt{\alpha_1}}\right) = A_{2n}\cos\left(\frac{\beta_n a}{\sqrt{\alpha_2}}\right) - B_{2n}\sin\left(\frac{\beta_n a}{\sqrt{\alpha_2}}\right)$$

$$\left[A_{2n}\cos\left(\frac{\beta_n b}{\sqrt{\alpha_2}}\right) - B_{2n}\sin\left(\frac{\beta_n b}{\sqrt{\alpha_2}}\right)\right] + \frac{h_3\sqrt{\alpha_2}}{k_2^*\beta_n}\left[A_{2n}\sin\left(\frac{\beta_n b}{\sqrt{\alpha_2}}\right) + B_{2n}\cos\left(\frac{\beta_n b}{\sqrt{\alpha_2}}\right)\right] = 0$$

and expressing in the form of a matrix

$$\begin{bmatrix} \sin\gamma & -\sin\left(\frac{a}{b}\eta\right) & -\cos\left(\frac{a}{b}\eta\right) \\ K\cos\gamma & -\cos\left(\frac{a}{b}\eta\right) & \sin\left(\frac{a}{b}\eta\right) \\ 0 & \frac{H}{\eta}\sin\eta+\cos\eta & \frac{H}{\eta}\cos\eta-\sin\eta \end{bmatrix} \begin{bmatrix} 1 \\ A_{2n} \\ B_{2n} \end{bmatrix} = \begin{bmatrix} 0 \\ 0 \\ 0 \end{bmatrix} \qquad (2)$$

where $\gamma \equiv \dfrac{a\beta_n}{\sqrt{\alpha_1}}$ $\quad \eta \equiv \dfrac{b\beta_n}{\sqrt{\alpha_2}}$ $\quad H \equiv \dfrac{bh_3}{k_2^*}$ $\quad K = \dfrac{k_1}{k_2}\sqrt{\dfrac{\alpha_2}{\alpha_1}}$

Now, there are three equations and two unknowns to be found. Therefore, taking first two equations, the matrix is reduced to

$$\begin{bmatrix} \sin\left(\frac{a}{b}\eta\right) & \cos\left(\frac{a}{n}\eta\right) \\ \cos\left(\frac{a}{b}\eta\right) & -\sin\left(\frac{a}{b}\eta\right) \end{bmatrix} \begin{bmatrix} A_{2n} \\ B_{2n} \end{bmatrix} = \begin{bmatrix} \sin\gamma \\ K\cos\gamma \end{bmatrix}$$

Thus, the coefficients A_{2n} and B_{2n} can be expressed as

$$A_{2n} = \frac{1}{\Delta}\left[-\sin\gamma \sin\left(\frac{a}{b}\eta\right) - K\cos\gamma\cos\left(\frac{a}{b}\eta\right)\right]$$

$$B_{2n} = \frac{1}{\Delta}\left[K\cos\gamma \sin\left(\frac{a}{b}\eta\right) - \sin\gamma\cos\left(\frac{a}{b}\eta\right)\right]$$

where

$$\Delta = -\sin^2\left(\frac{a}{b}\eta\right) - \cos^2\left(\frac{a}{b}\eta\right) = -1$$

To obtain the eigenvalues (β_n) the necessary condition is that the coefficients of the determinant equation (2) must vanish. This condition yields the transcendental equation for the determination of the eigenvalues (β_n)

$$\begin{vmatrix} \sin\gamma & -\sin\frac{a}{b}\eta & -\cos\frac{a}{b}\eta \\ K\cos\gamma & -\cos\frac{a}{b}\eta & \sin\frac{a}{b}\eta \\ 0 & \frac{H}{\eta}\sin\eta+\cos\eta & \frac{H}{\eta}\cos\eta-\sin\eta \end{vmatrix} = 0 \qquad (3)$$

Thus, knowing the coefficients and the eigenvalues (β_n) the eigenfunctions $\psi_{1n}(x)$ $\psi_{2n}(x)$ can be obtained. Then the temperature distribution in the slab can be determined from the equation

$$T_i(x,t) = \sum_{n=1}^{\infty} e^{-\beta_n^2 t}\frac{1}{N_n}\psi_{in}(x)\sum_{j=1}^{M}\frac{k_j}{\alpha_j}\int_{x'=x_j}^{x_{j+1}} x'^p\psi_{jn}(x')F_j(x')dx'$$

in $\quad x_i < x < x_{i+1}, \quad i = 1,2,\ldots,M$

where the norm N_n is defined as

$$N_n = \sum_{j=1}^{M}\frac{k_j}{\alpha_j}\int_{x_j}^{x_j+1} x'^p\psi_{jn}^2(x')dx'$$

where

$$p = \begin{cases} 0 & \text{slab} \\ 1 & \text{cylinder} \\ 2 & \text{sphere} \end{cases}$$

as

$$T_i(x,t) = \sum_{n=1}^{\infty}\frac{1}{N_n}e^{-\beta_n^2 t}\psi_{in}(x)\left[\frac{k_1}{\alpha_1}\int_{x'=0}^{a}\psi_{1n}(x')F_1(x')dx' + \frac{k_2}{\alpha_2}\int_a^b \psi_{2n}(x')F_2(x')dx'\right], \quad i = 1,2$$

where

$$N_n = \frac{k_1}{\alpha_1}\int_0^a \psi_{1n}^2(x')dx' + \frac{k_2}{\alpha_2}\int_a^b \psi_{2n}^2(x')dx'$$

$$\psi_{1n}(x) = \sin\left(\frac{\beta_n}{\sqrt{\alpha_1}}x\right)$$

$$\psi_{2n}(x) = A_{2n}\sin\left(\frac{\beta_n}{\sqrt{\alpha_2}}x\right) + B_{2n}\cos\left(\frac{\beta_n}{\sqrt{\alpha_2}}x\right)$$

• PROBLEM 3-8

A metal block in the shape of a right circular cone is initially heated in an oven to a temperature T_0. It is then removed and suspended in a cooling fluid which is at a temperature T_f. It has two different heat transfer coefficients: h_1 on the bottom face and h_2 on the curved surface. Find an expression for the cone temperature, as a function of time.

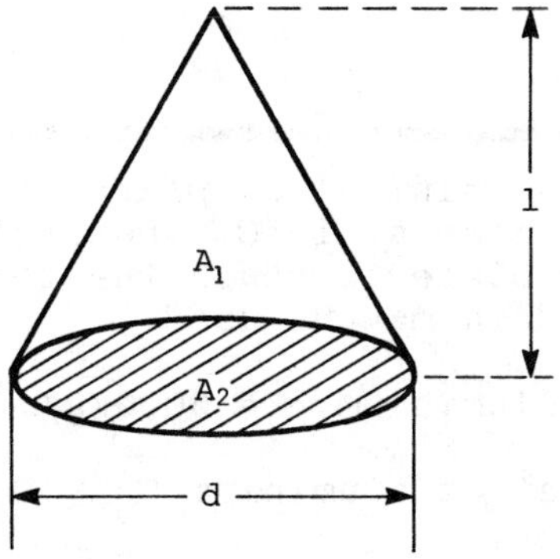

Solution: Assume the cone to represent the control volume and then apply the energy balance

$$Q_{in} + Q_{gen} = Q_{out} + Q_{stor} \tag{1}$$

The surface areas are as marked in the figure. The heat rate, out of the body, by Newton's cooling law is

$$Q_{out} = h_1A_1(T - T_f) + h_2A_2(T - T_f)$$

$$= (h_1A_1 + h_2A_2)(T - T_f) \tag{2}$$

Also
$$Q_{stored} = \rho C_p V_0 \frac{dT}{d\tau}, \quad Q_{gen} = Q_{in} = 0 \tag{3}$$

Substituting eqs.(2) and (3) into eq.(1)

$$0 = (h_1A_1 + h_2A_2)(T - T_f) + \rho\, C_p\, V_0 \frac{dT}{d\tau} \tag{4}$$

Defining
$$\psi = \frac{h_1A_1+h_2A_2}{\rho\, C_p\, V_0} + \theta = T - T_f \quad \text{such that} \quad \frac{d\theta}{d\tau} = \frac{dT}{d\tau}$$

eq.(4) becomes

$$\frac{d\theta}{d\tau} + \psi\theta = 0 \tag{5}$$

The solution to eq.(5) is

$$\theta = B_0\, e^{-\psi\tau} \tag{6}$$

At $\tau = 0$, $T = T_0$; therefore, $\theta = T_0 - T_f = \theta_0$ and from eq.(6) $B_0 = \theta_0$, hence eq.(6) becomes

$$\frac{T-T_f}{T_0-T_f} = e^{-\psi\tau}$$

• PROBLEM 3-9

A 3mm thick aluminium alloy plate at 200°C is suddenly quenched into water at 10°C. Determine the time required for the plate to reach the temperature of 30°C. The following data may be used.

Density of aluminium ρ = 3000kg/m^3

Specific heat of aluminium C_p = 0.2kcal/kg°C

Plate dimensions = 20 cm × 20 cm

Heat transfer coefficient h = 5000 kcal/m.^{2}hr.°C

Solution: The following relation is employed

$$\frac{t-t_s}{t_0-t_s} = e^{\frac{-hA}{\rho C_p Y}\tau}$$

where

t = temperature of the plate at any time τ

V = volume of the plate

t_s = temperature of the surrounding

t_0 = temperature of the plate at time $\tau = 0$

A = surface area of the plate.

For the given problem we have

$$t_0 = 200^0C, \quad t_s = 10^0C, \quad t = 30^0C$$

Neglecting the edge area, the area responsible for convective heat transfer

$$A = 2 \times 0.2 \times 0.2 = 0.08m^2$$

and $V = \frac{3}{1000} \times 0.2 \times 0.2 = 0.12 \times 10^{-3} m^2$

Hence

$$\frac{30 - 10}{200 - 10} = e^{\frac{-5000 \times 0.08}{3000 \times 0.2 \times 0.12 \times 10^{-3}} \tau}$$

or

$$\frac{20}{190} = e^{-5555 \tau}$$

or

$$\log_e\left(\frac{20}{190}\right) = -5555\tau$$

which gives $\tau = 4.052 \times 10^{-4} \times 3600 = 1.45$ seconds.

• PROBLEM 3-10

The initial uniform temperature of a steel rod, 3mm dia. and 10 cm. long, is 200^0C. Find the temperature distribution, after 0.01 seconds, of the rod immersed in a fluid medium at temperature 40^0C with heat transfer coefficient as $50W/m^2\ ^0C$. One end is maintained at the same initial temperature. Assume the properties of steel as follows:

density = $7800\ kg/m^3$

and

$C_p = 0.47\ J/kg^0C$.

Solution: The length of the rod is divided into four equal regions $\Delta x = \frac{0.1}{4} = 0.025$ meter. Each segment has a node at the end as shown in the figure.

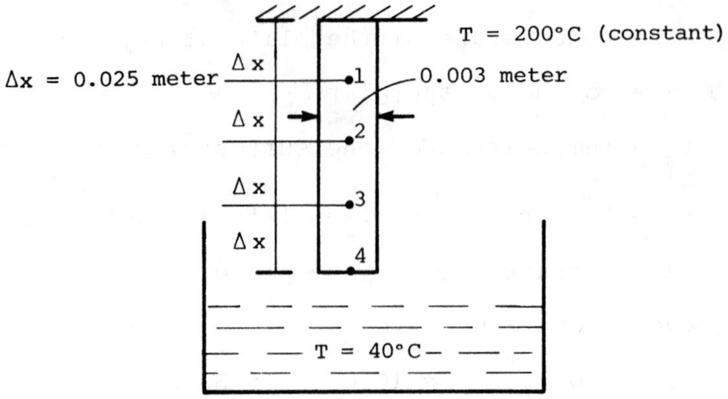

Fig. 1

The rod resistances are in series and each resistance is given by

$$R_{(m+1)} = R_{m-1} = \frac{\Delta x}{kA} = \frac{0.025}{(50)(0.003)^2 \frac{\pi}{4}}$$

$$= 70.73\,^0C/W$$

The subscript (m+1) represents the section of length Δx after the node m, similarly (m-1) represents the section of length Δx before the node m.

$$R_\infty = \frac{1}{h(\pi d \Delta x)} = \frac{1}{(50)\pi(0.003 \times 0.025)}$$

$$= 84.88\,^0C/W$$

where R_∞ is the thermal resistance offered by the boundary layer film.

$$C = \rho C_p \Delta V = 7800 \times 0.47 \times \left[(0.003)^2 \frac{\pi}{4} \times 0.025 \right]$$

$$= 6.48 \times 10^{-4}\ J/^0C$$

where C is the capacitance.

The quantities $R_{(m+1)}$, $R_{(m-1)}$, R_∞ and C for node 4 are

$$R_{(m+1)} = \frac{1}{hA} = 2829\,^0C/W$$

$$R_{(m-1)} = \frac{x}{kA} = 70.73\,^0C/W$$

$$R_\infty = \frac{2}{h}(\pi d \Delta x) = 169.77\,^0C/W$$

and

$$C = \frac{\rho C_p \Delta V}{2} = 3.24 \times 10^{-4}\ J/^0C.$$

Now, assuming the time increment to be 0.015 i.e. total number of time increments necessary $\frac{0.15}{0.015} = 10$.

The temperature at any node point i after an increment of time is given by the equation

$$T_i' = \left(\Sigma \frac{T_j}{R_{ij}}\right)\frac{\Delta\tau}{C_i} + \left(1 - \frac{\Delta\tau}{C_i}\sum_j \frac{1}{R_{ij}}\right)T_i$$

where j corresponds to the surrounding quantity. For i = 1, and substituting the respective values, yields

$$T_1' = \left(\frac{40}{84.88} + \frac{2(200)}{70.73}\right)\frac{0.01}{6.48\times10^{-4}} + 200\left(1 - \frac{0.01\times0.04006}{6.48\times10^{-4}}\right)$$

$$= 170.87^0\text{F}.$$

Similarly all the nodal temperatures for each time increment are calculated and put in the tabular form as given in table 2. Table 1 gives the

$\Sigma \frac{1}{R_{ij}}$, C_i and $\frac{C_i}{\Sigma 1/R_{ij}}$, values for each node point according to the condition of stability.

Table 1

Node	$\sum 1/R_{ij}$	C_i	$\frac{C_i}{\sum 1/R_{ij}}$, s
1	0.04006	6.479×10^{-4}	0.01617
2	0.04006	6.479×10^{-4}	0.01617
3	0.04006	6.479×10^{-4}	0.01617
4	0.02038	3.24×10^{-4}	0.0159

Table 2

Time increment	Node temperature			
	T_1	T_2	T_3	T_4
0	200	200	200	200
1	170.87	170.87	170.87	169.19
2	153.40	147.04	146.68	145.05
3	141.54	128.86	126.98	125.54
4	133.04	115.04	111.24	109.70
5	126.79	104.48	98.76	96.96
6	122.10	96.36	88.92	86.78
7	118.53	90.09	81.17	78.71
8	115.80	85.23	75.08	72.34
9	113.70	81.45	70.31	67.31
10	112.08	78.51	66.57	63.37

The rate of heat transfer at the end of 0.1 second is given by

$$q = \sum_i \frac{T_i - T_\infty}{R_{i\infty}}$$

and

$$q = \frac{200-40}{2(84.88)} + \frac{112.08 + 78.51 + 66.57 - 3 \times 40}{84.88}$$

$$+ \left(\frac{1}{169.77} + \frac{1}{2829} \right) (63.37-40)$$

$$= 2.704 \text{ W}$$

PERIODIC CONDUCTION

• PROBLEM 3-11

Determine the complete periodic solution for a semi-infinite slab using the inversion theorem. The oscillating surface temperature of the slab is $T_0 \cos\omega t$ with the initial uniform temperature as zero.

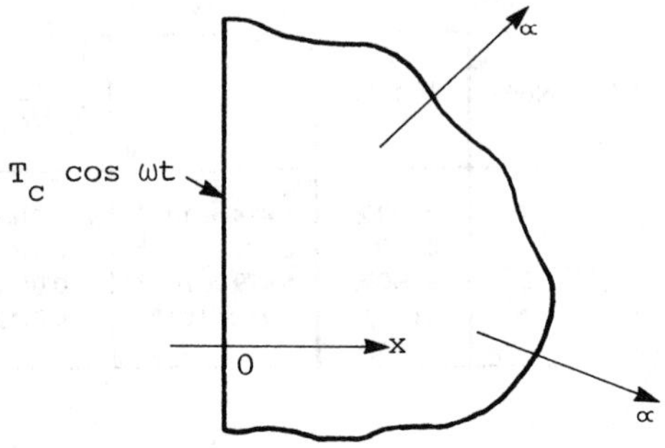

Fig. 1

Solution: The heat equation for the case can be represented as

$$\frac{\partial T}{\partial t} = \alpha \frac{\partial^2 T}{\partial x^2} \tag{1}$$

and boundary conditions

$$T(x,0) = 0$$

$$T(0,t) = T_0 \cos\omega t$$

$$T(\infty,t) = 0$$

The Fourier transformation of equation (1) is

$$\frac{\overline{T}(x,p)}{T_0} = \frac{pe^{-qx}}{(p^2+\omega^2)} \tag{2}$$

By inverse transformation, equation (2) becomes

$$\frac{T(x,t)}{T_0} = \frac{1}{2\pi i}\int_{(r-i\infty)}^{(r+i\infty)} \frac{ze^{tz-(z/a)^{\frac{1}{2}}x}}{(z^2+\omega^2)}\, dz \tag{3}$$

The expression under integral sign has a branch point at $z = 0$ and two poles at $z = \pm i\omega$ as plotted in Figure 2.

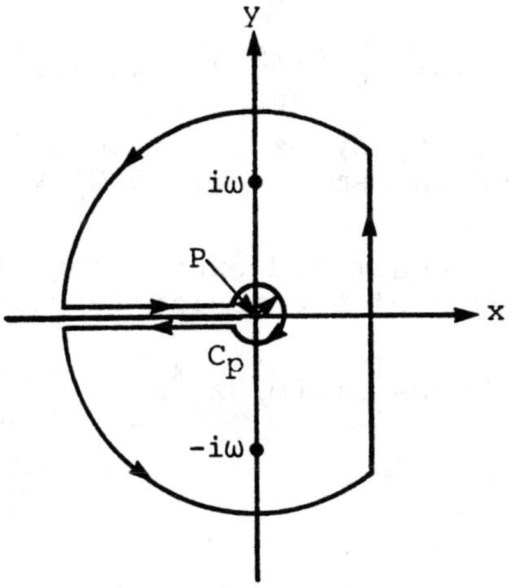

Fig. 2

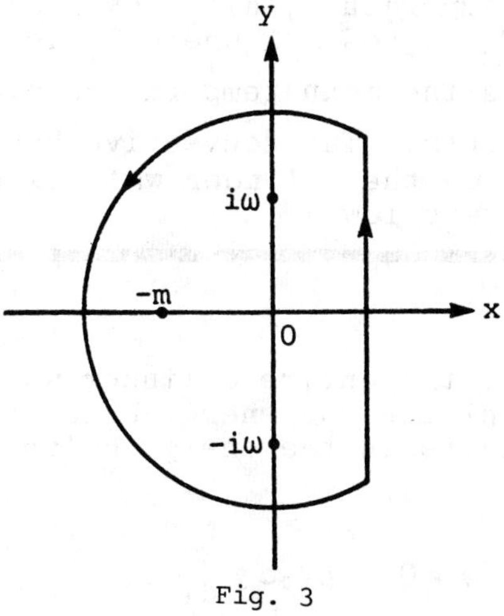

Fig. 3

Thus from equation (2) and contour plot II

$$\frac{T(x,t)}{T_0} = \text{Res}(\pm i\omega) + \lim_{\rho\to 0} \int_{C_p} \frac{ze^{\left[tz-(z/a)^{\frac{1}{2}}x\right]}}{(z^2+\omega^2)}\,dz$$

$$- \frac{1}{2\pi i}\int_0^\infty \rho e^{-t\rho}\left[\frac{e^{i(\rho/a)^{\frac{1}{2}}x} - e^{-i(\rho/a)^{\frac{1}{2}}x}}{(\rho^2+\omega^2)}\right] d\rho \qquad (4)$$

Now, residues at $z = \pm i\omega$ are

$$\text{Res}(i\omega) = \frac{1}{2}\, e^{i\omega t-(1+i)(\omega/2a)^{\frac{1}{2}}x}$$

$$\text{Res}(-i\omega) = \frac{1}{2}\, e^{-i\omega t-(1-i)(\omega/2a)^{\frac{1}{2}}x}$$

Adding the residues gives

$$\text{Res}(i\omega) + \text{Res}(-i\omega) = e^{-(\omega/2a)^{\frac{1}{2}}x} \cos\left[\omega t-(\omega/2a)^{\frac{1}{2}}x\right]$$

This gives the steady part of the solution. As $z \to 0$ the integral around the small circle C_p vanishes.

Applying this result and expressing the last integral of equation (4) in terms of a circular function yields

$$\frac{T(x,t)}{T_0} = e^{-(\omega/2a)^{\frac{1}{2}}x} \cos\left[\omega t-(\omega/2a)^{\frac{1}{2}}x\right] - \frac{1}{\pi}\int_0^\infty \frac{\rho e^{-t\rho}\sin(\rho/a)^{\frac{1}{2}}x}{(\rho^2+\omega^2)}\,d\rho$$

• PROBLEM 3-12

Determine the cylinder wall temperature, as a function of time, for a reciprocating internal combustion engine, in itially at temperature T_0, using the lumped parameter method. The gas temperature, after explosion, varies with time as $T_f = T_m + T_1 \cos \omega\tau$ where T_f is temperature at any time τ, T_m is the mean temperature of the gas and T_1 and ω are constants. The convective heat transfer coefficent (h) inside the cylinder wall is constant and on the outside is very low.

Solution: Consider the entire cylinder wall to be the control volume and applying the energy balance condition to this control volume results in the energy balance equation

$$hA_s(T_f - T) + 0 = \rho V_0 c_p \frac{dT}{d\tau}\,.$$

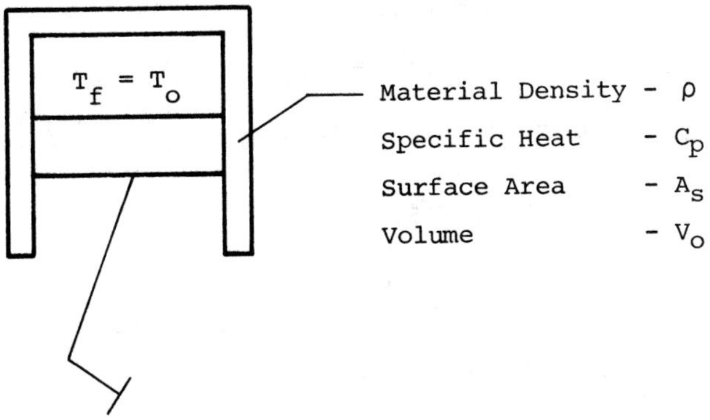

Substituting ϕ for $\frac{h\,A_s}{\rho C_p V_0}$ and rearranging the terms, the equation transforms into a linear, first-order, nonhomogeneous, ordinary differential equation, where ϕ is a constant.

$$\frac{dT}{d\tau} + \phi T = \phi T_f \,.$$

Substituting for T_f from the given time-temperature relation

$$\frac{dT}{d\tau} + \phi(T - T_m) = \phi T_1 \cos\omega\tau. \qquad (1)$$

Replacing $(T - T_m)$ by θ, then $\frac{d\theta}{d\tau} = \frac{dT}{d\tau}$.

Substituting the variable θ into equation (1) yields

$$\frac{d\theta}{d\tau} + \phi\theta = \phi T_1 \cos\omega\tau. \qquad (2)$$

Equation (2) can be solved by assuming the complementary and particular solutions. The complementary function θ_c is

$$\theta c = Be^{-\phi\tau} \qquad (3)$$

and the particular integral is

$$\theta p = a_1 \cos\omega\tau + a_2 \sin\omega\tau \qquad (4)$$

Introducing (3) and (4) into equation (2) yields

$$-\omega a_1 \sin\omega\tau + \omega a_2 \cos\omega\tau + \phi a_1 \cos\omega\tau + \phi a_2 \sin\omega\tau = \phi T_1 \cos\omega\tau,$$

or

$$(\phi a_2 - \omega a_1)\sin\omega\tau + (\phi a_1 + \omega a_2)\cos\omega\tau = \phi T_1 \cos\omega\tau. \qquad (5)$$

For this equation to hold for all values of time τ, the coefficient of sin ωt on the left must equal the coefficient of sin ωt on the right, which is zero. Similarly the coefficient of cos ωt on the left must equal the coefficient of cos ωt on the right.

$$\phi a_2 - \omega a_1 = 0$$

$$\phi a_1 + \omega a_2 = \phi T_1$$

Solving for constants, yields

$$a_1 = \frac{\phi^2 T_1}{\phi^2+\omega^2}$$

$$a_2 = \frac{\omega\phi T_1}{\phi^2+\omega^2}$$

The complete solution for θ is

$$\theta = Be^{-\phi\tau} + a_1\cos\omega\tau + a_2\sin\omega\tau \qquad (6)$$

Applying the initial condition i.e. at $\tau = 0$, $\theta = T_0 - T_m$, therefore the constant $B = (T_0 - T_m) - a_1$. Substituting the values of the constants into equation (6), yields

$$T = T_m + \left(T_0 - T_m - \frac{\phi^2 T_1}{\phi^2+\omega^2}\right) e^{-\phi\tau} + \frac{\phi T_1}{\phi^2+\omega^2}(\phi\cos\omega\tau + \omega\sin\omega\tau). \qquad (7)$$

The resulting equation (7) is a typical unsteady state equation, as it contains both regular periodic unsteady state and transient conduction terms. Further, the last term in equation (7) indicates that the cylinder wall cannot precisely follow the gas temperature variation and it lags behind the gas temperature.

• PROBLEM 3-13

Three sides of a flat plate are maintained at a constant temperature T_1 and the fourth side is subjected to

(1) a sinusoidal distribution of temperature

(2) a constant temperature $T = f(x)$.

Find the solution of the equation using the given boundary conditions.

The relevant Laplace equation is

$$\frac{\partial^2 T}{\partial x^2} + \frac{\partial^2 T}{\partial y^2} = 0$$

Solution: The solution of the Laplace's equation is assumed to be in product form

$$T = XY$$

where

$$X = X(x)$$

and

$$Y = Y(y)$$

The boundary conditions are applied to find the form of the X and Y functions.

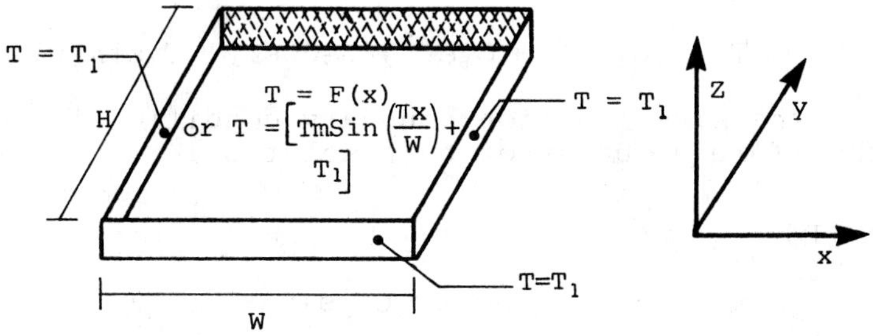

(1) Considering the sinusoidal temperature distribution at the fourth side, the boundary conditions are,

$$T = T_1 \quad \text{at} \quad y = 0$$
$$T = T_1 \quad \text{at} \quad x = 0$$
$$T = T_1 \quad \text{at} \quad x = W$$
$$T = T_m \sin\left(\frac{\pi x}{W}\right) + T_1 \quad \text{at} \quad y = H$$

where T_m is the amplitude of the sine function.

Substituting the boundary conditions in the given Laplace equation yields

$$-\frac{1}{X}\frac{d^2X}{dx^2} = \frac{1}{Y}\frac{d^2Y}{dy^2} = \lambda^2 \tag{1}$$

Equation (1) can also be rewritten in the form

$$\frac{d^2X}{dx^2} + \lambda^2 X = 0 \tag{2}$$

and

$$\frac{d^2Y}{dy^2} - \lambda^2 Y = 0 \tag{3}$$

λ^2 is the separation constant, the value of which depends on the boundary conditions. The form of the solution to equations (2) and (3) depends on the sign of λ^2; a different form would possibly result if λ^2 were zero. The correct form can be determined by the application of the boundary conditions.

Case I: For $\lambda^2 = 0$

$$X = C_1 + C_2 x$$
$$Y = C_3 + C_4 y$$
$$T = (C_1 + C_2 x)(C_3 + C_4 y)$$

This function does not fit the sine function boundary condition, so $\lambda^2 = 0$ cannot be a solution.

Case II: For $\lambda^2 < 0$

$$X = C_5 e^{-\lambda x} + C_6 e^{\lambda x}$$

$$Y = C_7 \cos\lambda y + C_8 \sin\lambda y$$

$$T = (C_5 e^{-\lambda x} + C_6 e^{\lambda x})(C_7 \cos\lambda y + C_8 \sin\lambda y)$$

Again, the sine-function boundary condition is not satisfied, therefore $\lambda < 0$ cannot be a solution.

Case III: For $\lambda^2 > 0$

$$X = C_9 \cos\lambda x + C_{10} \sin\lambda x$$
$$Y = C_{11} e^{-\lambda y} + C_{12} e^{\lambda y}$$
$$T = (C_9 \cos\lambda x + C_{10} \sin\lambda x)(C_{11} e^{-\lambda y} + C_{12} e^{\lambda y})$$

This function satisfies the sine-function boundary condition.

Substituting θ for $(T-T_1)$ transforms the boundary conditions to

$\theta = 0$ at $y = 0$

$\theta = 0$ at $x = 0$

$\theta = 0$ at $x = W$

and $\theta = T_m \sin\frac{\pi x}{W}$ at $y = H$

Using these boundary conditions

$$0 = (C_9 \cos\lambda x + C_{10} \sin\lambda x)(C_{11} + C_{12}) \quad (4)$$

$$0 = C_9 (C_{11} e^{-\lambda y} + C_{12} e^{\lambda y}) \quad (5)$$

$$0 = (C_9 \cos\lambda W + C_{10} \sin\lambda W)(C_{11} e^{-\lambda y} + C_{12} e^{\lambda y}) \quad (6)$$

$$T_m \sin\frac{\pi x}{W} = (C_9 \cos\lambda x + C_{10} \sin\lambda x)(C_{11} e^{-\lambda H} + C_{12} e^{\lambda H}) \quad (7)$$

Accordingly, $C_{11} = -C_{12}$

$C_9 = 0$

and from (6), $0 = C_{10} C_{12} \sin\lambda W (e^{\lambda y} - e^{-\lambda y})$

This requires that

$$\sin\lambda W = 0 \quad (8)$$

Now, the separation constant may be expressed as

$$\lambda = \frac{n\pi}{W} \quad \text{called eigenvalues.}$$

where n is an integer - 1,2,3...etc. The solution to the differential equation may thus be written as the sum of the solutions for each value of $n(1 \rightarrow \infty)$.

$$\theta = T - T_1 = \sum_{n=1}^{\infty} C_n \sin \frac{n\pi x}{W} \sinh \frac{n\pi y}{W} \tag{9}$$

All the constants are combined into one constant C_n and the exponential expressions are replaced by the hyperbolic function.

Applying the last boundary condition to the fourth side gives

$$T_m \sin \frac{\pi x}{W} = \sum_{n=1}^{\infty} C_n \sin \frac{n\pi x}{W} \sinh \frac{n\pi H}{W}$$

for $n > 1$, $C_n = 0$.

The final solution is

$$T = T_m \frac{\sinh(\pi y/W)}{\sinh(\pi H/W)} \sin\left(\frac{\pi x}{W}\right) + T_1$$

(2) A constant temperature distribution across the edge.

The boundary conditions are

$T = T_1$ at $y = 0$

$T = T_1$ at $x = 0$

$T = T_1$ at $x = W$

$T = T_2$ at $y = H$

On applying the first three boundary conditions, the solution takes the form given by equation (9).

Now, using the fourth boundary condition, eq.(9) becomes

$$T_2 - T_1 = \sum_{n=1}^{\infty} C_n \sin \frac{n\pi x}{W} \sinh \frac{n\pi H}{W} \tag{10}$$

The above equation is a Fourier sine series and the value of the constants can be determined by expanding the right-hand side expression in a Fourier series for the interval $0 < x < W$, therefore

$$T_2 - T_1 = (T_2 - T_1)\frac{2}{\pi} \sum_{n=1}^{\infty} \frac{(-1)^{n+1} + 1}{n} \sin \frac{n\pi x}{W} \tag{11}$$

The constants from equations (10) and (11) are expressed as

$$C_n = \frac{2}{\pi}(T_2 - T_1)\frac{1}{\sinh(n\pi H/W)}\frac{(-1)^{n+1}+1}{n}$$

Thus, the solution to Laplace's equation can be written as

$$\frac{T - T_1}{T_2 - T_1} = \frac{2}{\pi}\sum_{n=1}^{\infty}\frac{(-1)^{n+1}+1}{n}\sin\frac{n\pi x}{W}\frac{\sinh(n\pi y/W)}{\sinh(n\pi H/W)}$$

• PROBLEM 3-14

Estimate the temperature distribution in a large slab of thickness $L = 1$ft. The initial temperature distribution in the slab is a sine function $T = 300\sin\pi x + 150$ and is so maintained by an internal source of heat. The slab is suddenly exposed to a fluid at a temperature of $100\,^0$F. while the internal heat source is cut off. The sides of the slab are maintained at this temperature ($100\,^0$F). As-Assume the thermal diffusivity of the slab material as 1.138 ft^2/hr. Use M=3.

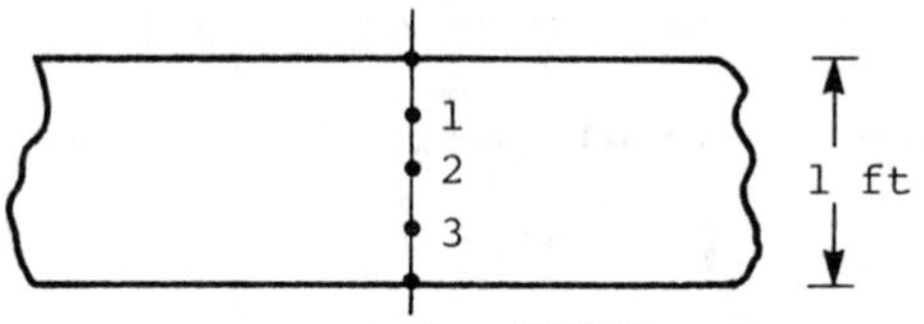

Solution: The slab thickness is divided into (say) 4 equal segments as shown in the figure. The temperature distribution at any point i in the slab with heat generation source present is (in time ΔT)

$$T_i' = \frac{1}{M}(T_{i-1} + T_{i+1}) + q_0\frac{(\Delta x)^2}{kM} + \left(1 - \frac{2}{M}\right)T_i$$

where

T_i' - is the temperature at point i

q_0 - rate of internal heat generation

Δx - thickness of the slab segment

k - thermal conductivity of the slab material

M is the distribution constant.

Now, at time $t = 0$, $q = 0$ and face temperature is 100^0F.

For points 1, 2 and 3 the temperatures can be written for any time $t > 0$ as

$$T_1' = \frac{1}{M}(100 + T_2) + \left(1 - \frac{2}{M}\right) T_1 \qquad (1)$$

$$T_2' = \frac{1}{M}(T_1 + T_3) + \left(1 - \frac{2}{M}\right) T_2 \qquad (2)$$

$$T_3' = \frac{1}{M}(T_2 + 100) + \left(1 - \frac{2}{M}\right) T_3$$

At instant $T_1' = T_3'$, both the sides of the slab are at the same temperature and points 1 and 3 are located at equal distance from the respective faces.

The term M is defined as

$$M = \frac{(\Delta x)^2}{\alpha \Delta t}$$

where Δt - is the time increment (during which the temperature distribution is calculated).

therefore $\Delta t = \frac{(\Delta x)^2 \times 60}{\alpha M} = \frac{(1/4)^2 \times 60}{1.138 \times 3} = 1.1$ min.

At time $t = 0$ the temperature distribution is

$$T = 300 \sin \pi x + 150$$

$$T_1 = 300 \sin(\pi/4) + 150 = 362^0\text{F}.$$

$$T_2 = 300 \sin(\pi/2) + 150 = 450^0\text{F}$$

Using equations (1) and (2)

$$T_1' = \frac{(100 + T_2)}{3} + \frac{T_1}{3}$$

and

$$T_2' = \frac{2}{3} T_1 + \frac{T_2}{3}$$

For time increment 1 Δt

$$T_1' = \frac{1}{3}(100 + 450) + \frac{362}{3} = 304$$

and

$$T_2' = \frac{2}{3}(362) + \frac{450}{3} = 391$$

For time increment 2 Δt

$$T_1' = \frac{1}{3}(100 + 391) + \frac{304}{3} = 265$$

$$T_2' = \frac{2}{3}(304) + \frac{391}{3} = 333.$$

The method can be repeated for any number of increments of time. The table below shows the calculations for 14 increments of time.

Finite Difference Temp	Calculation	Value(^{0}F)	Time(Δt)
T_1	$300 \sin(\pi/4) + 150$	362	0
T_2	$300 \sin(\pi/2) + 150$	450	
T_1'	$\frac{1}{3}(100 + 450) + \frac{362}{3}$	304	1
T_2'	$\frac{2}{3}(362) + \frac{450}{3}$	391	
T_1'	$\frac{1}{3}(100 + 391) + \frac{304}{3}$	265	2
T_2'	$\frac{2}{3}(304) + \frac{391}{3}$	333	
T_1'	$\frac{1}{3}(100 + 333) + \frac{265}{3}$	232	3
T_2'	$\frac{2}{3}(265) + \frac{333}{3}$	287	
T_1'	$\frac{1}{3}(100 + 287) + \frac{232}{3}$	206	4
T_2'	$\frac{2}{3}(232) + \frac{287}{3}$	250	
T_1'	$\frac{1}{3}(100 + 250) + \frac{206}{3}$	185	5
T_2'	$\frac{2}{3}(206) + \frac{250}{3}$	220	
T_1'	$\frac{1}{3}(100 + 220) + \frac{185}{3}$	168	6
T_2'	$\frac{2}{3}(185) + \frac{220}{3}$	196	
T_1'	$\frac{1}{3}(100 + 196) + \frac{168}{3}$	154	7
T_2'	$\frac{2}{3}(168) + \frac{196}{3}$	177	

T_1'	$\frac{1}{3}(100 + 177) + \frac{154}{3}$	143	8
T_2'	$\frac{2}{3}(154) + \frac{177}{3}$	161	
T_1'	$\frac{1}{3}(100 + 161) + \frac{143}{3}$	134	9
T_2'	$\frac{2}{3}(143) + \frac{161}{3}$	149	
T_1'	$\frac{1}{3}(100 + 149) + \frac{134}{3}$	127	10
T_2'	$\frac{2}{3}(134) + \frac{149}{3}$	139	
T_1'	$\frac{1}{3}(100 + 139) + \frac{127}{3}$	122	11
T_2'	$\frac{2}{3}(127) + \frac{139}{3}$	131	
T_1'	$\frac{1}{3}(100 + 131) + \frac{122}{3}$	117	12
T_2'	$\frac{2}{3}(122) + \frac{131}{3}$	125	
T_1'	$\frac{1}{3}(100 + 125) + \frac{117}{3}$	114	13
T_2'	$\frac{2}{3}(117) + \frac{125}{3}$	119	
T_1'	$\frac{1}{3}(100 + 119) + \frac{114}{3}$	111	14
T_2'	$\frac{2}{3}(114) + \frac{119}{3}$	115	

• PROBLEM 3-15

Two temperature-measuring devices, an iron-constantan thermocouple and a mercury thermometer, are used to measure the temperature-time history of a fluid flowing through a tube. The temperature of the fluid varies as $T = (100 + 50 \sin 2\pi\theta)\,^0F$. Assume the initial temperature of both the devices as $60\,^0F$ and both have the heat transfer coefficients of 5 Btu/hr.ft.$^2\,^0F$. The mercury thermometer is made of glass, 1/4 in. diameter and 1 in. long. The iron-constantan thermocouple is of 1/32 in. diameter and 2 in. long immersed in the fluid. Compare the temperature-time response of both the devices.

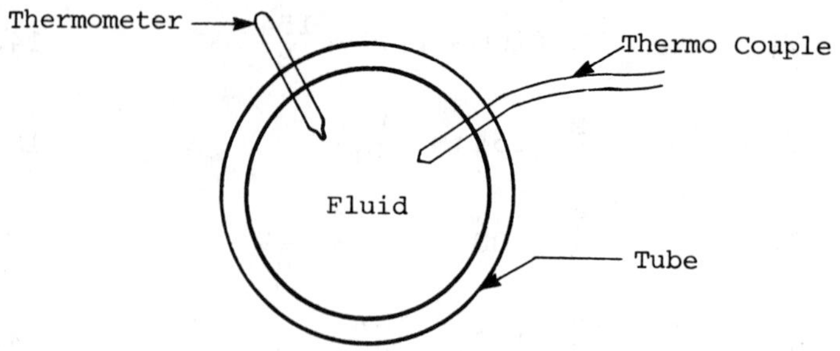

Fig. 1

Solution: The general differential equation relating time and temperature is

$$\frac{dT(\theta)}{d\theta} + \frac{\bar{h}A_s}{\rho cV} T(\theta) = \frac{\bar{h}A_s}{\rho cV} T_\infty(\theta) \qquad (1)$$

The transient response can be obtained by the complementary function of this equation. The complementary function is obtained by equating the left-hand side of the equation to zero,

$$\text{i.e.} \quad \frac{dT}{d\theta} + \frac{\bar{h}A_s}{c\rho V} T = 0$$

and the solution is

$$T_1 = C_1 e^{-(\bar{h}A_s/c\rho V)\theta}$$

Since the temperature of the fluid is sinusoidal, the response should also be sinusoidal. Further, the temperature contains a constant factor, therefore the response should also contain a constant factor. Also, when the solutions of linear differential equations are superposed (added), the sum is also a solution. Then the equation for temperature can be as follows:

$$T_{ss} = C_2 \sin 2\pi\theta + C_3 \cos 2\pi\theta + C_4 \ldots \qquad (2)$$

Taking the derivative of equation (2) with respect to θ

$$\frac{dT_{ss}}{d\theta} = 2\pi C_2 \cos 2\pi\theta - 2\pi C_3 \sin 2\pi\theta \qquad (3)$$

Inserting equations (2) and (3) in equation (1) and setting $m = \frac{\bar{h}A_s}{c\rho V}$, results in the following equations, after collecting the terms

$$(2\pi C_2 + mC_3)\cos 2\pi\theta + (mC_2 - 2\pi C_3)\sin 2\pi\theta + mC_4$$
$$= m100 + m50 \sin 2\pi\theta$$

The coefficients of like terms on each side of the equation are equal, therefore

$$2\pi C_2 + mC_3 = 0 \qquad \text{from the cosine terms}$$

$$mC_2 - 2\pi C_3 = 50m \qquad \text{from the sine terms}$$

$$C_4 = 100 \qquad \text{from the constant term}$$

Solving these equations for C_2 and C_3 simultaneously results in

$$C_2 = \frac{50}{1 + (2\pi/m)^2} \quad \text{and} \quad C_3 = -\frac{(2\pi/m)50}{1 + (2\pi/m)^2}$$

Under steady state condition the temperature response is therefore

$$T_{ss} = 100 + \frac{50}{1 + (2\pi/m)^2} \sin 2\pi\theta - \frac{(2\pi/m)50}{1 + (2\pi/m)^2} \cos 2\pi\theta \quad (4)$$

The sine and cosine difference terms can be put together by applying the equation

$$C_2 \sin 2\pi\theta - C_3 \cos 2\pi\theta$$

$$= \sqrt{C_2^2 + C_3^2}\left(\frac{C_2}{\sqrt{C_2^2 + C_3^2}} \sin 2\pi\theta - \frac{C_3}{\sqrt{C_2^2 + C_3^2}} \cos 2\pi\theta\right)$$

Constructing a right triangle with $\sqrt{C_2^2 + C_3^2}$ as the hypotenuse and C_2 and C_3 as the sides, gives

$$C_2/\sqrt{C_2^2 + C_3^2} = \cos\delta \text{ and } C_3/\sqrt{C_2^2 + C_3^2} = \sin\delta.$$

But since $\sin(A - B) = \sin A \cos B - \cos A \sin B$,

$$C_2 \sin 2\pi\theta - C_3 \cos 2\pi\theta = \sqrt{C_2^2 + C_3^2} \sin(2\pi\theta - \delta)$$

where δ is equal to the $\tan^{-1}(C_3/C_2)$.

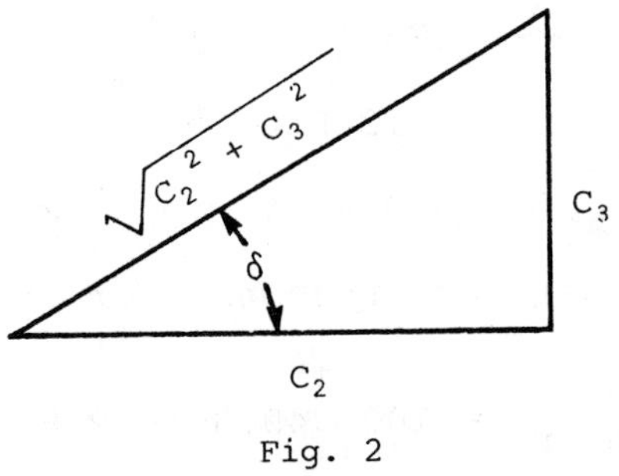

Fig. 2

Thus, combining the last two terms of Eq.(4) in this manner and adding T_1 yields

$$T = 100 + \frac{50}{1 + (2\pi/m)^2}\sqrt{1 + (2\pi/m)^2}\,\sin(2\pi\theta - \delta) + C_1 e^{-m\theta} \quad (5)$$

where $\delta = \tan^{-1}(2\pi/m)$ and represents the time lag in the temperature response of the instruments.

The constant of integration C_1 can now be evaluated from the initial condition, i.e., $T = 60\,^0F$ at $\theta=0$. Substituting this condition into Eq.(5) yields

$$T_{\theta=0} = 60 = 100 + \frac{50}{\sqrt{1 + (2\pi/m)^2}}\sin(-\delta) + C_1$$

Using the trigonometric identity for $\sin\delta$

$$C_1 = \frac{100\pi/m}{1 + (2\pi/m)^2} - 40$$

the expression for the temperature-time history of the instrument becomes

$$T = \left[\frac{100\pi/m}{1 + (2\pi/m)^2} - 40\right] e^{-m\theta} + \frac{50}{\sqrt{1 + (2\pi/m)^2}}\sin(2\pi\theta - \delta) + 100$$

The time required for the steady state response cycle is 1 hr as the time increases sinusoidally by 2π times. Therefore, the time lag δ in hours can be obtained by dividing the lag in radians by the number of radians corresponding to a unit increase of time, which is 2π radians. The numerical value of m for the instruments can be determined as

$$m = \frac{\bar{h}A_s}{c\rho V} = \frac{\bar{h}}{c\rho}\,\frac{\pi DL}{(\pi/4)D^2L} = \frac{4\bar{h}}{c\rho D}$$

The value of m, using physical properties of mercury

ρ = 849 lb/cu. ft.

c = 0.0325 Btu/lb.°F.

D = 0.021 ft. $\qquad$ $m = 35.2\ hr^{-1}$.

and, for the thermocouple, using properties of iron as an approximation

ρ = 475 lb/cu.ft.

c = 0.12 Btu/lb.°F.

D = 0.0026 ft. $\qquad$ $m = 135\ hr^{-1}$.

Thus, the time-temperature response for thermometer and thermocouple are

$$T_{meter} = 100 + 49.3\sin(2\pi\theta - 0.178) - 31.35e^{-35.2\theta}$$

$$T_{couple} = 100 + 50\sin(2\pi\theta - 0.0465) - 37.7e^{-135\theta}$$

respectively.

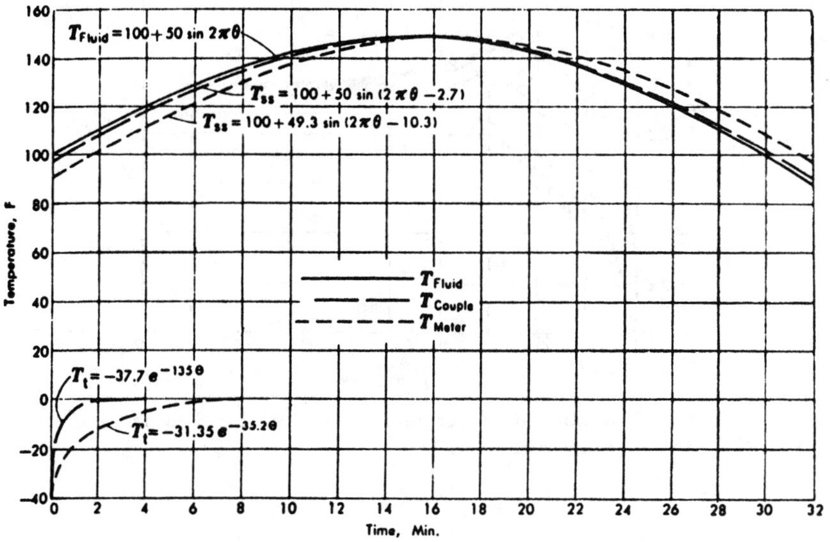

Fig. 3 Transient and steady-state response of the thermocouple and the thermometer.

The result is graphically represented in figure 3.

• PROBLEM 3-16

A two stroke internal combustion engine operates at a speed of 2500 r.p.m. Determine the depth of penetration of the temperature oscillations into the walls of the cylinder and the temperature fluctuations at the surface. Assume the cylinder block is made of steel, and

$h = 800\ W/m^2\,{}^0K \quad \alpha = 0.065\ m^2/hr \quad k = 60\ W/m\,{}^0K$

Solution: This is a case of periodic heat transfer. The engine is a two stroke internal combustion engine, and will therefore produce 2,500 peak temperatures per minute. The time period (τ_0) of one oscillation is

$$\tau_0 = \frac{1}{60(2500)} = \frac{1}{15 \times 10^4}\ hr.$$

The thermal diffusivity (α) of iron is $\alpha = 0.0595\ m^2/hr$. In the figure shown the values of t/t_{om} and $x/2\sqrt{\pi\tau_0\alpha}$ are re-

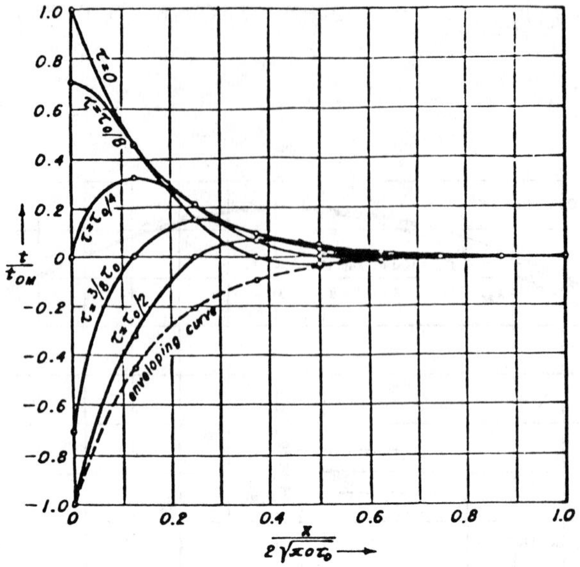

Penetration of a temperature oscillation into an infinitely thick wall.

quired. t_{om} is the maximum value of surface temperature variation. The penetration depth of the temperature oscillation

$$x/(2\sqrt{\pi\alpha\tau_0}) = 0.8$$

is taken, since t/t_{om} is fairly constant at this value. Substituting the numerical values of α and τ_0 and solving for x, the practical depth of penetration of temperature fluctuations, gives

$$x = 1.6\sqrt{\frac{\pi \times 0.065}{15 \times 10^4}} = 1.87 \times 10^{-3}\text{m} = 1.87\text{mm}$$

$$x = 1.87\text{mm}$$

The actual fluctuation of the surface temperature will be of a lesser amplitude than that of the gas in the cylinder. To determine this reduction in amplitude, the following factor is used.

$$f = 1/\sqrt{1 + 2\sqrt{\pi n k^2/\alpha\tau_0 h^2} + 2(\pi n k^2/\alpha\tau_0 h^2)}$$

where $n = 1$ for a fundamental wave

$k = 60\ \text{W/m}^0\text{K}$ for steel

$h = 800\ \text{W/m}^{2\,0}\text{K}$

Substituting the values in

$$\frac{\pi k^2}{\alpha \tau_0 h^2} = \frac{\pi \times (60)^2}{\left[\dfrac{0.065 \times (800)^2}{15 \times 10^4}\right]} = 40780$$

therefore $$f = \left[1 + 2\sqrt{40780} + 2(40780)\right]^{-1}$$

$$= 3.49 \times 10^{-3}$$

If the temperature oscillation of the gas has a double amplitude of 2200^0C, then the surface temperature of the cylinder has a double amplitude

$$= 2200 \times f$$

$$= 2200(3.49 \times 10^{-3}) = 7.68^0\text{C} \simeq 8^0\text{C}$$

Tests performed on an Otto cycle engine operating at 2,500rpm gave results of surface temperature fluctuations of 11^0C.

• PROBLEM 3-17

A thick brick wall has a surface which experiences a daily temperature variation from 50^0F to 130^0F. The properties of the brick are:

$$k = 0.4 \text{ Btu/hr. ft}^2(^0\text{F/ft})$$

$$C_p = 0.3 \text{ Btu/lb. }^0\text{F}$$

$$\rho = 120 \text{ lb/ft}^3$$

$$\alpha = \frac{k}{c\rho} = 0.01111 \text{ ft}^2\text{/hr}$$

Assume the variation to be cosine in nature, and the cement between the bricks has the same properties of of the bricks.

Find

(a) the temperature lag at a point 4 in. below the surface;

(b) the amplitude at this depth;

(c) the temperature deviation from the mean after 3 hours;

(d) the heat flow during the half period.

Solution: (a) The temperature lag 4 inches below the surface is determined, using the equation

$$\theta = \frac{x}{2}\sqrt{\frac{1}{\alpha\pi f}}$$

where x is the depth in feet $= \frac{4}{12} = 0.333$ ft

f is the frequency $= \frac{1}{24} = 0.041666$

Therefore

$$\theta = \frac{0.333}{2}\sqrt{\frac{1}{0.01111 \times \pi \times 1/24}} = 4.37 \text{ hrs.}$$

$$\theta = 4.37 \text{ hrs.}$$

(b) The amplitude can be calculated, using the equation

$$\Delta t_{x=x} = \Delta t_m \, e^{-x\sqrt{\pi f/\alpha}}$$

where $\Delta t_m = \frac{130 - 50}{2} = 40^\circ F$

$f = \text{frequency} = 1/24$

and, introducing the known values

$$\Delta t_{x=x} = 40 \times e^{-(0.333)\sqrt{\frac{\pi \times 0.04166}{0.01111}}}$$

gives $\Delta t_{x=x} = 12.74^\circ F$

(c) The temperature variation after 3 hours is

$$\Delta t_x = \Delta t_m \, e^{-x\sqrt{\pi f/\alpha}} \cos\left(2\pi f\theta - x\sqrt{\frac{\pi f}{\alpha}}\right)$$

$$= (\Delta t_{x=x})\cos\left(2\pi f\theta - x\sqrt{\frac{\pi f}{\alpha}}\right)$$

Substituting the values

$$\Delta t_x = (12.74)\cos\left(2\pi \times 0.04166 \times 3 - 4x\sqrt{\frac{\pi \times 0.041666}{0.0111}}\right)$$

$$= 11.84^0\text{F}$$

(d) Heat flow during the half period is determined from the equation

$$\frac{Q}{A} = k\ \Delta t_m\ \sqrt{2/\alpha\pi f}$$

$$= 0.4 \times 40\sqrt{\frac{2}{0.01111 \times \pi \times 0.041666}} = 594\ \text{Btu/hrft}^2$$

• PROBLEM 3-18

In the month of June, at a place in New York, the highest and the lowest temperature measured on a day was 98°F and 50°F respectively. Estimate for that day the temperature range and the temperature lag at depths 4 ft. and 7 ft. Assume the properties of sand at that place as

Thermal conductivity (k) = 1.45 Btu/hr.ft.°F

Density (ρ) = 200 lb_m/ft^3.

Heat capacity (C_p) = 0.2 Btu/lb_m °F.

Solution: For the calculation of the amplitude of variation, in temperature, is given by

$$\theta_{(x,\alpha)} = \theta_{(0,\alpha)}\ e^{-x\sqrt{\frac{\pi n}{\alpha}}}$$

where $\theta_{(0,\alpha)}$ → mean change in temperature at the surface $= \dfrac{98 - 50}{2} = 24^0\text{F}$

x → depth at which the amplitude of temperature is to be found.

$n \rightarrow$ periodic frequency for change in temperature per hour $= \frac{1}{24}$

and $\alpha \rightarrow$ thermal diffusivity

$$= \frac{k}{\rho c_p} = \frac{1.45}{200 \times 0.2} \text{ ft}^2/\text{hr.} = 0.03625 \text{ ft}^2/\text{hr.}$$

$t_m \rightarrow$ mean temperature at the surface $= \frac{98+50}{2} = 74^0$

and the time at any depth x is given by

$$\Delta\tau = \frac{x}{2}\sqrt{\frac{1}{\alpha\pi n}}$$

a) Therefore at a depth x = 4ft.

$$\theta_{(x,\alpha)} = \theta_{(0,\alpha)} e^{-x\sqrt{\frac{\pi n}{\alpha}}}$$

$$= 24 \times e^{-4\sqrt{\frac{\pi}{0.03625} \times \frac{1}{24}}}$$

$$= 0.012^0\text{F}$$

Therefore the temperature at that point would range between $\pm 0.012^0$F the surface temperature , i.e. 74.012^0F and 73.988^0F

and $$\Delta\tau = \frac{x}{2}\sqrt{\frac{1}{\alpha\pi n}}$$

$$= \frac{4}{2}\sqrt{\frac{1}{0.03625 \times \pi} \quad \frac{24}{1}} \simeq 29 \text{ hrs.}$$

Therefore time lag at the depth of 4 ft. is 29 hrs.

Similarly

b) At a depth of x = 7 ft.

$$\theta_{(x,\alpha)} = 24 \times e^{-7\sqrt{\frac{\pi}{0.03625} \times \frac{1}{24}}}$$

$$= 4 \times 10^{-5} = 0.00004^0\text{F}$$

This value is negligible and can be neglected. Therefore, there is no variation at this depth and

$$\Delta\tau = \frac{7}{2}\sqrt{\frac{1}{0.03625 \times \pi}}\ \frac{24}{1} \simeq 51 \text{ hrs.}$$

Time lag at this depth is 51 hrs.

• PROBLEM 3-19

An incompressible fluid at uniform velocity V flows steadily through a tube (figure 1). The thermal conductivity, density, specific heat, and cross-sectional area of the tube and fluid are (k',k), (ρ',ρ), (c',c), and (A',A), respectively. The inner and outer peripheries of the tube are P_i, P_0, and the inside and outside heat transfer coefficients h_i, h_0. Initially the system is at the ambient temperature, say zero; then the inlet temperature starts to oscillate as $T_0 \cos \omega t$. Using a radially lumped analysis, find the steady periodic temperature of the system.

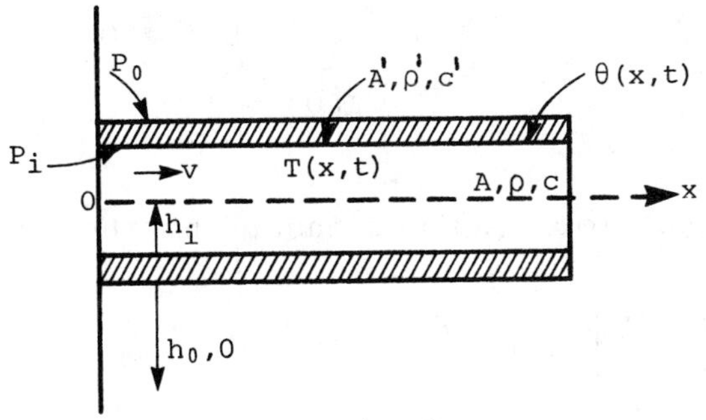

Fig. 1

Solution: The formulation of the problem is

$$\frac{\partial T}{\partial t} + V\frac{\partial T}{\partial x} + b_1(T-\theta) = 0,$$

$$\frac{\partial \theta}{\partial t} - b_2(T-\theta) + b_3\theta = 0,$$

$$T(x,0) = 0,$$

$$\theta(x,0) = 0,$$

$$T(0,t) = T_0 \cos \omega t,$$

where $b_1 = (h_i P_i/\rho c A)$, $b_2 = (h_i P_i/\rho' c' A')$, and $b_3 = (h_0 P_0/\rho' c' A')$.

The complex temperatures of the fluid and of the wall,

$$\psi = T + iT^*, \quad \phi = \theta + i\theta^*,$$

satisfy the equations

$$\frac{\partial\psi}{\partial t} + V\frac{\partial\psi}{\partial x} + b_1(\psi - \phi) = 0,$$

$$\frac{\partial\phi}{\partial t} - b_2\psi + b_4\phi = 0,$$

$$\psi(x,0) = 0,$$

$$\phi(x,0) = 0,$$

$$\psi(0,t) = T_0 e^{i\omega t},$$

where $b_4 = b_2 + b_3$. The complex steady periodic temperatures of the fluid and of the wall,

$$\psi(x,t) = \Lambda(x)e^{i\omega t}, \quad \phi(x,t) = \Gamma(x)e^{i\omega t}, \tag{1}$$

introduced into eqs. give

$$i\omega\Lambda + V\frac{d\Lambda}{dx} + b_1(\Lambda - \Gamma) = 0, \tag{2}$$

$$i\omega\Gamma - b_2\Lambda + b_4\Gamma = 0, \tag{3}$$

$$\Lambda(0) = T_0. \tag{4}$$

Solving eq.(3) for $\Gamma(x)$ and inserting the result into eq.(2) yields

$$\frac{d\Lambda}{dx} + \frac{1}{V}\left[i\omega + b_1\left(\frac{b_3 + i\omega}{b_4 + i\omega}\right)\right]\Lambda = 0. \tag{5}$$

The solution of eq.(5) which satisfies eq.(4) is

$$\Lambda(x) = T_0 \exp\left[-\left(\frac{b_3 b_4 + \omega^2}{b_4^2 + \omega^2}\right)\left(\frac{b_1 x}{V}\right)\right]\exp\left[-i\left(1 + \frac{b_1 b_2}{b_4^2 + \omega^2}\right)\left(\frac{\omega x}{V}\right)\right]. \tag{6}$$

Using this result and eq.(3), we obtain

$$\Gamma(x) = T_0\left(\frac{b_2}{b_4^2 + \omega^2}\right)\exp\left[-\left(\frac{b_3 b_4 + \omega^2}{b_4^2 + \omega^2}\right)\left(\frac{b_1 x}{V}\right)\right]$$
$$\times \exp\left\{-i\left[\left(1 + \frac{b_1 b_2}{b_4^2 + \omega^2}\right)\left(\frac{\omega x}{V}\right) + \beta\right]\right\}, \tag{7}$$

where $\beta = \tan^{-1}(\omega/b_4)$.

Thus from eqs.(1), (6) and (7) the complex steady periodic temperatures of the fluid and wall are readily obtained. The real parts of these temperatures,

$$\frac{T(x,t)}{T_0} = \exp\left[-\left(\frac{b_3 b_4+\omega^2}{b_4^2+\omega^2}\right)\left(\frac{b_1 x}{V}\right)\right]\cos\left[\omega t-\left(1+\frac{b_1 b_2}{b_4^2+\omega^2}\right)\left(\frac{\omega x}{V}\right)\right],$$

$$\frac{\theta(x,t)}{T_0} = \left(\frac{b_2}{b_4^2+\omega^2}\right)\exp\left[-\left(\frac{b_3 b_4+\omega^2}{b_4^2+\omega^2}\right)\left(\frac{b_1 x}{V}\right)\right]$$

$$\times\cos\left[\omega t-\left(1+\frac{b_1 b_2}{b_4^2+\omega^2}\right)\left(\frac{\omega x}{V}\right)-\beta\right]$$

are the steady periodic temperatures of the fluid and wall, respectively.

SEMI-INFINITE CONDUCTION

• PROBLEM 3-20

Consider a thick walled insulated duct to be a flat plate of thickness L, fixed to a semi-infinite slab. The initial uniform temperature of the system is T_0. Condensing vapor at temperature T_∞ rushes into the duct at time $t = 0$. The insulation has a large finite thermal resistivity. Determine the unsteady temperature of the system. Assume the contact resistance at the interface to be negligibly small.

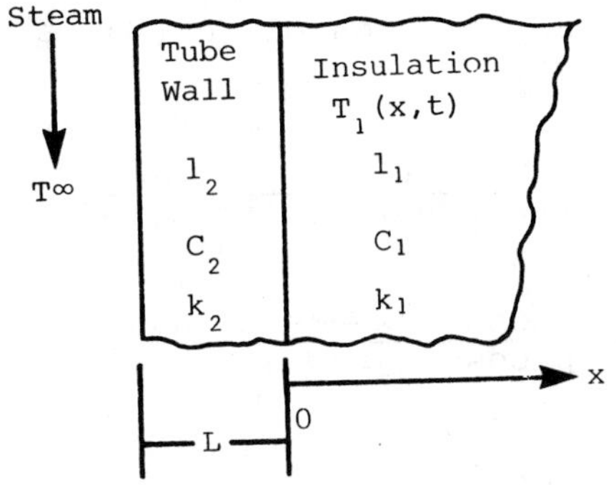

Fig. 1

Solution: Assuming negligible interface resistance the relevant differential equations are

$$\frac{\partial T_1}{\partial t} = \alpha_1 \frac{\partial^2 T_1}{\partial x^2}$$

$$\frac{\partial T_2}{\partial t} = \alpha_2 \frac{\partial^2 T_2}{\partial x^2}$$

$$T_1(x,0) = T_2(x,0) = T_0 \quad \text{(initially)} \tag{1}$$

Also, $T_1(0,t) = T_2(0,t)$,

$$k_1 \frac{\partial T_1(0,t)}{\partial x} = k_2 \frac{\partial T_2(0,t)}{\partial x}$$

$$\lim_{x\to\infty} T_1(x,t) \to T_0 \text{ and } T_2(-L,t) = T_\infty$$

x measured from the interface.

Substituting $\theta_1 = T_1 - T_0$, $\theta_2 = T_2 - T_0$,

$$q_1^2 = p/\alpha_1 \quad \text{and } q_2^2 = p/\alpha_2$$

then equation (1) takes the form

$$\frac{d^2\overline{\theta}_1}{dx^2} - q_1^2\overline{\theta}_1 = 0, \quad \frac{d^2\overline{\theta}_2}{dx^2} - q_2^2\overline{\theta}_2 = 0,$$

$$\overline{\theta}_1(0,p) = \overline{\theta}_2(0,p), \quad k_1 \frac{d\overline{\theta}_1(0,p)}{dx} = k_2 \frac{d\overline{\theta}_2(0,p)}{dx}, \tag{2}$$

$$\lim_{x\to\infty} \overline{\theta}_1(x,p) \to 0, \quad \overline{\theta}_2(-L,p) = \theta_\infty/p,$$

where $\theta_\infty = T_\infty - T_0$. The solution of eq.(2) is

$$\frac{\overline{\theta}_1(x,p)}{\theta_\infty} = \frac{e^{-q_1 x}}{p(\cosh q_2 L + \varepsilon \sinh q_2 L)}, \tag{3}$$

$$\frac{\overline{\theta}_2(x,p)}{\theta_\infty} = \frac{\cosh q_2 x - \varepsilon \sinh q_2 x}{p(\cosh q_2 L + \varepsilon \sinh q_2 L)}, \tag{4}$$

where $\varepsilon = (\rho_1 c_1 k_1/\rho_2 c_2 k_2)^{\frac{1}{2}}$. Inverting equation (3) yields

$$\frac{\theta_1(x,t)}{\theta_\infty} = \frac{1}{2\pi i}\int_{\gamma-i\infty}^{\gamma+i\infty} \frac{e^{tz-(z/a_1)^{\frac{1}{2}}x}}{z[\cosh(z/a_2)^{\frac{1}{2}}L + \varepsilon \sinh(z/a)^{\frac{1}{2}}L]}dz \tag{5}$$

The integrand of equation (5) has $e^{-(z/a_1)^{\frac{1}{2}}x}$ and $\sinh(z/a_2)^{\frac{1}{2}}L$, both are double valued functions and have offshoots at $z=0$, therefore contour II instead of contour I is appropriate to the evaluation of equation (5). The integrand possesses no poles, and integration around the small circle of loop L, as shown in Figure 2, gives $2\pi i$ within the limit, as the integrand approaches $i\theta$ when $\rho \to 0$.

The expression for $\dfrac{\theta_1(x,t)}{\theta_\infty}$ from equation

$$T(r,t) = \frac{1}{2\pi i}\int_{\gamma-i\infty}^{\gamma+i\infty} e^{tz}\bar{T}(r,z)dz = \sum_{n=1}^{\infty} \mathrm{Res}(a_n)$$

$$+ \frac{1}{2\pi i}\lim_{\rho\to o}\int_{C\rho} e^{tz}\bar{T}(r,z)dz + \frac{1}{2\pi i}\int_0^\infty e^{-t\rho}$$

$$\left[\bar{T}(r,\rho e^{-i\pi}) - \bar{T}(r,\rho e^{i\pi})\right]d\rho.$$

is

$$\frac{\theta_1(x,t)}{\theta_\infty}$$

$$= 1 + \frac{1}{2\pi i}\int_0^\infty e^{-t\rho} x \left\{\frac{e^{-x\left[(\rho/a_1)^{\frac{1}{2}} x\ e^{-i\pi/2}\right]}}{\cosh\left[(\rho/a_2)^{\frac{1}{2}}e^{-i\pi/2}L\right] + \varepsilon\sinh\left[(\rho/a_2)^{\frac{1}{2}}e^{-i\pi/2}L\right]}\right.$$

$$\left. - \frac{e^{-x\left[(\rho/a_1)^{\frac{1}{2}}xe^{i\pi/2}\right]}}{\cosh\left[(\rho/a_2)^{\frac{1}{2}}e^{i\pi/2}L\right] + \varepsilon\sinh\left[(\rho/a_2)^{\frac{1}{2}}e^{i\pi/2}L\right]}\right\}\frac{d\rho}{\rho} \tag{6}$$

Substituting

$\xi = (\rho/a_2)^{\frac{1}{2}}$, $\eta = (a_2/a_1)^{\frac{1}{2}}$, $e^{-i\pi/2} = -i$, and $e^{i\pi/2} = i$ in

equation (6) and reordering the terms, yields

$$\frac{\theta_1(x,t)}{\theta_\infty} = 1 - \frac{2}{\pi}\int_0^\infty e^{-a_2\xi^2 t}\left(\frac{\cos\xi L\sin\eta\xi x + \varepsilon\sin\xi L\cos\eta\xi x}{\cos^2\xi L + \varepsilon^2\sin^2\xi L}\right)\frac{d\xi}{\xi} \tag{7}$$

The function $\theta_2(x,t)$ can be similarly obtained as $\theta_1(x,t)$ and will be of the form

$$\frac{\theta_2(x,t)}{\theta_\infty} = 1 - \varepsilon\,\frac{2}{\pi}\int_0^\infty \frac{e^{-a_2\xi^2 t}\sin\xi(x+L)}{\cos^2\xi L + \varepsilon^2\sin^2\xi L}\,\frac{d\xi}{\xi} \tag{8}$$

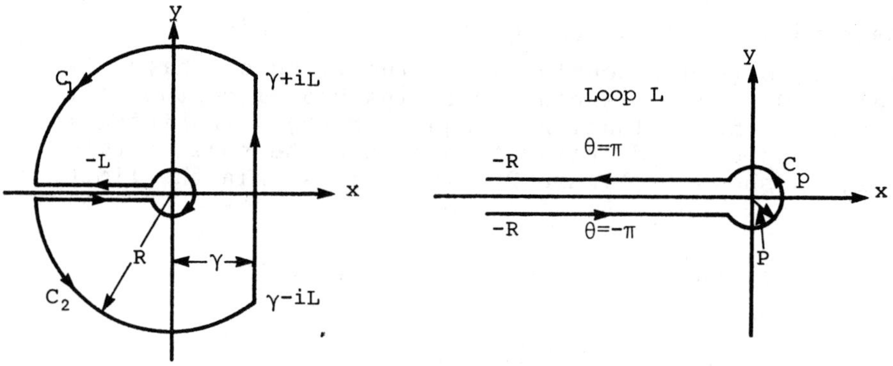

(a) Contour II (b)

Fig. 2

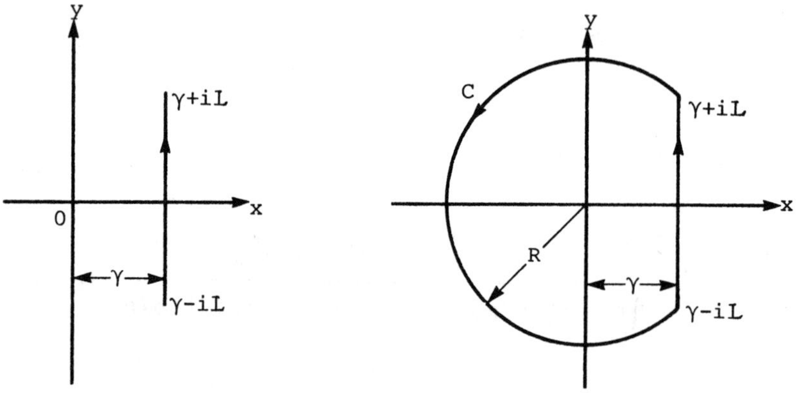

(a) (b) Contour I

Fig. 3

The integrals of equations (7) and (8) can be numerically evaluated.

• PROBLEM 3-21

The uniform temperature of the earth's surface at a location is 50^0F. At any instant of time the ambient air temperature suddenly drops down to 10^0F. Estimate the depth from the surface where the temperature would be 30^0F after 4 hours of time from that instant, given that $k = 0.2$, $h = 1.0$ and $\alpha = 0.02$. Use the given chart.

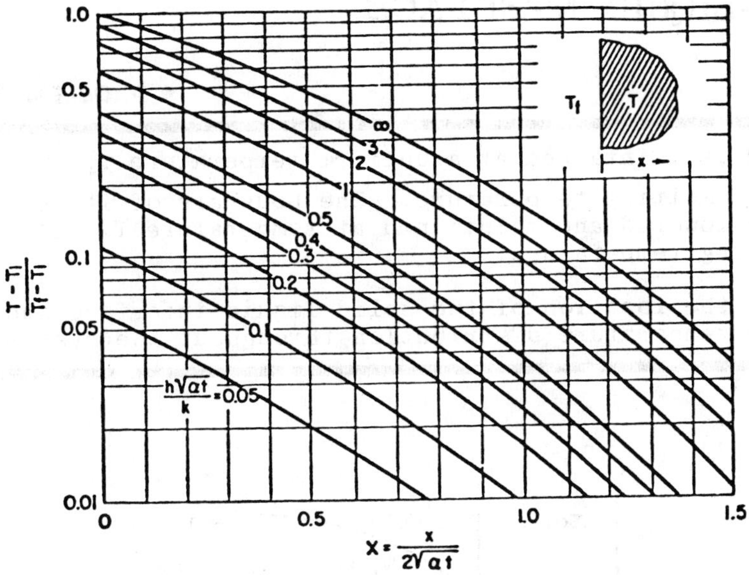

Fig. 1 Temperature history in a semi-infinite solid initially at a uniform temperature and suddenly exposed at its surface to a fluid at constant temperature.

<u>Solution</u>: The given chart indicates the temperature distribution in the earth due to sudden changes in temperature at the surface.

Now, $$\frac{T - T_1}{T - T_2} = \frac{50 - 30}{50 - 10} = \frac{20}{40} = 0.5$$

and

$$\frac{h\sqrt{\alpha t}}{k}$$

$$= \frac{1\sqrt{0.02 \times 4}}{0.2} = 1.414$$

For $\frac{T - T_1}{T - T_2} = 0.5$ and $\frac{h\sqrt{\alpha t}}{k} = 1.414$

from the chart $X = 0.17 = \frac{x}{2\sqrt{\alpha t}}$

or $$x = 0.17 \times 2\sqrt{0.02 \times 6} \text{ ft.}$$

$$= 0.117 \text{ ft.}$$

or 1.41 in.

SOLID-LIQUID INTERFACE

• PROBLEM 3-22

A liquid is maintained at a uniform temperature T_i, above its melting temperature. One boundary of it is suddenly lowered and maintained at temperature T_0 below its melting temperature.

Evaluate the location of the solid liquid interface and also the temperature distribution throughout the system.

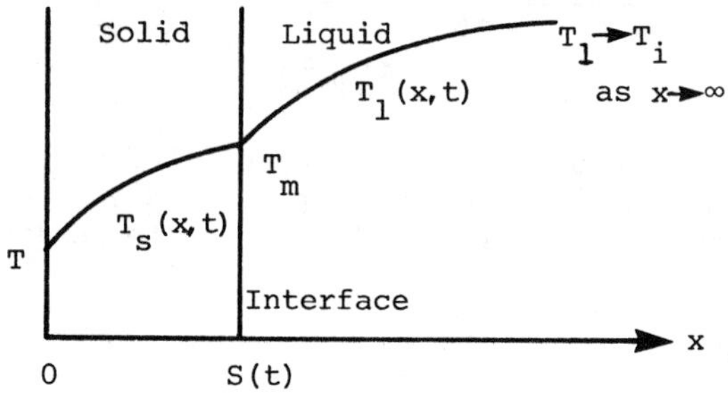

Solidification in a half-space.
Two phase problem.

Solution: The differential equation for the representation of the solid phase is

$$\frac{\partial^2 T_s}{\partial x^2} = \frac{1}{\alpha_s} \frac{\partial T_s(x,t)}{\partial t} \qquad \text{in } 0 < x < s(t),\ t > 0 \qquad (1)$$

$$T_s(x,t) = T_0 \qquad \text{at } x = 0,\ t > 0 \qquad (2)$$

and, for liquid phase

$$\frac{\partial^2 T_\ell}{\partial x^2} = \frac{1}{\alpha_l} \frac{\partial T_\ell(x,t)}{\partial t} \qquad \text{in } s(t) < x < \infty,\ t > 0 \qquad (3)$$

$$T_\ell(x,t) = T_i \qquad \text{for } t = 0,\ \text{in } x > 0 \qquad (4)$$

$$T_\ell(x,t) \to T_i \qquad \text{as } x \to \infty,\ t > 0 \qquad (5)$$

conditions for and at the interface $x = s(t)$

$$T_s(x,t) = T_\ell(x,t) = T_m \quad \text{at } x = s(t),\ t > 0 \qquad (6)$$

$$k_s \frac{\partial T_s}{\partial x} - k_\ell \frac{\partial T_\ell}{\partial x} = \rho L \frac{ds(t)}{dt} \text{ at } x = s(t),\ t > 0 \qquad (7)$$

Now, assume a solution of the form

$$T_s(x,t) = T_0 + A \operatorname{erf}\left[x/2(\alpha_s t)^{\frac{1}{2}}\right] \qquad (8)$$

The equation (1) for solid phase, and its boundary condition are satisfied.

Again, assume a solution for $T_\ell(x,t)$ as

$$T_\ell(x,t) = T_i + B \operatorname{erfc}\left[x/2(\alpha_1 t)^{\frac{1}{2}}\right] \qquad (9)$$

The differential equation (3) for liquid phase and initial and boundary conditions (5) and (4) are satisfied.

A and B are the constants. Substituting equations (8) and (9) into the interface equation (6) yields

$$T_0 + A \operatorname{erf}(\lambda) = T_i + B \operatorname{erfc}\left[\lambda\left(\frac{\alpha_s}{\alpha_\ell}\right)^{\frac{1}{2}}\right] = T_m \qquad (10)$$

where $$\lambda = \frac{s(t)}{2(\alpha_s t)^{\frac{1}{2}}} \quad \text{or} \quad s(t) = 2\lambda(\alpha_s t)^{\frac{1}{2}} \qquad (11)$$

The constants A and B found from equation (10) are

$$A = \frac{T_m - T_o}{\operatorname{erf}(\lambda)}, \quad B = \frac{T_m - T_i}{\operatorname{erfc}[\lambda(\alpha_s/\alpha_\ell)^{\frac{1}{2}}]}$$

Now, the temperature distribution in the solid and liquid phases can be obtained by substituting the values of constants A and B into equations (8) and (9).

Thus $$\frac{T_s(x,t) - T_o}{T_m - T_o} = \frac{\operatorname{erf}[x/2(\alpha_s t)^{\frac{1}{2}}]}{\operatorname{erf}(\lambda)} \qquad (12)$$

$$\frac{T_\ell(x,t) - T_i}{T_m - T_i} = \frac{\operatorname{erfc}[x/2(\alpha_\ell t)^{\frac{1}{2}}]}{\operatorname{erfc}[\lambda(\alpha_s/\alpha_\ell)^{\frac{1}{2}}]} \qquad (13)$$

In order to find an expression for the constant λ, expressions for $s(t)$, $T_s(x,t)$ and $T_\ell(x,t)$, obtained from equations (11), (12) and (13) respectively, are substituted into equation (7), which results in

$$\frac{e^{-\lambda^2}}{\text{erf}(\lambda)} + \frac{k_\ell}{k_s}\left(\frac{\alpha_s}{\alpha_\ell}\right)^{\frac{1}{2}} \frac{T_m - T_i}{T_m - T_o} \frac{e^{-\lambda(\alpha_s/\alpha_\ell)}}{\text{erfc}[\lambda(\alpha_s/\alpha_\ell)^{\frac{1}{2}}]} = \frac{\lambda L\sqrt{\pi}}{C_{ps}(T_m - T_o)} \quad (14)$$

Thus, after evaluating λ, the interface temperature s(t), the temperature distribution $T_s(x,t)$ in the solid phase and $T_\ell(x,t)$ in the liquid phase can be obtained from equations (11), (12) and (13) respectively.

• PROBLEM 3-23

Find the location for the solid-liquid interface as a function of time for a liquid initially at a temperature T_i, above its melting temperature T_m, which is in a thin container of thickness b as shown. The side at $x = 0$ is maintained at temperature T_o, while the side at $x = b$ is fully insulated. The solidification starts at $x = 0$ and the solid-liquid interface moves in the positive x-direction.

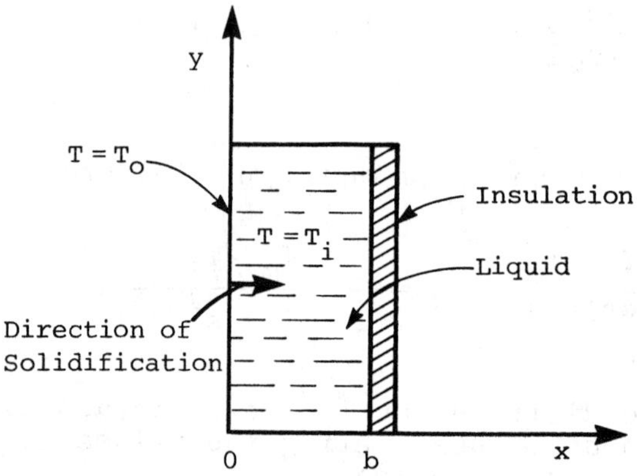

Fig. 1

Solution: The problem can be solved by exact integral method. The mathematical formulation is analogous to the Neumann's problem and can be written in dimensionless parameters.

The dimensionless parameters are defined as

$$\theta_j \equiv \frac{T_j - T_o}{T_i - T_o} \quad \text{with } j = s,\ \ell,\ \text{or } m,$$

$$X = \frac{x}{b},\quad S = \frac{s}{b},\quad \tau = \frac{\alpha_s t}{b^2}$$

where the subscripts ℓ, s, and m correspond to the liquid phase, solid phase and melting temperature, respectively.

Expression for solid phase can be written as

$$\frac{\partial^2 \theta_s}{\partial X^2} = \frac{\partial \theta_s(X,\tau)}{\partial \tau} \qquad \text{in } 0 < X < S(t),\ \tau > 0 \qquad (1)$$

$$\theta_s(X,\tau) = 0 \qquad \text{at } X = 0,\ \tau > 0 \qquad (2)$$

for the liquid phase as

$$\frac{\partial^2 \theta_\ell}{\partial X^2} = \frac{\alpha_s}{\alpha_\ell} \frac{\partial \theta_\ell(X,\tau)}{\partial \tau} \qquad \text{in } S(\tau) < X < 1,\ \tau > 0 \qquad (3)$$

$$\frac{\partial \theta_\ell}{\partial X} = 0 \qquad \text{at } X = 1,\ \tau > 0 \qquad (4)$$

$$\theta_\ell = 1 \qquad \text{for } \tau = 0 \text{ in } 0 < X < 1 \qquad (5)$$

and for the solid-liquid interface as

$$\theta_s(X,\tau) = \theta_\ell(X,\tau) = \theta_m \qquad \text{at } X = S(\tau),\ \tau > 0 \quad (6)$$

$$\frac{\partial \theta_s}{\partial X} - \frac{k_\ell}{k_s} \frac{\partial \theta_\ell}{\partial X} = \frac{L}{C_{ps}(T_i - T_o)} \frac{dS(\tau)}{d\tau} \quad \text{at } X=S(\tau),\ \tau > 0 \qquad (7)$$

The temperature distribution for the solid region is given by the equation

$$\frac{T_s(x,t) - T_o}{T_m - T_o} = \frac{\operatorname{erf}[x/2(\alpha_s t)^{\frac{1}{2}}]}{\operatorname{erf}(\lambda)}$$

or in terms of dimensionless parameters

$$\frac{\theta_s(X,\tau)}{\theta_m} = \frac{\operatorname{erf}[X/2\sqrt{\tau}]}{\operatorname{erf}(\lambda)}, \quad \tau > 0 \qquad (8)$$

where the parameter λ is unknown and is to be determined. The equation (8) satisfies the differential equation (1) and also the boundary condition equation (2).

The solid-liquid interface, as a function of time $S(\tau)$, can be assumed to be given by equation

$$S(\tau) = 2\lambda\sqrt{\tau} \qquad (9)$$

Now, applying integral method to find the temperature distribution in the liquid phase

$$\theta_\ell(X,\tau) = 1 \quad \text{at } X = \delta(\tau)$$

$$\frac{\partial \theta_\ell}{\partial X} = 0 \quad \text{at } X = \delta(\tau) \qquad (10)$$

where $\delta(\tau)$ is the thermal layer as indicated in figure 2. The thermal layer should be less than or equal to unity for the integral method to be valid. Thus, the heat conduction equation is integrated between the limits $X = S(\tau)$ and $X = \delta(\tau)$ and using the boundary conditions gives the energy equation

$$-\frac{\alpha_\ell}{\alpha_s}\frac{\partial\theta_\ell}{\partial X}\bigg|_{X=S(\tau)} + \frac{d\delta}{d\tau} - \theta_m\frac{dS}{d\tau} = \frac{d}{d\tau}\left[\int_{S(\tau)}^{\delta(\tau)} \theta_\ell(X,\tau)dX\right] \quad (11)$$

In order to solve equation (11) the form of $\theta_\ell(X,\tau)$ must be known, hence the form is assumed as

$$\theta_\ell(X,\tau) = 1 - (1-\theta_m)\left(\frac{\delta - X}{\delta - S}\right)^n, \quad n \geq 2 \quad (12)$$

where n is a dimensionless index. The assumed form satisfies the boundary condition equations (5) and (10).

Let $\delta(\tau) = 2\beta\sqrt{\tau}$ (13)

where β is unknown.

Now, introducing equations (12) and (13) into equation (11) and solving, yields

$$\beta - \lambda = \frac{n+1}{2}\left[-\lambda + \sqrt{\lambda^2 + \frac{2n}{n+1}\frac{\alpha_\ell}{\alpha_s}}\right] \quad (13)$$

In order to obtain λ the equations (8) and (12) are substituted into equation (6) which on simplification yields

$$\frac{e^{-\lambda^2}}{\mathrm{erf}(\lambda)} + \frac{k_\ell}{k_s}\left(\frac{\alpha_s}{\alpha_\ell}\right)^{\frac{1}{2}}\frac{\theta_m - 1}{\theta_m}\frac{1}{Z_n} = \frac{\lambda L\sqrt{\pi}}{C_{ps}(T_m - T_o)}$$

where

$$Z_n \quad \frac{n+1}{n\sqrt{\pi}}\left[-\gamma + \sqrt{\gamma^2 + \frac{2n}{n+1}}\right]$$

$$\gamma = \lambda\left(\frac{\alpha_s}{\alpha_\ell}\right)^{\frac{1}{2}}$$

$$\frac{\theta_m - 1}{\theta_m} = \frac{T_m - T_i}{T_m - T_o}$$

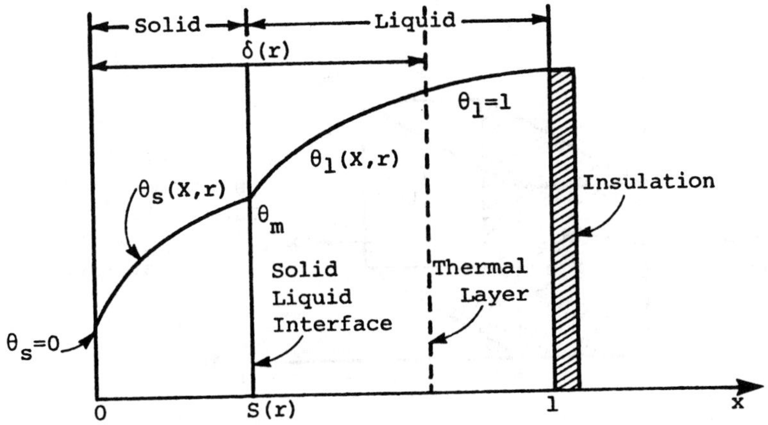

Fig. 2 Solidification of liquid of finite thickness.

Thus, after finding λ, β can be evaluated. Then substituting for λ and β into equations $S(\tau) = 2\lambda\sqrt{\tau}$ and $\delta(\tau) = 2\beta\sqrt{\tau}$ the solid-liquid interface $S(\tau)$ and the thickness of thermal layer $\delta(\lambda)$ can be found.

SOLID-GAS INTERFACE

• PROBLEM 3-24

Determine the temperature distribution in a rectangular plate $T(x,y,t)$ as a function of time $(0 \le t \le \infty)$. At one end the two sides of the plate $(x=0, y=0)$ are insulated while the other end $(x=a, y=b)$ convects heat to the surroundings at temperature $T=0$, and heat is generated at a constant rate of $q\ W/m^3$. The initial uniform temperature of the plate is $T=T_o$ and the orthogonal thermal conductivities of the plate are k_x and k_y.

Solution: The two-dimensional Fourier equation for heat conduction with internal heat source can be written as

$$\frac{\partial^2 T}{\partial x^2} + \frac{1}{\varepsilon^2}\frac{\partial^2 T}{\partial y^2} + \frac{q}{k_x} = \frac{1}{\alpha_x}\frac{\partial T}{\partial t} \qquad (1)$$

for $0 < x < a$, $0 < y < b$ at $t > 0$ with boundary conditions

$$\frac{\partial T}{\partial x} = 0 \qquad \text{at } x = 0,\ t > 0$$

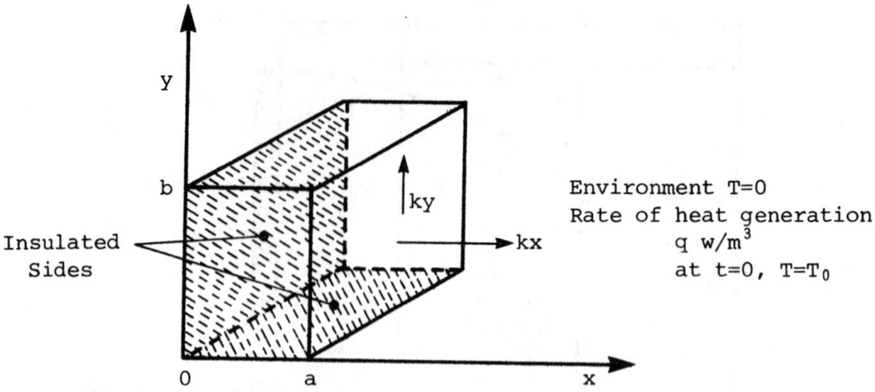

$$\frac{\partial T}{\partial x} + H_x T = 0 \qquad \text{at } x = a,\ t > 0$$

$$\frac{\partial T}{\partial y} = 0 \qquad \text{at } y = 0,\ t > 0$$

$$\frac{\partial T}{\partial y} + H_y T = 0 \qquad \text{at } y = b,\ t > 0$$

$$T = T_o \qquad \text{for } t = 0, \text{ in the region}$$

where

$$\varepsilon^2 \equiv \frac{k_x}{k_y}, \quad H_x = \frac{h_x}{k_x}, \quad H_y = \frac{h_y}{k_y}$$

and

$$\alpha_x = \frac{k_x}{\rho C_p}$$

The partial differential equation can be separated into steady temperature distribution $T_s(x,y)$ and transient temperature distribution $T_h(x,y,t)$.

Therefore $T(x,y,t) = T_s(x,y) + T_h(x,y,t)$

where for solving for steady state condition

$$\frac{\partial^2 T_s}{\partial x^2} + \frac{1}{\varepsilon^2}\frac{\partial^2 T_s}{\partial y^2} + \frac{q}{k_x} = 0 \qquad (2)$$

for $0 < x < a$ and $0 < y < b$ with boundary conditions

$$\frac{\partial T_s}{\partial x} = 0 \qquad \text{at } x = 0$$

$$\frac{\partial T_s}{\partial x} + H_x T_s = 0 \qquad \text{at } x = a$$

$$\frac{\partial T_s}{\partial y} = 0 \qquad \text{at } y = 0$$

$$\frac{\partial T_s}{\partial y} + H_y T_s = 0 \qquad \text{at } y = b$$

And, for transient temperature $T_h(x,y,t)$

$$\frac{\partial^2 T_h}{\partial x^2} + \frac{1}{\varepsilon^2}\frac{\partial^2 T_h}{\partial y^2} = \frac{1}{\alpha_x}\frac{\partial T_h}{\partial t} \quad \text{for } 0 < x < a,\ 0 < y < b,\ \text{at } t > 0 \tag{3}$$

with boundary conditions

$$\frac{\partial T_h}{\partial x} = 0 \qquad \text{at } x = 0,\ t > 0$$

$$\frac{\partial T_h}{\partial x} + H_x T_h = 0 \qquad \text{at } x = a,\ t > 0$$

$$\frac{\partial T_h}{\partial y} = 0 \qquad \text{at } y = 0,\ t > 0$$

$$\frac{\partial T_h}{\partial y} + H_y T_h = 0 \qquad \text{at } y = b,\ t > 0$$

$$T_h = T_o - T_s(x,y) \equiv F(x,y) \quad \text{for } t = 0, \text{ in the plate.}$$

Now, the integral-transform pair with respect to the x-variable is defined as

Transform: $$\bar{T}(\beta_m,y,t) = \int_0^a X(\beta_m,x')T(x',y,t)dx' \tag{4}$$

Inversion: $$T(x,y,t) = \sum_{m=1}^{\infty} \frac{1}{N(\beta_m)}X(\beta_m,x)\bar{T}(\beta_m,y,t) \tag{5}$$

where

$$X(\beta_m,x) = \cos\beta_m x, \quad \frac{1}{N(\beta_m)} = 2\,\frac{\beta_m^2 + H_x^2}{a(\beta_m^2+H_x^2)+H_x} \tag{6}$$

and β_m's are the roots of

$$\beta_m \tan\beta_m a = H_x$$

and, the integral transform pair with respect to y-variable is defined as

Transform: $$\tilde{\bar{T}}(\beta_m,\gamma_n,t) = \int_0^b Y(\gamma_n,y')\bar{T}(\beta_m,y',t)dy' \tag{7}$$

Inversion: $$\bar{T}(\beta_m,y,t) = \sum_{n=1}^{\infty} \frac{Y(\gamma_n,y)}{N(\gamma_n)}\tilde{\bar{T}}(\beta_m,\gamma_n,t) \tag{8}$$

where

$$Y(\gamma_n,y) = \cos\gamma_n y, \quad \frac{1}{N(\gamma_n)} = 2\,\frac{\gamma_n^2 + H_y^2}{b(\gamma_n^2+H_y^2)+H_y} \tag{9}$$

and γ_n's are the roots of

$$\gamma_n \tan\gamma_n b = H_y$$

Applying the transform equation (4) to equation (3), the integral transform of (4) with respect to variable x can be expressed as

$$-\beta_m^2\overline{T}_h(\beta_m,y,t) + \frac{1}{\varepsilon^2}\frac{\partial^2\overline{T}_h}{\partial y^2} = \frac{1}{\alpha_x}\frac{\partial\overline{T}}{\partial t} \qquad \text{for } 0 < y < b, \text{ at } t > 0 \tag{10}$$

$$\frac{\partial\overline{T}_h}{\partial y} = 0 \qquad \text{at } y = 0,\ t > 0$$

$$\frac{\partial\overline{T}_h}{\partial y} + H_y\overline{T}_h = 0 \qquad \text{at } y = b,\ t > 0$$

$$\overline{T}_h = \overline{F}(\beta_m,y) \qquad \text{for } t = 0, \text{ in } 0 \leq y \leq b$$

And, the integral transform of equation (10) with respect to variable y using the equation (7) yields

$$-\beta_m\tilde{\tilde{T}}_h(\beta_m,\gamma_n,t) - \frac{1}{\varepsilon^2}\gamma_n^2\tilde{\tilde{T}}_n = \frac{1}{\alpha_x}\frac{d\tilde{\tilde{T}}_h}{dt} \tag{11}$$

or

$$\frac{d\tilde{\tilde{T}}_h}{dt} + \alpha_x\gamma_{mn}^2\tilde{\tilde{T}}_h = 0 \qquad \text{for } t > 0 \tag{12}$$

$$\tilde{\tilde{T}}_h(\beta_m,\gamma_n,t) = \tilde{\tilde{F}}(\beta_m,\gamma_n)$$

where

$$\lambda_{mn}^2 = \beta_m^2 + \frac{1}{\varepsilon^2}\gamma_n^2 \tag{13}$$

The solution of equation (11) is

$$\tilde{\tilde{T}}_h(\beta_m,\gamma_n,t) = e^{-\alpha_x\lambda_{mn}^2 t}\,\tilde{\tilde{F}}(\beta_m,\gamma_n) \tag{14}$$

Taking inverse transformations successively on equation (14) yields the solution for transient temperature distribution $T_h(x,y,t)$

$$T_h(x,y,t) = \sum_{m=1}^{\infty}\sum_{n=1}^{\infty}\frac{e^{-\alpha_x\lambda_{mn}^2 t}}{N(\beta_m)N(\gamma_n)}\cos\beta_m x\cos\gamma_n y \tag{15}$$

$$\int_{x'=0}^{a}\int_{y'=0}^{b}\cos\beta_m x'\cos\gamma_n y' F(x',y')dx'dy'$$

The equation can be evaluated by numerical method.

• PROBLEM 3-25

Find the temperature distribution in a composite concentric cylindrical solid (refer figure 1) with initial temperatures of $T_1(r,t) = f_1(r)$ and $T_2(r,t) = f_2(r)$ in the inner and outer cylinder respectively. From the outer surface of the external cylinder, heat is convected to the surrounding air at constant temperature T_∞. Assume the interface to be in perfect thermal contact.

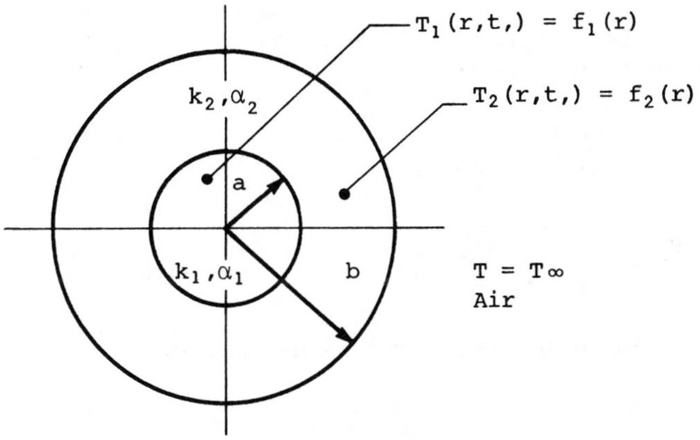

Fig. 1 Composite Concentric Cylinder

Solution: The initial temperature of the cylinders can be referred to as in excess of the ambient temperature T_∞

$$\theta_i(r,t) = T_i(r,t) - T_\infty$$

where subscript i assumes 1 and 2 referring to inner and outer cylinder respectively.

The differential equation for heat transfer can be expressed as at $t > 0$

$$\frac{\partial\theta_1(r,t)}{\partial t} = \frac{\alpha_1}{r}\frac{\partial}{\partial r}\left(r\frac{\partial\theta_1}{\partial r}\right)$$

for inner cylinder and,

$$\frac{\partial\theta_2(r,t)}{\partial t} = \frac{\alpha_2}{r}\frac{\partial}{\partial r}\left(r\frac{\partial\theta_2}{\partial r}\right)$$

for outer cylinder.

The boundary conditions expressed in terms of θ at $t > 0$ are

$$\theta_1(r,t) = \text{finite} \qquad \text{at } r = 0$$

$$\theta_1(r,t) = \theta_2(r,t) \qquad \text{at } r = a$$

$$k_1 \frac{\partial \theta_1}{\partial r} = k_2 \frac{\partial \theta_2}{\partial r} \qquad \text{at } r = a$$

$$k_2 \frac{\partial \theta}{\partial r} + h_3 \theta_2 = 0 \qquad \text{at } r = b$$

and the initial condition, at $t = 0$

$$\theta_1(r,t) = f_1(r) - T_\infty \equiv F_1(r) \quad \text{for } 0 < r < a$$

$$\theta_2(r,t) = f_2(r) - T_\infty \equiv F_2(r) \quad \text{for } a < r < b$$

Expressing the heat transfer equation in the form of eigenvalue problem as

$$\frac{1}{r}\frac{d}{dr}\left(r\frac{d\Psi_{1n}}{dr}\right) + \frac{\beta_n^2}{\alpha_1}\psi_{1n}(r) = 0 \qquad \text{in } 0 \le r < a$$

$$\frac{1}{r}\frac{d}{dr}\left(r\frac{d\psi_{2n}}{dr}\right) + \frac{\beta_n^2}{\alpha_2}\psi_{2n}(r) = 0 \qquad \text{in } a < r < b$$

and the boundary conditions transform as

$$\psi_{1n}(r) = \text{finite} \qquad \text{at } r = 0$$

$$\psi_{1n}(r) = \psi_{2n}(r) \qquad \text{at } r = a$$

$$k_1 \frac{d\psi_{1n}}{dr} = k_2 \frac{d\psi_{2n}}{dr} \qquad \text{at } r = a$$

$$k_2^* \frac{d\psi_{2n}}{dr} + h_3\psi_{2n} = 0 \qquad \text{at } r = b$$

From the table the solution of the eigenvalue problem can be written as

$$\psi_{in}(r) = A_{in}J_o\left(\frac{\beta_n}{\sqrt{\alpha_i}}r\right) + B_{in}Y_o\left(\frac{\beta_n}{\sqrt{\alpha_i}}r\right), \qquad i = 1,2$$

From the boundary conditions the value of constant $B_{1n} = 0$ and $A_{1n} = 1$ then the solutions for cylinder (1) and (2) respectively are

$$\psi_{1n}(r) = J_o\left(\frac{\beta_n}{\sqrt{\alpha_1}}r\right) \qquad \text{in } 0 \le r < a$$

$$\psi_{2n}(r) = A_{2n}J_o\left(\frac{\beta_n}{\sqrt{\alpha_2}}r\right) + B_{2n}Y_o\left(\frac{\beta_n}{\sqrt{\alpha_2}}r\right) \qquad \text{in } a < r < b \tag{1}$$

To evaluate coefficients A_{2n} and B_{2n} the remaining boundary conditions are applied. Now, at $r = a$ $\psi_{1n}(r) = \psi_{2n}(r)$

$$J_o\left(\frac{\beta_n a}{\sqrt{\alpha_1}}\right) = A_{2n} J_o\left(\frac{\beta_n a}{\sqrt{\alpha_2}}\right) + \beta_{2n} Y_o\left(\frac{\beta_n a}{\sqrt{\alpha_2}}\right)$$

$$\frac{k_1}{k_2}\sqrt{\frac{\alpha_2}{\alpha_1}}\, J_1\left(\frac{\beta_n a}{\sqrt{\alpha_1}}\right) = A_{2n} J_1\left(\frac{\beta_n a}{\sqrt{\alpha_2}}\right) + B_{2n} Y\left(\frac{\beta_n a}{\sqrt{\alpha_2}}\right)$$

$$-\left[A_{2n} J_1\left(\frac{\beta_n b}{\sqrt{\alpha_2}}\right) + B_{2n} Y_1\left(\frac{\beta_n b}{\sqrt{\alpha_2}}\right)\right] + \frac{h_3\sqrt{\alpha_2}}{k_2^*\beta_n}\left[A_{2n} J_o\left(\frac{\beta_n b}{\sqrt{\alpha_2}}\right)\right.$$

$$\left. + B_{2n} Y_o\left(\frac{\beta_n b}{\sqrt{\alpha_2}}\right)\right] = 0$$

writing in matrix form

$$\begin{bmatrix} J_o(\gamma) & -J_o\left(\frac{a}{b}\eta\right) & -Y_o\left(\frac{a}{b}\eta\right) \\ KJ_1(\gamma) & -J_1\left(\frac{a}{b}\eta\right) & -Y_1\left(\frac{a}{b}\eta\right) \\ 0 & \frac{H}{\eta}J_o(\eta)-J_1(\eta) & \frac{H}{\eta}Y_o(\eta)-Y_1(\eta) \end{bmatrix} \begin{bmatrix} 1 \\ A_{2n} \\ B_{2n} \end{bmatrix} = \begin{bmatrix} 0 \\ 0 \\ 0 \end{bmatrix} \qquad (2)$$

where γ, η, H and K are defined as

$$\gamma \equiv \frac{a\beta_n}{\sqrt{\alpha_1}} \qquad \eta \equiv \frac{b\beta_n}{\sqrt{\alpha_2}} \qquad H \equiv \frac{bh_3}{k_2^*} \qquad K = \frac{k_1}{k_2}\sqrt{\frac{\alpha_2}{\alpha_1}}$$

There are three equations and two unknowns A_{2n} and B_{2n}, considering first two equations only, the matrix reduces to

$$\begin{bmatrix} J_o\left(\frac{a}{b}\eta\right) & Y_o\left(\frac{a}{b}\eta\right) & A_{2n} \\ J_1\left(\frac{a}{b}\eta\right) & Y_1\left(\frac{a}{b}\eta\right) & B_{2n} \end{bmatrix} = \begin{bmatrix} J_o(\gamma) \\ KJ_1(\gamma) \end{bmatrix}$$

Therefore the constants can be written as

$$A_{2n} = \frac{1}{\Delta}\left[J_o(\gamma)Y_1\left(\frac{a}{b}\eta\right) - KJ_1(\gamma)Y_o\left(\frac{a}{b}\eta\right)\right]$$

$$B_{2n} = \frac{1}{\Delta}\left[KJ_1(\gamma)J_o\left(\frac{a}{b}\eta\right) - J_o(\gamma)J_1\left(\frac{a}{b}\eta\right)\right]$$

where

$$\Delta = J_o\left(\frac{a}{b}\eta\right)Y_1\left(\frac{a}{b}\eta\right) - J_1\left(\frac{a}{b}\eta\right)Y_o\left(\frac{a}{b}\eta\right)$$

To obtain the eigenvalues (β_n) the necessary condition is that the coefficient of the determinant equation (2) should vanish. Thus,

$$\begin{vmatrix} J_o(\gamma) & -J_o\left(\frac{a}{b}\eta\right) & -Y_o\left(\frac{a}{b}\eta\right) \\ KJ_1(\gamma) & -J_1\left(\frac{a}{b}\eta\right) & -Y_1\left(\frac{a}{b}\eta\right) \\ 0 & \frac{H}{\eta}J_o(\eta)-J_1(\eta) & \frac{H}{\eta}Y_o(\eta)-Y_1(\eta) \end{vmatrix} = 0$$

Now, knowing the coefficients and the eigenvalues, the eigenfunctions $\psi_{1n}(r)$ and $\psi_{2n}(r)$ can be obtained from the equation (1), then the solution for temperature distribution can be expressed as

$$T_i(x,t) = \sum_{n=1}^{\infty} e^{-\beta_n^2 t}\,\frac{1}{N_n}\,\psi_{in}(x)\sum_{j=1}^{M}\frac{k_j}{\alpha_j}\int_{x'=x_j}^{x_{j+1}} x'^p\psi_{jn}(x')F_j(x')dx'$$

$$\text{in } x_i < x < x_{i+1}, \quad i = 1,2,\ldots,M$$

where the norm N_n is defined as

$$N_n = \sum_{j=1}^{M}\frac{k_j}{\alpha_j}\int_{x_j}^{xj+1} x'^p\psi_{jn}^2(x')dx'$$

where

$$p = \begin{cases} 0 & \text{slab} \\ 1 & \text{cylinder} \\ 2 & \text{sphere} \end{cases}$$

$$\theta_i(r,t) = \sum_{n=1}^{\infty}\frac{1}{N_n}e^{-\beta_n^2 t}\psi_{in}(r)\left[\frac{k_1}{\alpha_1}\int_0^a r'\psi_{1n}(r')F_1(r')dr'\right.$$

$$\left. + \frac{k_2}{\alpha_2}\int_a^b r'\psi_{2n}(r')F_2(r')dr'\right], \qquad i = 1,2$$

where

$$N_n = \frac{k_1}{\alpha_1}\int_0^a r'\psi_{1n}^2(r')dr' + \frac{k_2}{\alpha_2}\int_a^b r'\psi_{2n}^2(r')dr'$$

$$\psi_{1n}(r) = J_o\left(\frac{\beta_n}{\sqrt{\alpha_1}}r\right)$$

$$\psi_{2n}(r) = A_{2n}J_o\left(\frac{\beta_n}{\sqrt{\alpha_2}}r\right) + B_{2n}Y_o\left(\frac{\beta_n}{\sqrt{\alpha_2}}r\right)$$

• PROBLEM 3-26

Develop a computer software to find the temperature distribution with time in a large 10 cm. thick wall initially at a uniform temperature distribution of 20°C and suddenly exposed to ambient air at 40°C on one side and at 200°C on the other side. The convective heat transfer coefficient on the low temperature side is 50 $W/m^2 {}^0K$. The properties of the wall refractory are $k = 0.5$ W/m^0K, $\rho = 1800$ kg/m^3, $C_p = 840$ J/kg^0K. Also, find the time required to reach the steady temperature distribution.

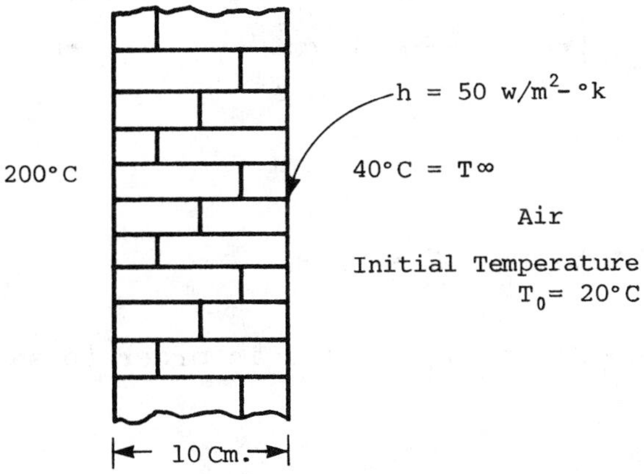

Fig. 1 Temperature distribution across a refractory wall.

Solution: The problem is one-dimensional unsteady state conduction. The 10 in. thick wall is divided into say 10 equal segments each of thickness Δx.

The value of α for the wall is

$$\alpha = \frac{k}{\rho C_p} = \frac{0.5}{840 \times 1800} = 0.33 \times 10^{-6} m^2/s.$$

The Fourier number is given by

$$\Delta F_0 = \frac{\alpha \Delta \tau}{(\Delta x)^2}$$

The temperature at any node i in any time increment n is given by

$$T_i^{(n+1)} = \Delta F_0 (T_{i-1}^n + T_{i+1}^n) + (1 - 2\Delta F_0) T_i^n$$

i varies from 1 to (M-1)

n=1 corresponds to the initial conditions i.e. when T=20°C uniform temperature of the wall

$$\Delta x = \frac{L}{M-1} = \frac{10}{11-1} = 1 \text{ cm.}$$

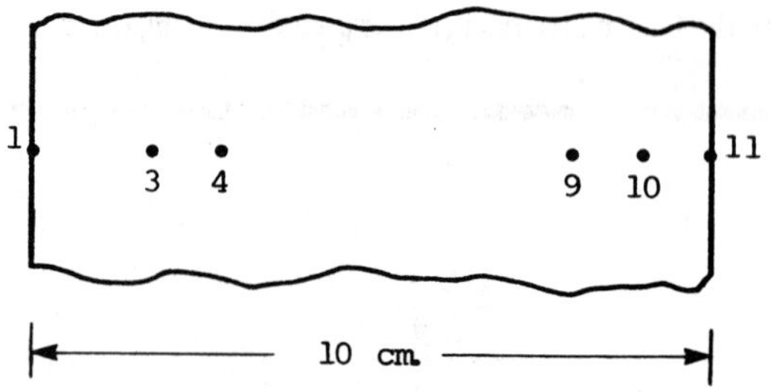

Fig. 2

M = 1 to 11, L = 10 cm.

The time increment $\Delta\tau$ is selected in order to satisfy the equation

$$\Delta\tau = \frac{0.99}{\dfrac{2\alpha}{(\Delta x)^2} + \dfrac{2h}{\rho C_p \Delta x}}$$

where α is the thermal diffusivity.

The temperature at the exposed face at 40°C is at node 11 is given by

$$T_M^{(n+1)} = 2\Delta F_0 \, T_{(M-1)}^n + 2\Delta F_0 \, \Delta B_i \, T_\infty + (1 - 2\Delta F_0 - 2\Delta B_i \Delta F_0) T_M^n$$

Further, steady state is reached when the heat conducted through the wall is equal to the heat convected from the face of the wall at temperature $T_\infty = 40°C$. The Do loop in the computer program will stop function once the $Q_{cond.} = Q_{conv.}$ the range of 1% of difference.

The complete calculations are performed in the following five steps.

```
      DIMENSION T(20),TT(20),TN(20),OTT(20)
      REAL L,K
      READ (5,10)L,K,CP,RHO
10    FORMAT(10F8.4)
      READ(5,10)TO,TS,H,TI
      WRITE (108,20)L,K,CP,RHO,TO,TS,H,TI
20    FORMAT(1H ,//,
     1'THICKNESS                 = ',F10.4,'   CMS',/,
     2'THERMAL CONDUCTIVITY      = ',F10.4,'   WATTS PER M.K',/,
     3'SPECIFIC HEAT             = ',F10.4,'   WATTA PER KG.K',/,
     4'DENSITY                   = ',F10.4,'   KG PER M**3',/,
     5'INITIAL TEMPERATURE       = ',F10.4,'   DEG CELSIUS',/,
     6'NEW TEMPERATURE           = ',F10.4,'   DEG CELSIUS',/,
     7'HEAT TRASFER COEFF.       = ',F10.4,'   WATTS PER M**2 .K',/,
     8'AMBIENT TEMPERATURE       = ',F10.4,'   DEG CELSIUS',//)
      M=11
      DX=(L/10.)/100.
      ALPHA=K/(CP*RHO)
      DT=DX*DX/ALPHA
      DT= 0.99/(  (2./DT) +   ((2.*H)/(RHO*CP*DX))  )
      WRITE (108,30)DT
30    FORMAT(1H ,/,
     1'TIME STEP                 = ',F10.6,' SECONDS')
15    INPUT ZC,ZT
      TIME=0
      B1=ALPHA*DT/(DX*DX)
      B2=2.*H*DT/(RHO*CP*DX)
      B3=1.-2*B1-B2
      DO 40 I=1,11
      TT(I)=TO
40    T(I)=TO
      T(1)=TS
      N=1
      NSTEP=0
      NN=ZT*60/DT
      NCOUNT=0
      TEZT=0
      WRITE(108,45)
45    FORMAT(///,
     1'TIME(MTS.) STEPS  TEMP1    TEMP2   TEMP3   TEMP4',
     2'   TEMP5   TEMP6    TEMP7    TEMP8   TEMP9',
     3'  TEMP10  TEMP11',//)
50    DO  60  I=1,M
      OTT(I)=T(I); OTT(1)=TS
60    TT(I)=T(I)
      DO 70  J=1,NN
      DO 80  I=2,10
80    TN(I)=B1*(T(I-1)       + T(I+1) +((1./B1)-2.)*T(I))
      TN(M)=2.*B1*T(M-1) + B2*TI  +B3*T(M)
      NSTEP=NSTEP + 1
      DO 90  I=2,11
90    T(I) = TN(I)
      N=N+1
70    CONTINUE
      N=0
      NCOUNT=NCOUNT+1
      TIME=NSTEP*DT/60
      QCOND=  (T(1)-T(M))/L*K*100
      QCONV=H*(T(M)-TI)
      Z=(QCOND-QCONV)/(.5*(QCOND+QCONV))
      IF(ABS(Z).GT.ZC)GO TO 117
      IF(TEZT.GT.999)GO TO 117
      NCOUNT=NCOUNT-1
      NSTEP=NSTEP-NN
      IF(TEZT.GT.600)ZT=1
      IF(TEZT.LT.600)ZT=10
      NN=ZT*60/DT
      DO 112 I=1,M
112   T(I)=OTT(I)
      IF(TEZT.GT.600)TEZT=1000
      IF(TEZT.LT.600)TEZT=900
      GOTO 50
117   WRITE(108,110)TIME,NSTEP,(T(I),I=1,M)
110   FORMAT(F10.2,I5,11F8.2)
      IF(ABS(Z).GT.ZC)GO TO 50
115   WRITE(108,120) TIME,QCOND,QCONV
120   FORMAT(1H ,//,
     1'STEADY STATE IS REACHED AFTER ',F10.4,' MINUTES',/,
     2'QCOND                          = ',F10.4,/,
     3'QCONV                          = ',F10.4,//)
      IF(ZC.LT.999)GO TO  15
      STOP
      END
```

Time step is calculated.

Steady-state criterion and the time interval for outputting temperatures are read.

The initial conditions are implemented.

T_i^n s are calculated.

Testing for steady state is done.

a. To read the input data-physical properties, initial and boundary conditions.

b. To compute the quantities Δx, α and $\Delta\tau$.

c. To compute the number of time increments necessary (N) for a time period of say 30 minutes.

d. To calculate the temperature T_i^n at each node for each increment of time.

e. To calculate the heat conducted through the wall and convected from the fall and find the difference and stop when the difference is less than or equal to 1%.

The program given needs the changes in input format and certain characters to match with the particular compiler of the computer. However, the basic algorithm remains unchanged.

The program is run and the results are printed as follows.

```
THICKNESS                =    10.0000  CMS
THERMAL CONDUCTIVITY     =      .5000  WATTS PER M.K
SPECIFIC HEAT            =   840.0000  WATTA PER KG.K
DENSITY                  =  1800.0000  KG PER M**3
INITIAL TEMPERATURE      =    20.0000  DEG CELSIUS
NEW TEMPERATURE          =   200.0000  DEG CELSIUS
HEAT TRASFER COEFF.      =    50.0000  WATTS PER M**2 .K
AMBIENT TEMPERATURE      =    40.0000  DEG CELSIUS

TIME STEP                =  74.843964 SECONDS
```

TIME(MTS.)	STEPS	TEMP1	TEMP2	TEMP3	TEMP4	TEMP5	TEMP6	TEMP7	TEMP8	TEMP9	TEMP10	TEMP11
29.94	24	200.00	159.46	122.13	90.54	66.07	48.93	38.37	33.15	31.89	33.34	36.47
59.88	48	200.00	171.48	144.23	119.36	97.73	79.86	65.91	55.71	48.79	44.50	42.02
89.81	72	200.00	177.13	154.96	134.12	115.12	98.32	83.89	71.80	61.85	53.64	46.66
119.75	96	200.00	180.39	161.19	142.79	125.50	109.54	95.03	81.97	70.22	59.57	49.58
149.69	120	200.00	182.35	164.96	148.05	131.82	116.40	101.88	88.24	75.41	63.25	51.56
179.63	144	200.00	183.55	167.26	151.27	135.69	120.61	106.08	92.08	78.59	65.51	52.72
209.56	168	200.00	184.29	168.67	153.24	138.06	123.19	108.65	94.44	80.54	66.89	53.42
239.50	192	200.00	184.74	169.54	154.45	139.51	124.77	110.23	95.89	81.74	67.74	53.86
269.44	216	200.00	185.02	170.07	155.19	140.40	125.73	111.19	96.77	82.47	68.26	54.12
299.38	240	200.00	185.18	170.39	155.64	140.95	126.33	111.78	97.32	82.92	68.58	54.29
329.31	264	200.00	185.29	170.59	155.92	141.28	126.69	112.15	97.65	83.20	68.78	54.39
339.29	272	200.00	185.31	170.64	155.98	141.36	126.78	112.23	97.73	83.26	68.82	54.41
340.54	273	200.00	185.32	170.64	155.99	141.37	126.79	112.24	97.74	83.27	68.83	54.41
341.79	274	200.00	185.32	170.65	156.00	141.38	126.80	112.25	97.75	83.28	68.84	54.42

```
STEADY STATE IS REACHED AFTER   341.7874 MINUTES
QCOND                       =   727.9243 WATTS
QCONV                       =   720.7563 WATTS
```

• PROBLEM 3-27

Estimate the heat loss from the above ground utilidor, as shown in the figure. A bare asbestos cement pipe is enclosed in an air filled duct. The relevant data is

Wind velocity	U_∞ = 5mph
Convective coefficient	k_u = 0.75 Btu/hr ft ^{0}F
Liner conductivity	k_ℓ = 0.02
Pipe conductivity	k_p = 0.375
Mean perimeter of the duct	P_u = 8.25 ft
Mean perimeter of the liner	P_ℓ = 7.0 ft
Equivalent circular diameters for the duct	D_i = 0.598 ft
	D_o = 1,273 ft
$\delta = \frac{D_o - D_i}{2}$	δ = 0.3375

Emissivities of the liner $E_\ell = 0.91$

The properties of the air in the duct are:

$\nu = 12.44 \times 10^{-5}$ ft²/sec

$k = 0.0136$ Btu/hr ft°F

$\rho = 0.717$

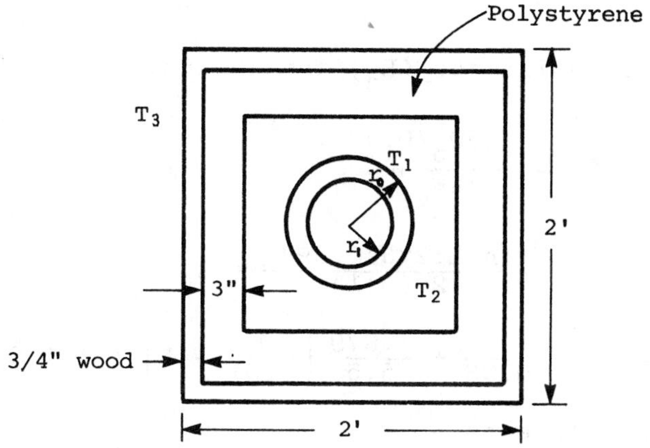

Solution: The first step in solving this problem is to make an approximate guess of the temperatures.

Let $T_1 = 45°F$

$T_2 = 27°F$

$T_3 = -49°F$

The fluid flowing inside the pipe, has a convective resistance (R_1), where

$$R_1 = \frac{1}{2\pi r_i h_i}$$

$$\simeq 0.033 \quad \text{This can be neglected.}$$

The conductive resistances of the pipe wall (R_2) and the insulation (R_3) are

$$R_2 = \frac{\ln(r_o/r_i)}{2\pi k_p} = \frac{\ln(0.299/0.244)}{2\pi(0.375)}$$

$$= 0.086$$

$$R_3 = \frac{\ln(r_1/r_0)}{2\pi k_i} = 0 \quad \text{no insulation}$$

The Grashof-Prandtt number is

$$N = \frac{32.2(45 - 27)(0.3375)^3}{496(12.44 \times 10^{-5})^2}$$

$$= 2.081 \times 10^6$$

Then from the equation

$$k_e/k = 0.4N^{0.2} \qquad 10^6 < N < 10^9$$

$$k_e = 0.100$$

The convective resistance (R_c) for an enclosed cylinder is given by

$$R_c = \frac{\ln(D_o/D_i)}{2\pi k_e}$$

$$= \frac{\ln(1.273/0.598)}{2\pi(0.1)} = 1.205$$

Therefore $P = \frac{1}{0.93} + \frac{1.879}{5.5}\left(\frac{1}{0.91} - 1\right) = 1.109$

The mean temperature is

$$\overline{T} = \left(\frac{45 + 27}{2}\right) + 460 = 496\,^0R$$

$$h_r = \frac{4(0.1714 \times 10^{-8})(496)^3}{1.109} = 0.754$$

The radiative resistance is $R_r = 0.7056$

The value of p is given by $P = \frac{1}{e_c} + \frac{A_1}{A_c}\left(\frac{1}{\varepsilon_\ell} - 1\right)$ where

$A_1 = 2\pi r$, outside area of inner cylinder

$A_c = 2(b+a)-4(t_\ell+t_u)$ inside area of duct liner

$\varepsilon_c, \varepsilon_\ell$ = emissivities of the insulation and liner

$$R_4 = \frac{1}{\frac{1}{R_c} + \frac{1}{R_r}}$$

$$= \frac{1}{\frac{1}{1.205} + \frac{1}{0.7056}} = 0.445$$

The resistances of the duct liner (R_5) and the duct (R_6) are

$$R_5 = \frac{t_\ell}{P_\ell k_\ell} = 1.786$$

$$R_6 = \frac{t_u}{P_u k_u} = 0.101$$

The outside convective coefficient can be estimated by the equation

$$h_\infty = 0.31 \left(\frac{T_u - T_\infty}{a+b} \right)^{0.25} \sqrt{12.5\, U_\infty + 1}$$

$$= 0.31 \left(\frac{1}{4} \right)^{0.25} \sqrt{12.5(5) + 1}$$

Therefore $h_\infty = 1.75$

The resistance (R_7) for the utilidor is given as

$$R_7 = \begin{cases} \dfrac{1}{k_g \left(5.7 + \dfrac{b}{2a}\right)} \ln \dfrac{3.5(H + H_0)}{a^{0.75}\, b^{0.25}} & \text{if the utilidor is buried (A)} \\[2ex] \dfrac{1}{2(a+b)h_\infty} & \text{if above ground (B)} \end{cases}$$

substituting the required values in equation B gives

$$R_7 = 0.72$$

Adding all the resistances gives the total resistance R_T

$$R_T = 2.523$$

$$q/L = \frac{100}{2.523} = 39.64 \text{ Btu/hr ft}$$

The initial temperatures that were assumed, can now be checked by calculating the temperature drop across each resistance. This calculation gives

$$T_1 = 45.3^0\text{F} \quad T_2 = 27.7 \quad T_3 = -47.2$$

The initial guess was chosen so as to avoid iteration, but initially any temperatures may be assumed and iterated until there is no significant variation between the temperatures assumed and calculated.

• PROBLEM 3-28

A thermocouple is welded to the bottom of a pocket made of copper tubing of 3/8 in. outside diameter. The pocket is fitted perpendicular to the copper pipe wall, through which gas flows. The pocket thickness is 0.036 in. and projects 3 in. into the pipe. The thermocouple gives a reading of 400°F, when the wall temperature is at 200°F. The heat transfer coefficient from the gas to the copper tube is 25 Btu/hr ft²°F. Calculate the corrected gas temperature. Conductivity for copper is 200 Btu/hr ft°F.

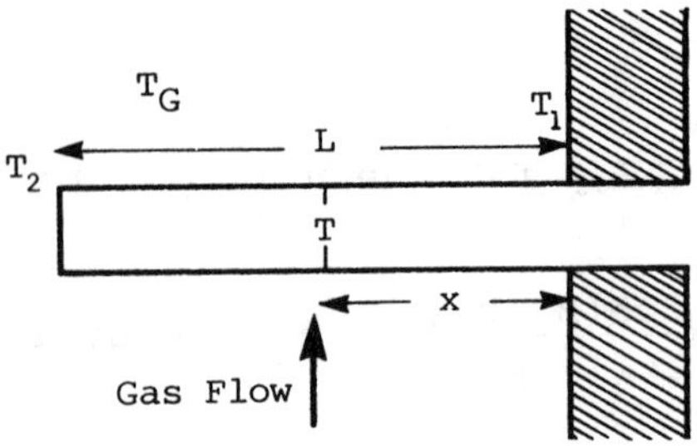

Fig. 1 Heat transfer to thermometer pocket.

Solution: If the temperature difference θ

$$= T - T_G$$

then for a fin (taking the pocket to be an extended surface, as the pocket and the tube are made of copper)

$$\theta = \frac{\theta_1 \cosh m(L - x)}{\cosh mL}$$

At $x = L$

$$\theta = \frac{\theta_1}{\cosh mL}$$

$$\text{or} \quad m^2 = \frac{25 \times 144 \times 0.375}{12 \times 200 \times 0.339 \times 0.036} = 46.2$$

$$\text{or} \quad m = 6.8$$

Also

$$\theta_1 = T_G - 200$$

$$\theta_2 = T_G - 400$$

and $$\frac{\theta_1}{\theta_2} = \cosh mL$$

Therefore substituting the value of θ_1 and θ_2 in this equation gives

$$\frac{T_G - 200}{T_G - 400} = \cosh mL$$

$$= \cosh(6.8 \times 3/12)$$

$$= \cosh 1.7$$

$$= 2.828$$

Therefore, $$T_G - 200 = 2.828(T_G - 400)$$

$$T_G = 2.828T_G - (2.828 \times 400) + 200$$

$$1.828T_G = 2.828 \times 400 - 200$$

Therefore the gas temperature is

$$T_G = \frac{2.828 \times 400 - 200}{1.828}$$

$$T_G \simeq 510^0 F$$

• PROBLEM 3-29

A thermocouple in a cylindrical well (see figure) is inserted in a gas stream. The estimated temperature of the gas stream is 500^0F. The length and wall thickness of the well are $L = 0.2$ ft. and $B = 0.08$ in., respectively. Take the value of $h = 120$ Btu/hr. ft.^{2} ^0F and $k = 60$ Btu/hr ft.^0F. The temperature of the pipe wall is 350^0F. Estimate the true temperature of the gas stream.

Solution: The thickness of the well as compared to its diameter is very small and since only one side of the well is in contact with the gas stream, the temperature distribution along this wall will be approximately the same as that along a bar of thickness 2B, in contact with the gas stream on both sides.

The temperature at the end of the well is determined by

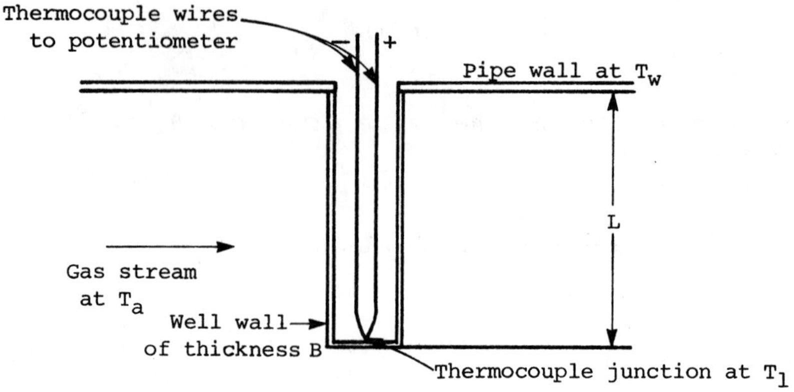

Fig. 1 Thermocouple in cylindrical well.

using the equation for dimensionless temperature

$$\theta = \frac{T - T_a}{T_w - T_a} = \frac{\cosh N(1 - \zeta)}{\cosh N}$$

where $\zeta = Z/L$ = dimensionless distance

$N = \sqrt{\frac{hL^2}{KB}}$ = dimensionless heat transfer coefficient

$$\frac{T_1 - T_a}{T_w - T_a} = \frac{\cosh 0}{\cosh N} = \frac{1}{\cosh\sqrt{hL^2/KB}}$$

$$= \frac{1}{\cosh\sqrt{(120)(0.2)^2/(60)(0.08/12)}}$$

$$= \frac{1}{\cosh 2\sqrt{3}} = \frac{1}{16}$$

The actual ambient gas temperature is obtained by solving the equation for T_a;

$$\frac{500 - T_a}{350 - T_a} = \frac{1}{16.0}$$

$$\text{or} \quad T_a = 510^0F$$

Therefore the estimated temperature of the gas stream is lower than the actual gas temperature by 10^0F. Note: If the variation in temperature along the x-axis had been taken into consideration, the actual gas temperature of 511^0F would have been obtained.

CHAPTER 4

UNSTEADY STATE CONDUCTION: USE OF CHARTS

Basic Attacks and Strategies for Solving Problems in this Chapter. **See pages 210 to 271 for step-by-step solutions to problems.**

There are several types of unsteady state conduction problems which have charts available to use in their solution. Problems using six types of charts are illustrated in this chapter.

The Gaussian error function chart is useful for problems involving a semi-infinite body (discussed in the introduction to chapter 3), which is initially at a uniform temperature and then has the surface changed to a different temperature. Problem 4.1 shows the analytical solution of the conduction equation for the above conditions to give the error function as the solution. The next several problems illustrate the use of the chart to solve particular semi-infinite body problems.

Three different types of charts are available which give a dimensionless temperature or dimensionless heat loss rate as a function of a Fourier Number and Biot Number. Each of these charts is for a particular shaped solid, such as a sphere, long cylinder, or large flat slab. These charts apply to the situation where the solid is initially at a uniform temperature and is then exposed to a fluid at a different temperature. One type of chart relates center temperature, time, and size of the solid. If two of those parameters are known, the other can be found from the chart. Another type of chart relates temperature within the solid at a given time to position and center temperature at that time. The third type of chart relates the heat loss rate from one of the solid shapes to time and size of the solid. In each case, values are needed for other parameters, such as the initial uniform temperature of the solid, the constant temperature to which the solid is exposed, the thermal conductivity of the solid, the thermal diffusivity, and the surface convection coefficient.

Gurnie-Lurie charts apply to the same situation and relate the same parameters as the Fourier-Biot charts, except that the heat loss rate is not included, and the temperature, time, position relationship are all condensed into a single chart for a given solid shape. These charts are available for the same shapes: sphere, long cylinder, and large slab. For each shape, the Gurnie-Lurie chart provides a means of finding any one of the following parameters if the other two are known: 1) position in the solid, ii) time, and iii) temperature at that time and position. In each case the size of the solid and values for the other parameters mentioned for the Fourier-Biot charts must be known also.

Although the Fourier-Biot and Gurnie-Lurie charts are the two types most widely available in references, there are a variety of other charts which have been prepared for similar purposes. A couple of them are illustrated in problems 4-16 to 4-18 in this chapter.

The thermal conductivity of any substance is, in general, dependent upon temperature and pressure. For a particular type of substance the thermal conductivity is sometimes expressed as a function of reduced temperature and reduced pressure. Reduced temperature is the ratio of actual temperature to critical temperature and reduced pressure is the ratio of actual pressure to critical pressure. Problem 4-19 illustrates the use of a chart relating thermal conductivity to reduced temperature and pressure.

There are several types of charts which can be used to estimate depth of freezing or depth of thawing in soil in terms of parameters such as surface temperature and properties of the soil. The last several problems in this chapter illustrate the use of such charts.

Step-by-Step Solutions to Problems in this Chapter, "Unsteady State Conduction: Use of Charts"

GAUSSIAN ERROR FUNCTION CHARTS

• PROBLEM 4-1

A three dimensional body occupies the half space ($0 \leq x \leq \infty$). The body is initially at the uniform temperature T_l. The surface at x=0 is instantaneously changed to and held at T_s for all time greater than $\tau=0$. Derive an expression for the time-dependent temperature profile for the heat transfer rate at a given location at a specified time.

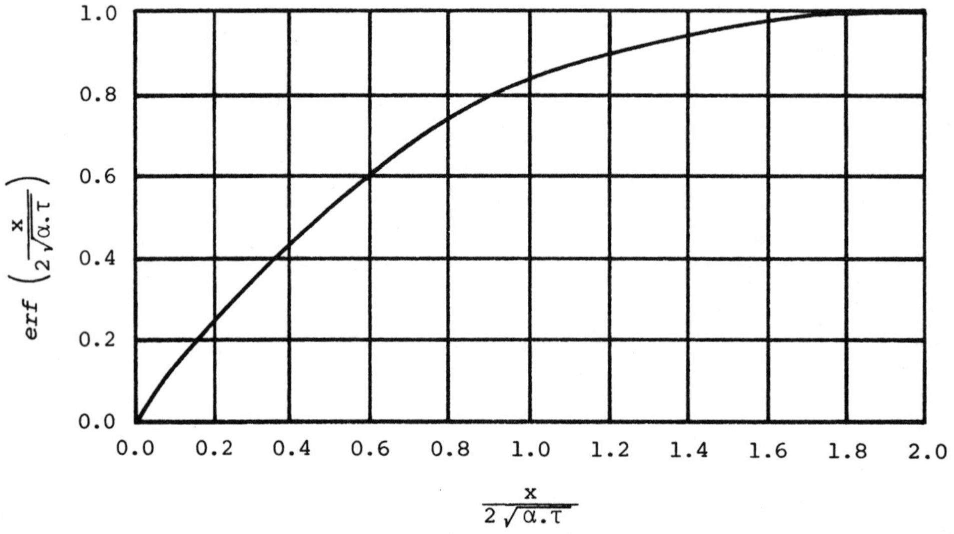

Gauss's error integral

Solution: The semi-infinite body for the given conditions obeys

$$\frac{\partial T}{\partial \tau} = \alpha \frac{\partial^2 T}{\partial x^2} \qquad (1)$$

subjected to the boundary conditions

$T(x,0) = T_i$ for $\tau=0$ -- Initial conditions

$T(0,\tau) = T_s$ at $x=0$

for $\tau>0$ -- Boundary conditions

$T(\infty,\tau) = T_i$ at $x=\infty$

Solving the differential equations by variables separable method and applying the boundary conditions gives the solution

$$\frac{T(x,\tau) - T_s}{T_i - T_s} = \frac{2}{\sqrt{\pi}} \int_0^{x/\sqrt{4\alpha\tau}} e^{-\eta^2} d\eta \qquad (2)$$

For heat transfer rate at a given location within the semi-infinite body at a specified, time differentiating by Leibnitz rule gives

$$\frac{\partial T}{\partial x} = \frac{T_i - T_s}{\sqrt{\pi\alpha\tau}} e^{-x^2/4\alpha\tau} \qquad (3)$$

Substituting this temperature gradient in equation (1) and simplifying gives the heat flux

$$\frac{q}{A} = \frac{k(T_s - T_i)}{\sqrt{\pi\alpha\tau}} e^{-x^2/4\alpha\tau} \qquad (4)$$

Thus the above equation gives the heat transfer rate at any location at a specified time.

Now heat transfer at the location

$x=0$

$$\left.\frac{q}{A}\right|_{x=0} = \frac{k}{\sqrt{\pi\alpha\tau}} (T_s - T_i)$$

The "thermal penetration thickness" is defined as

$$\delta_T = 4\sqrt{\alpha\tau}$$

Here the error function erf $\left(\frac{x}{2\sqrt{\alpha\tau}}\right)$

approaches the value 1.0 as the term $\frac{x}{2\sqrt{\alpha\tau}}$ approaches 2.

(Shown in figure).

• PROBLEM 4-2

A very thick wall, surface area 8 sq. ft, has a surface temperature 60°F, after 9 hours changes suddenly to 1400°F and remains constant thereafter. Estimate the temperature in a plane 8 in. from the surface and the total energy taken up by the wall in 9 hours. Take the average thermal conductivity and diffusivity values of the wall material to be 0.6 Btu/hr ft°F and 0.04 ft²/hr respectively.

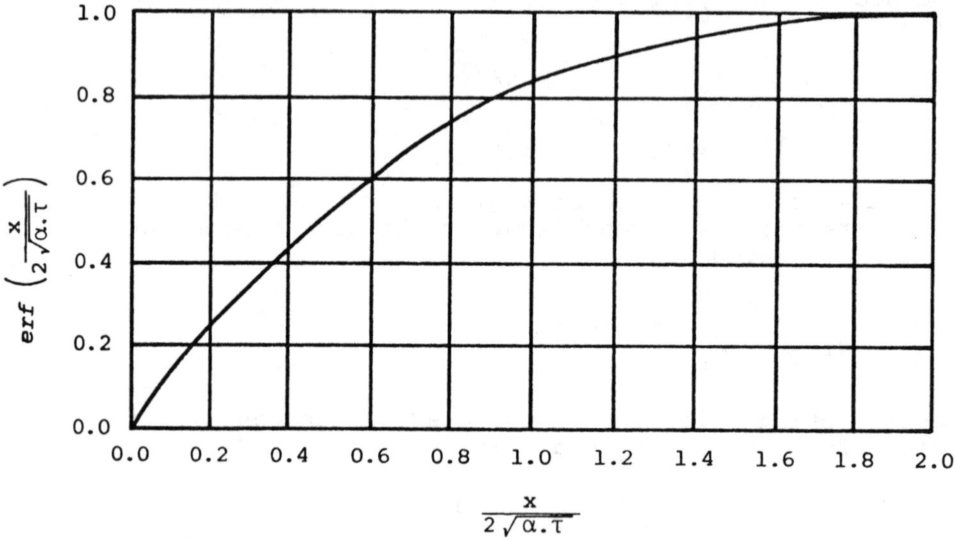

Gauss's error integral

Solution: The temperature at a finite distance from the surface is expressed by the equation

$$t = t_s + (t_i - t_s)\,\mathrm{erf}\left(\frac{x}{2\sqrt{\alpha\tau}}\right) \qquad (1)$$

where t is the temperature at a distance x from the surface

τ--time period in hours

t_s-surface temperature after change

t_i-initial surface temperature

α--thermal diffusivity of the material

$$\frac{x}{2\sqrt{\alpha\tau}} = \frac{8/12}{2\sqrt{0.04 \times 9}} = 0.555$$

From Gauss -error function graph corresponding to this value

$$\text{erf} \frac{x}{2\sqrt{\alpha\tau}} = 0.57$$

Introducing this value in equation (1) yields

$$t = 1500 + (70 - 1500)\ 0.57$$

$$\cong 685°F.$$

Now, the heat flow is expressed as

$$q = -kA\ (t_i - t_s)\ \frac{e^{-x^2/4\alpha\tau}}{\sqrt{\pi\alpha\tau}}$$

substituting values

$$q = -(0.6)(8)\ (60 - 1400) \times \frac{e^{-(0.555)^2}}{\sqrt{\pi \times 0.04 \times 9}}$$

$$\cong 4445\ \text{Btu}/_{\text{hr}}$$

Again, total heat energy absorbed by the wall is equal to the heat entered into the surface, and expressed as

$$Q = -kA(t_i - t_s)\ 2\sqrt{\frac{\tau}{\pi\alpha}}$$

$$= -(0.6)\ (8)\ (60 - 1400)\ 2\sqrt{\frac{8}{0.04\pi}}$$

$$= 102640\ \text{Btu}$$

• PROBLEM 4-3

A large rock, having a crevice 8 ft wide, is initially at 30°C. Molten lava enteres the crevice at time t=0 and at a temperature of 1200°C. Calculate the temperature at the center of the molten lava after 6 months, assuming the thermal diffusivity for both the lava and the rock to be 0.029 ft^2/hr.

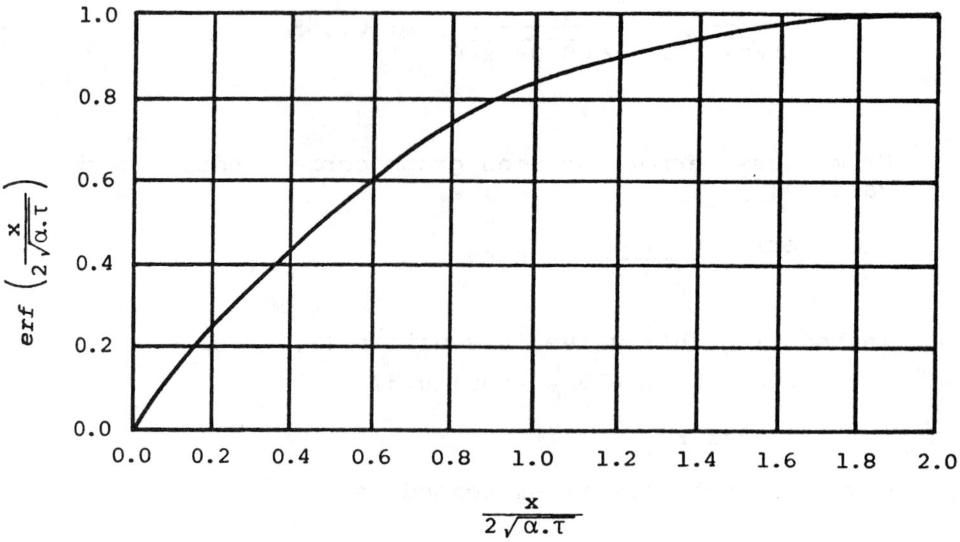

Gauss's error integral

Solution: The non-dimensional temperature variation through the semi-infinite media is

$$\frac{T - T_\infty}{T - T_\infty} = \frac{1}{2}\left[\text{erf}\left(\frac{b-x}{2\sqrt{\alpha\tau}}\right) - \text{erf}\left(\frac{a-x}{2\sqrt{\alpha\tau}}\right)\right]$$

Here a=0, b=8 ft, $\tau = \frac{365}{2} \times 24 = 4380$ hrs

for mid point x=4 ft. and $\alpha = 0.029\ \text{ft}^2/\text{hr}$

$$\text{erf}\ \frac{b-x}{2\sqrt{\alpha\tau}} = \text{erf}\ \frac{8-4}{2\sqrt{0.029 \times 4380}}$$

$$= \text{erf}\ (0.1775) = 0.18 \text{ (from graph)}$$

and

$$\text{erf}\ \frac{a-x}{2\sqrt{\alpha\tau}} = \text{erf}\ \frac{0-4}{2\sqrt{0.029 \times 4380}}$$

$$= \text{erf}\ (-0.1775)$$

$$= -\text{erf}\ (0.1775) = -0.18 \text{ (from graph)}$$

Substituting the values in

$$\frac{T - T_\infty}{T - T_\infty} = \frac{1}{2}\left[\text{erf}\left(\frac{b-x}{2\sqrt{\alpha\tau}}\right) - \text{erf}\left(\frac{a-x}{2\sqrt{\alpha\tau}}\right)\right]$$

$$= \tfrac{1}{2}\quad (0.18) - (-0.18) \quad = 0.18$$

i.e. $\dfrac{T - T_\infty}{T_i - T_\infty} = 0.18$

where T_∞ = Surrounding temperature = 30°C.

T_l = Temperature of lava = 1200°C.

Substituting the values

$$T = 30 + (1200 - 30)0.18 = 240.6°C.$$

• PROBLEM 4-4

A semi-infinite steel slab is initially at a homogeneous temperature of 40°C. Compute the temperature at a depth of 3 cm. from the surface after 40 seconds have elapsed under the following conditions:

(1) When the slab is suddenly exposed to a temperature of 300°C.

(2) When the slab is exposed to constant heat flux of 4×10^5 W/m².

Consider the steel thermal conductivity

$k = 60$ W/m°C and diffusivity

$\alpha = 1.6 \times 10^{-5}$ m²/s.

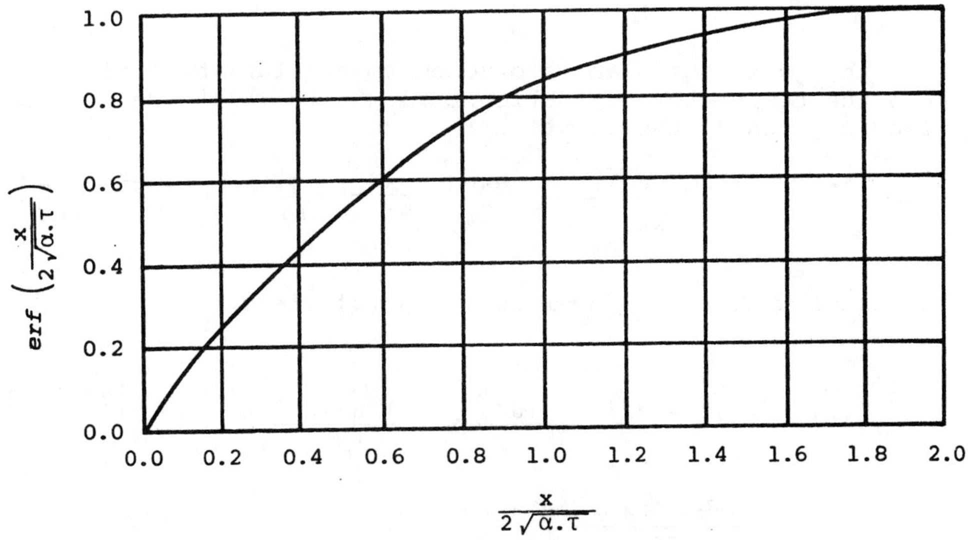

Gauss's error integral

Solution: The nondimensional temperature variation through the semi-infinite slab is expressed as

$$\frac{T(x,\tau) - T_o}{(T_i - T_o)} = \text{erf}\,\frac{x}{2\sqrt{\alpha\tau}} \tag{1}$$

where the expression $\frac{x}{\sqrt{\alpha\tau}}$, called Fourier number, represents the nondimensional time parameter.

$$\text{Now}\quad \frac{x}{2\sqrt{\alpha\tau}} = \frac{0.03}{2\sqrt{1.6 \times 10^{-5} \times 40}} = 0.593$$

The error function is determined, from the graph, as

$$\text{erf}\,\frac{x}{2\sqrt{\alpha\tau}} = \text{erf}\ 0.593 = 0.6$$

Since $T_i = 40°C$ and $T_o = 300°C$,

(a) required temperature at x = 3 cm from the surface can be evaluated using equation (1).

$$T(x,\tau) = T_o + (T_i - T_o)\ \text{erf}\,\frac{x}{2\sqrt{\alpha\tau}}$$

$$= 300 + (40 - 300)\ (0.6)$$

$$= 144°C$$

(b) when the slab is exposed to constant heat flux then the temperature variation through the thickness of the slab is given by the equation

$$T - T_i = \frac{2q_o\sqrt{\alpha\tau/\pi}}{kA}\ \exp\left(\frac{-x^2}{4\alpha\tau}\right) - \frac{q_o x}{kA}\left(1 - \text{erf}\,\frac{x}{2\sqrt{\alpha\tau}}\right) \tag{2}$$

substituting the values in the equation (2) yields

$$T(x,\tau) = 40 + \frac{2x(4 \times 10^5)}{60} \times \left[\frac{(1.6x10^{-5})40}{\pi}\right]^{\frac{1}{2}} x\ e^{-\left[(0.593)^2\right]}$$

$$- \frac{(0.03) \times (4 \times 10^5)}{60}\ (1 - 0.6)$$

$\therefore T(x, \tau) = 94°C$ at $x = 3$ cm and $\tau = 40s$

For constant heat flux after 40 seconds is evaluated at $x = 0$

$$\therefore T(x=0) = 40 + \frac{2x(4x10^5)}{60} \times \left[\frac{(1.6x10^{-5})40}{\pi}\right]^{\frac{1}{2}} \times e^{-0} - 0$$

$$= 230°C.$$

• PROBLEM 4-5

The initial uniform temperature of a thick and large aluminium slab is 120°F. Instantaneously only one of its surface temperatures is changed to and maintained at 1300°F. Calculate the following at a depth of 5 in. from the surface after 3 hrs:

(a) The temperature at that point

(b) The rate of heat flowing across that point at that time, and

(c) The amount of heat passed till that time.

Given properties of aluminium:

$C_p = 0.208$ Btu/lb°F, $k = 130$ Btu/hr. ft°F

and $\rho = 170$ lbm/ft^3.

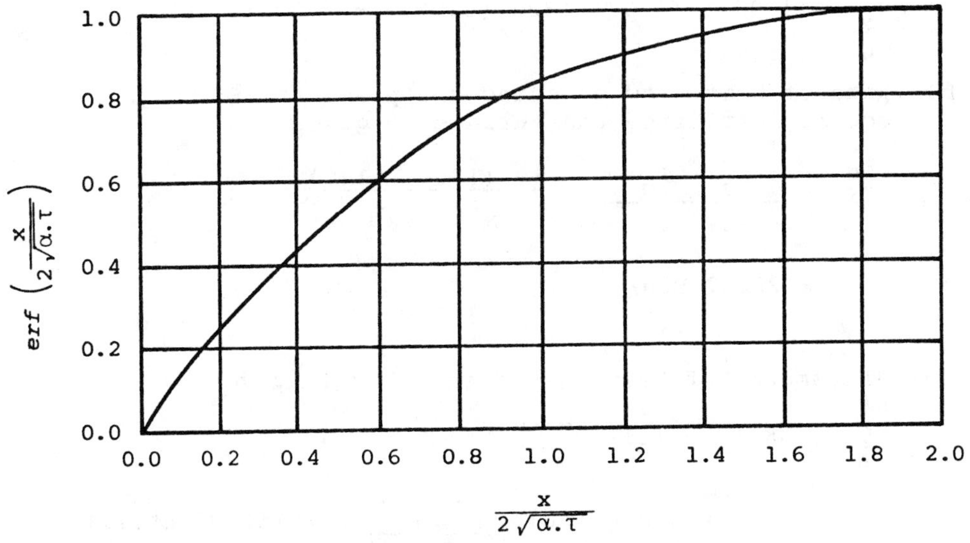

Gauss's error integral

Solution: The temperature variation through the wall can be expressed as

$$\frac{T_x - T_s}{T_o - T_s} = \text{erf} \frac{x}{2\sqrt{\alpha\tau}} \qquad (1)$$

where x -- distance from the surface = 5 in = $\frac{5}{12}$ ft

= 0.4166 ft

τ -- time after the temperature is raised = 3 hrs.

T_s -- Surface temperature (for $\tau>0$) = 1300°F.

T_o -- Surface temperature (for $\tau=0$) = 120°F.

T_x -- Temperature at depth (x) from the surface

$$\text{Now } \alpha = \frac{k}{C_p\rho} = \frac{130}{0.208 \times 170} = 3.68 \text{ ft}^2/\text{hr}$$

$$\therefore \frac{x}{2\sqrt{\alpha\tau}} = \frac{0.41666}{2\sqrt{3.68 \times 3}} = 0.063$$

For $\frac{x}{2\sqrt{\alpha\tau}} = 0.063$, from Gauss-error-function graph

$$\text{erf} \frac{x}{2\sqrt{\alpha\tau}} = 0.065$$

a) Substituting the values in equation (1)

$$T_x = 1300 + (120 - 1300)(0.065) = 1223°\text{F}.$$

b) Rate of heat flow crossing the plane at x=5 in, per square feet, from the surface is given by

$$\frac{q}{A} = \frac{k(T_s - T_o)}{\sqrt{\pi\alpha\tau}} = \frac{130(1300 - 120)}{\sqrt{\pi \times 3.68 \times 3}}$$

$$= 26047 \text{ Btu/hr. ft}^2$$

c) The amount of heat passed in 3 hrs is given by

$$\frac{Q}{A} = 2k(T_s - T_o)\sqrt{\frac{\tau}{\pi\alpha}}$$

$$= 2 \times 130(1300 - 120)\sqrt{\frac{3}{\pi \times 3.68}} = 156285 \text{ Btu/ft}^2$$

• PROBLEM 4-6

Estimate the depth to which a water pipe line is to be laid such that the water flowing in the pipe does not freeze due to low temperatures at the surface. The mean temperature during cold season is 5°C and minimum temperature goes down to -10°C and exists for a maximum period of 48 hours. Assume the properties of the earth as

$$k = 0.52\ W/m - °K,\quad \rho = 1840\ kg/m^3$$

$$C_p = 2050\ J/kg - °K.$$

Table 1: The Complimentary Error Function

$$\text{erfc}(\eta) = 1 - \frac{2}{\sqrt{\pi}} \int_0^{\eta} e^{-u^2} du$$

η	erfc(η)	η	erfc(η)
0.0	1.0000	1.1	0.11980
0.05	0.9436	1.2	0.08969
0.1	0.8875	1.3	0.06599
0.15	0.8320	1.4	0.04772
0.2	0.7773	1.5	0.03390
0.25	0.7237	1.6	0.02365
0.3	0.6714	1.7	0.01621
0.35	0.6206	1.8	0.01091
0.4	0.5716	1.9	0.00721
0.45	0.5245	2.0	0.00468
0.5	0.4795	2.1	0.00298
0.55	0.4367	2.2	0.00186
0.6	0.3961	2.3	0.001143
0.65	0.3580	2.4	0.000689
0.7	0.3222	2.5	0.000407
0.75	0.2889	2.6	0.000236
0.8	0.2579	2.7	0.000134
0.85	0.2293	2.8	0.000075
0.9	0.2031	2.9	0.000041
0.95	0.1791	3.0	0.000022
1.00	0.1573		

Solution: The thermal diffusivity of the earth is

$$\alpha = \frac{k}{\rho C p} = \frac{0.52}{(2050)(1840)}$$

$$= 1.378 \times 10^{-7}\ m^2/s$$

The depth of freezing is given by

$$x = \eta(4\alpha t)^{\frac{1}{2}}$$

where x - is depth of freezing

η - a positive number

t - time period in seconds

α - thermal diffusivity

The nondimensional temperature parameter is given as

$$\frac{T - T_o}{T_s - T_o} = 1 - \frac{2}{\pi} \int_0^{\eta} e^{-u^2} du$$

$$= \text{erfc}(\eta)$$

where erfc (η) is the complementary error function and it is obtained from the table.

Introducing the known values in the equation, results in

$$\frac{0 - 5}{-10 - 5} = 0.333$$

then from the table for erfc (η) = 0.333, η = 0.68.

Thus,

$$x = 0.68(4 \times 1.378 \times 10^{-7} \times 48 \times 3600)^{\frac{1}{2}}$$

$$= 0.20 \text{ meters} = 20 \text{ cm}$$

Therefore, to avoid freezing of water in the pipe line it is to be laid 20 cm below ground level.

• PROBLEM 4-7

Compute the earth's surface temperature after being exposed for $5\frac{1}{2}$ hours to cold air at -20°C, the initial uniform temperature of the earth's surface being 14°C. The convective heat transfer coefficient above the soil is 10.8W/m^2 °K. Ignore the effects due to latent heat. The physical properties of the soil are k=0.888W/m°K, α =4.7 x 10^{-7}m^2 /sec. Find the depth at which the temperature will be 0°C in a $5\frac{1}{2}$ hour period.

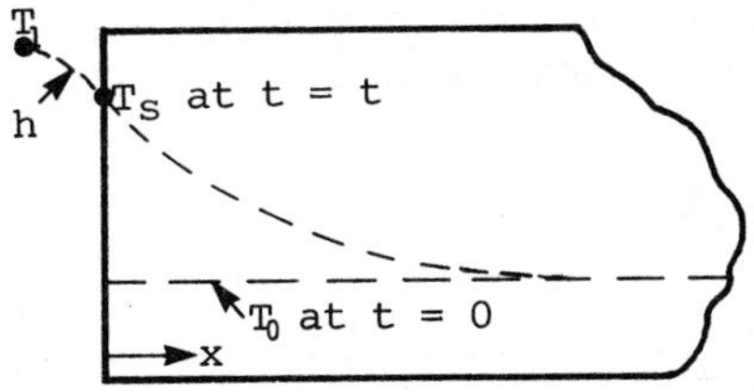

Fig. 1 Unsteady state conduction in a semi-infinite solid.

<u>Solution</u>: a) The surface temperature after 5½ hours of being exposed to a temperature of -20°C can be calculated using the chart shown in figure 2.

$$\frac{h\sqrt{\alpha t}}{k} = \frac{10.8\sqrt{4.7 \times 10^{-7} \times 5\frac{1}{2} \times 3600}}{0.888}$$

$$= 1.173$$

Again, $\frac{x}{2\sqrt{\alpha t}} = 0$

($\because$ at surface x=0)

From chart, corresponding to $\frac{x}{2\sqrt{\alpha t}} = 0$

and $\frac{h\sqrt{\alpha t}}{k} = 1.173$

$$1 - y = \frac{T - T_0}{T_1 - T_0} = 0.62$$

Substituting the values, $T_0 = 14 + 273.2 = 287.2°K$.

$T_1 = -20 + 273.2 = 253.2°K$.

Gives $T = (0.62(253.2 - 287.2) + 287.2)°K$

$= 266.12°K$

or -7.08°C

b) To find the depth at which the temperature will be 0°C or 273.2°K after $5\frac{1}{2}$ hours of being exposed to -20°C (or 253.2°K).

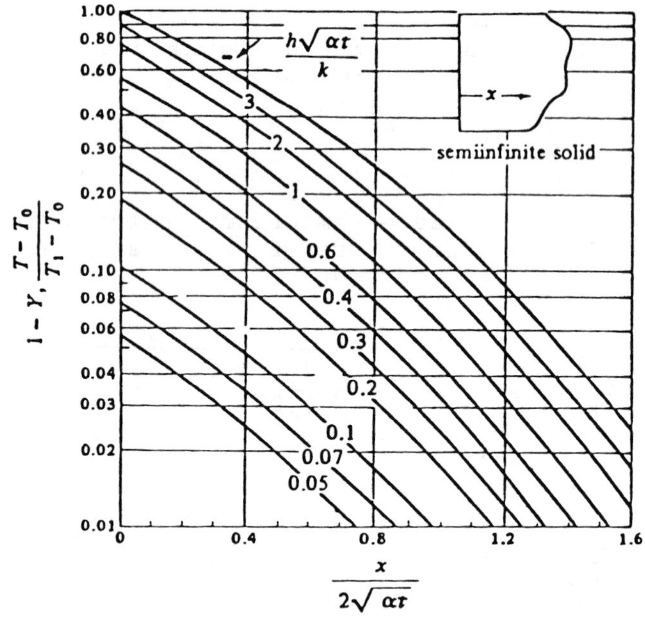

Fig. 2 Unsteady-state heat conduction in a semi-infinite solid with surface convention.

$$\therefore \quad 1 - y = \frac{T - T_0}{T_1 - T_0} = \frac{273.2 - 287.2}{253.2 - 287.2}$$

$$= 0.411$$

From figure 2 corresponding to (1-y) = 0.411 and

$$\frac{h\sqrt{\alpha t}}{k} = 1.173, \qquad \frac{x}{2\sqrt{\alpha t}} = 0.2$$

$$\text{or } x = 0.2 \times 2 \times \left[4.7 \times 10^{-7} \times 5.5 \times 3600\right]^{\frac{1}{2}}$$

$$= 0.0385 \text{ meter}$$

or 3.85 cm.

BIOT-FOURIER NUMBER CHARTS

• **PROBLEM 4-8**

A metal sleeve of negligible thickness and length 2ℓ rotates with a constant angular velocity ω, over a solid rod of radius R and length 2L (see figure). The ends of the rod are insulated and the pressure between the sleeve and the rod is P. Determine the steady temperature in this system for the following cases:

1) A single heat transfer coefficient for the entire system, $h_1 = h_2 = h$

2) Two heat transfer coefficients, h_1 for the rotating sleeve and h_2 for the stationary rod.

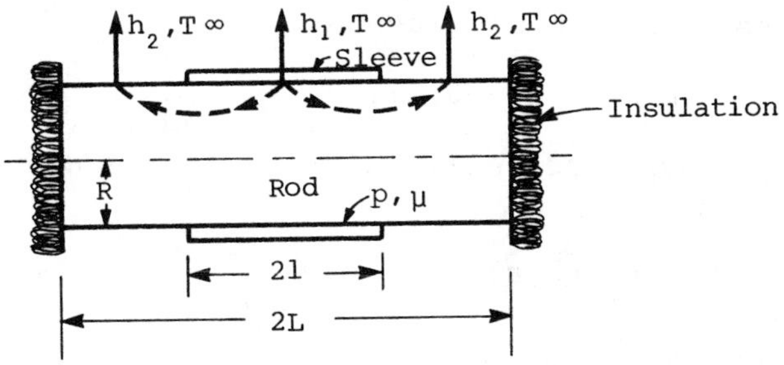

Fig. 1

Solution: The ends of the rod are insulated, but from the heat flux paths shown in the figure, it follows that the problem is two dimensional.

1. The formulation of the problem for the first case of a single heat transfer coefficient is

$$\frac{1}{r}\frac{\partial}{\partial r}\left(r\frac{\partial \theta}{\partial r}\right) + \frac{\partial^2 \theta}{\partial z^2} = 0,$$

$$\frac{\partial \theta\ (0,z)}{\partial r} = 0$$

$$-k\frac{\partial \theta\ (R,z)}{\partial r} + \begin{Bmatrix} \mu P \omega R, & 0 \le z < \ell \\ 0, & \ell < z \le L \end{Bmatrix} = h\theta(R,z)$$

$$\frac{\partial \theta(r,0)}{\partial z} = 0 \qquad \frac{\partial \theta(r,L)}{\partial z} = 0$$

Taking the r axis as the homogeneous direction, this problem is solved as a two domain problem. The peripheral boundary conditions of the two domains differ only in homogeneity and the characteristic values of both the domains are the same. Next, proceeding to a one domain solution by taking the z axis as the homogenous direction, in this case, the separation of variables gives

$$\frac{d^2Z}{dz^2} + \lambda^2 Z = 0,$$

$$\frac{dZ(0)}{dz} = 0, \qquad \frac{dZ(L)}{dz} = 0,$$

$$\frac{d}{dr}\left(r\frac{dR}{dr}\right) - \lambda^2 rR = 0$$

$$\frac{dR(0)}{dr} = 0$$

The characteristic functions in z are

$$\phi_n(z) = \cos \lambda_n z$$

and the characteristic values are

$$\lambda_n L = n\pi, \qquad n = 0, 1, 2, \ldots\ldots.$$

The appropriate particular solution in r which satisfies the centerline symmetry is

$$R_n(r) = I_o(\lambda_n r)$$

The product solution is

$$\theta(r,z) = a_o + \sum_{n=1}^{\infty} a_n I_o(\lambda_n r) \cos \lambda_n z \tag{1}$$

Using the non-homogeneous boundary conditions in r yields

$$-k\sum_{n=1}^{\infty} a_n \lambda_n I_1(\lambda_n R) \cos \lambda_n z + \mu P\omega R \begin{Bmatrix} 0 \le z < \ell \\ \ell < z \le L \end{Bmatrix} \tag{1}$$

$$= ha_0 + h\sum_{n=1}^{\infty} a_n I_0(\lambda_n R) \cos \lambda_n z,$$

from which the values of a_0 and a_n are

$$a_0 = \frac{\mu P\omega R\ell}{hL}$$

$$a_n = \frac{2(\mu P\omega R^2/k) \sin \lambda_n \ell}{\lambda_n L\left[Bi\ I_0(\lambda_n R) + (\lambda_n R)\ I_1(\lambda_n R)\right]}$$

where Bi is the Biot number and is given as

$$Bi = hR/k$$

Introducing these values into equation (1), the temperature distribution in the system is given by

$$\frac{\theta(r,z)}{\mu P\omega R^2/k} = \frac{\ell/L}{Bi} + 2\sum_{n=1}^{\infty} \frac{\sin \lambda_n \ell \, I_0(\lambda_n r)}{\lambda_n L\left[Bi\, I_0(\lambda_n R) + (\lambda_n R) I_1(\lambda_n R)\right]} \cos \lambda_n z$$

In the case of $h_1 \neq h_2$, there are two sets of characteristic values and the problem cannot be solved as a one-domain problem.

• PROBLEM 4-9

The initial temperature of a combustion chamber, of wall thickness ½ in, is 90°F. Estimate the span of time the chamber can operate with the wall temperature of 2100°F. The flue gas temperature immediately from the start of firing in the chamber is 4600°F. The flame-side heat transfer coefficient is 100 Btu/hr ft²°F. Assume the properties of the chamber wall material as $k = 20$ Btu/hr ft°F,

$\rho = 600$ lb/ft³

$C_p = 0.38$ Btu/lb°F.

Solution: The wall of the combustion chamber can be considered as a flat infinite plate as the thickness is very small compared to its radius. The outside wall, if exposed to atmosphere, would have a very low value of h and hence it would be convenient to assume the outer wall to be insulated.

At the outer surface $\frac{x}{x_0} = 1.0$

$$\alpha = \frac{k}{\rho C_p} = \frac{20}{600 \times 0.38} = 0.088 \text{ ft}^2/\text{hr}$$

$$\frac{k}{hx_0} = \frac{20}{1000\ (\frac{1}{2} \times 12)} = 0.48$$

$$Y_0 = \frac{4600 - 2100}{4600 - 90} = 0.55$$

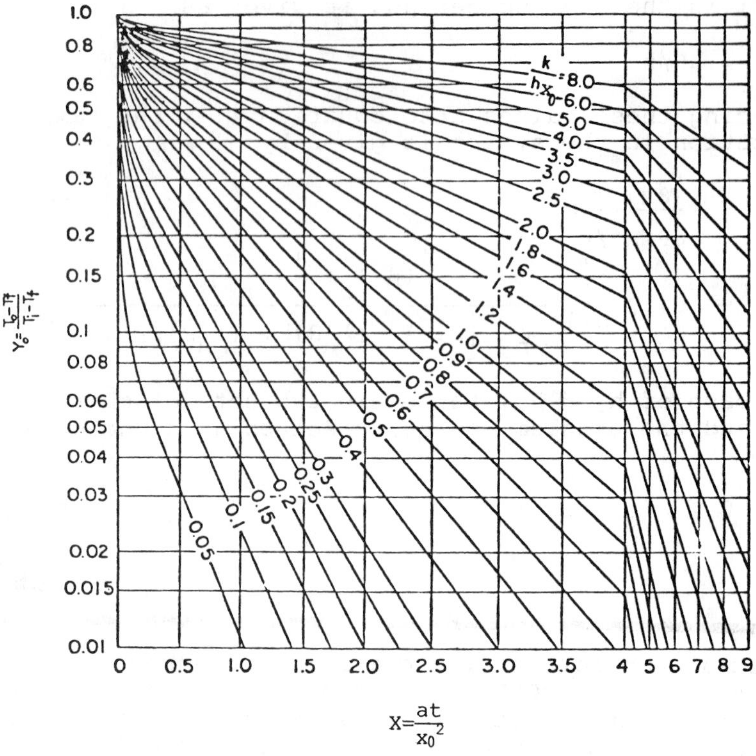

Fig. 1 Temperature history at the surface of an infinite slab.

From the chart corresponding to the calculated values of y_0 and $\frac{k}{hx_0}$,

$$x = 2.48 = \frac{\alpha t}{x_0^2}$$

or $$t = \frac{2.48 \times \left(\frac{1}{2 \times 12}\right)^2}{0.088}$$

$$= 0.049 \text{ hr}$$

or $$= 2.94 \text{ seconds.}$$

Find the temperatures at the center and at 18 mm from the center for a ball bearing of 40 mm od at 800°F, after it is dipped for 30 seconds in a pool of liquid at 40°F. Assume the average convective heat transfer coefficient as 1555 W/ m^2°K. Also, find the heat convected by the spherical ball in the first 30 seconds in the liquid. The physical properties of the bearing material are

$$\rho = 7625 \text{ kg/m}^3, \quad C_p = 460 \text{ J/Kg°K}$$
$$k = 19 \text{ W/m-°K}$$

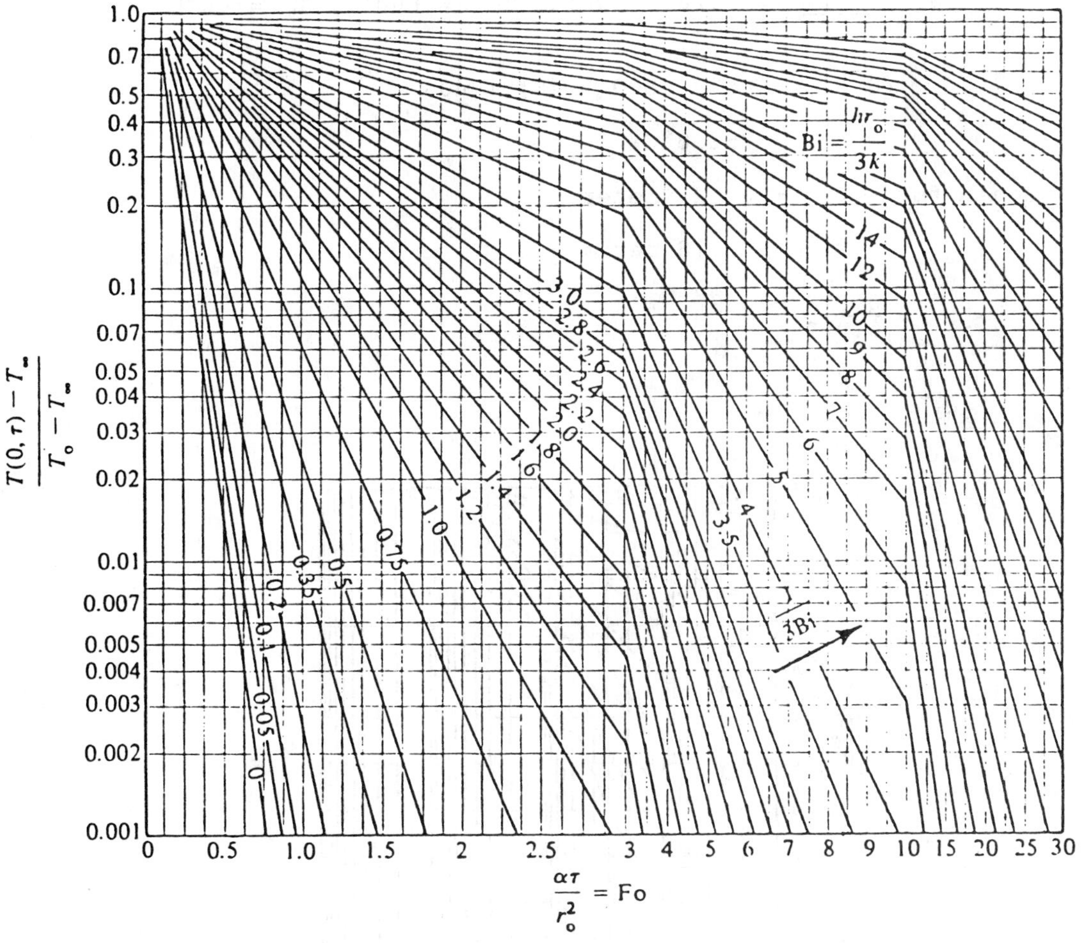

Fig. 1 Center temperature for a sphere of radius, r_0.

Solution: The Biot number for the ball bearing is expressed as

$$Bi = \frac{hr_0}{3k} = \frac{1555\ (20/1000)}{3 \times 19}$$

$$= 0.545$$

and $$\frac{1}{3Bi} = \frac{1}{3 \times 0.545} = 0.611$$

Fourier number $$F_0 = \frac{(\alpha)\tau}{r_0^{\,2}}$$

$$F_0 = \frac{k}{\rho C_p} \times \frac{\tau}{r_0^{\,2}} = \frac{19 \times (3600)}{7625 \times 460} \times \frac{\left(\frac{1}{120}\right)}{\left(\frac{20}{1000}\right)^2}$$

$$= 0.406$$

From figure (1)

$$\frac{T\ (0,\tau) - T_\infty}{T_0 - T_\infty} = 0.28 \qquad \text{where } T_\infty = 800°C,\ T_0 = 40°C$$

$$\therefore\ T\ (0,\tau) = 0.28\ (800 - 40) + 40$$

$$= 252.8°C$$

Again for $r_0 = 0.02$ meter

$r = 0.018$ meter

$$\frac{r}{r_0} = \frac{0.018}{0.020} = 0.9$$

$$\frac{1}{3Bi} = 0.611$$

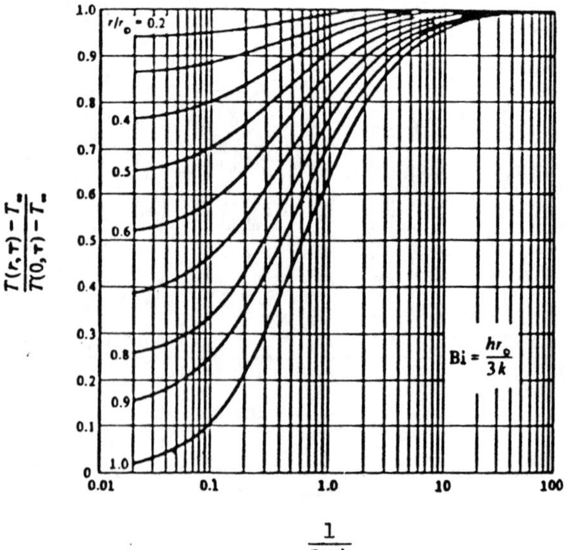

Fig. 2 Temperature as a function of center temperature for a sphere of radius, r_0.

From figure (2)

$$\frac{T(r,\tau) - T_0}{T(0,\tau) - T_\infty} = 0.6$$

or $T(r,\tau) = 0.6\ (252.8 - 40) + 40$

$= 167.68°C$

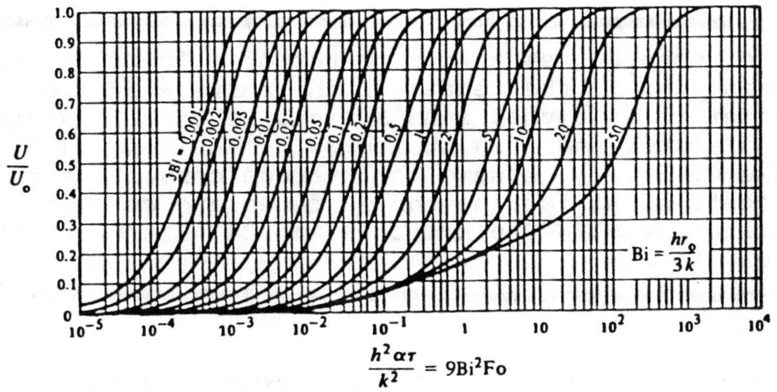

Fig. 3 Dimensionless heat loss U/U_0 of a sphere of radius, r_0, with time.

The overall heat convected by the spherical ball can be estimated from figure (3).

$$U_0 = \rho C_p V\ (T_0 - T_\infty)$$

$$U_0 = 7625 \times 460 \times \frac{4}{3}\pi\ (0.02)^3\ (800 - 40)$$

$$= 8.9328 \times 10^4\ J$$

and Bi = 0.545 or 3Bi = 3 x 0.545

$= 1.635$

$$\frac{h^2\alpha\tau}{k^2} = (1555)^2 \times \frac{k}{\rho C_p} \times \frac{1}{120} \times \frac{1}{k^2}$$

$$= \frac{(1555)^2}{460 \times 7625} \times \frac{1}{120} \times \frac{1}{19}$$

$$= 3.02 \times 10^{-4}$$

Corresponding to $\frac{h^2\alpha\tau}{k} = 3.03 \times 10^{-4}$

and $\quad 3Bi = 1.635, \quad \frac{U}{U_0} = 0.02$

$\therefore \quad U = 0.02 \times 8.9328 \times 10^4 J$

Thus heat lost by the ball bearing in 30 seconds is 1786.56 Joules.

• PROBLEM 4-11

> Compute the temperatures at the center of a 18 Cr 8Ni (V2A) circular disc and at the center of the surface at one end, 6 minutes after it is dipped into a fluid at 0°F. The disc is 14 in. in diameter and 3 in. thick with a uniform initial temperature of 510°F. Assume the average convective heat transfer coefficient for the end surfaces and cylindrical curved surface as 60.8 Btu/hr ft^2°F and 243 Btu/hr ft^2°F, respectively.

TABLE Property Values for Metals (English System of Units)

Metals	Properties at 68 F				k, thermal conductivity, Btu/hr ft F									
	ρ lb/ft^3	c_p $\frac{Btu}{lb F}$	k $\frac{Btu}{hr\ ft\ F}$	a $\frac{ft^2}{hr}$	−148 F −100 C	32 F 0 C	212 F 100 C	392 F 200 C	572 F 300 C	752 F 400 C	1112 F 600 C	1472 F 800 C	1832 F 1000 C	2192 F 1200 C
Cr-Ni (chrome-nickel): 15 Cr, 10 Ni.	491	0.11	11	0.204										
18 Cr, 8 Ni (V2A)...............	488	0.11	9.4	0.172	.....	9.4	10	10	11	11	13	15	18	
20 Cr, 15 Ni....................	489	0.11	8.7	0.161										
25 Cr, 20 Ni....................	491	0.11	7.4	0.140										

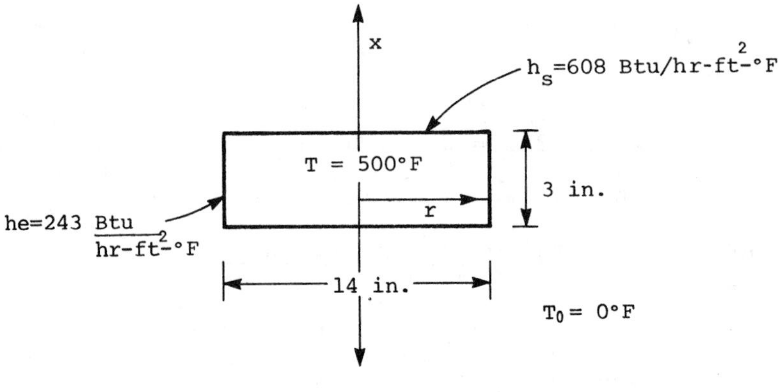

Fig. 1

Solution: The temperature distribution in a cylindrical body is expressed as

$$\frac{T(r,x,\tau) - T\infty}{T_0 - T\infty} = P(x,\tau)\ C(r,\tau)$$

where T_0 - is the initial uniform temperature

$T\infty$ - convective environment temperature

$P(x,\tau)$ - the temperature distribution in an infinite slab

and $C(r,\tau)$-the temperature distribution in an infinite cylinder.

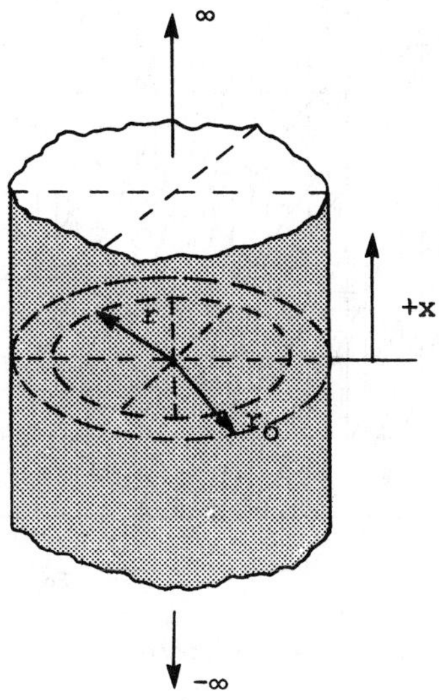

Fig. 2 Infinite cylinder

The Biot number relating the heat conduction in a body and convection from the surface is expressed as, for a slab,

$$Bi = \frac{hL}{k} \qquad \text{and Fourier number } F_0 = \frac{\alpha\tau}{L^2}$$

Now, from the table, the properties of the disc material are

ρ = 488 lb/ft^3, C_p = 0.11 Btu/lb °F

α = 0.172 ft^2/hr and k = 10 Btu/hr ft°F

then $$Bi = \frac{60.8 \times \left(\frac{3}{12 \times 2}\right)}{10} = 0.76$$

and $$\frac{1}{Bi} = 1.315$$

$$\text{Again, } F_0 = \frac{\alpha\tau}{L^2} = \frac{0.172 \times \left(\frac{6}{60}\right)}{\left(\frac{3}{12 \times 2}\right)^2} = 1.1$$

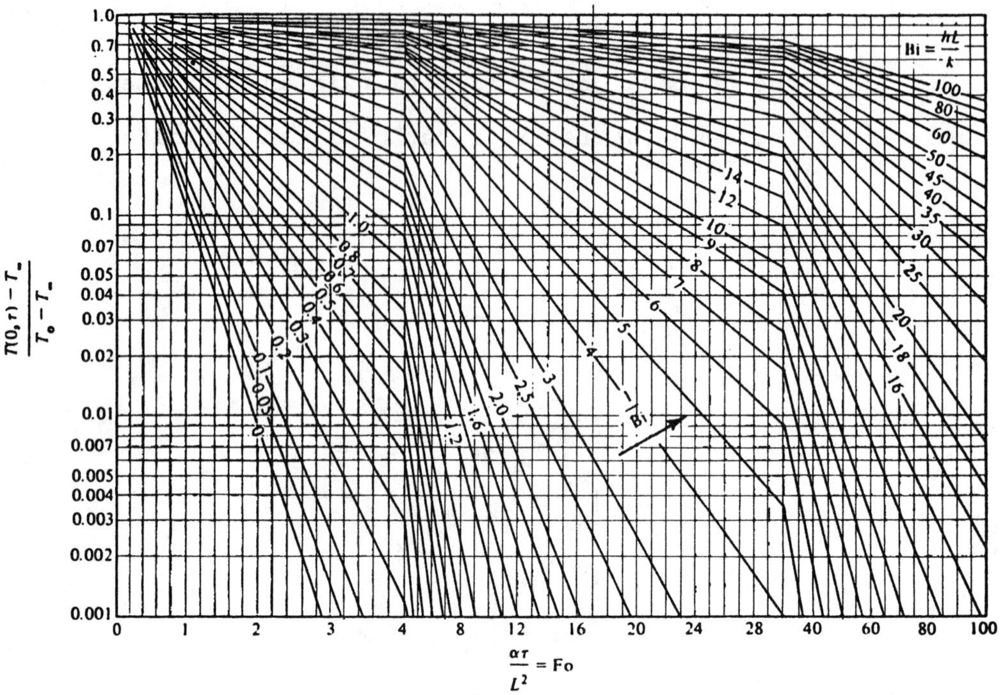

Fig. 3 Centerline temperature for a slab of thickness, $2L; Bi=\frac{hL}{k}$.

From figure 3 corresponding to $F_0 = 0.917$

and $\frac{1}{Bi} = 1.315, \quad P(0,\tau) = \frac{T(0,\tau) - T\infty}{T_0 - T\infty}$

$= 0.5$

$$\therefore \quad P(0,\tau) = \left[\frac{T(0,\tau) - T\infty}{T_0 - T\infty}\right] = 0.5$$

From the figure, corresponding to $\frac{1}{Bi} = 1.315$

and $\frac{x}{L} = \frac{1.5}{3/2} = 1,$

$$\left[\frac{P(1.5, 5 \text{ min}) - T\infty}{P(0,5\text{min}) - T\infty}\right]_{\text{slab}} = 0.87$$

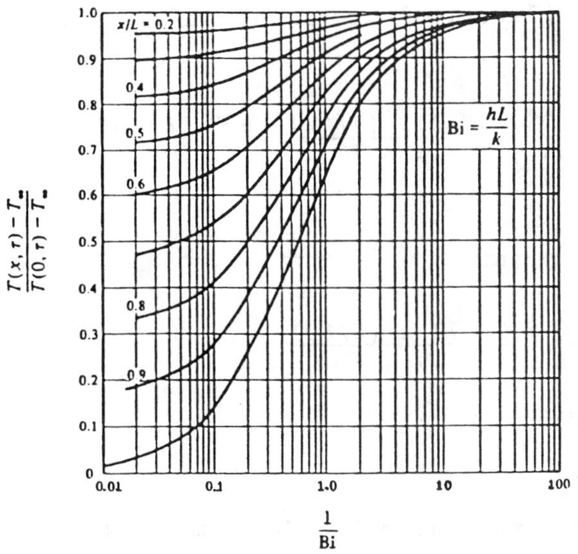

Fig. 4 Temperature as a function of centerline temperature in a slab of thickness, 2L.

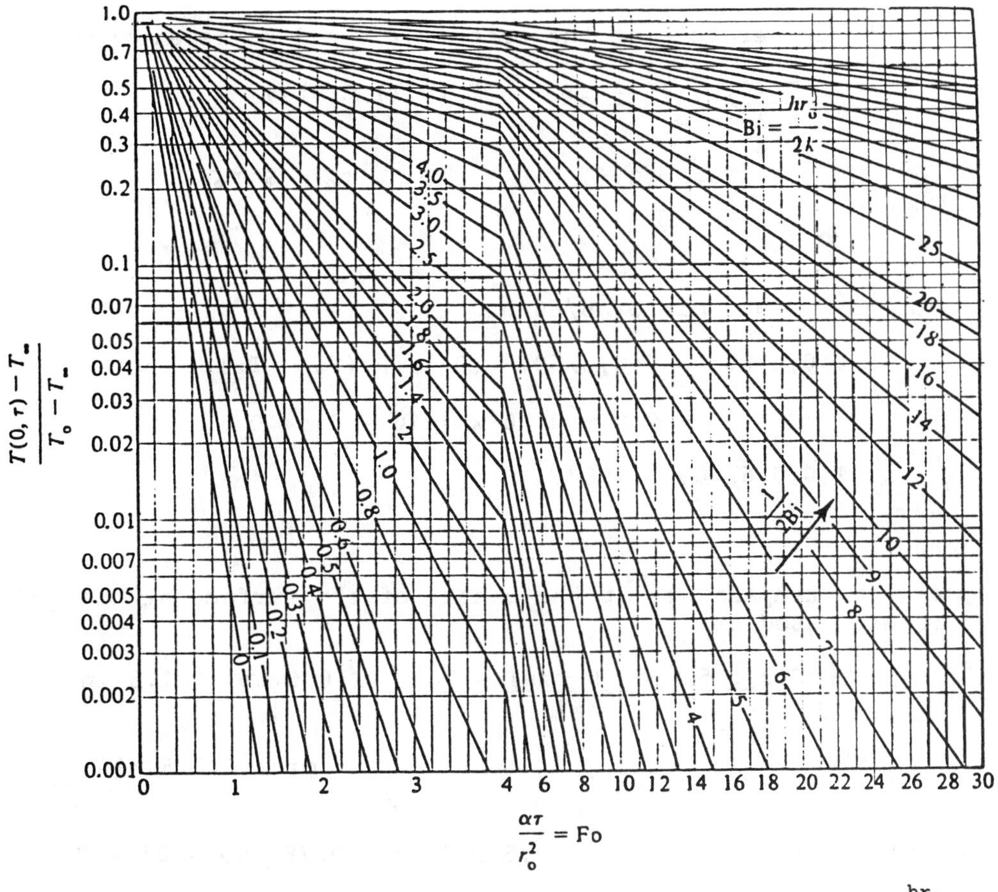

Fig. 5 Axis temperature for a long cylinder of radius, r_0; $\text{Bi}=\frac{hr_0}{2k}$.

Again, for cylindrical surface, the Biot number is defined as

$$Bi = \frac{hr_0}{2k} = \frac{243\ (7/12)}{2 \times 10} = 7.087$$

or $\frac{1}{Bi} = 0.141$

and $F_0 = \frac{\alpha\tau}{r_0^2} = \frac{0.172 \times (6/60)}{(7/12)^2}$

$= 0.05$

From figure 5 the value for C(0,6 min) is

$$C(0,6\ \text{min}) = \left[\frac{T(0,\tau) - T_\infty}{T_0 - T_\infty}\right]_{cyl} = 0.78$$

The temperature at the center of the disc is expressed as

$$\frac{T\ (0,0,\tau) - T_\infty}{T_0 - T_\infty} = P\ (0,\tau)\ C(0\ \tau)$$

or $T(0,0,\tau) = P(0,\tau)\ C(0,\tau)\ (T_0 - T_\infty) + T_\infty$

For $\tau = 6$ min or 6/60 hr

$$T(0,0,6/60) = P(0,6/60)\ C(0,6/60)(T_0 - T_\infty) + T_\infty$$

$$= 0.5 \times 0.78 \times (510 - 0) + 0$$

$\therefore T(0,0,6/60) = 195°F$

and the temperature at the center of the surface at one end is expressed as

$$\frac{T(0,1.5\ \text{in},\ 6\ \text{min}) - T_\infty}{(T_0 - T_\infty)} = P(0,6\text{min})\ P(1.5\ \text{in},\ 6\ \text{min})\ C(0,6\ \text{min})$$

or $T(0,1.5\ \text{in},\ 6\ \text{min}) = 0.5 \times 0.87 \times 0.78(500 - 0) + 0$

$= 169.65°F$

• PROBLEM 4-12

An ice cream block, 6 in.x6 in.x8 in., is initially at a uniform temperature of -12°F. It is suddenly exposed to air at 80°F. Determine the time period before the thawing occurs. The properties of ice cream block are k=1.26Btu/hr-ft-°F and the thermal diffusivity α=0.05ft^2/hr. Assume the heat transfer coefficient at the interface as 1.55Btu/hr-ft^2-°F.

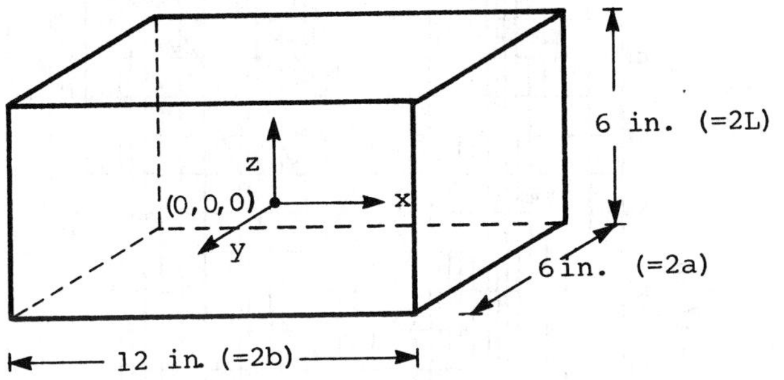

Fig. 1 Ice Cream Block

Solution: Thawing occurs initially at the corners of the ice-cream block since the surface area at the corners for heat transfer is larger than at any other part of the block. The problem involves a three dimensional transient heat transfer and hence it can be solved by product of solution in three orthogonal planes. The general solution in three dimensional plane can be expressed as

$$\psi = \psi_1(L)\ \psi_2(a)\ \ \psi_3(b) \qquad (1)$$

The thawing occurs, at the corners, at a temperature of 32°F (the melting temperature of ice). The temperature in excess of thawing at the corners is

$$\theta_e = 32-80 = -48°F$$

The temperature difference at time t=0 between ice and air θ_0 is

$$\theta_0 = -12-80 = -92°F$$

The solution for corner a=L=3 in.

$$\psi = \frac{\theta_e}{\theta_0} = \frac{-48}{-92} = 0.521$$

$$= \psi_2^2 \Big|_{\text{at } x/L=1.0}^{(L)} \times \psi_2 \Big|_{\text{at } y/b=1.0}^{(b)} \quad (2)$$

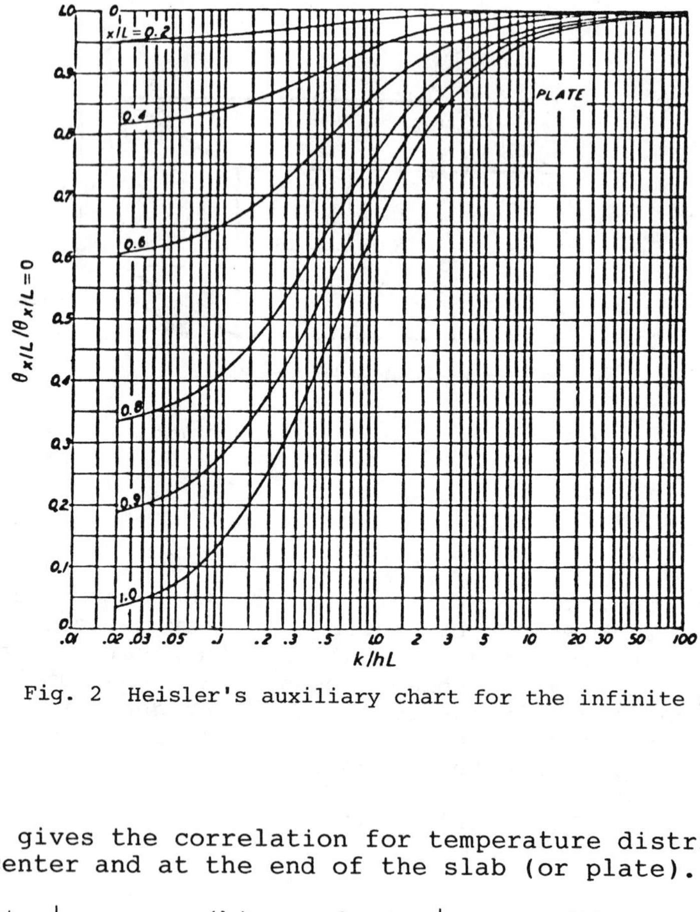

Fig. 2 Heisler's auxiliary chart for the infinite slab.

Figure 2 gives the correlation for temperature distribution at the center and at the end of the slab (or plate).

i.e. $\psi_2 \Big|_{\text{at } y/b=1.0}^{(b)}$ and $\psi_2 \Big|_{\text{at } y/b=0}^{(b)}$

For $L=b=\frac{12}{2}=6$ in, $\frac{k}{hL}=\frac{1.26}{1.55 \times (6/12)} = 1.62$

Now, for y/b = 1.0 from chart (fig 2) i.e. corresponding to

$\frac{k}{hL} = 1.62$

gives $$\frac{\theta_{x(L=1)}}{\theta_{x(L=0)}} = 0.75$$

or $$\frac{\psi_2\big|_{at\ y/b\ =\ 1.0}\ (b)}{\psi_2\big|_{at\ y/b\ =\ 0}\ (b)} = 0.75$$

or

$$\psi_2\Big|_{at\ y/b\ =\ 1.0}(b) = 0.75 \times \psi_2\Big|_{at\ y/b=0}(b) \qquad (3)$$

But $$\psi_2\Big|_{at\ y/b=0}(b) = \left(\frac{T - T_f}{T_0 - T_f}\right)\ \text{at } y/b=0$$

Similarly for other sides

$$\frac{k}{hL} = \frac{1.26}{1.55\left(\frac{3}{12}\right)} = 3.25$$

Again, from figure 2,

$$\theta_{x/L=1}/\theta_{x/L=0} = 0.86$$

$$\psi\Big|_{at\ x/L=1.0}(L) = 0.86\ \ \psi\Big|_{at\ x/L=0}(L) \qquad (4)$$

Also $$\psi\Big|_{at\ x/L=0}(L) = \left(\frac{T - T_f}{T_0 - T_f}\right)\ \text{at } x/L=0$$

Now, substituting equations(3) and (4) in equation (2) results in

$$\psi = 0.521 = 0.75\ \psi_2\Big|_{at\ y/b=0}(b) \times (0.86)^2\ \psi_2^2\Big|_{at\ x/L=0}(L)$$

or $$\frac{0.521}{(0.86)^2(0.75)} = 0.939 = \psi^2\Big|_{at\ x/L=0}(L)\ \ \psi\Big|_{at\ y/b=0}(b) \qquad (5)$$

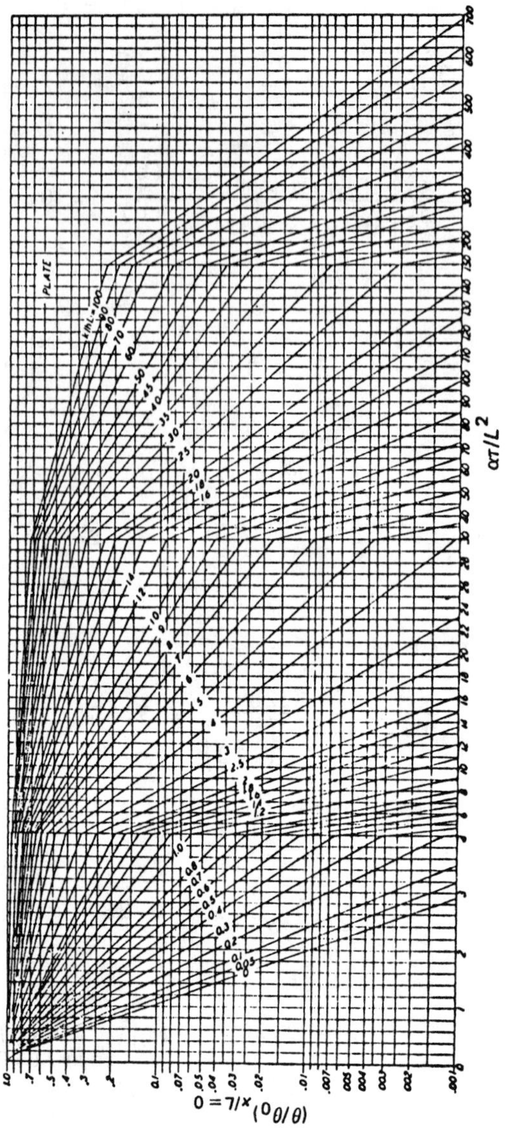

Fig. 3 Heisler's main chart for the infinite slab.

From figure 3 corresponding to

$$\frac{k}{hL} = 1.62 \text{ and } \frac{k}{hL} = 3.25$$

the Fourier numbers $F = \frac{\alpha\tau}{L^2}$ are

$$F_b = \frac{\alpha\tau}{b^2} = \frac{0.05\tau}{\left(\frac{6}{12}\right)^2} = 0.2\tau \tag{6}$$

and $$F_L = \frac{\alpha\tau}{L^2} = \frac{0.05\tau}{\left(\frac{6}{12} \times 2\right)^2} = 0.8\tau \qquad (7)$$

respectively. Thus, there are three equations and three unknowns, $\psi\big|_{at\ x/L=0}(L)$, $\psi\big|_{at\ y/b=0}(b)$, and τ.

The equations can only be solved by a trial and error method.

Let $$\psi\big|_{at\ y/b=0}(b) = 1$$

Substituting this value in equation (5) gives

$$\psi^2\big|_{at\ x/L=0}(L) = 0.939$$

or $$\psi\big|_{at\ x/L=0}(L) = 0.969$$

Corresponding to this value from figure 3 and $\frac{k}{hL} = 3.25$ the Fourier number F_L is $F_L = \frac{\alpha\tau}{(L)^2} = 0.6 = 0.8\tau$

or $$\tau = 3/4 \text{ hr.}$$

Introducing this value of τ in equation (6)

$$0.2 \times \tau = 0.2 \times 3/4 = 0.15$$

$$\therefore \quad F_b = 0.15$$

Again, corresponding to $F_b = 0.15$ and $\frac{k}{hb} = 1.62$,

$$\psi\big|_{at\ y/b=0} = 0.999 \simeq 1.0$$

which is the assumed value.

Thus, in time 3/4 hr the thawing is initiated.

GURNEY-LURIE CHARTS

• PROBLEM 4-13

Determine the time required for the temperature to reach 290°F at the center of a ½ in. thick rubber plate which is initially at a uniform temperature of 70°F. The surfaces of the plate are maintained at a temperature of 292°F. Assume the thermal diffusivity $\alpha = \frac{k}{\rho C_p}$ of the material as 0.0028 ft^2/hr.

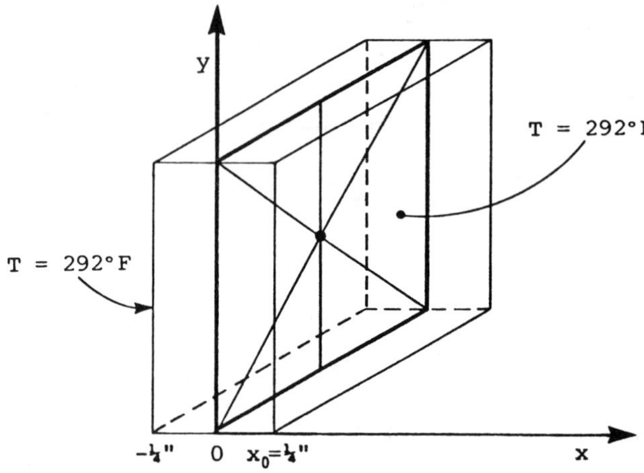

Solution: The unsteady state heat conduction in solids is expressed as

$$\frac{\partial T}{\partial \theta} = \frac{k}{C_p \rho} \frac{\partial^2 T}{\partial x^2}$$

Expressing the equation in dimensionless parameters

$$\frac{\partial \Delta}{\partial \tau} = \frac{\partial^2 \Delta}{\partial \eta^2} \qquad (1)$$

where $\Delta = \frac{292 - T}{292 - 70}$, $\eta = \frac{x}{x_0}$ and $\tau = \frac{k\theta}{C_p \rho x_0^2}$

The differential equation (1) can be solved by the method of separation of variables and the form of the solution can be assumed as

$$\Delta = \tau N \qquad (2)$$

where N is a function of η only. Differentiating equation (2) yields

$$N \frac{\partial \tau}{\partial \tau} = \tau \frac{\partial^2 N}{\partial \eta^2}$$

or

$$\frac{1}{\tau} \frac{\partial \tau}{\partial \tau} = \frac{1}{N} \frac{\partial^2 N}{\partial \eta^2}$$

Now, as the time and distance are independent variables, the left side of the equation is independent of distance and the right side is independent of time. Therefore each side must be constant. Let this constant be $-a^2$.

Then

$$\frac{d\tau}{d\tau} + a^2\tau = 0$$

and

$$\frac{d^2N}{d\eta^2} + a^2N = 0$$

Solutions of these equations are

$$\tau = C_1 e^{-a^2\tau}$$

and

$$N = C_2 \sin(a\eta) + C_3 \cos(a\eta)$$

Thus, the complete solution of equation (1) can be written as

$$\Delta = \tau N$$

$$= C_1 e^{-a^2\tau} (C_2 \sin(a\eta) + C_3 \cos(a\eta)) \qquad (3)$$

To evaluate the constants the following boundary conditions can be applied to the solution

I) The plate is symmetrical about the axis-y, so no heat is conducted across the center line (x=0). Therefore

$$\eta = 0, \quad \frac{\partial \Delta}{\partial \eta} = 0$$

II) At the surface, the temperature is constant at 292°F. Thus $\eta=1$, $\Delta=0$

III) The temperature of the plate at time t=0 is 70°F. Therefore $\tau=0$, $\Delta=1$

Differentiating equation (3) with respect to η

$$\frac{\partial \Delta}{\partial \eta} + C_1 e^{-a^2\tau}(C_2 a \cos(a\eta) - C_3 a \sin(a\eta))$$

and for the boundary condition (I)

$$\frac{\partial \Delta}{\partial \eta} = aC_1 C_2 e^{-a^2\tau} = 0$$

$\therefore$ constant $C_2 = 0$

$$\Delta = Ae^{-a^2\tau}\cos(a\eta)$$

where $A = C_1C_3$

From boundary condition II $0 = Ae^{-a^2\tau} \cos a$

The expression on the right hand side is equal to zero for the values equal to $\pi/2$, $3\pi/2$, $5\pi/2$, $7\pi/2$ and so on.

Next, applying the third boundary condition

i.e. $\tau = 0$, $\Delta = 1$, results in

$$1 = A_1 \cos\frac{\pi}{2} n + A_2\text{Cos}\frac{3\pi}{2} n + A_3\cos\frac{5\pi}{2} n + \ldots$$

$$+ A_\iota \cos\frac{2\iota - 1}{2}\pi n + \ldots$$

Now, the general solution to equation can be written as

$$\Delta = A_1 e^{-(\pi/2)2\tau}\cos\left(\frac{\pi}{2}\right)n + A_2 e^{-(3\pi/2)2\tau}\cos\left(\frac{3\pi}{2}\right)n$$
$$+ A_3 e^{-(5\pi/2)2\tau}\cos\left(\frac{5\pi}{2}\right)n + \quad \ldots \tag{4}$$
$$+ A_\iota e^{-[(2\iota-1)\pi/2]2\tau}\cos\left[\frac{(2\iota-1)\pi}{2}\right]n + \ldots$$

where ι assumes integer values only. Further, in order to evaluate the coefficients in equation (4), multiplying both sides by the expression $[(2\iota-1)/2]\,\pi n\,dn$ and integrating the resulting expression between the limits 0 to 1 yields for the left hand side

$$\int_0^1 \cos\left(\frac{2\iota-1}{2}\right)\pi n\,dn = \frac{2}{2\iota-1}\,\frac{1}{\pi}\left[\sin\left(\frac{2\iota-1}{2}\right)\pi n\right]_0^1$$

$$= -\frac{2}{2\iota-1}\,\frac{1}{\pi}\,(-1)^\iota$$

First term on the right hand side:

$$\int_0^1 A_1\cos\left(\frac{\pi}{2}\right)n\,\cos\left(\frac{2\iota-1}{2}\right)\pi n\,dn$$

$$= A_1 \left[\frac{\sin\ (\iota-1)\pi n}{(2\iota-2)\pi} + \frac{\sin\ \iota\pi n}{2\iota\pi} \right]_0^1$$

$$= A_1 \left[\frac{\sin\ (\iota-1)\pi}{(2\iota-2)\pi} + \frac{\sin\ \iota\pi}{2\iota\pi} \right]$$

$$= 0$$

Similarly, except the ιth term, all the terms are each equal to zero.

For ιth term $A_\iota \int_0^1 \cos^2\left(\frac{2\iota-1}{2}\right) \pi n\ dn$

$$= A_\iota \frac{2}{2\iota-1} \frac{1}{\pi} \left[\frac{\pi}{2} \frac{2\iota-1}{2} n + \tfrac{1}{4} \sin(2\pi) \frac{2\iota-1}{2} n \right]_0^1$$

$$= A_\iota \frac{2}{2\iota-1} \frac{1}{\pi} \frac{\pi}{2} \frac{2\iota-1}{2} + 0 - 0 - 0$$

$$= \frac{A_\iota}{2}$$

Writing L H S and R H S of the equation,

$$- \frac{2}{2\iota-1} \frac{1}{\pi} (-1)^\iota = \frac{A_\iota}{2}$$

or

$$A_\iota = \frac{-4(-1)^\iota}{\pi(2 - 1)}$$

Substituting the values of A in equation (4), yields

$$\Delta = \frac{4}{\pi} e^{-(\pi/2)2\tau} \cos\left(\frac{\pi n}{2}\right) - \frac{4e^{-(3\pi/2)2\tau}}{3\pi} \cos \frac{3\pi n}{2}$$

$$+ \frac{4e^{-(5\pi/2)2\tau}}{5\pi} \cos \frac{5\pi n}{2} \quad \ldots.$$

or

$$= \sum_{\iota=1}^{\iota=\infty} \frac{-2(-1)^\iota}{[(2\iota - 1)/2]\pi} e^{-[(2\iota-1)\pi/2]2\tau} \cos\left(\frac{2\iota-1}{2}\right) \pi n$$

Now substituting numerical values

$$\Delta = \frac{292 - 290}{292 - 70} = 0.009$$

and $\eta = \frac{x}{x_0} = \frac{0}{x_0} = 0,\ \tau = \frac{k\theta}{C_p \rho x_0^2}$

or $\tau = 0.0028\theta/x_0^2$

$$\therefore \quad 0.0090 = \frac{4}{\pi} e^{-(\pi/2)2\tau} - \frac{4}{3\pi} e^{-(3\pi/2)2\tau} + \frac{4}{5} e^{-(5\ /2)2} - \ldots$$

The only unknown is θ and can be determined by trial and error method.

Considering only the first term in the series to exist and all other terms to vanish, then

$$\tau = -\left(\frac{2}{\pi}\right)^2 \ell n \ \frac{\pi 0.009}{4}$$

$$= 2.01$$

$$\theta = [(C_p\rho)/k]x_0^2\tau$$

or $$\theta = \frac{2.01\left(\frac{1}{4} \times \frac{1}{12}\right)}{0.0028} = 0.313 \text{ hr}$$

or $$\theta = 18.7 \text{ min.}$$

• PROBLEM 4-14

Using Gurney-Lurie Graphs determine the time period required for the temperature to reach 290°F at the center of a ½ in. thick plate. The initial uniform temperature of the plate is 73°F. Then the surfaces of the plate are instantaneously raised to 300°F and maintained at that temperature. The thermal diffusivity of the material is 0.003 ft^2/hr.

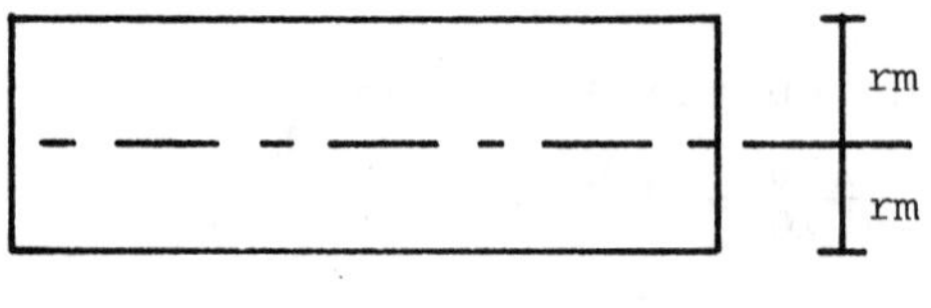

Fig. 1

<u>Solution:</u>

$$\Delta = \frac{t_s - t}{t_s - t_0} = \frac{300 - 290}{300 - 75} = 0.0444$$

where $t_s = 300°F$ $\qquad t = 290°F$

$$m = \frac{k}{hr_m} = 0$$

as the surface temperature is constant; in other words the coefficient of heat transfer is infinity.

again, $\quad n = \frac{r}{r_m} = 0 \quad$ at the center of the plate.

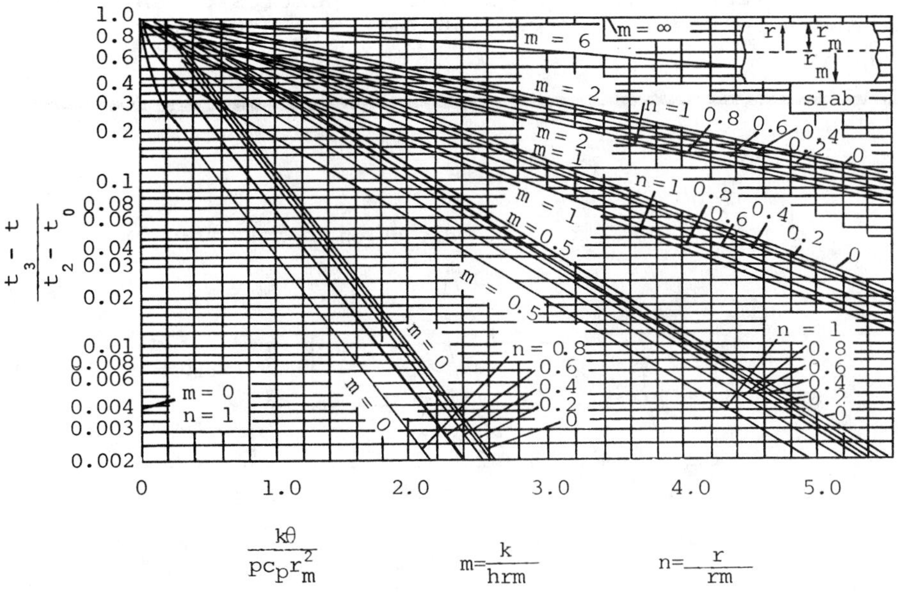

Fig. 2 Gurney-Lurie Chart

Now, from the G-L chart corresponding to $m = 0 = n$ and $\Delta = 0.0444$,

$$\frac{k\theta}{\rho C_p r_m^2} = 1.4 \qquad (1)$$

and by calculation

$$\frac{k\theta}{\rho C_p r_m^2} = \frac{0.003 \times \theta}{(\frac{1}{4} \times \frac{1}{12})^2} = 6.912\theta \qquad (2)$$

Now, from equations (1) and (2)

$$\theta = \frac{1.4}{6.912} = 0.020 \text{ hr}$$

or $\theta = 12.15$ min.

• PROBLEM 4-15

A margarine slab 50 mm thick with a uniform temperature of 5°C is placed in an environment which is at a temperature of 24°C. The heat transfer occurs only at the top surface as all other sides including the bottom are insulated. Using the given charts compute the temperature of the margarine at the top, at the bottom surface and at mid depth after a lapse of 5 hours. Assume the surface convective heat transfer coefficient as 9.0 $W/m^2{}^\circ K$. The properties of the margarine can be assumed as $k=0.166 W/m^\circ K$, $C_p=2300$ J/kg°K and $\rho=1000$ kg/m^3.

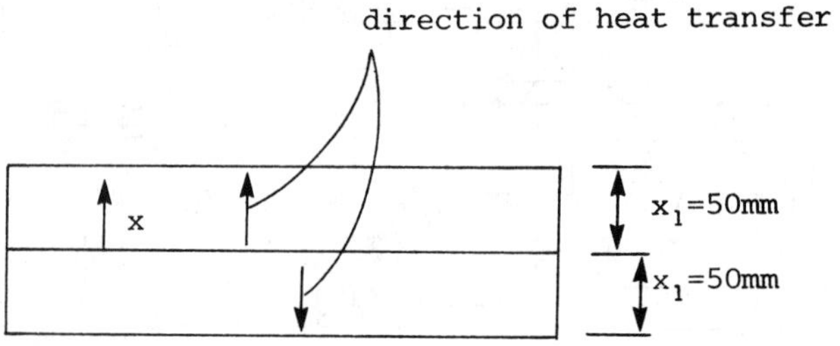

Fig. 1

Solution: The margarine slab can be analyzed taking two slabs taken together where the centre would act as an insulated surface.

α - thermal diffusivity

$$= \frac{k}{\rho C_p} = \frac{0.166}{2300 \times 1000} = 7.22 \times 10^{-8}$$

t = time in seconds

$$= 5 \times 3600 = 1.8 \times 10^4 \text{ seconds}$$

$x_1 = 0.05$ meter

$$\therefore \quad X = \frac{\alpha t}{x_1^2} = \frac{7.22 \times 10^{-8} \times 1.8 \times 10^4}{(0.05)^2}$$

$$= 0.52$$

$$n = \frac{x}{x_1} = \frac{0.05}{0.05} = 1 \qquad m = \frac{k}{hx_1} = \frac{0.225}{9.0 \times 0.05} = 0.500$$

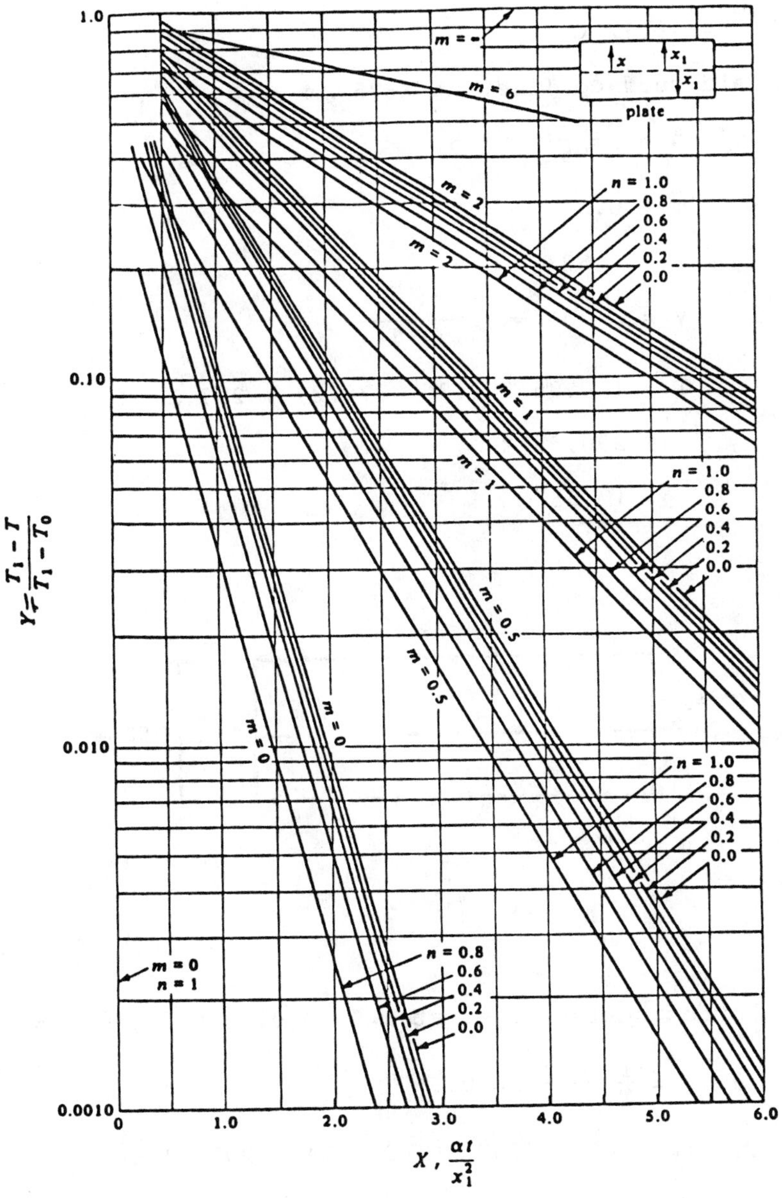

Fig. 2 Unsteady-state heat conduction in a large flat plate.

For X = 0.704, m = 0.5 and n=1 from chart (fig. 2)

$$y = \frac{T_1 - T}{T_1 - T_0} = 0.32$$

where $T_1 = 24 + 273.2 = 297.2°K$

$T_0 = 5 + 273.2 = 278.2°K$

$\therefore \quad T = (-0.32(297.2 - 278.2) + 297.2)°K$

$= 291.12°K$

or $= 18.12°C$

Again, for mid surface (i.e. at x=25 mm or 0.025 meter)

$$n = \frac{0.025}{0.05} = 0.5$$

$m = 0.5$

and $x = 0.52$

$$y = 0.35 = \frac{T_1 - T}{T_1 - T_0}$$

or $T = (-0.35(297.2 - 278.2) + 297.2)°K$

$= 290.55°K$

or $= 17.55°C$

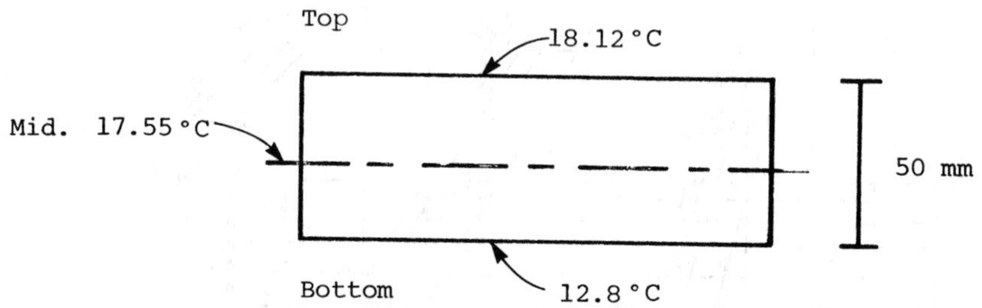

Fig. 3

Again, for bottom surface i.e. at x=0, n=0, x=0.52, m=0.5

$$y = \frac{T_1 - T}{T_1 - T_0} = 0.6$$

$\therefore \quad T = (-0.60(297.2 - 278.2) + 297.2)°K$

$= 285.8°K$

or $= 12.8°C$

TEMPERATURE DISTRIBUTION CHARTS FOR VARIOUS SOLIDS

• PROBLEM 4-16

The initial uniform temperature of a long aluminium rod, 8 in. dia., is 100°F. The rod is suddenly immersed, at time t=0, in medium at temperature 1000°F. Determine the temperature at the center line after 10 min. of its immersion. Use the chart given and the value $\alpha = 0.46$ ft^2/hr.

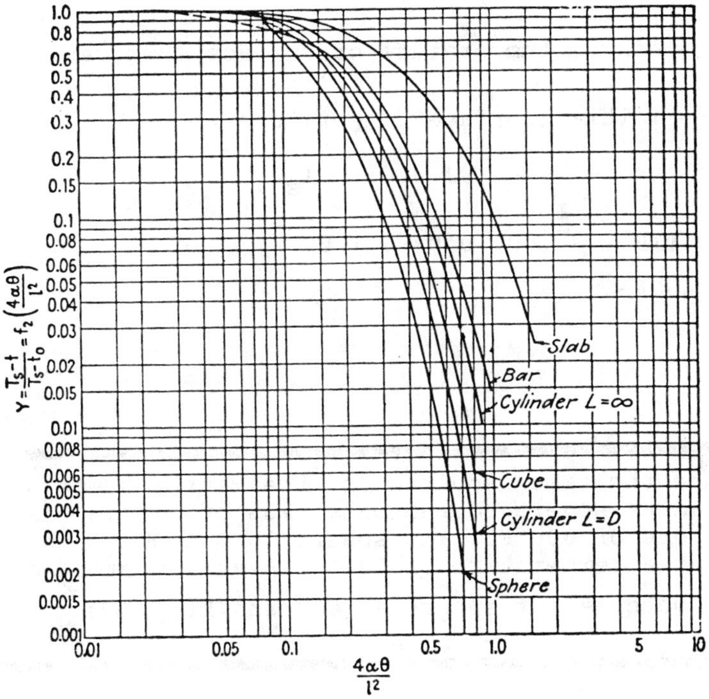

Fig. 1 Temperature distribution curves for different shapes.

Solution: Let the surface temperature of the rod be the same as the heat source temperature 1000°F. Also, the rod is of infinite length.

Corresponding to the value of $\frac{4\alpha\theta}{\ell^2}$ and infinite length of rod Δ can be read from the ordinate of the chart.

where θ is in hours $= \frac{10}{60} = 0.1666$ hr

and ℓ is in ft. $= \frac{8}{12} = 0.666$ ft

$$\frac{4\alpha\theta}{\ell^2} = \frac{4 \times 0.46 \times 0.1666}{(0.666)}$$

$$= 0.69$$

For $\frac{4\alpha\theta}{\ell^2} = 0.69$ and length $L = \infty$

from graph $\Delta = \frac{T_s - t}{T_s - t} = 0.03$

T_s - source temperature or surface temperature = 1000°F

t_0 - initial surface temperature = 100°F

$\therefore \quad t = -(T_s - t_0)\ 0.03 + T_s$

or $t = -(1000 - 100)\ 0.03 + 1000$

$= 973°F$

• PROBLEM 4-17

Calculate the time required for the center temperature of a 6 in. radius sphere to reduce to 25% of its initial uniform temperature of 800°F. The sphere is plunged into a pool of liquid at a temperature of 80°F. Assume thermal diffusivity for the material of sphere as 0.7 ft^2/hr. Use the given chart.

Solution: The flow of heat occurs in two stages i.e. conduction through the body of the sphere and then convection from the surface of the sphere. The conduction and convection resistances to heat flow are related by the Biot number defined as

$$Bi = \frac{hx}{k}$$

where x is the characteristic dimension of the sphere

$$x = \frac{\text{Volume}}{\text{surface area of sphere}} = \frac{\frac{4}{3}\pi r^3}{4\pi r^2}$$

$$= \frac{r}{3}$$

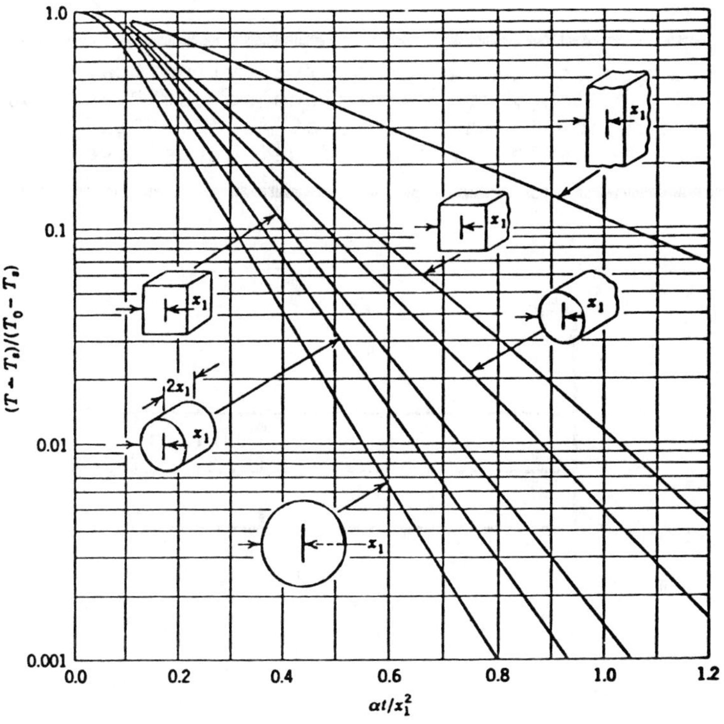

Fig. 1 Distribution of temperature within the solids of different configurations.

Considering Bi = 0.1 as the limiting factor, h is found to be in the range of about 20 Btu/hr ft^2 °F with thermal conductivity of k = 29 Btu/hr ft°F. This is very close to a reasonable value indicating that the assumption of no temperature gradient within the sphere is not an accurate one. In such a case, the accompanying figure is used for the central temperature history.

$$\frac{T_c - T_s}{T_0 - T_s} = \left(\frac{0.25 \times 800 - 80}{800 - 80} \right)$$

$$= 0.1667$$

where c and s subscripts stand for center and surrounding.

From the chart

$$\frac{\alpha t}{x_1^2} \simeq 0.12 \quad \text{or } t = \frac{\left(\frac{6}{12}\right)^2 \times 0.12}{0.7}$$

$$\therefore \quad t = 0.043 \text{ hrs.}$$

$$\simeq 2.6 \text{ min}$$

$$\simeq 155 \text{ sec}$$

• **PROBLEM 4-18**

The temperature at the two ends of a cylindrical rod, 8 in. dia. and 12 in. long, are maintained at 300°F and 600°F. The curved surface is maintained at 400°F. Find the temperature at the center of the rod using the given graph.

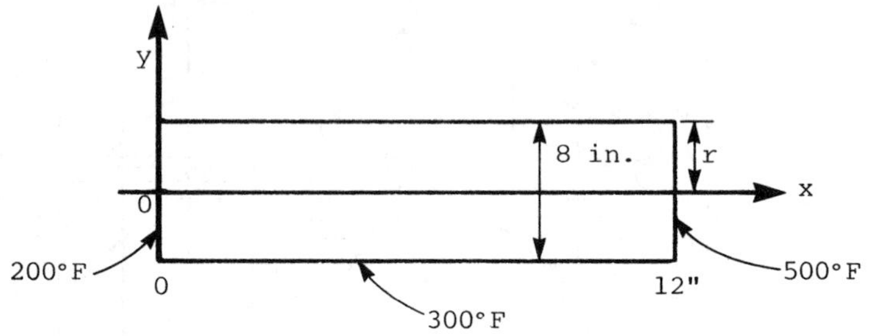

Fig. 1

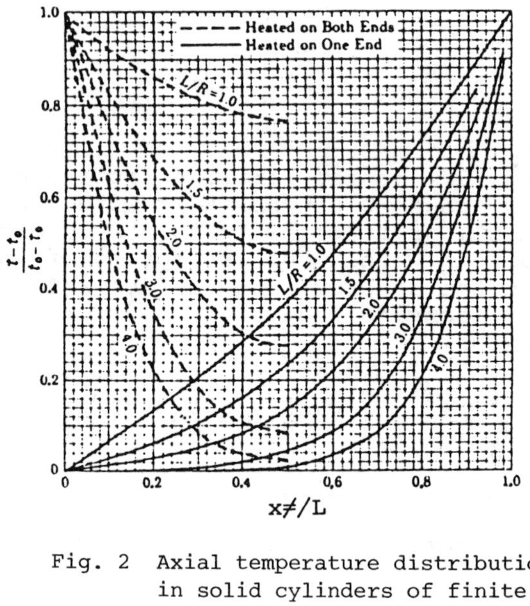

Fig. 2 Axial temperature distribution in solid cylinders of finite length.

Solution: The problem can be analyzed by the method of superposition taking the equivalent temperature distribution on the cylinder.

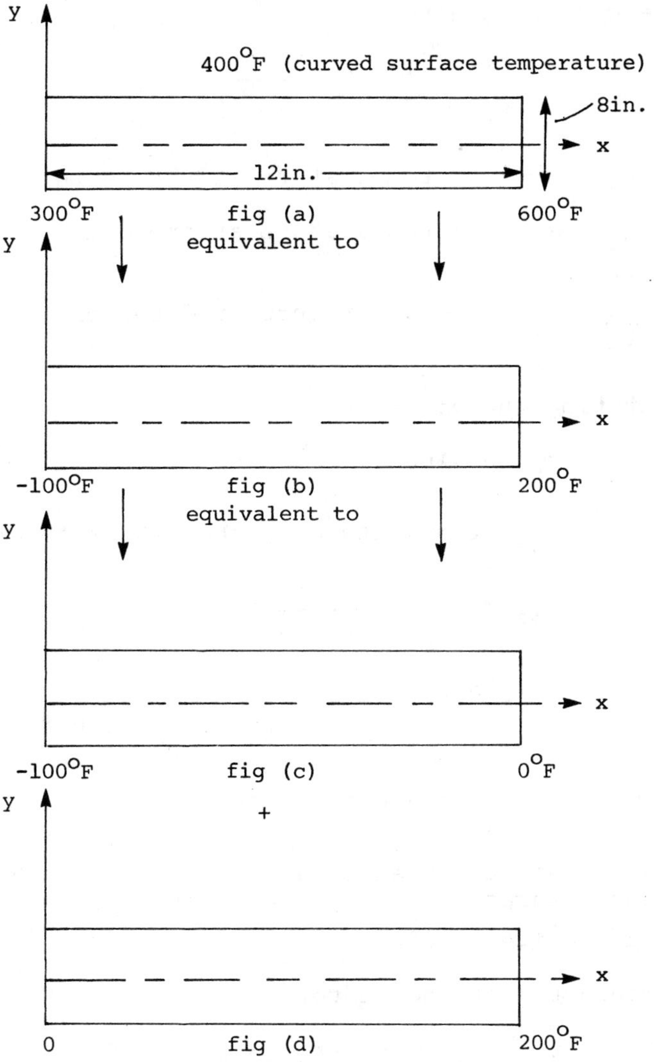

Equivalent temperature distribution
[fig (a) = fig (b) = fig (c) + fig (d)]

Fig. 3

The two temperature distributions in figure 3 are equivalent to that in figure 1.

The ratio $\frac{L}{R} = \frac{12}{4} = 3$

and at the mid section

$$\frac{x}{L} = \frac{6}{12} = 0.5$$

For temperature distribution in figure 3(a) the temperature at the center is given by

$$t_{x_1} = t \quad x \quad \frac{t - t_c}{t_0 - t_c}$$

where t_{x_1} ⟶ temperature at the center for figure 3(a)

t_1 ⟶ temperature at one end = -100°F

$\frac{t - t_c}{t_0 - t_c}$ ⟶ temperature gradient obtained from graph = 0.8

Substituting the values

$$t_{x_1} = -100 \times 0.8 = -80°F$$

Similarly t_{x_2}, temperature at the center for figure 3(b)

$$t_{x_2} = t_2 \times \left(\frac{t - t_c}{t_0 - t_c}\right)$$

where $t_2 = 200°F$ and $\frac{t - t_c}{t_0 - t_c} = 0.8$

$$\therefore t_{x_2} = 200 \times 0.8 = 160°F$$

Now the actual temperature at the center is the sum of the two temperatures t_{x_1} and t_{x_2} plus the initially subtracted temperature i.e. 400°F.

.. Temperature at the center is

$$t_{(x=6\ in)} = (t_{x_1} + t_{x_2} + 400)°F$$

$$= (-80°F + 160°F + 400°F) = 480°F$$

PRESSURE DISTRIBUTION CHARTS

• PROBLEM 4-19

The thermal conductivity of a monoatomic substance at 153°F and at atmospheric pressure is k_a = 0.016 Btu/hr ft°F. Find the thermal conductivity at a pressure of 191.8 atmospheres, given critical temperature and pressure are 305.4°F and 48.2 atm., respectively.

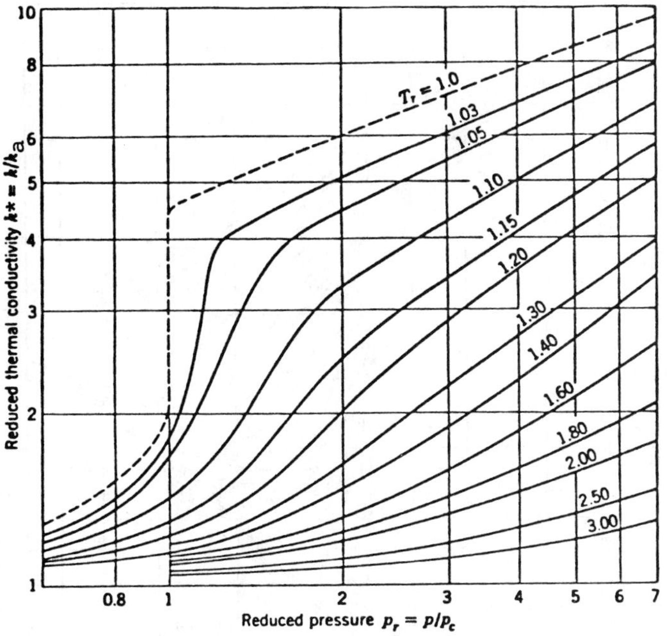

Fig. 1 Reduced thermal conductivity $k^* = k/k_a$ as a function of reduced temperature and reduced pressure.

Solution: The thermal conductivity corresponding to any temperature and pressure conditions can be read from figure 1 for reduced temperature and reduced pressure.

Reduced temperature is

$$T_r = \frac{T}{T_c} = \frac{153 + 460}{1.8 \times 305.4} = 1.115$$

and reduced pressure

$$P_r = \frac{p}{p_c} = \frac{191.8}{48.2} = 3.98$$

then from figure 1

$$k^* = \frac{k}{k_a} = 4.7$$

or $\quad k = k_a \times 4.7$

$$= 0.016 \times 4.7$$

$$= 0.0752 \text{ Btu/hr ft}^\circ\text{F}$$

This problem can also be solved by the chart (figure 2) showing the variation of thermal conductivity with reduced pressure and reduced temperature.

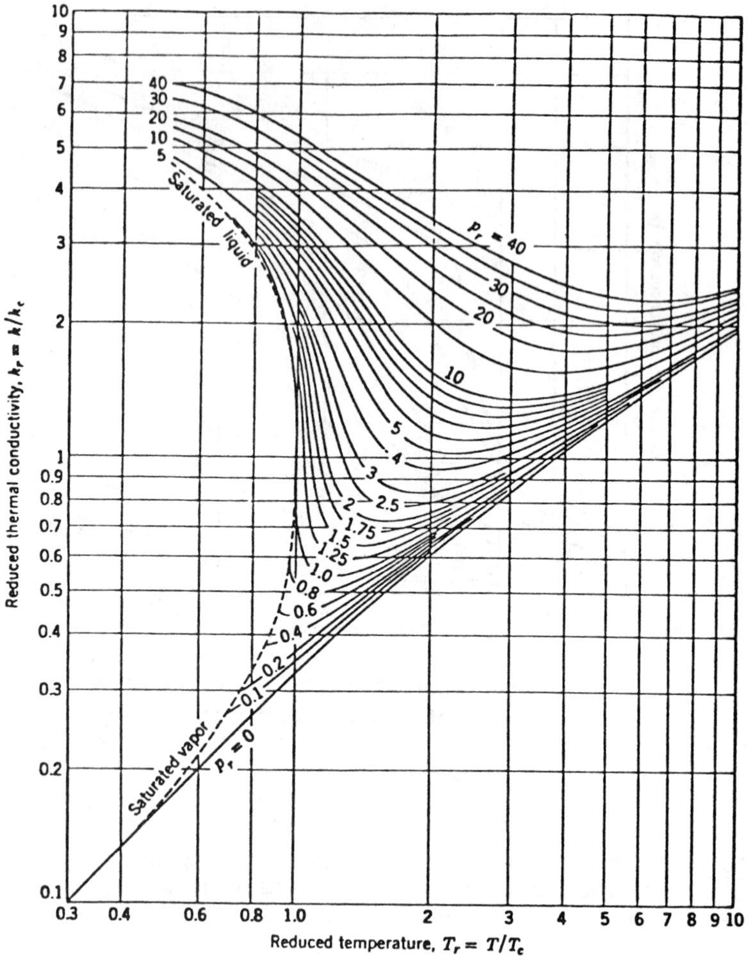

Fig. 2 Reduced thermal conductivity for monatomic substances as a function of reduced temperature and pressure.

Corresponding to reduced temperature T_r = 1.115 and pressure pr = 0, the k_r = 0.36.

Now $k_c = \frac{k}{k_r} = \frac{0.016}{0.36} = 0.0442$ Btu/hr ft°F

Again corresponding to Pr = 3.98 and T_r = 1.115 from figure 2, k_r = 2.07

$\therefore$ $k = k_r k_c = 2.07 \times 0.0442$

$= 0.0914$ Btu/hr ft°F

FREEZING INDEX

• PROBLEM 4-20

The freezing index at a place is 9050°F-day with θ_s= 255 days. The moisture percent of dry weight is 40%. The soil is homogeneous with density 1.2 gm/cm³. The mean annual air temperature is 55°C. Estimate the freezing depth at that place.

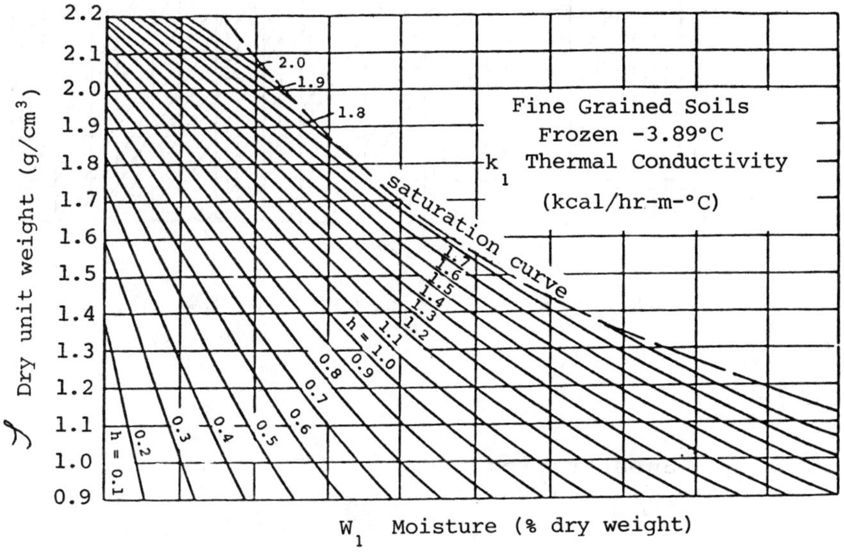

Fig. 1. Thermal Conductivity of Fine-Grained Soils

<u>Solution</u>: For values of W = 40%

and γ_d = 1.2 gm/cm³

from the graph in figure 1, k_f = 1.7

and from figure 2, C_f = 0.44

Now $T_f - T_s = \dfrac{(FI)}{\theta s}$

where FI ⟶ freezing index = 9050

and θ_s ⟶ 255 days

where T_f ⟶ phase change temperature = 32°F

T_s ⟶ the average surface temperature

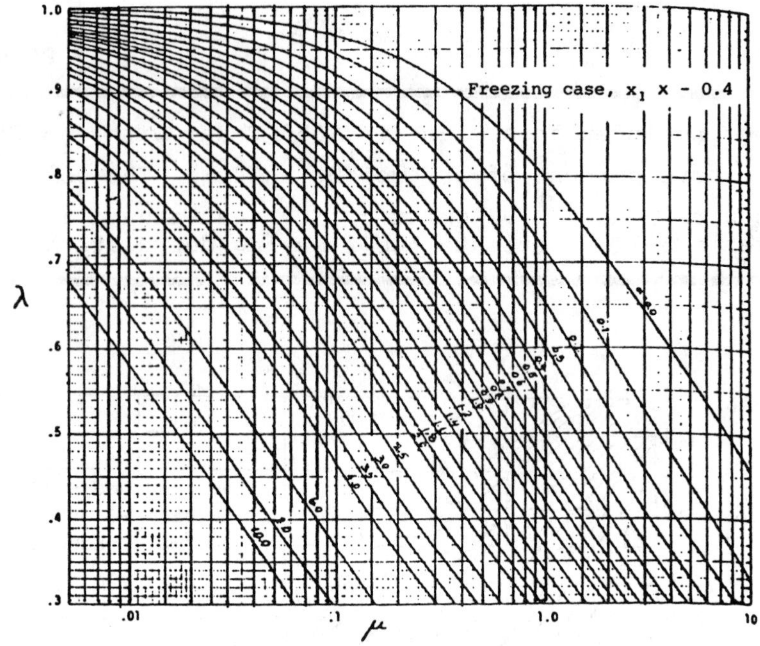

Fig. 2 Freezing Case, $x_1 = 0.4$.

Substituting the values

$$T_f - T_s = \frac{9050}{255} \cong 36°F$$

$\therefore$ The average surface temperature

$$T_s = 32°F - 36°F = -4°F$$

Now, thermal diffusivity

$$\alpha = \frac{T_0 - T_f}{T_f - T_s}$$

where $T_0 = 55°F$, $T_f = 32°F$

and $T_f - T_s = 36°F$

$$\therefore \alpha = \frac{55 - 32}{36} = 0.64$$

$$\text{Viscosity } \mu = \frac{C_f \times (T_f - T_s)}{2L}$$

where $C_f = 0.44$

$$T_f - T_s = 36°F$$

and $L = \frac{\rho \times W \times h_{if}}{100} = \frac{1.2 \times 40 \times 79.7}{100}$

$L = 38.26$

Substiting

$$\mu = \frac{0.44}{2 \times 38.26} \times \frac{36}{1.8} = 0.115$$

For $\mu = 0.115$ and $\alpha = 0.64$

From figure 3, $\lambda = 0.78$

∴ For the depth of freezing

$$x = 16.33\ \lambda \sqrt{\frac{R_f(FI)}{L}}$$

where $\lambda = 0.78$, $k_f = 1.7$, $FI = 9050$ and $L = 38.26$

$$\therefore \quad x = 16.33 \times 0.78 \times \sqrt{\frac{1.7 \times 9050}{38.26}} = 255.4 \text{ cm}$$

• PROBLEM 4-21

A building located in Chesterfield, NWT has the dimensions of 70 x 210 ft. The following data are given: $\gamma_a = 1.85$ g/cm³, W = 15%, $T_a = 11°F$, $T_F = 60°F$ (year-round), FI = 8750°F-day. (a) Calculate the centerline thaw after 8.7 years. (b) Calculate the temperature at y/a = 0, x/a = 0.5, and z/a = 0.5 and 1.5 for t = 8.7 years and also after an infinite time.

$k_1 = 2.23$ Btu/hr ft°F $\quad k_2 = 2.63 \quad k_{21} = 1.18$

$C_1 = 30$ Btu/ft³°F

$L = 2480$ Btu/ft³

Solution: (a) Method 1: Use the correction factor

$\nu_o = 32 - 11 = 21 \quad \nu_s = 28 \quad TI = (60 - 32)(365)(8.7)$
$= 89{,}000$

$\alpha = 0.75 \quad \mu = 0.34 \quad \lambda = 0.75$

$X = 46.5$ ft of thaw for an infinite system

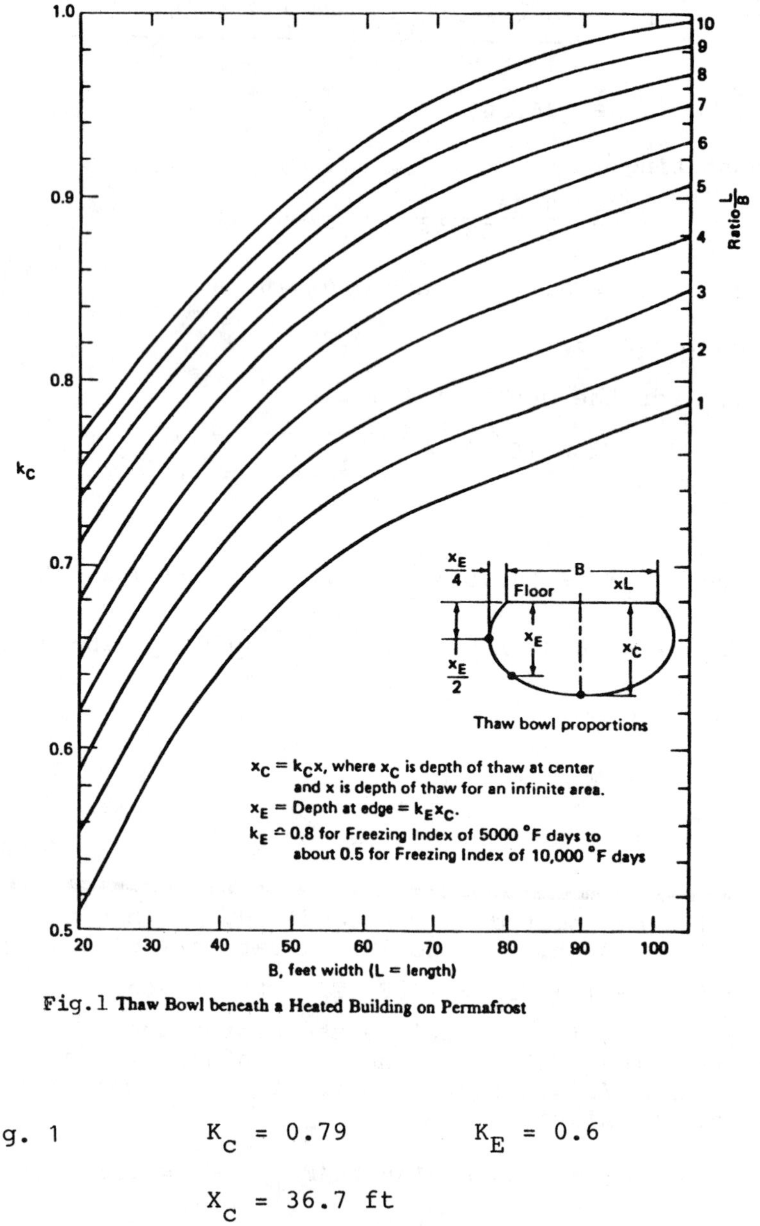

Fig. 1 Thaw Bowl beneath a Heated Building on Permafrost

From Fig. 1 $K_C = 0.79$ $K_E = 0.6$

$X_C = 36.7$ ft

$X_E = 22.0$ ft

Method 2: Use quasi-steady procedure

$$n = \frac{b}{a} = 3$$

$$I = k_1 \frac{(T_p - T_f)t}{4a^2L} = .392$$

$$\beta = -k_{21}\frac{T_o - T_f}{T_p - T_f} = .89$$

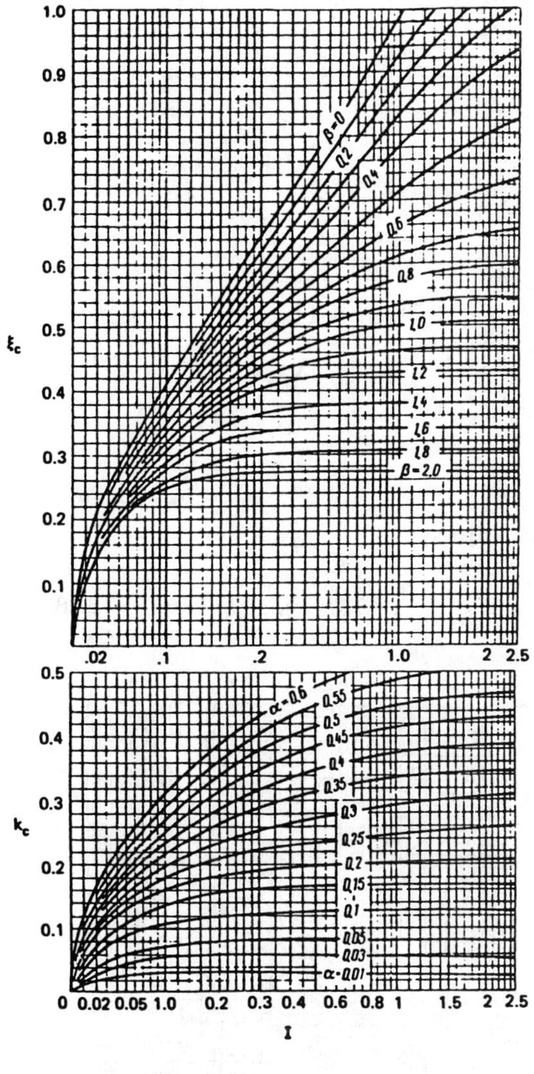

Fig. 2 **Ground Thaw Depth Coefficients for Center of Building**

From Fig. 2

$$\xi_c = .483$$

$$\xi_{oc} = X_c = .483(70) = 33.8 \text{ ft}$$

The two methods give very close results, but the quasi-steady method is preferred, especially as the permafrost temperature decreases.

A more realistic calculation of the thaw depth beneath the building would include the thermal resistance of the floor structure. Assume that the floor resistance R_i=13.5 ft²hr-°F/ Btu. Then using

$$X = kR_i\left(\sqrt{1 + \frac{48\lambda^2 I}{kLR_i^2}} - 1\right)$$

one obtains

$$X = 25.3 \text{ ft}$$

and

$$X_C = 20.0 \text{ ft}$$

$$X_E = 12.0 \text{ ft}$$

For this case, I = .392, β = .89, α = 0.43, and k_i = 1.0.
From Fig. 2, ξ_C = .483, k_C = .357

$$\xi_{oc} = (.483 - .357)(70) = 8.8 \text{ ft}$$

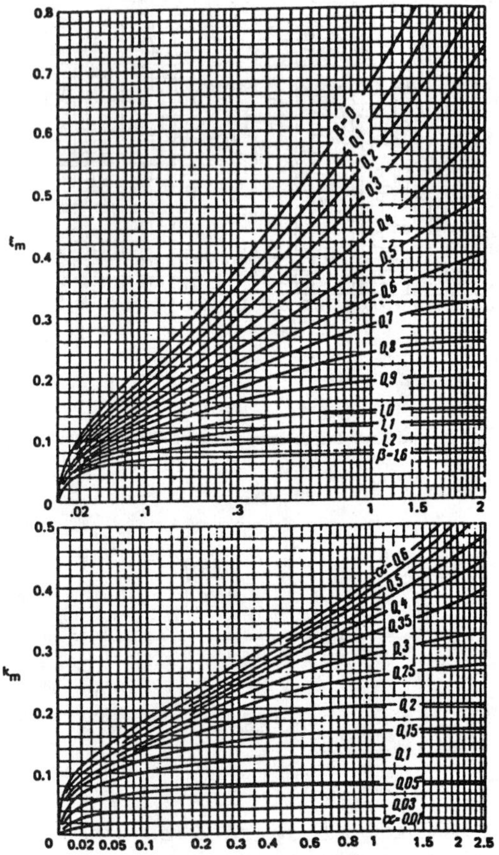

Fig. 3 Ground Thaw Depth Coefficients under Edge of Building

From Fig. 3, $k_m = .270, \quad \xi_m = .175$

$$\xi_e = [.175 - .270 - (.1)(.89)\sqrt{.392}]\ 70 < 0, \text{ implies}$$

no thawing.

(b) At t = 8.7 years $\quad \frac{\xi_o}{a} = .966$

21P2 Then from the eq.

$$f(\xi) = \frac{2}{\pi}\tan^{-1}\left[\frac{n}{\frac{\xi}{a}\left[\frac{\xi}{a}^2 + n^2 + 1\right]^{\frac{1}{2}}}\right]$$

$$f_o = f(\xi_o) = 0.481$$

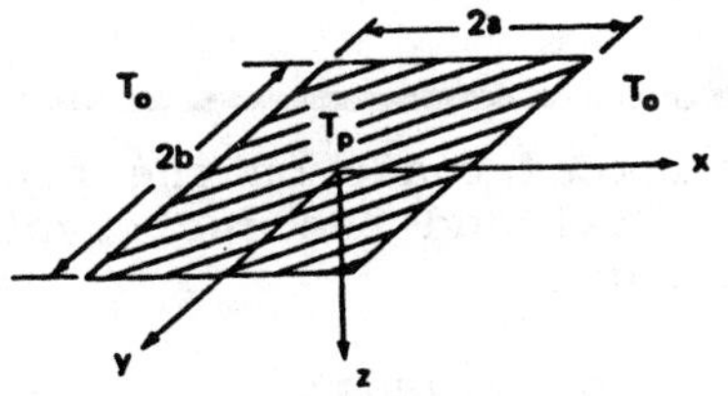

Fig. 4

From the eq.

$$(4) \qquad T - T_o = \frac{(T_p - T_o)}{2\pi} \tan^{-1}\frac{(x+a)(y+b)}{z\sqrt{z^2+(x+a)^2+(y+b)^2}}$$

$$-\tan^{-1}\frac{(x-a)(y+b)}{z\sqrt{z^2+(x-a)^2+(y+b)^2}}$$

$$-\tan^{-1}\frac{(x+a)(y-b)}{z\sqrt{z^2+(x+a)^2+(y-b)^2}}$$

$$+\tan^{-1}\frac{(x-a)(y-b)}{z\sqrt{z^2+(x-a)^2+(y-b)^2}}$$

$f(.5,0,.5) = 0.63$

$f(.5,0,1.5) = 0.32$

Then, from the eq.

$$(5) \qquad T_1 = T_p f + \frac{(T_p f_o - T_f)(1 - f)}{f_o - 1}$$

$$T_2 = \frac{(T_f - T_o)f}{f_o} + T_o$$

$$T(.5,0,.5) = (60)(.63) + \frac{(60)(.481)-32}{.481-1}(1-.63) = 40.1°F$$

(thawed)

$$T(.5,0,1.5) = (32 - 11)\frac{.32}{.481} + 11 = 25.0°F \text{ (frozen)}$$

After steady state is reached, $q_1 = .375$, and $\xi_\infty = 45.9$ ft. Then $f(\xi_o) = .375$

$$T(.5,0,.5) = 53.0°F$$

$$T(.5,0,1.5) = 28.9°F$$

• PROBLEM 4-22

Neglecting the mechanical loads on the foundation, determine the depth of gravel fill and the duct system, as well as the floor thermal resistance.

The following data is provided:

Length of duct, L=220 ft
Year-round temperature of the floor, $T_F = 60°F$

Properties of gravel pad: $\gamma_d = 2.0 g/cm^3$, $W = 2.5\%$, $x_\ell = 0.05$

$K_u = 1.10 \frac{Btu}{hr\text{-}ft°F}$, $C_u = 23.4 \frac{Btu}{ft^3 °F}$

$L_s = 450 \frac{Btu}{ft^3}$

FI=6400°F-days freezing index.
Length of freezing season, FS=228 days.
Length of thawing season, TS=137 days.
Conductivity of foam glass insulation, Ki=0.0333 Btu/hr-ft°F
Conductivity of concrete, $K_c = 0.47 — 0.81$
Conductivity of floor finish layer (wood), $K_F = 0.11$

Solution: During the thawing season, the gravel pad will thaw to a certain depth. During this time, the air ducts are closed to prevent warmer air from outside augmenting the thaw effect. The thawed layer of gravel must be refrozen during the freezing season, so as to keep the permafrost table stationary, thus the ducts are open allowing free circulation of cold outside air by free convection.

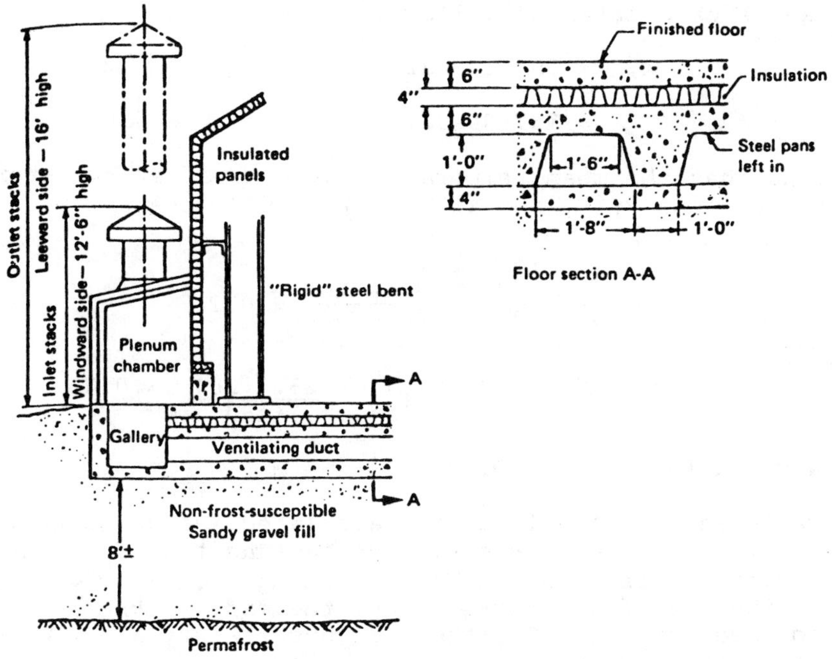

Fig. 1 **Ducted Foundation for Example 4 (adapted from Sanger 1969).**

In order to calculate the thaw depth, it is necessary to determine the thermal resistance of the floor. This is done by first assuming a certain thickness of insulation and then using the following empirical relation:

$$H = \frac{1}{140}\left[\frac{L}{\gamma_s(t+1)}\right]^2(L + 250)$$

where H = stack height (ft)

t = thickness of foamglass insulation

γ_s = FI/FS = 6400/228 = 28.1°F

This is the average daily temperature below freezing for the freezing season. During the freezing season the average air temperature is 3.9°F. Assuming the insulation thickness as 4 in., then from the above empirical relation, the initial stack height is 8.3 ft. The pad thickness is given by the relation:

$$X = K_u R\left[\sqrt{1 + \frac{48\lambda^2 I_f}{K_u L_s R^2}} - 1\right]$$

where I_f = thawing index at the floor

L_s = latent heat of soil

R = resistance of structure above pad

X = depth of thaw penetration

I_f can be calculated as follows:

$$I_f = TS\,(T_f - 32)$$
$$= 137(60-32) = 3836 \text{ degree-days}$$

Assuming that the mean temperature of the pad at the start of thawing will be 26°F, and if $T_s = T_F = 60°$, then

$$\alpha = \frac{T_f - T_0}{T_s - T_f} = \frac{32 - 26}{60 - 32} = 0.21$$

$$\mu = \frac{C_u}{2L_s}(T_s - T_f) = \frac{23.4(28)}{2(450)} = 0.73$$

The Newmann function $\lambda = 0.72$.

The resistance of the floor is calculated, assuming that the resistance of the air layer is the same as that for concrete. The resistance is $13.5 < R < 16.1$, using $R=13.5$ ft²-hr-°F/Btu and making use of the correlation; the pad thickness X=6.4 ft. Due to uncertainty in T_0, there is uncertainty about λ. If λ is 1.0, then the pad thickness would be 10.8 ft. Choosing an average value 8.6 ft, the temperature of the top of the gravel pad after complete thaw is given by the relation:

$$T_{d_1} = T_s + \frac{(T_f - T_F)d_1}{d_1 + K_1X/K_u}$$

where K_1 = conductivity of the floor = $\frac{d_1}{R}$

d = thickness of the floor = 2.67

The approximate mean temperature of the frozen and unfrozen pad is 26°F and 37.4°F. Fig 2 shows the temperature profiles of both.

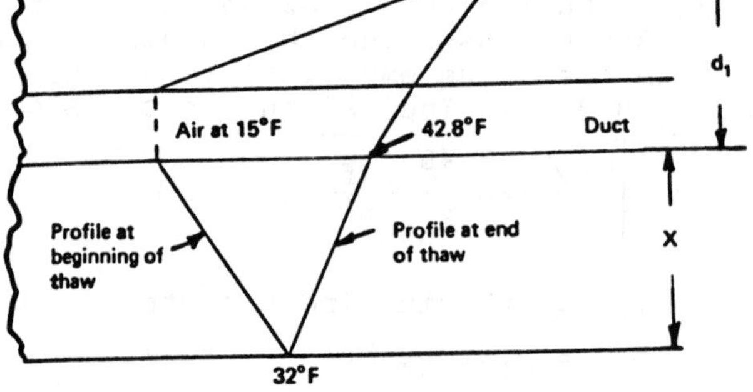

Fig. 2 **Temperature Profiles in Duct System.**

Heat added to the pad is the sum of sensible and latent heat.

$$Q_A = X(L_s + C\Delta T)$$

$$= 8.6\ [450 + 23.4(10.3)] = 5935\ \text{Btu/ft}^2$$

The ducts during the freezing season are opened and this amount of heat would be withdrawn from the pad to refreeze 8.6 ft of gravel.

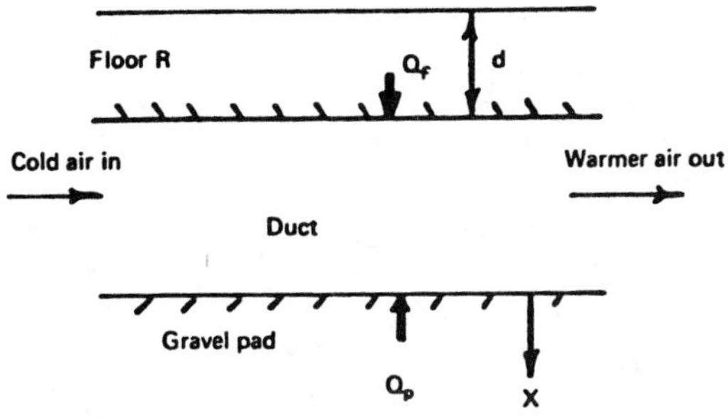

Fig. 3 **Heat Flows in Duct During Freeze Season.** Q_f **= heat flux rate into duct from the heated floor and** Q_p **= heat flux rate into duct from the pad.**

Figure 3 shows the heat transfer flow during the cooling season. The freezing season is 228 days and therefore the average rate of heat withdrawal is

$$Q_P = \frac{Q_A}{Fs}$$

$$= \frac{5935}{228(24)} = 1.09\ \frac{\text{Btu}}{\text{hr-ft}^2}$$

22P2 At the gravel-pad surface, the freezing index can be calculated from the relation

$$I_P = \frac{L_s\ X^2}{48\lambda^2 K_f} = 1602°\text{F-days}$$

where $K_f = 0.73$, $C_f = 21.8$, $T_0 = 36.7$

$$T_s = \frac{T_{st} + T_{sf}}{2} = \frac{40.5 + 3.9}{2} = 22°\text{F}$$

$\alpha = 0.4$, $\mu = 0.3$,, $\lambda = 0.77$

∴ the average surface temperature of the duct should be:

$$32 - T_{out} = \frac{1602}{228} = 7.0$$

$$T_{out} = 25°F$$

The mean temperature of air is 14.4°F, thus

$$Q_f = \frac{60 - 14.4}{R} = \frac{60 - 14.4}{12} = 3.80 \frac{Btu}{hr\text{-}ft^2}$$

∴ total heat-flow picked up by the air is

$$Q_T = Q_P + Q_f$$

$$= 1.09 + 3.80$$

$$= 4.89$$

Properties of air at 14.4°F are

$$\rho = 0.084 \text{ Btu/ft}^3$$

$$C_P = 0.24 \text{ Btu/lb}_m{}°F$$

Each duct covers 32 in. strip of floor,

$$Q_D + \left(\frac{32}{12}\right) (1)(220)(4.89) = 2867 \frac{Btu}{hr}$$

The cross-sectional area of the duct is 1.58 ft ^{2},

$$Q_d = \dot{m}C_p\Delta T = \rho AVC_p\Delta T$$

thus mean air velocity is

$$V = 71.1FPM = 1.19FPS$$

The design must now be checked to see if the stack height will deliver the required pressure differential. Assuming duct roughness ε=0.003, the equivalent length of duct bends can be found from reference tables as

$\frac{L_e'}{D}$	Right angle bend	Outlet	Inlet
	65	10	10

Therefore L_e'/D is 150 for each duct, which is the equivalent length of duct due to fittings and bends.

Assuming stack efficiency to be 80%,

$$L = 220 + 20 + 5 = 245 \text{ ft}$$

$$D_e = \frac{4A}{P} = 1.22 \text{ ft}$$

$$L_e' = 183 \text{ ft}$$

$$L_e = 428 \text{ ft}$$

Reynold number Re = 12020

$$\frac{\varepsilon}{D_e} = \frac{0.003}{1.22} = 0.0025$$

The friction factor can be found by using the relation

$$f = 0.0055 \left[1 + 20000 \frac{\varepsilon}{D_e} + \frac{10^6}{Re}^{1/3} \right]$$

then f = 0.035

Pressure drop in the duct is calculated by the following formula.

$$h_f = f \frac{L_e V^2 \rho_a}{D_e \, g_c \rho_w}$$

$$= 4.33 \times 10^{-3} \text{ in. } H_2O$$

For starting condition and for steady flow condition, the pressure rise for the stack may be calculated as follows:

For starting condition, the inside temperature is uniform such that the density $\rho_0 > \rho_i$

$$P_1 = P_0 + \rho_i h_i g - \rho_0 g (h_0 - h_i)$$

$$P_2 = P_0 + \rho_i g h_0$$

Pressure rise is

$$P_1 - P_2 = \frac{g}{g_c} (\rho_0 - \rho_i)(h_0 - h_i)$$

Starting pressure rise depends upon the inlet and outlet stacks (fig 4)

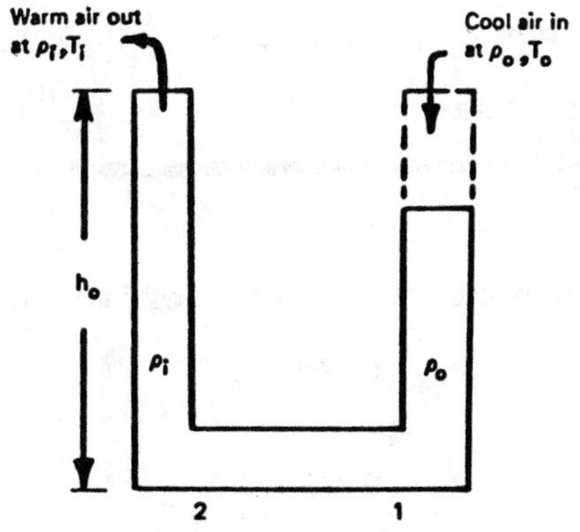

Fig. 4 **Pressure Rise in Stacks.**

For steady flow conditions, the pressure differential is given as follows:

$$P_1 - P_2 = \frac{g}{g} h\ (\rho_0 - \rho_i)$$

In this case the pressure rise does not depend upon the diference in the inlet and outlet stack heights.

$$h_D = H(\rho_0 - \rho_i) = H\frac{P}{R}\ \frac{1}{T_0} - \frac{1}{T_i}$$

$$H = 0.131\ h_D\ \frac{T_0}{1 - \frac{T_0}{T_i}}$$

but $h_d = h_f$

$T_0 = 463.9°R$

$T_i = 485°R$

$H = 6.05 \times 1.2 = 7.3ft$

A stack with a height of 8 ft would probably be suitable.

• PROBLEM 4-23

Hot air at 200°F is used to heat a layer of granitic type gravel consisting of 1 inch particles, which is initially at a temperature of 50°F. It is considered to be non-efficient if the exit air leaves at a temperature above 90°F, from an energy recovery standpoint. The air flow is 60 lb/hr ft² for 6 hours. If the bed has 45 percent voids, calculate the depth of the bed.

<u>Data</u>:	Specific heat of the gravel	= 0.25 Btu/lb°F
		= 41.3 Btu/ft³ °F
	Actual density of gravel	= 165 lb/ft³
	Specific heat of air	= 0.0191 Btu/ft³ °F

<u>Solution</u>: The heat transfer coefficient is given by

$$h = 0.79\ \frac{G}{D_e}^{0.7}$$

$$= 0.79\ \frac{60}{1/12}^{0.7}$$

$$= 79.4$$

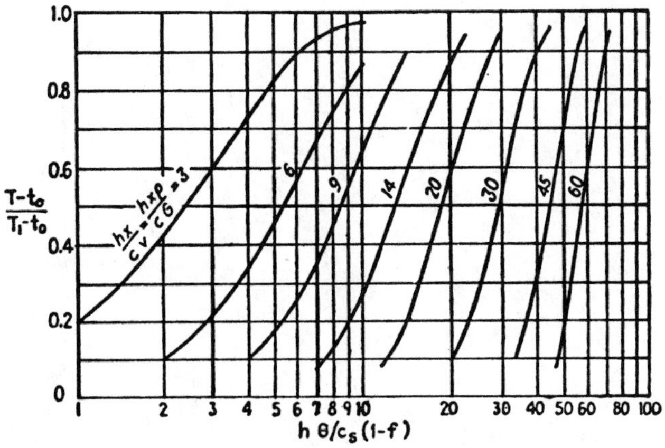

Fig. 1 Schumann curves.

In the correlation above, G is the mass flux and D_e is the diameter of the particles. The quantity (1 - f) is used to represent the ratio of the volume of the particles to the volume of the layer.

$$\frac{h\,\theta}{c_s(1 - f)} = \frac{79.4 \times 6}{41.3 \times 0.55}$$

$$= 21$$

The temperature difference ratio is given by

$$\frac{T - t_0}{T_1 - t_0} = \frac{90 - 50}{200 - 50}$$

$$= 0.267$$

Using this value in the figure shown, the value of

$$\frac{hx\rho}{cG} = 24.5$$

$$\rho = 0.0807 \text{ lb/ft}^3 \text{ of air}$$

$$x = 24.5 \times \frac{0.0191 \times 60}{79.4 \times 0.0807}$$

$$= 4.4 \text{ ft}$$

CHAPTER 5

UNSTEADY STATE HEAT CONDUCTION: NUMERICAL ANALYSIS AND 3-DIMENSIONAL PROBLEMS

Basic Attacks and Strategies for Solving Problems in this Chapter. **See pages 272 to 310 for step-by-step solutions to problems.**

The numerical solutions presented in this chapter are based on the "Schmidt plot" graphical method, which is illustrated in problem 5-1. The Schmidt numerical method is suitable for a flat wall or slab which initially has a known temperature distribution and then has each of its two faces held at a constant temperature. The two faces may be at the same temperature or they may be different. The initial temperature of the slab may be uniform or it may be nonuniform as long as it is known.

The Schmidt numerical method allows determination of the temperature at selected locations within the slab at different time intervals after the surface temperatures are fixed. The thickness of the solid is divided into equal space intervals with the thickness of each interval equal to x. *Then the appropriate time interval to use is determined from the equation:*

$$\Delta t = (\Delta x)^2 / \propto m \, .$$

In this equation, m *must be greater than one and is often set equal to 2.0 because this simplifies the calculation of temperatures. Once the space and time intervals are fixed, the temperature at any time and location node is calculated from the temperatures of adjacent nodes at the previous time interval. The use of this technique is illustrated in problems 5-2 to 5-5.*

Analytical solutions for temperature distribution as a function of time and position in a solid are possible for some 3-dimensional configurations. In each case the solution consists of solving the appropriate form of the 3-dimensional heat conduction differential equation. A suitable number of initial and boundary conditions are needed. The physical system typically is a solid 3-dimensional shape with a prescribed initial temperature distribution, which then has some or all of its surface held at some constant temperature. Problems 5-6 to 5-13 show solutions for several problems of this type. Knowledge of techniques for solution of partial differential equations is needed to handle this type of problem.

Step-by-Step Solutions to Problems in this Chapter, "Unsteady State Heat Conduction: Numerical Analysis and 3-Dimensional Problems"

SCHMIDT'S GRAPHICAL SOLUTIONS

• PROBLEM 5-1

A plane asbestos wall initially at a uniform temperature of 100^0F is required to have the temperature on one side to be below 300^0F, while the other side is to be maintained at 1500^0F, for a period of 1 hour. Determine the thickness of the wall.

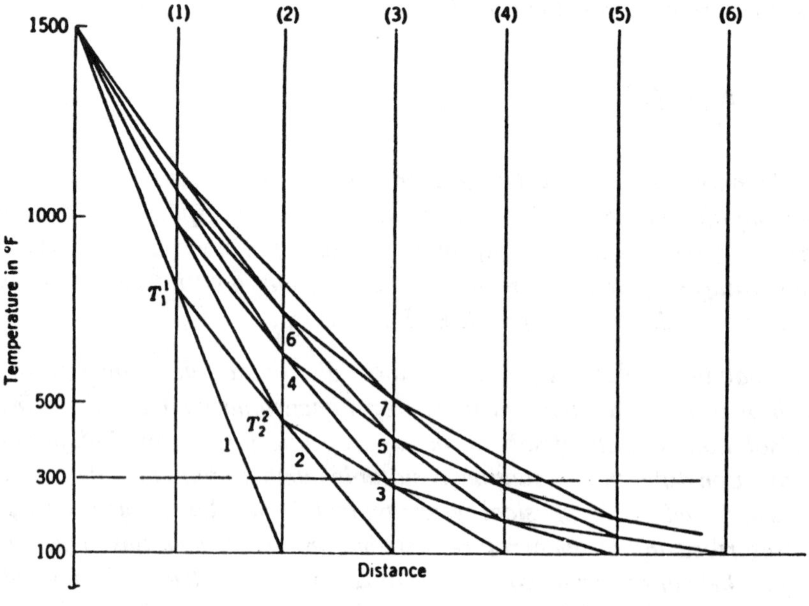

Fig. 1 Schmidt plot.

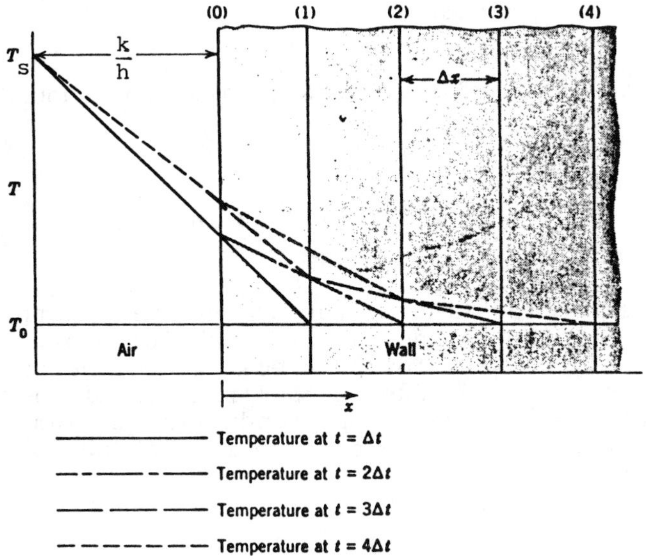

Fig.2 **Schmidt plot with surface resistance.**

<u>Solution</u>: Figure 1 is a temperature-thickness graph, with equal spatial intervals in the conducting medium. The procedure for drawing the graph is as follows: The temperature of plane (1), T_1^1, after the first time period is found by joining points, 1500°F on temperature reference plane (0) and 100°F on the temperature reference plane (2). This time period is designated by (1). For the second time period (2), the temperature T_2^2 is determined by joining the temperature points on the two adjacent temperature references. In the third time period (3), the temperature at the reference planes (1) and (2) are raised. This procedure can be repeated for n Δt time periods. For each time period, the temperature profile across the conducting medium is represented graphically by construction. The temperature 300°F, as determined from the graph is at a position 4.3ΔL from the heated surface, after the 8th time period.

The total time is 1 hour, therefore $\Delta t = 1/8$ hr. The thermal diffisivity of asbestos α is 0.01 ft^2/hr. The equation for maximum time or spatial increment, while maintaining stability is

$$\Delta t = \frac{(\Delta L)^2}{2\alpha}$$

substituting the numerical values into this equation yields,

$$(\Delta L)^2 = 2\ \alpha\ (\Delta t)$$

$$= 2\ (0.01)\ (1/8)$$

$$= 0.0025\ ft^2$$

or $\Delta L = 0.05$ ft

The minimum required thickness of asbestos should be at least

$$4.3\Delta L = 4.3\ (0.05)$$

$$= 0.215 \text{ ft or } 2.58 \text{ in}$$

This procedure for finding the temperature distribution in a conducting medium is known as the Schmidt plot. Schmidt's plot can also be used for problems where the surface temperature is not constant due to the transfer of energy by convection at the surface of the conducting medium. The rate of energy transfer at the surface for conduction and convection is:

$$(qx/A)_{surface} = k \left.\frac{\partial T}{\partial x}\right|_{surface} = h\ (Ts - To)$$

At the reference plane (0) the temperature gradient is

$$-\left.\frac{\partial T}{\partial x}\right|_{o} = \frac{(T_x - To)}{k/h} = \frac{Ts - To}{\Delta x^*}$$

where Δx^* is the fictitious wall thickness. This fictitious distance (k/h), is added to the actual wall thickness, because of the resistance at the surface to convective heat transfer and if h the surface conduction varies with time, the value of (k/h) can be corrected after each time period Δt. This is shown in figure 2.

SCHMIDT'S NUMERICAL SOLUTIONS

• **PROBLEM 5-2**

Find the temperatures 46.8 minutes later at different locations in a 9 in. thick wall as shown in the figure below. The variation of temperatures with time in the wall at different locations is given in the table. The thermal diffusivity of the wall is 0.02 ft^2/hr.

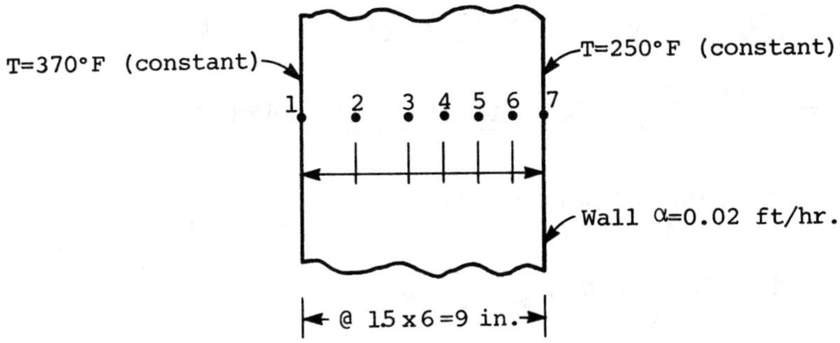

Fig. 1

Table 1: Temperature distribution in the wall with time.

Station	Time in Hr.				
	τ	τ + 0.195	τ + 0.390	τ + 0.585	τ + 0.780
1	370	370	370	370	370
2	435	430	425	420	415
3	480	470	461	452	444
4	485	473	462	452	442
5	440	431	422	413	404
6	360	352	346	341	336
7	250	250	250	250	250

Solution: Using the numerical analysis for temperature distribution in a solid , the value of M is defined as

$$M = \frac{(\Delta x)^2}{\alpha\ \Delta\tau}$$

$$\text{or } \Delta\tau = \frac{(\Delta x)^2}{\alpha M} = \frac{(1.5/12)^2}{0.02 \times 4}$$

($\because$ value of 4 is assumed for M)

Again,

$$t_2{}^1 = \frac{t_1 + t_3 + (M-2)t_2}{M}$$

$$\text{or} \quad t_2{}^1 = \frac{t_1 + t_3 + 2t_2}{4}$$

Now, at time τ the temperature distribution is given (see first column in the table) and the temperature at the outer surfaces is given as constant. At time ($\tau+\Delta\tau$), the temperature distribution is computed as follows

$$t_2{}^1 = \frac{t_1+t_3+2t_2}{4} = \frac{370+480+2(435)}{4} = 430^{\circ}F$$

$$t_3^{1} = \frac{t_2+t_4+2t_3}{4} = \frac{435+485+2(480)}{4} = 470^{\circ}F$$

$$t_4^{1} = \frac{t_3+t_5+2t_4}{4} = \frac{480+440+2(485)}{4} = 473^{\circ}F$$

$$t_5^{1} = \frac{485+360+2(440)}{4} = 431^{\circ}F$$

$$t_6^{1} = \frac{440+250+2(360)}{4} = 352^{\circ}F$$

The procedure requiring 4 $\Delta\tau$ intervals is shown in the figure. For each successive time interval the distribution is computed by allowing the present distribution to be the primed one, and the previously computed prime distribution becomes the unprimed one.

• PROBLEM 5-3

Compute the temperature at the center of a 12 in. thick slab 15 min. after the slab is put in a constant temperature medium at 100°F using

I the Schmidt tabular method based on m=2

II the initial temperature as $\frac{t_a+t_s}{2}=400^{\circ}F$

III the analytical method.

Assume the initial uniform temperature of the slab is 700°F. The properties of the slab material are: $c_p=0.13$ Btu/lb.°F, $\rho=490$ lb./ft.^{3}, $k=25$ Btu/hr. - ft.-°F.

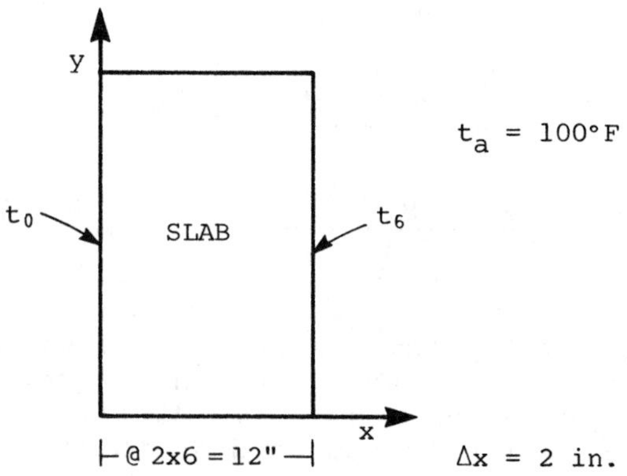

Fig. 1

Solution: (I) In Schmidt tabular method the space interval is assumed as 2 in. i.e. $\Delta x = 2$ in. and $m=2$. Using the relation

$$t_i^1 = \frac{t(i-1) + t(i+1)}{2}$$ the table of temperature distribution with time is written down

$$\Delta\tau = \frac{(\Delta x)^2}{\alpha m} = \frac{(1/6)^2}{0.392(2)} = 0.0354 \text{ hr.}$$

$$(\because \alpha = \frac{k}{\rho Cp} = \frac{25}{(490)(0.13)} = 0.392 \text{ ft}^2/\text{hr})$$

Total number of time steps required is $\frac{15}{60 \text{x} 0.0354} \simeq 7 = N\Delta\tau$

For $\Delta\tau = 1$

$$t_1^1 = \frac{t_0 + t_2}{2} = \frac{100 + 700}{2} = 400^\circ F$$

$$t_2^1 = \frac{700 + 700}{2} = 700^\circ F$$

$$t_3^1 = \frac{700 + 700}{2} = 700$$

As the slab is symmetric about the midplane $t_o = t_6$, $t_1 = t_5$, $t_2 = t_4$.

For $\Delta\tau = 2$

$$t_1^1 = \frac{100 + 700}{2} = 400^\circ F$$

$$t_2^1 = \frac{400 + 700}{2} = 550^\circ F$$

$$t_3^1 = \frac{700 + 700}{2} = 700^\circ F$$

The procedure is continued until $\Delta\tau = 7$. The temperature distribution so obtained is given in table 1.

Table 1:

Δt	t_0	t_1	t_2	t_3	t_4	t_5	t_6
0	100	700	700	700	700	700	100
1	100	400	700	700	700	400	100
2	100	400	550	700	550	400	100
3	100	325	550	550	550	325	100
4	100	325	438	550	438	325	100
5	100	269	438	438	438	269	100
6	100	269	354	438	354	269	100
7	100	227	354	354	354	227	100

The table shows discrete temperature distribution in the slab. The temperature changes with every increment of time. This is an idealization of the method, and does not necessarily occur in actual practice. In actual practice the variation of temperature continues with continuously changing time.

(II) With initial surface temperature of $\frac{t_a + t_s}{2} = 400^{\circ}F$, m=2 and $t_0 = 100^{\circ}F$ and employing the similar procedure as in (1), the table 2 is obtained.

For $\Delta\tau = 1$

$$t_1^{\,1} = \frac{400 + 700}{2} = 550^{\circ}F$$

$$t_2^{\,1} = \frac{700 + 700}{2} = 700^{\circ}F$$

$$t_3^{\,1} = \frac{700 + 700}{2} = 700^{\circ}F$$

For $\Delta\tau = 2$

$$t_1^{\,1} = \frac{100 + 700}{2} = 400^{\circ}F$$ and so on.

Table 2:

$\Delta\theta$	t_0	t_1	t_2	t_3	t_4	t_5	t_6
0	400	700	700	700	700	700	400
1	100	550	700	700	700	550	100
2	100	400	625	700	625	400	100
3	100	362	550	625	550	362	100
4	100	325	493	550	493	325	100
5	100	296	437	493	437	296	100
6	100	268	395	437	395	268	100
7	100	248	352	395	352	248	100

(III) The analytical solution is obtained by the rapidly converging infinite series

$$\frac{t_a - t}{t_a - t_b} = \frac{4}{\pi}\left(e^{-\alpha_1 x}\sin\frac{\pi x}{2r_m} + \frac{1}{3}\,e^{-9\alpha_1 x}\sin\frac{3\pi x}{2r_m} + \frac{1}{5}\,e^{-25\alpha_1 x}\sin\frac{5\pi x}{2r_m} + \ldots\right)$$

where α_1, equals $(\pi/2)^2$, X represents the dimensionless ratio $\alpha\tau/r_m^{\,2}$

$t = t_3$, $x_3 = r_m$, $t = t_a$ at $x = 0$ and
$x = 2r_m$ and $t = t_b$ at $\tau = 0$

$$\alpha_1 = \left(\frac{\pi}{2}\right)^2, \quad x = \frac{\alpha\tau}{r_m^2} = \frac{(0.392)(\frac{1}{4})}{(0.5)^2} = 0.392$$

$$\alpha_1 x = \left(\frac{\pi}{2}\right)^2 0.392 = 0.967$$

Substituting the values in the equation yields

$$\frac{t_3 - 100}{700 - 100} = \frac{4}{\pi}\left(e^{-0.967} \sin\frac{\pi}{2} + \frac{1}{3} e^{-9x0.967} \sin\frac{3\pi}{2} + \frac{1}{5} e^{-25x0.967} \sin\frac{5\pi}{2} + \ldots\right)$$

gives $\frac{t_3 - 100}{700 - 100} = \left(\frac{4}{\pi}\right) 0.384 = 0.484$

(Neglecting the higher order terms)

$$t_3 = 390.4^\circ F$$

It is found that the temperature at the center of the slab calculated by three methods are different. The difference in I and III is more than between II and III. This may be due to improper selection of initial temperature in method I, beside the inherent method's imperfection.

• PROBLEM 5-4

Compute the temperature distribution across a large 12 in. thick slab 36 minutes after one of its faces is exposed to an ambient temperature of 0°F. The other face of the slab is insulated. The coefficient of convective heat transfer for the exposed face is infinity. Assume the initial uniform temperature of the slab as 170°F. Use the Schmidt numerical method with M=2.0.

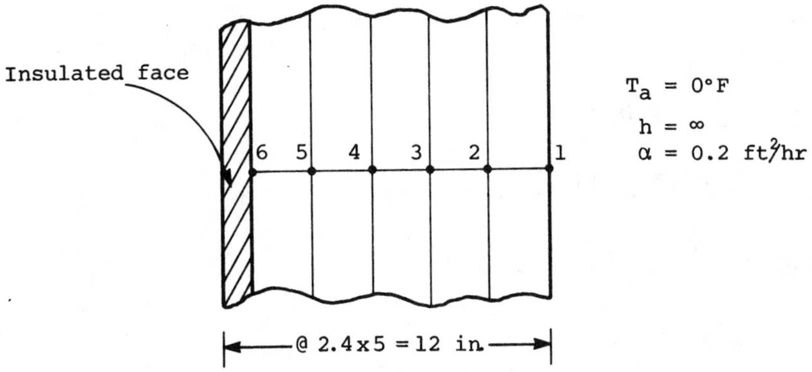

Solution: Using the numerical analysis for temperature distribution in a solid, the term M is defined as

$$M = \frac{(\Delta x)^2}{\alpha \Delta t}$$

where Δx is the linear increment in the direction of the temperature distribution desired.

Δt - time increment
α - thermal diffusivity of the material.

$$M = 2, \ \Delta x = 2.4 \text{ in. or } \frac{2.4}{12} \text{ ft.}$$

$$\alpha = 0.2 \text{ ft}^2/\text{hr}$$

$$\therefore \ \Delta t = \frac{(2.4/12)^2}{0.2x2} = 0.1 \text{ hr.}$$

Thus, the total number of time increment to be used are

$$\frac{t}{\Delta t} = \frac{36/60}{0.1} = 6$$

The temperature distribution in the slab is computed using the equation

$$t_i^1 = \frac{t_{i-1} + t_{i+1}}{2}$$

where i - is the slab segment number as identified in the figure.

i = 1 to 5

For exposed face the value of i = 1.

Now, for the first time increment

$$t_1^1 = \frac{0 + 170}{2} = 85^\circ F$$

$$t_2^1 = \frac{85 + 170}{2} = 127.5^\circ F$$

$$t_3^1 = t_4^1 = t_5^1 = 170^\circ F = t_6^1$$

For 2nd time increment i.e. $2\Delta t = 2x0.1 = 0.2$ hrs.

$$t_1^1 = 0^\circ F$$

$$t_2^1 = \frac{0 + 170}{2} = 85^\circ F$$

$$t_3^1 = \frac{127.5 + 170}{2} = 148.75^\circ F$$

$$t_4^1 = \frac{170 + 170}{2} = 170^\circ F$$

$$t_4^1 = t_5^1 = t_6^1$$

For the 3rd time increment i.e. $3\Delta t = 0.1x3 = 0.3$ hr.

$$t_1{}^1 = 0$$

$$t_2{}^1 = \frac{0 + 148.75}{2} = 74.37^{\circ}F$$

$$t_3{}^1 = \frac{85 + 170}{2} = 127.5^{\circ}F$$

$$t_4{}^1 = \frac{148.75 + 170}{2} = 159.37^{\circ}F$$

$$t_5{}^1 = \frac{170 + 170}{2} = 170^{\circ}F$$

$$t_5{}^1 = t_6{}^1$$

For the 4th time increment i.e. $4\Delta t = 0.1x4 = 0.4$ hr.

$$t_1{}^1 = 0$$

$$t_2{}^1 = \frac{0 + 127.5}{2} = 63.75^{\circ}F$$

$$t_3{}^1 = \frac{74.37 + 159.37}{2} = 116.87^{\circ}F$$

$$t_4{}^1 = \frac{127.5 + 170}{2} = 148.75^{\circ}F$$

$$t_5{}^1 = \frac{159.37 + 170}{2} = 164.68^{\circ}F$$

$$t_6{}^1 = 170^{\circ}F.$$

For the 5th increment of time i.e. $5\Delta t = 0.1x5 = 0.5$ hr.

$$t_1{}^1 = 0$$

$$t_2{}^1 = \frac{0 + 116.87}{2} = 58.44^{\circ}F$$

$$t_3{}^1 = \frac{63.75 + 148.75}{2} = 106.25^{\circ}F$$

$$t_4{}^1 = \frac{116.87 + 164.68}{2} = 140.77^{\circ}F$$

$$t_5{}^1 = \frac{148.75 + 170}{2} = 159.38^{\circ}F$$

$$t_6{}^1 = 164.68^{\circ}F$$

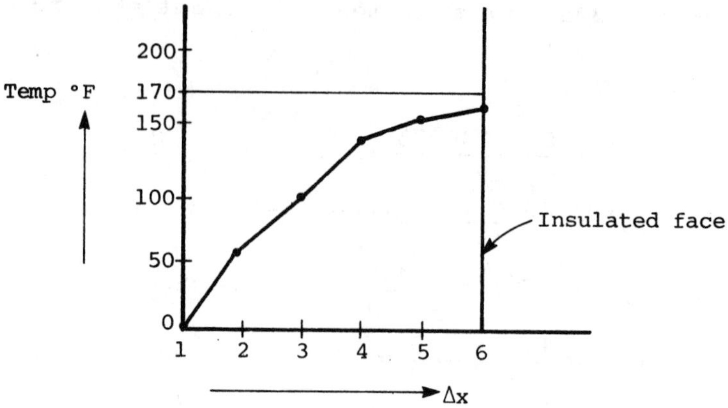

Temperature distribution at 0.6 hr.

For 6th increment of time i.e. $6\Delta t = 0.1x6 = 0.6$ hr.

$$t_1{}^1 = 0$$

$$t_2{}^1 = \frac{0 + 106.25}{2} = 53.13^{\circ}F$$

$$t_3{}^1 = \frac{58.44 + 140.77}{2} = 99.60^{\circ}F$$

$$t_4{}^1 = \frac{106.25 + 159.38}{2} = 132.81^{\circ}F$$

$$t_5{}^1 = \frac{140.77 + 164.68}{2} = 152.72^{\circ}F$$

$$t_6{}^1 = 159.38^{\circ}F$$

• PROBLEM 5-5

Compute the time-temperature distribution across the thickness of a wall 10 in. thick for the first 60 mins. The temperature of the side wall is suddenly raised and maintained at 500°F. The initial uniform temperature being 100°F, the properties of the wall material $k = 1.0$ Btu/hr - ft - °F, ρ = 144 lb/ft³, Cp = 0.2 Btu/lbm - °F. Use numerical method.

Solution: The wall is split into ten regions, each 1 in. thick. As the wall is symmetrical about the mid plate only half thickness can be analyzed. The properties are lumped at the center of each region and represented in an equivalent electrical circuit as shown in the figure 1.

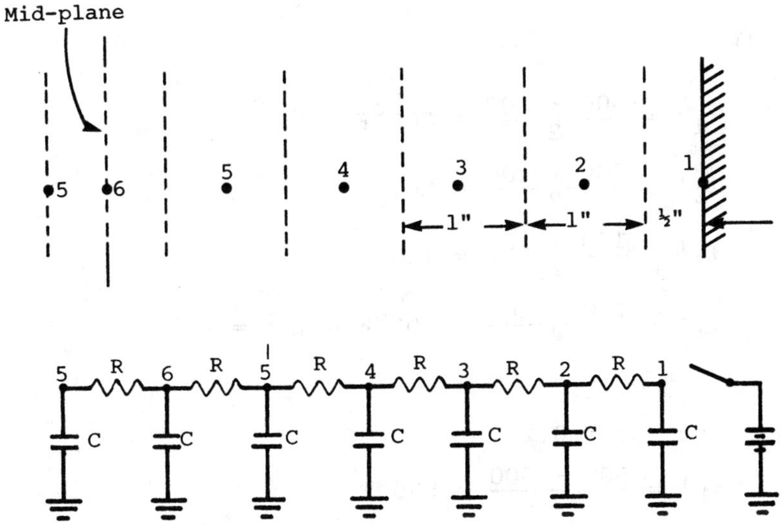

Fig. 1 The unsteady state circuit for plane wall.

The circuit stability requirement is given by

$$\delta\tau \leq \tfrac{1}{2}RC$$

$$\leq \frac{1}{2}\,\frac{\rho c_p}{k}(\delta x)^2$$

$$\delta\tau \leq \frac{1}{2}\,\frac{(\delta x)^2}{\alpha}$$

for the exterior nodes, where $\delta\tau$ is the time increment and δx is the distance between nodes.

The maximum permissible time increment is defined as

$$\delta\tau \leq \left(\sum_j \frac{1}{R_{ij}C_i}\right)^{-1} = \frac{RC}{2}$$

$$\left(\because \sum_j \frac{1}{R_{ij}C_i} = \frac{2}{RC}\right)$$

$$RC = \frac{(\delta x)^2}{\alpha} = \frac{(1/12)^2}{0.0347}$$

$$= 0.2 \text{ hr. (12 min.)}$$

$$\therefore\ \delta\tau \leq \frac{RC}{2} = \frac{0.2}{2} = 0.1 \text{ hr. (6 min.)}$$

Corresponding to this time increment, the nodal equations for temperature are

$$t_i{}^1 = \frac{t_{i-1} + t_{i+1}}{2}$$

Now at $\Delta\tau = 0.1$

$$t_1{}^1 = \frac{500 + 500}{2} = 500^{\circ}F$$

$$t_2{}^1 = \frac{300 + 100}{2} = 200^{\circ}F$$

$$t_3{}^1 = \frac{100 + 100}{2} = 100^{\circ}F$$

$$t_4{}^1 = \frac{100 + 100}{2} = 100^{\circ}F = t_5{}^1 = t_6{}^1$$

For $\Delta\tau = 0.2$

$$t_1{}^1 = \frac{500 + 500}{2} = 500^{\circ}F$$

$$t_2{}^1 = \frac{500 + 100}{2} = 300^{\circ}F$$

$$t_3{}^1 = \frac{200 + 100}{2} = 150^{\circ}F$$

$$t_4{}^1 = \frac{100 + 100}{2} = 100^{\circ}F = t_5{}^1 = t_6{}^1$$ and so on.

The procedure is repeated until the 1 hr. time period and the results can be tabulated as given in the table below.

Table 1: Numerical calculations for large wall.

Time. hr \ Node	1	2	3	4	5	6
0.0	300	100.0	100.0	100.0	100.0	100.0
0.1	500	200.0	100.0	100.0	100.0	100.0
0.2	500	300.0	150.0	100.0	100.0	100.0
0.3	500	325.0	200.0	125.0	100.0	100.0
0.4	500	350.0	225.0	150.0	112.5	100.0
0.5	500	362.5	250.0	168.8	125.0	112.5
0.6	500	375.0	265.7	188.5	140.7	125.0
0.7	500	382.9	281.8	203.2	156.8	140.7
0.8	500	390.0	293.1	219.3	172.0	156.8
0.9	500	396.6	305.1	232.6	188.1	172.0
1.0	500	402.6	314.6	246.6	202.3	188.1

The results of numerical calculations for time periods 0.2 hr. and 0.5 hr. are compared with the exact solution in figure 2 and found that for 1 hr. period the exact and numerical solution results are very close. The temperature variation with the time increment is graphically presented in figure 3. It is observed that the smaller the increment, the closer are the results of numerical solution to that of exact solution.

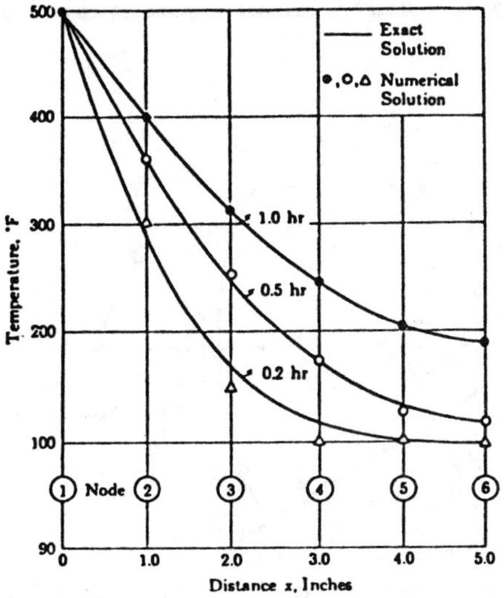

Fig. 2 Comparison of analytical and numerical results for a plane wall, showing the temperature distribution at various times when using the maximun time increment.

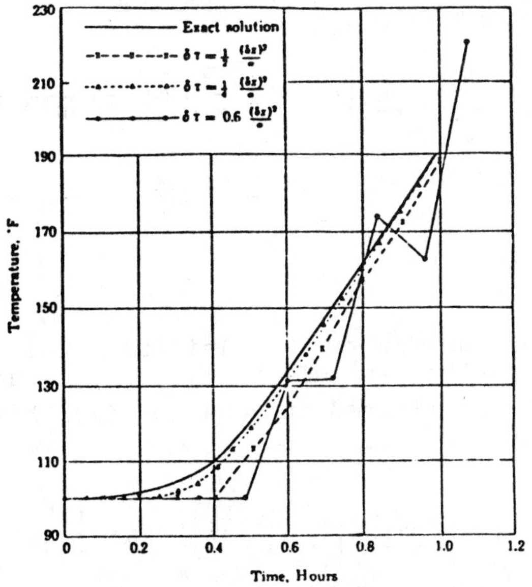

Fig. 3 Comparison of analytical and numerical results for a plane slab, showing the temperature-time history of the centerline for different time increments.

3-DIMENSIONAL PROBLEMS

• PROBLEM 5-6

Find the temperature distribution T (r, μ, ϕ, t) in a cone $\theta=\theta_0$ cut from a sphere as shown in the figure. The initial temperature (at t=0) of the cone is $T=F(r, \mu, \phi)$ and at any time $(t > 0)$ the surface temperature is $T = 0$.

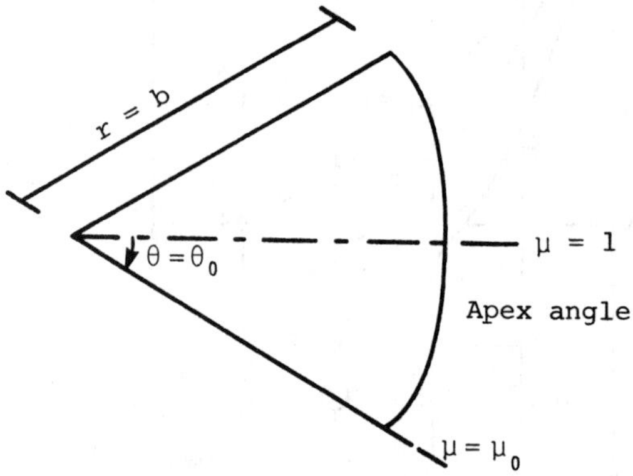

Fig 1. A cone cut out from a sphere of radius b.

Solution: The heat conduction equation for unsteady state in spherical coordinate system is

$$\frac{\partial^2 T}{\partial r^2} + \frac{2}{r}\frac{\partial T}{\partial r} + \frac{1}{r^2} + \frac{\partial}{\partial \mu}\left[(1-\mu^2)\frac{\partial T}{\partial \mu}\right] + \frac{1}{r^2(1-\mu^2)}\frac{\partial^2 T}{\partial \phi^2} = \frac{1}{\alpha}\frac{\partial T}{\partial t} \quad (1)$$

Boundary conditions for $0 \le t \le \infty$ are

$0 \le r < b, \ \mu_o < \mu \le 1$

$0 \le \phi \le 2\pi$

$T=0$ at $r=b$ and $\mu=\mu o$ for $t > 0$. Initial condition at $t=0$ is $T=F(r, \mu, \phi)$. Let a variable $V(r, \mu, \phi, t)$ defined as $V = \sqrt{r}\ T$ be substituted in equation (1). Then it is transformed to

$$\frac{\partial^2 V}{\partial r^2} + \frac{1}{r}\frac{\partial V}{\partial r} - \frac{1}{4}\frac{V}{r^2} + \frac{1}{r^2}\frac{\partial}{\partial \mu}\left[(1-\mu^2)\frac{\partial V}{\partial \mu}\right] + \frac{1}{r^2(1-\mu^2)}\frac{\partial^2 V}{\partial \phi^2} =$$

$$\frac{1}{\alpha}\frac{\partial V}{\partial t} \quad (2)$$

The corresponding boundary conditions are transformed to

for $0 \leq t \leq \infty$

$0 \leq r < b$, $\mu_o < \mu \leq 1$,

$0 \leq \phi \leq 2\pi$

$V=0$ at $r=b$ and $\mu=\mu_o$ at $t > 0$ and $V = \sqrt{r}\, F(r,\mu,\phi)$ at $t=0$.

Based on the physical situation of the problem, the primary solutions of equation (2) can be expressed as

$$e^{-\alpha\lambda^2 t}, \quad J_{n+1/2}(\lambda r), P_n^{-m}(\mu), \quad (A \cos m\phi + B \sin m\phi)$$

The expression $P_n^{-m}(\mu)$ is the generalized Legendre function of the first kind and is defined as

$$P_n^{-m}(\mu) = \frac{1}{\Gamma(m+1)}\left(\frac{1-\mu}{1+\mu}\right)^{m/2} F\left(-n,\ n+1,\ 1+m,\ \frac{1-\mu}{2}\right)$$

$$= \frac{1}{\Gamma(m+1)}\left(\frac{1-\mu}{1+\mu}\right)^{m/2}\left[1+\sum_{r=1}^{\infty}\frac{(-n)_r\ (n+1)_r}{r!(1+m)_r}\left(\frac{1-\mu}{2}\right)^r\right]$$

where m is a positive integer (m=0, 1, 2, 3 ....) and n's assumed values greater than $-\frac{1}{2}$. The positive value of m is consistent with the requirement that T(or V) is periodic with period 2π in the range of $0 \leq \phi \leq 2\pi$.

Now, the complete solution of the equation (2) can be expressed as

$$V(r, \mu, \phi, t) = \sum_n \sum_p \sum_{m=0}^{\infty} e^{-\alpha\lambda^2_{np} t} J_{n+\frac{1}{2}}(\lambda_{np} r) P_n^{-m}(\mu) \qquad (3)$$

$$\cdot(A_{mnp}\cos m\phi + B_{mnp}\sin m\phi)$$

If λ_{np}'s are the positive roots of $J_{(n+\frac{1}{2})}(\lambda_{np} b) = 0$ then the above solution satisfies the boundary condition V=0 at r = b.

Further, if the roots n are greater than $-\frac{1}{2}$ of equation

$$P_n^{-m}(\mu_0) = 0 \qquad \text{then}$$

the boundary condition V=0 at $\mu=\mu_0$ is satisfied.

The coefficients A_{mnp} and B_{mnp} can be obtained by applying the initial condition i.e. at t=0, $V=r^{\frac{1}{2}}F(r,\mu,\phi)$ in equation (3), yields

$$r^{\frac{1}{2}}F(r,\mu,\phi) = \sum_n \sum_p \sum_{m=0}^{\infty} J_{n+\frac{1}{2}}(\lambda_{np} r) P_n^{-m}(\mu)(A_{mnp}\cos m\phi + B_{mnp}\sin m\phi)$$

$$\text{in } \mu_0 < \mu \leq 1, 0 \leq \phi \leq 2\pi,\ 0 \leq r \leq b \qquad (4)$$

Multiplying both sides of equation (4) by

$$\int_0^b rJ_{n+\frac{1}{2}}(\lambda_{np}r)dr$$

and applying the orthogonality condition equation

$$\int_0^b rR_v(\beta_m,r)R_v(\beta_n,r)dr = \begin{cases} 0 & \text{for } m \neq n \\ N(\beta_m) & \text{for } m = n \end{cases}$$

where $R_v(\beta_m,r)$ are orthogonal in the interval $0 \leq r < b$

with respect to the weight function w=r, results in

$$f(\mu, \phi)=\sum_n \sum_{m=0}^{\infty} P_n^{-m}(\mu) \left[A_{mnp}\cos m\phi+ B_{mnp}\sin m\phi\right] N(\lambda_{np}) \quad (5)$$

$$\text{in } \mu_0 < \mu \leq 1,\ 0 \leq \phi \leq 2\pi$$

where

$$f(\mu, \phi) \equiv \int_0^b r^{3/2}F(r,\mu,\phi)J_{n+\frac{1}{2}}(\lambda_{np}r)dr$$

$$N(\lambda_{np}) \equiv \int_0^b rJ^2_{n+\frac{1}{2}}(\lambda_{np}r)dr = \tfrac{1}{2}b^2\left[J'_{n+\frac{1}{2}}(\lambda_{np}b)\right]^2$$

Again, applying the equation

$$J_{(n+\frac{1}{2})}(\lambda_{np}b) = 0 \quad \text{and}$$

$$\int_0^b rG_v^2(\beta r)dr=\tfrac{1}{2}r^2\left[G'^2_v(\beta r) + \left(1 - \frac{v^2}{\beta^2 r^2}\right)G^2_v(\beta r)\right] \quad \text{the}$$

norm $N(\lambda_{np})$ can be obtained from equation (5)

$$N(\lambda_{np}) \left[A_{mnp}\cos m\phi + B_{mnp}\sin m\phi\right] \equiv \frac{1}{\pi N(m,n)} \cdot \int_{\phi'=0}^{2\pi} \int_{\mu'=\mu_0}^{1} f(\mu',\phi')P_n^{-m}(\mu') \cdot \cos m(\phi-\phi')d\mu'd\phi'$$

where π should be replaced by 2π for m = 0, and the norm N(m,n) subject to the condition $P_n^{-m}(\mu_0) = 0$ is obtained from the equation

$$N(m,n) = -\frac{1 - \mu_0^2}{2n + 1}\left[\frac{d}{dn} P_n^{-m}(\mu)\frac{d}{d\mu}P_n^{-m}(\mu)\right]_{\mu = \mu_0}$$

as $$N(m,n) = -\frac{1-\mu_0^2}{2n+1}\frac{d}{dn}P_n^{-m}(\mu_0)\frac{d}{d\mu_0}P_n^{-m}(\mu_0) \tag{6}$$

Substituting for the coefficients from equation (6) into the equation (2) and transforming the equation into the variable temperature $T(r,\mu,\phi,t)$, yields

$$T(r,\mu,\phi,t) = \frac{1}{\pi}\sum_n\sum_p\sum_{m=0}^{\infty}\frac{e^{-\alpha\lambda_{np}^2 t}}{N(m,n)N(\lambda_{np})} r^{-\frac{1}{2}}J_{n+\frac{1}{2}}(\lambda_{np}r)P_n^{-m}(\mu)$$

$$\cdot\int_{r'=0}^{b}\int_{\mu'=\mu_0}^{1}\int_{\phi'=0}^{2\pi} r'^{3/2}J_{n+\frac{1}{2}}(\lambda_{np}r')P_n^{-m}(\mu')\cos m(\phi-\phi')$$

$$\cdot F(r',\mu',\phi')d\phi'\ d\mu'\ dr'$$

where π should be replaced by 2π for m=0, the eigenvalues λ_{np}'s are the positive roots of

$$J_{(n+\frac{1}{2})}(\lambda_{np}b) = 0$$

and n's are roots, greater than $-\frac{1}{2}$ of the equation

$$P_n^{-m}(\mu_0) = 0$$

• PROBLEM 5-7

Determine the steady state temperature of a finite solid rod that has one half of its circumferential surface maintained at a constant temperature θ_0 while the remainder of the surface area, including the ends, is at 0°. The radius of the rod is R and the length is l.

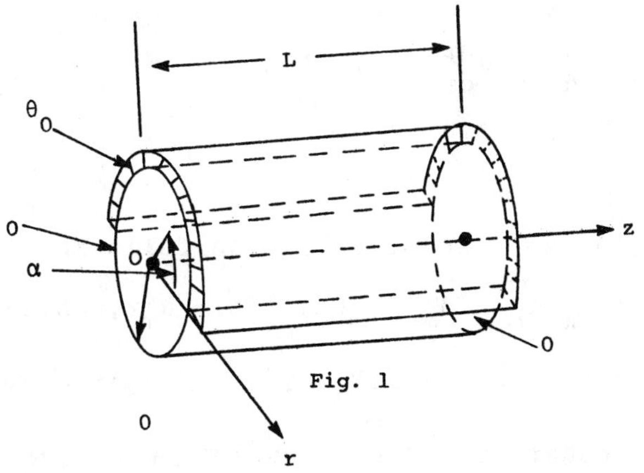

Fig. 1

Solution: The problem involves a 3 dimensional solution. The differential equation and the boundary conditions describing the problem are

$$\frac{1}{r}\frac{\partial}{\partial r}\left(\frac{r\partial\theta}{\partial r}\right) + \frac{1}{r^2}\frac{\partial^2\theta}{\partial\phi^2} + \frac{\partial^2\theta}{\partial z^2} = 0$$

Boundary conditions:

$$\left.\begin{aligned} \theta(r,\phi,0) &= 0 \\ \theta(r,\phi,1) &= 0 \end{aligned}\right\} \text{(i)}$$

$$\left.\begin{aligned} \theta(r,\phi,z) &= \theta(r,\phi+2\pi,z) \\ \frac{\partial\theta}{r\partial\phi}(r,\phi,z) &= \frac{\partial\theta(r,\phi+2\pi,z)}{r\partial\phi} \end{aligned}\right\} \text{(ii)}$$

(iii)

$$\left.\begin{aligned} \theta(0,\phi,z) &= \text{finite} \\ \theta = (R,\phi,z) &= \begin{cases} \theta_0 & \text{for } 0<\phi<\pi \\ 0 & \text{for } \pi<\phi<2\pi \end{cases} \end{aligned}\right\} \text{(iii)}$$

Employing the product solution $\theta(r,\phi,z)=P(r)\Phi(\phi)Z(z)$, the problem can be divided as follows

$$\left.\begin{aligned} &\frac{d^2Z}{dz^2} + \alpha^2 Z = 0 \\ &Z(0),\ Z(1) = 0 \end{aligned}\right\} \quad (1)$$

$$\left.\begin{aligned} &\frac{d^2\Phi}{d\phi^2} + \beta^2\Phi = 0 \\ &\Phi(\phi) = \Phi(\phi+2\pi) \end{aligned}\right\} \quad (2)$$

$$\left.\begin{aligned} &r\frac{d}{dr}\left(r\frac{dP}{dr}\right) - (\alpha^2 r^2 + \beta^2)P = 0 \\ &P(0) = \text{finite} \end{aligned}\right\} \quad (3)$$

Equation (1) has a solution of the following form

$Z_m(z) = A_m \phi_m(z)$, $\phi_m(z) = \sin \alpha_m z$, characteristic functions

and $\alpha_m l = m\pi$, $m = 1,2,3,4 \ldots$ characteristic values

Similarly, equation (2) has a solution of the following form

$$\Phi_\beta(\phi) = B_\beta \cos \beta\phi + C_\beta \sin \beta\phi$$

$\cos \beta\phi$, $\sin\beta\phi$ characteristic functions

and $\beta = 0, 1, 2, 3, 4 \ldots$ characteristic values

Finally, the solution of equation (3) is obtained

$$P_{m\beta}(r) = D_{m\beta} I_\beta(\alpha_m r)$$

The product solution may now be obtained and is given below

$$\theta(r,\phi,z) = \sum_{m=1}^{\infty} a_{mo} I_o(\alpha_m r) \sin \alpha_m z + \sum_{m=1}^{\infty} \sum_{\beta=1}^{\infty} (a_{m\beta} \cos\beta\phi + b_{m\beta} \sin\beta\phi) I_\beta(\alpha_m r) \sin\alpha_m z \qquad (4)$$

where,

$$a_{mo} = A_m B_o D_{mo},\ a_{m\beta} = A_m B_\beta D_{m\beta} \text{ and } b_{m\beta} = A_m C_\beta D_{m\beta}.$$

The use of the nonseparable boundary condition gives the following expression

$$\theta_o = \sum_{m=1}^{\infty} a_{mo} I_o(\alpha_m R) \sin\alpha_m z + \sum_{m=1}^{\infty} \sum_{\beta=1}^{\infty} (a_{m\beta} \cos\beta\phi + b_{m\beta} \sin\beta\phi) I_\beta (\alpha_m R) \sin\alpha_m z \qquad (5)$$

It is observed that equation (5) is complete in ϕ but not in z as it carries only the sine terms in z. This is a characteristic of a double Fourier series. Hence, we may say that equation (5) is a double Fourier series representation of θ_o. The coefficients of this series may be determined by taking equation (5) to be a simple Fourier series in ϕ and setting $\theta_o = 0$ for the interval $\pi<\phi<2\pi$. This gives,

$$\sum_{m=1}^{\infty} a_{mo} I_o(\alpha_m R) \sin \alpha_m z = \frac{\theta_o}{2\pi} \int_o^{\pi} d\phi$$

$$\sum_{m=1}^{\infty} a_{m\beta} I_\beta(\alpha_m R) \sin\alpha_m z = \frac{\theta_o}{\pi} \int_o^{\pi} \cos \beta\phi d\phi$$

$$\sum_{m=1}^{\infty} b_{m\beta} I_\beta(\alpha_m R) \sin \alpha_m z = \frac{\theta_o}{\pi} \int_o^{\pi} \sin \beta\phi d\phi$$

Simplifying the above three relations we get,

$$\sum_{m=1}^{\infty} a_{mo} I_o(\alpha_m R) \sin\alpha_m z = \frac{\theta_o}{2} \tag{6}$$

$$\sum_{m=1}^{\infty} a_{m\beta} I_\beta(\alpha_m R) \sin\alpha_m z = 0 \tag{7}$$

$$\sum_{m=1}^{\infty} b_{m\beta} I_\beta(\alpha_m R) \sin\alpha_m z = \frac{2\theta_o}{\beta\pi} \text{ where } \beta = 1, 3, 5, 7, \ldots \tag{8}$$

Each of the equations (6), (7) and (8) is a Fourier sine-series expansion in z and the three coefficients a_{mo}, $a_{m\beta}$ and $b_{m\beta}$ may be found by simple Fourier series analysis. The expressions for the three coefficients are given below

$$a_{mo} I_o\ (\alpha_m R) = \frac{2\theta_o}{m\pi} \quad \text{where} \quad m = 1, 3, 5, 7$$

$$a_{m\beta} = 0$$

$$b_{m\beta} I_\beta\ (\alpha_m R) = \frac{8\theta_o}{(\beta\pi)(m\pi)}$$

Substituting the values of all the coefficients in equation (4) we get,

$$\frac{\theta(r,\phi,z)}{\theta_o} = \sum_{m=1}^{\infty} \left(\frac{2}{m\pi}\right) \frac{I_o(\alpha_m r)}{I_o(\alpha_m R)} \sin \alpha_m z + 2 \sum_{m=1}^{\infty} \sum_{\beta=1}^{\infty} \left(\frac{2}{m\pi}\right)\left(\frac{2}{\beta\pi}\right)$$

$$\frac{I_\beta(\alpha_m r)}{I\ (\alpha_m R)} \sin \alpha_m z \sin \beta\phi$$

• PROBLEM 5-8

Obtain an expression for the differential energy balance for a sphere, in the spherical coordinate system, assuming that there is no variation of temperature with angular position.

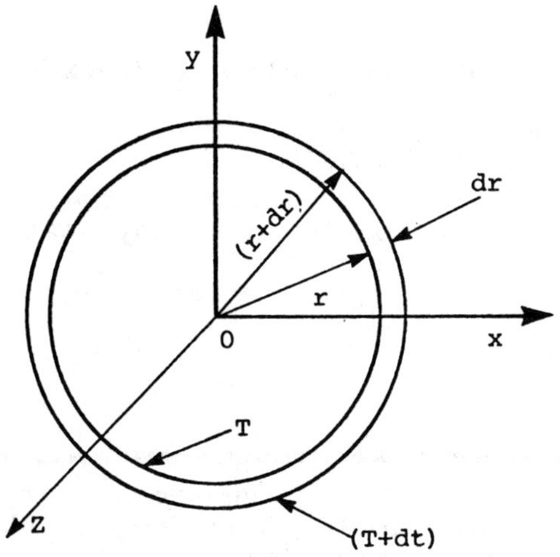

Fig. 1

Solution: Consider a sphere with its center at the origin. Consider two spherical surfaces at radii r and (r+dr) with uniform temperatures T and (T+dT) respectively--as shown in the diagram. The volume enclosed between the radii r and (r+dr) is the control volume. Then, the rate of heat inflow through the control volume is given by the Fourier Law for heat conduction

$$Q=-k(A_s)\frac{\partial T}{\partial r} \quad \text{where } A_s \rightarrow \text{ surface area of the}$$

sphere at radius r $\quad Qin = -k(4\pi r^2)\frac{\partial T}{\partial r}$ (1)

and rate of heat flow out of the control volume at radius (r+dr)

$$Qout=-k\left[4\pi(r+dr)^2\right]\left[\frac{\partial T}{\partial r}+d\left(\frac{\partial T}{\partial r}\right)\right]$$

or $\quad =-4\pi k(r^2+2rdr+(dr)^2]\left(\frac{\partial T}{\partial r}+\frac{\partial^2 T}{\partial r^2}dr\right)$ (2)

The net heat transfer, neglecting 2nd and 3rd order differentials, the product of small terms, can be written as (Qout-Qin)

$$4\pi k\left[r^2\frac{\partial^2 T}{\partial r^2}dr + 2r\frac{\partial T}{\partial r}dr\right] \quad (3)$$

The rate of heat flow into the control volume can be expressed as

$$(4\pi r^2 dr)\ \rho Cp\ \frac{\partial T}{\partial \theta} \quad (4)$$

For equilibrium conditions, equation (3) and (4) are equal.

Thus,

$$4\pi k\left(r^2\frac{\partial^2 T}{\partial r^2}dr + 2r\frac{\partial T}{\partial r}dr\right)=(4\pi r^2 dr)\rho x Cp\frac{\partial T}{\partial \theta}$$

or $$\frac{\partial T}{\partial \theta} = \frac{k}{\rho Cp}\left[\frac{\partial^2 T}{\partial r^2}+\frac{2}{r}\frac{\partial T}{\partial r}\right]$$

which is the equation for unsteady heat conduction in the spherical coordinate system.

• PROBLEM 5-9

Derive an expression for the temperature distribution in a segment of a hollow cylinder, as shown in the figure. The segment, initially, at time t=0, is at an uniform temperature T= F(r, ϕ, z). The boundary surfaces are maintained at a constant temperature T=0 (for all time t > 0).

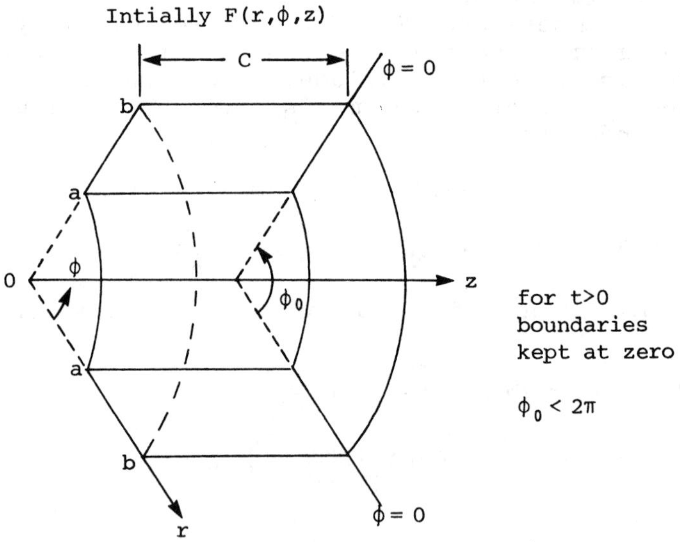

Fig. 1 Boundary and initial conditions for a segment of a hollow cylinder.

Solution: The Fourier equation for heat conduction in the cylindrical coordinate system can be written as

$$\frac{\partial^2 T}{\partial r^2}+\frac{1}{r}\frac{\partial T}{\partial r}+\frac{1}{r^2}\frac{\partial^2 T}{\partial \phi^2}+\frac{\partial^2 T}{\partial z^2} = \frac{1}{\alpha}\frac{\partial T}{\partial t} \quad \text{in } a \leq r < b,\ 0 < z < c,\ 0 < \phi < \phi_o(<2\pi),\ t > 0$$

$T=0$	at all boundaries, $t > 0$
$T=F(r, \phi, z)$	for $t=0$, in the region

Assuming the solutions for the separated variables as

$$e^{-\alpha(\beta_m^2+\eta_p^2)t}, R_\nu(\beta_m, r), \quad Z(\eta_p, z), \quad \Phi(\nu, \phi)$$

The solution for temperature distribution in the segment $T(r,\phi, z,t)$ can be obtained by combining the separated solutions and can be written as

$$T(r, \phi, z, t) = \sum_{m=1}^{\infty} \sum_{p=1}^{\infty} \sum_{\nu} c_{mp\nu} R_\nu (\beta_m, r) Z(\eta_p, z) \Phi(\nu, \phi) \, e^{-\alpha(\beta_m^2+\eta_p^2)t} \tag{1}$$

At $t=0$ the equation (1) reduces to

$$F(r, \phi, z) = \sum_{m=1}^{\infty} \sum_{p=1}^{\infty} \sum_{\nu} c_{mp\nu} R_\nu (\beta_m, r) Z(\eta_p, z) \Phi(\nu, \phi)$$

$$\text{in } a < r < b, \; 0 < z < c, \quad 0 < \phi < \phi_o \tag{2}$$

In order to determine the coefficient $C_{mp\nu}$ both sides of equation (2) are successively multiplied by

$$\int_a^b r R_\nu(\beta_{m'}, r) dr, \qquad \int_0^c Z(\eta_{p'}, z) dz, \text{ and } \qquad \int_0^{\phi_o} \Phi(\nu', \phi) d\phi$$

Simplifying the expression and applying the orthogonality property of eigenfunctions yields

$$C_{mn\nu} = \frac{1}{N(\beta_m) N(\eta_p) N(\nu)} \int_{r=a}^{b} \int_{z=0}^{c} \int_{\phi=0}^{\phi_o} r R_\nu(\beta_m, r) Z(\eta_p, z) \Phi(\nu, \phi)$$

$$F(r, \phi, z) \, dr \, dz \, d\phi$$

Now, substituting the value for $C_{mp\nu}$ into equation (1), yields

$$T(r, \phi, z, t) = \sum_{m=1}^{\infty} \sum_{n=1}^{\infty} \sum_{\nu}^{\infty} \frac{e^{-\alpha(\beta_m^2+\eta_p^2)t}}{N(\beta_m) N(\eta_p) N(\nu)} R_\nu(\beta_m, r) Z(\eta_p, z) \Phi(\nu, \phi)$$

$$\cdot \int_{r'=a}^{b} \int_{z'=0}^{c} \int_{\phi'=0}^{\phi_o} r' R_\nu(\beta_m, r') Z(\eta_p, z') \Phi(\nu, \phi') F(r', \phi', z')$$

$$d\phi' \; dz' \; dr' \tag{3}$$

Table 1 The Solution $R_\nu(\beta_m, r)$, the Norm $N(\beta_m)$ and the Eigenvalues β_m of the Differential Equation

$$\frac{d^2R(r)}{dr^2} + \frac{1}{r}\frac{dR(r)}{dr} + \left(\beta^2 - \frac{\nu^2}{r^2}\right)R(r) = 0 \qquad \text{in} \qquad a < r < b$$

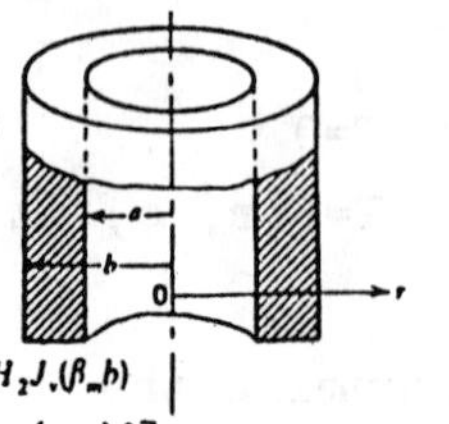

Subject to the Boundary Conditions Shown in the Table. The following notation is used in this Table.

$$S_\nu \equiv \beta_m Y_\nu'(\beta_m b) + H_2 Y_\nu(\beta_m b), \qquad U_\nu \equiv \beta_m J_\nu'(\beta_m a) - H_1 J_\nu(\beta_m a), \qquad V_\nu \equiv \beta_m J_\nu'(\beta_m b) + H_2 J_\nu(\beta_m b)$$

$$W_\nu \equiv \beta_m Y_\nu'(\beta_m a) - H_1 Y_\nu(\beta_m a), \qquad B_1 \equiv H_1^2 + \beta_m^2\left[1 - \left(\frac{\nu}{\beta_m a}\right)^2\right], \qquad B_2 \equiv H_2^2 + \beta_m^2\left[1 - \left(\frac{\nu}{\beta_m b}\right)^2\right]$$

No.	Boundary Condition at $r = a$	Boundary Condition at $r = b$	$R_\nu(\beta_m, r)$ and $1/N(\beta_m)$	β_m's are the positive roots of
1	$\frac{dR}{dr} = 0$	$R = 0$	$R_\nu(\beta_m, r) = J_\nu(\beta_m r)Y_\nu(\beta_m b) - J_\nu(\beta_m b)Y_\nu(\beta_m r)$ $\frac{1}{N(\beta_m)} = \frac{\pi^2}{2}\frac{\beta_m^2 J_\nu'^2(\beta_m a)}{J_\nu'^2(\beta_m a) - \left[1 - \left(\frac{\nu}{\beta_m a}\right)^2\right]J_\nu^2(\beta_m b)}$	$J_\nu'(\beta_m a)Y_\nu(\beta_m b)$ $- J_\nu(\beta_m b)Y_\nu'(\beta_m a) = 0$
2	$R = 0$	$\frac{dR}{dr} + H_2 R = 0$	$R_\nu(\beta_m, r) = S_\nu J_\nu(\beta_m r) - V_\nu Y_\nu(\beta_m r)$ $\frac{1}{N(\beta_m)} = \frac{\pi^2}{2}\frac{\beta_m^2 J_\nu^2(\beta_m a)}{B_2 J_\nu^2(\beta_m a) - V_\nu^2}$	$S_\nu J_\nu(\beta_m a) - V_\nu Y_\nu(\beta_m a) = 0$
3	$R = 0$	$\frac{dR}{dr} = 0$	$R_\nu(\beta_m, r) = J_\nu(\beta_m r)Y_\nu'(\beta_m b) - J_\nu'(\beta_m b)Y_\nu(\beta_m r)$ $\frac{1}{N(\beta_m)} = \frac{\pi^2}{2}\frac{\beta_m^2 J_\nu^2(\beta_m a)}{\left[1 - \left(\frac{\nu}{\beta_m b}\right)^2\right]J_\nu^2(\beta_m a) - J_\nu'^2(\beta_m b)}$	$J_\nu(\beta_m a)Y_\nu'(\beta_m b)$ $- J_\nu'(\beta_m b)Y_\nu(\beta_m a) = 0$
4	$R = 0$	$R = 0$	$R_\nu(\beta_m, r) = J_\nu(\beta_m r)Y_\nu(\beta_m b) - J_\nu(\beta_m b)Y_\nu(\beta_m r)$ $\frac{1}{N(\beta_m)} = \frac{\pi^2}{2}\frac{\beta_m^2 J_\nu^2(\beta_m a)}{J_\nu^2(\beta_m a) - J_\nu^2(\beta_m b)}$	$J_\nu(\beta_m a)Y_\nu(\beta_m b)$ $- J_\nu(\beta_m b)Y_\nu(\beta_m a) = 0$

Table 2 The Solution $X(\beta_m, x)$, the Norm $N(\beta_m)$ and the Eigenvalues β_m of the Differential Equation

$$\frac{d^2X(x)}{dx^2} + \beta^2 X(x) = 0 \qquad \text{in} \qquad 0 < x < L$$

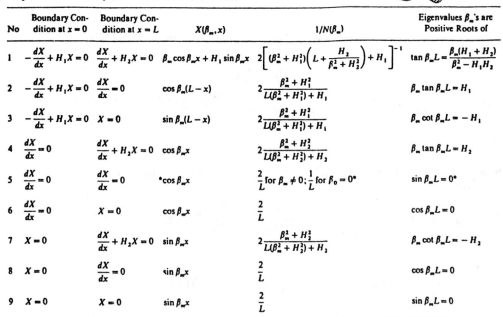

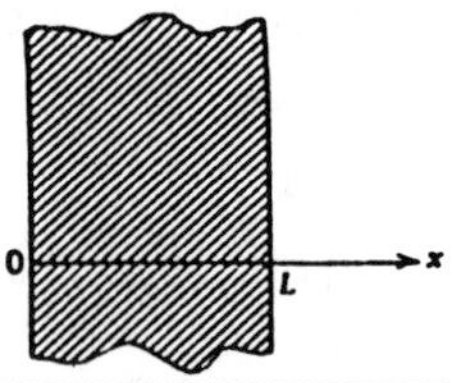

Subject to the Boundary Conditions Shown in the Table

No	Boundary Condition at $x = 0$	Boundary Condition at $x = L$	$X(\beta_m, x)$	$1/N(\beta_m)$	Eigenvalues β_m's are Positive Roots of
1	$-\frac{dX}{dx} + H_1 X = 0$	$\frac{dX}{dx} + H_2 X = 0$	$\beta_m \cos\beta_m x + H_1 \sin\beta_m x$	$2\left[(\beta_m^2 + H_1^2)\left(L + \frac{H_2}{\beta_m^2 + H_2^2}\right) + H_1\right]^{-1}$	$\tan\beta_m L = \frac{\beta_m(H_1 + H_2)}{\beta_m^2 - H_1 H_2}$
2	$-\frac{dX}{dx} + H_1 X = 0$	$\frac{dX}{dx} = 0$	$\cos\beta_m(L - x)$	$2\frac{\beta_m^2 + H_1^2}{L(\beta_m^2 + H_1^2) + H_1}$	$\beta_m \tan\beta_m L = H_1$
3	$-\frac{dX}{dx} + H_1 X = 0$	$X = 0$	$\sin\beta_m(L - x)$	$2\frac{\beta_m^2 + H_1^2}{L(\beta_m^2 + H_1^2) + H_1}$	$\beta_m \cot\beta_m L = -H_1$
4	$\frac{dX}{dx} = 0$	$\frac{dX}{dx} + H_2 X = 0$	$\cos\beta_m x$	$2\frac{\beta_m^2 + H_2^2}{L(\beta_m^2 + H_2^2) + H_2}$	$\beta_m \tan\beta_m L = H_2$
5	$\frac{dX}{dx} = 0$	$\frac{dX}{dx} = 0$	*$\cos\beta_m x$	$\frac{2}{L}$ for $\beta_m \neq 0$; $\frac{1}{L}$ for $\beta_0 = 0$*	$\sin\beta_m L = 0$*
6	$\frac{dX}{dx} = 0$	$X = 0$	$\cos\beta_m x$	$\frac{2}{L}$	$\cos\beta_m L = 0$
7	$X = 0$	$\frac{dX}{dx} + H_2 X = 0$	$\sin\beta_m x$	$2\frac{\beta_m^2 + H_2^2}{L(\beta_m^2 + H_2^2) + H_2}$	$\beta_m \cot\beta_m L = -H_2$
8	$X = 0$	$\frac{dX}{dx} = 0$	$\sin\beta_m x$	$\frac{2}{L}$	$\cos\beta_m L = 0$
9	$X = 0$	$X = 0$	$\sin\beta_m x$	$\frac{2}{L}$	$\sin\beta_m L = 0$

*For this particular case $\beta_0 = 0$ is also an eigenvalue corresponding to $X = 1$.

The solution $R_\nu(\beta_m,r)$, the norm $N(\beta_m)$ and the eigenvalues β_m are obtained from table 1 as

$$R_\nu(\beta_m,r)=J_\nu(\beta_m r)Y_\nu(\beta_m b)-J_\nu(\beta_m b)Y_\nu(\beta_m r)$$

$$\frac{1}{N(\beta_m)} = \frac{\pi^2}{2}\frac{\beta_m^2 J_\nu^2(\beta_m a)}{J_\nu^2(\beta_m a)-J_\nu^2(\beta_m b)} \tag{4}$$

and β_m's are the positive roots of

$$J_\nu(\beta_m a)Y_\nu(\beta_m b)-J_\nu(\beta_m b)Y_\nu(\beta_m a)=0$$

From table 2 terms defining $Z(\eta_p, z)$, $N(\eta_p)$ and η_p are obtained as follows,

$$Z(\eta_p,z)=\sin\eta_p z, \quad \frac{1}{N(\eta_p)} = \frac{2}{c}$$

and η_p's are the roots of

$$\sin\eta_p c=0 \text{ (or } \eta_p = \frac{p\pi}{c}, \ p = 1, 2, 3 \ldots) \tag{5}$$

Further, from table 2 terms defining $\Phi(\nu,\phi)$, $N(\nu)$ and ν are, also, obtained as follows

$$\Phi(\nu,\phi)=\sin \nu\phi, \quad \frac{1}{N(\nu)} = \frac{2}{\phi_o}$$

and ν's are the roots of

$$\sin \nu\phi_o=0 \tag{6}$$

The complete solution for temperature distribution can be obtained by substituting equations (4), (5) and (6) into equation (3).

• PROBLEM 5-10

A solid metallic block of width 2b is initially at a uniform temperature T_o. Instantaneously the temperature, at time t=0, at both the end surfaces is raised to and maintained at T_1.

Derive an expression for the temperature distribution along its width as a function of time.

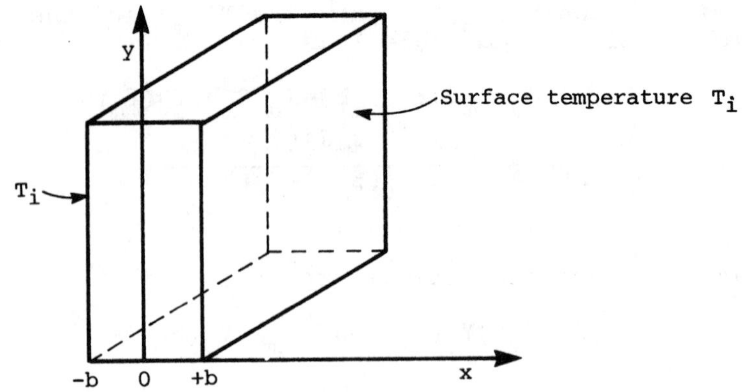

Fig. 1

Solution: The figure shows the cross section of the slate with temperature for time t > 0. The problem involves a three dimensional analysis.

For simplifying the problem non-dimensional parameters are introduced.

For temperature gradient

$$\theta = \frac{T-T_1}{T_o-T_1} \quad \text{non-dimensional temperature} \tag{1}$$

$$\ell = \frac{y}{b} \quad \text{non-dimensional width} \tag{2}$$

$$\tau = \frac{\alpha t}{b^2} \quad \text{non-dimensional time} \tag{3}$$

Now the problem reduces to the simpler partial differential equation

$$\frac{\partial\theta}{\partial\tau} = \frac{\partial^2\theta}{\partial\ell^2} \tag{4}$$

with boundary conditions

$\theta = 1$ for time $\tau = 0$ Initial condition

$\theta = 0$ at $\ell = \pm 1$ for $\tau > 0$.

Equation (4) can be solved by variables separation method and the solution can be written as

$$\theta = ae^{-\beta^2\tau}(c\sin\beta\ell + d\cos\beta\ell) \tag{5}$$

where a, c and d are some constants.

Since the problem is symmetric the above function should be even. This implies that $c = 0$ and hence the term $c\sin\beta\ell$ vanishes.

Substituting the boundary condition

$$\theta = 0 \text{ for } \tau > 0$$

$$0 = (ae^{-\beta^2\tau})d\cos\beta\ell$$

This implies that $d\cos\beta\ell = 0$

since $ae^{-\beta^2\tau} \neq 0$

and $\cos\beta\ell = 0 \quad \because d \neq 0$

$$\therefore \text{ the value of } \beta_n = (n + \tfrac{1}{2})\pi \tag{6}$$

where $n = 0, \pm 1, \pm 2,$

$\therefore$ The solution is

$$\theta = (ae^{-\beta^2\tau})\ (d\cos\beta\ell)$$

where $\beta = (n + \frac{1}{2})\pi$

The above equation can be written as

$$\theta_n = A_n \times e^{-\beta_n{}^2\tau} \times \cos\beta_n\ell$$

where A_n is another constant $= a \times d$. For different values of n, θ will be different, hence for all integrals from

$n = -\infty$ to $n = +\infty$

$$\theta = \sum_{n=0}^{\infty} A_n\ e^{-\beta_n{}^2\tau}\ \cos\beta_n\ell \tag{7}$$

Now to find the value of A_n substitute the initial condition

i.e. $\theta = 1$ for time $\tau = 0$

$$1 = \sum_{n=0}^{\infty} A_n \cos\beta_n 1 \tag{8}$$

Multiplying both sides by $\cos\beta_m$ and integrating between the limits -1 and +1

$$\int_{-1}^{+1} \cos\beta_m 1 = \int_{-1}^{+1} \sum_{n=0}^{\infty} A_n \cos\beta_n\ell \cos\beta_m\ell$$

On the right hand side, except for m = n, all the other terms when m ≠ n are each equal to zero. When m = n it is equal to $\frac{A_n}{2}$.

On the left hand side

$$\int_{-1}^{+1} \cos\beta_m l = \frac{(-1)^m}{(m+\frac{1}{2})\pi}$$

$$\therefore \quad A_n = \frac{2\,(-1)^m}{(m+\frac{1}{2})\pi} \qquad (9)$$

Substituting the value of A_n and β_n in equation (7)

$$\theta = 2\sum_{n=0}^{\infty} \frac{(-1)^m}{(m+\frac{1}{2})\pi} \times e^{-[(n+\frac{1}{2})\pi]^2\tau} \times \cos[(n+\tfrac{1}{2})\pi\ell] \qquad (10)$$

Now substituting the transformed parameters in equations(1), (2) and (3) and putting m = n

$$\frac{T-T_1}{T_o-T_1} = 2\sum_{n=0}^{\infty} \frac{(-1)^n}{(n+\frac{1}{2})\pi} \times e^{-(n\pi+\frac{\pi}{2})^2\frac{\alpha t}{b^2}} \times \cos(n\pi+\tfrac{\pi}{2})\frac{y}{b} \qquad (11)$$

This is the most generalized equation.
Solutions for heat conduction in rectangular parallelopipeds and finite cyclinders can be worked out similarly.

• PROBLEM 5-11

Determine the unsteady temperature of the empty frying pan initially at uniform ambient temperature T. Suddenly heat flux q, is applied at the bottom of the skillet as shown in the diagram.

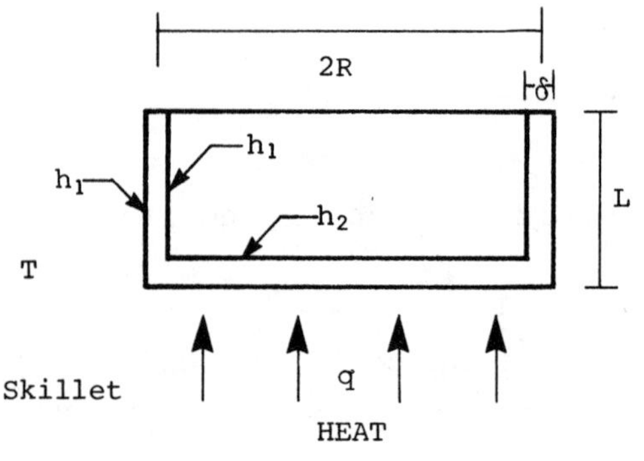

Fig. 1 Skillet

Solution: The lumped parameter analysis for the bottom of the frying pan, yields

$$\rho C_p A\delta \frac{dT_2}{dt} = qA - h_2 A (T_2 - T) \tag{1}$$

where A is the total surface area of the bottom.

Applying the initial condition to equation (1) $T_2(0) = T$ gives

$$\theta_2(t) = \frac{q}{h_2}(1 - e^{-bt}) \tag{2}$$

where $\theta_2(t) = T_2(t) - T$ and $b = h_2/\rho C_p \delta$.

Using equation (2), the differential equations for side walls of frying pan can be written as

$$\frac{\partial^2 \theta_1}{\partial x^2} - m_1^2 \theta_1 = \frac{1}{\alpha}\frac{\partial \theta_1}{\partial t}, \qquad m_1^2 = 2h_1/k\delta, \tag{3}$$

$$\theta_1(x,0) = 0,$$

$$\frac{\partial \theta_1(0,t)}{\partial x} = 0, \qquad \theta_1(L,t) = \frac{q}{h_2}(1 - e^{-bt}).$$

Consider the function $\psi(x,t)$, the construction of which is similar to equation (3) except for the last boundary condition which is replaced by the step disturbance

$$\theta_1(L,t) = q/h_2$$

By superposition

$$\psi(x,t) = \Phi(x,t) + \Phi(x)$$

where $\Phi(x)$ and $\Phi(x,t)$ comply with

$$\frac{d^2\phi}{dx^2} - m_1^2\phi = 0, \quad \frac{d\phi(0)}{dx} = 0, \quad \phi(L) = \frac{q}{h_2} \tag{4}$$

and

$$\frac{\partial^2 \Phi}{\partial x^2} - m_1^2 \Phi = \frac{1}{\alpha}\frac{\partial \Phi}{\partial t}, \qquad \Phi(x, 0) = -\phi(x), \tag{5}$$

$$\frac{\partial \Phi(0,t)}{\partial x} = 0, \qquad \Phi(L,t) = 0$$

The solution of equation (4) can be expressed as

$$\frac{\Phi(x)}{q/h_2} = \frac{\text{Cosh}(m_1 x)}{\text{Cosh}(m_1 L)} \tag{6}$$

By separation of variables equation 5 gives

$$\frac{d^2X}{dx^2} + \lambda^2 X = 0, \tag{7}$$

$$\frac{dX(0)}{dx} = 0, \quad X(L) = 0;$$

$$\frac{d\tau}{dt} + \alpha(\lambda^2 + m^2)\tau = 0. \tag{8}$$

Solution of equation 7 is

$X_n(x)=A_n \cos\lambda_n x$, $\phi_n(x)=\cos \lambda_n x$, characteristic functions

$\lambda_n L=(2n+1)\pi/2$, $n = 0, 1, 2, 3, \ldots.$, characteristic values

and the solution of Eq. (8) is

$$\tau_n(t) = C_n e^{-\alpha(\lambda_n^2+m^2)t}$$

Therefore, the product solution of Φ becomes

$$\Phi(x,t) = \sum_{n=0}^{\infty} a_n e^{-\alpha(\lambda_n^2+m^2)t} \cos \lambda_n x. \tag{9}$$

At t=0,

$$-\frac{q}{h_2}\frac{\cosh(m_1 x)}{\cosh(m_1 L)} = \sum_{n=0}^{\infty} a_n \cos \lambda_n x \tag{10}$$

where coefficient $a_n = -(-1)^n \frac{2}{L}\left(\frac{q}{h_2}\right)\frac{\lambda_n}{m_1^2+\lambda_n^2}$ (11)

Now, the solution for function $\psi(x,t)$ can be obtained by substituting equation (11) into (9) and summing equation (6), which yields

$$\frac{\psi(x,t)}{q/h_2} = \frac{\cosh m_1 x}{\cosh m_1 L} - 2\sum_{n=0}^{\infty}\frac{(-1)^n(\lambda_n L)e^{-\alpha(m_1^2+\lambda_n^2)t}}{(m_1^2+\lambda_n^2)L^2}\cos\lambda_n x. \tag{12}$$

Duhamel's superposition integral gives the solution $\phi(r,t)$ corresponding to the time-dependent disturbance D(t) in terms of the solution $\psi(r,t)$ corresponding to the stepwise disturbance of the same kind.

$$\Phi(r,t) = D(0)\,\psi(r,t) + \int_0^t \psi(r,t-s)\frac{dD(s)}{ds}\,ds \tag{13}$$

where $D(s) = 1 - e^{-bs}$

Applying Duhamel's superposition integral yields

$$\frac{\theta_1(x,t)}{q''/h_2} = b \int_0^t \left[\frac{\cosh m_1 x}{\cosh m_1 L} - 2 \sum_{n=0}^{\infty} \frac{(-1)^n (\lambda_n L) e^{-\alpha(m_1^2+\lambda_n^2)(t-s)}}{(m_1^2 + \lambda_n^2)L^2} \cos \lambda_n x \right] e^{-bs} ds \tag{14}$$

Simplifying the equation (14) gives the unsteady state temperature of the side walls as

$$\frac{\theta_1(x,t)}{q''/h_2} = \frac{\cosh m_1 x}{\cosh m_1 L}(1 - e^{-bt})$$

$$+ 2\left(\frac{b}{a}\right) \sum_{n=0}^{\infty} \frac{(-1)^n (\lambda_n L)\left[e^{-\alpha(m_1^2+\lambda_n^2)t} - e^{-bt}\right]}{(m_1^2+\lambda_n^2)L^2(m_1^2+\lambda_n^2 - b/a)} \cos \lambda_n x \tag{15}$$

Combining with equations (10) and (11), yields

$$\frac{\cosh m_1 x}{\cosh m_1 L} = 2 \sum_{n=0}^{\infty} \frac{(-1)^n (\lambda_n L)}{(m_1^2 + \lambda_n^2)L^2} \cos \lambda_n x \tag{16}$$

Substituting $(m_1^2 - b/a)^{\frac{1}{2}}$ for m, in equation (16) gives

$$\frac{\cosh (m_1^2 - b/a)^{\frac{1}{2}} x}{\cosh (m_1^2 - b/a)^{\frac{1}{2}} L} = 2 \sum_{n=0}^{\infty} \frac{(-1)^n (\lambda_n L)}{(m_1^2 + \lambda_n^2 - b/a)L^2} \cos \lambda_n x \tag{17}$$

From equation (16) and (17)

$$\frac{\cosh (m_1^2 - b/a)^{\frac{1}{2}} x}{\cosh (m_1^2 - b/a)^{\frac{1}{2}} L} - \frac{\cosh m_1 x}{\cosh m_1 L} = 2\left(\frac{b}{a}\right) \sum_{n=0}^{\infty} \frac{(-1)^n (\lambda_n L) \cos \lambda_n x}{(m_1^2+\lambda_n^2)L^2(m_1^2+\lambda_n^2-b/a)}. \tag{18}$$

Again, multiplying equation (18) by term e^{-bt} and substituting the expression into equation (15) the solution for $\theta_1(x,t)$ is obtained as

$$\frac{\theta_1(x,t)}{q''/h_2} = \frac{\cosh m_1 x}{\cosh m_1 L} - \frac{e^{-bt}\cosh(m_1^2 - b/a)^{\frac{1}{2}} x}{\cosh (m_1^2 - b/a)^{\frac{1}{2}} L}$$

$$+ 2\left(\frac{b}{a}\right) \sum_{n=0}^{\infty} \frac{(-1)^n (\lambda_n L) e^{-\alpha(m_1^2+\lambda_n^2)t}}{(m_1^2 + \lambda_n^2)L^2(m_1^2 + \lambda_n^2 - b/a)} \cos\lambda_n x.$$

Find the temperature distribution T(x,y,z,t) at any time (t > 0) for the solid shown in the figure. The initial temperature of the solid is T = F(x,y,z,t) and the boundary surfaces are maintained at temperature T = 0 (for all t > 0).

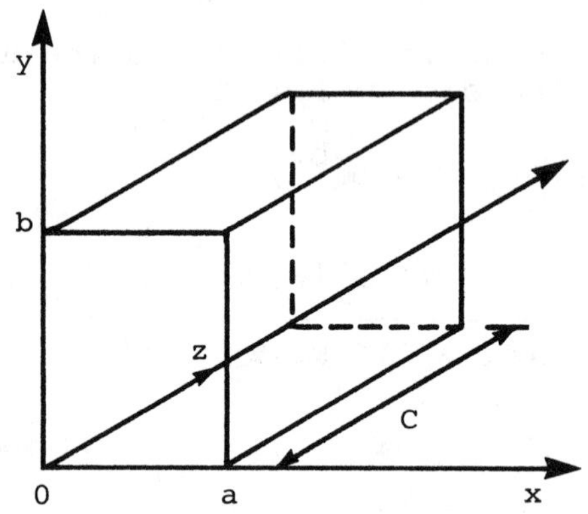

Fig. 1 A solid slab.

Solution: The unsteady state heat conduction equation for solids is

$$\frac{\partial^2 T}{\partial x^2} + \frac{\partial^2 T}{\partial y^2} + \frac{\partial^2 T}{\partial z^2} = \frac{1}{\alpha}\frac{\partial T}{\partial t} \qquad \text{for } 0 \le t \le \infty .$$

$$\text{Boundary conditions} \begin{cases} T=0 \text{ at all boundary surfaces } 0 \le t \le \infty \\ T=F(x,y,z) \text{ at } t = 0. \end{cases}$$

The differential equation can be solved by the separation of variables method. Assume the solution to the differential equation to take a product form as follows

$$T(x,y,z,t) = \Gamma(t)*X(x)*Y(y)*Z(z)$$

Subsequently, the solution in terms of these separated functions may be expressed as

$$T(x,y,z,t) = \sum_{m=1}^{\infty}\sum_{n=1}^{\infty}\sum_{p=1}^{\infty} C_{mnp}\, e^{-\alpha(\beta_m^2+\gamma_n^2+\eta_p^2)t} X(\beta_m,x)Y(\gamma_n,y) Z(\eta_p,z) \qquad (1)$$

Applying the initial condition

T = F(x,y,z) at t=0 to the equation (1), yields

$$F(x,y,z)=\sum_{m=1}^{\infty}\sum_{n=1}^{\infty}\sum_{p=1}^{\infty} C_{mnp}X(\beta_m,x)Y(\gamma_n,y)Z(\eta_p,z) \qquad (2)$$

Now, multiplying both sides of the equation (2) by the expressions

$$\int_0^a X(\beta_m,x)dx,\quad \int_0^b Y(\gamma_n,y)dy,\ \text{and}\ \int_0^c Z(\eta_p,z)dz \quad \text{in succession}$$

and using the condition of orthogonality of the eigenfunctions, the value of the coefficient C_{mnp} is obtained

$$C_{mnp}=\frac{1}{N(\beta_m)N(\gamma_n)N(\eta_p)}\int_{x'=0}^{a}\int_{y'=0}^{b}\int_{z'=0}^{c} X(\beta_m,x')Y(\gamma_n,y')Z(\eta_p,z')$$

$$\cdot F(x',y'z')\ dx'\ dy'\ dz'$$

where

$$N(\beta_m)\equiv\int_0^a X^2(\beta_m,x)dx,\quad N(\gamma_n)\equiv\int_0^b Y^2(\gamma_n,y)dy,\ \text{and}\ N(\eta_p)\equiv\int_0^c Z^2(\eta_p,z)dz$$

Next, substituting the value of the coefficient C_{mnp} in equation (1), the temperature distribution in the solid slab is obtained

$$T(x,y,z)=\sum_{m=1}^{\infty}\sum_{n=1}^{\infty}\sum_{p=1}^{\infty} e^{-\alpha(\beta_m^2+\gamma_n^2+\eta_p^2)t}\frac{1}{N(\beta_m)N(\gamma_n)N(\eta_p)}X(\beta_m,x)$$

$$Y(\gamma_n,y)Z(\eta_p,z)\int_{x=0}^{a}\int_{y=0}^{b}\int_{z=0}^{c} X(\lambda_m,x')Y(\eta_n,y')Z(\ _p,z')$$

$$F(x',y',z')dx'dy'dz' \qquad (3)$$

Table 1 The Solution $X(\beta_m, x)$, the Norm $N(\beta_m)$ and the Eigenvalues β_m of the Differential Equation

$$\frac{d^2X(x)}{dx^2}+\beta^2X(x)=0 \quad \text{in} \quad 0<x<L$$

Subject to the Boundary Conditions Shown in the Table

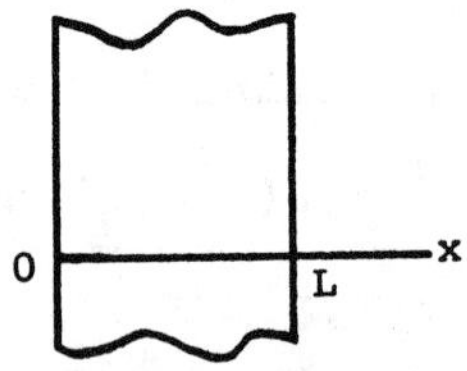

No	Boundary Condition at $x=0$	Boundary Condition at $x=L$	$X(\beta_m, x)$	$1/N(\beta_m)$	Eigenvalues β_m's are Positive Roots of
1	$-\frac{dX}{dx}+H_1X=0$	$\frac{dX}{dx}+H_2X=0$	$\beta_m\cos\beta_m x+H_1\sin\beta_m x$	$2\left[(\beta_m^2+H_1^2)\left(L+\frac{H_2}{\beta_m^2+H_2^2}\right)+H_1\right]^{-1}$	$\tan\beta_m L=\frac{\beta_m(H_1+H_2)}{\beta_m^2-H_1H_2}$
2	$-\frac{dX}{dx}+H_1X=0$	$\frac{dX}{dx}=0$	$\cos\beta_m(L-x)$	$2\frac{\beta_m^2+H_1^2}{L(\beta_m^2+H_1^2)+H_1}$	$\beta_m\tan\beta_m L=H_1$
3	$-\frac{dX}{dx}+H_1X=0$	$X=0$	$\sin\beta_m(L-x)$	$2\frac{\beta_m^2+H_1^2}{L(\beta_m^2+H_1^2)+H_1}$	$\beta_m\cot\beta_m L=-H_1$
4	$\frac{dX}{dx}=0$	$\frac{dX}{dx}+H_2X=0$	$\cos\beta_m x$	$2\frac{\beta_m^2+H_2^2}{L(\beta_m^2+H_2^2)+H_2}$	$\beta_m\tan\beta_m L=H_2$
5	$\frac{dX}{dx}=0$	$\frac{dX}{dx}=0$	$^*\cos\beta_m x$	$\frac{2}{L}$ for $\beta_m\neq0$; $\frac{1}{L}$ for $\beta_0=0^*$	$\sin\beta_m L=0^*$
6	$\frac{dX}{dx}=0$	$X=0$	$\cos\beta_m x$	$\frac{2}{L}$	$\cos\beta_m L=0$
7	$X=0$	$\frac{dX}{dx}+H_2X=0$	$\sin\beta_m x$	$2\frac{\beta_m^2+H_2^2}{L(\beta_m^2+H_2^2)+H_2}$	$\beta_m\cot\beta_m L=-H_2$
8	$X=0$	$\frac{dX}{dx}=0$	$\sin\beta_m x$	$\frac{2}{L}$	$\cos\beta_m L=0$
9	$X=0$	$X=0$	$\sin\beta_m x$	$\frac{2}{L}$	$\sin\beta_m L=0$

The solutions to eigenvalue problems satisfying the functions x,y,z are obtained from the table

$X(\beta_m,x)=\sin\beta_m x,\ \frac{1}{N(\beta_m)}=\frac{2}{a}$ and β_m's are roots of $\sin\beta_m a=0$

$Y(\gamma_n,y)=\sin\gamma_n Y,\ \frac{1}{N(\gamma_n)}=\frac{2}{b}$ and γ_n's are roots of $\sin\gamma_n b=0$

$Z(\eta_p,z)=\sin\eta_p z,\ \frac{1}{N(\eta_p)}=\frac{2}{c}$ and η_p's are roots of $\sin\eta_p c=0$

Substituting these in the equation (3) gives

$$T(x,y,z,t)=\frac{8}{abc}\sum_{m=1}^{\infty}\sum_{n=1}^{\infty}\sum_{p=1}^{\infty} e^{-\alpha(\beta_m^2+\gamma_n^2+\eta_p^2)t}\sin\beta_m x\ \sin\gamma_n y$$

$$\sin\eta_p z \int_{x'=0}^{a}\int_{y'=0}^{b}\int_{z'=0}^{c}\sin\beta_m x'\ \sin\gamma_n y'\ \sin\eta_p z'$$

$$F(x',y',z')\ dx'\ dy'\ dz'$$

where

$$\beta_m = \frac{m\pi}{a}, \quad m = 1,2,3, \ldots$$

$$\gamma_n = \frac{n\pi}{b}, \quad n = 1,2,3, \ldots$$

$$\eta_p = \frac{p\pi}{c}, \quad p = 1,2,3, \ldots$$

• PROBLEM 5-13

Derive an expression for the temperature distribution $T(r,z,t)$, at times $t > 0$, for an isotropic infinite cylinder of radius b. The initial temperature, at time $t=0$ is $F(r,z)$. Then at time $t > 0$ the boundary surface at radius $r=b$ is maintained at temperature $T=0$.

Solution: The appropriate general heat conduction equation in cyclindrical coordinate system is represented by

$$\varepsilon_{11} \frac{1}{r} \frac{\partial}{\partial r} \left(r\frac{\partial T}{\partial r}\right) + \varepsilon_{22} \frac{1}{r^2} \frac{\partial^2 T}{\partial \phi^2} + \varepsilon_{33} \frac{\partial^2 T}{\partial z^2} + (\varepsilon_{12}+\varepsilon_{21}) \frac{1}{r} \frac{\partial^2 T}{\partial \phi \partial z}$$

$$+ (\varepsilon_{13}+\varepsilon_{31}) \frac{\partial^2 T}{\partial r\, \partial z} + \frac{\varepsilon_{13}}{r} \frac{\partial T}{\partial z} + (\varepsilon_{23}+\varepsilon_{32}) \frac{1}{r} \frac{\partial^2 T}{\partial \phi \partial z} + g(r,\phi,z,t)$$

$$= \rho C p \frac{\partial T(r,\phi,z,t)}{\partial t} \qquad (1)$$

Here for the cylinder $\varepsilon_{ij} = \varepsilon_{ji}$ and since there is variation of temperature along the horizontal angular distance from the reference to the point $\frac{\partial T}{\partial \phi} = 0$

Applying these to equation (1) gives

$$\frac{1}{r} \frac{\partial}{\partial r} \left(r\frac{\partial T}{\partial r}\right) + \varepsilon_{33} \frac{\partial^2 T}{\partial z^2} + 2\varepsilon_{13} \frac{\partial^2 T}{\partial r \partial z} + \varepsilon_{13} \frac{1}{r} \frac{\partial T}{\partial z} = \frac{1}{\alpha_{11}} \frac{\partial T}{\partial t} \qquad (2)$$

for radius $0 \le r < b$,

for length $-\infty < z < \infty$, and time $t > 0$

The boundary conditions are

$$T = 0 \quad \text{at } r = b \quad \text{for } t > 0$$

and $T = F(r,z)$ for $0 \leq r < b$, $-\infty < z < \infty$ for $t = 0$

The partial differential equation (2) can be reduced by the application of Fourier integral transformations.

$$\text{Inversion: } -T(r,z,t) = \frac{1}{2\pi}\int_{-\infty}^{+\infty} e^{i\omega z_1}\hat{T}(r,\omega,t)\, d\omega$$

$$\text{Transform: } -\hat{T}(r,\omega,t) = \int_{z_1=-\infty}^{\infty} e^{i\omega z_1} T(r,z_1,t)\, dz_1 \tag{3}$$

Transforming equation (2) gives

$$\frac{\partial^2\hat{T}}{\partial r^2} + \frac{1}{r}\frac{\partial\hat{T}}{\partial r} - \omega^2\varepsilon_{33}\hat{T} - 2i\omega\varepsilon_{13}\frac{\partial\hat{T}}{\partial r} - i\omega\frac{\varepsilon_{13}}{r}\hat{T} = \frac{1}{\alpha_{11}}\frac{\partial\hat{T}}{\partial t} \tag{4}$$

in $0 \leq r < b$ for $t > 0$,

where $\alpha_{11} = \dfrac{k_{11}}{\rho Cp}$

$\hat{T} = 0$ at $r = b$ and $t > 0$

$\hat{T} = \hat{F}(r,\omega)$ at $0 \leq r < b$ and $t = 0$

where $\hat{T} = \hat{T}(r,\omega,t)$.

Defining a new variable

$$\hat{W}(r,\omega,t)\, e^{i\omega\varepsilon_{13}r} = \hat{T}(r,\omega,t) \tag{A}$$

the transformed equation is

$$\frac{\partial^2\hat{W}}{\partial r^2} + \frac{1}{r}\frac{\partial\hat{W}}{\partial r} - (\varepsilon_{33}-\varepsilon_{13}^2)\,\omega^2\hat{W} = \frac{1}{\alpha}\frac{\partial\hat{W}}{\partial t} \tag{5}$$

with $\hat{W} = 0$ at $r = b$ for $t > 0$

$\hat{W} = e^{-i\omega\varepsilon_{13}r}\,\hat{F}(r,\omega)$ at $0 \leq r \leq b$ for $t = 0$.

To remove the partial derivative of r, integral transforms are used, which are

$$\text{Inversion: } \hat{W}(r,\omega,t) = \sum_{m=1}^{\infty}\frac{1}{N(\beta_m)}\, J_o\, \hat{\hat{W}}(\beta_m,\omega,t) \tag{6}$$

Transform: $$\hat{\hat{W}}(\beta_m,\omega,t) = \int_0^b r\, J_o\,(\beta_m r)\, \hat{W}(r,\omega,t)\, dr$$

where
$$\frac{1}{N(\beta_m)} = \frac{2}{b^2}\,\frac{1}{J_o'^2(\beta_m b)} = \frac{2}{b^2 J_1^2(\beta_m b)}$$

and β_m's are the roots of (B)

$$J_o(\beta_m b) = 0$$

Applying transformations (6) to equation (5)

$$\frac{d\hat{\hat{W}}}{dt} + \alpha_{11}\lambda^2\hat{\hat{W}}(\beta_m,\omega,t) = 0 \quad \text{for } t > 0$$

$$\hat{\hat{W}}(\beta_m,\omega,t) = \hat{\hat{H}}(\beta,\omega) \quad \text{for } t = 0 \tag{7}$$

where $\lambda^2 = \beta_m^2 + \omega^2(\varepsilon_{33}-\varepsilon_{13}^2)$

$$\varepsilon_{33}-\varepsilon_{13}^2 > 0$$

$$\hat{\hat{H}}(\beta_m,\omega) = \int_0^b r_1 J_o(\beta_m,r_1)\, e^{-i\omega\varepsilon_{13}\, r_1}\, \hat{F}(r,\omega)\, dr_1$$

(8)

and $$\hat{F}(r_1,\omega) = \int_{z_1=-\infty}^{+\infty} e^{i\omega z_1} F(r_1,z_1)\, dz_1$$

The solution for equation (7) is

$$\hat{\hat{W}}(\beta_m,\omega,t) = e^{-\alpha_{11}\lambda^2 t}\,\hat{\hat{H}}\,(\beta_m,\omega) \tag{9}$$

The inversion of equation (9) gives

$$\bar{W}(r,\omega,t) = \sum_{m=1}^{\infty} \frac{1}{N(\beta_m)}\, J_o(\beta_m,r)\, e^{-\alpha_{11}\lambda^2 t}\, \hat{\hat{H}}(\beta_m,\omega) \tag{10}$$

Substituting the value of $\bar{W}(r,\omega,t)$ in equation (A)

$$\hat{T}(r,\omega,t) = \sum_{m=1}^{\infty} \frac{1}{N(\beta_m)}\, J_o(\beta_m r) \times e^{-\alpha_{11}\lambda^2 t + i\omega\varepsilon_{13} r}\, \hat{\hat{H}}(\beta_m\omega) \tag{11}$$

Inverting this equation using equation (3)

$$T(r,z,t) = \frac{1}{2\pi}\sum_{m=1}^{\infty} \int_{\omega=-\infty}^{\infty} \frac{1}{N(\beta_m)} \cdot J_o(\beta_m r)\ \hat{\bar{H}}(\beta_m,\omega)$$

$$x\ e^{-\left[\alpha_{11}\lambda^2 t + i\omega(z-\varepsilon_{13}r)\right]} dw \quad (12)$$

where $\lambda^2 = \beta_m^{\ 2} + \omega^2(\varepsilon_{33}-\varepsilon_{13}^{\ 2})$

Substituting the value of $\hat{\bar{H}}\ (\beta_m,\omega)$ from equation (8) and re-arranging

$$T(r,z,t) = \sum_{m=1}^{\infty} e^{-\alpha_{11}\beta_m^2 t}\ x\ \frac{1}{N(\beta_m)} J_o(\beta_m,\ r)$$

$$x \int_{z_1=-\infty}^{+\infty} \int_{r_1=0}^{b} r\ J_o(\beta_m,r)F(r,z)\ dr\ dz$$

$$x\left\{\frac{1}{2\pi}\int_{\omega=-\infty}^{\infty} e^{-\omega^2\alpha_{11}(\varepsilon_{33}-\varepsilon_{13}^{\ 2})t\ -\ i\omega(z-z_1)\ +\ \varepsilon_{13}(r_1-r)}\ d\,\omega\right\}$$

(13)

The integral with respect to ω is evaluated using equation (3) and the solution is

$$T(r,z,t) = \frac{1}{\sqrt{4\pi\alpha_{11}(\varepsilon_{33}-\varepsilon_{13}^{2})t}} \sum_{m=1}^{\infty} e^{-\alpha_{11}\beta_m^2 t} \cdot \frac{1}{N(\beta_m)} J_o(\beta_m r)$$

$$x \int_{z_1=-\infty}^{\infty} \int_{r_1=0}^{b} r_1\ J_o(\beta_m,r_1).\ F(r_1,z_1)$$

$$x\ e^{\left\{\frac{(z-z_1)\ +\ \varepsilon_{13}(r_1-r)^2}{4\alpha_{11}(\varepsilon_{33}-\varepsilon_{33}^{\ 2})}\right\}} dr_1\ dz_1$$

where $N(\beta_m)$ is given by equation (B) and β_m's are the roots of the equation (B).

CHAPTER 6

FREE CONVECTION HEAT TRANSFER

Basic Attacks and Strategies for Solving Problems in this Chapter. See pages 311 to 350 for step-by-step solutions to problems.

Convection heat transfer in general occurs due to movement of a fluid near a solid which is at a different temperature than the fluid. If the solid is higher in temperature than the fluid, there will be heat transfer to the fluid. If the solid is cooler than the fluid, there will be heat transfer from the fluid to the solid. Free convection, also sometimes called natural convection, occurs just due to the natural movement of the fluid in the vicinity of the solid because of the different densities of the fluid at different temperatures. For example, a cool fluid in contact with a hot solid will be heated just adjacent to the solid. The heating will cause it to expand and rise carrying heat with it.

Free convection problems generally center upon determination of a heat transfer coefficient, h, for free convection from some solid shape such as the vertical plates, horizontal plates, pipes, and other shapes considered in this chapter. The free convection heat transfer coefficient is usually calculated from an empirical equation which gives Nusselt number as a function of a Prandtl number and a Grashof number. The Nusselt number is given by

$$N_u = hL / k,$$

so if a value for N_u *can be determined and values are known for L and k, then the value of h can be calculated. The Prandtl number is*

$$P_r = C_p \mu / k.$$

It depends only on properties of the fluid and is often tabulated as a function of temperature for common fluids such as air. The Grashof number, G_r*, depends upon fluid properties, the size of the solid and the temperature difference between the solid and fluid. The Grashof number is defined in problems 6-5 and 6-6.*

The free convection empirical equations typically give N_u *as a function of the product* $G_r P_r$*, which is sometimes called the Rayleigh number,* R_a*. The equation for a given solid configuration will typically have different sets of constants for different ranges of values for* $G_r P_r$*.*

If the heat transfer rate between solid and fluid is to be calculated or is given in the problem, then the general convection heat transfer equation, $q = hA(T_s - T_f)$*, which was discussed in Chapter 1, must be used.*

Step-by-Step Solutions to Problems in this Chapter, "Free Convection Heat Transfer"

FREE CONVECTION HEAT TRANSFER FROM VERTICAL PLATES

• PROBLEM 6-1

Using energy equations, derive an expression for heat transfer in terms of the system variables for a plate at temperature T_0 dipped in a large volume of liquid which is at temperature T_1. Assume the physical properties of liquid to be constant and free convection conditions prevail.

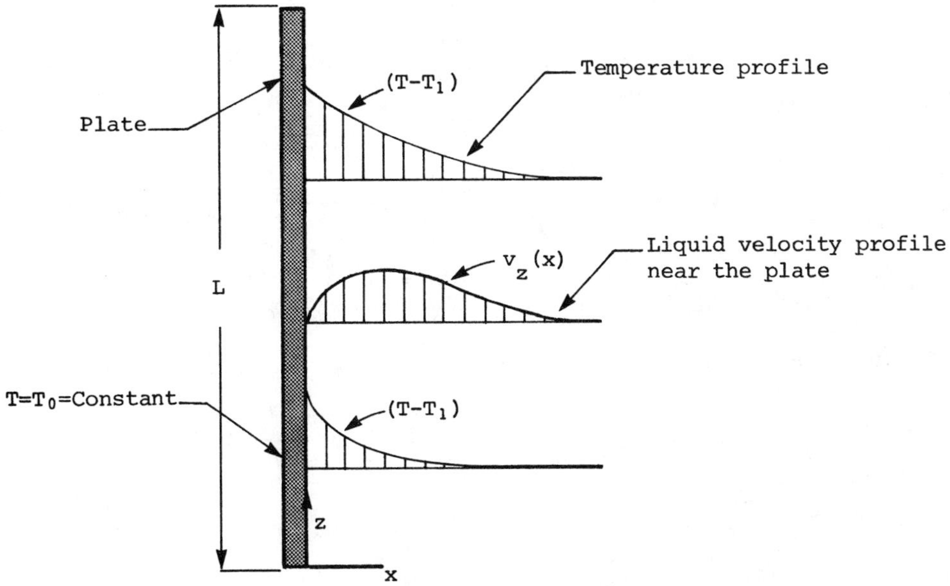

Solution: When a heated plate is dipped in a still fluid, the fluid around the plate is set in motion. In this analysis it is assumed that the plate dimension in y-direction is very large as compared to the dimensions in the x and z directions. Thus, the fluid velocity components v_x and v_z in directions x and z are independent of y. It is also assumed that the component v_x is very small compared to the z component.

Hence, the fluid moves in the vertical direction only.

Now, based on the above assumptions the equations for motion and energy are as follows.

Equation for motion:

$$\rho\, v_x\left(\frac{\partial}{\partial x} + v_z\frac{\partial}{\partial z}\right) v_z = \mu\left(\frac{\partial^2 v_z}{\partial x^2} + \frac{\partial^2 v_z}{\partial z^2}\right) + \rho g \beta\,(T - T_1) \qquad (1)$$

Equation for energy:

$$\rho C_p\left(v_x\frac{\partial}{\partial x} + v_z\frac{\partial}{\partial z}\right) T - T_1) = k\left(\frac{\partial^2}{\partial x^2} + \frac{\partial^2}{\partial z^2}\right)(T - T_1) \qquad (2)$$

The two-dimensional continuity equation is

$$\frac{\partial v_x}{\partial x} + \frac{\partial v_z}{\partial z} = 0 \qquad (3)$$

The boundary conditions are

$$\text{at } x = 0,\ v_x = v_z = 0 \text{ and } T = T_0 \qquad (4)$$

$$\text{at } x = \infty,\ v_z = 0 \quad \text{and } T = T_1 \qquad (5)$$

$$\text{at } z = -\infty,\ v_x = v_z = 0 \text{ and } T = T_1 \qquad (6)$$

For convenience, the equations (1), (2) and (3) are expressed in dimensionless parameters

equation (1) may be written as

$$\frac{1}{Pr}\left(\phi_x\frac{\partial}{\partial \eta} + \phi_z\frac{\partial}{\partial \xi}\right)\phi_z = \frac{\partial^2 \phi_z}{\partial \eta^2} + \theta \qquad (7)$$

equation (2) as

$$\phi_x\frac{\partial \theta}{\partial \eta} + \phi_z\frac{\partial \theta}{\partial \xi} = \frac{\partial^2 \theta}{\partial \eta^2} \qquad (8)$$

and finally, equation (3) as

$$\frac{\partial \phi_x}{\partial \eta} + \frac{\partial \phi_z}{\partial \xi} = 0 \tag{9}$$

where the dimensionless parameters are defined as follows

$$\theta = (T - T_1)/(T_0 - T_1) \tag{10}$$

$$\xi = Z/L \tag{11}$$

$$\eta = \left(\frac{B}{\mu \alpha L}\right)^{\frac{1}{2}} x \tag{12}$$

$$\phi_z = \left(\frac{\mu}{B \alpha L}\right)^{\frac{1}{4}} v_z$$

$$\phi_x = \left(\frac{\mu L}{\alpha^3 B}\right)^{\frac{1}{4}} v_x \ , \ \alpha = \frac{k}{\rho C p}$$

and $$B = \rho g \beta \ (T_0 - T_1)$$

The boundary conditions are given below

at $\eta = 0$; $\phi_x = \phi_z = 0$ and $\theta = 1$

at $\eta = \infty$; $\phi_x = \phi_z = 0$ and $\theta = 0$

at $\xi = -\infty$; $\phi_x = \phi_z = 0$ and $\theta = 0$

Observing equations (7) and (8), it can be inferred that the dimensionless velocity components ϕ_x and ϕ_z and the dimensionless temperature θ depend on η, ξ and Pr. The flow in free convection is very slow, therefore, the terms containing Pr will be very small and their dependence on Prandtl number would be much less compared to the other parameters.

The expression for average heat transfer from the plate can be written as

$$q_{av} = \frac{k}{L} \int_0^L - \left(\frac{\partial T}{\partial x} \bigg|_{\text{at } x = 0} \right) dz$$

and in terms of dimensionless parameters

$$q_{av} = k(T_0 - T_1) \left(\frac{B}{\mu \alpha L}\right)^{\frac{1}{4}} \int_0^1 \left(- \frac{\partial \theta}{\partial \eta} \bigg|_{\text{at } \eta=0} \right) d\xi$$

$$= k(T_0 - T_1) \left(\frac{B}{\mu \alpha L}\right)^{\frac{1}{4}} C$$

$$= C \frac{k}{L} (T_0 - T_1) (Gr \ Pr)^{\frac{1}{4}}$$

C is a weak function of the Prandtl number and it remains constant.

• PROBLEM 6-2

Water at 80° F is being heated by a vertical plate. Determine the temperature at which the plate should be maintained if the rate of heat transfer from the plate to water is 5000 Btu/hr. The values of the constants for the number may be taken as a = 0.13 and m = 1/3.

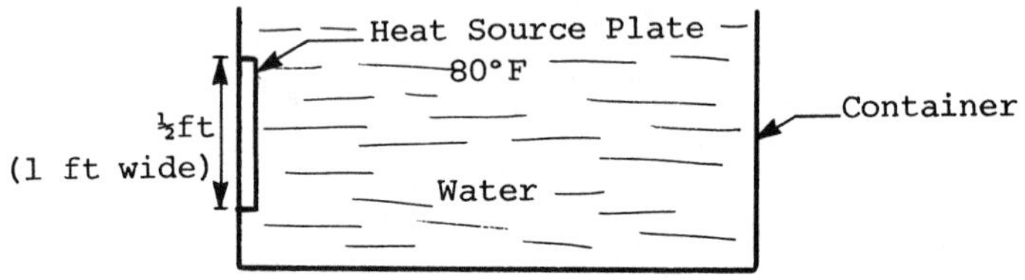

t(°F)	Saturation pressure (psia)	ρ (lbm/ft³)	μ (lb/hr ft)	k (Btu/hr ft °F)	c_p (Btu/lbm °F)	N_{Pr}	a† x 10⁻⁶ (1/ft³ °F)
40	0.122	62.43	3.74	0.326	1.0041	11.5	
50	0.178	62.41	3.16	0.334	1.0013	9.49	
60	0.256	62.36	2.72	0.341	0.9996	7.98	154
70	0.363	62.30	2.37	0.347	0.9987	6.81	238
80	0.507	62.22	2.08	0.353	0.9982	5.89	330
90	0.698	62.12	1.85	0.358	0.9980	5.15	435
100	0.949	62.00	1.65	0.363	0.9980	4.55	547
110	1.28	61.86	1.49	0.367	0.9982	4.06	670
120	1.69	61.71	1.35	0.371	0.9985	3.64	804
130	2.22	61.55	1.24	0.375	0.9989	3.29	949
140	2.89	61.38	1.13	0.378	0.9994	3.00	1100
150	3.72	61.20	1.05	0.381	1.0000	2.74	1270
160	4.74	61.00	0.968	0.384	1.0008	2.52	1450
170	5.99	60.80	0.900	0.386	1.0017	2.33	1640
180	7.51	60.58	0.839	0.388	1.0027	2.17	1840
190	9.34	60.36	0.785	0.390	1.0039	2.02	2050
200	11.53	60.12	0.738	0.392	1.0052	1.89	2270
212	14.696	59.83	0.686	0.394	1.0070	1.76	2550
220	17.19	59.63	0.655	0.394	1.0084	1.67	2750
240	24.97	59.11	0.588	0.396	1.0124	1.51	3270
260	35.43	58.86	0.534	0.397	1.0173	1.37	3830
280	49.20	57.96	0.487	0.397	1.0231	1.26	4420
300	67.01	57.32	0.449	0.396	1.0297	1.17	5040
320	89.66	56.65	0.418	0.395	1.0368	1.10	5660
340	118.01	55.94	0.393	0.393	1.0451	1.04	6300
360	153.0	55.19	0.371	0.391	1.0547	1.00	6950
380	195.8	54.38	0.351	0.388	1.0662	0.97	7610
400	247.3	53.51	0.333	0.384	1.0800	0.94	8320
420	308.8	52.61	0.316	0.379	1.0968	0.92	9070
440	381.6	51.68	0.301	0.374	1.1168	0.90	9900
460	466.9	50.70	0.285	0.368			
480	566.1	49.67	0.272	0.362			

Solution: An iterative method is adopted to solve the problem. A plate temperature is assumed and using the Nusselt number, the convective heat transfer coefficient is obtained Subsequently, the plate temperature is re-calculated. The above procedure is repeated until the value

$T_{p_{assumed}} \cong T_{p_{calculated}}$ is approached.

The heat transferred from the plate is equal to the heat supplied to the plate

Therefore, $q = hA\Delta T \quad (\Delta T = T_p - T_w)$

$$\text{or} \quad 5000 = h(\tfrac{1}{2} \times 1)(T_p - 80)$$

where subscripts p and w refer to plate and water, respectively

$$\text{or} \quad 10{,}000 = h(T_p - 80) \qquad (1)$$

The average film temperature at the plate-water interface is given by

$$T_f = \tfrac{1}{2}(T_p + T_w)$$

The turbulent (boundary layer type) free convection flow is given by the equation

$$\frac{h\ell}{k_f} = Nu = 0.13\,(Ra)^{1/3}$$

$$\text{or} \quad h = 0.13\,\frac{K_f}{l}\,(Ra)^{1/3} \qquad (2)$$

Also, $Ra = a_f L^3 \Delta T$

Hence, $$h = 0.13\,K_f\,a_f^{1/3}\,(T_p - T_w)^{1/3} \qquad (3)$$

Equating (1) and (2), we get

$$T_p - 80 = \left(\frac{77{,}000}{K_f a_f^{1/3}}\right)^{3/4} \qquad (4)$$

A value of T_p is assumed in order to calculate the film temperature, the values of a_f and k_f are obtained at this temperature (T_f), then T_p is calculated from eq. (3) and compared with the assumed value. If the two values differ substantially then this calculated value becomes the new assumed value and the procedure is repeated till we get nearly the same value.

Assume $T_p = 120°F$, then

$$T_f = \frac{1}{2}(120 + 80) = 100°F$$

From property tables of water at 100°F,

$$a_f = 547 \times 10^6$$

$$K_f = 0.363$$

Therefore, $a_f^{1/3} = 812.$

$$T_p - 80 = \left[\frac{77{,}000}{0.363(812)}\right]^{3/4}$$

or $\quad T_p = 144°F$

Assume $T_p = 144°F$

$$T_f = \frac{1}{2}(144 + 80) = 112°F.$$

Therefore, $a_f = 697 \times 10^6$

$$K_f = 0.367$$

$$a_f^{1/3} = 883$$

$$T_p - 80 = \left[\frac{77{,}000}{0.367(883)}\right]^{3/4}$$

or $\quad T_p = 140°F$ which is reasonably close to the assumed value.

Checking for the Rayleigh number

$$R_{a_f} = a_f L^3 \Delta T$$

$$= 697 \times 10^6 \left(\tfrac{1}{2}\right)^3 (140-80) = 5.24 \times 10^9$$

Since $R_{a_f} > 10^9$, use of equation (2) is justified, and the required plate surface temperature is 140°F.

• PROBLEM 6-3

Compute the film heat transfer coefficient and the heat flux from a vertical surface at 210°F. The surrounding air temperature is 70°F.

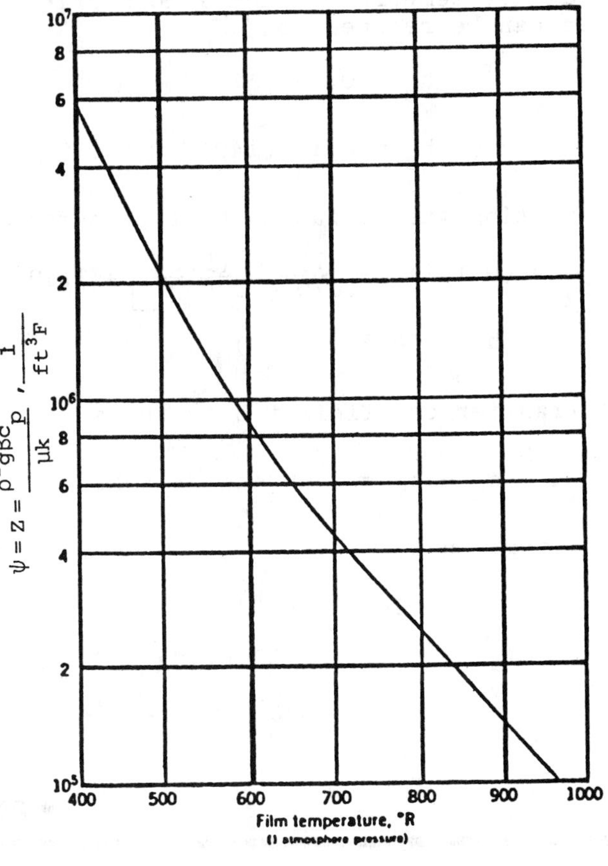

Solution: The average film temperature $t_f = \frac{210 + 70}{2} = 140°F$.
From the graph, corresponding to this temperature (140 + 460 = 600°R), the value of ψ (from ψ vs τ graph) $\psi = 0.88 \times 10^6$.

Let
$$X = \frac{L^3 \rho_f^2 g \beta_f \Delta t}{\mu_f^2} \left(\frac{C_p \mu}{k}\right)_f$$

$$= 0.88 \times 10^6 \ L^3 \ \Delta t$$

$$\Delta t = 210\text{-}70 = 140$$

Therefore,
$$X = 1.23 \times 10^8 \ L^3$$

For air flow in the range of $X = 10^9$ to 10^4, i.e., in the laminar flow range, the recommended equation for Nusselt number is

$$\frac{h_c L}{k_f} = 0.59 \left[\frac{L^3 \rho_f^2 g \beta_f \Delta t}{\mu_f^2} \left(\frac{C_p \mu}{k}\right)_f\right]^{\frac{1}{4}}$$

For air at ordinary temperature and atmospheric pressure, the above equation can be reduced to

$$h_c = 0.19\,[\Delta t]^{1/3}$$

Thus, $$h_c = 0.19\,(140)^{1/3} = 0.99$$

Further, calculating the radiation heat transfer coefficient h_r

$$h_{r_b} = (0.171)\left[\frac{(6.7)^4 - (5.3)^4}{(670 - 530)}\right] = 1.49$$

Therefore, $$h_r = (0.9 \times 1.49) = 1.34$$

The net heat transfer coefficient h is the sum of h_c and h_r

$$h = h_c + h_r = 0.99 + 1.34 = 2.33$$

The heat flux is given by

$$q/A = h(\Delta t)$$

$$= 2.33\,(140)$$

$$= 326 \text{ Btu/hr-ft}^2$$

• PROBLEM 6-4

Consider a heating chamber with a 1 ft high vertical wall. The wall surface temperature is 450°F and the ambient temperature is 100°F. Find the convective heat transfer coefficient and the rate of heat transfer per unit surface area of the wall. Neglect radiation heat transfer.

Table 1.

Physical Properties of Air at 101.325 kPa (1 Atm Abs) (English Units)

T (°F)	ρ $\left(\frac{lb_m}{ft^3}\right)$	c_p $\left(\frac{btu}{lb_m \cdot °F}\right)$	μ (centipoise)	k $\left(\frac{btu}{h \cdot ft \cdot °F}\right)$	Pr	$\beta \times 10^3$ $(1/°R)$	$g\beta\rho^2/\mu^2$ $(1/°R \cdot ft^3)$
0	0.0861	0.240	0.0162	0.0130	0.720	2.18	4.39×10^6
32	0.0807	0.240	0.0172	0.0140	0.715	2.03	3.21×10^6
50	0.0778	0.240	0.0178	0.0144	0.713	1.96	2.70×10^6
100	0.0710	0.240	0.0190	0.0156	0.705	1.79	1.76×10^6
150	0.0651	0.241	0.0203	0.0169	0.702	1.64	1.22×10^6
200	0.0602	0.241	0.0215	0.0180	0.694	1.52	0.840×10^6
250	0.0559	0.242	0.0227	0.0192	0.692	1.41	0.607×10^6
300	0.0523	0.243	0.0237	0.0204	0.689	1.32	0.454×10^6
350	0.0490	0.244	0.0250	0.0215	0.687	1.23	0.336×10^6
400	0.0462	0.245	0.0260	0.0225	0.686	1.16	0.264×10^6
450	0.0437	0.246	0.0271	0.0236	0.674	1.10	0.204×10^6
500	0.0413	0.247	0.0280	0.0246	0.680	1.04	0.163×10^6

Table 2.

Constants for use with Eq. (1) for Nusselt Number

Physical Geometry	(Gr Pr)	*a*	*m*	*Ref.*
Vertical planes and cylinders				
[vertical height $L < 1$ m (3 ft)]				
	$< 10^4$	1.36	$\frac{1}{5}$	(P1)
	10^4–10^9	0.59	$\frac{1}{4}$	(M1)
	$> 10^9$	0.13	$\frac{1}{3}$	(M1)
Horizontal cylinders				
[diameter D used for L and $D < 0.20$ m (0.66 ft)]				
	$< 10^{-5}$	0.49	0	(P1)
	10^{-5}–10^{-3}	0.71	$\frac{1}{25}$	(P1)
	10^{-3}–1	1.09	$\frac{1}{10}$	(P1)
	1–10^4	1.09	$\frac{1}{5}$	(P1)
	10^4–10^9	0.53	$\frac{1}{4}$	(M1)
	$> 10^9$	0.13	$\frac{1}{3}$	(P1)
Horizontal plates				
Upper surface of heated plates or lower surface of cooled plates	10^5–2×10^7	0.54	$\frac{1}{4}$	(M1)
	2×10^7–3×10^{10}	0.14	$\frac{1}{3}$	(M1)
Lower surface of heated plates or upper surface of cooled plates	10^5–10^{11}	0.58	$\frac{1}{5}$	(F1)

Solution: For a vertical plate at constant temperature and having a height of less than 3 ft, the average natural convective heat transfer coefficient is given by the following equation

$$Nu = \frac{hL}{k} = a\left(\frac{L^3\rho^2 g\,\beta\,\Delta T}{\mu^2}\;\frac{C_p\mu}{k}\right)^m$$

$$= a(GrPr)^m$$

where a and m are constants and can be found from the given table.

The physical properties of air are found at the average film temperature T_f

$$T_f = \frac{T_w + T_b}{2} = \frac{450 + 100}{2} = 275°F$$

Hence, from table 1, at 275°F

$$k = 0.0198 \text{ Btu/hr-ft-°F}$$

$$\rho = 0.0541 \text{ lbm/ft}^3$$

$$\mu = (0.0232 \text{ Cp})\ (2.4191) = 0.0562 \text{ lbm/ft-hr}$$

$$\beta = 1/(460 + 275) = 1.36 \times 10^{-3}/°R$$

$$Pr = 0.690 \text{ and } \Delta T = 450 - 100 = 350°F$$

Grashoff number is defined as

$$Gr = \frac{L^3 \rho^2 g \beta \Delta T}{\mu^2}$$

$$= \frac{(1.0)^3 (0.0541)^2 (32.174)(3600)^2 (1.36 \times 10^{-3})(350)}{(0.0562)^2}$$

$$= 1.84 \times 10^8$$

$$Gr\ Pr = (1.84 \times 10^8)(0.69)$$

$$= 1.27 \times 10^8$$

From table 2, for vertical plates, corresponding to $Gr = 1.27 \times 10^8$, the constants a and m are 0.59 and $\frac{1}{4}$, respectively. Therefore,

$$Nu = \frac{hL}{k} = 0.59\ (1.27 \times 10^8)^{\frac{1}{4}}$$

$$\text{or} \quad h = 0.59\ (1.27 \times 10^8)^{\frac{1}{4}} \left(\frac{0.0198}{1} \right)$$

$$\therefore \quad h = 1.24 \text{ Btu/hr-ft}^2\text{-°F}.$$

The rate of heat transfer per unit wall surface area is given by

$$q = hA(T_w - T_b)$$

$$= 1.24 \times 1 \times (450-100) = 434 \text{ Btu/hr}.$$

• PROBLEM 6-5

Compute the rate of heat transfer from a 3 ft × 2.5 ft front panel of an oven. The panel is at 105°F and the ambient air temperature is 70°F. Assume the heat transfer to take place by natural convection only.

TABLE PROPERTY VALUES OF AIR AT ATMOSPHERIC PRESSURE

T, F	ρ, lb/ft^3	c_p, Btu/lb F	μ lb/sec ft	v, ft^2/sec	k, Btu/hr ft F	α, ft^2/hr	Pr
-280	0.2248	0.2452	0.4653×10^{-5}	2.070×10^{-5}	0.005342	0.09691	0.770
-190	0.1478	0.2412	0.6910	4.675	0.007936	0.2226	0.753
-100	0.1104	0.2403	0.8930	8.062	0.01045	0.3939	0.739
-10	0.0882	0.2401	1.074	10.22	0.01287	0.5100	0.722
80	0.0735	0.2402	1.241	16.88	0.01516	0.8587	0.708
170	0.0623	0.2410	1.394	22.38	0.01735	1.156	0.697
260	0.0551	0.2422	1.536	27.88	0.01944	1.457	0.689

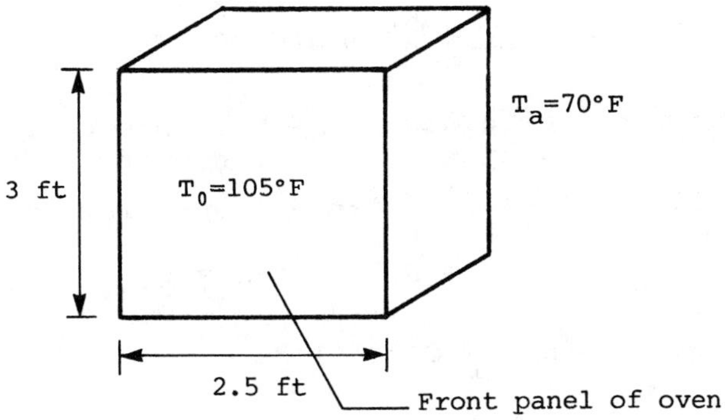

Solution: The characteristics of the air flow around the panel are determined by the product of Prandtl and Grashoff numbers.

The properties of air at mean temperature

$$T_f = \frac{T_o + T_a}{2} = \frac{105 + 70}{2} = 87.5°F \text{ are}$$

$$Cp = 0.24 \text{ Btu/lbm °F},$$

$$\nu = 0.174 \times 10^{-3} \text{ ft}^2\text{/sec.}$$

$$\beta = \frac{1}{(87.5 + 460)R} \quad \text{i.e. } \beta = 1.83 \times 10^{-3} \text{ °R}^{-1}$$

$$k = 0.0151 \text{ Btu/hr-ft-°F}$$

and $\quad Pr = 0.72$

The Grashoff number is defined as

$$Gr = \frac{L^3\rho^2 g\beta\Delta T}{\mu^2} \qquad (\nu = \frac{\mu}{\rho})$$

$$= \frac{L^3 g\beta\Delta T}{\nu^2}$$

$$= \frac{(3)^3(32.2)(1.83 \times 10^{-3}) \times (105\text{-}70)}{(0.174 \times 10^{-3})^2}$$

$$= 1.85 \times 10^9$$

Therefore, $Gr \times Pr = 1.85 \times 10^9 \times 0.72 = 1.32 \times 10^9$. Since the product $Gr\ Pr > 10^9$, the flow around the panel is turbulent.

The heat transfer coefficient can be obtained from the Nusselt number which is defined as

$$Nu = \frac{hL}{k} = a(Gr\ Pr)^m$$

where a and m are obtained from the table corresponding to the product of Grashof and Prandtl numbers. Hence for $Gr\ Pr = 1.32 \times 10^9$ (Refer to table in problem No. 4),

$$a = 0.13 \text{ and } m = 1/3$$

Therefore,
$$h = \frac{k}{L}\ 0.13\ (1.32 \times 10^9)^{1/3}$$

or
$$h = \frac{0.0151}{3}\ (0.13)\ (1.32 \times 10^9)^{1/3}$$

$$= 0.72 \text{ Btu/hr-ft}^2\text{-°F}$$

Therefore, the heat transfer rate

$$q = hA\Delta T$$

$$= 0.72 \times (3 \times 2.5) \times (105-70)$$

$$= 189 \text{ Btu/hr}$$

• PROBLEM 6-6

Compute the natural convective heat transfer coefficient for a vertical plate at a distance of 8 in. from the bottom of the plate. The plate is heated to a temperature of 200°F on one side and the other side is in contact with air at 68°F and at atmospheric pressure.

Solution: The convective heat transfer coefficient at any point, on a vertical wall, is given by

$$h = 2\frac{k}{\delta}$$

where δ - is the film or boundary layer thickness.

The Grashof number is defined as

$$Gr = \frac{L^3 \rho^2 g\beta\Delta T}{\mu^2} = \frac{L^3 g\beta\Delta T}{\nu^2} \qquad (\text{as } \nu = \frac{\mu}{\rho})$$

Average temperature $T = \frac{200 + 68}{2} = 134°F$. Properties of air at atmospheric pressure and at temperature 200°F are (refer to table in problem No. 5)

$$k = 0.0181 \text{ Btu/hr-ft-°F}$$

$$\nu = 24.21 \times 10^{-5} \text{ ft}^2\text{/sec}, \quad Pr = 0.694$$

and $$g = 32.2 \text{ ft/sec}^2$$

Temperature difference $\Delta T = 200\text{-}68 = 132°F$ and expansion coefficient $\beta = \frac{1}{528R}$

Substituting these values

$$Gr = 40.7 \times 10^6 \text{ (value is between } 10^4 - 10^9\text{)}$$

∴ The boundary layer is laminar

$$\therefore \frac{\delta}{x} = (3.93)\left(\frac{0.952 + Pr}{Gr \times Pr^2}\right)^{1/4}$$

$$= (3.93)\left[\frac{(0.952 + 0.694)}{(40.7 \times 10^6)(0.694)^2}\right]^{1/4}$$

$$\frac{\delta}{x} = 0.0670 \qquad x = 8 \text{ in}$$

$$\therefore \delta = 0.0670 \times 8 = 0.536 \text{ in}$$

$$\therefore h = 2\frac{k}{\delta} = 2 \times \frac{0.0181 \times 12}{0.536}$$

$$h = 0.81 \text{ Btu/hr-ft}^2\text{-}^0\text{F}.$$

The average heat transfer coefficient over the length, 8 in, of the plate is

$$\bar{h} = \frac{4}{3} h = \frac{4}{3} \times 0.81 = 1.08 \text{ Btu/hr-ft}^2\text{- }^0\text{F}.$$

• PROBLEM 6-7

A 3.5m × 2m vertical plate is subjected to a constant heat flux of 800 W/m². The back of the plate is insulated and the ambient air temperature is 30°C. Determine the average surface temperature of the plate, assuming that all the incident radiation is lost by free convection to the surrounding air.

Physical Properties of Air at Atmopheric Pressure (SI Units)

T (°C)	T (K)	ρ (kg/m³)	c_p (kJ/kg · K)	μ × 10⁵ (Pa·s, or kg/m·s)	k (W/m · K)	Pr	β × 10³ (1/K)	$g\beta\rho^2/\mu^2$ (1/K·m³)
-17.8	255.4	1.379	1.0048	1.62	0.02250	0.720	3.92	2.79×10^8
0	273.2	1.293	1.0048	1.72	0.02423	0.715	3.65	2.04×10^8
10.0	283.2	1.246	1.0048	1.78	0.02492	0.713	3.53	1.72×10^8
37.8	311.0	1.137	1.0048	1.90	0.02700	0.705	3.22	1.12×10^8
65.6	338.8	1.043	1.0090	2.03	0.02925	0.702	2.95	0.775×10^8
93.3	366.5	0.964	1.0090	2.15	0.03115	0.694	2.74	0.534×10^8
121.1	394.3	0.895	1.0132	2.27	0.03323	0.692	2.54	0.386×10^8
148.9	422.1	0.838	1.0174	2.37	0.03531	0.689	2.38	0.289×10^8
176.7	449.9	0.785	1.0216	2.50	0.03721	0.687	2.21	0.214×10^8
204.4	477.6	0.740	1.0258	2.60	0.03894	0.686	2.09	0.168×10^8
232.2	505.4	0.700	1.0300	2.71	0.04084	0.684	1.98	0.130×10^8
260.0	533.2	0.662	1.0341	2.80	0.04258	0.680	1.87	1.104×10^8

Solution: In this problem, the plate surface temperature and the heat transfer coefficient are not known. The analysis is initiated with an assumed heat transfer coefficient and using this value, the surface temperature is calculated. Subsequently, using the Nusselt number, a new heat transfer coefficient is evaluated. This procedure is repeated till the assumed value nearly equals the calculated Nusselt number.

Assuming a value of $h = 10\text{W/m}^2\text{°C}$, heat transfer per unit area is

$$q = h\Delta T$$

or
$$\Delta T = \frac{q}{h} = \frac{800}{10} = 80\text{°C}.$$

Therefore,
$$T_f = \frac{80}{2} + 30 = 70\text{°C or } 343\text{°K}.$$

The properties of air at 70°C are

$$\nu = 2.005 \times 10^{-5}\text{m}^2/\text{s} \;, \quad \beta = \frac{1}{T_f} = 2.92 \times 10^{-3}\text{°K}^{-1}$$

$$k = 0.0295 \text{ W/m°C} \quad \text{and} \quad Pr = 0.7$$

The Grashof number is defined as

$$Gr = \frac{g\beta q_w x^4}{k\nu^2}$$

$$Gr = \frac{9.8 \times (2.92 \times 10^{-3})(800)(3.5)^4}{(0.0295)(2.005 \times 10^{-5})^2}$$

$$= 2.9 \times 10^{14} \quad (\text{greater than } 10^9)$$

The flow of air around the plate is turbulent as the Grashof number is greater than 10^9. Hence, the Nusselt number is given by

$$Nu = 0.17\ (Gr\ Pr)^{1/4}$$

$$Nu = \frac{hx}{k} = 0.17\ (Gr\ Pr)^{1/4}$$

or $$h = \frac{0.0295}{3.5}\ (0.17)\ (2.9 \times 10^{14} \times 0.7)^{1/4}$$

$$= 5.41\ W/m^2\ °C.$$

Now, in a turbulent flow the value of h does not change appreciably, therefore it is reasonable to assume a constant value of h for further iterations.

$$\Delta T = \frac{q}{h} = \frac{800}{5.41} = 148°C$$

Therefore, $$T_f = \frac{148}{2} + 30 = 104°C$$

Properties of air at 104°C are

$$\nu = 2.354 \times 10^{-5}\ m^2/s\ ,\ \beta = \frac{1}{T_f} = 2.65 \times 10^{-3}$$

$$k = 0.0320\ W/m°C\ ,\ Pr = 0.695$$

Hence, $$Gr = \frac{(9.8)(2.65 \times 10^{-3})(800)(3.5)^4}{(0.0320)(2.354 \times 10^{-5})^2}$$

$$= 1.758 \times 10^{14}$$

or using the Nusselt number

$$h = \frac{k}{x}\ (0.17)(Gr\ Pr)^{1/4}$$

$$= \frac{(0.0320)(0.17)(1.758 \times 10^{14} \times 0.695)^{1/4}}{3.5}$$

$$= 5.17\ W/m^2°C$$

Thus, the new temperature difference is

$$\Delta T = \frac{800}{5.17} = 155°C$$

and the average wall temperature is

$$155 + 30 = 185°C.$$

FREE CONVECTION HEAT TRANSFER FROM HORIZONTAL PLATES

• PROBLEM 6-8

Compute the rate of heat transfer by natural convection from the surface of a horizontal, 7 in. square, plate at a temperature of 300°F. The ambient air temperature is 80°F.

EMPIRICAL EQUATIONS FOR NATURAL CONVECTION FROM HORIZONTAL PLATES TO ROOM TEMPERATURE AIR AT ATMOSPHERIC PRESSURE

Configuration	Equation	
	Turbulent Region Range of X $2 \times 10^7 - 3 \times 10^{10}$	Laminar Region Range of X $10^5 - 2 \times 10^7$ L is expressed in feet
Heated horizontal plates facing upward	$h = 0.22\Delta t^{1/3}$	$h = 0.27\left(\frac{\Delta t}{L}\right)^{1/4}$
Heated horizontal plates facing downward	...	$h = 0.12\left(\frac{\Delta t}{L}\right)^{1/4}$
Cooled horizontal plates facing upward	...	$h = 0.12\left(\frac{\Delta t}{L}\right)^{1/4}$
Cooled horizontal plates facing downward	$h = 0.22\Delta t^{1/3}$	$h = 0.27\left(\frac{\Delta t}{L}\right)^{1/4}$

Solution: The average film temperature at the interface of the air and plate is

$$T_f = \frac{300 + 80}{2} = 190°F$$

or $$460 + 190 = 650°R$$

From the graph, corresponding to 650°R (refer to graph in problem No. 3), the value of Z is 61×10^4. The Rayleigh number $Ra = ZL^3\Delta T$.

Hence, $$Ra = 61 \times 10^4\left(\frac{7}{12}\right)^3 (300-80)$$

$$= 2.66 \times 10^7$$

This value of Rayleigh number falls in the range for which the flow of air around the plate is turbulent, and hence the relation $h = 0.22(\Delta T)^{1/3}$ is applicable.

Therefore, $$h = 0.22\ (300-80)^{1/3}$$

$$= 1.32\ \text{Btu/hr-ft}^2\text{-°F}$$

Thus, the rate of heat transfer by natural convection is

$$q = 1.32 \times \frac{7 \times 7}{12 \times 12}(300-80)$$

$$= 9.88\ \text{Btu/hr.}$$

A 16 cm dia plate is suspended in a large reservoir of water at 70°C. Determine the rate of heat input required to the plate so as to maintain it at a constant surface temperature of 130°C.

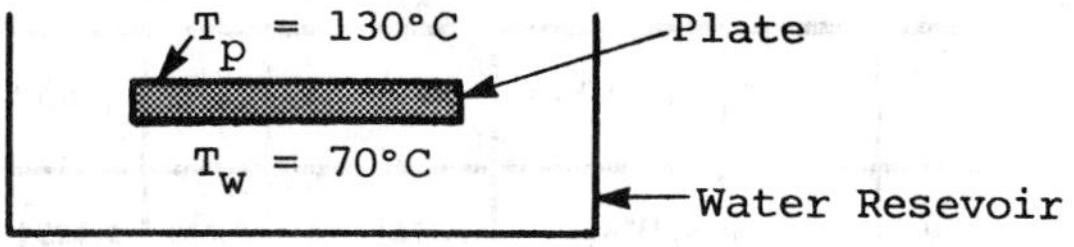

Solution: Since it is not specified, it is assumed that the plate convects heat from both its top and bottom surfaces.

Rate of heat input to the plate = Heat convected to water by natural convection

In order to estimate the heat loss, the heat transfer coefficient for the top (h_t) and the bottom (h_b) surfaces are determined. The flow characteristics of water depend on the value of the product of Grashof (Gr) and Prandtl number (Pr).

Grashof number $$Gr = \frac{g\beta\Delta T\, L_c^{\,3}}{\nu^2}$$

where L_c is the characteristic length of the circular plate

$$= \frac{(\text{surface area of plate})}{\text{perimeter of plate}} = \frac{(\pi/4)D^2}{\pi D}$$

$$= \frac{D}{4} = \frac{16}{4} = 4 \text{ cm. or } 0.04 \text{ m.}$$

g is the acceleration due to gravity

$= 9.81 \text{ m/sec}^2$

β is the coefficient of expansion per °K

$= 0.75 \times 10^{-3}$ 1/°K

ΔT is the temperature difference

$= (130-70) = 60°C$

ν is the viscosity of the water $= \frac{\mu}{\rho}$

(from the table, $\frac{\mu}{\rho}$ at the average film temperature $\frac{130+70}{2} = 100°C$ is 0.294×10^{-6} m²/sec.)

$$\text{Therefore, } Gr = \frac{(9.81) \times (0.75 \times 10^{-3}) \times (60) \times (0.04)^3}{(0.294 \times 10^{-6})^2}$$

$$= 3.27 \times 10^8$$

Property Values for Water in a Saturated State
(SI System of Units)

t, C	ρ, kg/m³	cp, j/kg K ($\times 10^3$)	ν, m²/s ($\times 10^{-6}$)	k, W/m K	α, m²/s ($\times 10^{-7}$)	Pr	β, K⁻¹
0	1,002.28	4.2178	1.788	0.552	1.308	13.6	
20	1,000.52	4.1818	1.006	0.597	1.430	7.02	0.18×10^{-3}
40	994.59	4.1784	0.658	0.628	1.512	4.34	
60	985.46	4.1843	0.478	0.651	1.554	3.02	
80	974.08	4.1964	0.364	0.668	1.636	2.22	
100	960.63	4.2161	0.294	0.680	1.680	1.74	
120	945.25	4.250	0.247	0.685	1.708	1.446	
140	928.27	4.283	0.214	0.684	1.724	1.241	
160	909.69	4.342	0.190	0.680	1.729	1.099	
180	889.03	4.417	0.173	0.675	1.724	1.004	
200	866.76	4.505	0.160	0.665	1.706	0.937	
220	842.41	4.610	0.150	0.652	1.680	0.891	
240	815.66	4.756	0.143	0.635	1.639	0.871	
260	785.87	4.949	0.137	0.611	1.577	0.874	
280.6	752.55	5.208	0.135	0.580	1.481	0.910	
300	714.26	5.728	0.135	0.540	1.324	1.019	

Corresponding to a film temperature of 100°C the Prandtl number from the table is 1.75.

Hence, $Ra = 3.27 \times 10^8 \times 1.75 = 5.72 \times 10^8$

The relation for the Nusselt number for convective heat transfer, from the top surface, when $Ra > 10^7$ is given by

$$Nu_t = 0.15\,(Ra)^{1/3}$$

Also, the relation for the Nusselt number for convective heat transfer from the bottom surface for $3 \times 10^5 < Ra < 3 \times 10^{10}$ is given by

$$Nu_b = 0.27\,(Ra)^{1/4}$$

The Nusselt number is also expressed as

$$Nu = \frac{h\,L_c}{k}$$

For the top surface,

$$Nu_t = \frac{h_t\,L_c}{k} = 0.15\,(Ra)^{1/3}$$

or $$h_t = \frac{k}{L_c} 0.15 (Re)^{1/3}$$

$$= \frac{0.68}{0.04} \times 0.15 \times (5.72 \times 10^8)^{1/3}$$

$$= 2117 \text{ W/m}^2 \text{ °k.}$$

From the table, k = 0.68 W/m°K at 100°C.

Similarly, for the bottom surface,

$$Nu_b = \frac{h_b L_c}{k} = 0.27 (Re)^{1/4}$$

or $$h_b = \frac{k}{L_c} 0.27 (Re)^{1/4}$$

$$= \frac{(0.68)}{0.04} 0.27 (5.72 \times 10^8)^{1/4}$$

$$= 710 \text{ W/m}^2 \text{ °k}$$

Heat loss from the top surface

$$Q_t = A h_t (T_p - T_w)$$

and from the bottom surface

$$Q_b = A h_b (T_p - T_w)$$

where subscripts p and w stand for plate and water, respectively.

The total heat loss is given by the sum of the heat losses from the two surfaces,

$$Q = A (T_p - T_w) (h_t + h_f)$$

$$= \frac{\pi}{4} (0.26)^2 (130-70)(2117 + 710)$$

$$= 3410 \text{ W}$$

Thus, the rate of heat input to the plate to maintain its temperature at 130°C is 3410 W.

FREE CONVECTION HEAT TRANSFER FROM PIPES

• PROBLEM 6-10

Saturated steam at 30 psig and 50°F is flowing through a 2-in OD pipe. Calculate the rate of heat transfer when

(a) the pipe is surrounded by air at 55°F

(b) the pipe is surrounded by water at 55°F.

Assume the emissivity of the pipe surface as 0.9.

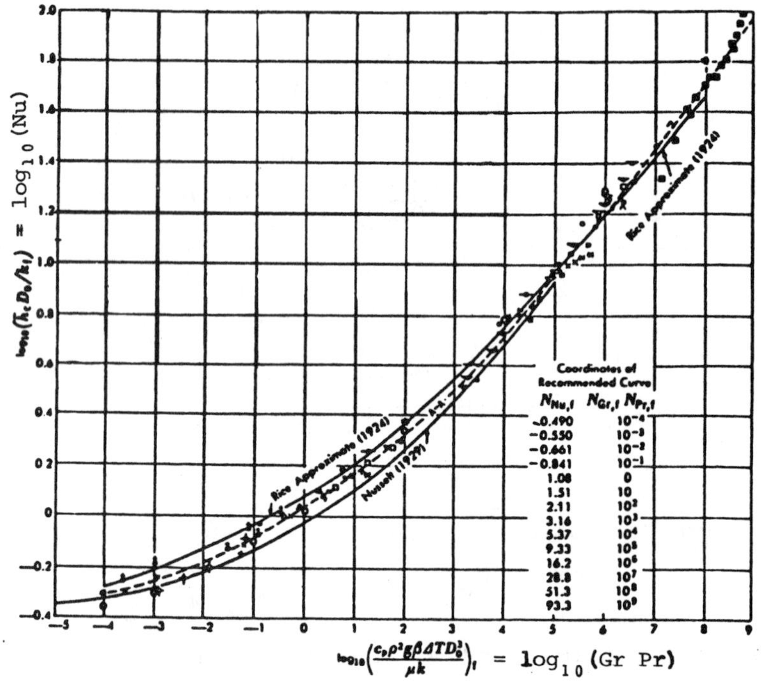

Free-convection heat transfer correlation for horizontal cylindrical surfaces.

Solution: Transfer of heat takes place by convection only when the pipe is surrounded by water, and when surrounded by air, heat transfer is by convection and radiation.

From the table, corresponding to an absolute pressure of 30 + 14.7 ≅ 45 psia the saturation temperature of steam is 274.4°F. Therefore, the average film temperature at the pipe-fluid interface is

$$T_f = \frac{274.4 + 55}{2} = 164.7°F$$

Physical Properties of Air at 1 Atm Abs

T (°F)	ρ $\left(\frac{lb_m}{ft^3}\right)$	c_p $\left(\frac{btu}{lb_m \cdot °F}\right)$	μ (centipoise)	k $\left(\frac{btu}{h \cdot ft \cdot °F}\right)$	Pr	$\beta \times 10^3$ (1/°R)	$g\beta\rho^2/\mu^2$ (1/°R·ft³)
0	0.0861	0.240	0.0162	0.0130	0.720	2.18	4.39×10^6
32	0.0807	0.240	0.0172	0.0140	0.715	2.03	3.21×10^6
50	0.0778	0.240	0.0178	0.0144	0.713	1.96	2.70×10^6
100	0.0710	0.240	0.0190	0.0156	0.705	1.79	1.76×10^6
150	0.0651	0.241	0.0203	0.0169	0.702	1.64	1.22×10^6
200	0.0602	0.241	0.0215	0.0180	0.694	1.52	0.840×10^6
250	0.0559	0.242	0.0227	0.0192	0.692	1.41	0.607×10^6
300	0.0523	0.243	0.0237	0.0204	0.689	1.32	0.454×10^6
350	0.0490	0.244	0.0250	0.0215	0.687	1.23	0.336×10^6
400	0.0462	0.245	0.0260	0.0225	0.686	1.16	0.264×10^6
450	0.0437	0.246	0.0271	0.0236	0.674	1.10	0.204×10^6
500	0.0413	0.247	0.0280	0.0246	0.680	1.04	0.163×10^6

The physical properties of air and water are obtained from the tables.

For air, refer to table in problem No. 4.

$$\mu = 0.04943 \text{ lbm/hr-ft}$$

$$k = 0.0171 \text{ Btu/hr-ft-°F}$$

$$C_p = 0.241 \text{ Btu/lbm °F}$$

$$\beta = \frac{1}{(164.7 + 460)} = 1.6 \times 10^{-3} \quad \frac{1}{°R}$$

$$Pr = 0.697, \ \rho = 0.0636 \text{ lbm/ft}^3$$

For water, refer to table in problem No. 2.

$$Pr = 2.43, \ \mu = 0.936 \text{ lb/hr ft}$$

$$\rho = 60.906 \text{ lbm/ft}^3, \ k = 0.385 \text{ Btu/hr ft°F}$$

$$C_p = 1.0012 \text{ Btu/lbm°F}, \ \beta = 1.6 \times 10^{-3} \frac{1}{°R}$$

Rayleigh number is the product of Grashof and Prandtl numbers.

$$Ra = (Gr \cdot Pr)$$

For air, $Gr = \dfrac{g\rho^2\beta\Delta TL^3}{\mu^2}$

$$= \frac{32.2 \times (0.0636)^2 \ 1.6 \times 10^{-3}}{(0.04943)^2} \times \left(\frac{2}{12}\right)^3 \times (3600)^2 \times$$

$$\times (274.4-55) = 1.1228 \times 10^6$$

For water, $Gr = \frac{32.2 \times (60.906)^2 \times 1.6 \times 10^{-3}}{(0.936)^2} \left(\frac{2}{12}\right)^3 (3600)^2$

$\times (274.4-55)$

$= 2.87 \times 10^9$

Therefore, $Ra = 1.1228 \times 10^6 \times 0.697 = 0.782 \times 10^6$ for air

$= 2.87 \times 10^9 \times 2.43 = 6.974 \times 10^9$ for water.

Knowing the values of the Rayleigh number for air and water, the corresponding values of the Nusselt number can be obtained from the graph. For air,

$$Ra = 0.782 \times 10^6$$

$\log (0.782 \times 10^6) = 5.893$ and corresponding to this value, $\log (Nu) = 1.2$

hence $Nu = 15.84$

For water

$$Ra = 6.974 \times 10^9$$

$\log (6.974 \times 10^9) = 9.843$ and corresponding to this value, $\log (Nu) = 1.98$ which gives

$$Nu = 95.5$$

Next, the convective heat transfer coefficient for water and air are calculated.

$$Nu = \frac{h_c D_o}{k}$$

For air, $h_c = \frac{15.84 \times 0.0171}{2/12} = 1.625$ Btu/hr-ft^2-°F

For water, $h_c = \frac{95.5 \times 0.385}{2/12} = 220.6$ Btu/hr-ft^2-°F.

The rate of heat transfer per unit length of pipe - when submerged in water:

$$q_w = h_c A \Delta T$$

$$= \frac{220.6 \times \pi \times 2 \times (274.4-55)}{12}$$

$$= 25{,}342 \text{ Btu/hr-ft}$$

When exposed to air:

$$\text{by convection, } q_1 = \frac{1.625 \times \pi \times 2 \times (274.4-55)}{12}$$

$$= 186.6 \text{ Btu/hr-ft}$$

$$\text{by radiation, } q_2 = \varepsilon A \sigma (T_1^4 - T_2^4)$$

where,

σ = Stefan-Boltzman constant

$= 0.171 \times 10^8$ Btu/hr $ft^2 °R^4$

$T_1 = 274.4 + 460 = 734.4°R$

$T_2 = 55 + 460 = 515°R$

ε = emissivity of the pipe surface = 0.9

$$q_2 = 0.9 \times \frac{\pi \times 2 \times 1}{12} \times 0.171 \left[\left(\frac{734.4}{100}\right)^4 - \left(\frac{515}{100}\right)^4 \right]$$

$$= 177.2 \text{ Btu/hr-ft}$$

The total heat transferred in air

$$= q_1 + q_2$$

$$= 186.6 + 177.2$$

$$= 363.8 \text{ Btu/hr-ft}$$

It is interesting to note that the rate of heat transfer in air is only 1.44 percent of the rate of heat transfer when the pipe is submerged in water.

• PROBLEM 6-11

The cylindrical housing of an electrical device is 48 in high and 30 in dia. The base is perfectly insulated. Determine the surface temperature of the housing when it is losing heat to air at 69°F by free convection, at a rate of 1500 watts.

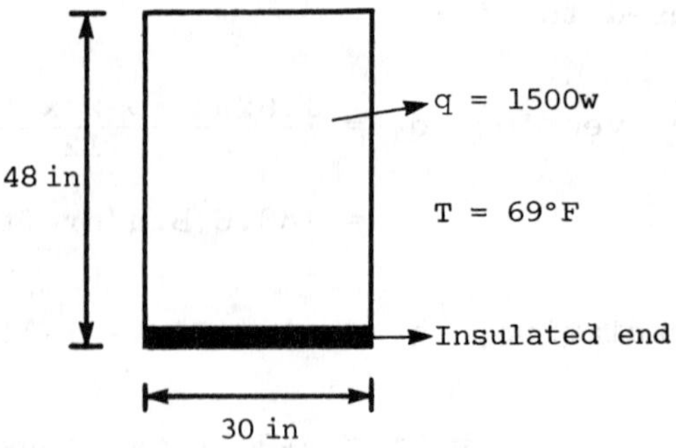

Geometry	Range of application	C	n	L
Vertical planes and cylinders	$10^4 < Gr_L Pr < 10^9$	0.29	$\frac{1}{4}$	height
	$10^9 < Gr_L Pr < 10^{12}$	0.19	$\frac{1}{3}$	1
Horizontal cylinders	$10^3 < Gr_D Pr < 10^9$	0.27	$\frac{1}{4}$	diameter
	$10^9 < Gr_D Pr < 10^{12}$	0.18	$\frac{1}{3}$	1
Horizontal plates—heated plates facing up or cooled plates facing down	$10^5 < Gr_L Pr < 2 \times 10^7$	0.27	$\frac{1}{4}$	length of side
	$2 \times 10^7 < Gr_L Pr < 3 \times 10^{10}$	0.22	$\frac{1}{3}$	1
Cooled plates facing up or heated plates facing down	$3 \times 10^5 < Gr_L Pr < 3 \times 10^{10}$	0.12	$\frac{1}{4}$	length of side

Solution: The heat transfer rate is given by the equation

$$q = hA\Delta T$$

where, A is the surface area

ΔT is the temperature difference

and h is the coefficient of heat transfer

$$= C\left(\frac{\Delta T}{L}\right)^n$$

C and n are obtained from the table.

It is assumed that the flow of air, around the heat convecting surface is laminar. Then, the values of C and n for the horizontal surface (top end) are ≅ 0.27 and $\frac{1}{4}$, respectively and for the vertical curved surface C and n are ≅ 0.29 and $\frac{1}{4}$, respectively

$$\therefore\ h_1 = 0.27\left(\frac{T-69}{30/12}\right)^{\frac{1}{4}} \quad \text{for top end}$$

$$h_2 = 0.29\left(\frac{T-69}{48/12}\right)^{\frac{1}{4}} \quad \text{for vertical surface}$$

Therefore, $q_1 = \frac{\pi}{4}\left[\left(\frac{30}{12}\right)^2\right] \times \left[0.27\left(\frac{T-69}{30/12}\right)^{\frac{1}{4}}\right](T-69)$

$$= 1.08\ (T-69)^{5/4}\ \text{Btu/hr} \quad \text{for top end}$$

and $q_2 = \left[\pi\left(\frac{30}{12}\right)\left(\frac{48}{12}\right)\right]\left[0.29\left(\frac{T-69}{30/12}\right)^{1/4}\right](T-69)$

$$= 6.44\ (T-69)^{5/4}\ \text{Btu/hr} \quad \text{for vertical surface.}$$

The total heat transfer is given by

$$q = 1500 \text{ watt} = q_1 + q_2$$

$$= (1.08 + 6.44)(T-69)^{5/4} \text{ Btu/hr}$$

$$[1500 \text{ watts} = (1500 \times 3.413) \text{ Btu/hr}]$$

or $1500 \times 3.413 = 7.52\ (T-69)^{5/4}$ Btu/hr

or $T = 184.67 + 69$

$$= 253.67°F = 254°F$$

• PROBLEM 6-12

The surface temperature of a long circular duct, 8 in OD, is 115°F and the ambient air temperature is 65°F. Determine the rate of heat loss from the surface of the duct if it is placed horizontally.

Solution: The mean film temperature at the interface of duct surface and air is

$$T_f = \frac{115 + 65}{2} = 90°F$$

From table 1, properties of air at 90°F are (refer to table in problem No. 4)

$$a = 1.948 \times 10^6/°R\ ft^3$$

$$k = 0.0152 \text{ Btu/hr-ft-°F}$$

and $Pr = 0.7026$

Raleigh number Ra is defined as

$$Ra = Gr * Pr$$

where $Gr = a\Delta T L^3$

Substituting values $a = 1.948 \times 10^6/°R\ ft^3$

$$\Delta T = 115-65 = 50°F$$

and $$L = \frac{8}{12} = 0.666\ ft.$$

$$Gr = (1.948 \times 10^6)(50) * (0.666)^3$$

$$= 2.88 \times 10^7$$

and $$Pr = 0.7026$$

$$\therefore\ Ra = (2.88 \times 10^7)(0.7026)$$

$$= 2.03 \times 10^7$$

For $Ra = 2.03 \times 10^7$ from table 2 constants for Nusselt number are $a = 0.53$ and $m = \frac{1}{4}$

$$\therefore\ \text{Nusselt Number } Nu = a \times (Ra)^m$$

$$Nu = 0.53 \times (2.03 \times 10^7)^{\frac{1}{4}} = 39.22$$

But $$Nu = \frac{hL}{k} = 39.22$$

$$\therefore\ h = \frac{Nu \times k}{L} = \frac{39.22 \times 0.0152}{8/12}$$

$$= 0.894\ Btu/hr\text{-}ft^2\text{-}°F$$

For the rate of heat transfer per unit length of the duct,

$$q = hA\Delta T$$

$$= 0.894 \quad (\pi \times \frac{8}{12}) \times 50$$

$$= 93.64\ Btu/hr\text{-}ft$$

• PROBLEM 6-13

A mercury thermometer, $\frac{1}{4}$ in dia, at 76°F is placed in an oven, which is at 400°F, at time $t = 0$. The thermometer behaves as an infinitely long cylinder with negligible resistance to heat transfer. Estimate the time required for the thermometer to reach a temperature of 275°F. The convective heat transfer coefficient between the thermometer surface and the hot air in the oven is 2 $Btu/hr\text{-}ft^2\,°F$.

Physical Properties of Mercury

Temperature (°F)	Thermal conductivity (Btu/hr ft °F)	Density (lbm/ft³)	Heat capacity (Btu/lbm °F)	Absolute† viscosity $\left(\frac{lbm}{ft\ sec} \times 10^3\right)$	Kinematic viscosity $\left(\frac{ft^2}{sec} \times 10^6\right)$	Thermal diffusivity (ft²/hr)	c_p/k $\left(\frac{hr\ ft}{lbm} \times 10^3\right)$	Pr
50	4.7	847	0.033	1.07	1.2	0.17	7.1	0.027
200	6.0	834	0.033	0.84	1.0	0.22	5.5	0.016
300	6.7	826	0.033	0.74	0.90	0.25	4.9	0.012
400	7.2	817	0.032	0.67	0.82	0.27	4.5	0.011
600	8.1	802	0.032	0.58	0.72	0.31	4.0	0.0084

Solution:

$$\begin{pmatrix}\text{Heat convected to}\\ \text{the thermometer}\end{pmatrix} = \begin{pmatrix}\text{Heat absorbed by}\\ \text{the mercury}\end{pmatrix}$$

$$hA(T_o - T_x) = \rho C_p V \frac{dT}{d\theta}$$

On rearrangement,

$$\frac{dT}{T_o - T} = \frac{hA}{\rho C_p V} d\theta$$

Assuming that h, ρ and C_p behave as constants, the solution for the differential equation is

$$\log\left(\frac{T_o - T_x}{T_o - T_m}\right) = -\frac{hA}{\rho C_p V}\theta$$

where $\frac{A}{V} = \frac{\pi D}{\frac{\pi}{4} D^2} = \frac{4}{D}$ and $T_x = 275°F$

$T_o = 400°F$

$T_m = 76°F$

For the mean temperature $\frac{275+76}{2} = 175.5°F$,

from tables, $C_p = 0.033$ Btu/lbm°F

$\rho = 836.1$ lbm/ft³

Substituting the values,

$$\theta = \frac{836.1 \times 0.033}{2} \times \left(\frac{0.25}{4\times 12}\right) \ln\left(\frac{400-76}{400-275}\right)$$

$= 0.0684$ hrs

$= 4.1$ min.

• PROBLEM 6-14

Consider a 0.521 ft dia pipe having a surface temperature of 950°F. The ambient air temperature is 86°F. Determine the heat loss by free convection per unit pipe length.

Properties of air at atmospheric pressure.

t(°F)	μ(lbm/hr ft)	k(Btu/hr ft °F)	c_p(Btu/lbm °F)	Pr	$a^* \times 10^{-6}$ (1/ft³°F)
-100	0.0319	0.0104	0.239	0.739	10.22
-50	0.0358	0.0118	0.239	0.729	5.4
0	0.0394	0.0131	0.240	0.718	3.13
50	0.0427	0.0143	0.240	0.712	1.94
100	0.0459	0.0157	0.240	0.706	1.26
150	0.0484	0.0167	0.241	0.699	0.86
200	0.0519	0.0181	0.241	0.693	0.59
250	0.0547	0.0192	0.242	0.690	0.42
300	0.0574	0.0203	0.243	0.686	0.312
400	0.0626	0.0225	0.245	0.681	0.180
500	0.0675	0.0246	0.248	0.680	0.111
600	0.0721	0.0265	0.250	0.680	0.072
700	0.0765	0.0284	0.254	0.682	0.049
800	0.0806	0.0303	0.257	0.684	0.0346
900	0.0846	0.0320	0.260	0.687	0.0251
1000	0.0884	0.0337	0.263	0.690	0.0187

$a^* \quad \dfrac{g\beta\rho^2 c_p}{\mu k}$

Solution: The average film temperature at the air-pipe (wall) interface is

$$T_f = \frac{950 + 86}{2} = 518°F.$$

The properties of air corresponding to this temperature, as read from the table, are:

$$\rho = 0.0405 \text{ lbm/ft}^3, \quad \mu = 0.0685 \text{ lbm/ft-hr.}$$

$$k = 0.0250 \text{ Btu/hr-ft-°F}, \qquad Pr = 0.68$$

$$\beta = \frac{1}{80+460} = \frac{1}{540} \quad \frac{1}{°R}$$

The Rayleigh number is defined as

$$Ra = Gr\ Pr$$

$$R_a = \left(\frac{g\beta L^3\rho^2\Delta T}{\mu^2}\right)Pr$$

$$= \frac{(32.2)(0.521)^3(0.0405)^2(3600)^2(950-86)}{(0.0685)^2 \times 546} \times 0.68$$

$$= 3.26 \times 10^7 \times 0.68 =$$

$$= 2.22 \times 10^7$$

The Nusselt number is defined as $Nu = a(Ra)^m$

Corresponding to a value of $Ra = 2.22 \times 10^7$, $a = 0.525$ and $m = \frac{1}{4}$.

Therefore, $$Nu = 0.525\ (2.22 \times 10^7)^{\frac{1}{4}}$$

$$= 36.1$$

Also, $$Nu = \frac{hL}{k} = \frac{h \times 0.521}{0.025} = 36.1$$

or $$h = 1.73\ \text{Btu/hr-ft}^2\text{-°F}$$

Hence, heat loss $$\frac{q}{L} = hA\ (T_p - T_a)$$

$$= 1.73 \times \pi(0.521)(950-86)$$

$$= 2440\ \text{Btu/hr-ft}$$

• PROBLEM 6-15

Derive the dimensionless equation for the heat transfer coefficient for natural convection in a horizontal pipe. The controlling physical variables, given below, are to be represented in terms of the fundamental dimensions length (L), mass (M), temperature (T), and time (θ).

Solution: For determining the heat transfer coefficient, the following factors are considered.

$$h = f\ (\rho,\ L,\ v,\ \mu,\ C_p,\ k)$$

$$\therefore \qquad f\ (h,\ \rho,\ L,\ v,\ \mu,\ C_p,\ k) = 0 \qquad (1)$$

This problem is controlled by 7 physical quantities containing 4 fundamental dimensions, so according to Buckingham theorem, this problem can be controlled by (7-4) = 3 non-dimensional factors which are known as π-terms.

$$\therefore \qquad f\ (\pi_1,\ \pi_2,\ \pi_3) = 0$$

Quantity	Dimensions	Dimensional Formula in M, L, T, θ system	Symbol
Absolute viscosity	kg_m/m-sec	$ML^{-1}T^{-1}$	μ
Velocity	m/sec	LT^{-1}	v
Specific heat	kcal/kg_m-°C	$L^2T^{-2}\theta^{-1}$	C_p
Thermal conductivity	kcal/m-hr-°C	$MLT^{-3}\theta^{-1}$	K
Heat transfer coefficient	kcal/m^2-hr-°C	$MT^{-3}\theta^{-1}$	h
Density	kg_m/m^3	ML^{-3}	ρ

According to Buckingham theorem, terms marked in equation (1) which contain all four fundamental dimensions must form a non-dimensional group with each remaining term.

Taking four factors as a group from equation (1) such that the group contains all the fundamental dimensions controlling the physical phenomenon, and substituting the dimensional formulae,

$$\therefore \quad \pi_1 = \mu(h)^a(\rho)^b(L)^c(v)^d = M^oL^oT^o\theta^o$$

$$\therefore \quad (ML^{-1}T^{-1})(MT^{-3}\theta^{-1})^a(ML^{-3})^b(L)^c(LT^{-1})^d = M^0L^0T^0\theta^0$$

Equating the powers of M, L, T, and θ of both sides, we get the following equations:

$$1 + a + b = 0$$

$$-1 - 3b + c + d = 0$$

$$-1 - 3a - d = 0$$

$$-a = 0$$

Solving the above equations, we get the following values :

$$a = 0, \quad b = -1, \quad c = -1, \quad d = -1$$

$$\pi_1 = \frac{\mu}{\rho l v} \qquad (2)$$

Similarly,

$$\pi_2 = C_p(h)^{a_1}(\rho)^{b_1}(L)^{c_1}(v)^{d_1} = M^oL^oT^o\theta^o$$

$$(L^2T^{-2}\theta^1)(MT^{-3}\theta^{-1})^{a_1}(ML^{-3})^{b_1}(L)^{c_1}(LT^{-1})^{d_1} = M^oL^oT^o\theta^o$$

Equating the powers of M, L, T, and θ of both sides, we get the following equations:

$$a_1 + b_1 = 0$$

$$2 - 3b_1 + c_1 + d_1 = 0$$

$$-2 - d_1 - 3a_1 = 0$$

$$-1 - a_1 = 0$$

Solving the above equations, we get the following values:

$$a_1 = -1, \quad b_1 = 1, \quad c_1 = 0, \quad \text{and} \quad d_1 = 1$$

$$\pi_2 = \frac{C_p \rho v}{h}$$

As dimensions of h and k/L are the same,

$$\pi_2 = \frac{C_p \rho v L}{k}$$

Similarly,

$$\pi_2 = k(h)^{a_2} (\rho)^{b_2} (L)^{c_2} (v)^{d_2} = M^\circ L^\circ T^\circ \theta^\circ$$

$$\therefore (MLT^{-3}\theta^{-1})(MT^{-3}\theta^{-1})^{a_2} (ML^{-3})^{b_2} (L)^{c_2} (LT^{-1})^{d_2} = M^\circ L^\circ T^\circ \theta^\circ$$

Equating the powers of M, L, T and θ of both sides, we get the following equations:

$$1 + a_2 + b_2 = 0$$

$$1 - 3b_2 + c_2 + d_2 = 0$$

$$-3 - 3a_2 - d_2 = 0$$

$$-1 - a_2 = 0$$

Solving the above equations, we get the following values.

$$a_2 = -1, \; b_2 = 0, \; c_2 = -1, \; d_2 = 0$$

$$\therefore \qquad \pi_3 = \frac{k}{hL}$$

According to Buckingham theorem,

$$\pi_3 = f(\pi_1, \pi_2)$$

$$\therefore \qquad \frac{k}{hL} = f\left(\frac{\mu}{\rho Lv}\right)^{n'}\left(\frac{Cp\rho Lv}{k}\right)^{m'}$$

where n' and m' are constants.

Assuming n' > m', we can write the above equations in the following form:

$$\frac{hL}{k} = f\left(\frac{\mu}{\rho Lv}\right)^{m'}\left(\frac{\mu}{\rho Lv}\right)^{n'-m'}\left(\frac{Cp\rho Lv}{k}\right)^{m'}$$

$$= f\left(\frac{\mu}{\rho Lv}\right)^{n'-m'} \cdot \left(\frac{\mu}{\rho LV}\frac{Cp\rho Lv}{k}\right)^{m'}$$

$$= f\left(\frac{\mu}{\rho Lv}\right)^{n'-m'}\left(\frac{\mu Cp}{k}\right)^{m'}$$

$$= f\left(\frac{\mu}{\rho Lv}\right)^{n}\left(\frac{\mu Cp}{k}\right)^{m}$$

Here n is substituted for n' - m' and m for m'.

$$Nu = f(Re)^n\,(Pr)^m = C\,(Re)^n\,(Pr)^m \qquad (5)$$

where C, n and m are constants and Nu, Re and Pr are known as Nusselt, Reynold and Prandtl numbers.

The constants are calculated from experiments.

• PROBLEM 6-16

Air at 5 atm and 230° F is flowing through a 3 in diameter pipe. The pipe wall temperature is 70°F. Using the given nomograph, determine the heat transfer coefficient for air.

Solution: The average film temperature at the interface of air and the pipe wall is

$$T_f = \frac{230 + 70}{2} = 150°F$$

From the table for gas, the number for air is 7. Join the points 7 and 150°F on the T_f scale (extreme left).

Now, consider the term $\frac{P^2 \Delta T}{D}$,

where P is the pressure in atm. = 5 atm

ΔT is the difference in temperature = 160°

and D is the pipe diameter = 3 in.

$$\frac{P^2 \Delta T}{D} = \frac{(5)^2(230-70)}{3} = 1333.33$$

The scale representing this factor is shown at the extreme right of the nomograph.

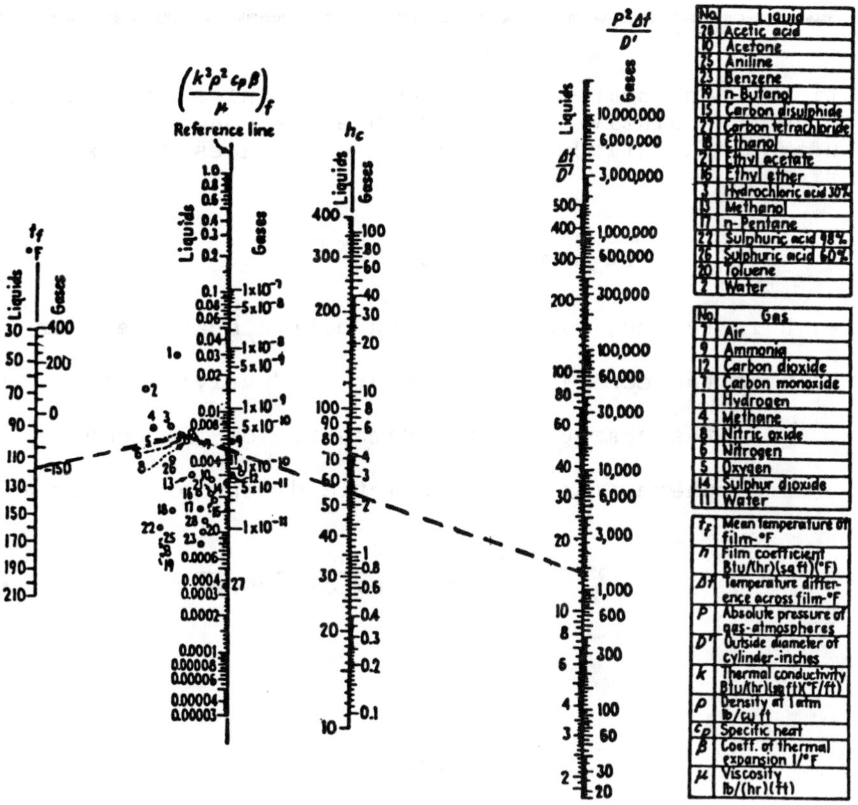

Nomograph for finding h with a laminar boundary layer on horizonal cylinder.

The value 1333.33 is indicated on the scale and a line joining this point and point 7 is drawn. The intersection of this line with the line for h_c gives the required value of the heat transfer coefficient, and can be read as ←2.7 Btu/hr-ft^2-°F.

FREE CONVECTION HEAT TRANSFER FROM ADDITIONAL SHAPES

• PROBLEM 6-17

A spherical ball, 1 in. in radius and at a temperature of 800°F, is suddenly quenched in a large pool of liquid which is maintained at 280°F. Estimate the surface temperature of the ball after one hour. The average convective heat transfer coefficient h = 2 Btu/hr ft^2°F. The properties of the ball material are:

k = 25 Btu/hr ft°F $\qquad$ ρ = 492 lbm/ft^3

Cp = 0.11 Btu

Solution: Biot number (Bi), the dimensionless parameter, compares the relative values of internal conduction resistance and surface convective resistances to heat transfer.

$$\text{Biot number - Bi} = \frac{hx}{k} < 0.1$$

where h is the coefficient of convective heat transfer

k is the coefficient of heat conductivity

x is the characteristic dimension of the body.

For a sphere, the characteristic dimension is given by

$$x = \frac{\text{volume of sphere}}{\text{surface area of sphere}}$$

$$= \frac{4\pi r^3/3}{4\pi r^2}$$

$$= \frac{r}{3} = \frac{1}{3} \text{ in.}$$

Now, substituting values in the expression for the Biot number,

$$\text{Bi} = \frac{(2) \times \left(\frac{1}{3\times 12}\right)}{(25)} = 0.00222$$

Since Bi is less than 0.1, the problem can be solved by using the lumped parameter analysis.

$$\text{Hence, } \frac{T - T_\infty}{T_o - T_\infty} = e^{-\left(\frac{hA}{Cp\rho v}\right)t}$$

$$\frac{hA}{Cp\rho v} = \frac{2}{0.11 \times 492 \times (1/36)} \text{ per hr}$$

$$= 1.333 \text{ per hr}$$

Therefore,

$$\frac{T-280}{800-280} = e^{-(1.333)1.0}$$

or $$T = 417°F$$

• PROBLEM 6-18

> It is required to keep a liquid above its freezing point in a large tank. Establish a relation to determine the surface temperature of the heating element in terms of energy input rate (Q) and the fluid temperature (T_0). The diameter and the uniform surface temperature of the heating element are D and T_1, respectively. Neglect the physical properties except density (ρ).

Solution: The general expression for the rate of heat energy input to the heating element is

$$Q = -k \int_s \left.\frac{\partial T}{\partial r}\right|_c ds$$

where s represents the surface

c - conduction in the fluid in the vicinity of the heating coil

and r - the outward normal distance from the surface.

The equation, expressed in terms of non-dimensional parameters, is

$$Q = -k\ (T_1-T_0)D \int_\eta \left.\frac{\partial \theta}{\partial \phi}\right|_c d\eta$$

or $$\frac{Q}{k(T_1-T_0)D} = \psi\left[\left(\frac{Cp\mu}{k}\right),\left(\frac{\rho^2\beta(T_1-T_0)gD^3}{\mu^2}\right)\right] \qquad (1)$$

where θ, ϕ and η are non-dimensional parameters that correspond to temperature, diameter and surface area, respectively.

$$Q = \frac{T-T_0}{T_1-T_0}\ ,\ \phi = \frac{r}{D}\ ,\ \eta = \frac{A}{D^2},$$

and ψ is a function of Grashof and Prandtl numbers.

Multiplying both sides of equation (1) by Grashof number

$$\frac{Q}{k(T_1-T_0)D} * \frac{\rho^2 \beta g D^3(T_1-T_0)}{\mu^2} = \psi\left[\left(\frac{Cp\mu}{k}\right)\left(\frac{\rho^2 \beta g D^3(T_1-T_0)}{\mu^2}\right)\right] * Gr$$

where $\frac{C_p\mu}{k}$ = Prandtl number = Pr, and

$$\frac{\rho^2 \beta g D^3(T_1-T_0)}{\mu^2} = \text{Grashof number} = Gr.$$

On simplification

$$\frac{\rho^2 \beta g D^2 Q}{k\mu^2} = (Gr)\ \psi\left[(Pr)(Gr)\right] \qquad (2)$$

Since the properties of the fluid, except density, are independent of temperature variations, the Prandtl number $\left(Pr = \frac{C_p \mu}{k}\right)$ can be considered to be constant.

Solving equation (2) for the Grashof number,

$$T_1-T_0 = \left(\frac{\mu^2}{\rho^2 \beta g D^3}\right) * \xi\left(\frac{D^2 Q \rho^2 \beta g}{k\mu^2}\right)$$

The expression $\xi\left(\frac{D^2 Q \rho^2 \beta g}{k\mu^2}\right)$ is determined experimentally by a model test.

• PROBLEM 6-19

An electrically heated wire, 20 ft long, 1/600 in dia, is maintained at 180°F. Compute the rate of heat loss by natural convection if the surrounding air temperature is 63°F.

Solution: The properties of air at the mean film temperature are found from the table.

$$T_f = \frac{180 + 63}{2} = 121.5°F$$

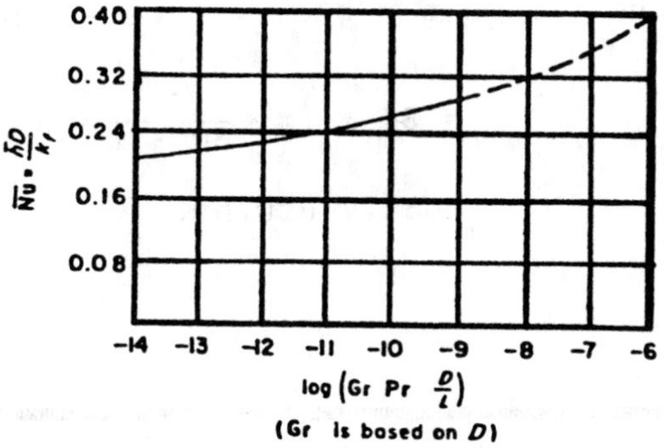

Free-convective heat transfer from small vertical wire.

From the table,

$$g\beta\rho^2/\mu^2 = 1.49 \times 10^6 \text{ per } {}^\circ\text{R-ft}^3$$

$$k = 0.0163 \text{ Btu/hr-ft-}{}^\circ\text{F}$$

$$C_p = 0.245 \text{ Btu/lbm-}{}^\circ\text{F}$$

$$Pr = 0.7035$$

Now, the Grashof number can be expressed as

$$Gr = \left(\frac{g\beta\rho^2}{\mu^2}\right) * \Delta T * L^3$$

Substituting the values,

$$Gr = (1.49 \times 10^6) * (180-63) * \left(\frac{1}{600\times12}\right)^3$$

$$= 4.67 \times 10^{-4}$$

$$\therefore \quad Gr * Pr * \frac{D}{L} = 4.67 \times 10^{-4} * 0.7035 * \frac{1}{600\times12} * \frac{1}{20}$$

$$= 2.28 \times 10^{-9}$$

and $\log\left(GrPr\frac{D}{L}\right) = -8.642$

from the graph, $Nu = \frac{hD}{k} \cong 0.29$

$$\therefore \quad h = \frac{0.29 \times 0.0163}{\left(\frac{1}{600\times12}\right)}$$

$$= 34 \text{ Btu/hr-ft}^2\,{}^\circ\text{F}$$

And the rate of heat loss q = hAΔT   (A = πDL)

$$= 34 * (\pi \times \frac{1}{600\times 12} \times 20) * (180-63)$$

$$= 34.7 \text{ Btu/hr.}$$

• PROBLEM 6-20

A 1/12 in. dia metal wire is placed in still air at 66°F. Compute the rate of heat generation required to maintain the surface temperature of the wire at 1734°F. The heat loss due to radiation may be neglected. Use the given chart.

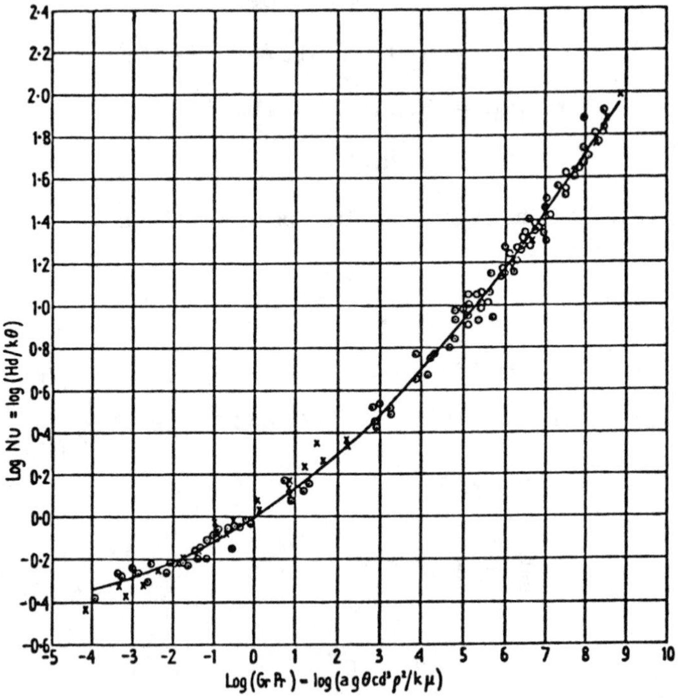

Natural convection for horizontal cylinders. ⊙s are experimental points for gases, Xs are experimental points for liquids.

Solution: The mean film temperature at the air-wire interface is

$$T_f = \frac{1734 + 66}{2} = 900°F$$

Properties of air at 900°F are: (refer to the table in problem No. 14)

$$k = 0.032 \text{ Btu/hr-ft-°F}$$

$$a = \frac{g\beta\rho^2 C_p}{\mu^2 k} = 0.0251 \times 10^6 \; (1/\text{ft}^3\text{-°F})$$

$$Gr \times Pr = \frac{\rho^2 \beta g \Delta T D^3}{\mu^2} \times \frac{C_p \mu}{k} = \frac{g\beta\rho^2 C_p}{\mu k} \times (\Delta T) \times (D^3)$$

$$= 0.0251 \times 10^6 \times (1734-66) \times \left(\frac{1}{12 \times 12}\right)^3$$

$$= 14.02$$

$$\log\,(Gr \times Pr) = \log\,(14.02) = 1.146$$

From the graph, log Nu = log(0.15)·

$$Nu = \frac{hD}{k} = 1.412$$

or $$h = \frac{1.412k}{D} = \frac{1.412 \times 0.032}{\left(\frac{1}{12\times12}\right)} = 6.51$$

Hence,

$$q = h \times A \times \Delta T$$

$$= 6.51 \times \left(\pi \times \frac{1}{12\times12}\right) \times \left(1734-66\right)$$

$$\cong 237 \text{ Btu/hr ft}$$

• PROBLEM 6-21

Compute the surface temperature of a 1/12 in. diameter resistance wire generating heat at the rate of 236.77 Btu/hr per foot length. The ambient air temperature is 66°F. Neglect the radiation heat loss.

Solution: The solution is approached by using the trial and error method. An initial value of the surface temperature is assumed and the heat rate required to maintain this temperature is calculated. This procedure is repeated until the calculated value of the heat rate equals the given value.

As a first estimate, let the surface temperature of the wire be 1700°F. The properties of air at the mean film temperature

$$T_f = \frac{1700 + 66}{2} = 883°F$$

are: (refer to the table in problem No. 14)

$$k = 0.0317 \text{ Btu/ft-hr-°F}$$

$$\frac{g\beta\rho^2 C_p}{\mu k} = 0.0267$$

Therefore,

$$Gr \times Pr = 0.0267 \times 10^6 \times (1700\text{-}66)\left(\frac{1}{12\times 12}\right)^3$$

$$= 14.61$$

or $$\log (Gr \times Pr) = \log (14.61) = 1.165$$

From the graph, (refer to graph in problem No. 20) corresponding to log (14.61) = 1.165, the value of log Nu is 0.015

or $$Nu = 1.412$$

Also, $$Nu = \frac{hD}{k} = 1.412$$

or $$h = 1.412 \frac{k}{D}$$

The rate of heat generated

$$q = hA\Delta T$$

$$= \frac{1.412 \times 0.0317 \times (1700\text{-}66) \times \pi \times D}{D}$$

$$= 229.77 \text{ Btu/hr per ft. length}$$

This value of heat generated is quite close to the given value. Hence, the surface temperature of the wire ≅ 1700°F.

CHAPTER 7

FORCED CONVECTION HEAT TRANSFER

Basic Attacks and Strategies for Solving Problems in this Chapter. **See pages 351 to 453 for step-by-step solutions to problems.**

Forced convection heat transfer, like free convection, occurs due to the movement of a fluid near a solid which is at a different temperature than the fluid. With forced convection, however, the fluid is pumped or blown past the solid or else the solid is moved through the fluid. The movement of the fluid is forced rather than due to natural buoyancy created by heating it. A typical example is a fluid being pumped through a pipe which is at a higher or lower temperature than the fluid. Heat transfer will thus take place between the pipe wall and the fluid.

As with free convection, the problems involving forced convection usually center on determination of a heat transfer coefficient, h. The heat transfer coefficient is typically determined from an empirical equation for the Nusselt number, N_u. *With forced convection,* N_u *is typically given as a function of the Prandtl number,* P_r, *the Reynolds number,* R_e, *and perhaps other parameters such as the length to diameter ratio for a pipe.*

The actual equation to use for a given situation depends upon the configuration of the solid and the value of R_e. *Common solid configurations for forced convection are flow through a pipe, flow past a flat plate, and external flow perpendicular to a pipe or a bank of pipes. Empirical equations are available for each of these configurations as illustrated in the problems in this chapter. For a given solid configuration there will typically be two or more equations, each for a different range of values for the Reynolds number. There is always at least one equation for laminar flow and another for turbulent flow. Sometimes there are additional equations for particular ranges of values for* R_e. *Some of the equations have restrictions on the range of values for* P_r *also.*

In calculating heat transfer rate to or from a flat plate or a pipe, the average heat transfer coefficient over the entire surface is needed, but the heat transfer coefficient is not constant over the surface. Some problems involve determination of an equation for local heat transfer coefficient as a function of position on the solid, and then integration over the surface to find the average coefficient. Many of the empirical equations give average heat transfer coefficient directly.

For problems involving flow past a flat plate, calculations relating to the boundary layer which forms along the plate may be required. There will always be a laminar boundary layer at the leading edge of the plate and it will become turbulent at some point if the plate is long enough. The plate length at which the boundary layer becomes turbulent is called the critical length. There are equations available to calculate i) the thickness of the boundary layer at any length along the plate, ii) the frictional drag coefficient, and iii) the drag force on the plate. These equations are available for both the laminar portion and the turbulent portion of the boundary layer.

Step-by-Step Solutions to Problems in this Chapter, "Forced Convection Heat Transfer"

INTRODUCTION TO FORCED CONVECTION

• PROBLEM 7-1

Consider the heat transfer involved in turbulent fluid flow in a pipe. It is known that the dependent variables are:

k - thermal conductivity G - mass fluid velocity

d - pipe diameter μ - viscosity

c_p - specific heat

Use dimensional analysis to discover the physical relationship involved, and from this, define the heat transfer coefficient.

Solution: The basic dimensional system to be used is the $ML\theta$ (mass, length, time) system. First, each dependent variable must be non-dimensionalized. So, if T = temperature and H = heat, then

$$k - \left(\frac{H}{\theta LT}\right) \qquad G - \left(\frac{M}{\theta L^2}\right)$$

$$d - (L) \qquad \mu - \left(\frac{M}{L\theta}\right)$$

$$c_p - \left(\frac{H}{MT}\right)$$

For purposes of simplification, the constant $K_H - \left(\frac{ML^2}{H\theta^2}\right)$ is included, since M and H appear in it. Now the heat transfer coefficient may be expressed as

$$h = f(k, G, d, \mu, c_p, K_H) \tag{1}$$

Taking an infinite series of this,

$$h = (C_1 k^{a_1} G^{b_1} d^{e_1} \mu^{f_1} c_p^{\ i_1} K^{m_1}) + (C_2 k^{a_2} G^{b_2} d^{e_2} \mu^{f_2} c_p^{\ i_2} K^{m_2}) + \ldots \quad (2)$$

The law of dimensional homogeneity states that each grouping must have the same overall dimensions. Therefore, the first group is taken as a complete representation. Since $h - \left(\frac{H}{\theta L^2 T}\right)$, substituting all the dimensional formulae into eq.(2) gives

$$\frac{H}{\theta L^2 T} = C_1 \left(\frac{H}{\theta L T}\right)^a \left(\frac{M}{\theta L^2}\right)^b (L)^e \left(\frac{M}{L\theta}\right)^f \left(\frac{H}{MT}\right)^i \left(\frac{ML^2}{H\theta^2}\right)^m \quad (3)$$

Taking each dimension separately and collecting exponents,

ΣH: $1 = a + i - m$

$\Sigma \theta$: $-1 = -a - b - f - 2m$

ΣL: $-2 = -a - 2b + e - f + 2m$

ΣT: $-1 = -a - i$

ΣM: $0 = b + f - i + m$

Solving these five simultaneous equations yields:

$m = 0$, $b = e+1$, $a = 1-i$, and $a + e + f = 0$

Now, if the variables as shown in eq.(2) are given these exponents and rearranged, the result is

$$\frac{hd}{k} = C_1 \left(\frac{c_p \mu}{k}\right)^i \left(\frac{dG}{\mu}\right)^e$$

or

$$\frac{hd}{k} = f\left[\left(\frac{c_p \mu}{k}\right), \left(\frac{dG}{\mu}\right)\right]$$

$\left(\frac{hd}{k}\right)$ is called the Nusselt number (Nu), $\left(\frac{c_p \mu}{k}\right)$ is the Prandtl number (Pr), and $\left(\frac{dG}{\mu}\right)$ is the Reynolds number (Re).

Thus, in terms of the non-dimensional parameters is

$$Nu = f\left[Pr,\ Re\right].$$

Hydrogen gas is to be heated by means of forced convection in a pipe. This experiment will first be modeled on a smaller scale, using air as the flow component. The prototype specifications are:

$G_H = 4210$ lbm/ft^2-hr $\qquad d_{iH} = 0.1727$ ft

$L_H = 18.5$ ft $\qquad \mu_H = 0.0242$ lbm/ft-hr

$C_{pH} = 3.4$ Btu/lbm-^{0}F

while the model will use

$d_{iA} = 0.0358$ ft $\qquad \mu_A = 0.047$ lbm/ft-hr

$C_{pA} = 0.24$ Btu/lbm-^{0}F

Using appropriate relations of similitude, find the required values of L_A, G_A and the ratio h_H/h_A.

Solution: When considering a flow subjected to the transfer of heat by forced convection, the relevant dimensionless groups are known to be the Stanton number, the Prandtl number, the Reynolds number, and the ratio of the length to the diameter. The Stanton number is a function of the other three, or

$$St = f(Pr,\ Re,\ L/d) \tag{1}$$

Substituting in the parameters, this is written as

$$\frac{h}{\rho U c_p} = f\left(\frac{c_p \mu}{k},\ \frac{\rho U d}{\mu},\ L/d\right) \tag{2}$$

If the two flows are to be similar, the following must be true:

$$\left(\frac{c_p \mu}{k}\right)_H = \left(\frac{c_p \mu}{k}\right)_A , \tag{3}$$

$$\left(\frac{\rho U d}{\mu}\right)_H = \left(\frac{\rho U d}{\mu}\right)_A , \tag{4}$$

$$(L/d)_H = (L/d)_A \tag{5}$$

and consequently

$$\left(\frac{h}{\rho U c_p}\right)_H = \left(\frac{h}{\rho U c_p}\right)_A \qquad (6)$$

Now, the pipe length of the model is found by substituting into eq.(5) and rearranging:

$$L_A = (L/d)_H \, d_A$$

$$= \frac{(18.5)(0.0358)}{0.1727}$$

$$= 3.84 \text{ ft.}$$

If the mass velocity (denoted by G) of the air is desired, it may be found by using eq.(4). Since the mass velocity is a product of the density and the flow velocity, this can be written as

$$\left(\frac{Gd}{\mu}\right)_H = \left(\frac{Gd}{\mu}\right)_A$$

Substituting and rearranging yields

$$G_A = \left(\frac{Gd}{\mu}\right)_H \left(\frac{\mu}{d}\right)_A$$

$$= \frac{(4210)(0.1727)(0.047)}{(0.0242)(0.0358)}$$

$$= 39{,}444 \text{ lbm/ft}^2\text{-hr}$$

Finally, if the Stanton number similarity is to be maintained, from eq.(6), the ratio of the heat transfer coefficients must be

$$\frac{h_H}{h_A} = \frac{(Gc_p)_H}{(Gc_p)_A}$$

$$= \frac{(4210)(3.4)}{(39{,}444)(0.24)}$$

$$= 1.51$$

• PROBLEM 7-3

A fluid flowing through a long pipe is transferring heat by forced convection. The wall surface temperature and the rate of heat transfer per unit length of pipe are constant. Assuming that the flow is rod-like, show that the local Nusselt number is 8.

Solution: The thermal energy equation across the boundary layer is represented by

$$U_b \frac{\partial T}{\partial x} = \frac{\alpha}{r}\frac{\partial}{\partial r}\left(r\frac{\partial T}{\partial r}\right) \quad (1)$$

where r and x are the radial and axial distances, and U_b, T and α are the flow velocity, temperature and thermal diffusivity, respectively.

Now, rod-like (or slug-like) flow means that the velocity is independent of radial distance. Eq.(1) may be written as

$$\frac{U_b}{\alpha}\frac{\partial T}{\partial x} = \frac{\partial^2 T}{\partial r^2} + \frac{1}{r}\frac{\partial T}{\partial r} \quad (2)$$

Since $\frac{\partial T}{\partial x}$ is constant everywhere along the axial and radial directions, and α is defined as

$$\alpha = \frac{k}{\rho c_p},$$

it can be said that

$$\frac{U_b \rho c_p}{k}\frac{\partial T}{\partial x} = a$$

where a is an arbitrarily chosen constant. Writing in terms of bulk parameters and rearranging,

$$\frac{dT_b}{dx} = \frac{ka}{\rho c_p U_b} \quad (3)$$

By conservation of energy, the heat lost from the pipe section is equal to that gained by the fluid contained within. This is expressed as

$$q = h_x \pi d(dx)(T_s - T_b) = mc_p(dT_b) \quad (4)$$

The mass flow rate is

$$m = \rho U_b \frac{\pi}{4} d^2$$

Solving for h_x and substituting yields

$$h_x = \frac{\rho U_b d c_p}{4(T_s - T_b)}\left(\frac{dT_b}{dx}\right) \quad (5)$$

Substituting eq.(3) into eq.(5) gives

$$h_x = \frac{dka}{4(T_s - T_b)} \quad (6)$$

Since the local Nusselt number is

$$Nu_x = \frac{h_x d}{k},$$

eq.(6) becomes

$$Nu_x = \frac{h_x d}{k} = \frac{ad^2}{4(T_s - T_b)} \tag{7}$$

Now, eq.(2) is a second-order differential equation for temperature distribution at any radial distance. Let $\frac{\partial T}{\partial r}$ equal some variable, which converts it to a first-order equation. On integration, this gives

$$T = \frac{ar^2}{4} + C_1 \ln(r) + C_2 \tag{8}$$

Substituting the boundary conditions,

$$\frac{\partial T}{\partial r} = 0 \quad \text{at } r = 0$$

and

$$T = T_s \quad \text{at } r = \frac{d}{2},$$

the solution is

$$T = a\left(\frac{r^2}{4} - \frac{d^2}{16}\right) + T_s \tag{9}$$

Now, the bulk temperature is defined as

$$T_b = \frac{2}{r_i^2}\int_0^{r_i} Trdr \tag{10}$$

Substituting eq.(9) in eq.(10) and integrating yields

$$T_b = T_s - \frac{ad^2}{32} \tag{11}$$

Rearranging for the term $(T_s - T_b)$ and substituting into eq. (7) yields

$$Nu_x = \frac{ad^2}{4\left(\frac{ad^2}{32}\right)}$$

$$Nu_x = 8.$$

• PROBLEM 7-4

A stream of air is flowing over a flat plate with a velocity of 33 ft/sec, at atmospheric pressure and 60°F. Calculate the boundary layer thickness on the plate at 4 in. from the leading edge.

Solution: The conservation of momentum applies to the boundary layer region. From this principle, boundary layer thickness is derived as a function of distance from the edge.

$$\frac{\delta}{x} = \frac{4.64}{\sqrt{Re_x}}$$

where

δ = boundary layer thickness

x = distance from the leading edge

Re_x = local length Reynolds number = $\frac{Ux}{\nu}$

where

U = free stream velocity = 33 ft/sec

ν = kinematic viscosity

x = distance from the leading edge = $\frac{4}{12}$ ft.

From the air tables at the given temperature and pressure,

$$\nu = 1.6 \times 10^{-4} \text{ ft}^2/\text{sec}.$$

Then,

$$Re_x = \frac{(33)(4/12)}{1.6 \times 10^{-4}} = 68,750$$

and

$$\frac{\delta}{x} = \frac{4.64}{\sqrt{68,750}} = 0.0177$$

Therefore, the boundary layer thickness at 4 in. is

$$\delta = 4(0.0177) = 0.0708 \text{ in.}$$

• PROBLEM 7-5

A sailboat, 100 ft. long, is moving at a speed of 10 ft per sec. Assume it to act as a flat plate, and the water as the medium through which the transfer of heat occurs. Find the maximum thickness of the boundary layer. Take the density of water as 62.4 lbm/ft^3 and the viscosity as 3.0 lbm/hr-ft.

Solution: The flow must be categorized and hence the Reynolds number is to be calculated. The magnitude of the boundary layer thickness increases with the distance from the leading edge. The maximum occurs at the end of the plate, i.e.,at L = 100 ft. The Reynolds number is calculated over the entire length.

$$Re = \frac{\rho UL}{\mu} = \frac{(62.4)(10)(3600)(100)}{3} = 7.49 \times 10^7$$

where the conversion factor is 3600 sec/hr. Since the critical Reynolds number is 3×10^5 this is definitely a turbulent flow. The boundary layer thickness is a function of length and is given by

$$\frac{\delta}{L} = \frac{0.371}{(Re_L)^{1/5}}$$

Substituting the values,

$$\delta = \frac{0.371L}{(Re_L)^{1/5}} = \frac{0.371(100)}{(7.49 \times 10^7)^{1/5}} \simeq 0.987 \text{ ft.}$$

• PROBLEM 7-6

A plate loses heat to the surrounding atmosphere by forced convection. It is initially at 140°F, and the air stream is at 60°F and a pressure of 1 atm. The air flow velocity over the plate is 10 ft/sec. At the critical length and at one foot from the leading edge, calculate

(a) the velocity boundary layer thickness,
(b) the thermal boundary layer thickness,
(c) the local friction coefficient,
(d) the average friction coefficient,
(e) the local drag force,
(f) the local coefficient of heat transfer,
(g) the average coefficient of heat transfer, and
(h) the rate of heat transfer.

Assume unit (1 ft.) width.

Solution: The mean temperature is

$$T_m = \frac{60 + 140}{2} = 100^\circ F$$

The properties of air at this temperature are

$$\rho = 0.071 \text{ lbm/ft}^3 \qquad \mu = 1.285 \times 10^{-5} \text{ lbm/ft-sec}$$

$$k = 0.0154 \text{ Btu/hr-ft-}^\circ F \quad Pr = 0.72$$

The Reynolds number at any distance from the leading edge is given by

$$Re_x = \frac{\rho Ux}{\mu}$$

therefore at x = 1 ft.

$$Re_{(x=1)} = \frac{(0.071)(10)(1)}{1.285 \times 10^{-5}} \simeq 55{,}250$$

Conversely, at the critical Reynolds number, taken as 5×10^5, the critical length may be solved for

$$x_{cr} = \frac{Re_{cr}\mu}{\rho U} = \frac{(5\times10^5)(1.285\times10^{-5})}{(0.071)(10)}$$

$$= 9 \text{ ft.}$$

(a) The velocity boundary layer thickness δ is given by

$$\delta = \frac{-5x}{Re_x^{1/2}}$$

Then

$$\delta\Big|_{x=1} = \frac{5.0(1)}{(55{,}250)^{1/2}} = 0.0213 \text{ ft.}$$

and

$$\delta\Big|_{x=9} = \frac{5.0(9)}{(5\text{x}10^5)^{1/2}} = 0.064 \text{ ft.}$$

(b) The thermal boundary layer thickness may be calculated from

$$\delta_{th} = \frac{\delta}{Pr^{1/3}}$$

At x = 1 ft:

$$\delta_{th}\Big|_{x=1} = \frac{0.0213}{(0.72)^{1/3}} = 0.0238 \text{ ft.}$$

and at x = 9 ft:

$$\delta th\Big|_{x=9} = \frac{0.064}{(0.72)^{1/3}} = 0.0714 \text{ ft.}$$

(c) The local friction coefficient or skin friction coefficient is defined as

$$C_f = \frac{0.664}{Re^{\frac{1}{2}}}$$

therefore at x = 1 ft.

$$C_f\Big|_{x=1} = \frac{0.664}{(55{,}250)^{\frac{1}{2}}} = 0.0028$$

and at x = 9 ft:

$$C_f\Big|_{x=9} = \frac{0.664}{(5\times10^5)^{\frac{1}{2}}} = 0.00094$$

(d) The average friction coefficient is found by integrating the local coefficient over the entire length, or

$$\overline{C}_f = \frac{1}{L}\int_0^L C_f dx = \frac{1.33}{Re^{\frac{1}{2}}} = 2\ C_f$$

Therefore,

$$\overline{C}_f\bigg|_{x=1} = 2(0.0028) = 0.0056$$

$$\overline{C}_f\bigg|_{x=9} = 2(0.00094) = 0.00188$$

(e) The shear stress or drag at the plate surface is defined as

$$\tau = 0.332\ \frac{\mu U}{g_c x}\ (Re)^{\frac{1}{2}}$$

Substituting the values,

$$\tau\bigg|_{x=1} = \frac{0.332(1.285\times10^{-5})(10)}{(32.2)(1)}\ (55{,}250)^{\frac{1}{2}}$$

$$= 3.11\times 10^{-4}\ \text{lbf/ft}^2$$

and

$$\tau\bigg|_{x=9} = \frac{0.332(1.285\times10^{-5})(10)}{(32.2)(9)}\ (5\times 10^5)^{\frac{1}{2}}$$

$$= 1.04\times 10^{-4}\ \text{lbf/ft}^2$$

(f) For a flat plate in laminar flow, the local Nusselt number is expressed as

$$Nu_x = 0.332\ Pr^{1/3}\ Re_x^{\frac{1}{2}}$$

where

$$Nu_x = \frac{h_x x}{k}\ .$$

Therefore,

$$h_x = 0.332\ Pr^{1/3}\ Re_x^{\frac{1}{2}}\left(\frac{k}{x}\right)$$

Substituting the values,

$$h_x\bigg|_{x=1} = 0.332(0.72)^{1/3}(55{,}250)^{\frac{1}{2}}\left(\frac{0.0154}{1}\right)$$

$$= 1.077\ \text{Btu/hr-ft}^2\text{-}^0\text{F}$$

and

$$h_x\bigg|_{x=9} = 0.332(0.72)^{1/3}(5\times10^5)^{\frac{1}{2}}\left(\frac{0.0154}{9}\right)$$

$$= 0.36\ \text{Btu/hr-ft}^2\text{-}^0\text{F}.$$

(g) The average Nusselt number (over the plate length) is

$$Nu_L = 0.664\ Pr^{1/3}\ Re_L^{\frac{1}{2}}$$

or

$$Nu_L = 2Nu_x$$

Similarly,

$$\bar{h}_L = 2\ h_x$$

so

$$\bar{h}_L\Big|_{x=1} = 2(1.077) = 2.154\ \text{Btu/hr-ft}^2\text{-}^0\text{F}$$

and

$$\bar{h}_L\Big|_{x=9} = 2(0.36) = 0.72\ \text{Btu/hr-ft}^2\text{-}^0\text{F}$$

(h) The rate of heat transfer is given by

$$q = hA(T_s - T_\infty)$$

where

$$T_s = \text{surface temperature}$$

and

$$T_\infty = \text{fluid temperature}$$

The plate area is bL (b=width). The total rate of heat transfer is obtained by integrating the above equation between the limits x=0 to x=L, giving

$$q = \bar{h}_L\ bL(T_s - T_\infty)$$

Since the width is unity,

$$q = \bar{h}_L L(T_s - T_\infty)$$

Substituting the values,

$$q\Big|_{x=1} = 2.154(1)(140 - 60)$$

$$= 172.3\ \text{Btu/hr}$$

and

$$q\Big|_{x=9} = 0.72(9)(140 - 60)$$

$$= 518.4\ \text{Btu/hr}$$

A plate loses heat to a stream of air by convection. The plate is 2 ft. in length, and it is initially at a surface temperature of 500°F. The ambient air temperature is 100°F, and the air flow rate is observed to be 150 ft/sec. Assuming a unit (1 ft.) width, calculate

(a) the heat transferred from the laminar boundary layer region; and

(b) the heat transferred from the turbulent boundary layer region.

Then assume that the boundary layer is completely turbulent.

Find

(c) the heat transfer involved; and

(d) the error this assumption results in.

Solution: The mean film temperature is

$$T_m = \frac{100 + 500}{2} = 300^\circ F$$

At this temperature, the properties of air are

$\nu = 1.1026\ ft^2/hr \qquad k = 0.01995\ Btu/hr\text{-}ft\text{-}^\circ F$

and

$Pr = 0.701$

(a) The Reynolds number over the entire length of the plate is

$$Re_L = \frac{UL}{\nu} = \frac{(150)(3600)(2)}{1.1026}$$

$$= 9.8 \times 10^5$$

If the Reynolds number is used for computing the Nusselt number the resulting coefficient of heat transfer will be an average value, or one that is for the entire length. Similarly, the heat transfer calculated using the average coefficient will be the heat transfer over the entire length, or the total heat transfer. This is equal to the sum of the heat transfer from both the laminar and turbulent boundary layer regions, or

$$q_T = q_{Lam} + q_{turb} \tag{1}$$

Now, the length of the laminar boundary layer L_{cr} can be found, knowing the critical Reynolds number.

$$Re_{cr} = \frac{UL_{cr}}{\nu} \tag{2}$$

If the Nusselt number is found over the laminar boundary

layer region, then the coefficient of heat transfer of that region may be found, as

$$Nu_{Lam} = \frac{h_{Lam} L_{cr}}{k} \tag{3}$$

Then the heat transfer of that region is easily determined. Proceeding to that end, first consider the plate as a whole. For mixed flow (where there are both laminar and turbulent regions) the Nusselt number is

$$Nu_L = \left[0.037\ Re_L^{0.8} - 872\right] Pr^{1/3}$$

with the constraint

$$5 \times 10^5 < Re_L < 10^7$$

As this is satisfied,

$$Nu_L = \left[0.037(9.8\times10^5)^{0.8} - 872\right](0.701)^{1/3}$$

$$= 1265.97$$

Then the average heat transfer coefficient is

$$h_{av} = \frac{Nu_L k}{L} = \frac{(1265.97)(0.01995)}{2}$$

$$= 12.63 \text{ Btu/hr-ft}^2\text{-}^0\text{F}$$

This is used for the calculation of the total heat transferred, or

$$q_T = h_{av} A(T_s - T_\infty)$$

$$= 12.63(2)(1)(500 - 100)$$

$$= 10{,}104 \text{ Btu/hr}$$

(a) The length of the laminar boundary layer is found from eq.(2).

$$L_{cr} = \frac{Re_{cr}\nu}{U} = \frac{(5\times10^5)(1.1026)}{(150)(3600)}$$

$$= 1.021 \text{ ft.}$$

Now, for laminar flow, the Nusselt number is found as

$$Nu_L = 0.664\ Re_L^{\frac{1}{2}}\ Pr^{1/3}$$

where the subscript L refers to the length of the laminar boundary layer. The condition

$$0.6 \le Pr \le 50$$

is satisfied. Then

$$Nu_L = 0.664(5\times10^5)^{\frac{1}{2}}(0.701)^{1/3}$$

$$= 417.09$$

Then the heat transfer coefficient in the laminar region is found using eq.(3).

$$h_{Lam} = \frac{Nu_{Lam}k}{L_{cr}} = \frac{(417.09)(0.01995)}{1.021}$$

$$= 8.15 \text{ Btu/hr-ft}^2\text{-}^0\text{F}.$$

Finally, the heat transfer out of the laminar region is

$$q_{Lam} = h_{Lam}A_{Lam}(T_s - T_\infty)$$

$$= (8.15)(1)(0.817)(500 - 100)$$

$$= 2{,}663.3 \text{ Btu/hr}.$$

(b) Referring to eq.(1), the heat transferred from the turbulent boundary layer region is easily found as

$$q_{turb} = q_T - q_{Lam}$$

$$= 10{,}104 - 2{,}663.3$$

$$= 7440.7 \text{ Btu/hr}$$

(c) If the boundary layer is turbulent from the leading edge on the Nusellt number is given by

$$Nu_L = 0.037 \, Re_L^{0.8} \, Pr^{1/3}$$

Substituting,

$$Nu_L = 0.037(9.8\times10^5)^{0.8}(0.701)^{1/3}$$

$$= 2040.6$$

The heat transfer coefficient is then

$$h_L = \frac{Nu_L k}{L} = \frac{(2040.6)(0.01995)}{2}$$

$$= 20.36 \text{ Btu/hr-ft}^2\text{-}^0\text{F}$$

Therefore, the total heat transferred

$$q_T = (20.36)(2\times1)(500 - 100)$$

$$= 16{,}288 \text{ Btu/hr}$$

(d) The difference between the two values of the total heat transfer may be expressed as a percent error:

$$\% \text{ error} = \frac{16{,}288 - 10{,}104}{10{,}104} \times 100\%$$

$\simeq 61\%$

The overall turbulence assumed results in erroneous results and hence is to be avoided.

• PROBLEM 7-8

Consider an example of forced convection within a pipe. The local heat transfer coefficient is valid at the point it is evaluated at, and in its immediate vicinity. When the length of pipe as a whole is of concern, the average heat transfer coefficient applies. The average temperature difference is defined as

$$(T_s - T)_{av} = \frac{(T_s - T_{in}) + (T_s - T_{out})}{2}$$

where

T_s = wall surface temperature

T_{in} = bulk fluid temperature at entrance

T_{out} = bulk fluid temperature at exit.

Derive the average heat transfer coefficient h_{av} in terms of the local coefficient h_x.

Solution: The heat transferred over the length of pipe is given as

$$Q = h_{av} A (T_s - T)_{av} \quad (1)$$

As the length of pipe is taken as a whole, the average values may be used.

Taking only a differential length of pipe, the heat transferred is

$$q_x = h_x (T_s - T) 2\pi R \, dx$$

If this is integrated, the elements sum up to give the total heat transferred over the length of pipe.

$$Q = \int_0^L h_x (T_s - T) 2\pi R \, dx \quad (2)$$

Bringing the constants outside and integrating yields

$$Q = 2\pi R \int_0^L h_x(T_s - T)dx \tag{3}$$

$$= 2\pi R h_a(T_s - T)L$$

Eqs.(1) and (3) may be equated as

$$2\pi R \int_0^L h_x(T_s - T)dx = h_{av} 2\pi R L(T_s - T)_{av}$$

which gives

$$h_{av} = \frac{\int_0^L h_x(T_s - T)dx}{(T_s - T)_{av}}$$

LAMINAR FLOW OF PLATES AND PIPES

• PROBLEM 7-9

A flat plate loses heat to the atmosphere by forced convection. Air at 70^0F flows over the plate at the rate of 50 ft/sec. The plate is initially at 212°F. Find

(a) the length of the laminar boundary layer thickness on the plate;

(b) the local coefficient of heat transfer;

(c) the thickness of the velocity boundary layer; and

(d) the thickness of the thermal boundary layer.

Also find

(e) the average coefficient of heat transfer over the laminar region.

Solution: (a) The Reynolds number is defined as

$$Re = \frac{Ud}{\nu}$$

The laminar boundary layer extends to the point where the Reynolds number is transitional (the critical point). So, if the critical Reynolds number of 5×10^5 is used, the length x of the laminar boundary layer may be solved for. First, the properties must be evaluated at the mean temperature.

$$T_m = \frac{212 + 70}{2} = 141^0F.$$

The table for the properties of air at this temperature gives

$$\nu = 0.21 \times 10^{-3} \text{ ft}^2/\text{sec} \qquad k = 0.0164 \text{ Btu/hr-ft-}^0\text{F}$$

$$Pr = 0.72$$

Now, for the critical length,

$$Re_x = \frac{Ux_{(cr)}}{\nu}$$

or

$$x_{(cr)} = \frac{Re_x \nu}{U} = \frac{(5\times10^5)(0.21\times10^{-3})}{50} = 2.1 \text{ ft.}$$

(b) The local coefficient of heat transfer may be found from the appropriate Nusselt number relation. For laminar forced convection over a flat plate, this is given as

$$Nu_x = 0.332 \; Pr^{1/3} \; Re_x^{\frac{1}{2}}$$

Since

$$Nu_x = \frac{hx}{k}$$

the local heat transfer coefficient is

$$h = 0.332 \; Pr^{1/3} \left(Re_x\right)^{\frac{1}{2}}\left(\frac{k}{x}\right)$$

$$= 0.332(0.72)^{1/3}(5\times10^5)^{\frac{1}{2}}\left(\frac{0.0164}{2.1}\right)$$

$$= 1.64 \text{ Btu/hr-ft}^2\text{-}^0\text{F}$$

(c) The thickness δ of the velocity boundary layer at the critical length may be found from the expression

$$\frac{\delta}{x} = \frac{5.0}{Re_x^{\frac{1}{2}}}$$

Solving for the thickness and substituting,

$$\delta = \frac{5.0x}{Re_x^{\frac{1}{2}}} = \frac{5.0(2.1)}{(5\times10^5)^{\frac{1}{2}}}$$

$$= 0.0148 \text{ ft.}$$

(d) A comparison of the velocity and thermal boundary layers shows that

$$\frac{\delta}{\delta_{th}} = Pr^{1/3}$$

Solving for the thermal boundary layer thickness yields

$$\delta_{th} = \frac{\delta}{Pr^{1/3}} = \frac{0.0148}{(0.72)^{1/3}}$$

$$= 0.0165 \text{ ft.}$$

(e) The average Nusselt number over that length of plate is given by

$$Nu_L = 0.664 \, Pr^{1/3} \, Re_L^{\frac{1}{2}}$$

which is noted to be

$$Nu_L = 2Nu_x .$$

Consequently, the mean heat transfer coefficient is also

$$h_L = 2h_x$$

or

$$h_L = 2(1.64) = 3.28 \text{ Btu/hr-ft}^2\text{-}^0\text{F}.$$

• PROBLEM 7-10

A plate at 60 °C, is cooled by forced convection from an airstream. The steady flow velocity of the air stream is observed to be 2m/sec with the temperature and pressure as 27 °C and 1 atm., respectively. Find the heat transfer per unit width from the leading edge to a distance of (a) 20 cm, and (b) 40 cm. Also find (c) the drag force over the 40 cm distance.

Solution: When referring to a length of plate, the heat transfer must be calculated using a coefficient of heat transfer that applies over the length, or the average coefficient. Therefore, the average Nusselt number correlation is used.

(a) First, the Reynolds number must be calculated. The mean film temperature is

$$T_m = \frac{27 + 60}{2} = 43.5\,^0\text{C}$$

At this temperature, the air tables give

$$\nu = 17.36 \times 10^{-6} \text{ m}^2\text{/sec} \qquad k = 0.02749 \text{ W/m-}^0\text{C}$$

$$Pr = 0.7 \text{ and } C_p = 1.006 \text{ KJ/kg}^0\text{C}.$$

(a) Then at a length of 20 cm,

$$Re_{(x)} = \frac{UL}{\nu} = \frac{(2)(0.2)}{17.36 \times 10^{-6}}$$

$$\simeq 23{,}041$$

This is a laminar flow. As the condition

$$0.6 \leq Pr \leq 50$$

is satisfied, the average Nusselt number may be calculated as

$$Nu_L = 0.664\ Re_L^{1/2}\ Pr^{1/3}$$

$$= 0.664(23{,}041)^{1/2}\ (0.7)^{1/3}$$

$$\simeq 89.49$$

From this, the average heat transfer coefficient is found to be

$$h_L = \frac{Nu_L k}{L} = \frac{(89.49)(0.02749)}{0.2}$$

$$\simeq 12.3\ W/m^2\text{-}^0C$$

Then the heat transferred from this region is

$$q = h_L A(T_s - T_\infty)$$

$$= 12.3(0.2)(1)(60-27)$$

$$\simeq 81.19\ W$$

(b) The same process is repeated, but the plate length is now taken to be 40 cm. Thus

$$Re_L = \frac{UL}{\nu} = \frac{(2)(0.4)}{17.36\times10^{-6}}$$

$$= 46{,}083$$

The flow is still laminar, so the same Nusselt number expression is used.

$$Nu_L = 0.664\ Re_L^{1/2}\ Pr^{1/3}$$

$$= 0.664(46{,}083)^{1/2}(0.7)^{1/3}$$

$$= 126.56$$

This gives an average heat transfer coefficient of

$$h_L = \frac{Nu_L k}{L} = \frac{(126.56)(0.02749)}{0.4}$$

$$= 8.7\ W/m^2\text{-}^0C$$

Finally, the heat transferred is

$$q = h_L A(T_s - T_\infty)$$

$$= 8.7(0.4)(1)(60-27)$$

$$= 114.8 \text{ W}$$

(c) The drag force F_D is

$$F_D = \tau_0 A$$

where τ_0 is the wall shear stress. This may be evaluated knowing the relation

$$\tau_0 = \frac{1}{2} C_D \rho U^2$$

where C_D is the coefficient of drag. This is given by

$$C_D = \frac{1.328}{Re_L^{1/2}}.$$

For a length of 40 cm, the coefficient of drag is

$$C_D = \frac{1.328}{(46,083)^{1/2}} = 0.00619$$

At the mean temperature, the air tables give

$$\rho = 1.115 \text{ kg/m}^3$$

Then

$$\tau_0 = \frac{1}{2}(0.00619)(1.115)(2)^2$$

$$= 0.0138 \text{ N/m}^2$$

The drag force is, then,

$$F_D = (0.0138)(0.4)(1)$$

$$= 0.0055 \text{ N}$$

• PROBLEM 7-11

A flat plate is cooled by a flow of air, with the transfer of heat accomplished by means of forced convection. The air is measured at a speed of 33 ft/sec, and has a temperature and pressure of 68 °F and 1 atm, respectively. The temperature distribution across the plate is as follows:

from 0 - 2 in. T = 122 °F

2 - 3.2 in. 176 °F

3.2 - 4.75 in. 212 °F

At a point 4 in. from the leading edge, calculate the heat flow per unit time and area. Also, find the local heat transfer coefficient.

Solution: Properties must be evaluated at the mean temperature. At x = 4 in., the temperature is 212°F, so

$$T_m = \frac{68 + 212}{2} = 140\,^\circ F.$$

From the air tables, properties at this temperature and pressure are

$\nu = 0.7416\ ft^2/hr$ $\qquad$ $k = 0.0166\ Btu/hr\text{-}ft\text{-}^\circ F$

$Pr = 0.701$

Then the Reynolds number is

$$Re_x = \frac{Ux}{\nu} = \frac{(33)(3600)(4/12)}{0.7416}$$

$$= 53{,}398$$

As this is less than the critical value, the flow is laminar.

Now, for laminar flow over an unheated starting length of plate, the local Nusselt number is

$$Nu_x = \frac{h_x x}{k} = \frac{0.332 Pr^{1/3} Re_x^{\ 1/2}}{\left[1 - \left(\frac{x_0}{x}\right)^{3/4}\right]^{1/3}} \tag{1}$$

Since

$$q = h_x A \Delta T$$

or

$$q/A = h_x \Delta T,$$

eq.(1) becomes

$$\frac{q}{A} = \frac{0.332 Pr^{1/3} Re_x^{\ 1/2} k \Delta T}{\left[1 - \left(\frac{x_0}{x}\right)^{3/4}\right]^{1/3} x} \tag{2}$$

Since every region encountered when traveling from the leading edge to x = 4 in., must be accounted for, eq.(2) is written as:

$$\frac{q}{A} = \frac{0.332 Pr^{1/3} Re_x^{\ 1/2} k}{x} \left[\frac{(\Delta T)_{x=0}}{\left[1 - \left(\frac{x_0}{x}\right)^{3/4}\right]^{1/3}\Big|_{x=0}} + \frac{(\Delta T)_{x=2}}{\left[1 - \left(\frac{x_0}{x}\right)^{3/4}\right]^{1/3}\Big|_{x=2}} \right.$$

$$\left. + \frac{(\Delta T)_{x=3.2}}{\left[1 - \left(\frac{x_0}{x}\right)^{3/4}\right]^{1/3}\Big|_{x=3.2}} \right] \tag{3}$$

At $x = 0$,

$$\frac{\Delta T}{[1 - \left(\frac{x_0}{x}\right)^{3/4}]^{1/3}} = \frac{(122 - 68)}{[1 - \left(\frac{0}{4}\right)^{3/4}]^{1/3}}$$

$$= 54\,^0F$$

At $x = 2$ in,

$$\frac{\Delta T}{[1 - \left(\frac{x_0}{x}\right)^{3/4}]^{1/3}} = \frac{(176 - 68)}{[1 - \left(\frac{2}{4}\right)^{3/4}]^{1/3}}$$

$$= 145.93\,^0F$$

At $x = 3.2$ in,

$$\frac{\Delta T}{[1 - \left(\frac{x_0}{x}\right)^{3/4}]^{1/3}} = \frac{(212 - 68)}{[1 - \left(\frac{3.2}{4}\right)^{3/4}]^{1/3}}$$

$$= 268.59\,^0F$$

Then substituting into eq.(3) yields

$$q/A = \frac{0.332(0.701)^{1/3}(53,398)^{1/2}(0.0166)}{4/12}\left[54+145.93+268.59\right]$$

$$\simeq 1590 \text{ Btu/hr-ft}^2$$

Finally, at $x = 4$ in. the heat transfer coefficient is found, as

$$\frac{q}{A} = h_x(T_s - T)$$

Then

$$h_x = \frac{q/A}{(T_s - T)}$$

$$= \frac{1590}{(212 - 68)}$$

$$= 11.04 \text{ Btu/hr-ft}^2\text{-}^0F$$

• **PROBLEM 7-12**

Oil at 320 ^0F is used to heat the inside of a metalic tube by forced convection. The mass flow rate of oil inside the tube is 3000 lbm/hr. The tube is 50 ft. long with 3 in. internal diameter. If the tube surface temperature is 210 ^0F, calculate and compare the coefficient of heat transfer using the formulae of (a) Hausen and (b) Sieder-Tate.

Solution: The properties of air at 320 ^0F are

$\rho = 50.3$ lbm/ft^3 $\qquad \mu = 10.9$ lbm/ft-hr

$k = 0.076$ Btu/hr-ft-^{0}F $\qquad \nu = 0.216$ ft^2/hr

$Pr = 84$

The flow velocity can be found from $U = \frac{\dot{m}}{\rho A}$. Substituting the values, the flow velocity is

$$U = \frac{3000}{50.3\left[\pi \frac{(3/12)^2}{4}\right]} = 1215 \text{ ft/hr}$$

Now the Reynolds number is calculated.

$$Re_{(d)} = \frac{Ud}{\nu} = \frac{(1215)(3/12)}{0.216} = 1406$$

This is definitely a laminar flow.

(a) For laminar flow in a tube, Hausen's relation is

$$Nu_d = 3.66 + \frac{0.0668(d/L)Re_{(d)}Pr}{1+0.04\left[(d/L)(Re_{(d)})Pr\right]^{2/3}}$$

Since

$$Nu_d = \frac{hd}{k},$$

the heat transfer coefficient is

$$h = \left[3.66 + \frac{0.0668(d/L)Re_d Pr}{1+0.04\left[(d/L)Re_d Pr\right]^{2/3}}\right]\left(\frac{k}{d}\right)$$

$$= \left[3.66 + \frac{0.0668\left(\frac{3/12}{50}\right)(1406)(84)}{1+0.04\left[\left(\frac{3/12}{50}\right)(1406)(84)\right]^{2/3}}\right]\left(\frac{0.076}{3/12}\right)$$

$$= 4.26 \text{ Btu/hr-ft}^2\text{-}^0\text{F}$$

(b) Sieder and Tate give the Nusselt number correlation as

$$Nu_d = 1.86\left[\left(\frac{d}{L}\right)Re_d Pr\right]^{1/3}\left(\frac{\mu}{\mu_s}\right)^{0.14}$$

where μ_s is the viscosity at the wall surface temperature. The restrictions on this equation are as follows:

1. $0.48 < Pr < 16{,}700$

2. $\left(\frac{d}{L}\right) R_{ed} Pr > 10$

The first restriction is valid, as $Pr = 84$. The second one must be checked, so

$$\left[\left(\frac{d}{L}\right)(Re_d)Pr\right] = \left(\frac{3/12}{50}\right)(1406)(84) = 590.52$$

which is also valid. The oil tables, at 210°F, gives

$$\mu_s = 41.3 \text{ lbm/ft-hr.}$$

Then the Nusselt number is

$$Nu_d = 1.86(590.52)^{1/3}\left(\frac{10.9}{41.3}\right)^{0.14}$$

$$= 12.95$$

Finally, the heat transfer coefficient is solved for as

$$h = \frac{Nu_d k}{d} = \frac{(12.95)(0.076)}{(3/12)}$$

$$= 3.94 \text{ Btu/hr-ft}^2\text{-}^0\text{F.}$$

The value obtained by Hausen's relation may be considered more accurate, but under the specific constrains mentioned, the Sieder-Tate relation also gives accurate results.

• PROBLEM 7-13

Forced convective heat transfer occurs when water is flowing at a rate of 1.27 ft³/hr through a pipe, 0.0492 ft dia and 3.28 ft long, at 140°F. The temperature of water at the inlet and outlet are 68°F and 86°F, respectively. Assuming laminar flow, calculate the heat transfer coefficient by a) Sieder and Tate's formula, and b) Hausen's formula.

Solution: The bulk mean temperature is

$$T_m = \frac{68 + 86}{2} = 77\,^0\text{F.}$$

At this temperature, the properties of water are

$\rho = 62.2 \text{ lbm/ft}^3$ $\quad\quad$ $\mu = 2.16 \text{ lbm/ft-hr}$

$Pr = 6.13$ $\quad\quad$ $k = 0.353 \text{ Btu/hr-ft-}^0\text{F}$

The volume flow rate is given by

$$U = \frac{Q}{A} = \frac{1.27/3600}{\frac{\pi}{4}(0.0492)^2} = 0.186 \text{ ft/sec}$$

Then the Reynolds number is

$$Re = \frac{\rho U d}{\mu}$$

$$= \frac{(62.2)(0.186)(3600)(0.0492)}{2.16}$$

$$= 949$$

(a) The Sieder-Tate correlation is given by

$$Nu_d = 1.86\left[\left(\frac{d}{L}\right)Re_d Pr\right]^{1/3}\left(\frac{\mu}{\mu_s}\right)^{0.14} \qquad (1)$$

where μ_s is the viscosity at the wall surface temperature. The constraints are:

1. $0.48 < Pr < 16,700$

2. $\left(\frac{d}{L}\right) Re_d Pr > 10$

The first condition is valid, and the second is checked as

$$\left(\frac{d}{L}\right) Re_d Pr > 10$$

$$\left(\frac{0.0492}{3.28}\right)(949)(6.13) = 87.26 > 10.$$

From the table, $\mu_s = 1.14$ lbm/ft-hr at 140°F.

Substituting the values in eq.(1),

$$Nu_d = 1.86(87.26)^{1/3}\left(\frac{2.16}{1.14}\right)^{0.14}$$

$$= 9.022$$

Then the average heat transfer coefficient over the pipe length is

$$h = \frac{Nu_d k}{d} = \frac{(9.022)(0.353)}{0.0492}$$

$$= 64.73 \text{ Btu/hr-ft}^2\text{-}^0\text{F}$$

(b) The Hausen correlation is

$$Nu_d = 3.66 + \frac{0.0668(d/L)Re_d Pr}{1 + 0.04\left[(d/L)Re_d Pr\right]^{2/3}}$$

which is

$$Nu_d = 3.66 + \frac{0.0668(0.0492/3.28)(949)(6.13)}{1+0.04\left[(0.0492/3.28)(949)(6.13)\right]^{2/3}}$$

$$= 6.922$$

Therefore the average heat transfer coefficient

$$h = \frac{Nu_d k}{d} = \frac{(6.922)(0.353)}{0.0492}$$

$$= 49.66 \text{ Btu/hr-ft}^2\text{-}^0\text{F}.$$

• PROBLEM 7-14

Air at 200 ^{0}F is flowing at a rate of 3 lbm/hr through a tube, 0.5 in ID, maintained at a uniform temperature of 212 ^{0}F from x = 0 up to a selected length. The temperature of air, at the outlet, is noted (shown in table) for every length. Plot the graphs for h_l, h_a and h_{in} as functions of the rediced length L/D.

Experiment	A	B	C	D	E	F	G
L (in.)	1.5	3.0	6.0	12.0	24.0	48.0	96.0
T_{b2} (°F)	201.4	202.2	203.1	204.6	206.6	209.0	211.0

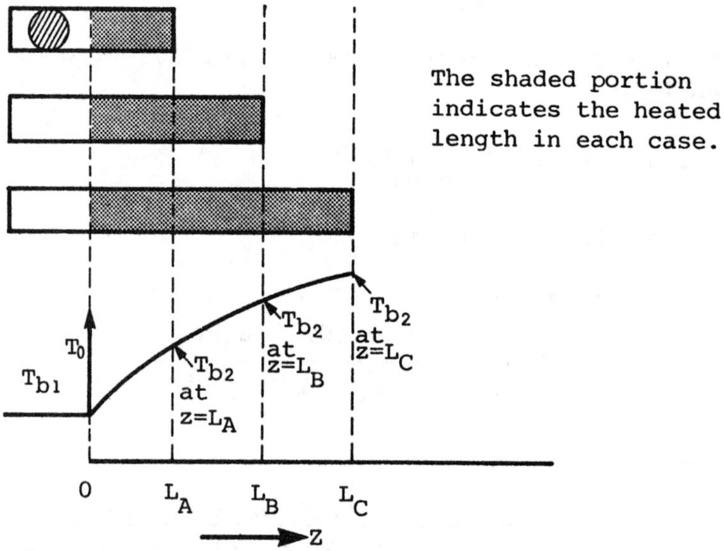

Fig. 1 Experimental determination of heat transfer coefficients.

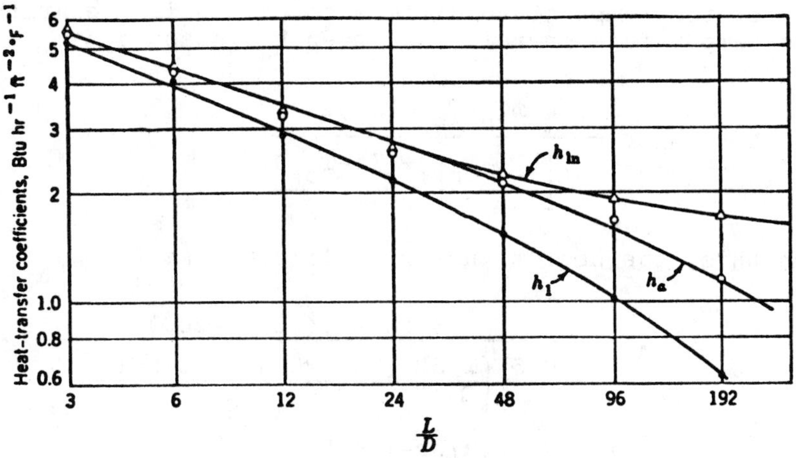

Fig. 2 Calculated heat-transfer coefficients.

Solution: The heat absorbed by the fluid in the selected length is given by

$$\dot{m}\, C_p(T_{b_2}-T_{b_1}) = h_1 \pi d L(\Delta T)_b \tag{1}$$

where T_{b_1} = water temperature at the inlet = 200°F and the temperature difference on the right-hand side is found according to the chosen method.

(a) If the temperature difference is to be based on T_{b_1}, then h_1 is solved for as

$$h_1 = \frac{\dot{m}\, C_p(T_{b_2}-T_{b_1})}{\pi d L(T_0-T_{b_1})} \tag{2}$$

The air tables give the value of C_p = 0.24 Btu/lbm-°F at the mean bulk temperature. Then at point A,

$$h_{1A} = \frac{(3)(0.24)(201.4-200)}{\pi\left(\frac{0.5}{12}\right)\left(\frac{1.5}{12}\right)(212-200)}$$

$$= 5.13 \text{ Btu/hr-ft}^2\text{-}^0\text{F}$$

Similarly at points B through G, the heat transfer coefficients are:

h_{1B} = 4.03 Btu/hr-ft²-°F h_{1C} = 2.84 Btu/hr-ft²-°F

h_{1D} = 2.11 Btu/hr-ft²-°F h_{1E} = 1.51 Btu/hr-ft²-°F

h_{1F} = 1.03 Btu/hr-ft²-°F h_{1G} = 0.63 Btu/hr-ft²-°F

(b) If the temperature difference chosen is the arithmetic bulk temperature correlation, then h_a is found as

$$h_a = \frac{\dot{m}C_p(T_{b_2} - T_{b_1})}{\pi dL\left[\frac{(T_0 - T_{b_1}) + (T_0 - T_{b_2})}{2}\right]} \tag{3}$$

From this, the heat transfer coefficient at A is

$$h_{aA} = \frac{(3)(0.24)(201.4-200)}{\pi\left(\frac{0.5}{12}\right)\left(\frac{1.5}{12}\right)\left[\frac{(212-200)+(212-201.4)}{2}\right]}$$

$$= 5.45 \text{ Btu/hr-ft}^2\text{-}^0\text{F}$$

Similarly,

$h_{aB} = 4.44$ Btu/hr-ft^2-^{0}F $h_{aC} = 3.26$ Btu/hr-ft^2-^{0}F

$h_{aD} = 2.61$ Btu/hr-ft^2-^{0}F $h_{aE} = 2.09$ Btu/hr-ft^2-^{0}F

$h_{aF} = 1.65$ Btu/hr-ft^2-^{0}F $h_{aG} = 1.16$ Btu/hr-ft^2-^{0}F

(c) Now, if the logarithmic mean temperature difference is used, the heat transfer coefficient is

$$h_{ln} = \frac{\dot{m}C_p(T_{b2} - T_{b1})}{\pi dL\left[\frac{(T_0 - T_{b1}) - (T_0 - T_{b2})}{\ln\left(\frac{T_0 - T_{b1}}{T_0 - T_{b2}}\right)}\right]} \tag{4}$$

Substituting in the parameters at A,

$$h_{lnA} = \frac{(3)(0.24)(201.4-200)}{\pi\left(\frac{0.5}{12}\right)\left(\frac{1.5}{12}\right)\left[\frac{(212-200)-(212-201.4)}{\ln\left(\frac{212-200}{212-201.4}\right)}\right]}$$

$$= 5.46 \text{ Btu/hr-ft}^2\text{-}^0\text{F}$$

and

$h_{lnB} = 4.46$ Btu/hr-ft^2-^{0}F $h_{lnC} = 3.28$ Btu/hr-ft^2-^{0}F

$h_{lnD} = 2.66$ Btu/hr-ft^2-^{0}F $h_{lnE} = 2.2$ Btu/hr-ft^2-^{0}F

$h_{lnF} = 1.91$ Btu/hr-ft^2-^{0}F $h_{lnG} = 1.71$ Btu/hr-ft^2-^{0}F

The above data is plotted in figure 2.

• PROBLEM 7-15

> Water enters a reactor through a length of pipe in which the necessary preheating is accomplished. The reactor has a maximum power rating of 150 W per meter of pipe, and its operating temperature is 350 °K. The flow is measured at 5 kg/hr, and it enters the pipe at a mean bulk temperature of 290 °K. If the internal diameter of the pipe is 0.005 m, calculate (a) the length of pipe over which the necessary heat transfer may be accomplished, and (b) the maximum exit temperature of the water.

Solution: (a) The heat balance over the length of pipe L is given by

$$q\pi dL = \dot{m}C_p \; \Delta T_b \tag{1}$$

Solving for L,

$$L = \frac{\dot{m}C_p \Delta T_b}{q\pi d} \tag{2}$$

Now, the mean bulk temperature of the water is

$$T_m = \frac{290 + 350}{2} = 320\,^0\text{K}$$

Properties of water at this temperature are

$\rho = 989 \text{ kg/m}^3$ $\qquad \nu = 0.59 \times 10^{-6} \text{ m}^2/\text{sec}$

$C_p = 4.174 \text{ kJ/kg}^0\text{K}$ $\qquad k = 0.641 \text{ W/m}^2\text{-}^0\text{K}$

The heat flow per meter of pipe is 150 watts. It is desired to express this per unit area of pipe, so

$$q = \frac{150 \text{ W/m}}{\pi d} = \frac{150}{\pi(0.005)}$$

$$= 9549.3 \text{ W/m}^2$$

Substituting into eq.(2) gives the required length as

$$L = \frac{\left(\frac{5}{3600}\right)(4.174)(1000)(350-290)}{(9549.3)\pi(0.005)}$$

$$= 2.32 \text{ m}$$

(b) The Reynolds number is

$$Re = \frac{Ud}{\nu}$$

Since

$$\dot{m} = \rho U \frac{\pi}{4} d^2$$

or

$$Ud = \frac{4\dot{m}}{\rho\pi d}$$

the Reynolds number may be calculated as

$$Re = \frac{4\dot{m}}{\rho\pi d\nu}$$

$$= \frac{4\left(\frac{5}{3600}\right)}{(989)\pi(0.005)(0.59\times10^{-6})}$$

$$= 606$$

This is a laminar flow. Now, if the flow is assumed to be fully developed, the Nusselt number will approach the value 4.364 by an asymptotic curve. Then

$$h = \frac{kNu}{d} = \frac{4.364k}{d} \tag{3}$$

Since the heat flow per unit area in a pipe is defined as

$$q/A = h\Delta T_b,$$

the change in bulk temperature along a length is

$$\Delta T_b = \frac{(q/A)}{h} \tag{4}$$

Substituting eq.(3) into eq.(4) yields

$$\Delta T_b = \frac{(q/A)d}{4.364k}$$

With the given parameters, this becomes

$$\Delta T_b = \frac{(9549.3)(0.005)}{4.364(0.641)}$$

$$= 17.07^\circ K$$

Then the exit temperature is

$$T_{b_2} = 350 + 17.07$$

$$= 367.07^\circ K$$

TURBULENT FLOW OF PLATES AND PIPES

• PROBLEM 7-16

Water cools the inside of a pipe, with the transfer of heat accomplished by the mechanism of forced convection. The mass flow rate is 200 lbm/hr. The pipe is 10 ft. long and has a 4 in. internal diameter. If the water has an entrance temperature of 40 °F and an exit temperature of 100 °F, what is the temperature of the inside wall surface?

Solution: As the flow must be categorized, the Reynolds number is calculated. The mean bulk water temperature is

$$T_b = \frac{40 + 100}{2} = 70\ {}^0\mathrm{F}$$

At this temperature, the water tables list the properties as

$\mu = 2.37$ lbm/hr-ft $\qquad k = 0.347$ Btu/hr-ft-°F

$Pr = 6.81 \qquad C_p = 1.0$ Btu/lbm-°F

Since the flow velocity is

$$U = \frac{\dot{m}}{\rho A},$$

the Reynolds number is found as

$$\mathrm{Re}_d = \frac{\rho U d}{\mu} = \frac{\rho \dot{m} d}{\rho \frac{\pi}{4} d^2 \mu} = \frac{4\dot{m}}{\pi d \mu}$$

$$= \frac{4(200)}{\pi(4/12)(2.37)}$$

$$= 322$$

This is a laminar flow. The Nusselt number may then be calculated using the Sieder-Tate relation

$$\mathrm{Nu}_d = 1.86\left[\left(\frac{d}{L}\right)\mathrm{Re}_d\ \mathrm{Pr}\right]^{1/3}\left(\frac{\mu}{\mu_s}\right)^{0.14} \qquad (1)$$

where μ_s is the viscosity at the unknown wall surface temperature. This relation may be used if the following conditions are satisfied:

(i) $0.48 < \mathrm{Pr} < 16{,}700$ and

(ii) $\left(\frac{d}{L}\right)\mathrm{Re}_d\mathrm{Pr} > 10$

The Prandtl number is within the given range.

For the second condition,

$$\left(\frac{d}{L}\right)Re_d Pr = \left(\frac{4/12}{10}\right)(322)(6.81) = 73.09$$

so the criterion is satisfied. Now, eq.(1) cannot be solved, as the value of μ_s is dependent on the unknown wall surface temperature. Substituting known values into this equation will give a relation between the coefficient of heat transfer h and μ_s. Equating heat lost by the pipe to the heat gained by the water contained within the specified length,

$$q = h\pi dL(T_s - T_b) = \dot{m}C_p(T_o - T_i) \tag{2}$$

where the subscripts o and i denote outlet and inlet, respectively. This equation directly relates the heat transfer coefficient and the wall surface temperature.

An iterative method is used to find the value of T_s. An initial value is assumed and the corresponding μ_s used to find h from eq.(1). Eq.(2) is now solved for T_s. The new value may be used for the second estimate, and the process repeated until convergence is achieved.

To begin with, assume $T_s = 200^0F$. The tables give a value of $\mu_s = 0.738$ lbm/hr-ft. Substituting into eq.(1) yields

$$Nu_d = \frac{hd}{k} = 1.86(73.09)^{1/3}\left(\frac{2.37}{0.738}\right)^{0.14}$$

$$= 9.157$$

which gives

$$h = \frac{Nu_d k}{d} = \frac{(9.157)(0.347)}{(4/12)}$$

$$= 9.53 \text{ Btu/hr-ft}^2\text{-}^0F$$

Using this in eq.(2),

$$(9.53)\pi(4/12)(10)(T_s - 70) = (200)(1)(100 - 40)$$

which gives

$$T_s = 190.2^0F$$

For the second iteration, $T_s = 190.2^0F$ is assumed. The tables give $\mu_s = 0.782$ lbm/hr-ft. Again, eq.(1) gives a heat coefficient of

$$h = 9.46 \text{ Btu/hr-ft}^2\text{-}^0F$$

Then eq.(2) gives a surface temperature of

$$T_s = 191.2^0F$$

This value of T_s agrees very closely to the assumed value, and thus further iterations are not needed.

• PROBLEM 7-17

Water, used as a cooling agent, is flowing at a speed of 0.02 m/sec through a pipe of 3.0 m length and 2.54 cm diameter. The wall surface temperature is maintained at 80°C. If the water enters at 60°C, calculate the temperature at the exit.

Solution: The heat lost by the pipe is equal to the heat gained by the water.

$$q = h\pi dL(T_s - T_b) = \dot{m}C_p(T_o - T_i) \qquad (1)$$

where

$$T_b = \frac{T_i + T_o}{2} \qquad (2)$$

and

$$\dot{m} = \rho U \frac{\pi}{4} d^2 \qquad (3)$$

Substituting eqs.(2) and (3) into (1),

$$h\pi dL\left[T_s - \left(\frac{T_i+T_o}{2}\right)\right] = \rho U \frac{\pi}{4} d^2C_p(T_o-T_i) \qquad (4)$$

Since T_o is unknown, an iterative procedure is adopted.

To begin with, the bulk temperature is assumed to be 60°C. Then the water tables at $T_b = \frac{60+60}{2} = 60°C$ give

$\rho = 985\ kg/m^3$ $\qquad \mu = 4.71 \times 10^{-4}$ kg/m-sec

$k = 0.651$ W/m-°C $\qquad Pr = 3.02$

$C_p = 4.18$ kJ/kg-°C

The Reynolds number is calculated as

$$Re_d = \frac{\rho Ud}{\mu} = \frac{(985)(0.02)(0.0254)}{4.71\times10^{-4}}$$

$$= 1062\ .$$

As this is a laminar flow, the Nusselt number may be calculated from Sieder and Tate's relation

$$Nu_d = 1.86\left[\left(\frac{d}{L}\right)Re_d Pr\right]^{1/3}\left(\frac{\mu}{\mu_s}\right)^{0.14} \tag{5}$$

where μ_s is the viscosity at the wall surface temperature. Eq.(5) may be used if the following conditions are satisfied:

1. $0.48 < Pr < 16{,}700$

2. $\left(\frac{d}{L}\right)Re_d Pr > 10$

The first condition is obviously satisfied. The second one is now tested:

$$\left(\frac{d}{L}\right)Re_d Pr = \left(\frac{0.0254}{3}\right)(1062)(3.02)$$

$$= 27.16$$

This is also satisfied.

The water tables give a value of

$$\mu_s = 3.55 \times 10^{-4} \text{ kg/m-sec} \quad \text{at } 80\,^0C.$$

Now, substituting into eq.(5) yields

$$Nu_d = 1.86(27.16)^{1/3}\left(\frac{4.71}{3.55}\right)^{0.14}$$

$$= 5.816$$

Since

$$Nu_d = \frac{hd}{k}$$

the coefficient of heat transfer is found to be

$$h = \frac{Nu_d k}{d} = \frac{(5.816)(0.651)}{0.0254}$$

$$= 149.1 \text{ W/m}^2\text{-}^0C$$

Now eq.(4) may be used. Substituting and rearranging yields

$$T_o = 71.98\,^0C$$

Assuming this value for the exit temperature, the second assumed bulk temperature will be

$$T_b = \frac{60 + 71.98}{2} = 66\,^0C$$

At this temperature, the water tables give

$\rho = 982 \text{ kg/m}^3$ $\qquad\qquad$ $\mu = 4.36 \times 10^{-4}$ kg/m-sec

$k = 0.656 \text{ W/m-}^0\text{C} \qquad Pr = 2.78$

$C_p = 4.18 \text{ kJ/kg-}^0\text{C}$

Again, the Reynolds number is calculated.

$$Re_d = \frac{\rho U d}{\mu} = \frac{(982)(0.02)(0.0254)}{4.36\times10^{-4}}$$

$$= 1144$$

Then

$$\left(\frac{d}{L}\right) Re_d Pr = \left(\frac{0.0254}{3}\right)(1144)(2.78)$$

$$= 26.93$$

The Nusselt number is

$$Nu_d = 1.86(26.93)^{1/3}\left(\frac{4.36}{3.55}\right)^{0.14}$$

$$= 5.74$$

and the coefficient of heat transfer is

$$h = \frac{Nu_d k}{d} = \frac{(5.74)(0.656)}{0.0254}$$

$$= 148.2 \text{ W/m}^2\text{-}^0\text{C}$$

Substituting into eq.(4) gives the new value of T_o as 71.92^0C. As the difference is quite small, an average exit temperature is taken as 71.95^0C.

• PROBLEM 7-18

Steam condenses on the outside surface of a pipe, and its coefficient of heat transfer is known to be 2000 Btu/hr-ft²-°F. Oil is used as a coolant inside, and it flows at the rate of 10 ft/min. The pipe is 2 ft. long and has an inside diameter of 1 in. If the oil entrance temperature is 70°F, what is the rate at which heat is transferred to the oil? Assume that the flow is fully developed, and that the resistance of the pipe itself is small enough to be neglected.

Solution: When considering an elemental length of pipe dx, the energy balance equation is

$$\rho U \frac{\pi}{4} d^2 C_p (T|_{x+\Delta x} - T|_x) = h\pi d\Delta x(T_s - T_m)$$

Integrating this over the limit as $\Delta x \to 0$ and rearranging yields

$$\ln\left(\frac{T_o - T_s}{T_i - T_s}\right) + \frac{h}{\rho U c_p}\frac{4L}{d} = 0 \qquad (1)$$

The outlet temperature is not known, and the coefficient of heat transfer must be calculated based on properties as evaluated at the mean bulk temperature. Thus, an iterative procedure is used to find the solution. This procedure will be as follows: the temperature at the outlet is assumed. Knowing the properties of oil at the bulk temperature, the Reynolds number is calculated. The appropriate Nusselt number correlation is then obtained. The coefficient of heat transfer may be found, and eq.(1) used to verify the assumed temperature. This process is repeated until convergence is achieved.

A mean bulk temperature of 100 °F is chosen. The appropriate tables yield

$\nu = 10.7 \times 10^{-5}$ ft²/sec $\qquad k = 0.069$ Btu/hr-ft-°F

$\mu = 5.56$ lbm/ft-hr $\qquad Pr = 136$

$\rho = 52$ lbm/ft³ $\qquad C_p = 0.467$ Btu/lbm-°F

The Reynolds number is calculated as

$$Re_d = \frac{Ud}{\nu} = \frac{(10/60)(1/12)}{10.7 \times 10^{-5}}$$

$$= 129.8$$

This is a laminar flow. The chosen Nusselt number correlation is that of Sieder and Tate:

$$Nu_d = 1.86\left[\left(\frac{d}{L}\right)Re_d Pr\right]^{1/3}\left(\frac{\mu}{\mu_s}\right)^{0.14}$$

Now, the wall surface temperature is not known. The assumed value is 210 °F. The tables give

$$\mu_s = 2.5 \text{ lbm/ft-hr}$$

The coefficient of heat transfer is then calculated as

$$h = 1.86\left[\left(\frac{d}{L}\right)Re_d Pr\right]^{1/3}\left(\frac{\mu}{\mu_s}\right)^{0.14}\left(\frac{k}{d}\right)$$

$$= 1.86\left[\left(\frac{1/12}{2.5}\right)(129.8)(136)\right]^{1/3}\left(\frac{5.56}{2.5}\right)^{0.14}\left(\frac{0.069}{1/12}\right)$$

$$= 14.43 \text{ Btu/hr-ft}^2\text{-}^0\text{F}$$

Now, substituting in eq.(1),

$$\ln\left(\frac{T_o - 210}{70-210}\right) + \frac{14.43}{(52)(10/60)(3600)(0.467)}\,\frac{(4)(2)}{(1/12)} = 0$$

or

$$T_o = 82.7^0\text{F}$$

As there is a great difference between the assumed and the calculated temperatures, the calculated temperature 82.7°F is taken for the second iteration. For this outlet temperature, the mean bulk temperature is

$$T_m = \frac{70 + 82.7}{2} = 76.4^0\text{F}$$

The heat transfer coefficient is found to be

$$h = 12.38 \text{ Btu/hr-ft}^2\text{-}^0\text{F}$$

Then eq.(1) gives

$$T_o = 81^0\text{F} \quad \text{(This value is nearer to the assumed)}$$

The mean bulk temperature is 75.5°F. Finally, the rate at which heat is transferred to the oil is

$$q = \dot{m}C_p\Delta T = \rho U \frac{\pi}{4} d^2 C_p(T_o - T_i)$$

$$= (52)(10)(60)\frac{\pi}{4}(1/12)^2(0.443)(81-70)$$

$$= 829 \text{ Btu/hr}$$

Oil flows through a tube, and heat is transferred to it by forced convection. The tube is 15 ft. long, and its internal diameter measures 0.0303 ft. The flow is measured at 80 lbm/hr, and the wall surface temperature is constant at 350°F. Properties of the oil are:

$$C_p = 0.5 \text{ Btu/lbm-}^\circ\text{F} \quad k = 0.083 \text{ Btu/hr-ft-}^\circ\text{F}$$

The viscosity - temperature profile is given by the following data:

T(°F)	(lbm/ft-hr)
150	15.73
200	12.22
250	9.2
300	6.82
350	4.72

Calculate the heat transfer coefficient and the mean exit temperature of the oil, assuming that is enters the tube at a temperature of 150°F.

Solution: An iterative procedure is followed to determine the outlet temperature T_o, which will first be assumed as 250°F. Then the bulk mean temperature is

$$T_m = \frac{150 + 250}{2} = 200^\circ\text{F}$$

At this temperature, the viscosity is given as 12.22 lbm/ft-hr. Since

$$Re_d = \frac{\rho U d}{\mu}$$

and

$$\dot{m} = \rho U \frac{\pi}{4} d^2,$$

the Reynolds number may be found as

$$Re_d = \frac{4\dot{m}}{\pi d \mu} = \frac{4(80)}{\pi(0.0303)(12.22)}$$

$$= 275$$

This is a laminar flow. Then the relationship of Sieder and Tate may be used:

$$Nu_d = \frac{hd}{k} = 1.86\left[\left(\frac{d}{L}\right) Re_d Pr\right]^{1/3}\left(\frac{\mu}{\mu_s}\right)^{0.14}$$

This applies if

1. $0.48 < Pr < 16{,}700$

2. $\left(\frac{d}{L}\right) Re_d Pr > 10$

The Prandtl number is found as

$$Pr = \frac{\mu C_p}{k} = \frac{(12.22)(0.5)}{0.083}$$

$$= 73.6$$

The first condition holds. To check the second one,

$$\left(\frac{d}{L}\right) Re_d Pr = \left(\frac{0.0303}{15}\right)(275)(73.6)$$

$$= 40.89 > 10$$

At 350°F, μ_s = 4.72 lbm/ft-hr. Then h is solved for, as

$$h = 1.86(40.89)^{1/3}\left(\frac{12.22}{4.72}\right)^{0.14}\left(\frac{0.083}{0.0303}\right)$$

$$= 20.05 \text{ Btu/hr-ft}^2\text{-}^0\text{F}$$

By energy balance,

$$q = h\pi dL\Delta T = \dot{m}C_p(T_o - T_i)$$

where

$$\Delta T = \frac{(T_s - T_i) + (T_s - T_o)}{2}$$

Substituting,

$$(20.05)\pi(0.0303)(15)\left[\frac{(350-150)+(350-T_o)}{2}\right] = (80)(0.5)(T_o - 150)$$

or

$$T_o = 255^0\text{F}$$

The value differs from the original by 5°F. Hence, this value is taken without any further iterations. Thus, the value of the heat transfer coefficient is 20.05 Btu/hr-ft²-°F, and the average outlet temperature is 252.5°F.

• PROBLEM 7-20

A car travels at a speed of 88 ft/sec, and the crankcase loses heat to the atmosphere by forced convection. The crankcase is 28 in. long and has a surface area of 7.06 ft². The surface temperature is 160°F, and the surrounding air is at 40°F. Assuming turbulent flow, calculate the rate of cooling of the crankcase.

Solution: From the Reynolds number, the appropriate Nusselt number may be found. From that, the heat transfer coefficient may be calculated, and the heat flow from the crankcase to the atmosphere can be found by

$$q = hA(T_s - T_\infty) \tag{1}$$

where $T_s \rightarrow$ surface temperature = 160°F

and $T_\infty \rightarrow$ air temperature = 40°F.

The mean bulk temperature is

$$T_m = \frac{40 + 160}{2} = 100\,^\circ F.$$

The properties of air at this temperature are

$\rho = 0.071$ lbm/ft³ $\qquad \mu = 1.285 \times 10^{-5}$ lbm/ft-sec

$Pr = 0.719$ $\qquad k = 0.0154$ Btu/hr-ft-°F

Then the Reynolds number is

$$Re_L = \frac{\rho UL}{\mu} = \frac{(0.071)(88)(28/12)}{1.285\times 10^{-5}}$$

$$= 1.13 \times 10^6 \quad (>5\times 10^5)$$

This is a turbulent flow. Since the crankcase is at a constant temperature and the flow is classified as "low speed", the average Nusselt number is found from

$$\overline{Nu}_L = 0.036\, Pr^{1/3} Re_L^{0.8}$$

Then

$$\overline{Nu}_L = 0.036(0.719)^{1/3}(1.13 \times 10^6)^{0.8}$$

$$\simeq 2244$$

From this, the average heat transfer coefficient is

$$\bar{h} = \overline{Nu}_L \frac{k}{L} = \frac{(2244)(0.0154)}{28/12}$$

$$= 14.8 \text{ Btu/hr-ft}^2\text{-}^\circ F$$

Substituting into eq.(1) gives the heat flow, or

$$q = (14.8)(7.06)(160-40)$$

$$= 12539 \text{ Btu/hr}$$

Water at 57°C is flowing at a rate of 12 litres/min through a standard ½ in. schedule 40 pipe. Calculate the heat transfer coefficient.

Solution: Properties of water at 57°C are

$\nu = 0.497 \times 10^{-6}$ m²/sec $\qquad$ $k = 0.648$ W/m-°C

Pr = 3.16

For a ½ in. schedule 40 pipe, the inside diameter is 0.622 in. or 0.0158 m. Then

$$U = \frac{12(0.001)}{\frac{\pi}{4}(0.0158)^2(60)}$$

$$= 1.02 \text{ m/sec}$$

The Reynolds number is found as

$$Re_d = \frac{Ud}{\nu} = \frac{(1.02)(0.0158)}{0.497\times 10^{-6}}$$

$$\simeq 32{,}426$$

This is greater than the critical value of 2300, so the flow is turbulent. Since

$$3 \times 10^4 < Re_d < 10^6,$$

the Colburn analogy is applicable. The Nusselt number is then calculated as

$$Nu_d = 0.023\, Re_d^{0.8}\, Pr^{1/3}$$

$$= 0.023(32{,}426)^{0.8}(3.16)^{1/3}$$

$$= 137.1$$

Then the coefficient of heat transfer is

$$h = \frac{Nu_d k}{d} = \frac{137.1(0.648)}{0.0158}$$

$$= 5622.8 \text{ W/m}^2\text{-}^\circ\text{C}$$

A lightweight oil is used to cool a heated pipe. The flow is measured at 30 ft/sec, and it moves through a diameter of 3 in. The oil enters at 60 °F and exits at 100 °F. If the wall surface temperature is 200 °F, determine the minimum pipe length over which the necessary heat transfer may be accomplished.

Solution: The Reynolds number must be calculated and the properties evaluated. Then the appropriate Nusselt number relation can be chosen.

The mean bulk temperature is

$$T_b = \frac{60 + 100}{2} = 80\,°F.$$

Tables for oil at this temperature yield

$\nu = 49 \times 10^{-5}$ ft²/sec $k = 0.077$ Btu/hr-ft-°F

$Pr = 570$ $\mu_o = 0.0025$ lbm/ft-sec $C_p = 0.44$ Btu/lbm-°F

$\rho = 56.8$ lbm/ft³

The Prandtl number is extremely large. The Sieder-Tate Nusselt number correlation can be used:

$$Nu_d = 0.023 Re_d^{0.8}\, Pr^{1/3} \left(\frac{\mu_o}{\mu_s}\right)^{0.14} \qquad (1)$$

where μ_s is the viscosity at the wall surface temperature and μ_o is the viscosity of the oil. For the application of eq.(1), the following should be satisfied:

1. $Re_d > 10{,}000$

2. $L/d > 60$

To check the first constraint, the Reynolds number is calculated.

$$Re_d = \frac{Ud}{\nu} = \frac{(30)(3/12)}{49 \times 10^{-5}}$$

$$= 15{,}306 \quad \text{(greater than 10,000)}$$

From the oil tables,

$\mu = 0.0278$ lbm/ft-sec at 80°F.

Substituting the values into eq.(1), the Nusselt number is given as

$$Nu_d = 0.023(15,306)^{0.8}(570)^{1/3}\left(\frac{0.0278}{0.0025}\right)^{0.14}$$

$$= 595.25$$

The coefficient of heat transfer is found as

$$h = \frac{Nu_d k}{d} = \frac{(595.25)(0.077)}{3/12}$$

$$= 183.3 \text{ Btu/hr-ft}^2\text{-}^0\text{F}.$$

The heat given off by the tube length is equal to the heat energy gained by the oil. The amount of heat transfer from the tube is

$$q = h\pi dL(T_s - T_b)$$

The heat energy gained by the oil is given by

$$q = \rho U\pi \frac{d^2}{4} C_p \Delta T_b$$

Equating these two and solving for the length gives

$$L = \frac{\rho U d C_p \Delta T_b}{4h(T_s - T_b)}$$

Substituting the values,

$$L = \frac{(56.8)(30)(3600)(3/12)(0.44)(100-60)}{4(183.3)(200-80)}$$

$$= 306.8 \text{ ft.}$$

Now the second constraint

$$L/d > 60$$

may be checked. As

$$L/d = 306.8/(3/12) = 1227,$$

the use of the Sieder-Tate analogy is justified.

• PROBLEM 7-23

Calculate the film coefficient at the inside surface of a standard 1 in. 18 gage tube, through which water at 80°F is flowing at a rate of 1.5 ft/sec.

The properties of water at 80°F are

$k = 0.355$ Btu/hr.ft.°F, $\nu = 0.0334$ ft²/hr

and $Pr = 5.85$.

Solution: The inside diameter of standard 1 in. 18 gage tube is 0.902 in. Then

$$Re_d = \frac{Ud}{\nu} = \frac{(1.5)(3600)(0.902/12)}{0.0334} = 1.22 \times 10^4$$

For liquid flow in a pipe, the critical Reynolds number is 2300. This is obviously a turbulent flow.

Since $0.7 < Pr < 160$ and $10^4 < Re_d < 10^6$, the Dittus-Boelter relation may be used to calculate the Nusselt number as

$$Nu_d = 0.023\, Re_d^{0.8}\, Pr^{0.4}$$

where

$$Nu_d = \frac{hd}{k}\ .$$

Solving for the heat transfer coefficient,

$$h = 0.023\, Re_d^{0.8}\, Pr^{0.4}\left(\frac{k}{d}\right)$$

$$= 0.023(1.22\times10^4)^{0.8}(5.85)^{0.4}\left(\frac{0.355}{0.902/12}\right)$$

$$= 409 \text{ Btu/hr-ft}^2\text{-}^0\text{F}.$$

• PROBLEM 7-24

A pipe is cooled by means of forced convection from a stream of water. The flow moves at 3 ft/sec, and is measured at a mean bulk temperature of 140°F. If the pipe is 2 ft. long with a diameter of 0.24 in., what is the film coefficient of heat transfer?

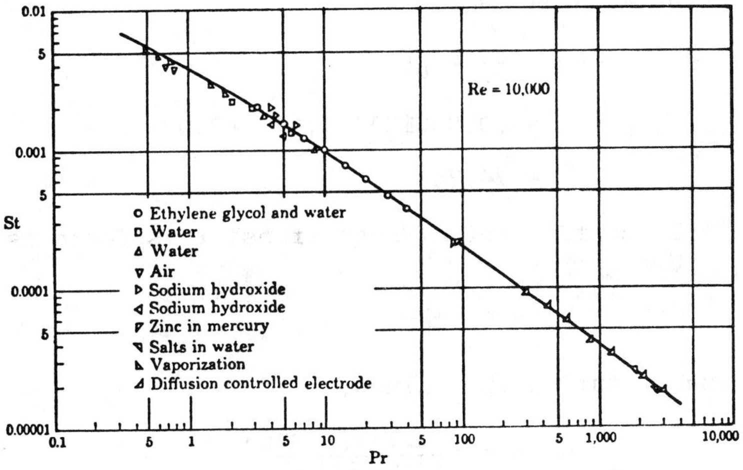

Fig. 1 Stanton number as function of Prandtl number for fully developed turbulent flow in a pipe.

Solution: The figure shown is an experimental plot of the Stanton number versus the Prandtl number. It is valid only for fully developed, turbulent flow.

The first step is to check that the flow fits the criteria. The Reynolds number must be calculated. At 140°F, the water tables give

$$\nu = 0.514 \times 10^{-5} \mathrm{ft}^2/\mathrm{sec}$$

Then

$$Re_d = \frac{Ud}{\nu} = \frac{(3)(0.24/12)}{0.514 \times 10^{-5}}$$

$$= 11,673$$

This is obviously a turbulent flow. Now to check that the flow is fully developed, the accepted rule is

$$L/d > 10$$

Then

$$2/(0.24/12) \simeq 100$$

Therefore, the given plot may be used.

The Stanton number is a ratio of dimensionless numbers, or

$$St = \frac{Nu_d}{Re_d Pr}$$

Knowing the Prandtl number, the plot gives the corresponding Stanton number.

For Pr = 3.02 the Stanton number from the graph is

$$St = 0.00212$$

Then the Nusselt number is

$$Nu_d = St\ Re_d Pr$$

$$= (0.00212)(11{,}673)(3.02)$$

$$= 74.74$$

The film coefficient of heat transfer is found as

$$h = \frac{Nu_d k}{d}$$

Substituting the values,

$$h = \frac{(74.74)(0.376)}{(0.24/12)}$$

$$= 1405 \text{ Btu/hr-ft}^2\text{-}^0\text{F}.$$

• PROBLEM 7-25

> Water flows through a tube with a mass flow rate of 2,000 lbm/hr. The tube has an inside diameter of 1 in., and the wall surface temperature is 200 °F. Water enters at 50 °F, and it is desired to heat it to 100 °F. Calculate the heat transfer coefficient and also the minimum length of tube required for the heat to be transferred.

Solution: The heat transfer coefficient can be determined from the Nusselt number. The properties of water at the average temperature of

$$T_{av} = \frac{50 + 100}{2} = 75\,^0\text{F} \quad \text{are:}$$

$\mu = 2.24$ lbm/ft-hr $\qquad$ $Pr = 6.4$

$k = 0.35$ Btu/hr-ft-°F $\qquad$ $C_p = 1$ Btu/lbm-°F

Since the mass flow rate is given and

$$\dot{m} = \rho VA = \rho V \frac{\pi}{4} d^2,$$

the Reynolds number is found from

$$Re_d = \frac{\rho Vd}{\mu} = \frac{4\dot{m}}{\mu\pi d}$$

$$= \frac{4(2000)}{2.24\pi(1/12)} = 13{,}642.$$

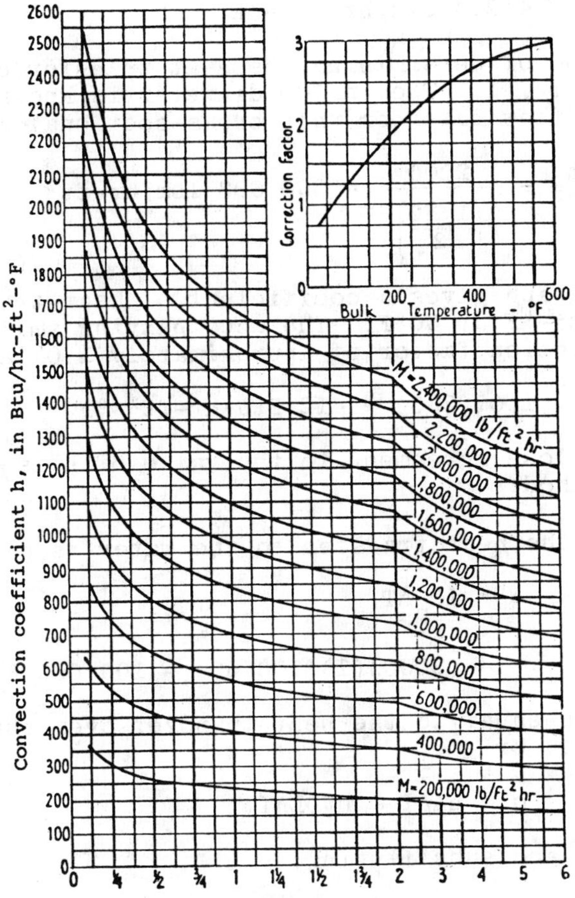

Inner diameter of tube - inches

Fig. 1 Heat transfer coefficients for different weights of water per unit area of pipe cross-section per unit time flowing through pipes of different diameters, with k and μ taken at 60°F.

Hence, the flow is turbulent.

Applying the Dittus-Boelter relation, the Nusselt number is

$$Nu_d = 0.023Re_d^{0.8} Pr^{0.4}$$

Since

$$Nu_d = \frac{hd}{k}$$

the heat transfer coefficient may be solved for as

$$h = 0.023 \, Re_d^{0.8} \, Pr^{0.4} \left(\frac{k}{d}\right)$$

$$= 0.023(13,642)^{0.8}(6.4)^{0.4}\left(\frac{0.35}{1/12}\right)$$

$$= 412.4 \text{ Btu/hr-ft}^2\text{-}^0\text{F}$$

Another method of determining the heat transfer coefficient is by using Fig. 1. The inside diameter is known to be 1 in. and the mass flow rate per unit cross-sectional area is

$$\frac{\dot{m}}{\pi r^2} = \frac{2000}{\pi\left(\frac{1/12}{2}\right)^2} = 366{,}693 \text{ lbm/hr-ft}^2$$

Reading the graph gives a coefficient of heat transfer of 385 Btu/hr-ft^2-^{0}F at 60^{0}F. The accompanying smaller graph gives a correction factor of approximately 1.07 at T_{av}=75^{0}F. Then

$$h \simeq 1.07(385) \simeq 412 \text{ Btu/hr-ft}^2\text{-}^0\text{F}$$

which is in close agreement to the value of h as obtained by the first method.

Finally, the heat transfer to the water is given by

$$q = \dot{m}C_p\Delta T$$

$$= (2000)(1)(100\text{-}50) = 100{,}000 \text{ Btu/hr.}$$

This energy is equal to the heat transfer over the entire length of pipe, or

$$q = hA\Delta T = h\pi DL\Delta T$$

Solving for L and substituting,

$$L = \frac{q}{\pi Dh\Delta T} = \frac{100{,}000}{\pi(1/12)(412.4)(200-75)} = 7.41 \text{ ft.}$$

• PROBLEM 7-26

Steam condenses on the outside surface of a pipe, maintaining an inside wall surface temperature of 488.7^{0}K. Air flows inside the pipe, and is heated by means of forced convection. The pipe is 2m long, with an inside diameter of 2.54 cm. The air velocity is 7.62 m/sec, and its pressure is 206.8 kPa. If the mean bulk temperature is 477.6^{0}K, deduce (a) the coefficient of heat transfer, and (b) the rate of heat transfer per unit area.

Solution: If the Reynolds number is calculated, the flow may be categorized, and the heat transfer coefficient found from the appropriate Nusselt number relation. Then the heat transfer rate is easily found.

(a) Substituting the values

$\mu = 2.6 \times 10^{-5}$ kg/m-sec $\qquad$ $k = 0.03894$ W/m-°K

$Pr = 0.686$,

the density may be found as

$$\rho = \frac{(MW)P}{RT} = \frac{(28.97)(206.8/101.33)}{(22.414)(477.6/273.2)}$$

$$= 1.509 \text{ kg/m}^3$$

The Reynolds number may be calculated as

$$Re = \frac{\rho U d}{\mu} = \frac{(1.509)(7.62)(0.0254)}{2.6 \times 10^{-5}}$$

$$= 11,233$$

This is a turbulent flow.

For the calculation of the Nusselt number, the relation of Sieder and Tate can be used,

$$Nu_d = 0.023 \, Re^{0.8} \, Pr^{1/3} \left(\frac{\mu}{\mu_s}\right)^{0.14}$$

as the ratio

$$L/d = 2/0.0254 = 78.74 > 60.$$

Since

$$Nu_d = \frac{hd}{k}$$

the heat transfer coefficient is

$$h = 0.023 \, Re^{0.8} \, Pr^{1/3} \left(\frac{\mu}{\mu_s}\right)^{0.14} \left(\frac{k}{d}\right)$$

At 488.70°K, $\mu_s = 2.64 \times 10^{-5}$ kg/m-sec.

Then

$$h = 0.023(11,233)^{0.8}(0.686)^{1/3}\left(\frac{2.6}{2.64}\right)^{0.14}\left(\frac{0.03894}{0.0254}\right)$$

$$= 53.98 \text{ W/m}^2\text{-°K}$$

(b) The rate of heat transfer per unit area is

$$\frac{q}{A} = h(T_s - T_m)$$

$$= 53.98(488.7 - 477.6)$$

$$= 599.2 \text{ W/m}^2$$

COMBINED FLOW

• PROBLEM 7-27

A tube is heated by means of forced convection from an airstream. The tube has a length of 1000 ft., and the wall surface temperature is 60°F. The flow is measured at a speed of 30 ft/sec, and it moves through an area measuring 3 in. across. If the air is at a mean bulk temperature of 70°F and a pressure of 29.4 psia, calculate

(a) the pressure drop over the tube length,

(b) the rate of heat flow from the air, and

(c) the coefficient of heat transfer by which this occurs.

Neglect the effects of any disturbance at the pipe entrance.

Solution: Since the temperature difference is small ($\Delta T = 70-60 = 10°F$), the properties of air may be evaluated at the mean bulk temperature. The air tables list the properties as

$$\mu = 0.0441 \text{ lbm/ft-hr} \qquad C_p = 0.24 \text{ Btu/lbm-°F}$$

$$Pr = 0.71 \qquad k = 0.0149 \text{ Btu/hr-ft-°F}$$

Density may be found from the equation of state as

$$\rho = \frac{P}{RT}$$

$$= \frac{(29.4)(144)}{(53.3)(70+460)}$$

$$= 0.15 \text{ lbm/ft}^3$$

Then the Reynolds number is given by

$$Re = \frac{\rho Ud}{\mu}$$

Substituting the values,

$$Re = \frac{(0.15)(30)(3600)(3/12)}{0.0441}$$

$$= 91,837$$

This is a turbulent flow.

(a) For this case, the pressure drop is defined as

$$\Delta p = 4f \frac{L}{d} \rho \frac{U^2}{2}$$

where the friction factor for a turbulent flow is

$$f = \frac{0.08}{Re^{1/4}}$$

$$= \frac{0.08}{(91,837)^{1/4}} = 0.0046$$

and

$$\Delta p = 4(0.0046)\left(\frac{1000}{3/12}\right)\left(\frac{0.15}{32.2}\right) \times \frac{(30/12)^2}{2}$$

$$= 1.071 \text{ psi}$$

(b) For calculating the rate of heat transfer, the following relation may be used:

$$q = \frac{QfC_pAU\Delta T}{2Pr}$$

where $C_p = \nu \times C = (0.15)(0.24) = 0.036 \text{ Btu/ft}^3\text{-}^0\text{F}$

and

$$Q = \frac{Pr^{1/2}}{1 + \frac{750(Pr)^{1/2}}{Re} + \frac{7.5(Pr)^{1/4}}{Re^{1/2}}}$$

Substituting,

$$Q = \frac{(0.71)^{1/2}}{1 + \frac{750(0.71)^{1/2}}{91,837} + \frac{7.5(0.71)^{1/4}}{(91,837)^{1/2}}}$$

$$= 0.818$$

and

$$A = \pi dL = \pi\left(\frac{3}{12}\right)(1000) = 785.4 \text{ ft}^2.$$

Substituting the values,

$$q = \frac{(0.818)(0.0046)(0.036)(785.4)(30)(70-60)}{2(0.71)}$$

$$\simeq 22.48 \text{ Btu/sec}$$

$$= 80928 \text{ Btu/hr.}$$

(c) Since $q = hA\Delta T$,

$$h = \frac{q}{A\Delta T}$$

$$= \frac{80928}{(785.4)(70-60)}$$

$$= 10.3 \text{ Btu/hr-ft}^2\text{-}^0\text{F}$$

Note: the heat flux can also be found from the correlation of Dittus-Boelter:

$$Nu_d = 0.023\ Re_d^{0.8}\ Pr^{0.4}$$

where as the process involves heating.

This is applicable as the difference of the pipe surface temperature and the bulk fluid temperature is small.

With this method, values of

$$q = 87{,}493 \text{ Btu/hr}$$

and

$$h = 11.14 \text{ Btu/hr-ft}^2\text{-}^0\text{F}$$

are obtained.

• PROBLEM 7-28

A pipe is cooled by forced convection from a stream of water. The flow is turbulent and moves at 10 ft/sec. It is a standard pipe of 1 in. diameter and 16-gage, and the wall surface temperature is 180°F. Consider the heat flow and find the coefficient of heat transfer using the analogies of (a) Dittus-Boelter, (b) Colburn, and (c) Sieder-Tate. Assume the entrance and exit water temperatures are 70°F and 130°F, respectively.

Solution: The properties of the water are evaluated at the average flow temperature,

$$T_{av} = \frac{70 + 130}{2} = 100^0\text{F}$$

At this temperature, the water tables give

$$\rho = 62.0 \text{ lbm/ft}^3 \qquad \mu = 1.648 \text{ lbm/ft-hr}$$

$$C_p = 0.999 \text{ Btu/lbm-}^0\text{F} \qquad k = 0.363 \text{ Btu/hr-ft-}^0\text{F}$$

The internal diameter of the pipe is 0.87 in.

The Reynolds number is given by

$$Re_d = \frac{\rho U d}{\mu} = \frac{(62.0)(10)(3600)(0.87/12)}{1.648}$$

$$= 98{,}192 \quad \text{This is turbulent flow.}$$

The Prandtl number may be calculated as

$$Pr = \frac{\mu C_p}{k} = \frac{(1.648)(0.999)}{0.363}$$

$$= 4.535$$

(a) The Dittus-Boelter relation gives the Nusselt number as

$$Nu_d = 0.023\ Re_d^{0.8}\ Pr^{0.4}$$

Since the Nusselt number is

$$Nu_d = \frac{hd}{k}$$

the heat transfer coefficient is

$$h = 0.023\ Re_d^{0.8}\ Pr^{0.4} \left(\frac{k}{d}\right)$$

$$= 0.023(98,192)^{0.8}\ (4.535)^{0.4} \left(\frac{0.363}{0.87/12}\right)$$

$$= 2078\ \text{Btu/hr-ft}^2\text{-}^\circ\text{F}.$$

(b) The Colburn relation is expressed as

$$Nu_d = 0.023\ Re_d^{0.8}\ Pr^{1/3}$$

Then solving for the heat transfer coefficient in a similar way gives

$$h = 0.023\ Re_d^{0.8}\ Pr^{1/3} \left(\frac{k}{d}\right)$$

$$= 0.023(98,192)^{0.8}\ (4.535)^{1/3} \left(\frac{0.363}{0.87/12}\right)$$

$$= 1879\ \text{Btu/hr-ft}^2\text{-}^\circ\text{F}.$$

(c) The equation of Sieder-Tate is

$$Nu_d = 0.023\ Re_d^{0.8}\ Pr^{1/3} \left(\frac{\mu}{\mu_s}\right)^{0.14}$$

where μ_s is the viscosity at the wall surface temperature. In this case, at 180°F μ=0.835 lbm/ft-hr. Then the heat transfer coefficient is found as

$$h = 0.023\ Re_d^{0.8}\ Pr^{1/3} \left(\frac{\mu}{\mu_s}\right)^{0.14} \left(\frac{k}{d}\right)$$

$$= 0.023(98,192)^{0.8}(4.535)^{1/3} \left(\frac{1.648}{0.835}\right)^{0.14} \left(\frac{0.363}{0.87/12}\right)$$

$$= 2066\ \text{Btu/hr-ft}^2\text{-}^\circ\text{F}.$$

The differences may or may not be significant, depending on the constraints of the application.

ITERATION OF LAMINAR AND TURBULENT PROBLEMS

• PROBLEM 7-29

A pipe is cooled by forced convection from a stream of water flowing at the rate of 20 gal/min. The pipe is 10 ft. long and has an inner diameter of 1 in. The wall surface temperature of the pipe is maintained at 210 °F. If the water has an entrance temperature of 50 °F, calculate the exit temperature. Use analogies of

(a) Reynolds,
(b) Colburn,
(c) Prandtl, and
(d) von Kármán.

Assume the flow is fully developed.

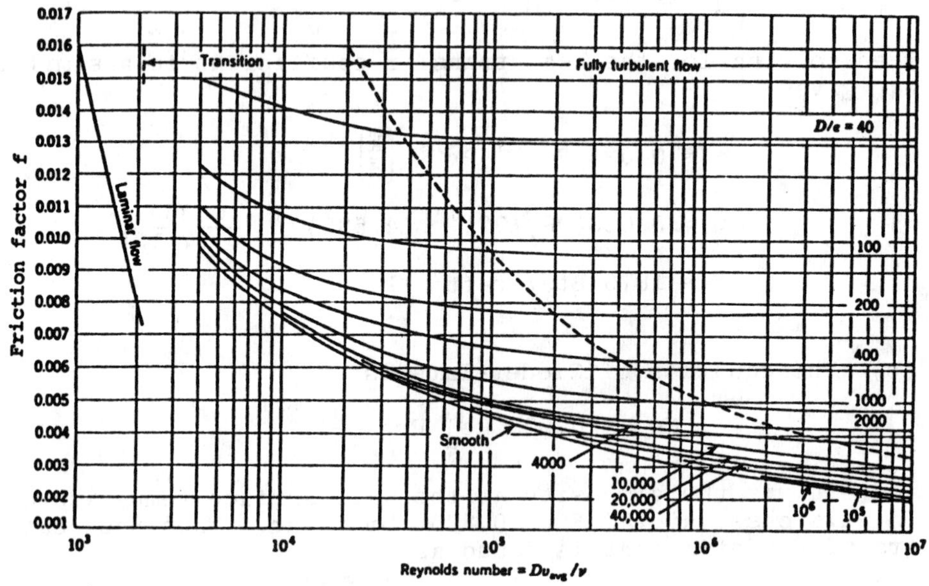

Fig. 1 Moody Diagram

Solution: Consider a pipe length dx. The heat balance between the pipe and the water is

$$\rho U \frac{\pi}{4} d^2 c_p (T|_{x+\Delta x} - T|_x) = h\pi d\Delta x(T_s - T)$$

where T is the mean bulk temperature. Letting $\Delta x \to 0$ and rearranging yields

$$\frac{dT}{dx} + \frac{h}{\rho U c_p} \frac{4}{d} (T-T_s) = 0$$

The group $\frac{h}{\rho U c_p}$ represents the Stanton number. This equation may be solved by the separation of variables into

$$\frac{dT}{T-T_s} + St \frac{4}{d} dx = 0$$

Integrating over a given length L yields

$$\int_{T_o}^{T_L} \frac{dT}{T-T_s} + St \frac{4}{d} \int_o^L dx = 0$$

or

$$\ln\left[\frac{T_L-T_s}{T_o-T_s}\right] + St \frac{4L}{d} = 0 \qquad (1)$$

If the Stanton number can be found, this equation may be used to find the exit temperature T_L.

For a given Reynolds number, the associated friction factor may be found from the accompanying Moody diagram (see Figure 1). Then the Stanton number may be found from an appropriate analogy. The exit temperature is solved for from eq.(1).

Assume a mean temperature of 70^0F. At this temperature, the water tables give values of

$$\nu = 1.06 \times 10^{-5} \text{ ft}^2/\text{sec}.$$

$$Pr = 6.82$$

The velocity is found as

$$U = \frac{(20\text{gal/min})(1\text{min}/60\text{sec})(1\text{ft}^3/7.48\text{gal})}{\frac{\pi}{4}(\frac{1}{12}\text{ft})^2}$$

$$= 8.17 \text{ ft/sec}$$

Then the Reynolds number is

$$Re = \frac{Ud}{\nu} = \frac{(8.17)(1/12)}{1.06\times10^{-5}}$$

$$= 64,230$$

Using the smooth pipe line on the Moody diagram, the friction factor is read as 0.005.

(a) Reynolds relates the Stanton number to the turbulent friction coefficient as

$$St = \frac{C_f}{2} \tag{2}$$

Then

$$St = \frac{0.005}{2}$$

$$= 0.0025$$

Substituting in eq.(1) gives

$$\ln\left(\frac{T_L - 210}{50 - 210}\right) + \frac{(0.0025)(4)(10)}{1/12} = 0$$

or

$$\frac{T_L - 210}{-160} = e^{-1.2}$$

Solving for T_L gives

$$T_L = 161.8^0F.$$

(b) Colburn expresses the Stanton number as

$$St = \frac{C_f}{2}(Pr)^{-2/3} \tag{3}$$

which gives

$$St = \frac{0.005}{2}(6.82)^{-2/3}$$

$$= 0.000695$$

Substituting in eq.(1) and solving gives

$$T_L = 210 - 160e^{-\left[\frac{(0.000695)(4)(10)}{1/12}\right]}$$

$$= 95.4^0F.$$

(c) The analogy of Prandtl is

$$St = \frac{C_f/2}{1+5\sqrt{C_f/2}(Pr-1)}$$

$$= \frac{0.005/2}{1+5\sqrt{0.005/2}(6\cdot 82-1)}$$

$$= 0.00102$$

Then eq.(1) yields

$$T_L = 210 - 160e^{-\left[\frac{(0.00102)(4)(10)}{1/12}\right]}$$

$$= 111.9^0F$$

(d) Finally, the von Karman relation is given by

$$St = \frac{C_f/2}{1+5\sqrt{C_f/2}\{(Pr-1)+\ln[1+\frac{5}{6}(Pr-1)]\}}$$

$$= \frac{0.005/2}{1+5\sqrt{0.005/2}\{(6.82-1)+\ln[1+\frac{5}{6}(6.82-1)]\}}$$

$$= 0.000863$$

and so

$$T_L = 210 - 160e^{-\left[\frac{(0.000863)(4)(10)}{1/12}\right]}$$

$$= 104.3^0F.$$

These values of T_L are summarized below.

Reynolds - $T_L = 161.8^0F$

Colburn - $T_L = 95.4^0F$

Prandtl - $T_L = 111.9^0F$

von Kármán - $T_L = 104.3^0F$

Only the Reynolds analogy proved to show a great discrepancy from the others. Based on the other three, the average outlet temperature is

$$T_L = \frac{95.4+111.9+104.3}{3} = 103.87^0F$$

Then the mean bulk fluid temperature is

$$T_m = \frac{50+103.87}{2} \cong 77^0F.$$

• PROBLEM 7-30

Steam condenses on the outer surface of a pipe. The coefficient of heat transfer is known to be 10,500 $W/m^2-{}^0K$ and the temperature is 107.8°C. Meanwhile, the interior is cooled by a flow of water, measured at a velocity of 2.44 m/sec and a mean bulk temperature of 65.6°C. It is a 2.54 cm schedule 40 steel pipe, with a length of 0.305 m. Find

(a) the coefficient of heat transfer from forced convection of the water,

(b) the overall heat transfer coefficient based on the inside surface area, and

(c) the rate of heat transfer to the water.

Solution: (a) At the mean bulk temperature, the water tables give the properties as

$\rho = 980 \text{ kg/m}^3$ $\qquad$ $\mu = 4.32 \times 10^{-4}$ kg/m-sec

$k = 0.633 \text{ W/m-}{}^0K$ $\qquad$ $Pr = 2.72$

Also, the standard pipe tables give values of

$$d_i = 0.0266\text{m}$$

and

$$d_o = 0.0334\text{m}$$

for the pipe. Now, to calculate the Reynolds number,

$$Re_d = \frac{\rho U d_i}{\mu} = \frac{(980)(2.44)(0.0266)}{4.32\times10^{-4}}$$

$$= 147,236$$

This is a turbulent flow. An appropriate Nusselt number correlation is that of Sieder and Tate:

$$Nu_d = \frac{hd_i}{k} = 0.023\ Re_d^{0.8}\ Pr^{1/3} \left(\frac{\mu}{\mu_s}\right)^{0.14}$$

where μ_s is the viscosity at the inside wall surface temperature. As this is an unknown quantity, it must be assumed. The temperature is chosen as 80°C. Again, the water tables give

$$\mu_s = 3.56 \times 10^{-4} \text{ kg/m-sec}$$

Then h is found as

$$h = 0.023(147{,}236)^{0.8}(2.72)^{1/3}\left(\frac{4.32}{3.56}\right)^{0.14}\left(\frac{0.633}{0.0266}\right)$$

$$= 10{,}697 \text{ W/m}^2\text{-}^0\text{K}.$$

(b) The overall heat transfer coefficient is defined as

$$U_i = \frac{1}{A_i \Sigma R}$$

where the sum of the resistances is

$$\Sigma R = R_i + R_m + R_o$$

$$= \frac{1}{h_i A_i} + \frac{r_o - r_i}{kA_{\ell m}} + \frac{1}{h_o A_o}$$

For the inside surface area,

$$A_i = \pi d_i L = \pi(0.0266)(0.305) = 0.0255\text{m}^2$$

For the bulk,

$$A_{\ell m} = \pi\left(\frac{d_i + d_o}{2}\right)L = \pi\left(\frac{0.0266+0.0334}{2}\right)(0.305) = 0.0287\text{m}^2$$

For the outer surface,

$$A_o = \pi d_o L = \pi(0.0334)(0.305) = 0.032\text{m}^2$$

Then, the resistance of the inside is

$$R_i = \frac{1}{h_i A_i} = \frac{1}{(12{,}558)(0.0255)}$$

$$= 0.00312 \text{ }^0\text{K/W}$$

The resistance of the steel pipe (based on the average thermal conductivity of 45 W/m-^{0}K) is

$$R_{\ell m} = \frac{r_o - r_i}{kA_m} = \frac{(0.0334 - 0.0266)/2}{(45)(0.0287)}$$

$$= 0.00263 \text{ }^0\text{K/W}$$

The outside resistance is

$$R_o = \frac{1}{h_o A_o} = \frac{1}{(10{,}500)(0.032)}$$

$$= 0.00298 \text{ }^0\text{K/W}$$

Now, the sum is

$$\Sigma R = 0.00312 + 0.00263 + 0.00298$$
$$= 0.00873 \ ^0K/W$$

Then the overall heat transfer coefficient is

$$U_i = \frac{1}{A_i \Sigma R} = \frac{1}{(0.0255)(0.00873)}$$
$$= 4492 \ W/m^2 - {}^0K$$

Also, the wall surface temperature may be verified, now that the resistances are known. The temperature drop may be related to the resistances as

$$\frac{T_s - T_m}{\Delta T_{Total}} = \frac{R_i}{\Sigma R}$$

Solving for T_s yields

$$T_s = T_m + \left(\frac{R_i}{\Sigma R}\right) \Delta T_{Total}$$
$$= 65.6 + \left(\frac{0.00312}{0.00873}\right)(107.8 - 65.6)$$
$$= 80.68^0C.$$

This is close enough to the original estimate.

(c) The rate of heat transfer to the water is given by

$$q = U_i A_i (T_o - T_m)$$
$$= (4492)(0.0255)(107.8 - 65.6)$$
$$= 4834 \ W$$

• PROBLEM 7-31

A stream of water cools a heated tube, serving as the medium for heat transfer by means of forced convection. The flow rate is measured at 2000 lbm/hr, entering and exiting at temperatures of 60 ^{0}F and 120 ^{0}F, respectively. The wall surface temperature is 150 ^{0}F. If the tube is 15 ft. long, what is the diameter? Assume a turbulent flow, and evaluate properties at the average fluid temperature.

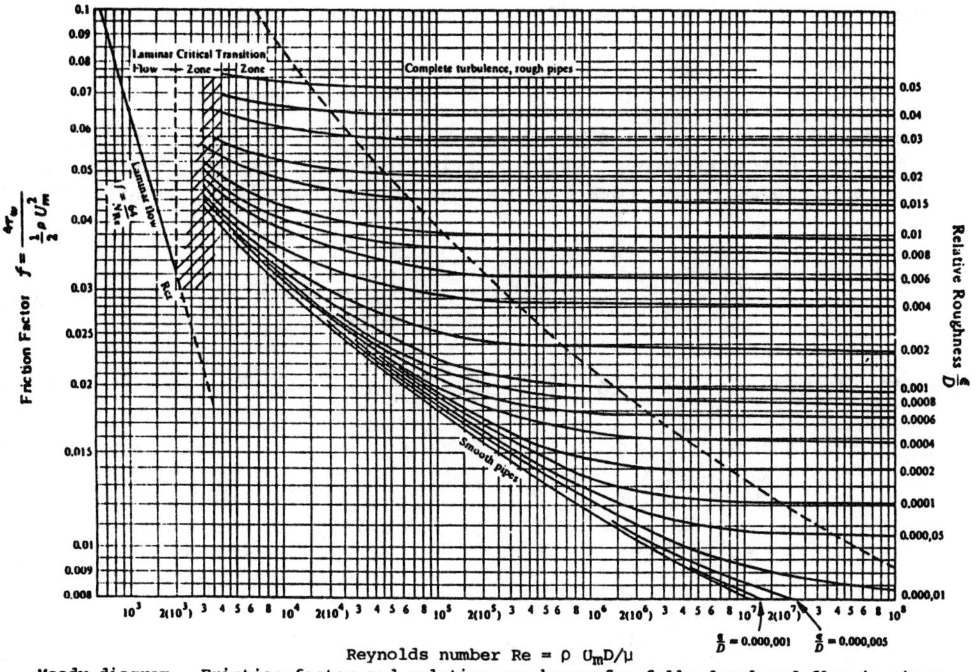

Moody diagram. Friction factor and relative roughness for fully-developed flow in pipes.

Solution: The average fluid temperature is

$$T_{av} = \frac{60 + 120}{2} = 90\,^0F$$

At this temperature, the water tables yield

μ = 1.35 lbm/hr-ft　　　　k = 0.371 Btu/hr-ft-^{0}F

Pr = 3.64　　　　ρ = 61.7 lbm/ft^3

c_p = 1.0 Btu/lbm-^{0}F

The energy balance equation states that the heat lost from the tube equal the heat gained by the water contained in the section, or

$$h\pi dL\Delta T_{LMTD} = \dot{m}c_p(T_o - T_i) \tag{1}$$

where o and i designate outlet and inlet temperatures, and LMTD denotes the logarithmic mean temperature difference, given by

$$\Delta T_{(LMTD)} = \frac{T_o - T_i}{\ln\left(\frac{T_s - T_i}{T_s - T_o}\right)} \tag{2}$$

where T_s is the wall surface temperature. Since

$$\dot{m} = \rho UA = \rho U \frac{\pi}{4} d^2, \tag{3}$$

the flow velocity is also unknown, as it is expressed in terms of the diameter. The third unknown is the coefficient of heat transfer. The von Kármán relation for turbulent flow is

$$Nu_d = \frac{(f/8)Re_d Pr}{1+5\sqrt{f/8}\{Pr-1+\ln[1+\frac{5}{6}(Pr-1)]\}} \tag{4}$$

where f is the friction factor. Since

$$Nu_d = \frac{hd}{k},$$

$$Re = \frac{\rho Ud}{\mu},$$

and

$$Pr = \frac{\mu c_p}{k},$$

eq.(4) becomes

$$h = \frac{\rho c_p U(f/8)}{1+5\sqrt{f/8}\{Pr-1+\ln[1+\frac{5}{6}(pr-1)]\}} \tag{5}$$

The friction factor f is found for any given Reynolds number from the Moody diagram (see figure).

Now, the variables d, h, U and f are unknown. Because of their interdependence, an iterative approach must be employed. First, the useful equations can be simplified for ease in calculation.

From eq.(3), substitution yields

$$2000 = 61.7\ U\left(\frac{\pi}{4}\right)d^2$$

or

$$U = 41.3/d^2 \tag{6}$$

Substituting eq.(6) and the various properties into eq.(5) yields

$$h = \frac{(61.7)(1.0)(41.3/d^2)f/8}{1+5\sqrt{f/8}\{3.64-1+\ln[1+\frac{5}{6}(3.64-1)]\}}$$

$$= \frac{318.5f/d^2}{1+6.72\sqrt{f}} \tag{7}$$

From eq.(2),

$$\Delta T_{LMTD} = \frac{120-60}{\ln\left(\frac{150-60}{150-120}\right)}$$

$$= 54.6\,^0F$$

Then eq.(1) becomes

$$h\pi d(15)(54.6) = 2000(1.0)(120-60)$$

or

$$h = 46.6/d$$

Then h can be eliminated from eq.(7), giving

$$\frac{46.6}{d} = \frac{318.5f/d^2}{1+6.72\sqrt{f}}$$

or

$$d = \frac{6.83f}{1+6.72\sqrt{f}} \qquad (8)$$

To begin, assume a value of d. The Reynolds number may then be found, as

$$Re = \frac{\rho Ud}{\mu} = \frac{(61.7)(41.3/d^2)d}{1.35}$$

$$= \frac{1887.6}{d} \qquad (9)$$

Then the corresponding value of f may be found from the Moody diagram. Substituting this value into eq.(8) gives a new value of d. This may be used as the second estimate, and the process repeated until convergence is achieved.

Proceeding as outlined above, the first estimate of d is chosen as 1/6 ft = 0.167 ft. Then from eq.(9),

$$Re = \frac{1887.6}{0.167} = 11,303$$

For this value, the Moody diagram gives f = 0.03. Then the new diameter as given by eq.(8),

$$d = \frac{6.83(0.03)}{1+6.72\sqrt{0.03}}$$

$$= 0.095 \text{ ft.}$$

Repeating the process,

$$Re = \frac{1887.6}{0.095} = 19,870$$

Using $f = 0.026$,

$$d = \frac{6.83(0.026)}{1 + 6.72\sqrt{0.026}}$$

$$= 0.085 \text{ ft.}$$

Then

$$Re = \frac{1887.6}{0.085} = 22,207$$

and $$f = 0.0255, \text{ giving}$$

$$d = \frac{6.83(0.0255)}{1 + 6.72\sqrt{0.0255}}$$

$$= 0.084 \text{ ft} = 1.008 \text{ in.}$$

As this value agrees very closely to the initial estimated value (0.085 ft), the process is terminated here.

• PROBLEM 7-32

Air flows through a pipe 5 ft. long with an internal diameter of 0.5 in. The air is flowing at a uniform speed of 80 ft/sec. The wall surface temperature is maintained at 210 °F. If air enters at a temperature of 60 °F and at atmospheric pressure, calculate the coefficient of heat transfer.

Solution: The energy balance equation for the pipe and the air is (heat absorbed by air) = (heat lost by pipe). The relation for the elemental length of the pipe is given by

$$\rho U \frac{\pi}{4} d^2 c_p (T|_{x+\Delta x} - T|_x) = h\pi d\Delta x(T_s - T) \qquad (1)$$

where T_s is the wall surface temperature and T is the mean bulk temperature. Integrating over a length L, the result is

$$\ln\left(\frac{T_o - T_s}{T_i - T_s}\right) + \frac{4Lh}{d\rho U c_p} = 0 \qquad (2)$$

where T_o and T_i refer to the outlet and inlet temperatures. Now, the outlet temperature is unknown, so is the heat transfer coefficient. The latter may be found from appropriate Nusselt number correlation. This can be chosen when the flow is categorized by calculation of the Reynolds number.

For an approximation, at the inlet condition of 60 °F, the air tables give

$$\nu = 1.58 \times 10^{-4}\ ft^2/sec$$

so the Reynolds number is

$$Re_d = \frac{Ud}{\nu}$$

$$= \frac{80\left(\frac{0.5}{12}\right)}{1.58 \times 10^{-4}} = 21{,}097$$

This is a turbulent flow. An appropriate Nusselt number relation is

$$Nu_d = \frac{hd}{k} = 0.023\ Re_d^{0.8}\ Pr^{1/3} \qquad (3)$$

This may be used if

1. $0.5 < Pr < 100$

and

2. $L/d > 60$

and if the properties are evaluated at the mean film temperature

$$T_m = \frac{T_s + T_b}{2}$$

This problem is solved by iteration, and is done as follows: to begin with, an exit temperature of 190°F is chosen. The mean bulk temperature is

$T_b = \frac{60 + 190}{2} = 125$ F and the mean film temperature is

$$T_m = \frac{210 + 125}{2} = 167.5°F$$

From the air tables, at 167.5°F ,

$c_p = 0.24$ Btu/lbm-°F, $\rho = 0.0633$ lbm/ft^3

$\nu = 2.2 \times 10^{-4}$ ft^2/sec $k = 0.0171$ Btu/hr-ft-°F

and

$Pr = 0.707$

The first condition of eq.(3) is satisfied. For the second,

$$L/d = 5/65x12) = 120.$$

Then from eq.(3), the heat transfer coefficient is

$$h = 0.023\ Re_d^{0.8}\ Pr^{1/3}\left(\frac{k}{d}\right)$$

$$= 0.023 \left[\frac{(80)\left(\frac{0.5}{12}\right)}{2.2 \times 10^{-4}} \right]^{0.8} (0.707)^{1/3} \left(\frac{0.0171}{0.5/12}\right)$$

$$= 18.6 \text{ Btu/hr-ft}^2\text{-}^0\text{F}$$

Substituting in eq.(2),

$$\ln\left(\frac{T_o - 210}{60 - 210}\right) + \frac{4(5)(18.6)}{\left(\frac{0.5}{12}\right)(0.0633)(80)(3600)(0.24)} = 0$$

Then

$$T_o = 190.5^0\text{F}.$$

This is close enough to the original assumed value, so no further iterations are required. The value of the heat transfer coefficient is, then, 18.6 Btu/hr-ft^2-^{0}F.

• PROBLEM 7-33

Air is heated as it flows through a pipe. It enters at a temperature of 75 ^{0}F and leaves at 125 ^{0}F. The air flow rate is 220 lbm/hr. If the pipe is 18 ft. long with diameter 2.75 in., what is the wall surface temperature? Assume turbulent flow.

Solution: A heat balance over the pipe length is

$$q = h\pi dL\Delta T_{\ell n} = \dot{m}c_p(T_o - T_i) \tag{1}$$

where the logarithmic mean temperature difference referred to is

$$\Delta T_{\ell n} = \frac{T_o - T_i}{\ln\left(\frac{T_s - T_i}{T_s - T_o}\right)} \tag{2}$$

The wall surface temperature is unknown, and hence the coefficient of heat transfer. Assuming a fully developed turbulent flow, the following relation is used:

$$Nu_d = \frac{hd}{k} = 0.023 \, Re^{0.8} \, Pr^{1/3} \tag{3}$$

as the conditions (1) $0.5 < Pr < 100$ and (2) $\frac{L}{d} > 60$ are satisfied. The problem is solved by a series of iterations by assuming values of T_s (the wall surface temperature).

A value of $T_s = 200^\circ F$ is assumed to begin. The mean bulk temperature is

$$T_b = \frac{75 + 125}{2} = 100\,^0F,$$

and the mean film temperature is

$$T_f = \frac{100 + 200}{2} = 150\,^0F$$

At this temperature, the air tables list

$c_p = 0.24$ Btu/lbm-^{0}F $\qquad \rho = 0.06508$ lbm/ft^3

$\mu = 0.04905$ lbm/ft-hr $\qquad \nu = 0.754$ ft^2/hr

$k = 0.01671$ Btu/hr-ft-^{0}F Pr= 0.7075

Since

$$Re = \frac{\rho U d}{\mu}$$

and

$$\dot{m} = \rho U \frac{\pi}{4} d^2,$$

$$Re = \frac{4\dot{m}}{\pi d \mu}$$

Substituting the values in eq.(3)

$$h = 0.023 \left[\frac{4}{\pi} \; \frac{220}{\left(\frac{2.75}{12}\right)^2} \; \frac{\left(\frac{2.75}{12}\right)}{0.04905} \right]^{0.8} (0.7075)^{1/3} \left(\frac{0.01671}{\frac{2.75}{12}}\right)$$

$\simeq 4.92$ Btu/hr-ft^2-^{0}F.

Now, from eq.(1),

$$\Delta T_{ln} = \frac{\dot{m} c_p (T_o - T_i)}{h \pi d L}$$

$$= \frac{(220)(0.24)(125-75)}{(4.92)\pi\left(\frac{2.75}{12}\right)(18)}$$

$$= 41.4\,^0F$$

Substituting into eq.(2),

$$41.4 = \frac{125-75}{\ln\left(\frac{T_s - 75}{T_s - 125}\right)}$$

or

$$T_s = 146.3^0F \quad \text{This is different from the assumed value of } 200^0F.$$

$T_s = 146.3^0F$ is assumed for the second iteration, giving

$$T_f = \frac{100 + 146.3}{2} = 123.15^0F$$

At this temperature, the air tables give

$c_p = 0.24$ Btu/lbm-^{0}F $\qquad \rho = 0.0681$ lbm/ft^3

$\mu = 0.0474$ lbm/ft-hr $\qquad \nu = 0.697$ ft^2/hr

$k = 0.0161$ Btu/hr-ft-^{0}F $\quad Pr = 0.709$

Then $\qquad h = 4.87$ Btu/hr-ft^2-^{0}F,

$$\Delta T_{ln} = 41.81^0F,$$

and $\qquad T_s = 146.7^0F.$

As the variation is small enough, no further iterations are required.

• PROBLEM 7-34

On a pipe, heat transfer is accomplished through a straight fin extending into an air stream. The fin may be idealized as a flat square plate with a 5.1 cm side. The flow is measured at 12.2m/s, and is at a temperature and pressure of 15.6 °C and 1 atm abs., respectively. If the plate surface temperature is 82.2 °C, calculate the heat transfer coefficient assuming (a) laminar flow, and (b) a rough leading edge, making the adjacent boundary layer completely turbulent.

Solution: The Reynolds number will be required. Thus, the properties must be evaluated. The mean film temperature is

$$T_m = \frac{15.6 + 82.2}{2} = 48.9\,^0C$$

The air tables at this temperature yield

$\rho = 1.097$ kg/m^3 $\qquad \mu = 1.95 \times 10^{-5}$ kg/m-sec

$Pr = 0.704$ $\qquad k = 0.028$ W/m-^{0}K

The Reynolds number is

$$Re_L = \frac{\rho UL}{\mu} = \frac{(1.097)(12.2)(0.051)}{1.95\times10^{-5}}$$

$$= 3.5 \times 10^4$$

which is indeed a laminar flow.

(a) For a laminar flow over a plane surface, the relation for the average Nusselt number is

$$Nu_L = 0.664\ Re_L^{1/2}\ Pr^{1/3}$$

(Since the constraint $0.6 \leq Pr \leq 50$ is satisfied)

where $$Nu_L = \frac{h_L L}{K}$$

The average heat transfer coefficient is

$$h_L = 0.664\ Re_L^{1/2}\ Pr^{1/3} \left(\frac{k}{L}\right)$$

$$= 0.664(3.5 \times 10^4)^{1/2}(0.704)^{1/3} \left(\frac{0.028}{0.051}\right)$$

$$= 60.67\ W/m\ -{}^0K$$

(b) Now, for turbulent flow (which may start at even smaller Reynolds numbers if the plate surface is rough enough),

$$Nu_L = 0.0366\ Re^{0.8}\ Pr^{1/3}$$

The constraint $Pr > 0.7$

is satisfied. The heat transfer coefficient is solved for as

$$h_L = 0.0366\ Re^{0.8}\ Pr^{1/3} \left(\frac{k}{L}\right)$$

$$= 0.0366(3.5 \times 10^4)^{0.8}(0.704)^{1/3} \left(\frac{0.028}{0.051}\right)$$

$$= 77.18\ W/m^2 -{}^0K$$

A steel cylinder contains liquid at a mean bulk temperature of 80 °F. Steam condensing at 212 °F on the outside surface is used for heating the liquid. The coefficient of heat transfer on the steam side is 1,000 Btu/hr-ft²-°F. The liquid is agitated by the stirring action of a turbine impeller. Its diameter is 2 ft., and it moves at an angular velocity of 100 rpm. The cylinder is 6 ft. long, with a diameter of 6 ft. and a wall thickness of 1/8 in. The thermal conductivity of steel may be taken as 9.4 Btu/hr-ft²-°F. Properties of the liquid, taken as constant, are:

$c_p = 0.6$ Btu/lbm-°F $\qquad$ $k = 0.1$ Btu/hr-ft-°F

$\rho = 60$ lbm/ft³

The viscosity at 130 °F is 653.4 lbm/ft-hr, and at 212 °F is 113.74 lbm/ft-hr. Calculate the time required to raise the mean bulk temperature of the liquid to 180 °F.

Solution: If an energy balance is taken over the cylinder, the result is

$$q = UA\Delta T_L = \mathring{m} c_p \Delta T_b$$

where

$$\Delta T_L = \frac{T_{b_2} - T_{b_1}}{\ln\left(\dfrac{T_s - T_{b_1}}{T_s - T_{b_2}}\right)}$$

where T_s is the steam temperature. Rearranging and substituting,

$$\frac{1}{\mathring{m}} = \frac{c_p}{UA} \ln\left(\frac{T_s - T_{b_1}}{T_s - T_{b_2}}\right) \tag{1}$$

Now, the overall heat transfer coefficient (based on the inside surface area) is defined as

$$\frac{1}{U_i} = \frac{1}{h_i} + \frac{A_i x}{A_m k} + \frac{A_i}{A_o h_s}$$

where m denotes a mean value, and x is the wall thickness. However, since the walls are relatively thin, the approximation

$$A_i \simeq A_o \simeq A_m$$

may be used, giving

$$\frac{1}{U_i} = \frac{1}{h_i} + \frac{x}{k} + \frac{1}{h_s} \quad (2)$$

The unknown quantity here is the inside heat transfer coefficient h_i. This may be found from the appropriate Nusselt number configuartion, which is

$$Nu = \frac{h_i d_T}{k} = 0.73 \; Re^{0.65} \; Pr^{1/3} \left(\frac{\mu_b}{\mu_w}\right)^{0.24} \quad (3)$$

where w refers to the inside wall surface temperature value and d_T is the total hydraulic diameter.

Now, if the Reynolds and Prandtl numbers are known, h_i may be found from eq.(3). With this, U_i may be found from eq.(2). Finally, if eq.(1) is used, the required length of time will be

$$\Delta t(hr) = \frac{1}{\dot{m}} \left(\frac{hr}{lbm}\right) \times weight(lbm)$$

so rewriting eq.(1) yields

$$\Delta t = \frac{(wt)c_p}{UA} \ln\left(\frac{T_s - T_{b_1}}{T_s - T_{b_2}}\right) \quad (4)$$

Calculations will be done at the mean temperature of

$$T_m = \frac{80 + 180}{2} = 130\,^0F$$

The Reynolds number is

$$Re = \frac{\rho U d}{\mu}$$

but given the angular velocity, $U = \omega d$ is used. Then

$$Re = \frac{(60)(100)(60)(2)^2}{653.4}$$

$$= 2204.$$

The Prandtl number is

$$Pr = \frac{\mu c_p}{k}$$

$$= \frac{(653.4)(0.6)}{0.1}$$

$$= 3920.4$$

From eq.(3),

$$h_i = 0.73(2204)^{0.65}(3920.4)^{1/3}\left(\frac{653.4}{113.74}\right)^{0.24}\left(\frac{0.1}{6}\right)$$

$$= 43.5 \text{ Btu/hr-ft}^2\text{-}^0\text{F}$$

Then, from eq.(2),

$$U_i = \frac{1}{\frac{1}{43.5} + \frac{(1/8)/12}{9.4} + \frac{1}{1,000}}$$

$$= 39.85 \text{ Btu/hr-ft}^2\text{-}^0\text{F}$$

Now, since $\rho = \text{wt/vol}$,

$$\text{wt} = (\rho)(\text{vol})$$

$$= (60)\pi\frac{(6)^2}{4}(6)$$

$$= 10,179 \text{ lbm}$$

The wall surface temperature is taken to be approximately equal to that of the condensing stream.

Substituting the values in eq.(4),

$$\Delta t = \frac{(10,179)(0.6)}{(39.85)\pi(6)(6)} \ln\left(\frac{212 - 80}{212 - 180}\right)$$

$$= 1.92 \text{ hr}$$

EXTERNAL FLOW

• PROBLEM 7-36

Air is flowing perpendicularly to a pipe, transferring heat by forced convection. The pipe has a diameter 4 in. and a surface temperature of 90^{0}F. The cooling air with flow velocity of 15 ft/sec, has a bulk temperature of 75^{0}F. Calculate the rate of heat transfer per unit length of pipe.

<u>Solution</u>: For external flow over a pipe, a common Nusselt number correlation is that of Churchill and Bernstein:

$$Nu_d = \left[0.3 + \frac{0.62 Re_d^{1/2} Pr^{1/3}}{[1+(0.4/Pr)^{2/3}]^{1/4}}\right]\left[1 + \left(\frac{Re_d}{2.82\times10^5}\right)^{5/8}\right]^{4/5}$$

This may be used if

$$Re_d Pr > 0.2$$

and the properties are evaluated at the mean temperature. This is

$$T_m = \frac{75 + 90}{2} = 82.5\,^0F$$

At this temperature, the air tables list

$\nu = 0.6136\ ft^2/hr$ $\quad$ $k = 0.01517\ Btu/hr\text{-}ft\text{-}^0F$

$Pr = 0.711$

The Reynolds number may be calculated as

$$Re_d = \frac{Ud}{\nu} = \frac{15(3600)(4/12)}{0.6136}$$

$$= 29335$$

Then

$$Re_d Pr = (29335)(0.711)$$

$$= 20857 > 0.2.$$

Substituting the values in eq.(1), yields

$$Nu_d = \left[0.3 + \frac{0.62(29335)^{1/2}(0.711)^{1/3}}{[1+(0.4/0.711)^{2/3}]^{1/4}}\right]\left[1 + \left(\frac{29335}{2.82\times10^5}\right)^{5/8}\right]^{4/5}$$

$$\simeq 99.41$$

The coefficient of heat transfer is found as

$$h = \frac{Nu_d k}{d} = \frac{99.41(0.01517)}{4/12}$$

$$= 4.52\ Btu/hr\text{-}ft^2\text{-}^0F$$

Since

$$q = hA\Delta T,$$

$$q/L = h\pi d\Delta T$$

$$= (4.52)\pi(4/12)(90 - 75)$$

$$\simeq 71\ Btu/hr\text{-}ft$$

• PROBLEM 7-37

Cooling water flows over a tube bank, transferring heat by forced convection. The flow velocity of water is 0.8 ft/sec, and the mean bulk temperature is 50 °F. Each tube has a diameter of 1 in. and a length of 10 ft., and the arrangement is staggered. In the longitudinal (flow) direction, the tube spacing is 2 in., with 6 rows. In the transverse direction, the tube spacing is 2 in., with 60 tubes. Inside the tubes, steam condenses, keeping the outside wall temperature at 210 °F. If the latent heat of vaporization of the steam h_{fg} is 950 Btu/lbm, calculate the rate of steam condensation.

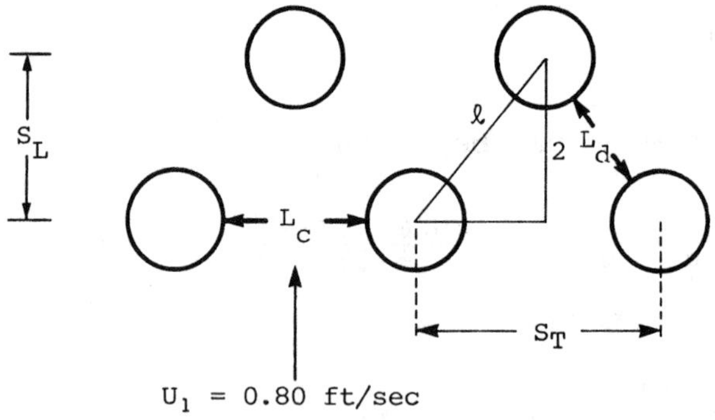

Fig. 1 Schematic arrangement of the tube bank.

Solution: By an energy balance, the heat lost by the steam is equal to the heat gained by the water,

$$hA(T_s - T) = \dot{m}h_{fg} \tag{1}$$

The mean film temperature is

$$T_m = \frac{50 + 210}{2} = 130\,^0F.$$

At this temperature, the water tables list the properties as

$k = 0.3728$ Btu/hr-ft-^{0}F $\quad \mu = 1.231$ lbm/ft-hr

Pr = 3.3

and at 50 ^{0}F, $\rho = 62.41$ lbm/ft^3.

The Reynolds number is calculated as

$$Re_d = \frac{\rho U d}{\mu}$$

Table 1: Constants B and m for use in Eq. 3 for tube banks of 10 or more.

Bank design	Longitudinal spacing ratio S_L/d	Transverse spacing ratio S_T/d							
		1.25		1.5		2.0		3.0	
		B	m	B	m	B	m	B	m
Staggered	0.600	...	...	...	...	...	...	0.213	0.636
	0.900	...	...	...	...	0.446	0.571	0.401	0.581
	1.000	...	...	0.497	0.558	...	...	...	...
	1.125	...	...	...	...	0.478	0.565	0.518	0.560
	1.250	0.518	0.556	0.505	0.554	0.519	0.556	0.522	0.562
	1.500	0.451	0.568	0.460	0.562	0.452	0.568	0.488	0.568
	2.000	0.404	0.572	0.416	0.568	0.482	0.556	0.449	0.570
	3.000	0.310	0.592	0.356	0.580	0.440	0.562	0.421	0.574
In-line	1.250	0.348	0.592	0.275	0.608	0.100	0.704	0.0633	0.752
	1.500	0.367	0.586	0.250	0.620	0.101	0.702	0.0678	0.744
	2.000	0.418	0.570	0.299	0.602	0.229	0.632	0.198	0.648
	3.000	0.290	0.601	0.357	0.584	0.374	0.581	0.286	0.608

Table 2: Ratio of the average heat transfer coefficient for the number of rows shown to that for 10 rows for staggered tube banks.

Number of Rows	4	5	6	7	8	9	10
h/h_{10}	0.88	0.92	0.94	0.97	0.98	0.99	1.0

The product ρU must be a maximum, and this will occur at the minimum area opening, whether transverse (L_c) or diagonal (L_d).

$$L_c = S_T - d = 2 - 1 = 1 \text{ in.}$$

$$L_d = \left[S_L^2 + \left(\frac{S_T}{2}\right)^2 \right]^{1/2} - d = \sqrt{2^2 + 1^2} - 1 = 1.236 \text{ in.}$$

Since $L_c < L_d$, the transverse opening provides the minimum area. By continuity,

$$\rho_1 U_1 S_T = (\rho U)_m L_c \tag{2}$$

Solving for the maximum product $(\rho U)_m$ yields

$$(\rho U)_m = \frac{\rho_1 U_1 S_T}{L_c} = \frac{(62.41)(0.8)(3600)(2)}{1}$$

$$\simeq 359482 \text{ lbm/ft}^2\text{-hr}$$

Then

$$Re_d = \frac{(359482)\left(\frac{1}{12}\right)}{(1.231)}$$

$$\simeq 24,335$$

Since $2000 < Re_d < 40,000$, the Nusselt number relation

$$Nu_d = \frac{hd}{k} = 1.13\ BRe_d^{\ m} Pr^{1/3} \tag{3}$$

is used. This holds for a bank at least 10 rows deep. Table 1 may be used to find the constants B and m. Since the transverse spacing ratio S_T/d is 2/1=2 and the longitudinal spacing ratio S_L/d is 2/1=2, the values of B and m are 0.482 and 0.556, respectively. Substituting into eq.(3) and solving for h yields

$$h = 1.13(0.482)(24,335)^{0.556}(3.3)^{1/3}\ \frac{0.3728}{1/12}$$

$$\simeq 996\ \text{Btu/hr-ft}^2\text{-}^0\text{F}.$$

Table 2 gives the appropriate modifications for a row depth of less than 10. Since there are 6 rows, the value is 94% of the computed value, or

$$h = 996(0.94)$$

$$= 936.5\ \text{Btu/hr-ft}^2\text{-}^0\text{F}.$$

The surface area involved must be multiplied by 360, the total number of tubes. Then substituting the values in eq. (1),

$$\dot{m} = \frac{hA(T_s - T)}{h_{fg}}$$

$$= \frac{(936.5)(360)\pi(1/12)(10)(210-50)}{950}$$

$$\simeq 148654\ \text{lbm/hr}$$

• PROBLEM 7-38

Air flows over a tube bank at a speed of 24 ft/sec. The bank is four rows deep, and the tubes are aligned as shown in Fig. 1. If the temperature of the air is 1100 °F and the outside wall temperature is at 600 °F, determine (a) the coefficient of heat transfer, and (b) the pressure drop.

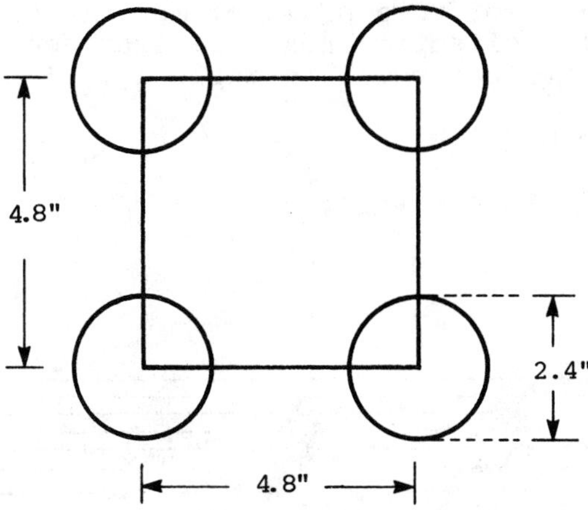

$$a = \frac{4.8}{2.4} = b = 2.0$$

Fig. 1

<u>Solution</u>: (a) The mean film temperature is

$$T_m = \frac{600 + 1100}{2} = 850\,{}^0F$$

At this temperature from the air tables,

$\nu = 2.744\ ft^2/hr$ $\qquad$ $k = 0.03054\ Btu/hr\text{-}ft\text{-}{}^0F$

$Pr = 0.703$

The Reynolds number is

$$Re_d = \frac{Ud}{\nu} = \frac{(24)(3600)\left(\frac{2.4}{12}\right)}{2.744}$$

$$= 6297.4$$

Since $\quad 2000 < Re_d < 40{,}000,\quad$ the expression

$$Nu_d = \frac{hd}{k} = 1.13\ B\ Re_d^{\ m} Pr^{1/3} \tag{1}$$

is used. The constants B and m are found from standard tube bank tables. Since both the longitudinal spacing ratio a and the transverse spacing ratio b are S/d = 4.8/2.4 = 2.0, constants B and m are listed as 0.229 and 0.632, respectively. Then eq.(1) becomes

$$\frac{hd}{k} = 0.2588\ Re_d^{\ 0.632} Pr^{\ 1/3} \tag{2}$$

and substituting the values,

$$h = 0.2588(6297.4)^{0.632}(0.703)^{1/3}\left(\frac{0.03504}{2.4/12}\right)$$

$$= 8.85\ Btu/hr\text{-}ft^2\text{-}{}^0F$$

However, eq.(1) holds for rows of 10 or more tubes. For a depth of 4 tubes, the effective heat transfer coefficient is 88% of the computed value (refer to table 1). Therefore,

$$h = 8.85(0.88)$$

$$= 7.79 \text{ Btu/hr-ft}^2\text{-}^0\text{F}$$

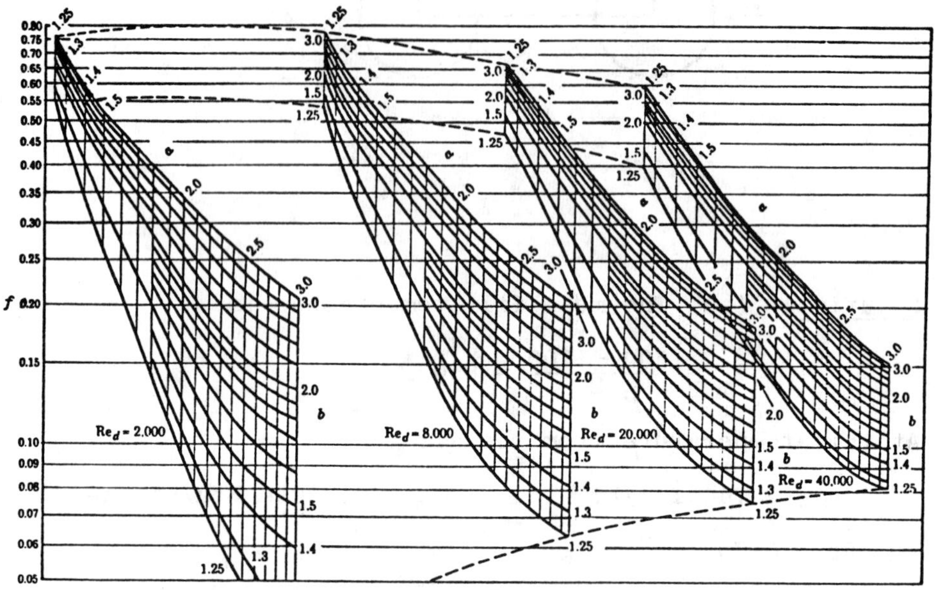

Fig. 2 Resistance coefficient for tube bundles - tubes aligned.

Table Ratio of the average heat transfer coefficient for the number of rows shown to that for 10 rows for staggered tube banks.

Number of Rows	4	5	6	7	8	9	10
h/h_{10}	0.88	0.92	0.94	0.97	0.98	0.99	1.0

(b) The pressure drop is given by

$$\Delta p = nf\rho \frac{U^2}{2g} \tag{3}$$

where n is the number of tubes, f the resistance coefficient of the tube bundles, and ρ the fluid density. Referring to

Fig. 2, the value of f is found to be approximately 0.225. Also, at the fluid temperature of 1100°F, the air tables give a value of

$$\rho = 0.02543 \text{ lbm/ft}^3$$

Substituting into eq.(3), the pressure drop is

$$\Delta p = 4(0.225)(0.02543)\frac{(24)^2}{2(32.2)}$$

$$\simeq 0.205 \text{ lb/ft}^2$$

• PROBLEM 7-39

Air is flowing over a 5-row tube bank. The air enters the enclosing duct at a temperature of 60 °F, and it leaves at 100 °F. The air flow is 500 ft³/min, and is at atmospheric pressure. The tubes have a diameter of 1 in. and a length of 4 ft. They are staggered, with the longitudinal and transverse spacing ratio of Z. The outside wall surface temperature is 227 °F. If the minimum free flow cross-sectional area is 1.667 ft², calculate the coefficient of heat transfer.

Table 1: Ratio of mean heat-transfer coefficient for bank of tubes N rows deep to the coefficient for tubes in a single row.

	N									
	1	2	3	4	5	6	7	8	9	10
Triangular grid	1	1.10	1.22	1.31	1.35	1.40	1.42	1.44	1.46	1.47
Square grid	1	1.25	1.36	1.41	1.44	1.47	1.50	1.53	1.55	1.56

Solution: The coefficient of heat transfer may be calculated from Grimison's correlation for air flowing over a tube bank 10 rows deep:

$$Nu_d = \frac{hd}{k} = B\ Re_d^{\ m} \tag{1}$$

The constants B and m are found from standard tables, and correspond directly to the spacing ratios. With the given values, the tables give values of B = 0.482 and m = 0.556. Eq.(1) becomes

$$\frac{hd}{k} = 0.482\ Re_d^{\ 0.556} \tag{2}$$

Reynolds number is given by

$$Re_d = \frac{\rho U d}{\mu}$$

The product ρU must be a maximum. This will occur at the minimum flow area, i.e. at 1.667 ft^2, and since the volume flow rate Q is given, the substitution

$$U = Q/A$$

can be made, giving

$$Re_d = \frac{\rho Q d}{\mu A}$$

The mean bulk air temperature is

$$T_b = \frac{60 + 100}{2} = 80^0F$$

Then the mean film temperature is

$$T_m = \frac{80 + 227}{2} = 153.5^0F$$

At this temperature, the air tables list

$k = 0.0164$ Btu/hr-ft-^{0}F $\quad \mu = 0.049$ lbm/ft-hr

and at the air entrance temperature (60^0F),

$\rho = 0.076$ lbm/ft^3

Then

$$Re_d = \frac{(0.076)(500)(60)(1/12)}{(0.049)(1.667)}$$

$$= 2326$$

Then from eq.(2), the coefficient of heat transfer is

$$h = 0.482(2326)^{0.556}\left(\frac{0.0164}{1/12}\right)$$

$$= 7.062 \text{ Btu/hr-ft}^2\text{-}^0\text{F}$$

Now, the bank is only 5 rows deep. Referring to Table 1, in a staggered arrangement (or triangular grid), the ratio of heat transfer coefficients having 5 and 10 rows is

$$\frac{h_5}{h_{10}} = \frac{1.35}{1.47}$$

Therefore,

$$h_5 = 0.9184\, h_{10} = 0.9184(7.062)$$

$$= 6.49 \text{ Btu/hr-ft}^2\text{-}^0\text{F}.$$

A plane flies at 500 ft/sec, at an elevation of 25,000 ft. above sea level. The atmosphere is at a temperature of approximately $-30^\circ F$. It is desired to keep the surface temperature of the wing at $32^\circ F$, in order that ice may not form on it. If the wing's leading edge contour may be idealized as a half of a 1 ft. diameter cylinder, calculate the coefficient of heat transfer across it.

Solution: In the case of external flow over a cylinder, a specific relation giving the coefficient of heat transfer as a function of the radial measurement θ is

$$h_\theta = 0.194\, T_f^{0.49} \left(\frac{\rho U}{d}\right)^{1/2} \left[1 - \left(\frac{\theta}{90^\circ}\right)^3\right] \tag{1}$$

where T_f is the mean film temperature, which is

$$T_f = \frac{T_s + T_b}{2} = \frac{32 - 30}{2} = 1^\circ F + 460 = 461^\circ R.$$

Now, the standard-atmosphere table lists a value of

$$\rho/\rho_0 = 0.4481$$

at an altitude of 25,000 ft. Since

$$\rho_0 = 0.07657 \text{ lbm/ft}^3, \text{ then}$$

$$\rho = 0.0343 \text{ lbm/ft}^3.$$

At $\theta = 0^\circ$ (called the stagnation point), h_θ is found from eq. (1) as

$$h_\theta = 0.194(461)^{0.49}\left[\frac{(0.0343)(500)}{1}\right]^{1/2} [1 - 0]$$

$$= 16.22 \text{ Btu/hr-ft}^2\text{-}^\circ F.$$

The variation in the coefficient of heat transfer is shown when calculations are performed in a similar manner at spacings of 15°. Results are:

At 15°, $h_\theta = 16.15$ Btu/hr-ft^2-$^\circ$F.

At 30°, $h_\theta = 15.62$ Btu/hr-ft^2-$^\circ$F.

At 45°, $h_\theta = 14.19$ Btu/hr-ft^2-$^\circ$F.

At 60°, $h_\theta = 11.41$ Btu/hr-ft^2-$^\circ$F.

At 75°, $h_\theta = 6.83$ Btu/hr-ft^2-$^\circ$F.

Finally, at $\theta = 90^\circ$, the heat transfer coefficient goes to zero, as the final term becomes one.

• PROBLEM 7-41

Air flows through a pipe, and cooling is achieved by means of forced convection. Air flowing at a speed of 100 ft/sec enters the pipe at a temperature and pressure of 68^0F and 14.2 psi, respectively. The pipe is 3 ft. in length, with a 0.8 in. diameter. The wall surface temperature is 200^0F. If the pressure drop over the length is 3.2 in. of water, calculate the heat transferred from the pipe to the air.

Solution: The energy balance is obtained by equating the heat lost by the pipe and the heat gained by the air.

$$\frac{Rc_p}{U}(T_s - T_b) = \dot{m}c_p(T_o - T_i) \qquad (1)$$

Now, R is the pipe resistance, defined as

$$R = \Delta pA = \Delta p \frac{\pi}{4} d^2$$

Also, since $\dot{m} = \rho UA = \rho U \frac{\pi}{4} d^2$

and

$$T_b = \frac{T_i + T_o}{2},$$

eq.(1) becomes, after substitution and simplification,

$$\frac{\Delta p}{U}\left[T_s - \left(\frac{T_i+T_o}{2}\right)\right] = \rho U(T_o-T_i) \qquad (2)$$

The pressure is given in inches of water. This may be converted to psi using the factor

$$27.72 \text{ in. water} = 1 \text{ psi}$$

The standard air density value is 0.0752 lbm/ft^3 at the entrance conditions. Substituting into eq.(2),

$$\frac{\left(\frac{3.2}{27.72}\right)(144)(32.2)}{(100)}\left[200 - \left(\frac{68 + T_o}{2}\right)\right] = (0.0752)(100)(T_o-68)$$

Solving this for T_o yields

$$T_o = 137.3^\circ F$$

Knowing this, either side of eq.(1) may be used to calculate the heat lost. If the left side is chosen,

$$q = \frac{Rc_p}{U}(T_s - T_b)$$

The average specific heat value is

$$c_p = 0.24 \text{ Btu/lbm-}^0\text{F}$$

Then

$$q = \Delta p \; \pi \frac{d^2}{4} \frac{c_p}{U}\left[T_s - \left(\frac{T_i + T_o}{2}\right)\right]$$

$$= \left(\frac{3.2}{27.72}\right) \frac{\pi}{4} (0.8)^2 (0.24) \frac{(32.2)}{(100)}\left[200 - \left(\frac{68+137.3}{2}\right)\right]$$

$$= 0.4365 \text{ Btu/sec } (3600 \text{ sec/hr})$$

$$= 1571.56 \text{ Btu/hr}$$

• PROBLEM 7-42

Air flows through a tube, causing the transfer of heat by forced convection. The tube has an internal diameter of 1 in., and the wall surface temperature is 212 ^{0}F. The air is flowing at a rate of 1570 ft^3/hr, and it enters at a temperature of 60 ^{0}F. If it leaves at 100 ^{0}F, calculate the length of the tube, and also the pressure drop over the length.

Solution: The energy balance equation is:

heat lost by the tube = heat gained by the air,

or

$$hA(T_s - T_m) = \dot{m}c_p(T_o - T_i) \qquad (1)$$

The volume flow rate Q is given and the mass flow rate is to be calculated. Since it is the product UA, and

$$\dot{m} = \rho UA,$$

then

$$\dot{m} = \rho Q$$

At the entrance conditions, the air tables give values of

$$\rho = 0.07633 \text{ lbm/ft}^3 \qquad c_p = 0.24 \text{ Btu/lbm-}^0\text{F}$$

Then

$$\dot{m} = (0.07633)(1570)$$

$$= 119.84 \text{ lbm/hr}$$

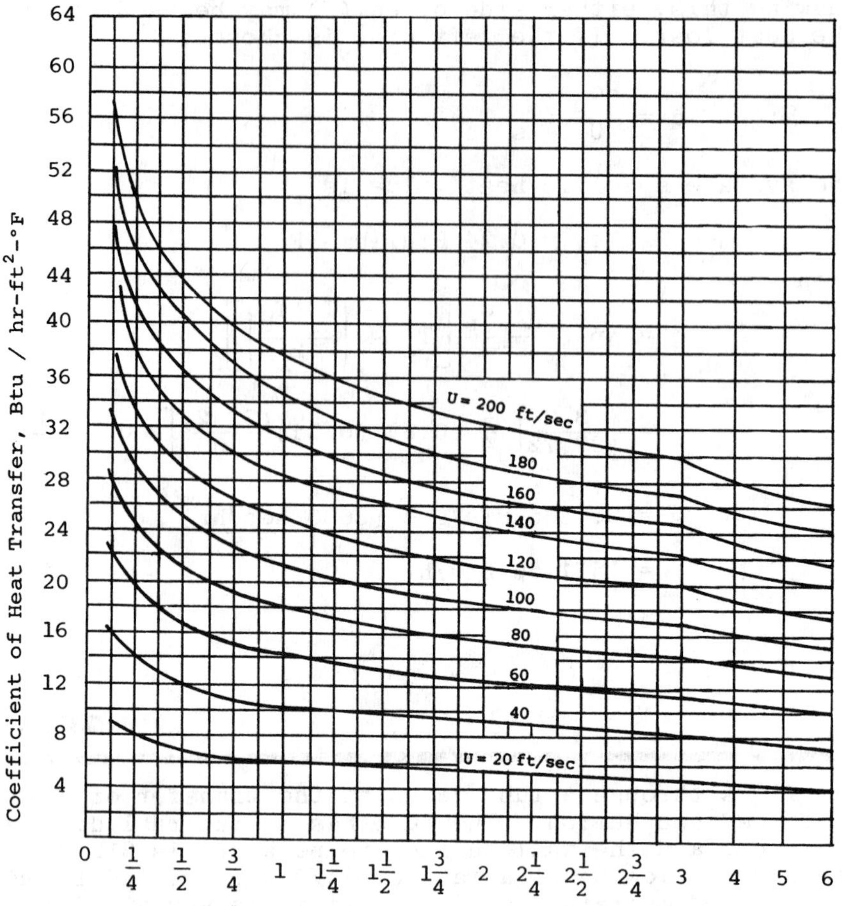

Fig. 1 Inner Diameter of Tube - Inches

The mean temperature is

$$T_m = \frac{60 + 100}{2} = 80\,^0\mathrm{F}$$

Referring to Fig. 1, the coefficient of heat may be read off the graph. Since

$$U = Q/A = \frac{(1570)}{\frac{\pi}{4}\left(\frac{1}{12}\right)^2 (3600)}$$

$$= 80 \text{ ft/sec}$$

and the inner diameter is 1 in., the coefficient of heat transfer is 17.8 Btu/hr-ft^2-^{0}F.

Now, as $A = \pi dL$, from eq.(1),

$$L = \frac{\dot{m}c_p(T_o - T_i)}{h\pi d(T_s - T_m)} \tag{2}$$

$$= \frac{(119.84)(0.24)(100-60)}{(17.8)\pi\left(\frac{1}{12}\right)(212-80)}$$

$$= 1.87 \text{ ft.}$$

For the pressure drop, consider the left-hand side of eq.(1). It can be written as

$$\frac{Rc_p}{U}(T_s - T_m)$$

Since R (the resistance) is

$$R = \Delta pA = \Delta p \frac{\pi}{4} d^2,$$

the pressure drop may be found as

$$\Delta p = \frac{4UhL}{dc_p}$$

$$= \frac{4(80)(17.8)(1.87)}{(1/12)(0.24)(3600)(32.2)}$$

$$= 4.59 \text{ lb/ft}^2$$

Since 1 psi = 2.31 ft. of water,

$$\Delta p = \frac{(4.59)(2.31)}{12}$$

$$= 0.884 \text{ in. of water}$$

• PROBLEM 7-43

Air flows through a tube, and heat is transferred by means of forced convection. The tube is 1 ft. long, with an internal diameter of 2 in. The wall surface temperature is constant at 80^0F, and the pressure drop over the length is 0.02 in. of water. The air flow velocity is 20 ft/sec, and the entrance temperature is 300^0F. Calculate the exit temperature, given the density of air to be 0.08 lbm/ft^3.

Solution: A heat balance taken over the tube length is

$$\frac{Rc_p}{U}(T_s - T_m) = \dot{m}c_p(T_o - T_i) \qquad (1)$$

where

$$R = \Delta p \frac{\pi}{4} d^2,$$

$$T_m = \frac{T_o + T_i}{2} ,$$

and

$$\dot{m} = \rho U \frac{\pi}{4} d^2 .$$

Substituting into eq.(1) and simplifying yields

$$\frac{\Delta p}{U}\left[T_s - \left(\frac{T_o + T_i}{2}\right)\right] = \rho U (T_o - T_i) \qquad (2)$$

Now, the pressure drop is converted to lb/ft^2. Using the conversion factor 1 psi = 2.31 ft. H_2O,

$$0.02 \text{ in } H_2O\left(\frac{12}{2.31}\right) = 0.1039 \text{ lb/ft}^2$$

Substituting the given values into eq.(2),

$$\frac{(0.1039)}{20}(32.2)\left[80 - \left(\frac{T_o + 300}{2}\right)\right] = (0.08)(20)(T_o - 300)$$

Solving for T_o,

$$T_o = 278^0\text{F}.$$

• PROBLEM 7-44

A plate is cooled by a stream of air, which moves at the speed of 655 ft/sec. The plate length is 1 in., and the static temperature is 65 ^{0}F. Considering (a) a surface temperature of 95 ^{0}F, calculate the adiabatic wall temperature (b) a surface temperature of 300 ^{0}F, calculate the rate of heat transfer at the plate.

Solution: (a) For high-speed flow over a flat plate, the adiabatic wall temperature is given by

$$T_w = T_\infty + \frac{r_c U^2}{2c_p} \qquad (1)$$

The recovery factor r_c is found by

$$r_c = Pr^{1/2} \qquad (2)$$

if the flow is laminar, and

$$r_c = Pr^{1/3} \qquad (3)$$

if it is turbulent. Properties are evaluated at the reference temperature T*, found as

$$T^* = T_\infty + 0.5(T_s - T_\infty) + 0.22(T_w - T_\infty) \qquad (4)$$

However, for this part, properties are evaluated at the mean temperature, since T_w is unknown. Then

$$T_m = \frac{65 + 95}{2} = 80\,^0\text{F}$$

At this temperature, the air tables give values of

$\nu = 0.6086 \text{ ft}^2/\text{hr}$ $\qquad$ Pr = 0.711

$c_p = 0.2403 \text{ Btu/lbm-}^0\text{F}$

Then the Reynolds number is

$$Re_L = \frac{UL}{\nu} = \frac{(655)(3600)(1/12)}{0.6086}$$

$$= 322{,}872$$

As this is less than the critical value of 5×10^5, the flow is laminar. Using eq.(2),

$$r_c = (0.711)^{1/2}$$

$$= 0.843$$

Then from eq.(1),

$$T_w = 65 + \frac{(0.843)(655)^2}{2(0.24)(778)(32.2)}$$

$$\simeq 95\,^0\text{F}$$

If this value is inserted into eq.(4), the reference temperature becomes

$$T^* = 65 + 0.5(95-65) + 0.22(95-65)$$

$$= 86.6\,^0\text{F}$$

If the properties are evaluated at this temperature instead of at the mean temperature, the properties change so little that the overall effect is negligible. Since the adiabatic wall temperature is equal to the surface temperature, there would not be any appreciable heat transfer occurring.

(b) The mean temperature is

$$T_m = \frac{65 + 300}{2} = 182.5\,^0\text{F}$$

Properties are

$\nu = 0.8253 \text{ ft}^2/\text{hr}$ $\qquad$ Pr = 0.706

$$c_p = 0.2412 \text{ Btu/lbm-}^0\text{F}$$

The Reynolds number is

$$Re = \frac{(655)(3600)(1/12)}{0.8253}$$

$$= 238{,}095$$

Since the flow is still laminar, eq.(2) again applies.

$$r_c = (0.706)^{1/2}$$

$$= 0.84$$

From eq.(1)

$$T_w = 65 + \frac{(0.84)(655)^2}{2(0.2412)(778)(32.2)}$$

$$= 94.8^0\text{F}$$

Using eq.(4),

$$T^* = 65 + 0.5(300-65) + 0.22(94.8-65)$$

$$= 189^0\text{F}.$$

Again, the difference between the reference temperature and the mean value is not large enough to cause a great discrepancy.

The heat flow per unit area at the plate is

$$\frac{q}{A} = h(T_s - T_w) \tag{5}$$

The coefficient of heat transfer is needed. This can be found from the Nusselt number relation describing laminar flow over a flat plate:

$$Nu_L = 0.664\ Re_L^{1/2}\ Pr^{1/3} \tag{6}$$

At 182.5^0F,

$$k = 0.0174 \text{ Btu/hr-ft-}^0\text{F}$$

Then the heat transfer coefficient is

$$h = 0.664(238{,}095)^{1/2}(0.706)^{1/3}\left(\frac{0.0174}{1/12}\right)$$

$$= 60.24 \text{ Btu/hr-ft}^2\text{-}^0\text{F}.$$

Finally, the heat flow is found from eq.(5) as

$$q/A = (60.24)(300 - 94.8)$$

$$= 12{,}361 \text{ Btu/hr-ft}^2$$

• PROBLEM 7-45

A plate is cooled by a high-speed stream of air, moving at 5867 ft/sec (4000 mph). The plate is 4 in. long and has a surface temperature of 196°F. If the air is at a temperature and pressure of 79°F and 14.2 psi, respectively, calculate the rate of heat transfer per unit area at the plate. Note: Viscous dissipation effects cannot be ignored.

Solution: The heat flow, when expressed in terms of enthalpy (e) changes, per unit area is

$$q/A = h_{av}(e_r - e_s) \qquad (1)$$

where r denotes a recovery property. To obtain the recovery enthalpy, the following relation is appropriate:

$$e_r = e_\infty + \frac{r_c U_\infty^2}{2g} \qquad (2)$$

The recovery factor r_c is also unknown. For a laminar flow, it is found as

$$r_c = Pr^{1/2} \qquad (3)$$

and for turbulent flow,

$$r_c = Pr^{1/3} \qquad (4)$$

Property values must be evaluated at the reference temperature, which corresponds to the reference enthalpy,

$$e^* = e_\infty + 0.5(e_s - e_\infty) + 0.22(e_r - e_\infty) \qquad (5)$$

but assuming first that $e_s = e_r$,

$$e^* = e_\infty + 0.72(e_r - e_\infty) \qquad (6)$$

The problem is solved by iteration by assuming a value for r_c.

Let $r_c = 1$. From the air tables, at 79°F

$$e_\infty = 129.06 \text{ Btu/lbm}$$

From eq.(2), substituting the value of r_c gives

$$e_r = 129.06 + \frac{(1)(5867)^2}{2(32.2)(778)}$$

$$= 816.1 \text{ Btu/lbm}$$

Using eq.(6),

$$e^* = 129.06 + 0.72(816.1 - 129.06)$$
$$= 623.7 \text{ Btu/lbm}$$

At this temperature, the tables list

$$\nu = 0.002038 \text{ ft}^2/\text{sec} \qquad Pr = 0.72$$

Then the Reynolds number is

$$Re = \frac{UL}{\nu} = \frac{(5867)(4/12)}{0.002038}$$
$$= 9.59 \times 10^5$$

As this is a turbulent flow, eq.(4) is used.

$$r_c = (0.72)^{1/3}$$
$$= 0.897$$

This is appreciably smaller than the assumed value. The value assumed for the second iteration is

$$r_c = \frac{1 + 0.897}{2}$$
$$= 0.948$$

Substituting in eq.(2),

$$e_r = 129.06 + \frac{(0.948)(5867)^2}{2(32.2)(778)}$$
$$= 780.4 \text{ Btu/lbm}$$

From eq.(6),

$$e^* = 129.06 + 0.72(780.4 - 129.06)$$
$$= 598.1 \text{ Btu/lbm}$$

If this is used to find the reference temperature, the tables give

$$Pr = 0.723$$

Then, from eq.(4),

$$r_c = (0.723)^{1/3}$$
$$= 0.897$$

The assumed value for the third iteration is

$$r_c = \frac{0.949 + 0.897}{2}$$

$$= 0.923$$

From eq.(2),

$$e_r = 129.06 + \frac{(0.923)(5867)^2}{2(32.2)(778)}$$

$$= 763.2 \text{ Btu/lbm}$$

This value may be used. Now, as $e_s \neq e_r$, eq.(5) is used. At the surface temperature of 196^0F, the tables give

$$e_s = 156.95 \text{ Btu/lbm.}$$

Then

$$e^* = 129.06 + 0.5(156.95 - 129.06) + 0.22(763.2 - 129.06)$$

$$= 282.52 \text{ Btu/lbm}$$

This corresponds to a reference temperature of 705.5^0F. At this temperature, the tables yield

$$\rho = 0.034 \text{ lbm/ft}^3 \qquad \text{Pr} = 0.682$$

$$\nu = 6.26 \times 10^{-4} \text{ ft}^2\text{/sec}$$

Then the Reynolds number is

$$\text{Re} = \frac{UL}{\nu} = \frac{(5867)(4/12)}{6.26 \times 10^{-4}}$$

$$= 3.13 \times 10^6$$

The coefficient of heat transfer is given by

$$h = \frac{\rho U c_p}{2(\text{Pr})^{1/3}} \quad \text{where} \quad c_f = \frac{0.074}{\text{Re}^{1/5}} - \frac{1700}{\text{Re}}$$

Substituting the values in c_f,

$$c_f = \frac{0.074}{(3.13\times10^6)^{1/5}} - \frac{1700}{3.13\times10^6}$$

$$= 0.00317$$

The coefficient of heat transfer is

$$h = \frac{(0.034)(5867)(3600)(0.00317)}{2(0.682)^{1/3}}$$

$$= 1293.1 \text{ lbm/hr-ft}^2$$

The average value is desired. This is found by multiplying by 1.25, or

$$\bar{h} = 1.25(1293.1)$$

$$= 1616.4 \text{ lbm/hr-ft}^2$$

Finally, from eq.(1),

$$q/A = 1616.4(763.2 - 129.06)$$

$$\simeq 1,025,024 \text{ Btu/hr-ft}^2$$

• PROBLEM 7-46

A compressible gas flows over a flat plate. Properties of the gas are closely similar to those of air. The flow is at a temperature and pressure of 700°F and 30 psia, respectively. The plate is 1 in. in length and is assumed to be perfectly insulated. If the gas is moving at a speed of 500 ft/sec, calculate the surface temperature of the plate. (Note: the speed is too great to neglect the effects of viscous dissipation.)

Solution: Heat transfer takes place by forced convection, and the high-speed flow results in viscous drag which cannot be neglected. This is similar to the aerodynamic heating effects on high-speed aircrafts, rockets and missiles where mechanical energy is transformed to thermal energy. For the given problem, properties must be evaluated at a reference temperature which is given by

$$T^* = 0.5(T_s + T_\infty) + 0.22(T_r - T_\infty) \tag{1}$$

where T_r is the recovery temperature of the surface that would be attained if it is allowed to come to equilibrium with the flow.

A related variable is the recovery factor. This is defined as

$$r = \frac{T_r - T_\infty}{U^2/2g_c c_p} \tag{2}$$

For laminar flow, it is given as

$$r = Pr^{*1/2} \tag{3}$$

while for turbulent flow, it is

$$r = Pr^{*1/3} \tag{4}$$

Now, for this problem, the plate is perfectly insulated. This simplifies the problem, as it can be assumed that $T_s = T_r$. Substituting this into eq.(1) yields

$$T^* = 0.72\, T_r + 0.28\, T_\infty \tag{5}$$

An iterative procedure is used to calculate the recovery temperature as follows:

to begin with, a value of $T_r = 800°F$ is assumed.

Substituting in eq.(5) yields

$$T^* = 0.72(800) + 0.28(700)$$

$$= 772°F.$$

At this temperature, the properties of air are:

$\mu^* = 0.0795$ lbm/ft-hr $Pr^* = 0.68$

$C_p^* = 0.256$ Btu/lbm-°F

Since $$\rho^* = \frac{P}{RT^*} = \frac{(30)(144)}{(53.3)(772+460)}$$

$$= 0.0657 \text{ lbm/ft}^3$$

The Reynolds number is

$$Re = \frac{\rho UL}{\mu} = \frac{(0.0657)(500)(3600)(1/12)}{0.0795}$$

$$= 123,962$$

which is a laminar flow, and hence eq.(3) is used.

$$r = (0.68)^{1/2}$$
$$= 0.825$$

Substituting this into eq.(2) and rearranging gives a new value of the recovery temperature as

$$T_r = T_\infty + \frac{U^2 r}{2g_c c_p}$$

$$= 700 + \frac{(500)^2(0.825)}{2(32.2)(778)(0.256)}$$

$$= 716°F.$$

The value of 778 ft-lb/Btu is used as a conversion factor.

There is a difference between the assumed and the calculated T_r. Hence, 716°F is assumed for the second iteration for the recovery temperature. Substituting in eq.(5) gives

$$T^* = 711.5°F$$

The air tables give corresponding values of

$\mu^* = 0.0773$ lbm/ft-hr $Pr^* = 0.7016$

$c_p^* = 0.254$ Btu/lbm-°F

Then

$$\rho^* = 0.069 \text{ lbm/ft}^3$$

and

$$Re = 134,254$$

As the flow is still laminar, eq.(3) is used, giving

$$r = 0.838$$

Then eq.(2) gives

$$T_r = 716.5^0F$$

As there is not much difference between the assumed and the calculated recovery temperature, this is acceptable. As the surface temperature is equal to the recovery temperature,

$$T_s = 716.25^0F.$$

This is the average value.

• PROBLEM 7-47

A pipe is cooled internally by forced convection from the flow of a non-Newtonian fluid. The pipe is 5 ft. long, with a diameter of 1 in. The flow is measured at 600 lbm/hr, and the wall surface temperature is 200°F. Assume it is a laminar flow of a power-law fluid. Physical and flow properties are:

$\rho = 65 \text{ lbm/ft}^3$ $\qquad c_p = 0.5 \text{ Btu/lbm-}^0F$

$k = 0.7 \text{ Btu/hr-ft-}^0F$

$n = n' = 0.4$ (constant)

$K = 94 \text{ lbm-sec}^{n-2}\text{/ft at } 100^0F$

$\quad = 42 \text{ lbm-sec}^{n-2}\text{/ft at } 200^0F$

A plot of log K vs. T^0F is a straight line. If the fluid enters the section at 100°F, at what temperature does it leave?

Solution: Newton's law of viscosity states that for the Newtonian fluid, the shearing stress of a fluid subjected to a laminar flow is linearly dependent on the velocity gradient, i.e. the proportionality constant known as the coefficient of dynamic viscosity $\mu$is constant. The fluid is said to be non-Newtonian if the dynamic viscosity is varying. Its behavior may be graphed with coordinates τ (shear) vs. $\frac{8v}{d}$, and the equation of the flow line is

$$\tau = \frac{d\Delta p}{4L} = K'\left(\frac{8v}{d}\right)^{n'} \tag{1}$$

The flow properties may be assumed constant over a given range (as given). Then

$$n' = n \tag{2}$$

and

$$K' = K\left(\frac{3n'+1}{4n'}\right)^{n'} \tag{3}$$

where n' is the slope of the line from a logarithmic plot, n is the flow behavior index, and K is the consistency index. For a Newtonian fluid, $K' = \mu$.

Flow parameters for a non-Newtonian fluid may be found in tables. Often they are related in the form of the generalized viscosity coefficient γ, which is

$$\gamma = K'8^{n'-1} \tag{4}$$

These parameters are classified according to temperature, and may be related in ratio form as

$$\frac{\gamma_m}{\gamma_s} = \frac{K'_m 8^{n'-1}}{K'_s 8^{n'-1}} = \frac{K'_m}{K'_s} = \frac{K_m}{K_s} \tag{5}$$

where the subscript m refers to the mean bulk temperature, and s to the wall surface temperature.

An energy balance for the pipe section can be written as

$$q = h\pi dL\Delta T = \dot{m}c_p(T_o-T_i) \tag{6}$$

where

$$\Delta T = \frac{(T_s-T_i)+(T_s-T_o)}{2} \tag{7}$$

The heat transfer coefficient may be found from the correlation of Metzner and Gluck:

$$Nu = \frac{hd}{k} = 1.75\left(\frac{3n'+1}{4n'}\right)^{1/3} G_z^{1/3}\left[\frac{\gamma_m}{\gamma_s}\right]^{0.14} \tag{8}$$

where G_z is the Graetz number. This may be used, provided that

1. $n' > 0.1$

2. $G_z > 20$

$$G_z = \frac{\dot{m}c_p}{KL} = \frac{(600)(0.5)}{(0.7)(5.0)}$$

$$= 85.8 > 20$$

and hence both conditions are satisfied. This problem is solved by iterations as follows: An outlet temperature of 130°F is assumed. Then the mean bulk temperature is

$$T_m = \frac{100 + 130}{2} = 115^\circ F.$$

Since the slope of a straight line is

$$m = \frac{y_2 - y_1}{x_2 - x_1},$$

the value of K at 115°F may be found from

$$m = \frac{\log 94 - \log 42}{100 - 200} = \frac{\log 94 - \log K}{100 - 115}$$

or

$$K = 83.3.$$

From eq.(5),

$$\frac{\gamma_m}{\gamma_s} = \frac{K_m}{K_s} = \frac{83.3}{42} = 1.983$$

Then the coefficient of heat transfer is

$$h = 1.75\left[\frac{3(0.4)+1}{4(0.4)}\right]^{1/3} (85.8)^{1/3}(1.983)^{0.14}\left(\frac{0.7}{1/12}\right)$$

$$= 79.35 \text{ Btu/hr-ft}^2\text{-}^\circ F.$$

Substituting into eqs.(6) and (7) yields

$$(79.35)\pi(1/12)(5)\left[\frac{(200-100)+(200-T_o)}{2}\right] = (600)(0.5)(T_o - 100)$$

or

$$T_o = 129.5^\circ F.$$

This is close to the assumed value of 130°F, and hence no further iterations are required. The fluid temperature at the exit ≃ 130°F.

• PROBLEM 7-48

Heat transfer by transpiration takes place when air flows through a flat porous plate, 0.7m long and 1m wide, with free stream velocity 918m/sec, at 0.05 atm and 233 °K. The surface temperature of the plate is 652 °K. At a point the Reynolds number is 5×10^4 and the injection parameter is 0.2. Assuming the properties are evaluated at the average between the recovery and the free stream temperatures, calculate at that point

a) the percent reduction in heat transfer, and

b) the required mass flow of coolant air per unit area.

(Given: Pr = 0.7)

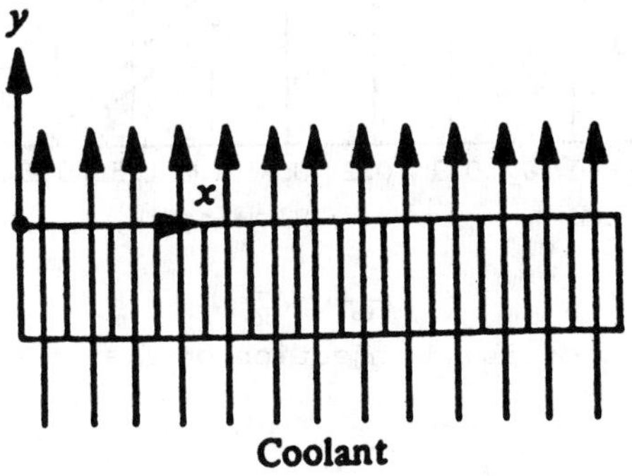

Fig. 1 Transpiration cooling

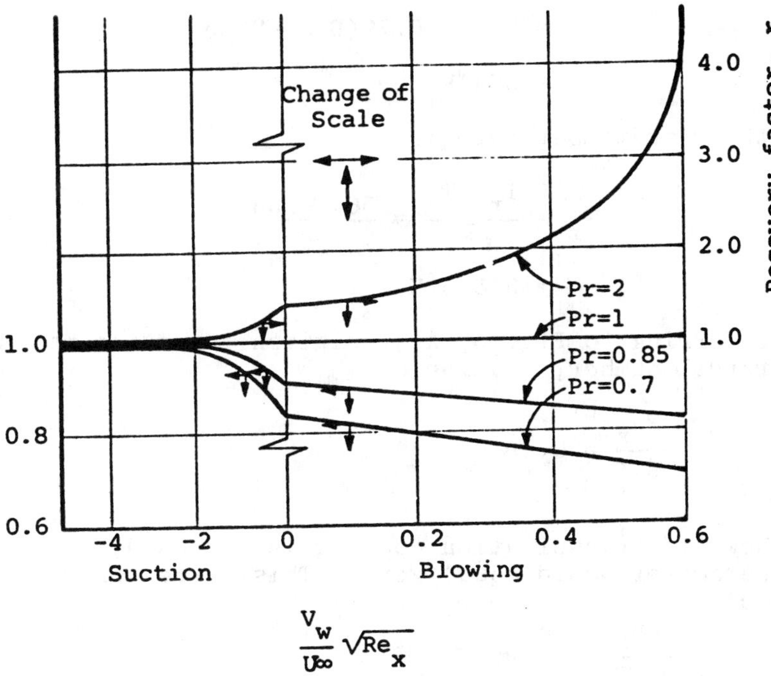

Fig. 2 Effects of fluid injection on recovery factor for flow over a flat plate.

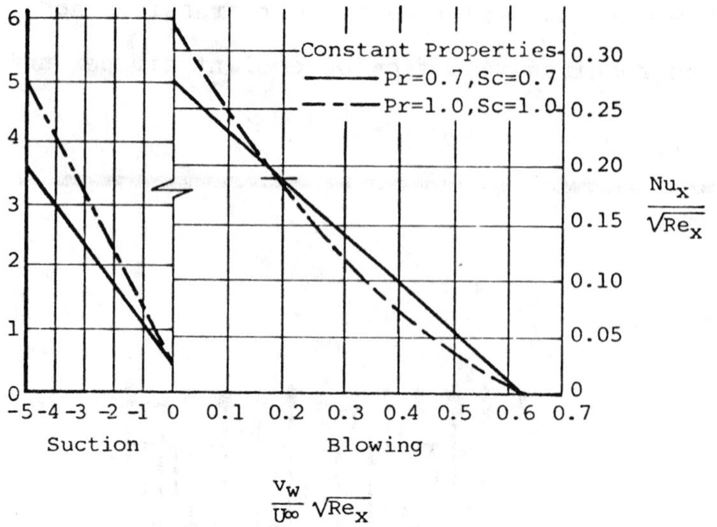

Fig. 3 Effect of fluid injection on flat-plate heat transfer.

Solution: Figure 1 shows the transpiration cooling process.

(a) At IP = 0.2 and Pr = 0.7, Fig. 2 gives a recovery factor r of 0.79. Since the recovery temperature is

$$T_r = T_\infty + r(T_o - T_\infty)$$

$$= 233 + 0.79(652 - 233)$$

$$= 564\,^0K$$

Then the mean temperature is

$$T_m = \frac{T_r + T_\infty}{2} = \frac{564 + 233}{2}$$

$$= 398.5\,^0K$$

Now, referring to Fig. 3, at the given injection parameter and Prandtl number, a value of

$$\frac{Nu_x}{\sqrt{Re_x}} = 0.19$$

Now, if transpiration cooling is not used, the injection parameter would equal zero. This corresponds to a value of

$$\frac{Nu_x}{\sqrt{Re_x}} = 0.29$$

A comparison of these numbers gives the per cent reduction in heat transfer:

$$\% \text{ reduction} = \frac{0.29 - 0.19}{0.29} \times 100\%$$

$$= 34.5\%$$

(b) The mass flow rate per unit area is defined by

$$\frac{\dot{m}}{A} = \rho v$$

For the coolant, this is

$$\frac{\dot{m}_w}{A} = \rho_w V_w$$

The density is found from the equation of state as

$$\rho_w = \frac{P}{RT} = \frac{(101.32\times10^3)(0.05)}{(287)(398.5)}$$

$$= 0.0443 \text{ kg/m}^3$$

The relation for the injection parameter is given by

$I \cdot P = \frac{V_w}{U_\infty}\sqrt{Re_{(x)}}$. Rearranging and substituting,

$$V_w = \frac{(I\cdot P)U_\infty}{\sqrt{Re_x}} = \frac{(0.2)(918)}{\sqrt{5\times10^4}}$$

$$= 0.82 \text{ m/s}$$

and therefore

$$\frac{\dot{m}_w}{A} = (0.0443)(0.82) = 0.0363 \text{ kg/m}^2\text{-sec}$$

• PROBLEM 7-49

Transpiration cooling occurs in a two concentric porous shells as shown in the figure. The outer surface of the inner shell is maintained at temperature T_K by the flow of air into the shell at temperature T_K. The air flows radially outward from the inner shell to the surroundings, passing through the intervening space and the outer shell at temperature T_1. Assuming steady laminar flow and low gas velocity, derive an expression for the required rate of heat transfer from the inner sphere as a function of the gas mass flow rate.

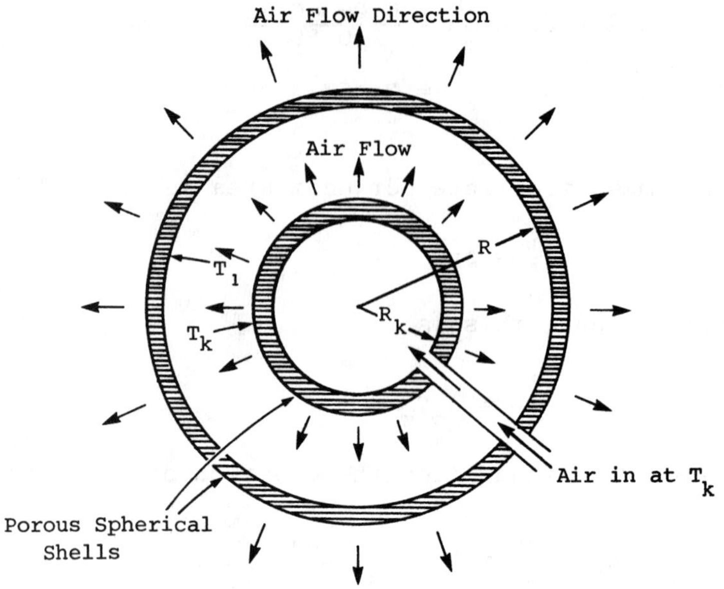

Fig. 1 Transpiration Cooling. The inner sphere is being cooled by means of the refrigeration coil to maintain its temperature at T_k. When air is blown outward, less refrigeration is required.

Solution: For any given system, the equation of continuity is

$$\frac{\partial \rho}{\partial t} + \vec{\nabla} \cdot (\rho \vec{V}) = 0 \tag{1}$$

For this system, in spherical coordinates, eq.(1) becomes

$$\frac{\partial \rho}{\partial t} + \frac{1}{r^2}\frac{\partial}{\partial r}(\rho r^2 V_r) + \frac{1}{r\sin\theta}\frac{\partial}{\partial \theta}(\rho V_\theta \sin\theta) + \frac{1}{r\sin\theta}\frac{\partial}{\partial \phi}(\rho V_\phi) = 0 \tag{2}$$

The only velocity is in the radial direction, so eq.(2) reduces to

$$\frac{1}{r^2}\frac{d}{dr}(r^2 \rho V_r) = 0 \tag{3}$$

for steady, incompressible flow. Integrating this,

$$r^2 \rho V_r = \text{constant}. \tag{4}$$

Mass flow rate is

$$\dot{m} = \rho V A,$$

and since

$$A = 4\pi r^2,$$

eq.(4) becomes

$$r^2 \rho V_r = \dot{m} \tag{5}$$

If the pressure may be assumed to be constant, the energy equation of the system may be written as

$$\rho c_p V_r \frac{dT}{dr} = k_{th} \frac{1}{(r^2)} \frac{d}{dr}\left(r^2 \frac{dT}{dr}\right) \tag{6}$$

assuming a constant thermal conductivity k_{th}. Substituting in eq.(5) and rearranging yields

$$\frac{dT}{dr} = \frac{4\pi k_{th}}{\dot{m} c_p} \frac{d}{dr}\left(r^2 \frac{dT}{dr}\right)$$

If

$$R_o = \frac{\dot{m} c_p}{4\pi k_{th}}$$

is used, and the equation is integrated twice to the limits

$$T(R_K) = T_K$$

$$T(R) = T_1,$$

the result is

$$\frac{T-T_1}{T_K-T_1} = \frac{e^{-R_o/r} - e^{-R_o/R}}{e^{-R_o/R_K} - e^{-R_o/R}} \tag{7}$$

This is the temperature distribution T(r) at any radius. For a small mass flow rate,

$$\frac{T-T_1}{T_K-T_1} = \frac{1/r - 1/R}{1/R_K - 1/R}$$

In addition, if the system is considered large enough, R may be assumed to be infinite, while R_K is a finite parameter. Then

$$\frac{T-T_\infty}{T_K-T_\infty} = \frac{R_K}{r} \tag{8}$$

Now, the refrigerant is removing heat from the system at the rate of

$$Q = -4\pi R_K^2 \, q_r \Big|_{r=R_K} \tag{9}$$

where q_r is the energy flux in the radial direction. The rate of heat flow for various regular surfaces is shown in table 1. Referring to table 1, for a sphere,

$$q_r = -k_{th} \frac{\partial T}{\partial r}$$

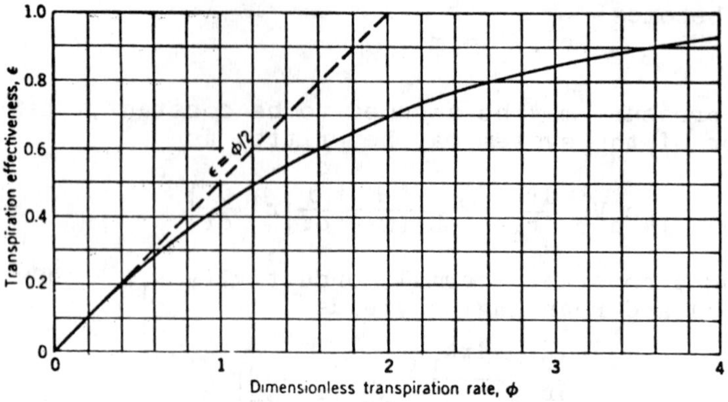

Fig. 2 Effect of transpiration cooling.

Substituting this into eq.(9) yields

$$Q = 4\pi k_{th} R_K^2 \left. \frac{dT}{dr} \right|_{r=R_K}$$

Using eq.(7) to find $\left. \frac{dT}{dr} \right|_{r=R_K}$ gives

$$Q = \frac{4\pi k_{th} R_o (T_1 - T_K)}{e^{(R_o/kR_K)\left(1 - \frac{R_K}{R_o}\right)} - 1} \tag{10}$$

Since transpiration effectiveness ε is defined as

$$\varepsilon = \frac{Q_o - Q}{Q_o} \tag{11}$$

where Q_o is the rate heat, is being removed at the gas flow rate of zero, eq.(10) must be evaluated at $R_o = 0$. Then

$$Q_o = \frac{4\pi k_{th} R_K (T_1 - T_K)}{1 - \frac{R_K}{R}} \tag{12}$$

To define a dimensionless transpiration rate φ,

$$\phi = \frac{R_o(1 - R_K/R_o)}{R_K}$$

From eq.(11), the effectiveness is then

$$\varepsilon = 1 - \frac{\phi}{e^{\phi} - 1} \tag{13}$$

A plot of ε vs. ϕ is shown in Fig. 2. It may be seen that for small ϕ, $\varepsilon \to \phi/2$.

Table 1: Components of the energy flux q.

Rectangular		Cylindrical		Spherical	
$q_x = -k \frac{\partial T}{\partial x}$	(A)	$q_r = -k \frac{\partial T}{\partial r}$	(D)	$q_r = -k \frac{\partial T}{\partial r}$	(G)
$q_y = -k \frac{\partial T}{\partial y}$	(B)	$q_\theta = -k \frac{1}{r} \frac{\partial T}{\partial \theta}$	(E)	$q_\theta = -k \frac{1}{r} \frac{\partial T}{\partial \theta}$	(H)
$q_z = -k \frac{\partial T}{\partial z}$	(C)	$q_z = -k \frac{\partial T}{\partial z}$	(F)	$q_\phi = -k \frac{1}{r \sin \theta} \frac{\partial T}{\partial \phi}$	(I)

CHAPTER 8

RADIATION HEAT TRANSFER

Basic Attacks and Strategies for Solving Problems in this Chapter. See pages 454 to 572 for step-by-step solutions to problems.

As with conduction and convection, heat transfer by radiation also always takes place from a higher temperature object or region to a lower temperature object or region. In this case, however, the heat is transferred by electromagnetic waves. Any object which is above absolute zero in temperature gives off electromagnetic waves (also called radiation) simply due to its temperature. As the temperature of an object increases the amount of energy given off in the radiation increases and the nature of the radiation changes. At any temperature, radiation over a spectrum of wavelengths is emitted. As the temperature increases, a larger fraction of radiation at shorter wavelengths and a smaller fraction at longer wavelengths is emitted.

The term "black body" is used to refer to an object which emits the maximum radiant energy possible at any given temperature. The total amount of radiant energy emitted by a black body can be calculated just from its temperature. It is given by the Stefan-Botzmann constant times the absolute temperature raised to the fourth power. The amount of radiant energy given off within any given range of wavelengths by a black body can be determined using a table or graph of Planck Radiation Functions as illustrated by problems 8-1, 8-3, & 8-4. For problems involving the wavelength of maximum emission at a given temperature or vice-versa, Wien's displacement law should be used as illustrated in problem 8-2.

The emissivity of any surface is the ratio of the radiant energy emitted by that surface to that which would be emitted by a black body at the same temperature. In general, the emissivity will be different for different wavelengths of radiation. If the emissivity can be considered to remain constant for a given surface, that surface is called a "gray" surface.

Calculations of the net rate of radiant heat transfer between two bodies often involves use of a shape factor. The shape factor, F_{12}, *is simply the fraction of the radiation leaving object number 1 which strikes object number 2. It is a purely geometric factor, independent of the temperatures of the objects. Shape factors may be determined by integration over the surfaces, by use of graphs for particular shapes and arrangements, and by use of shape factor algebra. A useful, fundamental equation for finding* F_{12} *when* F_{21} *is known or vice-versa is* $A_1F_{12} = A_2F_{21}$.

For two black surfaces at different temperatures, only the two temperatures and the shape factors and areas are needed to calculate the net rate of radiant heat transfer between them. The rate of radiant heat transfer from surface 1 to surface 2 is given by:

$$Q_{12} = A_1 F_{12} \sigma (T_1^4 - T_2^4) .$$

If the two surfaces at different temperatures are not black, but can be considered to be gray, then the rate of radiant heat transfer depends upon the emissivities of each as well as the variables in the previous paragraph. There is no general equation in this case, but equations for rate of radiant heat transfer between two gray surfaces have been derived for some fairly simple geometric arrangements such as large parallel planes or one surface completely enclosed within another.

For more complicated systems involving gray surfaces, two general approaches are the network method and the Gebhart method. The network method uses an electrical analogy to set up an electrical circuit. Equations can then be written from the circuit and solved for the desired unknowns. The Gebhart method involves writing a series of equations for the radiant energy emitted by each surface of an enclosure and solving the resulting set of equations to find the desired unknowns.

Solar radiation from the sun to the earth is of particular interest. The general principles discussed above apply to solar radiations as well as to other sources of radiation. Factors such as the angle of the sun's rays striking a particular surface may be required in problems involving solar collectors.

Step-by-Step Solutions to Problems in this Chapter, "Radiation Heat Transfer"

WAVELENGTH DISTRIBUTIONS OF EMITTED RADIANT ENERGY

• PROBLEM 8-1

Compute the fraction of the radiation emitted from the surface of a black object, maintained at a temperature of 500°F, in the wavelength band of 1.0 and 4.0 microns. Also compute, when the surface temperature of the black object is maintained at 1500°F.

Solution: The total emissive power of a black body is a function of its wavelength of radiation and the temperature.

$$W_b = f(\lambda, T)$$

where λ is the wavelength of radiation

T is the absolute temperature of the surface of the the body °R

and W_b is the black body emissive power.

$$W_b = \sigma T^4$$

σ is the Stefan-Boltzmann constant

$$= 0.1714\ 10^{-8}$$

$$T = 500 + 460 + 960°R$$

$$\therefore\ W_b = 0.1714 \times 10^{-8} \times (960)^4 = 1456\ \text{Btu/hr-ft}^2$$

The emissive power of a black body at any wavelength can be found from the Plank's Radiation Functions table shown, corresponding to the product of the wavelength λ and the absolute temperature of the black body.

Planck Radiation Functions

λT	$\frac{W_{b\lambda}}{\sigma T^5} \times 10^5$	$\frac{W_{b(0-\lambda)}}{\sigma T^4}$	λT	$\frac{W_{b\lambda}}{\sigma T^5} \times 10^5$	$\frac{W_{b(0-\lambda)}}{\sigma T^4}$	λT	$\frac{W_{b\lambda}}{\sigma T^5} \times 10^5$	$\frac{W_{b(0-\lambda)}}{\sigma T^4}$
1000	0.0000394	0	7200	10.089	0.4809	13400	2.714	0.8317
1200	0.001184	0	7400	9.723	0.5007	13600	2.605	0.8370
1400	0.01194	0	7600	9.357	0.5199	13800	2.502	0.8421
1600	0.0618	0.0001	7800	8.997	0.5381	14000	2.416	0.8470
1800	0.2070	0.0003	8000	8.642	0.5558	14200	2.309	0.8517
2000	0.5151	0.0009	8200	8.293	0.5727	14400	2.219	0.8563
2200	1.0384	0.0025	8400	7.954	0.5890	14600	2.134	0.8606
2400	1.791	0.0053	8600	7.624	0.6045	14800	2.052	0.8648
2600	2.753	0.0098	8800	7.304	0.6195	15000	1.972	0.8688
2800	3.872	0.0164	9000	6.995	0.6337	16000	1.633	0.8868
3000	5.081	0.0254	9200	6.697	0.6474	17000	1.360	0.9017
3200	6.312	0.0368	9400	6.411	0.6606	18000	1.140	0.9142
3400	7.506	0.0506	9600	6.136	0.6731	19000	0.962	0.9247
3600	8.613	0.0667	9800	5.872	0.6851	20000	0.817	0.9335
3800	9.601	0.0850	10000	5.619	0.6966	21000	0.702	0.9411
4000	10.450	0.1051	10200	5.378	0.7076	22000	0.599	0.9475
4200	11.151	0.1267	10400	5.146	0.7181	23000	0.516	0.9531
4400	11.704	0.1496	10600	4.925	0.7282	24000	0.448	0.9589
4600	12.114	0.1734	10800	4.714	0.7378	25000	0.390	0.9621
4800	12.392	0.1979	11000	4.512	0.7474	26000	0.341	0.9657
5000	12.556	0.2229	11200	4.320	0.7559	27000	0.300	0.9689
5200	12.607	0.2481	11400	4.137	0.7643	28000	0.265	0.9718
5400	12.571	0.2733	11600	3.962	0.7724	29000	0.234	0.9742
5600	12.458	0.2983	11800	3.795	0.7802	30000	0.208	0.9765
5800	12.282	0.3230	12000	3.637	0.7876	40000	0.0741	0.9881
6000	12.053	0.3474	12200	3.485	0.7947	50000	0.0326	0.9941
6200	11.783	0.3712	12400	3.341	0.8015	60000	0.0165	0.9963
6400	11.480	0.3945	12600	3.203	0.8081	70000	0.0092	0.9981
6600	11.152	0.4171	12800	3.071	0.8144	80000	0.0055	0.9987
6800	10.808	0.4391	13000	2.947	0.8204	90000	0.0035	0.9990
7000	10.451	0.4604	13200	2.827	0.8262	100000	0.0023	0.9992
							0	1.0000

Thus, for 1.0 micron of wavelength

$$T = 1.0 \times 960 = 960 \text{ micron-}^\circ R$$

From the table,

$$\frac{W_{b(0-1.0)}}{W_b} \approx 0$$

Again, for $\lambda = 4.0$ micron

$$\lambda T = 4.0 \times 960 = 3840 \text{ micron-}^\circ R$$

and

$$\frac{W_{b(0-4.0)}}{W_b} = 0.0890$$

$\therefore$ The difference is

$$\frac{W_{b(0-4.0)}}{W_b} - \frac{W_{b(0-1.0)}}{W_b} = \frac{W_{b(1.0-4.0)}}{W_b} = 0.089$$

$$\therefore \qquad W_{b(1.0-4.0)} = 0.089 \times W_b$$

$$= 0.089 \times 1456$$

or $$W_{b(1.0-4.0)} = 129.6 \text{ Btu/hr-ft}^2$$

Again, when the body temperature is 1500°F or

$$1500+460=1960°R$$

$$\lambda T = 1.0 \times 1960 = 1960 \text{ micron-°R}$$

and $$\frac{W_{b(0-1.0)}}{W_b} = 0.0008$$

Also, $$\lambda T = 4.0 \times 1960 = 7840 \text{ micron-°R}$$

and $$\frac{W_{b(0-4.0)}}{W_b} = 0.5416$$

$$\therefore \qquad \frac{W_{b(1.0-4.0)}}{W_b} = 0.5408$$

or $$W_{b(1.0-4.0)} = 0.5408 \times 0.1714 \times 10^{-8} \times (1960)^4$$

$$= 13{,}680 \text{ Btu/hr-ft}^2$$

• PROBLEM 8-2

Estimate the surface temperature of a star that is radiating energy with extreme concentration at a wavelength of $\lambda = 0.4$ microns. The star is assumed to be a black body. Also, determine the energy flow at the surface of the star.

Solution: (a) Wien's displacement law states that the concentration of the radiation occurs at wavelengths larger than the visible ones. The wavelength distribution of the energy emitted by a bright star corresponds to black body radiation at a temperature of 6500°K. At this temperature, 1/3 of the total heat flux is in the visible range. Or the intensity can be described as:

$$\lambda T = 0.2884$$

and $$T = \frac{0.2884}{0.4 \times 10^{-4} \text{cm}} = 7210°K$$

The energy radiated per unit time and per unit area by the star can be estimated using the Stefan-Boltzmann law:

$$E = \sigma T^4$$

$$= (0.1714 \times 10^{-8})(12978)^4 = 4.862 \times 10^7 \text{ Btu/hr-ft}^2$$

• PROBLEM 8-3

Compute the heat absorbed by a body which is exposed to an electric heater releasing energy at the rate of 1850 W/m. The absorbing body accepts 95% of the radiation falling above 2.7μm and 30% of radiation below 2.7μm. The diameter of the heating element is 25 mm while the two bodies are 10 cm apart.

Solution: The irradiation upon the surface is given by the equation

$$q^- = [I^-][2 \sin \gamma_1][\pi/2]$$

The intensity $I = [\dot{Q}/\pi DL][1/\pi]$

$$= [1850/\pi(0.025)1][1/\pi]$$

and $2 \sin \gamma_1 = 2.5/10$

Therefore $q^- = [1850/0.025\pi][1/\pi][2.5/10][\pi/2]$

$$= 2944 \text{ W/m}^2$$

Using the Stefan-Boltzmann Law

$$\sigma T^4 = \dot{Q}/\pi DL$$

$$T = \left[\frac{\dot{Q}}{(\pi DL)(\sigma)}\right]^{\frac{1}{4}}$$

$$= \left[\frac{1850}{(\pi 0.025)(1)(5.6697 \times 10^{-8})}\right]^{\frac{1}{4}}$$

$$= 803\text{K}$$

Accordingly, $\lambda T = (2.7\mu\text{m})(803\text{K})$

$$= 2168 \ \mu\text{mK}$$

and $f_e = 0.095$

Thus the absorbed flux αq^- is

$$\alpha q^- = [(0.95)(0.30) + (0.905)(0.95)][2944]$$

$$= 2615 \text{ W/m}^2$$

• PROBLEM 8-4

A glass window allows 95% of radiant energy with band between 0.32 and 2.6μ. Any wavelength out of this range will not pass through the glass. Calculate the energy rate passed through the glass window if the incident energy rate on the glass is 1200 W/m^2.

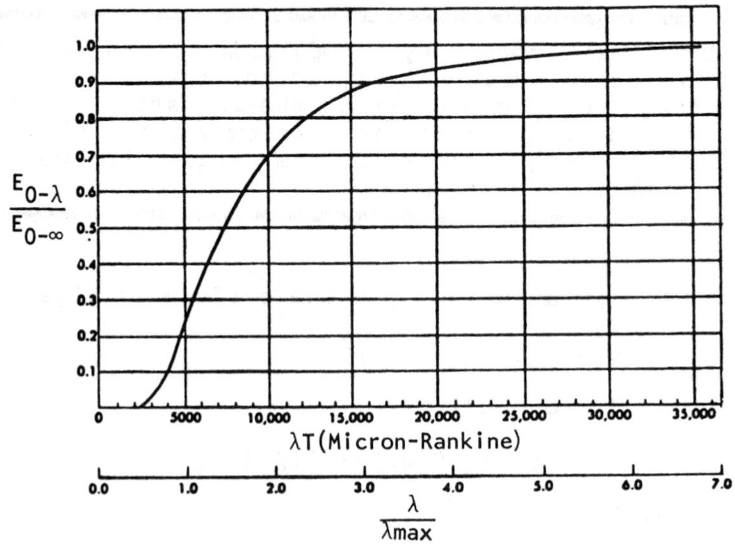

Fig. 1 Fraction of tatal emissive power in spectral region between λ = 0 and λ as a function of λT.

Solution: Referring to fig. 1 for λ=0.32μ or 3200μm, we obtain

$$\frac{\int_0^{0.32} E_{b\lambda} d\lambda}{\int_0^{\infty} E_{b\lambda} d\lambda} = \frac{E_{0-0.32}}{E_{0-\infty}} = 5\%$$

and for λ=2.6μ or 26000μm, we obtain

$$\frac{\int_0^{2.6} E_b d\lambda}{\int_0^{\infty} E_{b\lambda} d\lambda} = \frac{E_{0-2.6}}{E_{0-\infty}} = 96\%$$

Therefore, the percent of radiation incident upon the glass window from the heat source between the wavelength range of 0.32 and 26μ is 96%-5%=91%. The amount of energy allowed through the glass is (91%)(0.95)=85.5%.

∴ Energy rate passed through the window

$$=0.855 \times 1200 W/m^2 = 1026 W/m^2$$

• PROBLEM 8-5

Estimate the mean emissivity of a surface at temperatures of 2040°F and 3540°F. The variation of monochromatic emissivity of the surface is as shown in the figure below.

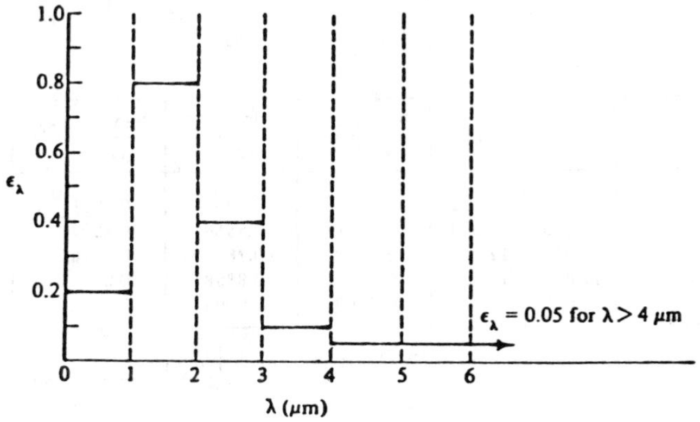

Solution: The emissivity of a surface considering the wavelength-dependent radiation, can be expressed as

$$\varepsilon = \frac{\int_0^\infty \varepsilon_\lambda(T) e_{b\lambda}(T) d\lambda}{\sigma T^4}$$

The expression with the integral sign can be approximated by considering discretised wavelength variation (0-1, 1-2, 2-3, 3-4, 4-∞ instead of continuous variation.

$$\int_0^\infty \varepsilon_\lambda(T) e_{b\lambda}(T) d\lambda \simeq \varepsilon_{0-1} e_{b0-1} + \varepsilon_{1-2} e_{b1-2} + \varepsilon_{2-3} e_{b2-3}$$

$$+ \varepsilon_{3-4} e_{b3-4} + \varepsilon_{4-\infty} e_{b4-\infty}$$

$$= \Sigma\ [\varepsilon_{(\lambda 1-\lambda 2)}\ e_{b(\lambda 1-\lambda 2)}]$$

where $\lambda 1-\lambda 2$ assumes $0 \longrightarrow 1$, $1 \longrightarrow 2$, $2 \longrightarrow 3$, $3 \longrightarrow 4$ and $4 \longrightarrow \infty$

$$\therefore \quad \varepsilon = \frac{\Sigma \varepsilon_{(\lambda 1-\lambda 2)} e_{b(\lambda 1-\lambda 2)}}{\sigma T^4}$$

$\sigma = 0.171 \times 10^{-8}$ and T is the absolute temperature in Rankine.

At temperature 2500° R

$\lambda_1 \to \lambda_2$	$\epsilon_{\lambda_1 - \lambda_2}$	$\lambda_1 T$	$\lambda_2 T$	$\frac{e_{b(0-\lambda_1)}}{\sigma T^4}$	$\frac{e_{b(0-\lambda_2)}}{\sigma T^4}$	$\frac{e_{b(\lambda_1-\lambda_2)}}{\sigma T^4}$	$\frac{[\epsilon_{\lambda_1-\lambda_2}][e_{b(\lambda_1-\lambda_2)}]}{\sigma T^4}$
0 → 1	0.2	0	2500	0	0.0075	0.0075	0.0015
1 → 2	0.8	2500	5000	0.0075	0.2229	0.2154	0.1723
2 → 3	0.4	5000	7500	0.2229	0.5103	0.2874	0.1149
3 → 4	0.1	7500	10,000	0.5103	0.6966	0.1863	0.0186
4 → -	0.05	10,000	-	0.6966	1.0000	0.3034	0.0152

$$\epsilon(2500) = \sum_{T = 2500} \frac{[\epsilon_{\lambda_1 - \lambda_2}]\,[e_{b(\lambda_1 - \lambda_2)}]}{\sigma T^4} = 0.32$$

At temperature 4000° R

$\lambda_1 \to \lambda_2$	$\epsilon_{\lambda_1 - \lambda_2}$	$\lambda_1 T$	$\lambda_2 T$	$\frac{e_{b(0-\lambda_1)}}{\sigma T^4}$	$\frac{e_{b(0-\lambda_2)}}{\sigma T^4}$	$\frac{e_{b(\lambda_1-\lambda_2)}}{\sigma T^4}$	$\frac{[\epsilon_{\lambda_1-\lambda_2}][e_{b(\lambda_1-\lambda_2)}]}{\sigma T^4}$
0 → 1	0.2	0	4000	0	0.1051	0.1051	0.0210
1 → 2	0.8	4000	8000	0.1051	0.5558	0.4507	0.3601
2 → 3	0.4	8000	12,000	0.5558	0.7876	0.2318	0.0927
3 → 4	0.1	12,000	16,000	0.7876	0.8868	0.0992	0.0099
4 → -	0.05	16,000	-	0.8868	1.0000	0.1132	0.0057

$$\epsilon(4000) = \sum_{T = 4000} \frac{[\epsilon_{\lambda_1 - \lambda_2}]\,[e_{b(\lambda_1 - \lambda_2)}]}{\sigma T^4} = 0.49$$

TABLE Radiation Functions

λT (μm·R)	λT (μm·K)	$\frac{e_b(0\to\lambda T)}{\sigma T^4}$	λT (μm·R)	λT (μm·K)	$\frac{e_b(0\to\lambda T)}{\sigma T^4}$	λT (μm·R)	λT (μm·K)	$\frac{e_b(0\to\lambda T)}{\sigma T^4}$
1000	556	0	7200	4000	.4809	13400	7445	.8317
1200	667	0	7400	4111	.5007	13600	7556	.8370
1400	778	0	7600	4222	.5199	13800	7667	.8421
1600	889	.0001	7800	4333	.5381	14000	7778	.8470
1800	1000	.0003	8000	4445	.5558	14200	7889	.8517
2000	1111	.0009	8200	4556	.5727	14400	8000	.8563
2200	1222	.0025	8400	4667	.5890	14600	8111	.8606
2400	1333	.0053	8600	4778	.6045	14800	8222	.8648
2600	1445	.0098	8800	4889	.6194	15000	8333	.8688
2800	1556	.0164	9000	5000	.6337	16000	8889	.8868
3000	1667	.0254	9200	5111	.6474	17000	9445	.9017
3200	1778	.0368	9400	5222	.6606	18000	10000	.9142
3400	1889	.0506	9600	5333	.6731	19000	10556	.9247
3600	2000	.0667	9800	5444	.6851	20000	11111	.9335
3800	2111	.0850	10000	5556	.6966	21000	11667	.9411
4000	2222	.1051	10200	5667	.7076	22000	12222	.9475
4200	2333	.1267	10400	5778	.7181	23000	12778	.9531
4400	2445	.1496	10600	5889	.7282	24000	13333	.9589
4600	2556	.1734	10800	6000	.7378	25000	13889	.9621
4800	2667	.1979	11000	6111	.7474	26000	14445	.9657
5000	2778	.2229	11200	6222	.7559	27000	15000	.9689
5200	2889	.2481	11400	6333	.7643	28000	15556	.9718
5400	3000	.2733	11600	6445	.7724	29000	16111	.9742
5600	3111	.2983	11800	6556	.7802	30000	16667	.9765
5800	3222	.3230	12000	6667	.7876	40000	22222	.9881
6000	3333	.3474	12200	6778	.7947	50000	27778	.9941
6200	3445	.3712	12400	6889	.8015	60000	33333	.9963
6400	3556	.3945	12600	7000	.8081	70000	38887	.9981
6600	3667	.4171	12800	7111	.8144	80000	44445	.9987
6800	3778	.4391	13000	7222	.8204	90000	50000	.9990
7000	3889	.4604	13200	7333	.8262	100000	55556	.9992
						-	-	1.0000

The emissivities of the surface at temperatures (2040+460=) 2500°R and (3540+460=) 4000°R are calculated in tabular form using the Radiation Function table.

• PROBLEM 8-6

A prototype oil burner, used for evaporations of water, is built in such a way that it will use a cylindrical flame 5 ft in diameter and 14 ft long. The prototype is a small scale model which produces a flame 1 ft in diameter. The 1 ft diameter flame has a temperature of T_{red}=1800K and T_{green}=1860K. Approximate the energy radiated by the flame in the actual burner.

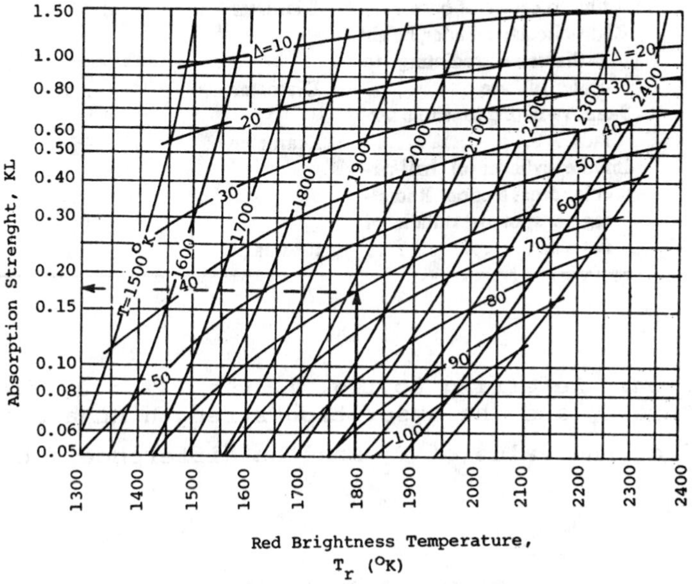

Fig. 1 Absorption Strength KL of Luminous Flames.

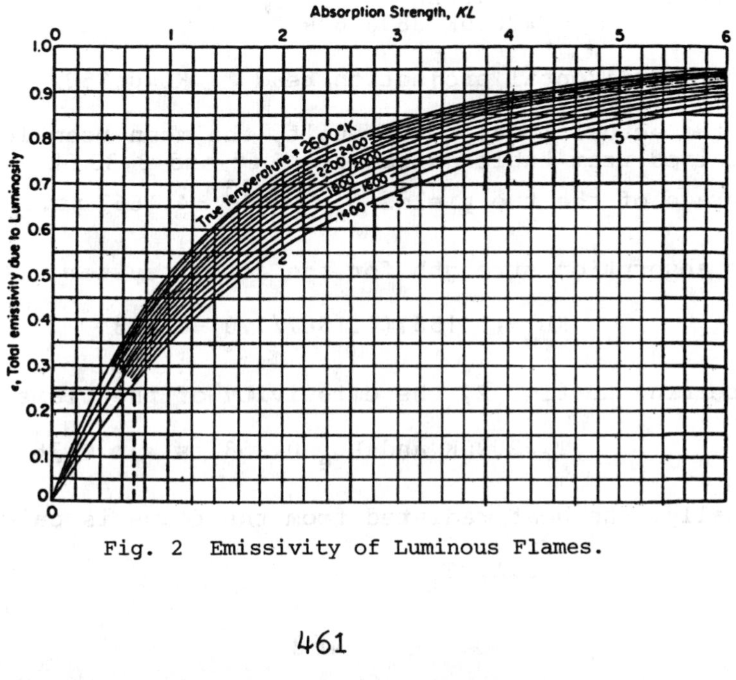

Fig. 2 Emissivity of Luminous Flames.

TABLE
Geometric Mean Beam Lengths *L* for Isothermal Gas Radiation

Shape	Characterizing dimension *X*	Factor by which *X* is multiplied to obtain mean beam length *L*	
		When $P_aL = 0$	For average values of P_aL
Sphere	Diameter	⅔	0.60
Infinite cylinder	Diameter	1	0.90
Semi-infinite cylinder, radiating to center of base	Diameter	....	0.90
Right-circular cylinder, height = diameter, radiating to center of base	Diameter	....	0.77
Same, radiating to whole surface	Diameter	⅔	0.60
Infinite cylinder of half-circular cross section. Radiating to spot on middle of flat side	Radius	....	1.26

<u>Solution</u>: Referring to fig. 1, the 1 ft dia. flame temperature T_f and the absorption strength KL are obtained based on the value of T_r and the difference Δ between T_r and T_g.

$$T_r = 1800K$$

$$\Delta = T_g - T_r = 1860K - 1800K = 60K$$

The point of intersection of 1800K and $\Delta = 60$ lies between 2000K and 2100K. Taking

$$T_f = 2020K = 3636.6°R$$

and by horizontal projection, read KL=0.158.

In the actual model to be built, the mean beam length is obtained from table as between 5 ft and 3.33 ft. Taking the average of the two yields $L_e = \frac{5+3.33}{2} = 4.1667$ ft.

The absorption strength for the actual system will be

$$KL_e = (0.158)(4.1667/1.0) = 0.658$$

Referring to fig. 2, the emissivity of the flame based on

$$T_f = 2020K \text{ and } KL_e = 0.658 \text{ is } \varepsilon_f \cong 0.36$$

Finally, the heat radiated from the flame is calculated by

$$q = \varepsilon_f A_f \sigma T_f^4$$

where A_f = surface area of flame $=(5)(14)\pi=219.9$ ft^2

σ = Stefan-Boltzmann constant $=0.1714\times10^{-8}$ Btu/h.ft^2°R

$q=(0.36)(219.9\text{ft}^2)(0.1714\times10^{-8}\text{Btu/h.ft}^2\text{°R})(3636.6\text{°R})^4$

$=2.373\times10^7$ Btu/hr

CALCULATION OF RADIANT ENERGY BY USING SHAPE FACTORS

• PROBLEM 8-7

Derive an expression for the shape factor between two parallel axially located circular surfaces separated by a distance ℓ, as shown in the figure.

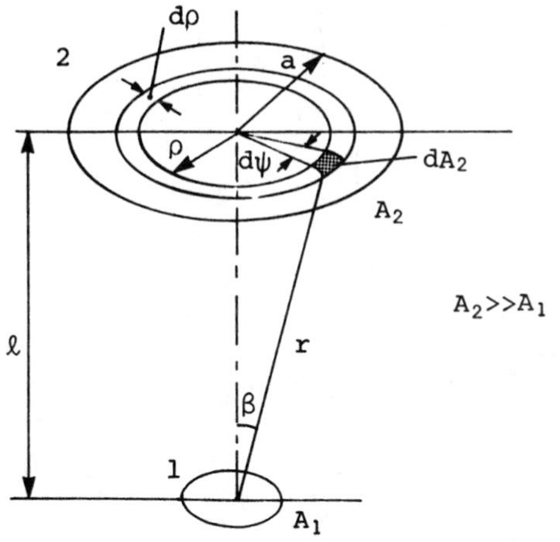

Fig. Two parallel concentric surfaces.

Solution: The shape factor from a constant temperature surface 1 to surface 2 is the fraction of emitted radiant energy from surface 1 that directly strikes surface 2.

The general expression for the shape factor F_{12} is

$$F_{12} = \frac{1}{A_1}\int_{A_2}\int_{A_1}\frac{dA_1\cos\beta_1 \; dA_2\cos\beta_2}{\pi r^2}$$

Introducing the conditions of the given problem, i.e., $A_1 << A_2$ and $\beta_1 = \beta_2 = \beta$, the expression for F_{12} simplifies to

$$F_{12} = \int_{A_2} \frac{\cos^2\beta \, dA_2}{\pi r^2}$$

dA_2 is the small elemental area, as shown in figure by the shaded portion of surface 2 and can be expressed as

$$dA_2 = \rho d\psi d\rho$$

Also $\cos\beta = \frac{\ell}{r}$. Then

$$F_{12} = \int_0^a \int_0^{2\pi} \left(\frac{\ell}{r}\right)^2 \frac{\rho d\psi \, d\rho}{\pi r^2}$$

$$= \int_0^a \int_0^{2\pi} d\psi \left[\left(\frac{\ell}{r}\right)^2 \rho d\rho\right] \frac{1}{\pi r^2}$$

$$= \int_0^a \left(\frac{2\pi}{\pi}\right) \left[\frac{\ell^2}{r^4} \rho d\rho\right]$$

$$= 2\ell^2 \int_0^a \frac{\rho}{r^4} d\rho$$

$$\therefore \quad r^2 = \rho^2 + \ell^2$$

$$\therefore \quad F_{12} = \int_0^a \frac{2\ell^2}{(\rho^2 + \ell^2)^2} \rho d\rho$$

$$= \frac{a^2}{a^2 + \ell^2}$$

$$\therefore \quad F_{12} = \frac{a^2}{(a^2 + \ell^2)}$$

• PROBLEM 8-8

Find the expression for the view factors F_{11}, F_{12}, and F_{21} for a hemisphere and a plane, as shown in figure 1. Also, derive an expression for the shape factor $F_{(dA_1)A_2}$ for planes shown in figure 2.

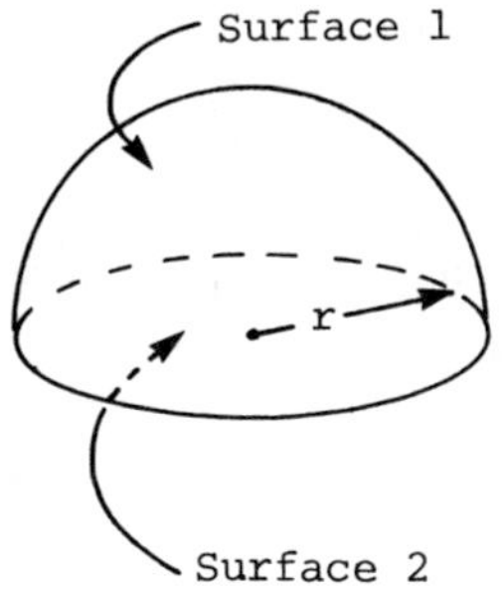

Figure 1 Hemisphere and plane

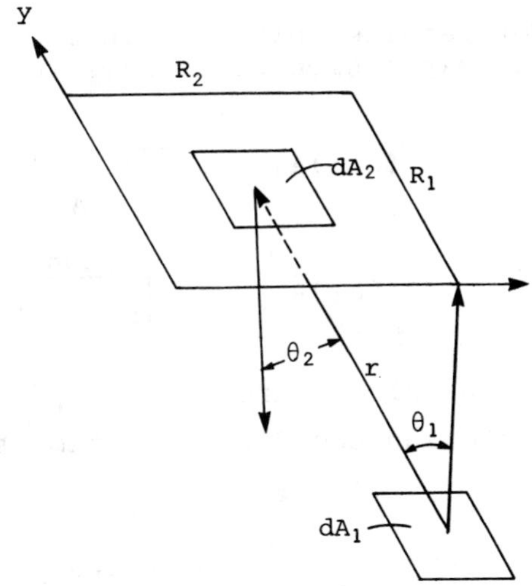

Fig. 2 Parallel Planes

Solution: a) In the case of a hemisphere placed on a plane, the shape factor for surface 2 to 1, i.e., F_{12}, is unity, as surface 2 sees nothing but the inside of hemisphere surface 1.

Again, $$F_{11} + F_{12} = 1$$

and $$A_1 F_{12} = A_2 F_{21}$$

or $$F_{12} = \frac{A_2}{A_1} F_{21}$$

$$= \frac{A_2}{A_1} = \frac{\pi r^2}{2\pi r^2} = \frac{1}{2}$$

$$\therefore F_{12} = \frac{1}{2}$$

and $$F_{11} = 1 - F_{12} = 1 - \frac{1}{2} = \frac{1}{2}$$

$$\therefore F_{12} = F_{11} = \frac{1}{2} \text{ and } F_{21} = 1$$

b) Considering figure 2, the smaller plane has a differential area (dA_1). The shape factor between two parallel planes will be obtained by analytical method. The shape factor between two surfaces is defined as

$$F_{12} = \frac{1}{A_1} \int_{A_1} \int_{A_2} \frac{\cos\theta_1 \cos\theta_2}{\pi r^2} (dA_1)(dA_2)$$

For parallel planes with areas (dA_1) and A_2, the shape factor expression transforms to

$$F_{(dA_1)A_2} = \frac{1}{(dA_1)} \int_{A_1} \int_{A_2} \frac{\cos\theta_1 \ \cos\theta_2}{\pi r^2} (dA_2)(dA_1)$$

or $$F_{(dA_1)A_2} = \frac{1}{\pi} \int_{A} \frac{\cos\theta_1 \ \cos\theta_2}{r^2} dA_2$$

As the planes are parallel, $\theta_1 = \theta_2 = \theta$ and $\cos\theta = \frac{D}{r}$, where r, in cartesian coordinate system, is defined as

$$r^2 = x^2 + y^2 + D^2$$

$$\therefore F_{(dA_1)A_2} = \frac{1}{\pi} \int_0^{L_1} \int_0^{L_2} \frac{D^2 dx \ dy}{(D^2 + x^2 + y^2)^2}$$

By integrating between the respective limits,

$$F_{(dA_1)A_2} = \frac{1}{2\pi} \left[\frac{L_1}{D^2 + L_1^2} \tan^{-1} \frac{L_2}{D^2 + L_1^2} + \right.$$

$$\left. \frac{L_2}{D^2 + L_2^2} \tan^{-1} \frac{L_1}{D^2 + L_2^2} \right]$$

• PROBLEM 8-9

A circular duct of diameter D is enclosed in a square chamber with sides 2S. Compute the shape factor for the section indicated in the figure below, if the ratio $\frac{D}{2S}$ is equal to 0.5. Describe the application of the method.

Solution: The heat flow lines and the isothermal lines, as shown in the figure, are drawn by a trial and error process. The following points are considered during the development of the grid:

(1) random number of squares can be drawn. The more the number of squares, the more accuracy of the result.

(2) The heat flow lines and the isothermal lines are always orthogonal to each other.

(3) Once the shape factor (f) is known, the heat transfer between two interactive surfaces can be computed by using the equation

$$q = fk(t_1 - t_2) \text{ Btu/hr-ft}^2$$

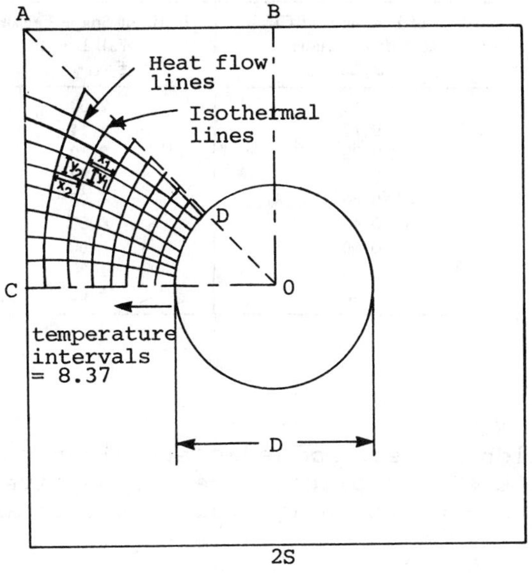

Fig. Isothermal lines and Heat flow lines in the cross-section of a wall.

The total number of flow channels for the portion indicated in the figure is

M= (Number of temperature intervals) x (Number of Heat flow lines)

$= 8 \times 9 = 72$

The shape factor is defined as

$$f = \frac{ML}{N}$$

where L is linear dimension

= length of the wall.

N - number of equal temperature intervals

$= 8.37$

$$f = \frac{72 \times L}{8.37} = 8.6L$$

Similarly, the shape factor f for different ratios of the radius of the circle to the half side of the square can be found out, and the results are given in the table.

Shape Factor for Different Ratios 2r/2s

Ratio of Diameter of Circle to Side of Square 2r/2s	Ratio of Shape Factor to Wall Length f/L
0.10	2.60
0.15	3.14
0.20	3.70
0.25	4.30
0.30	4.95
0.40	6.50
0.50	8.60
0.60	11.30
0.70	15.00

The application of the method is better suited to complex shapes, as it can be easily applied. The method gives less accurate results as compared to analytical or numerical methods.

• PROBLEM 8-10

The surface of a space ship, at temperature 40°F and emissivity 0.1, is exposed to space. Consider the space as a perfect black body at temperature 0°R, and the space ship as a gray infinite cylinder. Calculate the net radiant heat loss per square foot.

Solution: The radiation loss per unit area is:

$$\frac{Q_{12}}{A} = \sigma F_{12}\ (T_1^4 - T_2^4)$$

where

$$F_{12} = \frac{1}{1/\bar{F}_{12} + (1/\varepsilon_1 - 1) + A_1/A_2(1/\varepsilon_2 - 1)}$$

$\frac{A_1}{A_2} = 0$, because the area seen by the cylinder is relatively large compared to the surface area of the cylinder.

$\bar{F}_{12} = F_{12}$, because there is no heat re-emitted by other surfaces.

$F_{12} = 1.0$, because the cylinder only sees wide open area.

Hence, $$F_{12} = \frac{1}{1+(1/\varepsilon_1 - 1)+0} = 0.1$$

Introducing the values into eq. (1) yields:

$$\frac{Q}{A} = 0.1714\text{x}10^{-8}\ \text{Btu/hr-ft}^2\text{-°R}(0.1)(500^4 - 0^4\ \text{°R})$$

$$= 10.71\ \text{Btu/hr-ft}^2$$

• **PROBLEM 8-11**

A spherical metal part, 4 in. dia., is initially at a temperature of 75°F. It is plunged into a hot cubical chamber, 48 in. side, which is at a temperature of 540°F. Assuming all the surfaces as black, estimate the instantaneous heat transfer to the spherical part in the hot chamber.

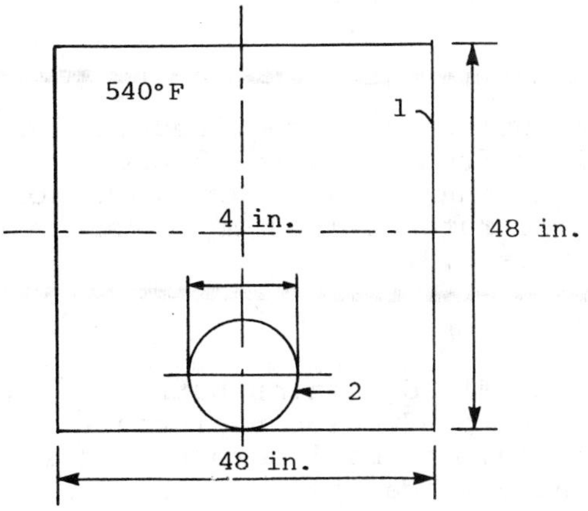

Fig. Hot Chamber

Solution: The shape factor or view factor for the sphere and the inside walls of the hot chamber, i.e., F_{2-1} is one, as all the points on the sphere can view the walls of the chamber. The shape factor for chamber walls to sphere can be found from the relation

$$F_{1-2} = \frac{A_2}{A_1} F_{2-1}$$

where A is the surface area and subscripts 1 and 2 correspond to chamber walls and sphere, respectively.

$$A = 6\text{x}48\text{x}48\ \text{sq in}$$

$$= \frac{6\text{x}48\text{x}48}{12\text{x}12} = 96\ \text{ft}^2$$

$$A_2 = 4\pi r^2 = 4\pi\left(\frac{2}{12}\right)^2 \text{ ft}^2$$

$$= 0.35 \text{ ft}$$

$$\therefore F_{1-2} = \frac{A}{A_1} F_{2-1} = \frac{0.35}{96} \times 1 = 3.63 \times 10^{-3}$$

The net heat transfer rate can be computed from the Stefan-Boltzmann equation

$$q = A_1 F_{1-2} \sigma(T_1^4 - T_2^4)$$

$$6 \times \frac{48}{12} \times \frac{48}{12} \times 3.63\times10^{-3} \times 0.171\times10^{-8}[(540+460)^4$$

$$-(75+460)^4]$$

$$= 547.1 \text{ Btu/hr}$$

• PROBLEM 8-12

Compute the surface area required in a heat exchanger to cool the fluid from 177°C to 77°C, carrying heat at the rate of 100 W/K. The heat exchanger gains solar energy by irradiation at the rate of 100 W/m^2. Assume the value of transfer factor f as 0.9.

Solution: The fluid remains sufficiently at a uniform temperature if the conduction resistance (or convection if the fluid is liquid) is small compared to surface radiation resistance. Mathematically,

$$\frac{}{KA} \quad \frac{1}{Af4\sigma T^4}$$

where k thermal conductivity of fluid

A surface area

f transfer factor

σ Stefan-Boltzmann constant = 5.67×10^{-8}

T absolute temperature in Kelvin

δ appropriate length for internal conduction

Then $$mC_p \frac{dT}{dt} = Q - Af\,\sigma(T_1^4 - T_2^4)(T_1 > T_2) \qquad (T \quad T)$$

The equation can be expressed in dimensionless parameters as

$$T_e = [T + \dot{Q}/Af\sigma]^{\frac{1}{4}}, \text{ where } \beta = \frac{T_e}{T_0}$$

$$T^* = T/T_0 \ , \ t^* = t/t_c$$

$$t_c = \frac{mc}{(Af\sigma T_0^3)}$$

then $$\frac{dT^*}{dt^*} = -(T^{*4} - \beta^4)$$

Solving the differential equation for t* vs. T*,

$$t^* = \frac{1}{2\beta^3} \ \frac{1}{2} \ \ell n \frac{(\beta+T^*)(\beta-1)}{(\beta-T^*)(\beta+1)} + \tan^{-1} \frac{T^* - 1}{\beta+(T^*/\beta)}$$

Now, total heat transferred per hour

$$\dot{m}C_p = 100 \text{ W/K}$$

and heat transferred, according to the Stefan-Boltzmann law

$$\sigma T_e^4 = \dot{m}C_p = 100 \text{ W/K}.$$

or $$T_e = [100/5.67\text{x}10^{-8}]^{\frac{1}{4}}$$

or $$T_e = 205 \text{ K}$$

Thus, $$\beta = \frac{T_e}{T_0} = \frac{205}{450} = 0.455$$

The required value of T* is

$$T^* = \frac{350}{450} = 0.778$$

and $t^* = Z^* = 0.407$

Now, the surface area of the tubes in the heat exchanger required can be computed using the relation

$$\text{Area} = Z^*\dot{m}C_p/f\sigma T_0^3$$

or $$\text{Area} = [(0.407)(100)]/[0.9\ (5.67\text{x}10^{-8})(450)^3]$$

$$= 8.76\text{m}^2$$

• PROBLEM 8-13

Find the heat loss, by leakage, per meter ($\dot{Q}/W$) through a 3 mm crack in a 6 cm thick insulation layer. The pipe inside temperature is $T_1 = 600°K$ and outer temperature is $T_2 = 300°K$.

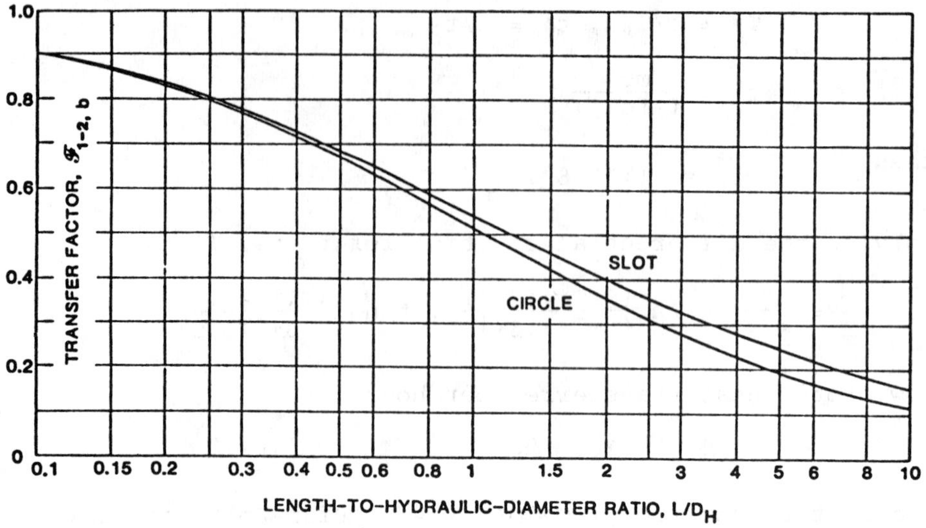

FIGURE TRANSFER FACTOR FOR DIFFUSE-REFRACTORY-WALLED PASSAGES

Solution: $D_H = 4A/P = 4WH/2H$

$= 2W = 2 \times 0.3 = 0.6$ cm.

$\therefore L/D_H = 6/0.6 = 10$

From the figure, corresponding to $L/D_H = 10$, the transfer factor $F_{1-2,b} = 0.16$.

The heat flow is

$$\dot{Q}/W = qWH/W = Hq = HF_{1-2,b}\sigma(T_1^4 - T_2^4)$$

$$\dot{Q}/W = (0.006)(0.16)(5.6697 \times 10^{-8})(600^4 - 300^4)$$

$$\dot{Q}/W = 6.6 \text{ W/m}$$

HEAT RADIATIONS BETWEEN PARALLEL AND ADJACENT SIDES

• PROBLEM 8-14

Determine the radiant energy transfer between two parallel gray plates. The temperatures and emissivities are $T_1 = 1000°R$, $T_2 = 1500°R$, and $\varepsilon_1 = 0.75$, $\varepsilon_2 = 0.65$ respectively. Also calculate the radiant energy transfer if the plates were black.

Solution: (a) The radiation per unit area is expressed as

$$\frac{q_{12}}{A} = \sigma \frac{\varepsilon_1 \varepsilon_2 T_1^4 - \varepsilon_1 \varepsilon_2 T_2^4}{\varepsilon_1 + \varepsilon_2 - \varepsilon_1 \varepsilon_2}$$

Simplifying, we obtain:

$$\frac{q_{12}}{A} = \sigma \frac{(T_1^4 - T_2^4)}{(1/\varepsilon_1)+(1/\varepsilon_2)-1.0}$$

$$= (0.1714\text{x}10^{-8}\ \text{Btu/hr.ft}^2.°\text{R}) \frac{(1500^4\ °\text{R} - 1000^4\ °\text{R})}{(1/0.75+1/0.65=1.0)}$$

$$= 3720\ \text{Btu/hr.ft}^2.$$

(b) If the plates were black bodies, the energy transfer would be almost twice of that of the gray body radiant heat transfer. The plates are black bodies, i.e. $\varepsilon_1 = \varepsilon_2 = 1.0$.

Eq. (1) will reduce to the relation for a black body:

$$\frac{q_{12}}{A} = \sigma(T_1^4 - T_2^4)$$

$$= (0.1714\text{x}10^{-8})\text{x}[1500^4 - 1000^4]$$

$$= 6963\ \text{Btu/hr.ft}^2.$$

• PROBLEM 8-15

The composite wall of a furnace consists of steel plate and 12 in. thick refractory wall separated by 24 in. air gap. Estimate the inside temperature of the refractory wall when the steel plate and ambient temperatures are 1000°F and 100°F, respectively. The thermal conductivity of the refactory wall is 0.05 Btu/hr.ft^2-(°F/ft). Assume there is no convective heat transfer in the air gap.

Solution: The heat transfer to refractory wall from the steel plate occurs by radiation only, and this heat is further transferred through the refractory wall by conduction to the outside air.

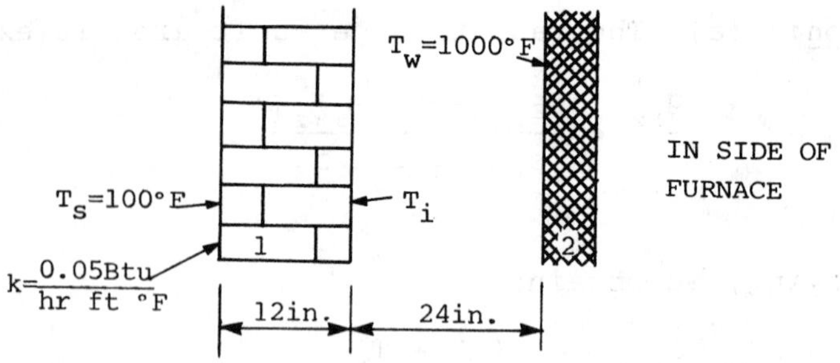

The radiation flux from steel plate to refractory wall is expressed as

$$q_R = \sigma A \varepsilon (T_w^4 - T_i^4)$$

where A is the surface area of the wall involved

- σ - Stefan-Boltzmann constant 0.171×10^{-8} Btu/hr.ft^2-°R^4
- T - absolute temperature in Rankine
- ε - emissivity of the surface = 1.0.

subscript R refers to radiation

T_w = 1000 + 1460°R.

T_i can be found from the equation of heat transfer by conduction, i.e.

$$q_c = kA \frac{T_i - T_s}{L}$$

where L is the thickness of the wall

subscript c refers to conduction. For energy balance $q_R = q_c$

$$\therefore q_c = 0.05A \frac{T_i - (100 + 460)}{1}$$

$$q_R = 0.171 \times 10^{-8} \times A \times 1 \; (1460^4 - T_i^4)$$

$$\because q_c = q_R$$

$$\therefore 0.05(T_i - 560) = 0.171 \times 10^{-8}(1460^4 - T_i^4)$$

or $\quad 0.171 \times 10^{-8} T_i^4 + 0.05 T_i = 0.171 \times 10^{-8} \times (1460)^4 + 0.05 \times 560$

By trial and error

$$T_i = 1458°R \text{ or } T_i = 998°F$$

• PROBLEM 8-16

Derive an expression for steady state heat radiation between two gray plates, placed parallel, with a radiation shield between them.

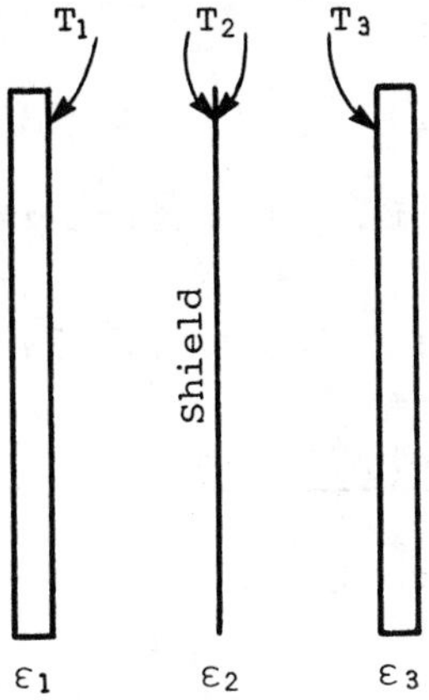

Fig. 1 Radiation between parallel plates with a radiation shield.

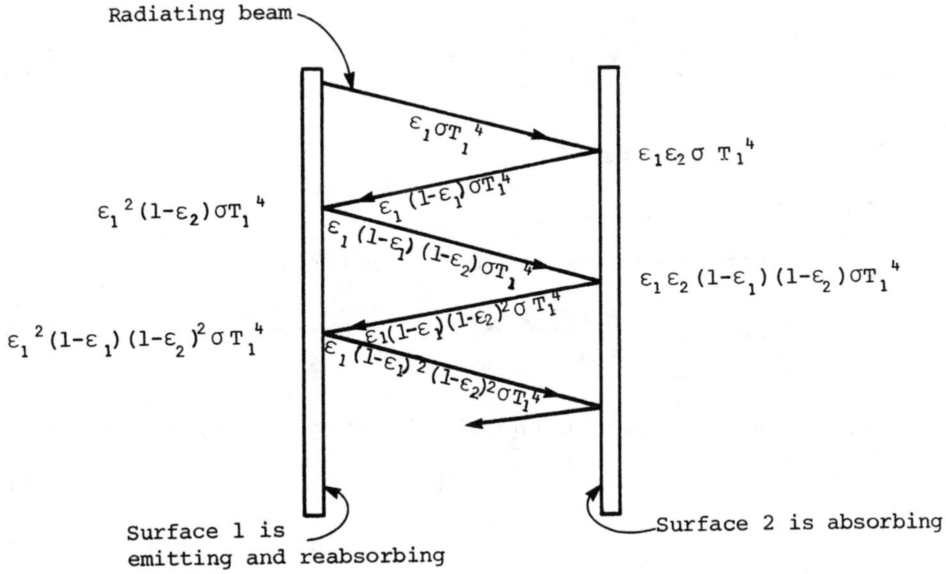

Fig. 2

Solution: Figure 1 shows the layout for the steady state radiation between the two parallel plates with a radiation shield inbetween.

Figure 2 shows the radiation balance between each pair of adjacent surfaces.

Referring to figure 2 the radiant heat transfer between the plate (1) and the radiation shield can be written as

$$Q_{12} = A\sigma(T_1^4 - T_2^4)[\varepsilon_1\varepsilon_2 + \varepsilon_1\varepsilon_2(1 - \varepsilon_2)(1 - \varepsilon_1) + \varepsilon_1\varepsilon_2(1 - \varepsilon_2)^2(1 - \varepsilon_1)^2 + \ldots]$$

$$= A\sigma\varepsilon_1\varepsilon_2(T_1^4 - T_2^4)\sum_{i=0}^{\infty}[(1 - \varepsilon_2)(1 - \varepsilon_1)]^i \qquad (1)$$

It may be shown that the summation reduces to

$$\sum_{i=0}^{\infty}[(1 - \varepsilon_2)(1 - \varepsilon_1)]^i = \frac{1}{1 - (1 - \varepsilon_2)(1 - \varepsilon_1)}$$

such that
$$Q_{12} = \frac{A\sigma(T_1^4 - T_2^4)}{\frac{1}{\varepsilon_1} + \frac{1}{\varepsilon_2} - 1} \qquad (2)$$

and similarly
$$Q_{23} = \frac{A\sigma(T_2^4 - T_3^4)}{\frac{1}{\varepsilon_2} + \frac{1}{\varepsilon_3} - 1} \qquad (3)$$

For steady state conditions

$$Q_{12} = Q_{23}$$

$$\therefore \quad \frac{A\sigma[T_1^4 - T_2^4]}{\frac{1}{\varepsilon_1} + \frac{1}{\varepsilon_2} - 1} = \frac{A\sigma[T_2^4 - T_3^4]}{\frac{1}{\varepsilon_2} + \frac{1}{\varepsilon_3} - 1} \qquad (4)$$

If $\varepsilon_1 = \varepsilon_2 = \varepsilon_3 = \varepsilon$

then $T_2^4 = \frac{1}{2}[T_1^4 + T_3^4]$

Substituting the values in equation (2),

$$Q_{12} = \frac{A\sigma[T_1^4 - T_3^4] \times \frac{1}{2}}{\frac{2}{\varepsilon} - 1} = \frac{A\ [T_1^4 - T_3^4]}{\frac{2}{\varepsilon} - 1} \times \frac{1}{2} \qquad (5)$$

Heat radiation without the shield is given by

$$Q_{13} = \frac{A\sigma[T_1^4 - T_2^4]}{\frac{2}{\varepsilon} - 1} \qquad (6)$$

Comparing equations (5) and (6)

Q_{13} (with shield) $= [\frac{1}{2}]Q_{13}$ (without shield)

It can be shown for n radiation shields as

Q (with n shields) $= \left[\frac{1}{n+1}\right]Q$ (without shield)

• PROBLEM 8-17

Compute the rate of radiation heat transfer between two large parallel plates having the temperatures and emissivities as shown in the figure. Also, compute the rate of radiation heat transfer, if a radiation shield, emissivity =0.06, is provided inbetween the walls.

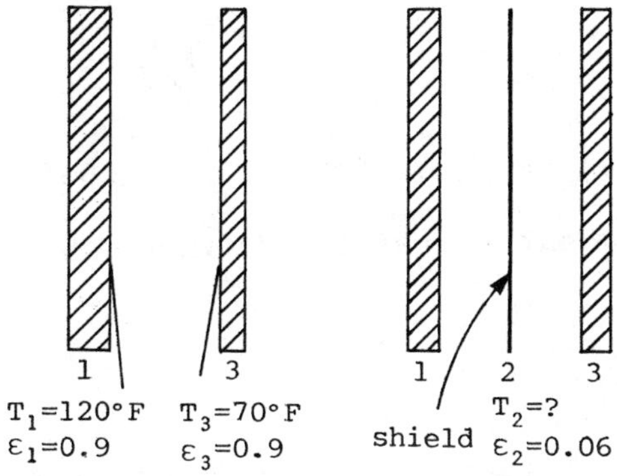

Fig. (a) Fig. (b)

Parallel plates.

Solution: a) The radiation heat flux is expressed as

$$q = Af\sigma(T_1^4 - T_3^4)\ \text{Btu/hr}$$

where A is the surface

σ Stefan-Boltzmann constant -- 0.171×10^{-8}

T-absolute temperature in Rankine

$$f = 1 \quad \frac{1}{\varepsilon_1} + \frac{1}{\varepsilon_3} - 1$$

$$= \frac{1}{\frac{1}{0.9} + \frac{1}{0.4} - 1} = 0.383$$

$$\therefore \frac{q}{A} = 0.383 \times 0.171 \times 10^{-8}\ [(120 + 460)^4 - (70 + 460)^4]$$

$$= 22.49\ \text{Btu/hr-ft}^2$$

Again, with a shield present between the surfaces, for equilibrium conditions,

$$\frac{q}{A} = \frac{\sigma(T_1^4 - T_2^4)}{\frac{1}{\varepsilon_1} + \frac{1}{\varepsilon_2} - 1} = \frac{\sigma(T_2^4 - T_3^4)}{\frac{1}{\varepsilon_2} + \frac{1}{\varepsilon_3} - 1}$$

Introducing the known values, the temperature of the shield can be found.

$$\frac{T_1^4 - T_2^4}{T_2^4 - T_3^4} = \frac{\frac{1}{\varepsilon_1} + \frac{1}{\varepsilon_2} - 1}{\frac{1}{\varepsilon_2} + \frac{1}{\varepsilon_3} - 1} = \frac{\frac{1}{0.9} + \frac{1}{0.06} - 1}{\frac{1}{0.06} + \frac{1}{0.4} - 1}$$

$$T_1^4 - T_2^4 = 0.92(T_2)^4 - 0.92(T_3)^4$$

or

$$1.92\, T_2^4 = T_1^4 + 0.92\,(T_3)^4$$

$$(T_2)^4 = \frac{[(120 + 460)^4 + 0.92(70+460)^4]}{1.92}$$

$$(T_2)^4 = 9.67\text{x}10^{10} \quad \therefore\; T_2 = 558°R(98°C)$$

The radiation heat transfer rate with a shield present is

$$\frac{q}{A} = \frac{(T_1^4 - T_2^4)}{\frac{1}{\varepsilon_1} + \frac{1}{\varepsilon_2} - 1}$$

$$= \frac{0.171\text{x}10^{-8}\;[(580)^4 - 9.7\text{x}10^{10}]}{\frac{1}{0.9} + \frac{1}{0.06} - 1}$$

$$= 1.65 \text{ Btu/hr-ft}^2$$

The presence of a radiation shield drastically reduced the net rate of radiation heat transfer.

• PROBLEM 8-18

Find the net radiant heat flux absorbed by the floor of an inert-gas-filled furnace, which has a floor 4 x 4 m and ceiling 2 m high.

The table below shows the average results of the observations made using a radiometer. The radiometer is calibrated to read an apparent temperature $T_{1(a)} = (q_1^+/\sigma)^{\frac{1}{4}}$, in which case

$$q_1^+ = \sigma T_{1(a)}^4$$

Furnace Side i	T_i,app[K]	q_i^+ [W/m²]
1	500	3544
2	1500	287000
3	1200	117600
4	1200	117600
5	1200	117600
6	1000	56700

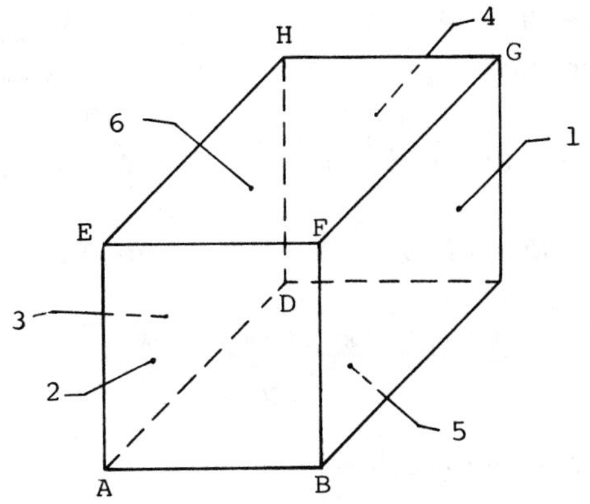

TABLE

CLOSED-FORM EXPRESSIONS OF SELECTED SHAPE FACTORS

Geometry	Expression
1. Opposite rectangles 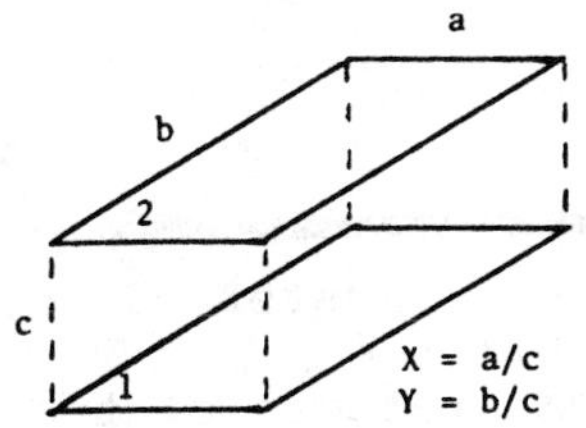 X = a/c Y = b/c	$F_{1-2} = \frac{2}{\pi XY}\left\{ \ln\sqrt{\frac{(1+X^2)(1+Y^2)}{1+X^2+Y^2}} + X\sqrt{1+Y^2}\tan^{-1}\frac{X}{\sqrt{1+Y^2}} + Y\sqrt{1+X^2}\tan^{-1}\frac{Y}{\sqrt{1+X^2}} - X\tan^{-1}X - Y\tan^{-1}Y \right\}$
2. Adjacent rectangles 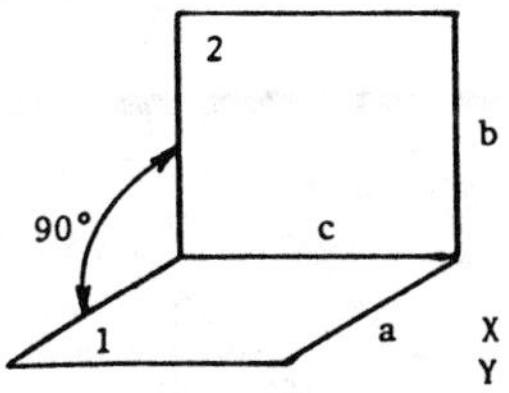 X = a/c Y = b/c	$F_{1-2} = \frac{1}{4\pi X}\left\{ 4X\tan^{-1}\frac{1}{X} + 4Y\tan^{-1}\frac{1}{Y} - 4\sqrt{X^2+Y^2}\tan^{-1}\frac{1}{\sqrt{X^2+Y^2}} + \ln\frac{(1+X^2)(1+Y^2)}{1+X^2+Y^2} + X^2\ln\frac{X^2(1+X^2+Y^2)}{(1+X^2)(X^2+Y^2)} + Y^2\ln\frac{Y^2(1+X^2+Y^2)}{(1+Y^2)(X^2+Y^2)} \right\}$

<u>Solution:</u> The irradiation q_1^- is given by

$$q_1^- = q_1^+F_{1-1} + q_2^+F_{1-2} + q_3^+F_{1-3} + q_4^+F_{1-4} + q_5^+F_{1-5} + q_6^+F_{1-6}$$

F_{1-1} is clearly zero as no solid-angle-cosine product is subtended by surface 1 itself, and from any point upon surface 1, the surface can not "see" itself.

The table below gives one relation for F_{i-j} where j = 2, 3, 4 or 5, and another for F_{1-2}. For the former

$$F_{1-2} = \frac{1}{4\pi x} \quad 4x\tan^{-1}\frac{1}{x} + \ldots \qquad = 0.14619$$

where x=4/4=1 and y=2/4=1/2. For the latter

$$F_{1-6} = \frac{2}{\pi xy} \quad \log[\frac{(1+x^2)(1+y^2)}{1+x^2+y^2}]^{\frac{1}{2}} + \ldots \qquad = 0.41525$$

where x=4/2=2 and y=4/2=2.

Therefore $q_1^- = (0.14619)(287000)+(3)(0.14619)(117600)$

$+(0.41515)(56700) = 117070 \text{ W/m}^2$

The net flux into the floor is $-q_1 = q_1^- - q_1^+$

$$-q_1 + 117070-3540 = 113500 \text{ W/m}^2$$

$$-\dot{Q}_1 = -q_1A_1 = (113500)(16) = 1816 \text{ kW}(6.2\text{x}10^6 \text{ Btu/hr},$$

• PROBLEM 8-19

Determine the radiant heat flux transferred between two parallel planes of dimension h=1.5 ft, L=2.5 ft, W=3.0 ft as shown in the figure. The temperatures of the planes are T_2=900°R and T_1=1100°R. The surroundings will be considered the third surface with T_3=0°R. The emissivities of the planes are ε_1=0.4 and ε_2=0.7.

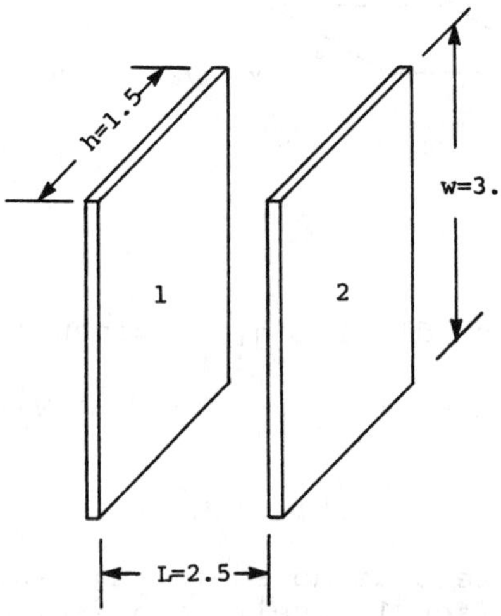

Fig. 1 Two parallel planes

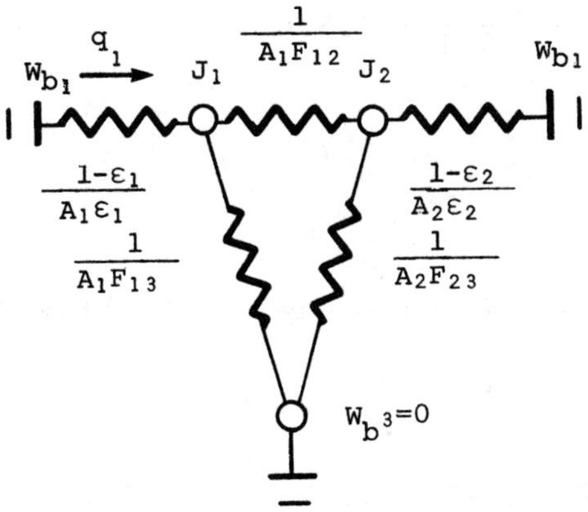

Figure 2

Solution: Considering three surfaces, gray surface 1, gray surface 2 and space, black surface 3, assumed to be at 0°R, interacting. Figure 2 shows the appropriate electrical network. Setting the currents into node(1) equal to zero to determine q_1 leads to

$$\frac{W_{b1} - J_1}{(1-\varepsilon_1)/A_1\varepsilon_1} + \frac{0 - J_1}{1/A_1F_{13}} + \frac{J_2 - J_1}{1/A_1F_{12}} = 0 \qquad (1)$$

Similarly at node 2

$$\frac{W_{b2} - J_2}{(1-\varepsilon_2)/A_2\varepsilon_2} + \frac{0 - J_2}{1/A_2F_{23}} + \frac{J_1 - J_2}{1/A_1F_{12}} = 0$$

Now $A_1 = A_2 = 1.5 \times 3 = 4.5 ft^2$

$$W_{b1} = \sigma T_1^4 = 0.1714 \times 10^{-8} \times (1100)^4$$

$$= 2509.5 \text{ Btu/hr.ft}^2$$

$$W_{b2} = \sigma T_2^4 = 0.1714 \times 10^{-8} \times (900)^4$$

$$= 1125 \text{ Btu/hr.ft}^2$$

$$\frac{1-\varepsilon_1}{A_1\varepsilon_1} = \frac{1-0.4}{4.5 \times 0.4} = 0.333$$

$$\frac{1-\varepsilon_2}{A_2\varepsilon_2} = \frac{1-0.7}{4.5 \times 0.7} = 0.0952$$

Since $F_{12} = 0.4$ the rest of emission from surface 1 would strike the space (surface 3) $\therefore F_{13} = 1-0.4=0.6$ and similarly for $F_{21} = 0.7$, $F_{23} = 0.3$.

$$\therefore \frac{1}{A_1F_{13}} = \frac{1}{4.5 \times 0.6} = 0.37$$

$$\frac{1}{A_1F_{12}} = \frac{1}{4.5x0.4} = 0.556$$

$$\frac{1}{A_2F_{23}} = \frac{1}{4.5x0.3} = 0.741$$

Substituting the values in equations (1) and (2)

$$\frac{(2509.5 - J_1)}{0.333} - \frac{J_1}{0.37} + \frac{J_2 - J_1}{0.556} = 0$$

gives $7529.3 - 3J_1 - 2.7J_1 + 1.8J_2 - 1.8J_1 = 0$

i.e. $7.5J_1 - 1.8J_2 = 7529.3$ (3)

Similarly $\frac{(1125 - J_2)}{0.0952} - \frac{J_2}{0.741} + \frac{J_1 - J_2}{0.556} = 0$

gives $11817 - 10.5J_2 - 1.35J_2 + 1.8J_1 - 1.8J_2 = 0$

i.e. $13.65J_2 - 1.8J_1 = 11817$ (4)

Solving equations (3) and (4)

$J_1 = 1251.3$ Btu/hr.ft^2 and $J_2 = 1030.7$ Btu/hr.ft^2

$\therefore$ for q_1

$$q_1 = \frac{W_{b_1} - J_1}{(1-\varepsilon_1)/A_1\varepsilon_1}$$

$$= \frac{2509.5 - 1251.3}{0.333}$$

$$\cong 3775 \text{ Btu/hr.}$$

• PROBLEM 8-20

Two circular discs, 10 in. diameter, at temperatures of 1960°R and 1460°R, respectively, are arranged as shown in the figure. Considering the discs as black bodies, estimate the heat transfer between them. Assume sides of the discs are linked with a reradiating surface and are 5 in. apart. The emissivities of the discs are 0.8 and 0.6, respectively.

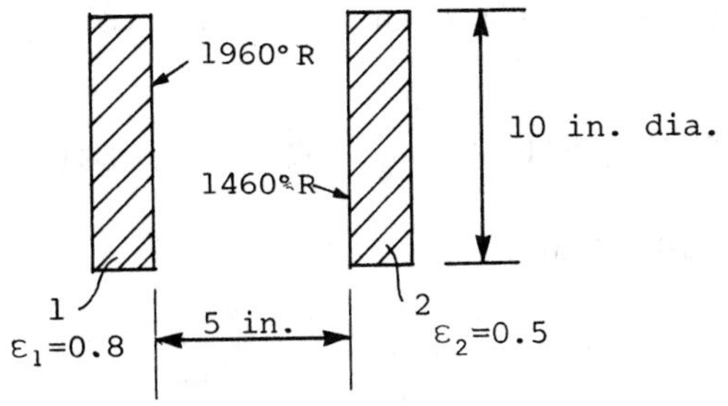

Fig. Parallel plates

Solution: The net heat exchange between two parallel discs placed at a finite distance apart can be expressed as

$$q = \sigma A f(T_1^4 - T_2^4)$$

where σ Stefan-Boltzmann constant = 0.1714×10^{-8}

A surface area of plate involved in heat transfer

$$= \frac{\pi}{4}(10)^2 \text{in}^2 \text{ or } \frac{\pi}{4}\left(\frac{10}{12}\right)^2 \text{ft}^2$$

T is the absolute temperature in Rankine.

f- the shape factor, can be found from the table corresponding to the D/L ratio.

Table: Shape factors for parallel squares and discs., directly opposed.

Ratio of Side or Diameter to Distance between Planes	Planes not Connected by Walls		Planes Connected by Non-conducting but Reradiating Walls	
	Squares	Discs	Squares	Discs
0	0	0	0	0
0.2	0.018	0.010	0.200	0.186
0.4	0.045	0.038	0.325	0.309
0.6	0.091	0.075	0.415	0.394
0.8	0.142	0.125	0.480	0.461
1.0	0.194	0.175	0.534	0.512
1.2	0.242	0.226	0.575	0.552
1.4	0.288	0.269	0.610	0.590
1.6	0.332	0.308	0.641	0.620
1.8	0.372	0.345	0.669	0.646
2.0	0.408	0.380	0.692	0.669
2.2	0.440	0.411	0.712	0.689
2.4	0.469	0.440	0.730	0.708
2.6	0.495	0.467	0.746	0.724
2.8	0.519	0.492	0.760	0.739
3.0	0.542	0.515	0.772	0.753

$\frac{D}{L} = \frac{10}{5} = 2.0$ and for surfaces non-conducting but re-radiating, the shape factor is 0.669.

Introducing the values in the equation for heat transfer yields

$$q = 0.1714\times10^{-8} \times \frac{\pi}{4}\left(\frac{10}{12}\right)^2 \times 0.669[(1960)^4 - (1460)^4]$$

$$= 6388 \text{ Btu/hr}$$

Again, if the emissivities of the plates are given then the shape factor can be computed as follows.

$$f_{1-2} = \frac{1}{\frac{1}{F_{1-2}} + \frac{1}{\varepsilon_1} - 1 + \frac{1}{\varepsilon_2} - 1}$$

$$= \frac{1}{\frac{1}{0.669} + \frac{1}{0.8} - 1 + \frac{1}{0.6} - 1}$$

or $f_{1-2} = 0.415$

Now, the heat transfer can be calculated using this value for the shape factor.

$$q = 0.1714\text{X}10^{-8} \times \frac{\pi}{4}\left(\frac{10}{12}\right)^2 \times 0.415[(1960)^4 - (1460)^4]$$

$$= 3962 \text{ Btu/hr}$$

• PROBLEM 8-21

(a) A cabin (see fig. 1 for dimensions) with a large viewport is to be heated by using the platform as the radiating heat source at 100°F. The glass viewport at 40°F will behave like a black body, radiating all the heat through it to the outside surroundings. The other three partitions and roof will behave like black bodies at 80°F. Calculate the net heat transfer between the platform and the glass port, and also the heat radiated by the platform.

(b) If all the partitions except the glass viewport and the platform specified in part (a) are regarded as one re-emitting plane, and if the platform and glass port are at identical temperatures of 80°F, determine the net heat transfer between the platform and viewport, and also the steady-state temperature of the re-emitting surface.

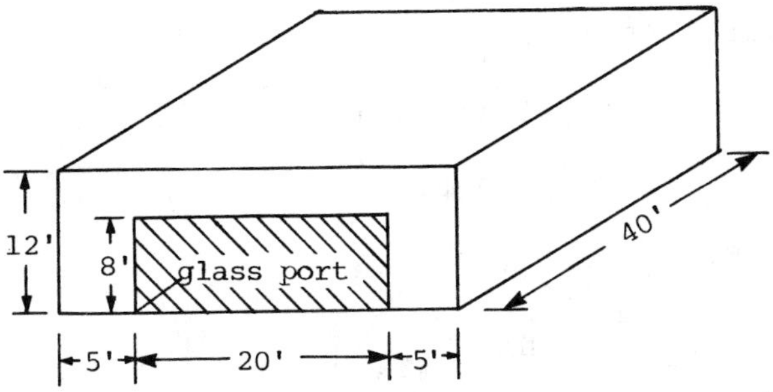

Fig. 1

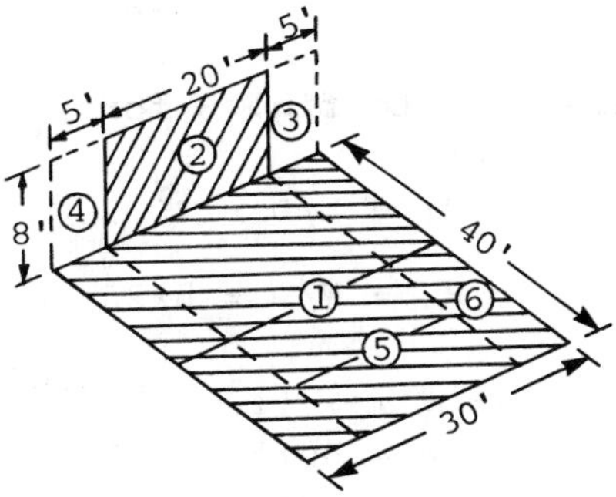

Fig. 2

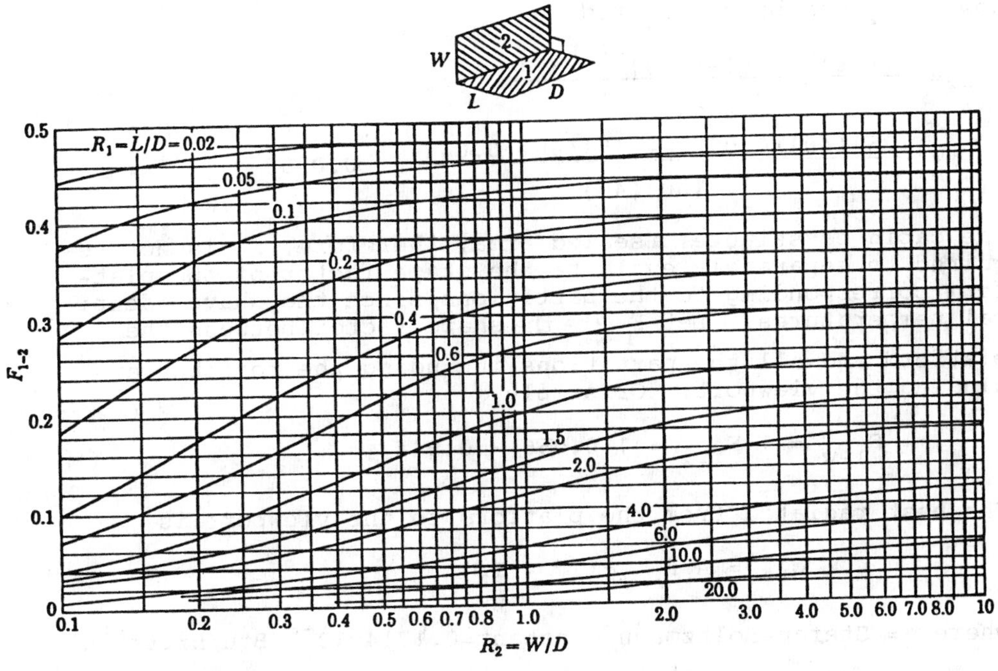

The radiation shape factor for perpendicular rectangles with a common edge.

Solution: (a) Refer to fig. 2 for the arrangement of the viewport and platform. The platform is represented by 1 and the viewport by 2. Numbers 3 through 6 are components to determine F_{1-2}. Applying the concept of shape factor yields

$$A_1F_{1-2} = A_5F_{5-2} + 2A_6F_{6-2}$$

$$= A_5F_{5-2} + [A_{(5,6)}F_{(5,6)-(2,3)}$$

$$- A_5F_{5-2} - A_6F_{6-3}]$$

$$= A_{(5,6)}F_{(5,6)-(2,3)} - A_6F_{6-3}$$

Referring to fig. 3 for $F_{(5,6)-(2,3)}$ and F_{6-3},

$$R_1 = L/D = 40/25 = 1.6$$

$$R_2 = W/D = 8/25 = 0.32$$

$$F_{(5,6)-(2,3)} = 0.075$$

$$R_1 = 40/5 = 8.0$$

$$R_2 = 8/5 = 1.6$$

$$F_{6-3} = 0.04$$

Now, F_{1-2} can be calculated as

$$F_{1-2} = \frac{A_{(5,6)}F_{(5,6)-(2,3)} - A_6F_{6-3}}{A_1}$$

$$= \frac{(25)(40)(0.075) - (5)(40)(0.04)}{(30)(40)} = 0.0558$$

The cabin constitutes a sealed area, therefore, it is not required to separately evaluate the shape factor of the platform corresponding to the partitions, since they have identical temperatures. Let F_{1-W} be the shape factor between the platform and all the partitions including the roof but excluding the viewport. Or simply,

$$F_{1-W} = 1 - F_{1-2} = 1 - 0.056 = 0.944$$

The heat radiated from the platform to the viewport is

$$q_{12} = A_1F_{1\ 2}\sigma(T_1^4 - T_2^4)$$

where σ = Stefan-Boltzmann constant $= 0.1714 \times 10^{-8}$ Btu/hr.ft^2 °R^4

$$q_{12} = (30)(40)(0.056)(0.1714 \mathrm{X} 10^{-8}\ \mathrm{Btu/hr.ft^2.°R^4})$$

$$(560^4 - 500^4\ °\mathrm{R})$$

$$= 4128\ \mathrm{Btu/hr}$$

The resulting heat emitted by the platform is:

$$q_1 = A_1\sigma[T_1^4 - (F_{1-2}T_2^4 + F_{1-W}T_W^4)]$$

$$= (30)(40)(0.1714 \mathrm{X} 10^{-8})[560^4 - (0.056 \mathrm{X} 500^4 + 0.944 \mathrm{x} 560^4)]$$

$$= 29980\ \mathrm{Btu/hr}$$

(b) The area of the platform is represented as A_1 and the glass port as A_2, and from part a, $F_{1-2}=0.056$. In this case, the prevailing conditions are one re-emitting plane and two emitting surfaces. Therefore, the following relation may be used for $\overline{F}_{1-2}$:

$$\overline{F}_{1-2} = \frac{A_2 - A_1 F_{1-2}^2}{A_1 + A_2 - 2A_1 F_{1-2}}$$

$$= \frac{(20)(8)-(30)(40)(0.056)^2}{(30)(40)+(8)(20)- 2(30)(40)(0.056} = 0.127$$

Under these conditions, the heat radiated from the platform to the viewport is

$$q_{12} = A_1\overline{F}_{1-2}\sigma(T_1^4 - T_2^4)$$

$$= (30)(40)(0.127)(0.1714\text{x}10^{-8})(560^4-500^4) = 9363 \text{ Btu/hr}$$

The steady-state temperature of the partitions is determined from the principle that the partitions transfer identical levels of heat with A_1 as with A_2. Hence, balancing the energy equation yields

$$\sigma A_1 F_{1-r}(T_1^4 - T_r^4) = \sigma A_r F_{r-2}(T_r^4 - T_2^4) \quad (1)$$

$$= \sigma A_2 F_{2-r}(T_r^4 - T_2^4)$$

Rearranging eq. (1) and solving for the steady-state temperature gives:

$$A_1(1-F_{1-2})(T_1^4 - T_r^4) = A_2(1-F_{2-1})(T_r^4 - T_2^4)$$

$$= (A_2 - A_1F_{1-2})(T_r^4 - T_2^4)$$

$$\frac{T_1^4 - T_r^4}{T_r^4 - T_2^4} = \frac{(A_2/A_1)-F_{1-2}}{1-F_{1-2}}$$

$$\frac{560^4 - T_r^4}{T_r^4 - 500^4} = \frac{(2/15)- 0.056}{0.944} = 0.0819$$

$$T_r^4 = \frac{560^4 + 5.125\text{x}10^9}{1.082} = 9.56\text{x}10^{10}$$

$$T_r = 556°\text{R} \quad \text{or } 96°\text{F}$$

• PROBLEM 8-22

One side of a cubical box, side 1m, is generating heat at 1227°C. The adjacent heat-absorbing side is at 227°C. The emissivity for both sides is 0.5. Except the generating and the absorbing black sides, all the sides are gray surfaces. Compute the temperature at the gray surfaces, also the net radiant heat exchange.

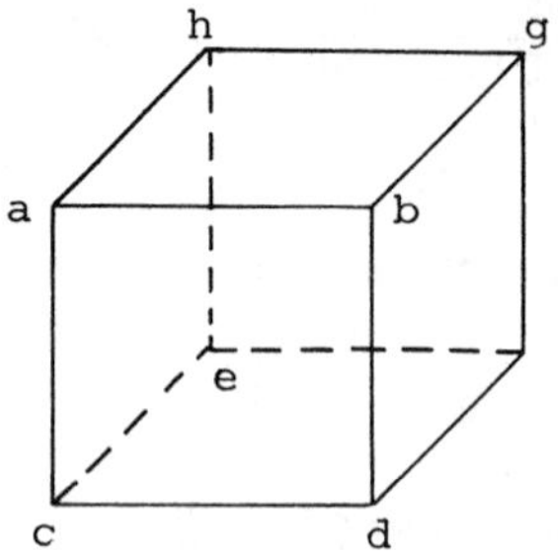

Fig. 1

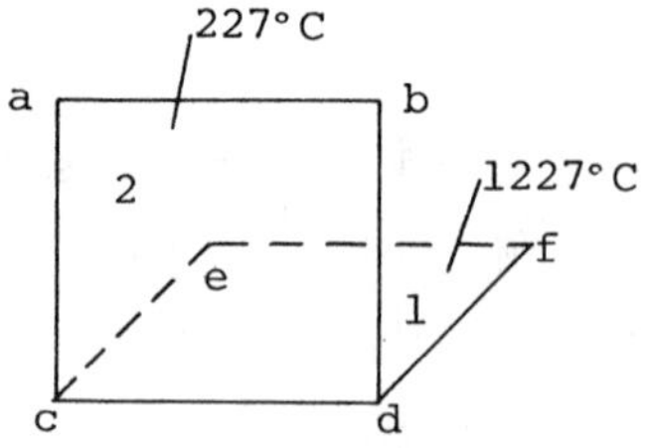

side 1 cdef heat generating
side 2 abcd heat absorbing

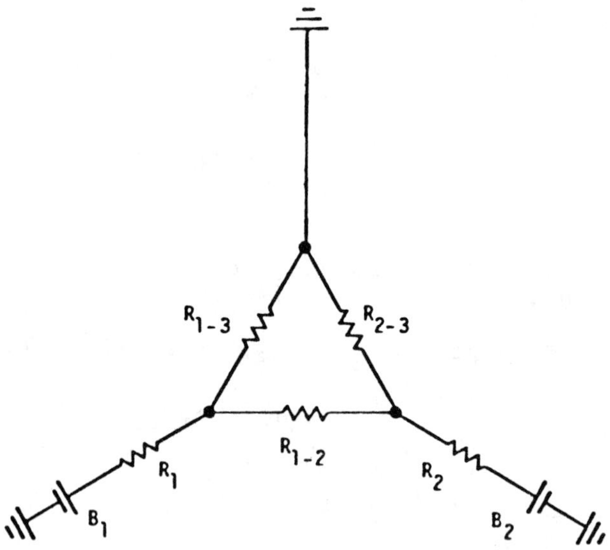

Fig. 2 Radiation network for 3 surfaces.

Solution: Figure 2 is the network representation of the problem. There are three surfaces in the radiation exchange--one is the heat-generating face (1), another is the heat-absorbing face (2) and third is the gray surface(s) (3).

Considering the symmetricity of the network, the radiosity of the surface 3 can be computed as the average of the radiosities of surfaces 1 and 2, and then from the radiosity of surface 3 the temperature of the gray surfaces can be determined.

Radiosity of surface 1 $\quad B_1 = \sigma (T_1)^4 = 5.67\text{x}10^{-8}(1227 + 273)^4$

$$= 5.67\text{x}10^{-8}(1500)^4$$

$$\cong 287040 \text{ W/m}^2$$

$$B_2 = \sigma T_2^4 = 5.67\text{x}10^{-8}\ (227 + 273)^4$$

$$= 5.67\text{x}10^{-8}\ (500)^4$$

$$= 3540 \text{ W/m}^2$$

$$\therefore\ B_3 = \frac{B_1 + B_2}{2} = \frac{287040 + 3540}{2}$$

$$= 145290 \text{ W/m}^2$$

or $\quad B_3 = T_3^4$

$$\therefore\ T_3 = \left(\frac{145290}{5.67\text{x}10^{-8}}\right)^{\frac{1}{4}}$$

$$T_3 \cong 1265°\text{K}\ (992°\text{C})$$

CLOSED-FORM EXPRESSIONS OF SELECTED SHAPE FACTORS

Geometry	Expression
1. Opposite rectangles 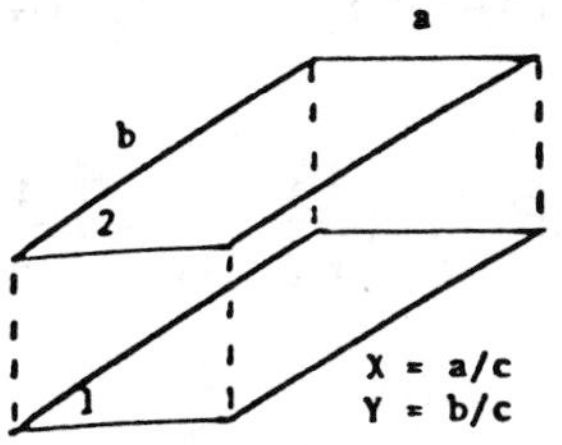	$F_{1-2} = \frac{2}{\pi XY}\left\{\ln\sqrt{\frac{(1+X^2)(1+Y^2)}{1+X^2+Y^2}} + X\sqrt{1+Y^2}\tan^{-1}\frac{X}{\sqrt{1+Y^2}} + Y\sqrt{1+X^2}\tan^{-1}\frac{Y}{\sqrt{1+X^2}} - X\tan^{-1}X - Y\tan^{-1}Y\right\}$
2. Adjacent rectangles 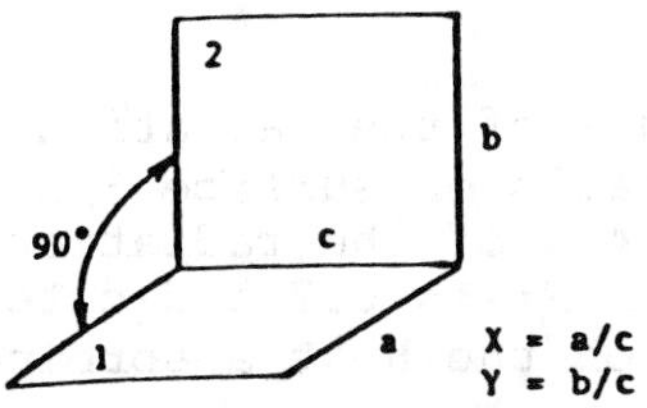	$F_{1-2} = \frac{1}{4\pi X}\left\{4X\tan^{-1}\frac{1}{X} + 4Y\tan^{-1}\frac{1}{Y} - 4\sqrt{X^2+Y^2}\tan^{-1}\frac{1}{\sqrt{X^2+Y^2}} + \ln\frac{(1+X^2)(1+Y^2)}{1+X^2+Y^2} + X^2\ln\frac{X^2(1+X^2+Y^2)}{(1+X^2)(X^2+Y^2)} + Y^2\ln\frac{Y^2(1+X^2+Y^2)}{(1+Y^2)(X^2+Y^2)}\right\}$

Again, from the table, the shape factor F_{1-2} for the adjacent faces is expressed as

$$F_{1-2} = \frac{1}{4\pi x}\ 4x\tan^{-1}\frac{1}{x} + 4y\tan^{-1}\frac{1}{y}$$

$$-4\ x^2 + y^2\ \tan^{-1}\frac{1}{x^2 + y^2}$$

$$+ \ln\frac{(1+x^2)(1+y^2)}{1+x^2+y^2}$$

$$+ x^2\ln\frac{x^2(1+x^2+y^2)}{(1+x^2)(x^2+y^2)}$$

$$+ y^2\ln\frac{y^2(1+x^2+y^2)}{(1+y^2)(x^2+y^2)}$$

where x and y are the ratios of the sides of the faces

$$x = \frac{a}{c} = \frac{1}{1} = 1$$

$$y = \frac{b}{c} = \frac{1}{1} = 1$$

$x = y = 1$, then,

$$F_{1-2} = \frac{1}{4\pi(1)}\ 4\tan^{-1}(1) + 4\tan^{-1}(1) - 4\ (1)^2 + (1)^2\ x$$

$$\tan^{-1}\frac{1}{1^2+1^2} + \ln\frac{(1+1^2)(1+1^2)}{1+1+1}$$

$$+ 1^2\ \ln\frac{1^2(1+1^2+1^2)}{(1+1^2)(1^2+1^2)} + 1^2\ln\frac{1^2(1+1^2+1^2)}{(1+1^2)(1^2+1^2)}$$

Simplifying,

$$F_{1-2} = \frac{1}{4\pi}\left[8\tan^{-1}(1) - 4\sqrt{2}\tan^{-1}\left(\frac{1}{\sqrt{2}}\right) + \ln\frac{4}{3} + 2\ln\frac{3}{4}\right]$$

$$\cong 0.2$$

$$F_{1-3} = F_{2-3} = (1-0.2) = 0.8 \text{ and}$$

$$F_{1-1} = F_{2-2} = 0$$

$F_{1-3} = 0.8$ shows that the 80% of the radiation from the heat generating surface 1 falls on surface 3, i.e., gray surface. 50% of this 80%, i.e., 40% of the radiation leaving surface 1 falls back to it. $F_{1-2} = 0.2$ indicate that only 20% of the radiation falls on the heat absorbing surface 2.

The equivalent conductance of the gray surfaces can be evaluated as from the expression

$$A_1 f_{1-2,b} = A_1F_{1-2} + \cfrac{1}{\cfrac{1}{A_1F_{1-3}} + \cfrac{1}{A_2F_{2-3}}}$$

$$A_1 = A_2 = 1\,m^2$$

$$\therefore\ f_{1-2,b} = 0.2 + \cfrac{1}{\cfrac{1}{0.8} + \cfrac{1}{0.8}}$$

$$= 0.6$$

The subscript b refers to black body.

Now, $$f_{1-2} = \cfrac{1}{\cfrac{1}{f_{1-2,b}} + \cfrac{1-\varepsilon_1}{\varepsilon_1} + \cfrac{1-\varepsilon_2}{\varepsilon_2}}$$

$$= \cfrac{1}{\cfrac{1}{0.6} + \cfrac{1-0.5}{0.5} + \cfrac{1-0.5}{0.5}}$$

$$= 0.273$$

The net exchange of heat flux is given by

$$Q = A_1 f_{1-2}\sigma(T_1^4 - T_2^4)$$

$$= 1 \times 0.273 \times 5.67 \times 10^{-8}(1500^4 - 500^4)$$

$$= 77396 \text{ W or } \cong 774 \text{ kW}$$

HEAT RADIATION BETWEEN REGULAR SURFACES

• PROBLEM 8-23

A duct, 4 in. dia., at 800°F is enclosed in a large refractory chamber at 1800°F as shown in the figure. Compute the heat transfer by radiation when a) the refractory chamber is very large compared to the duct diameter and b) the refractory chamber is a gray surface, 8 in. square, having emissivity as 0.8.

Emissivity of the duct = 0.43 and absorptivity = 0.58.

Solution: As the enclosure is much larger than the duct size, therefore, the emissivity from the enclosure to the duct is ignored and the equation for radiation heat transfer can be expressed as

$$q = A\sigma(\varepsilon_1 T_1^4 - \alpha T_2^4)$$

where A is the surface area of the duct

$$= \pi D = \pi \times \frac{4}{12} = 1.05 \text{ ft}^2$$

σ- Stefan-Boltzmann constant = 0.171×10^{-8}

T_1-absolute temperature of the duct surface - °R

$$= (800 + 460)°R$$

T_2-absolute temperature of the chamber surface - °R

$$= (1800 + 460)°R$$

ε_1-emissivity of the duct surface

α -thermal diffusivity of the enclosure lining.

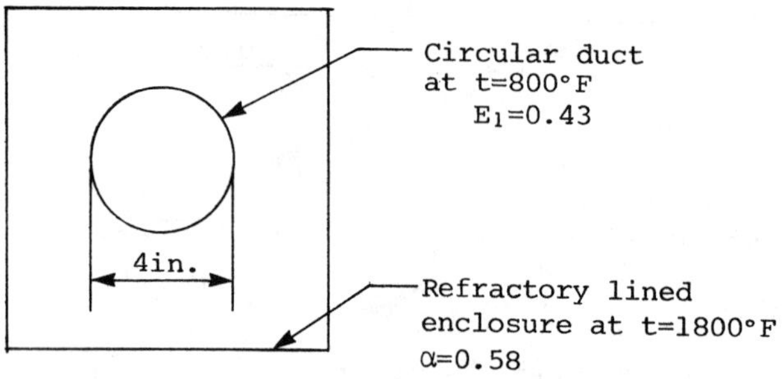

Introducing the known values in the equation yields

$$q = 1.05 \times 0.171 \times 10^{-8}[0.43(800 + 460)^4 - 0.58(1800 + 460)^4]$$

$$= -25,220 \text{ Btu/hr-ft length of the duct}$$

$$= -25,220 \text{ Btu/hr-ft of tube.}$$

(b) When the enclosure is small then the emissivity of the enclosure surface can not be neglected. Therefore, the heat transfer expression is modified and can be written as

$$q = A_1 \sigma f_{12} [T_1^4 - T_2^4]$$

where f_{12} is overall view factor

$$f_{12} = \frac{1}{\frac{1}{\varepsilon_1} - 1 + \frac{A_1}{A_2}\frac{1}{\varepsilon_2} - 1 + \frac{1}{F_{12}}}$$

$F_{12} = 1$, $A_1 = 1.05 \text{ ft}^2$, $A_2 = \frac{8 \times 4}{12} = 2.67 \text{ ft}^2$

$\varepsilon_1 = 0.58$, $\varepsilon_2 = 0.8$

$$f_{12} = \frac{1}{\frac{1}{0.58} - 1 + \frac{1.05}{2.67}\frac{1}{0.8} - 1 + 1}$$

$$= 0.549$$

$\therefore\ q = 1.05 \times 0.171 \times 10^{-8} \times 0.549[(800 + 460)^4 - (1800 + 460)^4]$

$= -23{,}230$ Btu/hr per foot length of the duct.

• PROBLEM 8-24

A 2 in. diameter pipe supplying steam at 300°F is enclosed in a 1 ft square duct at 70°F. The outside of the duct is perfectly insulated. Estimate the radiation heat transfer. Assume the emissivity of pipe surface and duct surfaces as 0.79 and 0.276, respectively.

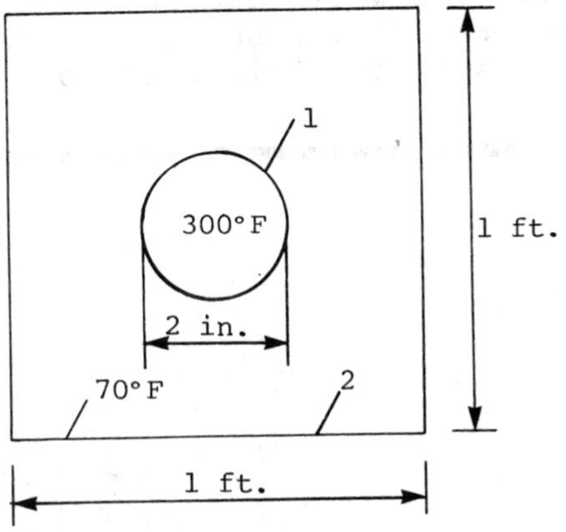

Solution: The radiant heat transfer between the pipe surface and duct can be expressed as

$$Q_{12} = A_1 F_{12}\ \sigma (T_1^4 - T_2^4)$$

where subscripts 1 and 2 refer to pipe and duct, respectively.

A_1 is the surface area per ft length of the pipe

T - absolute temperature in Rankine

f_{12} is the effective emissivity and expressed as

$$f_{12} = \frac{1}{\frac{1}{\varepsilon_1} \quad \frac{A_1}{A_1} \quad \frac{1}{\varepsilon_2} - 1}$$

$$\text{or } f_{12} = \frac{1}{\frac{1}{0.79} + \frac{\pi D1}{4 \text{ x(side) x } 1} \quad \frac{1}{0.276} - 1}$$

$$f_{12} = \frac{1}{\frac{1}{0.79} + \frac{\pi \times 2/12}{4.0} \quad \frac{1}{0.276} - 1}$$

$$= 0.621$$

$\therefore$ Heat transfer rate is

$$Q = \frac{\pi \times 2}{12} \times 0.621 \times 0.171 \times 10^{-8}[(300+460)^4 - (70+460)^4]$$

$$\simeq 142 \text{ Btu/hr per foot length.}$$

• PROBLEM 8-25

> Compute the rate of heat transfer to 2 in. diameter 4 ft long pipes fixed in the 4 ft x 6 ft furnace, as shown in the figure. The temperature of the furnace is 2000°F with emissivity of 0.7. The walls and roof are insulated and are perfectly radiating. The surface temperature of the pipes is 500°F. Assume the emissivity of the pipe surface as 0.8.

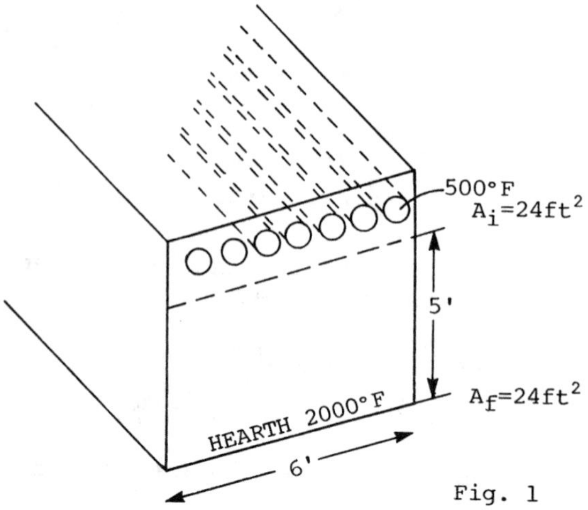

Fig. 1

Solution: The radiant heat transfer between the two surfaces is expressed as

$$q_{12} = A_1 f_{12} \sigma (T_1^4 - T_2^4)$$

where subscripts 1 and 2 refer to furnace floor and pipe surfaces, respectively.

T - the absolute temperature in Rankine

σ is the Stefan-Boltzmann constant $= 0.171 \times 10^{-8}$

A - surface area

f_{12} is the effective emissivity between the furnace floor and pipe surfaces.

The emissivity between floor and pipes f_{12} can be obtained from the equation

$$\frac{1}{A_1 f_{12}} = \frac{1}{A_1}\;\frac{1-\varepsilon_1}{\varepsilon_1} + \frac{1}{A_1 F_{12}} + \frac{1}{A_2}\;\frac{1-\varepsilon_2}{\varepsilon_2}$$

The tubes are idealized as an imaginary plane. Then the emissivity for the imaginary plane and the floor is expressed as

$$f_{if} = \frac{1}{\frac{1}{F_{if}} + \frac{1}{f_{it}} - 1 + 1\;\frac{1}{\varepsilon_f} - 1}$$

The emissivity between the imaginary plane and the tube is given as

$$f_{it} = \frac{1}{\frac{1}{\overline{\overline{F}}_{it}} + \frac{1}{1} - 1 + \frac{2}{\pi}\;\frac{1}{\varepsilon_{wall}} - 1}$$

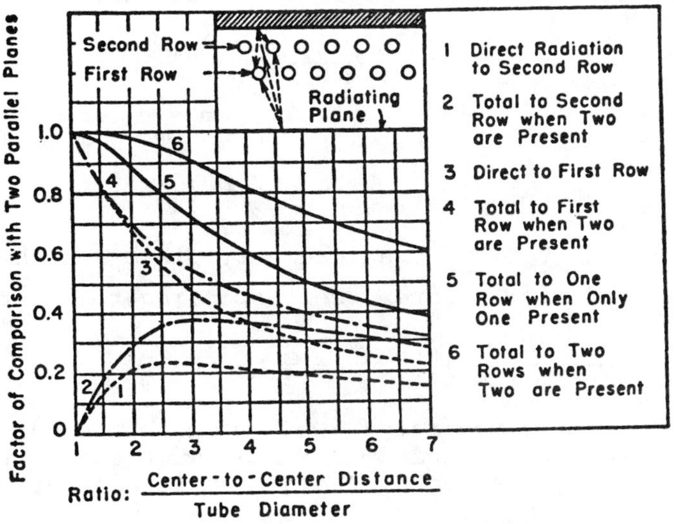

Fig. 2 View factor F and interchange factor F for radiation from plane to one or two rows of tubes above and parallel to the plane (6).

$\overline{\overline{F}}_{it}$ from figure 2 corresponding to

$$\frac{\text{center to center distance}}{\text{tube diameter}} = \frac{4}{2} = 2$$

and case 5, i.e., total to one row when only one present is 0.88

$$f_{it} = \frac{1}{\frac{1}{0.88} + 0 + \frac{2}{\pi}\;\frac{1}{0.8} - 1} = 0.836$$

Introducing this value of f_{it} in the expression for f_{if} yields

$$f_{if} = \frac{1}{\frac{1}{F_{if}} + \frac{1}{0.836} - 1 + 1 \quad \frac{1}{0.7} - 1}$$

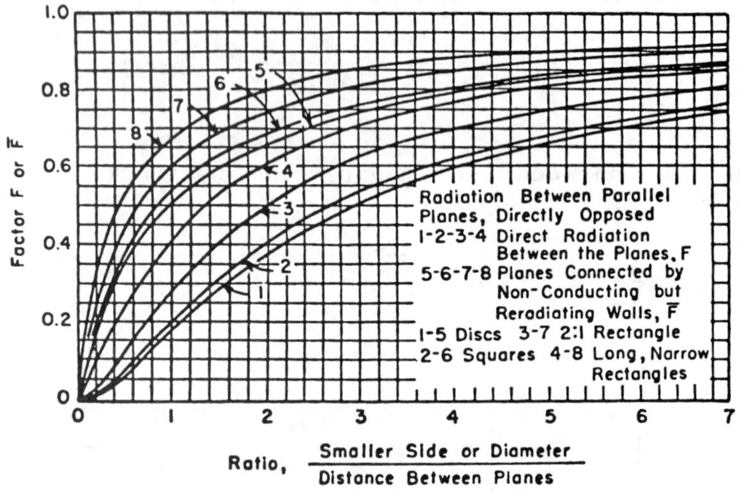

Fig. 3 View factor F and interchange factor F for radiation between parallel planes directly opposed.

F_{if} (view factor between the imaginary plane and the floor) can be otbained from figure 3 corresponding to the ratio

$$\frac{\text{smaller side}}{\text{Distance between planes}} = \frac{4}{5} \quad \text{and} \quad \frac{3}{5}$$

as the rectangle is separated into a 4 ft x 4 ft square and a 3 ft x 6 ft rectangle. And referred to lines 6 and 7, the value for $\bar{F}_{if}$ is obtained by the geometric mean of two values as

$$\bar{F}_{if} = \sqrt{0.48 \times 0.48} = 0.48$$

$$\therefore \quad f_{if} = \frac{1}{\frac{1}{0.48} + \frac{1}{0.836} - 1 + \frac{1}{0.7} - 1}$$

$$= 0.369$$

$$f_{if} = f_{12} = 0.369$$

Now

$$q_{12} = 4 \times 6 \times 0.369 \times 0.171 \times 10^{-8}[(2000 + 460)^4 - (500 + 460)^4]$$

$$= 541730 \text{ Btu/hr}$$

• **PROBLEM 8-26**

Determine the surface area of a radiant heating panel, operating at 80°C, fixed to the ceiling of a room such that the object with dry, non-heat-transferring black surface 200 cm below the ceiling is maintained at a temperature of 32°C. The **ambient air temperature in the room** is 16°C. Assume the heating panel to be black. Use $h_c = 40 W/m^2\text{-}°K$.

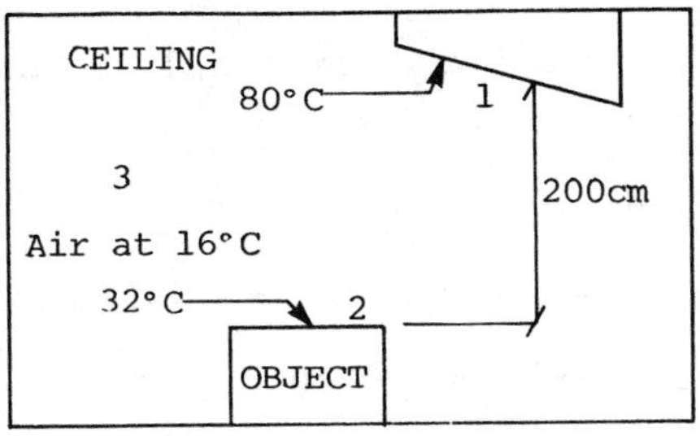

Fig. Heating by radiant heating panel.

Solution: The numbers 1, 2, 3 in the figure correspond to radiant heating panel, object, and room, respectively.

The heat balance equation for the object is rate of heat input is equal to rate of heat output at a steady temperature of 32°C of the object.

$$\dot{Q}_{in} = \dot{Q}_{out}$$

$$\dot{Q}_{in} = \sigma T_1^4 A_1 F_{1-2} + \sigma T_3^4 A_3 F_{3-2}$$

and, $$\dot{Q}_{out} = \sigma T_2^4 A_2 + h_c A_2 (T_2 - T_3)$$

as the rate of heat transfer from the surface of the object of radiation and convection processes.

The surfaces are facing each other and hence

$$A_1 F_{1-2} = A_2 F_{2-1}$$

Also, $$A_3 F_{3-2} = A_2 F_{2-3}$$

Again, $$F_{2-1} + F_{2-3} = 1$$

Introducing the above equations in the heat balance equation and solving for F_{2-1} yields

$$F_{2-1} = \frac{\sigma T_2^4 - \sigma T_3^4 + h_c (T_2 - T_3)}{\sigma T_1^4 - \sigma T_3^4}$$

Substituting the known values in the above equation gives

$$F_{2-1} = \frac{(492 - 396) + 4 \times (32 - 16)}{(882 - 396)}$$

$$= 0.328$$

Now, assuming the shape of the panel to be circular, then the shape factor F_{2-1} is the sine of angle at the panel

i.e., $\sin^2\theta = F_{2-1} = 0.328$

or $\theta = 34.9°$

is the angle subtended at the panel. The panel radius, therefore, must be equal to

$$R = 200 \tan \theta$$

$$= 200 \tan(34.9) = 140 \text{ cm.} = 1.40 \text{ m.}$$

and the panel area $\pi r^2 = \pi \times (140)^2$

$$= 6.157 \times 10^4 \text{ cm}^2$$

$$= 6.157 \text{ m}^2$$

• **PROBLEM 8-27**

Air flows through a conduit having a wall temperature of 425°C and a glass thermometer, as shown in the figure. The diameter of the thermometer bulb is 6.5 mm. and emissivity 0.96. Assuming the heat transfer coefficient between the thermometer and air as $h=100 W/m^2.K$, compute the air temperature, in the duct, if the thermometer reads 150°C. Also, estimate the thermometer reading if the bulb is covered with a foil having emissivity as 0.03.

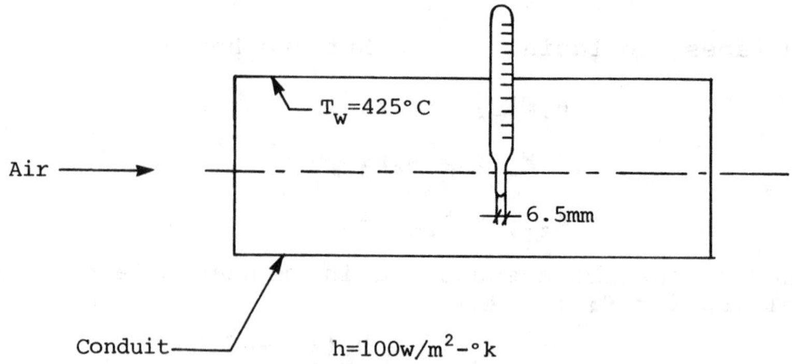

Fig. 1 Air Conduit

Solution: The heat transfer between thermometer bulb and the air in the conduit occurs by radiation. The view factor between the thermometer bulb and the conduit is unity.

The general equation for view factor is expressed as

$$f_{12} = \frac{1}{\frac{1}{F_{12}} + \left[\frac{1}{\varepsilon_1} - 1\right] + \left[\frac{A_1}{A_2}\right]\left[\frac{1}{\varepsilon_2} - 1\right]} \quad (1)$$

where f_{12} is the overall view factor

F_{12} is the view factor for surfaces 1 and 2

ε is the emissivity of the surface

A is the surface area

Subscripts 1 and 2 correspond to thermometer bulb and duct, respectively.

As $A_1 << A_2$, the term containing $\frac{A_1}{A_2}$ is very small, and hence reduces to zero. Substituting the known values in the term yields

$$f_{12} = \frac{1}{\frac{1}{1} + \left[\frac{1}{\varepsilon_1} - 1\right] + 0}$$

or $$f_{12} = \frac{1}{1 + \frac{1}{0.96} - 1} = 0.96$$

The air temperature is given by

$$T_a = T_1 - \frac{h_r f_{12} (T_2 - T_1)}{h_a}$$

where subscript 'a' refers to air.

h_r is the radiation heat transfer coefficient and can be obtained from the figure (2) corresponding to the temperatures 425°C and 150°C.

$$h_r = 43 \text{ W/m}^2.\text{K}.$$

$$\therefore \quad T_a = 150 - \frac{43 \times 0.96 (425 - 150)}{100}$$

$$= 36°\text{C}$$

Thus, the air temperature is 36°C.

Radiant heat transfer coeficient.
Temparature of other surfaces:

a = 538°c
b = 427°c
c = 316°c
d = 204°c
e = 93°c
f = -18°c

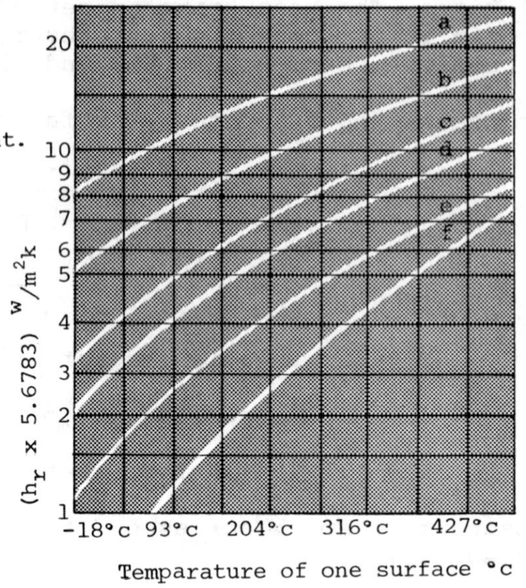

Again, when the thermometer bulb is covered by a foil having emissivity of 0.03, then

$$f_{12} = 0.03$$

Introducing the known values in equation (1) yields

$$36 = T_1 - \frac{h_r(0.03)}{100}(425 - T_1)$$

Let $T_1 = 40°C$, then from figure (2)

$$h_c = 34 \text{ W/m}^2\text{K.}$$

or

$$36 = 40 - \frac{(34)(0.03)(425 - 40)}{100}$$

$$= 40 - 3.9$$

$$= 36.1$$

Thus, left hand side and right hand side are equal, i.e., the assumed value of thermometer reading is correct, i.e., 40°C.

The interior surfaces of a long rectangular duct, used for heating, are gray. Three surfaces are diffusive surfaces while the fourth is a specular (regular) surface. The emissivities, shape factors and the temperatures of the surfaces are shown in the figure. Calculate the rate of radiant heat transfer at the specular surface.

System with side 4 as specular diffusion surface

sides 1, 2 & 3 as diffusive diffusion surfaces.

Dashed lines represent mirror images of the surfaces 1, 2 and 3 in surface 4.

$T_1 = 800°F$ $\varepsilon_1 = 0.8$ $F_{23} = F_{12} = F_{21} = F_{32} = F_{34} = F_{43} = F_{41} = F_{14} = 0.3$

$T_2 = 600°F$ $\varepsilon_2 = 0.7$ $F_{12(4)} = F_{21(4)} = F_{22(4)} = F_{23(4)}$

$= F_{32(4)} = 0.1$

$T_3 = 400°F$ $\varepsilon_3 = 0.6$ $F_{13(4)} = F_{31(4)} = 0.2$

$T_4 = 200°F$ $\varepsilon_4 = 0.1$ $F_{24} = 0.41 = F_{42}$

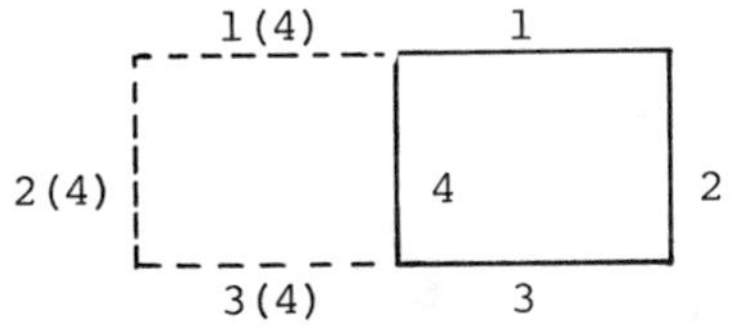

System with side 4 as specular diffusion surface
Sides 1,2, and 3 as diffusive surfaces.
Dashed lines represent mirror images of the surfaces 1,2, and 3 in surface 4.

$T_1 = 800°F$ $\varepsilon_1 = 0.8$ $F_{23} = F_{12} = F_{21} = F_{32} = F_{34} = F_{43} = F_{41} = F_{14} = 0.3$

$T_2 = 600°F$ $\varepsilon_2 = 0.7$ $F_{12(4)} = F_{21(4)} = F_{22(4)} = F_{23(4)} = F_{32(4)} = 0.1$

$T_3 = 400°F$ $\varepsilon_3 = 0.6$ $F_{13(4)} = F_{31(4)} = 0.2$

$T_4 = 200°F$ $\varepsilon_4 = 0.1$ $F_{24} = 0.41 = F_{42}$

Solution: If the radiant energy which is reflected from the surface leaves at the same angle to the surface as the incident radiation made with the surface, the surface is called a specular(regular) diffusion surface.

If the reflected radiant energy leaves the surface in all directions with no preferential direction, the surface is called a <u>diffusive-diffusion surface.</u>

The radiation shape factors for the system are given as:

$$F_{1-2} = F_{2-1} = F_{2-3} = F_{3-2} = F_{3-4} = F_{1-4} = 0.30$$

$$F_{1-2(4)} = F_{2-1(4)} = F_{2-2(4)} = F_{2-3(4)} = 0.10$$

$$F_{1-3(4)} = F_{3-1(4)} = 0.20$$

$$F_{2-4} = F_{3-1} = F_{1-3} = 0.41$$

The energy radiated per unit time and per unit area by each surface is calculated by:

$$E = \sigma T^4$$

where σ = Stefan-Boltzmann constant = 0.1714×10^{-8} Btu/hr.ft^2°R^4

$E_1 = (0.1714 \times 10^{-8})(800°F + 460°R)^4 = 4320$ Btu/hr.ft^2

$E_2 = (0.1714 \times 10^{-8})(600°F + 460°R)^4 = 2164$ Btu/hr.ft^2

$E_3 = (0.1714 \times 10^{-8})(400°F + 460°R)^4 = 938$ Btu/hr. ft^2

$E_4 = (0.1714 \times 10^{-8})(200°F + 460°R)^4 = 325$ Btu/hr. ft^2

The total radiation incident upon surfaces 1, 2 and 3 per unit time and per unit area can be represented by the following term:

$$G_i = \sum_{j=1}^{n} J_j A_j \quad F_{i-j} + \rho_4 F_{i-j(4)} \quad + E_{b4} F_{i-4} \qquad (1)$$

where G = irradiation

J = radiosity

ρ = reflectivity

The reflectivity of each surface may be obtained by

$$\rho = 1 - \alpha$$

where α is the absorptivity and is equal to the emissivity ε in this case.

$\varepsilon = \alpha$, this equation is called the Kirchoff's identity, and is valid only for the following conditions:

(1) The temperature T_i of the incident source is equal to the temperature T_s of the absorbing or emitting surface.

(2) When ε and α do not depend on wavelength (for non-black bodies radiation).

(3) Both the absorptance at T_s and the emittance at T_i of the same surface are equal, given that the incident beam is proportional to the wavelength range as a black body and the emittance does not depend on temperature.

(4) For cases involving metal surfaces, the absorptance at T_s is equal to the emittance at $T = \sqrt{T_i T_s}$

Condition (2) is valid for this problem, therefore,

$\alpha_1 = \varepsilon_1 = 0.8$, $\alpha_2 = \varepsilon_2 = 0.7$, $\alpha_3 = \varepsilon_3 = 0.6$, $\alpha_4 = \varepsilon_4 = 0.1$

and $\rho_1 = 1 - \alpha_1 = 1 - 0.8 = 0.2$, $\rho_2 = 1 - \alpha_2 = 1 - 0.7 = 0.3$

$\rho_3 = 1 - \alpha_3 = 1 - 0.6 = 0.4$, $\rho_4 = 1 - \alpha_4 = 1 - 0.1 = 0.9$

Using equation (1), the irradiation for surfaces 1, 2, and 3 is:

$$G_1 = J_1(F_{1-2} + \rho_4 F_{1-1(4)}) + J_2(F_{1-2} + \rho_4 F_{1-2(4)}) + J_3(F_{1-3} + \rho_4 F_{1-3(4)}) + E_4 F_{1-4}$$

$$G_2 = J_1(F_{2-1} + \rho_4 F_{2-1(4)}) + J_2(F_{2-2} + \rho_4 F_{2-2(4)} + J_3(F_{2-3} + \rho_4 F_{2-3(4)}) + E_4 F_{2-4}$$

$$G_3 = J_1(F_{3-1} + \rho_4 F_{3-1(4)}) + J_2(F_{3-2} + \rho_4 F_{3-2(4)}) + J_3(F_{3-3} + \rho_4 F_{3-3(4)}) + E_4 F_{3-4}$$

Note that the area terms dropped out, since all four surfaces are equal in size.

Substituting in the corresponding values yields:

$$G_1 = J_1[0 + 0.9(0)] + J_2[0.30 + 0.9(0.10)] + J_3[0.41 + 0.9(0.20)] + (325)(0.30)$$

$$G_2 = J_1[0.30 + 0.9(0.10)] + J_2[0 + 0.9(0.10)] + J_3[0.30 + 0.9(0.10)] + (325)(0.41)$$

$$G_3 = J_1[0.41 + 0.9(0.20)] + J_2[0.30 + 0.9(0.10)] + J_3[0 + 0.9(0)] + (325)(0.30)$$

Next, replace the irradiation in the radiosity equations with the irradiation relations:

$$J_1 - E_1 = \rho_1 G_1, \quad J_2 - E_2 = \rho_2 G_2, \quad J_3 - E_3 = \rho_3 G_3$$

Simplifying the substitutions, we obtain:

$$J_1 - 4320 = 0.2(0.39J_2 + 0.59J_3 + 97.5)$$

$$J_2 - 2164 = 0.3(0.39J_1 + 0.09J_2 + 0.39J_3 + 133.25)$$

$$J_3 - 938 = 0.4(0.59J_1 + 0.39J_2 + 97.5)$$

Simplifying again yields:

$$J_1 - 0.078J_2 - 0.118J_3 = 4339.5$$

$$-0.117J_1 + 0.973J_2 - 0.117J_3 = 2204$$

$$-0.236J_1 - 0.156J_2 + J_3 = 977$$

Solving the three equations simultaneously for J_1, J_2 and J_3 yields:

$$J_1 = 4896 \text{ Btu/hr. ft}^2$$

$$J_2 = 3168 \text{ Btu/hr. ft}^2$$

$$J_3 = 2626 \text{ Btu/hr. ft}^2$$

Substituting the radiosity values into the radiosity equations and solving for the irradiations yields:

$$G_1 = \frac{J_1 - E_1}{\rho_1} = \frac{4896 - 4320}{0.2} = 2880 \text{ Btu/hr. ft}^2$$

$$G_2 = \frac{J_2 - E_2}{\rho_2} = \frac{3168 - 2164}{0.3} \cong 3347 \text{ Btu/hr. ft}^2$$

$$G_3 = \frac{J_3 - E_3}{\rho_3} = \frac{2626 - 938}{0.4} = 4220 \text{ Btu/hr. ft}^2$$

The energies radiated to surfaces 1, 2, and 3 are

$$q_1'' = G_1 - J_1 = 2880 - 4896 = -2016 \text{ Btu/hr. ft}^2$$

$$q_2'' = G_2 - J_2 = 3347 - 3168 = +179 \text{ Btu/hr. ft}^2$$

$$q_3'' = G_3 - J_3 = 4220 - 2626 = 1594 \text{ Btu/hr. ft}^2$$

The energy radiated to surface 4 is given by

$$q_4'' = \alpha_4 G_4 - E_4$$

where $G_4 = F_{4-1}J_1 + F_{4-2}J_2 + F_{4-3}J_3$

Substituting the values,

$$G_4 = (0.3)\,4896 + (0.41)\,3168 + (0.3)\,2626$$

$$= 3555 \text{ Btu/hr. ft}^2$$

$$\therefore q_4^{11} = (0.1)\,3555 - 325$$

$$= 30.5 \text{ Btu/hr. ft}^2$$

HEAT RADIATION BETWEEN GASES AND REGULAR SURFACES

• PROBLEM 8-29

Compute the temperature of the filler metal point which is enveloped by an inert gas at 80°F. The area of the point and its emissivity are 0.014 ft^2 and 0.88, respectively. Assume the convective heat transfer coefficient between point and the inert gas as 2.0 Btu/hr^2 ft °F. The rate of heat generation at the point is 68.2 Btu/hr^2

Solution: The heat generated at the tip of the electrode is dissipated by convection and radiation processes. By drawing a heat balance, assuming steady state heat transfer, the following equation results:

Heat Generated = Heat dissipated

$$\begin{bmatrix}\text{Heat}\\\text{generated}\end{bmatrix} = \begin{bmatrix}\text{Radiant} & + & \text{convective}\\\text{heat transfer} & & \text{heat transfer}\end{bmatrix}$$

i.e. $$q_g = q_r + q_c$$

The radiation heat transfer is expressed as, by Stefan-Boltzmann Law,

$$q_r = \varepsilon_1 A_1 \sigma(T_1^4 - T_2^4)$$

where subscript 1 correspond to the tip of the electrode and 2 refers to surroundings or inert gas.

The convection heat transfer is given as

$$q_c = hA_1(T_1 - T_2)$$

Rewriting the heat balance equation,

$$68.2 = \varepsilon_1 A_1 \sigma(T_1^4 - T_2^4) + hA_1(T_1 - T_2)$$

$$\text{or } 68.2 = 0.88 \times 0.014 \times 0.171 \times 10^{-8}\ [T_1^4 - (80 + 460)^4]$$

$$+ 2.0 \times 0.014\ [T_1 - (80 + 480)]$$

$$= 2.1 \times 10^{-11}[T_1^4 - (540)^4] + 0.028\ [T_1 - 540]$$

$$2.1 \times 10^{-11} T_1^4 + 0.028\, T_1 - 16.9 = 68.2$$

$$\text{or} \quad 2.1 \times 10^{-11} T_1^4 + 0.028\, T_1 - 85.1 = 0$$

$$\text{or} \quad T_1^4 + 1.33 \times 10^9 T_1 - 4.05 \times 10^{12} = 0$$

Solving the equation by trial and error, the value of T_1 is obtained as

$$T_1 \simeq 1244°R$$

or temperature of the tip of the electrode is 1244 - 460 = 784°F.

• PROBLEM 8-30

A ventilating duct with non-gray interior surfaces, at temperature T_w=600°K, has a uniform thickness of 12 mm and absorption coefficient 120/m. A vapor at 1100°K is flowing through the channel. The black body radiosity of the vapor increases linearly, at each wall over the boundary layer thickness of 2 mm, and reaches a constant value at the outlet of the duct. Calculate the resultant radiant heat transfer at the wall.

Solution: The equation which represents the radiant energy flow from an elemental surface area is:

$$q(t) = \left[q_0 - J(0)\right]2E_3(t) - \left[q_L - J(t_L)\right] 2E_3(t_L - t) + \int_0^{t_L} 2E_3(|t - t'|) - \frac{dJ}{dt'}\, dt' \qquad (1)$$

where q_0 and q_L are at L=0 and L=12 mm, respectively.

$q(t)$ = net radiation per unit time

$J(0)$ - radiaiton leaving the surface initially

$J(t_L)$ = radiation leaving the surface at L=.012m per unit time t

$\frac{dJ}{dt'}$ = the change in radiosity with respect to film thickness at time t

$$= \frac{J_r - J_w}{t_\delta}$$

Since the interior surfaces will absorb all the radiant energy and no sudden temperature change occurs at the wall, we obtained:

$$q_0 - J(0) = 0 \text{ and } q_L - J(t_L) = 0$$

Therefore, the radiation at t=0 is:

$$q(0) = \int_0^{t_\delta} 2E_3(t') \left[- \frac{J_r - J_w}{t_\delta}\right] dt' \int_{t_\delta}^{t_L - t_\delta} 2E_3(t')(0)dt'$$

$$+ \int_{t_L - t}^{t_L} 2E_3(t') \left[- \frac{J_r - J_w}{t}\right] dt'$$

Integrating, we obtained:

$$- q(0) = \varepsilon_{duct} (J_r - J_w) \qquad (2)$$

where ε_{duct} is the emissivity of the duct and is represented as:

$$\varepsilon_{duct} = \frac{1}{t_L \alpha} \frac{2}{3} \left[1 - 3E_4(t_\delta) \right] - 2\left[E_4 \; t_L - t_\delta - E_4(t_L) \right]$$

and E = the emissive power of the duct

$$= \sigma T^4 = (5.67 \times 10^{-8} W/m^2.)(600K)^4 = 7.35 \times 10^3 W/m^2$$

$$\varepsilon_{duct} = \frac{1}{(120 \times 0.012)} \frac{2}{3} \left[1 - 3(7.35 \times 10^3)(.002) \right]$$

$$- 2 \left[7.35 \times 10^3 (.012 - .002) - 7.35 \times 10^3 (.012) \right]$$

$$= 0.463$$

Substituting the values into eq. (2) yields

$$- q(0) = 0.463 \left[(1100K^4 - 600K^4)(5.67 \times 10^{-8}) \right]$$

$$= 3.5 \times 10^4 W/m^2$$

• PROBLEM 8-31

Estimate the velicity of flue gases flowing through a 6 in. diameter and 36 in. long duct such that its (flue gas) temperature drops by 200°F from 1200°F. The outside of the duct is perfectly insulated, the average temperature and emissivity of the inside surface are 900°F and 0.8, respectively. The contents of the flue gas and the specific heat of the gas are given below. Assume heat exchange occurs by radiation alone.

$CO_2 \longrightarrow$ 10%

(Vapor) $H_2O \longrightarrow$ 20%

$C_p \longrightarrow$ 0.007 Btu/ft³-°F.

Solution: The radiant heat transfer depends upon the rate of flow for a given set of temperatures for the tube.

The net radiation heat flux from gas to duct wall is given by

$$A\varepsilon_{av}\sigma[E_{(c+w)1}T_1^4 - E_{(c+w)2}T_2^4]$$

where ε_{av} is the average emissivity $= \frac{0.8 + 1}{2} = 0.9$

σ - Stefan Boltzmann constant $= 0.1714 \times 10^{-8}$

T - Absolute temperature in Rankine

$E_{(c+w)}$ - effective emissivity of the flue gas at a temperature

$$p_w \ell = 0.2 \times 0.4 = 0.08 \text{ ft-atm.}$$

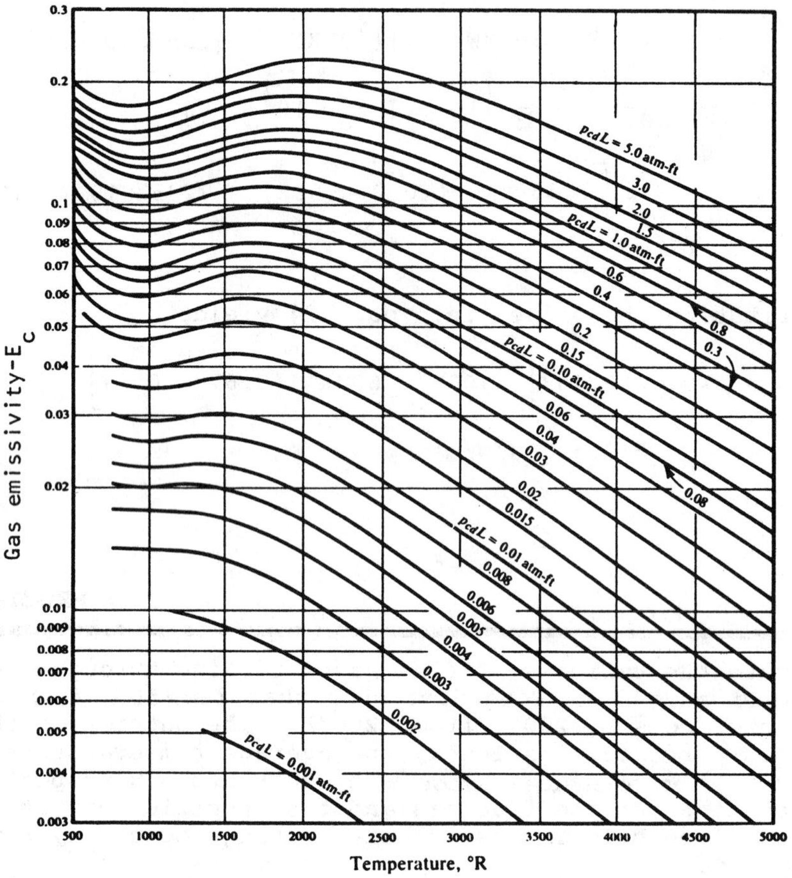

Fig. 1 Emissivity of carbon dioxide for total pressure 1 atmosphere.

From figure 1, the emissivity of carbon dioxide at a mean temperature of gas $\frac{1200 + 1000}{2} = 1100°F$ and of duct 900°F can be read as

$$E_c(1100) = 0.057 \text{ and}$$

$$E_c(900) = 0.055$$

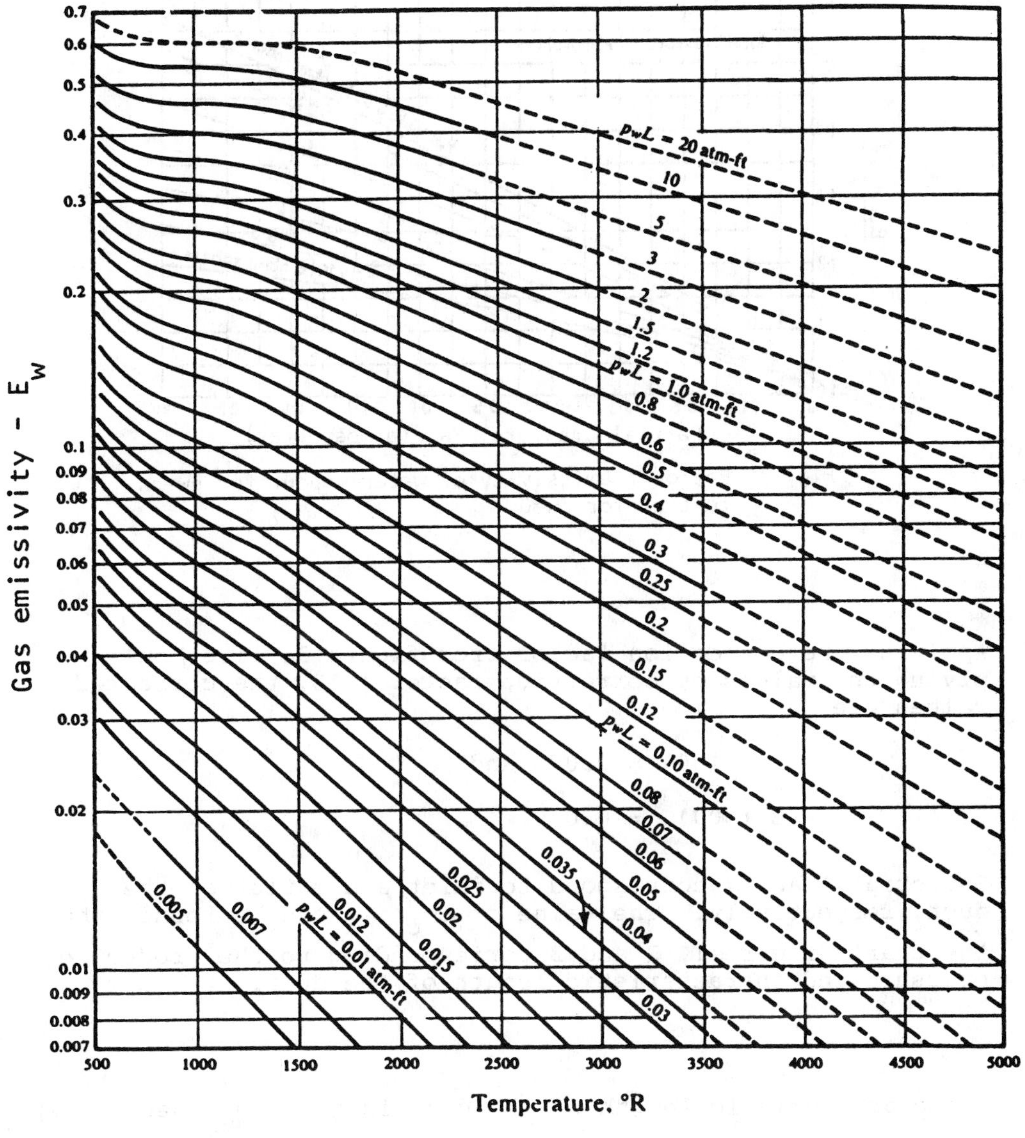

Fig. 2 Emissivity of water vapour for total pressure 1 atmosphere.

Similarly, for water vapor, from figure 2.

$$E_W(1100) = 0.061 \text{ and}$$

$$E_W(900) = 0.069$$

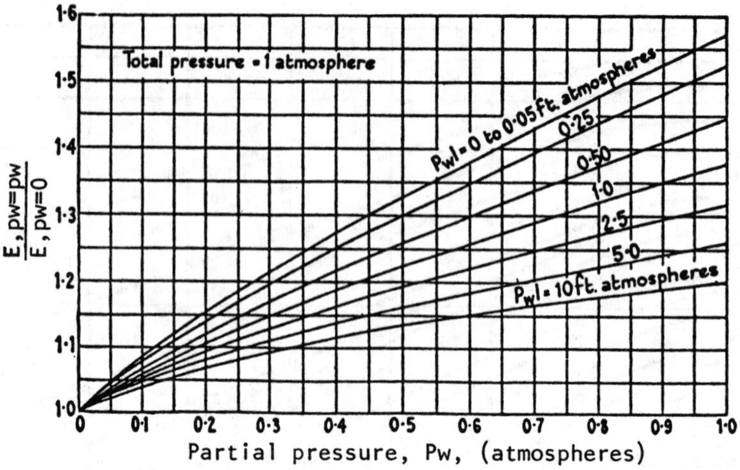

Fig. 3 Ratio of emissivity of water vapour for pw=pw to that for pw=0

Applying the correction factor from figure 3., i.e., multiplying the values by correction factor 1.15, the corrected values are

$$E_w(1100) = 0.07 \quad \text{and}$$

$$E_w(900) = 0.079$$

Subscripts 1 and 2 correspond to inlet and outlet of the duct, respectively. The value of $E_{(c+w)}$ can be found from the charts figures 1, 2 and 3 corresponding to the product of pressure and characteristic length of the duct.

The characteristic length of the duct is given by (see table)

$$\frac{3.4 \text{ x volume}}{\text{surface area}} = \frac{3.4 \times \pi r^2 \ell}{2\pi r\ell + 2\pi r^2}$$

$$= \frac{1.7 r\ell}{(\ell + r)} = \frac{1.7 \times 6 \times 36}{(36 + 6)} = 4.8 \text{ in.}$$

or 0.4 ft

Now, the product of the partial pressure of the gas and the characteristic length can be evaluated.

$$p_c = 0.1 \times 0.4 = 0.04 \text{ ft atm.}$$

Equivalent Thickness l for Non-Luminous Gas Radiation Layers of Different Shapes

Shape	Characteristic dimension Z	Factor by which Z is to be multiplied to give equivalent l for hemispherical radiation: Calculated by various workers	3·4 × (volume/area)
Sphere	Diameter	0·60	0·57
Cube	Side	0·60	0·57
Infinite cylinder radiating to walls	Diameter	0·90	0·85
Ditto, radiating to centre of base	Diameter	0·90	0·85
Cylinder, height = diameter, radiating to whole surface .	Diameter	0·60	0·57
Ditto, radiating to centre of base	Diameter	0·77	0·57
Space between infinite parallel planes	Distance apart	1·80	1·70
Space outside infinite bank of tubes with centres on equilateral triangles, tube diameter = clearance . .	Clearance	2·80	2·89
Ditto, but tube diameter = one-half clearance . .	Clearance	3·80	3·78
Ditto, with tube centres on squares, and tube diameter = clearance	Clearance	3·50	3·49
Rectangular parallelepiped, 1×2×6 radiating to:			
2×6 face . . .	Shortest edge	1·06	1·01
1×6 face	,,	1·06	1·05
1×2 face	,,	1·06	1·01
all faces	,,	1·06	1·02
Infinite cylinder of semicircular cross-section radiating to centre of flat side . .	Diameter	0·63	0·52

The total emissivity of the gas at 1100°F and 900°F can be obtained by combining the individual constituent of gas emissivities,

$$E_{(c+w)}(1100) = 0.055 + 0.07 = 0.127$$

$$E_{(c+w)}(900) = 0.057 + 0.079 = 0.134$$

Introducing the values in the equation for net radiant heat transfer yields

$$q = 0.9 \times 0.173 \times 10^{-8} \times \pi \frac{6}{12} \times \frac{36}{12} \times \left[0.127 \times (1560)^4 - 0.134(1360)^4\right]$$

$$= 2155 \text{ Btu/hr.}$$

Let v be the velocity of flow of the flue gas. Then heat transfered from the gas for a temperature drop of 200°F is

$$C_p Q \times \Delta T$$

$$= 0.007 \times \pi r^2 v \times 200$$

which should be equal to the net radiant heat transfer

$$\therefore\ 2140 = 0.007 \times \pi r^2 v \times 200$$

or
$$v = \frac{2140}{200 \times 0.007 \times (0.4)^2}$$

$$v = 3040.9 \text{ ft/hr}^2$$

• PROBLEM 8-32

A 6 mm thick material at a temperature of 627°C having thermal conductivity of 2 W/m-°K and surface emissivity of 0.48 is being heated up in a 6 m diameter furnace. The roof of the furnace is covered with insulating bricks, of area $50m^2$. The furnace is heated by air flowing at the rate of 2.7 kg/sec. and fuel flowing at the rate of 0.15 kg/sec. Both air and fuel are injected into the furnace at a temperature of 327°C. Given the emissivity of the hot gases in the furnace as 0.25, estimate the temperature of the gas and the material. Also, determine the net exchange of heat flux. The specific heat of hot gases is 1200 J/kg-°K.

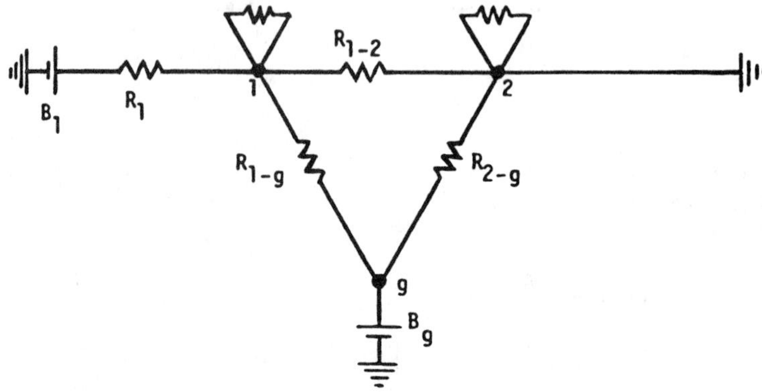

Fig. 1 Radiation network for n = 2 with isothermal gas g.

Solution: The hot gas is the source of the heat, and the material being heated up is the sink. The problem is represented in the form of a radiation network as shown in the figure. The source is represented with subscript g and sink with subscript 1. The refractory surface is represented with subscript 2.

The equivalent conductance, with refractory wall as radiatively adiabatic, can be expressed as

$$A\, f_{1-g} = \frac{1}{\dfrac{1-\varepsilon_1}{\varepsilon_1 A_1} + \left[\dfrac{1}{\dfrac{1}{R_{1-g}} + \dfrac{1}{R_{1-2} + R_{2-g}}}\right]}$$

where f_{1-g} is the equivalent conductance.

A surface area $= \frac{\pi}{4} \times 6^2 = 28.3\ m^2$

ε emissivity of the surface

$$R_{1-g} = \frac{1}{A_1 \varepsilon_{g,1-2}}$$

$$R_{2-g} = \frac{1}{A_1 F_{2-1} \varepsilon_{g,2-1} + A_1 F_{2-2} \varepsilon_{g,2-2}}$$

$$R_{1\ 2} = \frac{1}{A_1 F_{1-2}\ \tau_{g,1-2}}$$

where $\tau_{g,1-2}$ is the transmissivity of the gas to the surfaces 1 and 2.

$$= 1-\varepsilon_{g,1-2} = 1 - 0.25 = 0.75$$

shape factor $F_{1-2} = 1,\ F_{2-1} = 1,\ F_{2-2} = 0$

Introducing the known values in the equation yields

$$A\, f_{1-g} = \frac{1}{\dfrac{1-0.48}{(0.48)(28.3)} + \dfrac{1}{\dfrac{1}{(28.3)(0.25)} + \dfrac{1}{\dfrac{1}{(28.3)(0.75)} + \dfrac{1}{(50)(0.25)}}}}$$

$$= 6.13\ m^2$$

Now, the heat input to the furnace is equal to the heat transfer to the gas and the material. The heat input rate is given by

$$\dot{m}\, C_p\ (T_{gf} - T_g)$$

where $\dot{m}$ is net mass rate of flow

$$(2.7 + 0.15) = 2.85\ kg/sec$$

C_p - specific heat of gas = 1200 J/Kg-°K.

T_{gf}- temperature of the gaseous fuel when fired

$$= 500 + \frac{0.15 \times 5 \times 10^7}{\dot{m}C_p}$$

$$= 500 + \frac{0.15 \times 5 \times 10^7}{28.5 \times 1200} = 2693°K$$

Heat absorbed by the material

$$A_1 f_{1-g} (\sigma T_g^4 - \sigma T_1^4)$$

Thus, by drawing a heat balance,

$$\dot{m}C_p(T_{gf} - T_g) = A_1 f_{1-g}(\sigma T_g^4 - \sigma T_1^4)$$

or $\quad 3420(2693 - T_g) = A_1 f_{1-g} \sigma (T_g^4 - T_1^4) \qquad (1)$

Again, by drawing a heat balance for the material, i.e., heat radiated to the surface of the material is equal to heat conducted through it,

$$A_1 f_{1-g} \sigma (T_g^4 - T_1^4) = \frac{kA_1}{\delta} (T_1 - T_0)$$

where k is thermal conductivity of the material

$$k = 2 \text{ W/m-°K}$$

δ - thickness of the material = 6 mm or 0.006

$$\frac{kA_1}{\delta} = \frac{2 \times 28.3}{0.006} = 9433 \text{ W/K}$$

$A_1 f_{1-g} \sigma (T_g^4 - T_1^4) = 9433(T_1 - T_0) \qquad (2)$

where T_0 = 900°K

Equations (1) and (2) can be solved for temperature of the gas T_g and temperature of the surface of the material T . The equations are solved by iteration method by assuming the value of radiation heat transfer from gas to material surface $q_{g-1} = A_1 f_{1-g} \sigma (T_g^4 - T_1^4)$ and then using equation (1), the value of T_g is found. This value is then plugged in equation (2), and T_1 is evaluated. Then q_1 is evaluated from the calculated values of T_1 and T_g. The process is repeated until the calculated value of q_1 equals the assumed value of q_1. The values of iterations are as shown in the table.

<u>Table for Values of Iteration</u>

q_a (assumed)	Temperature of the gas T_g	Temperature of the wall T_1	q_e (calculated)	Difference $(q_c - q_a)$
3×10^6	1815.8	1218	3.0135×10^6	+13500
4×10^6	1523.4	1324	8.0391×10^6	$+4.039 \times 10^6$
3.2×10^6	1757.3	1239.2	2.4994×10^6	-0.7006×10^6
3.003×10^6	1814.9	1218.34	3.0052×10^6	+2200
3.004×10^6	1814.6	1218.44	3.0024×10^6	-1600
3.0035×10^6	1814.78	1218.39	3.0041×10^6	+600

Each time the T_g is calculated, the corresponding value of gas emissivity t_g is used in the subsequent calculation.

• PROBLEM 8-33

A gas mixture, 15% water vapor and 65% N_2 by volume, is introduced, at a pressure of 3 kg/cm and temperature 1273°K, between two parallel plates 0.7m apart. The temperatures of the plates are 473°K and 773°K. Calculate the radiant heat transfer at each plate.

473°K = T_1

Gas mixture at 3 Kg/cm² and 1273°K = T_g 70cm

Fig. 1 773°K = T_2

<u>Solution</u>: The rate of heat transfer by each flat surface is given by

$$q = GA - E_b A$$

or

$$q_1 = G_1 A_1 - E_{b_1} A_1 \text{ and}$$

$$q_2 = G_2 A_2 - E_{b_2} A_2$$

where subscripts 1 and 2 refer to the parallel surfaces.

G is the total heat energy available at the flat surface, i.e., sum of gas radiation energy plus radiation energy from the parallel opposite surface

A is the surface area

E_b is the radiant energy of the flat surface

The rate of heat transfer per unit area is

$$\frac{q_1}{A_1} = G_1 - E_{b_1} \text{ and}$$

$$\frac{q_2}{A_2} = G_2 - E_{b_2}$$

The radiation energy of the gas and flat surfaces can be computed using the Stefan-Boltzmann law

$$E_b = \sigma T^4$$

For the gas, $= \sigma T_g^4 = 5.67 \times 10^{-8}(1273)^4$

or $E_{bg} = 148.9 \text{ kW/m}^2$

For surface 1 $E_b = 5.67 \times 10^{-8} \ (473)^4$

$= 2.84 \text{ kW/m}^2$

and for surface 2 $E_{b_2} = 5.67 \times 10^{-8} \times (773)^4$

$= 20.24 \text{ kW/m}^2$

The equivalent beam length is given by the relation

$$L_e = 1.8L$$

where L is the gap between the flat surfaces = 0.7 meter

$$\therefore L_e = 1.8 \times 0.7 = 1.26 \text{ meter}$$

The partial pressure of individual gases in the flue gas mixture is evaluated as

for $CO_2 \longrightarrow p_c = (0.2) \times 3 \times (1.0132 \times 10^5)$

$= 60.8 \text{ kN/m}^2$

and for H O $p_w = 0.15 \times 3 \times (1.0132 \times 10^5)$

$= 45.6 \text{ kN/m}^2$

Partial pressure - beam length parameter for CO_2 is
$p_c L_e = 60.8 \times 1.26 = 76.6 \text{ kN/m}$ and

for H_2O is $p_w L_e = 45.6 \times 1.26 = 57.5 \text{ kN/m}$

It may be mentioned that the nitrogen and other non-polar, symmetrical molecules are essentially transparent at these temperatures, while CO_2 and H_2O and various hydrocarbon gases radiate to an appreciable extent.

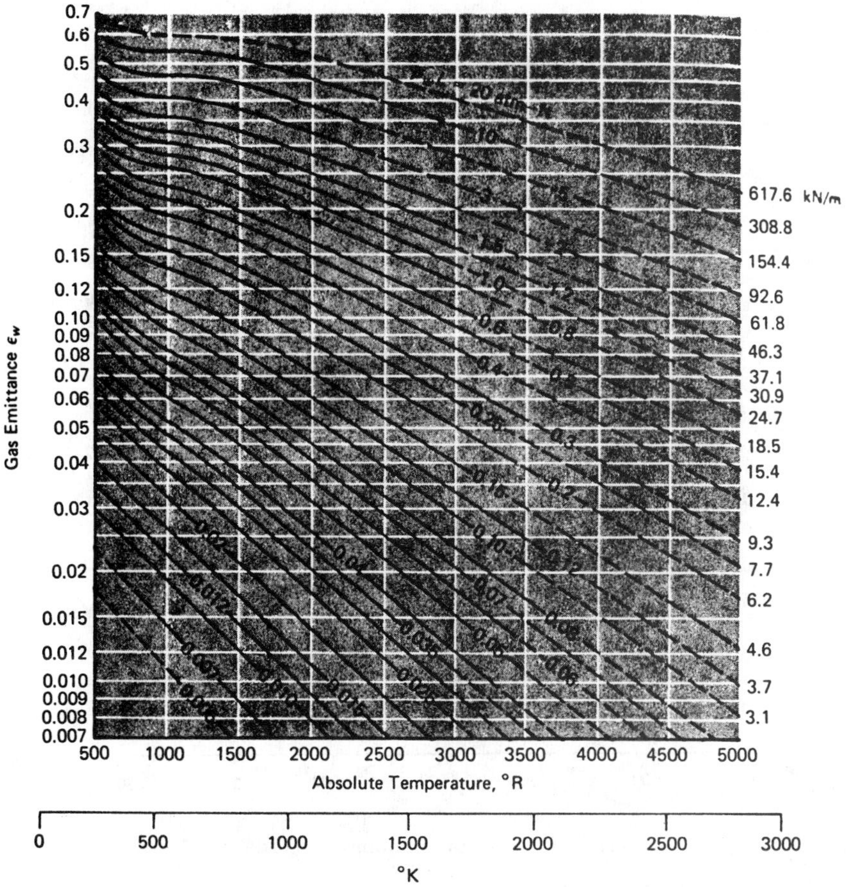

Fig. 2 Emissivity for water vapor at 1 atm.

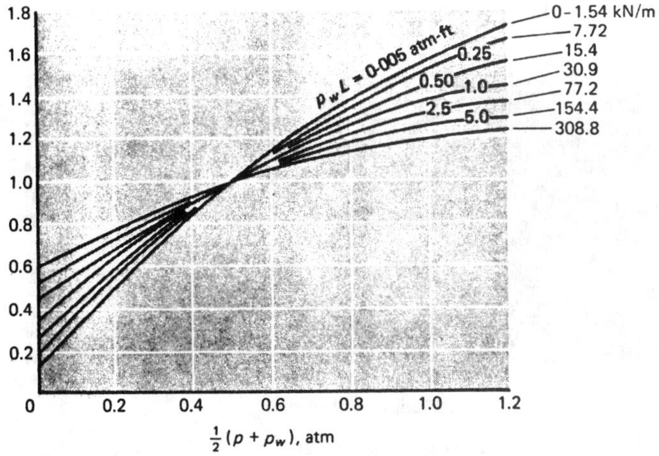

Fig. 3 Correction factor for emissivity of water vapor.

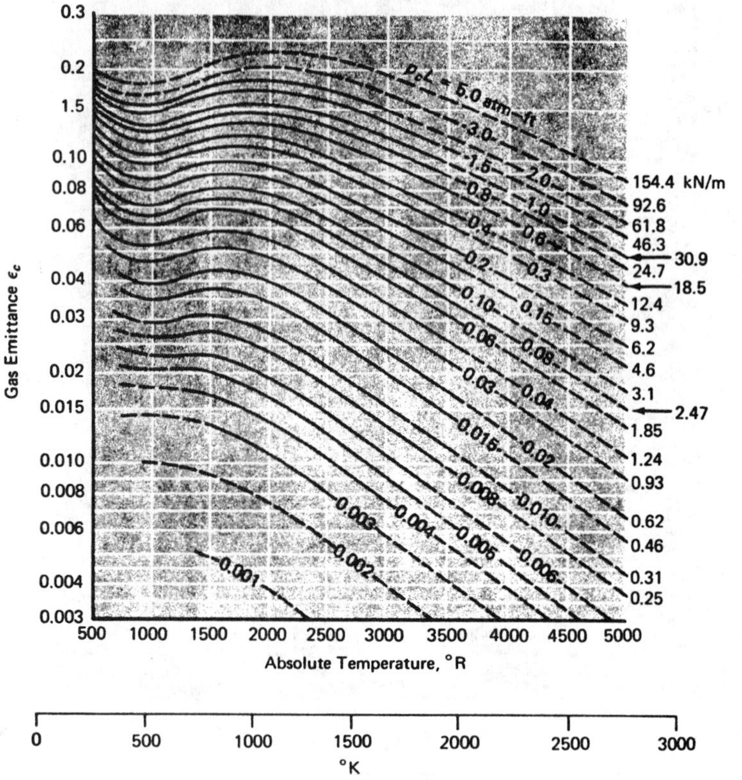

Fig. 4 Emissivity for carbon dioxide at 1 atm.

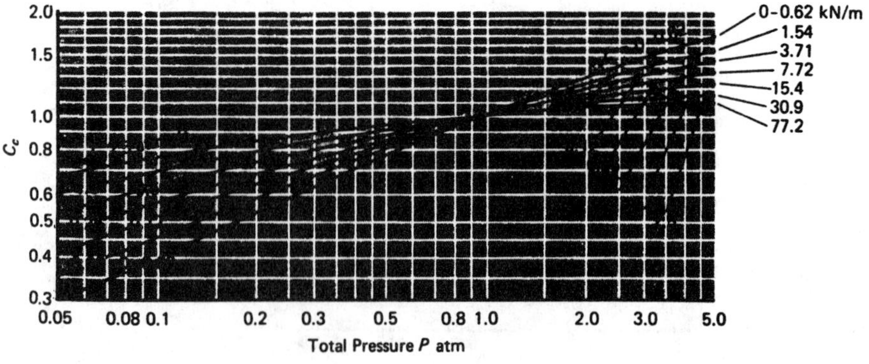

Fig. 5 Correction factor for emissivity of CO_2.

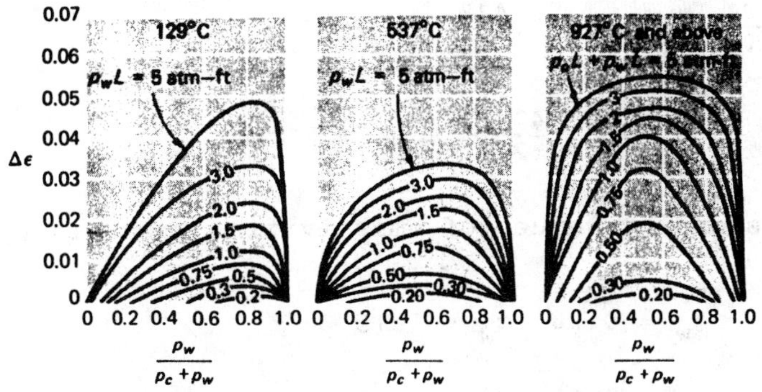

Fig. 6 Correction factor when CO_2 and H_2O are present in an enclosure at 30.881 kN/m.

From figure 2, the emissivity of water vapor is $E_w = 0.22$, and from figure 4, the emissivity of CO_2 is $\varepsilon_c = 0.17$. The correction factor for water vapor emissivity for pressure $\frac{1}{2}(p + p_w)$, i.e., $0.5(3 + 0.45) = 1.725$ atm, is $c_w = 1.4$ from figure 3. Similarly, the correction factor for emissivity of CO_2 is $C_c = 1.1$ from figure 5.

And, the correction factor for CO_2 and H_2O when both are present in a gaseous mixture corresponding to

$$\frac{p_w}{p_c + p_w} = \frac{0.45}{0.6 + 0.45} = 0.429 \text{ and } \Delta\varepsilon = 0.055$$

from figure 6.

Now, the gas emissivity ε_g can be computed using the relation

$$\varepsilon_g = C_c\varepsilon_c + C_w\varepsilon_w - \Delta\varepsilon$$

$$= (1.1 \times 0.17) + (1.4 \times 0.22) - 0.055$$

$$= 0.44$$

Now, the absorptivity of the gas at temperatures T_1 and T_2 can be determined.

The absorptivity of the gas α_g is expressed as

$$\alpha_g = \alpha_c + \alpha_w - \Delta\alpha \qquad \text{at a temperature}$$

The parameter at $T_1 = T_1 = 473°K$

$$p_cL_e \frac{T_1}{T_g} = 76.6 \frac{473}{1273} = 28.5 \text{ kN/m}$$

and $p_wL_e \frac{T_1}{T_g} = 57.5 \frac{473}{1273} = 21.4$ kN/m.

The corresponding values of emissivities from figures 2 to 6 are

$$\varepsilon_w = 0.26 \quad c_w = 1.15$$

$$\varepsilon_c = 0.13 \quad C_c = 1.15 \qquad \Delta\varepsilon = \Delta\alpha = 0.02$$

$$\alpha_c = C_c \times \varepsilon'_c \left(\frac{T_g}{T_c}\right)^{0.65}$$

$$= 1.15 \times 0.13 \left(\frac{1273}{473}\right)^{0.65}$$

$$= 0.285$$

$$\alpha_w = C_w \times \varepsilon'_w \left(\frac{T_g}{T_c}\right)^{0.45}$$

$$= 1.5 \times 0.26 \left(\frac{1273}{473}\right)^{0.45}$$

$$= 0.608$$

Thus, for $T = T_1$

$$\alpha_g (T_1) = 0.285 + 0.608 - 0.02$$

$$= 0.873$$

The transmissivity of gas at temperature T_1 is expressed as

$$\tau_g(T_1) = 1 - \alpha_g(T_1)$$

$$= 1 - 0.873$$

$$= 0.127$$

Again, at temperature T_2, i.e., $T=T_2 = 773°K$

$$p_cL_e \frac{T_2}{T_g} = 76.6 \frac{773}{1273} = 46.5 \text{ kN/m}$$

and $$p_wL_e \frac{T_2}{T_g} = 57.5 \frac{773}{1273} = 34.9 \text{kN/m}$$

From figures 2 to 6 the emissivities and correction factor are obtained as

$$\varepsilon''_w = 0.24, \qquad C_w = 1.45$$

$$\varepsilon''_c = 0.17, \qquad C_c = 1.13$$

$$\Delta\varepsilon = \Delta\alpha = 0.028$$

and absorptivity

$$\alpha_c = C_c\, \varepsilon''_c \left(\frac{T_g}{T_2}\right)^{0.65}$$

$$= 1.13 \times 0.17 \left(\frac{1273}{773}\right)^{0.65}$$

$$= 0.266$$

Also

$$\alpha_w = C_w\, \varepsilon''_w \left(\frac{T_g}{T_2}\right)^{0.45}$$

$$= 1.45 \times 0.24 \left(\frac{1273}{773}\right)^{0.45}$$

$$= 0.436$$

Again, the gas absorptivity at temperature T_2 is

$$\alpha_g(T_2) = \alpha_c + \alpha_w - \Delta\alpha$$

$$= 0.266 + 0.436 - 0.028$$

$$= 0.674$$

and the transmissivity of gas at temperature T_2 is

$$\tau_g(T_2) = 1 - \alpha_g(T_2) = 1 - 0.674 = 0.326$$

The heat energy at surface 1 is expressed as

$$G_1 = \varepsilon_g(T_g)E_{bg} + \tau_g(T_2)E_{b2}$$

introducing the values for known quantities,

$$G = 0.44 \times 148.9 + 0.326 \times 20.24$$

$$= 72.1 \text{ kW/m}^2$$

Also, for surface 2

$$G_2 = \varepsilon_g(T_g)E_{bg} + \tau_g(T_1)E_{b1}$$

$$= 0.44 \times 148.9 + 0.126 \times 2.84$$

$$= 65.9 \text{ kW/m}^2$$

Now the rate of heat transfer per unit area of the flat surfaces can be computed using the values obtained.

For surface 1

$$\frac{q_1}{A_1} = G_1 - E_{b1}$$

$$= (72.1 - 2.84)\ \text{kW/m}^2$$

$$= 69.3\ \text{kW/m}^2$$

and for surface 2

$$\frac{q_2}{A_2} = G_2 - E_{b2}$$

$$= (65.9 - 20.24)\ \text{kW/m}^2$$

$$= 45.7\ \text{kW/m}^2$$

Now, the total heat transferred is the sum of the heat transferred from the individual surfaces.

$$q = 69.3 + 45.7 = 115\ \text{kW/m}^2$$

• PROBLEM 8-34

A gas at atmospheric pressure is introduced between two long parallel surfaces, 3 inches apart. The surface temperatures are 1660°R and 1260°R, respectively. Compute the net radiant heat exchange between the two surfaces, if the emissivity of gas is 0.6.

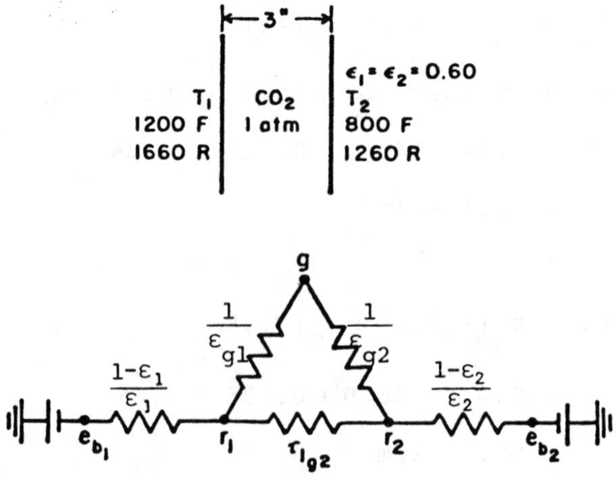

Fig. 1

Solution: Figure 1 shows the equivalent electrical network. The net radiant heat transfer between two parallel surfaces can be expressed as

$$q = f_{12} A \sigma (T_1^4 - T_2^4) \qquad (1)$$

where σ is Stefan-Boltzmann constant = 0.1714×10^{-8}

T - absolute temperature of the surfaces (or gas) in Rankine scale. $T_1 = 1660°R$ and $T_2 = 1260°R$.

A is surface area of the surface

f_{12} is the configuration factor where

$$f_{12} = \frac{1 - \varepsilon_1}{\varepsilon_1} + \frac{1 - \varepsilon_2}{\varepsilon_2} + \frac{1}{F_{12}\tau_{1g2} + \cfrac{1}{\cfrac{1}{\varepsilon_{g1}} + \cfrac{1}{\varepsilon_{g2}}}} \qquad (2)$$

where ε is emissivity and

τ is transmissivity.

The equivalent beam length or characteristic length L is given by

$$L = \frac{4 \text{ Volume } (V)}{\text{surface area } (A)}$$

$$= 0.85 \times 4 \times \frac{0.25}{2}$$

$$= 0.425 \text{ ft}$$

Now, let the temperature of the gas be 1000°F, then corresponding to $L \times p = 0.425 \times 1 = 0.425$, from figure 2, $\varepsilon_c = 0.118$, and hence $\tau_{1g_2} = 0.882$ ($\because \tau_{1g_2} = 1 - \varepsilon_c$).

Substituting the values in the expression for f_{12}, equation (2), yields

$$f_{12} = \frac{0.4}{0.6} + \frac{0.4}{0.6} + \frac{1}{1 \times 0.882 + \cfrac{1}{\cfrac{1}{0.118} + \cfrac{1}{0.118}}}$$

($\because F_{12} = 1$ for parallel surface)

or $f_{12} = 2.395$

$$\therefore q = f_{12} A \sigma (T_1^4 - T_2^4) \quad \text{-- equation (1)}$$

or $$\frac{q}{A} = (\tfrac{1}{2}.395) \times 0.1714 \times 10^{-8}[(1660)^4 - (1260)^4]$$

$$= 3630 \text{ Btu/hr-ft}^2$$

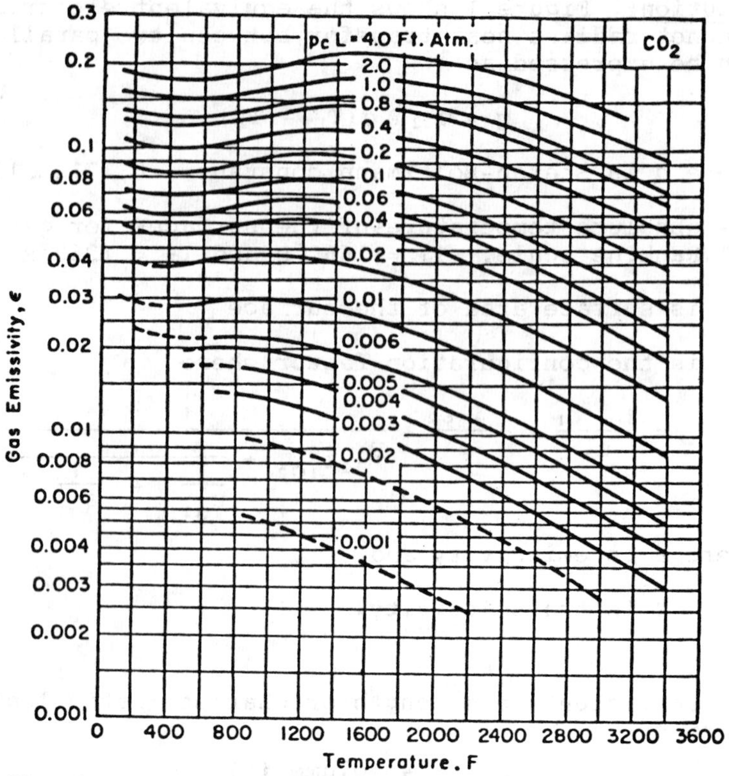

Fig. 2 Emissivity of carbon dioxide at 1 atmosphere pressure.

Now, the radiation from surface 1 is given as

$$e_{b1} = \sigma T_1^4 = 0.1714 \times 10^{-8} (1660)^4$$

$$\cong 13015 \text{ Btu/hr}$$

$$e_{b2} = \sigma T_2^4 = 0.1714 \times 10^{-8} (1260)^4$$

$$= 4320 \text{ Btu/hr}$$

and $$r_1 = e_{b1} - \frac{q/A}{\varepsilon_1/(1-\varepsilon_1)}$$

$$= 13000 - \frac{3630}{(0.6/0.4)} = 10580$$

Again, $$r_2 = e_{b2} + \frac{q/A}{\varepsilon_2/(1-\varepsilon_2)}$$

$$= 4320 + \frac{3630}{0.6/0.4} = 6740$$

$$e_{bg} = \frac{r_1 + r_2}{2} = \frac{10580 + 6740}{2}$$

$$= 8660$$

Thus, the gas temperature can be calculated, as

$$T_g = \frac{e_{bg}}{\sigma}$$

$$= \frac{8600}{0.1714 \times 10^{-8}}$$

or $$T_g = 1037°F$$

The difference between the assumed value of the gas temperature (1000°F) and the calculated one (1030°F) is very little and needs no further iteration. Thus, the net radiant heat exchange between the parallel plates is approximately 3630 Btu/hr-ft^2

• PROBLEM 8-35

A 20 cm cubical fuel burner (see Fig. 1) is injected with carbon dioxide gas at a pressure of 14.7 psi (1 atom) and temp. 1000°R. The temperatures and emissivities the door, floor, and rest of the walls are T_1=1200°R, ε_1=0.4, T_2= 1500°R, ε_2=0.8 and T_3=1400°R, ε_3=0.8, respectively. All three interior surfaces are black bodies. Evaluate the energy exchange rate and the heat absorbed by the gas.

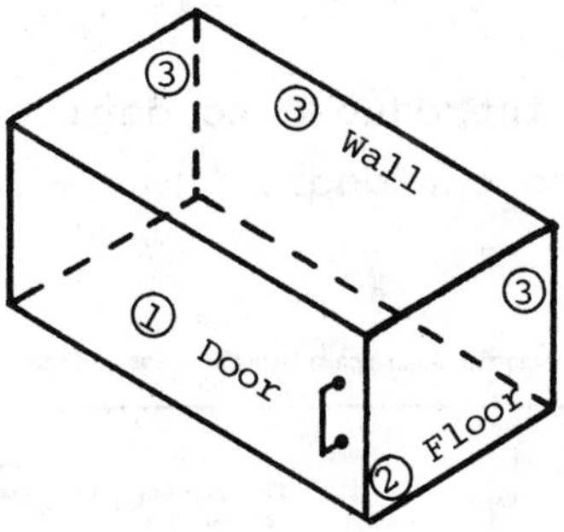

Fig. 1 Fuel burner with 2.7 μ bands wavelength.

Solution: The total radiation incident upon a surface per unit time and per unit area can be expressed as:

Irradiation on surface 1 = irradiation from surface i transmitted through gas + irradiation from gas or, for the three surfaces considered, we have:

$$\bar{G}_{v1}A_1\Delta v = \bar{J}_{v2}A_2F_{21}\bar{\tau}_{v21}\Delta v + \bar{J}_{v3}A_3F_{31}\bar{\tau}_{v31}\Delta v + \bar{E}_{bvg}A_2F_{21}\bar{\varepsilon}_{v21}\Delta v + \bar{E}_{bvg}A_3F_{31}\bar{\varepsilon}_{v31}\Delta v \qquad (1)$$

$$\bar{G}_{v2}A_2\Delta v = \bar{J}_{v1}A_1F_{12}\bar{\tau}_{v12}\Delta v + \bar{J}_{v3}A_3F_{32}\bar{\tau}_{v32}\Delta v + \bar{E}_{bvg}A_1F_{12}\bar{\varepsilon}_{v12}\Delta v + \bar{E}_{bvg}A_3F_{32}\bar{\varepsilon}_{v32}\Delta v \qquad (2)$$

$$\bar{G}_{v3}A_3\Delta v = \bar{J}_{v1}A_1F_{13}\bar{\tau}_{v13}\Delta v + \bar{J}_{v2}A_2F_{23}\bar{\varepsilon}_{v23}\Delta v + \bar{J}_{v3}A_3F_{33}\bar{\tau}_{v33}\Delta v + \bar{E}_{bvg}A_1F_{13}\bar{\varepsilon}_{v13}\Delta v + \bar{E}_{bvg}A_2F_{23}\bar{\varepsilon}_{v23}\Delta v + \bar{E}_{bvg}A_3F_{33}\bar{\varepsilon}_{v33}\Delta v \qquad (3)$$

where G = irradiation

A = surface area

J = radiosity

τ = transmissivity

ε = emissivity F = shape factor

For simplification, eq. (1) can also be written as:

$$\bar{G}_{v1}A_1\Delta v + \bar{J}_{v2}A_2F_{21}\bar{\tau}_{v21}\Delta v + \bar{J}_{v3}A_3F_{31}\bar{\tau}_{v31}\Delta v + \bar{E}_{bvg}A_1\left[F_{12}\bar{\varepsilon}_{v12} + F_{13}\bar{\varepsilon}_{v13}\right]\Delta v$$

or $\bar{G}_{v1}A_1\Delta v = \bar{J}_{v2}A_2F_{21}\bar{\tau}_{v21}\Delta v + \bar{J}_{v3}A_3F_{31}\bar{\tau}_{v31}\Delta v + \bar{E}_{bvg}A_1\left[\bar{\varepsilon}_{v\,encl-1}\right]\Delta v$

where $\bar{\varepsilon}_{v\ encl-1}$ is introduced to depict the radiation of the enclosed gas body to the door.

TABLE 1
Geometric Mean Beam Lengths *L* for Isothermal Gas Radiation

Shape	Characterizing dimension *X*	Factor by which *X* is multiplied to obtain mean beam length *L* — When $P_aL = 0$
Rectangular parallelepipeds:		
1 : 1 : 1 (cube)...........	Edge	$\frac{2}{3}$
1 : 1 : 4, radiating to 1 × 4 face............		0.90
radiating to 1 × 1 face............	Shortest edge	0.86
radiating to all faces...........		0.89
1 : 2 : 6, radiating to 2 × 6 face............		1.18
radiating to 1 × 6 face............	Shortest edge	1.24
radiating to 1 × 2 face............		1.18
radiating to all faces...........		1.20
1 : ∞ : ∞ (infinite parallel planes).........	Distance between planes	2

TABLE 2

Geometric Mean Beam Length Ratios and Configuration Factors for Parallel Equal Rectangles

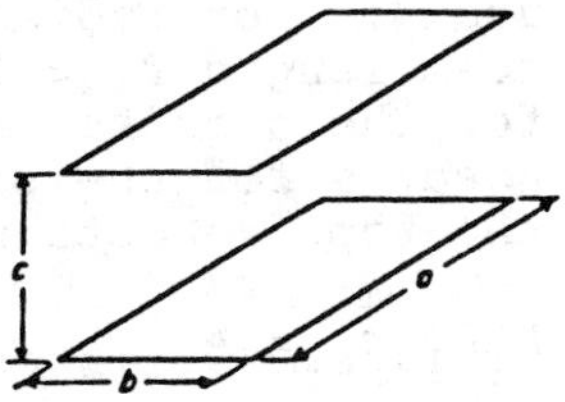

$\beta = b/c$, $\eta = a/c$, F = Configuration Factor, r = Geometric Mean Beam Length

η \ β:		0	0.1	0.2	0.4	0.6	1.0
0	r/c	1.000	1.001	1.003	1.012	1.025	1.055
	F	—	—	—	—	—	—
0.1	r/c	1.001	1.002	1.004	1.013	1.026	1.056
	F		0.00316	0.00626	0.01207	0.01715	0.02492
0.2	r/c	1.003	1.004	1.006	1.015	1.028	1.058
	F		0.00626	0.01240	0.02391	0.03398	0.04941
0.4	r/c	1.012	1.013	1.015	1.024	1.037	1.067
	F		0.01207	0.02392	0.04614	0.06560	0.09554
0.6	r/c	1.025	1.026	1.028	1.037	1.050	1.080
	F		0.01715	0.03398	0.06560	0.09336	0.13627
1.0	r/c	1.055	1.056	1.058	1.067	1.080	1.110
	F		0.02492	0.04941	0.09554	0.13627	0.19982
2.0	r/c	1.116	1.117	1.120	1.129	1.143	1.175
	F		0.03514	0.06971	0.13513	0.19342	0.28588

TABLE 3

Geometric Mean Beam Lengths for Rectangles at Right Angles

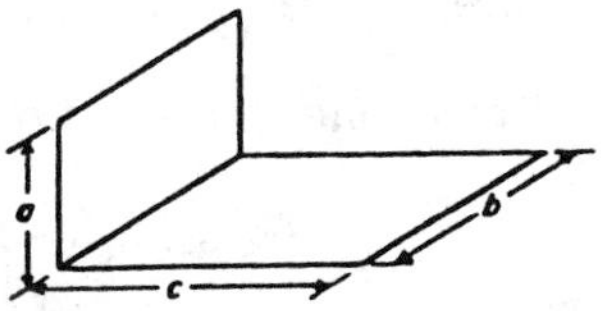

$\alpha = a/b$, $\gamma = c/b$, F = Configuration Factor

$\phi = FA/b^2$, $Z_r = FAR/abc$, R = Geometric Mean Beam Length

α \ γ:		0.05	0.10	0.20	0.4	0.6	1.0	2.0
0.02	ϕ	0.007982	0.008875	0.009323	0.009545	0.009589	0.009628	0.009648
	Z_r	0.17840	0.12903	0.08298	0.04995	0.03587	0.02291	0.01263
0.05	ϕ	0.014269	0.018601	0.02117	0.02243	0.02279	0.02304	0.02316
	Z_r	0.21146	0.18756	0.13834	0.08953	0.06627	0.04372	0.02364
0.10	ϕ		0.02819	0.03622	0.04086	0.04229	0.04325	0.04376
	Z_r		0.20379	0.17742	0.12737	0.09795	0.06659	0.03676
0.20	ϕ			0.05421	0.06859	0.07377	0.07744	0.07942
	Z_r			0.18854	0.15900	0.13028	0.09337	0.05356
0.40	ϕ				0.10013	0.114254	0.12770	0.13514
	Z_r				0.16255	0.14686	0.11517	0.07088
0.60	ϕ					0.13888	0.16138	0.17657
	Z_r					0.14164	0.11940	0.07830
1.0	ϕ						0.20004	0.23285
	Z_r						0.11121	0.08137
2.0	ϕ							0.29860
	Z_r							0.07086

Nest, the mean beam length can be obtained from tables 1, 2 and 3. For radiation from the door to the floor, the mean beam length r_{12}= 0.835 ft. Use the flux algebra techniques for the door to surface 3. The following subscripts will be used: s for the side panel, b for the rear panel, and t for the top. We obtain

$$A_1F_{13}\bar{r}_{13} = A_1F_{1s}\bar{r}_{1s} + A_1F_{1s}\bar{r}_{1s} + A_1F_{1b}\bar{r}_{1b} + A_1F_{1t}\bar{r}_{1t}$$

Inserting the values, we have $\bar{r}_{13}$ = 1.04 ft.

By symmetry, the mean beam length from the floor to surface 3 is:

$$\bar{r}_{23} = \bar{r}_{13} = 1.04 \text{ ft}$$

Applying the concept of reciprocity, we have:

$$\bar{r}_{21} = \bar{r}_{12} = 0.835 \text{ ft}$$

$$\bar{r}_{31} = \bar{r}_{13} = 1.04 \text{ ft}$$

$$\bar{r}_{32} = \bar{r}_{23} = 1.04 \text{ ft}$$

The radiation of surface 3 to itself, using flux algebra,

$$A_3F_{33}\bar{r}_{33} = A_3F_{3s}\bar{r}_{3s} + A_3F_{3t}\bar{r}_{3t} + A_3F_{3s}\bar{r}_{3s}$$

then $\bar{r}_{33}$ = 1.12 ft

From one panel to the rest of the walls, we have

$$A_sF_{s\text{-encl}}\bar{r}_{s\text{-encl}} = 4A_3F_{ss}\bar{r}_{ss} + A_bF_{bs}\bar{r}_{bs}$$

Then $r_{s\text{-encl}}$ = 1.00 ft

Now that the mean beam lengths have been determined, the diverse mass path lengths are then calculated.

$$(xPL)_{12} = 0.835 \text{ atm-ft}$$

$$(xPL)_{13} = 1.04 \text{ atm-ft}$$

$$(xPL)_{23} = 1.04 \text{ atm-ft}$$

$$(xPL)_{33} = 1.11 \text{ atm-ft}$$

$$(xPL)_{s\text{-encl}} = 1.00 \text{ atm-ft}$$

Referring to figs. 2 and 3, and applying eq. (4), we obtain

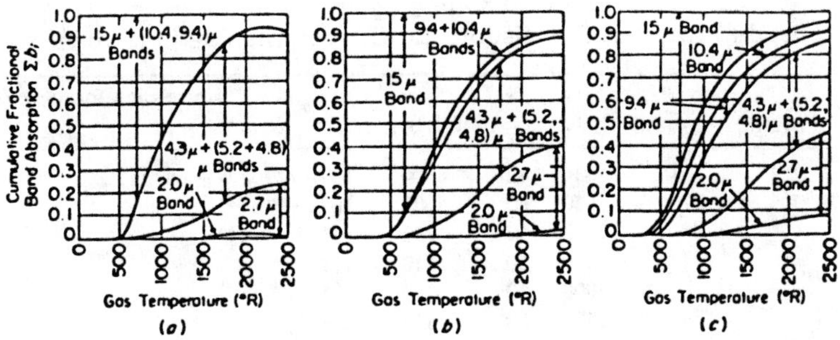

Fig. 2 Fractional band absorption for Carbon Dioxide (a) $p_cL=0.1$ atm-ft (b) $p_cL=1.0$ atm-ft (c) $p_cL=10$ atm-ft (5)

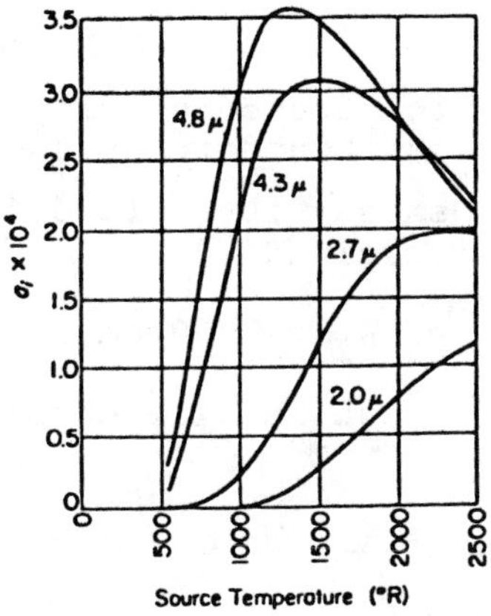

Fig. 3 Weighting Coefficient a_i for calculation of total absorption 4.8, 4.3, 2.7, and 2.0 μ bands of CO_2.

$$A_i\ T_g, W_g\ \bar{E}_{bvwi} = b_i\ T_g, W_g\ \varepsilon_g\ T_g, W_g\ \frac{a_i\ T_w}{a_i\ T_g}\ \sigma T_w^4 \tag{4}$$

$$\bar{\alpha}_{vjk}\Delta v = \frac{b_i\ T_g, W_g\ \varepsilon_g\ T_g, W_g}{a_i\ T_g}$$

and, $\alpha_{v12}\Delta v = 246/cm$ $\qquad$ $\bar{\varepsilon}_{v12}\ \Delta v = 246/cm$

$\alpha_{v13}\Delta v = 258/cm$ $\qquad$ $\bar{\varepsilon}_{v13}\Delta v = 258/cm$

$\alpha_{v23}\Delta v = 258/cm$ $\qquad$ $\bar{\varepsilon}_{v23}\Delta v = 258/cm$

$\alpha_{v33}\Delta v = 260/cm$ $\qquad$ $\bar{\varepsilon}_{v33}\Delta v = 260/cm$

and $\bar{\varepsilon}_{v\text{-}encl\text{-}s}\Delta v = 256/cm$

TABLE 4
Band-absorption Correlations for Carbon Dioxide in Nitrogen

Band	Fit	P_{E_L} (atm)	P_{E_U} (atm)	T_g (°R)	w_L (mlb/ft²)	w_U (mlb/ft²)	Correlation for $\mathcal{A}$ (cm⁻¹)	$\pm \Delta\mathcal{A}$ (cm⁻¹)
2.7μ	*S*	0.1	1.0	530	40	6500	$15 + 77 \log w + 68 \log P_E$	
		1	70	530	60	5000	$107w^{0.12}P_E^{0.06}$	22
				1000	50	750	$143w^{0.12}P_E^{0.02}$	22
				1500	50	510	$179w^{0.12}$	17
				2000	50	390	$205w^{0.12}$	17
				2500	50	310	$239w^{0.12}$	17
	M	0.1	1.0	530		3	$28.3w^{0.5}P_E^{0.43}$	
		0.5	1.0	530	13	60	$24w^{0.50}P_E^{0.4}$	22
		1	10	1000	4	50	$30w^{0.50}P_E^{0.06}$	22

Referring to table 4, the transmissivity for the 2.7μ band is $\overline{\tau}_v = 0.12$ with a maximum boundary at the band of 3830/cm.

The values of the transmissivity will differ due to the various band absorption received.

Let the maximum value of A_i be 260/cm, then $\Delta v = 296$/cm.

Inserting the value, we can readily obtain the transmittance and the emissivity:

$$\overline{\tau}_{v12} = 0.167, \qquad \overline{\varepsilon}_{v12} = 1 - \overline{\tau}_{v12} = 1 - 0.167 = 0.833$$

$$\overline{\tau}_{v13} = 0.126, \qquad \overline{\varepsilon}_{v13} = 1 - \overline{\tau}_{v13} = 1 - 0.126 = 0.874$$

$$\overline{\tau}_{v23} = 0.126, \qquad \overline{\varepsilon}_{v23} = 1 - \overline{\tau}_{v13} = 1 - 0.126 = 0.874$$

$$\overline{\tau}_{v33} = 0.120, \qquad \overline{\varepsilon}_{v33} = 1 - \overline{\tau}_{v33} = 1 - 0.12 = 0.880$$

and $\overline{\varepsilon}_{v\,encl-1} = F_{12}\overline{\varepsilon}_{v12} + F_{13}\overline{\varepsilon}_{v13} = 0.867$

These values will be substituted into the radiosity equations:

$$\overline{J}_{v1} = \overline{E}_{bv1} + \rho_1 \left[\overline{J}_{v2} F_{12} \overline{\tau}_{v12} + \overline{J}_{v3} F_{13} \overline{\tau}_{v13} + \overline{E}_{bvg} \varepsilon_{v\,encl-1} \right]$$

$$\overline{J}_{v2} = \overline{E}_{bv2} + \rho_2 \left[\overline{J}_{v1} F_{21} \overline{\tau}_{v21} + \overline{J}_{v3} F_{23} \overline{\tau}_{v23} + \overline{E}_{bvg} \overline{\varepsilon}_{v\,encl-2} \right]$$

$$\overline{J}_{v3} = \overline{E}_{bv3} + \rho_3 \quad \overline{J}_{v1} F_{31} \overline{\tau}_{v31} + \overline{J}_{v2} F_{32} \overline{\tau}_{v32} + \overline{J}_{v3} F_{33} \overline{\tau}_{v33}$$
$$+ \overline{E}_{bvg} \overline{\varepsilon}_{v\,encl-3}$$

The emissive power $\overline{E}_{bvg}$ is obtained from fig 3 as 0.041. The unknowns are $\overline{J}_{v1}$, $\overline{J}_{v2}$, $\overline{J}_{v3}$ and $\overline{\varepsilon}_{v\,encl-3}$, but $\overline{\varepsilon}_{v\,encl-3}$ can be determined from:

$$\overline{\varepsilon}_{v\ encl-3} = F_{31}\overline{\varepsilon}_{v31} + F_{32}\overline{\varepsilon}_{v32} + F_{33}\overline{\varepsilon}_{v33}$$

$$= 0.878$$

The three equations can now be solved simultaneously to yield:

$$\overline{J}_{v1} = 0.267 \text{ Btu per hr. ft}^2\text{.cm}^{-1}$$

$$\overline{J}_{v2} = 1.065 \text{ Btu per hr. ft}^2\text{.cm}^{-1}$$

$$\overline{J}_{v3} = 0.782 \text{ Btu per hr. ft}^2\text{.cm}^{-1}$$

The total radiation incident upon each surface is determined by

$$\overline{G}_{v1} = F_{12}\overline{\tau}_{v12}\overline{J}_{v2} + F_{13}\overline{\tau}_{v13}\overline{J}_{v3} + \overline{\varepsilon}_{v\ encl-1}\overline{E}_{bvg}$$

$$\overline{G}_{v1} = (0.2)(0.167)(1.065) + (0.8)(0.126)(0.782)+(0.867)(0.041)$$

$$= 0.150 \text{ Btu/hr. ft}^2\text{.cm}^{-1}$$

$$\overline{G}_{v2} = 0.123 \text{ Btu/hr. ft}^2\text{.cm}^{-1}$$

$$\overline{G}_{v3} = 0.126 \text{ Btu/hr. ft .cm}^{-1}$$

TABLE 1

Geometric Mean Beam Lengths L for Isothermal Gas Radiation

Shape	Characterizing dimension X	Factor by which X is multiplied to obtain mean beam length L When $PcL = 0$
Rectangular parallelepipeds:		
1 : 1 : 1 (cube)	Edge	⅔
1 : 1 : 4, radiating to 1 × 4 face		0.90
radiating to 1 × 1 face	Shortest edge	0.86
radiating to all faces		0.89
1 : 2 : 6, radiating to 2 × 6 face		1.18
radiating to 1 × 6 face	Shortest edge	1.24
radiating to 1 × 2 face		1.18
radiating to all faces		1.20
1 : ∞ : ∞ (infinite parallel planes)	Distance between planes	2

The net energy exchanges between the surfaces are:

$$q_{net_1} = \overline{G}_{v_1} - \overline{J}_{v_1} \quad \Delta v = (0.150 - 0.267)(296) = -34.6\ Btu/hr.ft^2$$

$$q_{net_2} = \overline{G}_{v_2} - \overline{J}_{2} \quad \Delta v = (0.123 - 1.065)(296) = -279\ Btu/hr.ft^2$$

$$q_{net_3} = \overline{G}_{v_3} - \overline{J}_{v_3} \quad \Delta v = (0.126 - 0.782)(296) = -194\ Btu/hr.ft^2$$

The negative sign means that the surfaces are losing heat to the carbon dioxide gas, thus the energy absorbed by the gas is:

$$q_{gas} = q_{net_1}A_1 + q_{net_2}A_2 + q_{net_3}A_3$$

where $A_1 = A_2 = A_3 = 2.78\ ft^2$

$$q_{gas} = (34.6)(2.78) + (279)(2.78) + (194)(2.78)(4\ surfaces)$$

$$= 3029\ Btu/hr$$

A more direct equation may be used to determine the heat absorbed by the carbon dioxide gas:

$$q_{gas} = \left[\overline{J}_{v_1} - \overline{E}_{bvg} \quad \overline{\varepsilon}_{v\,encl-1}\ A_1 + \overline{J}_{v_2} - \overline{E}_{bvg} \quad \overline{\varepsilon}_{v\,encl-2}\ A_2 + \overline{J}_{v_3} - \overline{E}_{bvg} \quad \overline{\varepsilon}_{v\,encl-3}\ A_3 \right] \Delta v$$

$$= [(0.267 - 0.041)(0.867)(2.78)$$
$$+ (1.065 - 0.041)(0.867)(2.78)$$
$$+ (0.782 - 0.041)(0.878)(11.12)] \times (296)$$

$$= 3033\ Btu/hr$$

SOLAR RADIATION

• PROBLEM 8-36

Calculate the energy radiated from the sun to the earth. The diameter of the sun is $D_{sun} = 8.6 \times 10^5$ miles, the distance between the sun and earth $r_{12} = 9.29 \times 10^7$ miles, and the heat flow radiated by the sun is $E_{b1} = 2.0 \times 10^7$ Btu/hr.ft^2

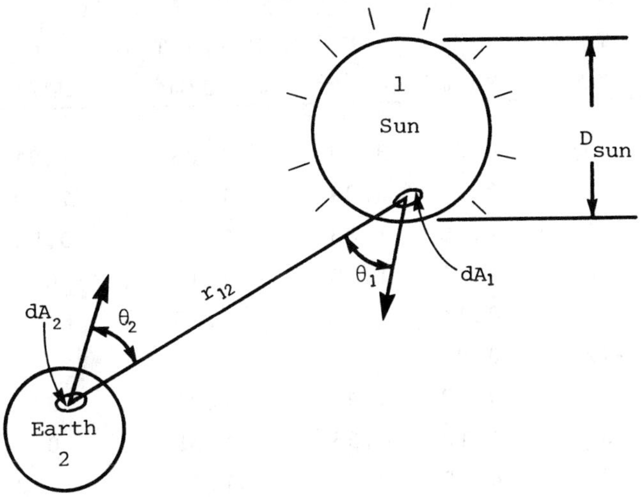

Fig. 1

Solution: The solar radiation incident on the earth's surface or the solar constant is approximated by:

$$\text{solar constant} = \frac{dQ_{12}}{\cos\theta_2 dA_2} = \frac{\sigma T_1^4}{\pi r_{12}^2} \int \cos\theta_1 dA_1$$

Integrating, we obtain:

$$\text{s. c.} = \frac{\sigma T_1^4}{\pi} \; \frac{\pi D^2_{sun}}{4r_{12}^2} = \frac{E_{b_1} D^2_{sun}}{4r_{12}^2}$$

$$= \frac{(2.0 \times 10^7 \text{Btu/hr.ft})(8.6 \times 10^5 \text{ miles})^2}{4(9.29 \times 10^7 \text{ miles})^2} = 428 \text{ Btu/hr.ft}^2$$

Not all the energy represented by the solar constant reaches the surface of the earth, because of strong absorption by CO_2 and water vapor in the atmosphere.

A series of solar collectors (each $3\,m^2$) are mounted at an angle of 61° and the surface azimuth angle is 0°. Table 1 gives the energy gained per hour by the solar panel K, the incident heat flux on the solar panel M, and the surrounding temperature each hour T_{amb}. The overall heat transfer coefficient is $U=9.0\ W/m^2{}^\circ C$ and the effective panel coefficient $F=0.852$. Water to be heated will enter the solar receiver at $T_i=45°C$ and a flow rate of $\dot{m}=0.04\,kg/s$. If 12 solar collectors are used, determine the applicable energy achieved each day and the performance.

Time	T_a °C	K MJ/m^2	M MJ/m^2	$U(T_i - T_a)$ MJ/m^2	q MJ/m^2	η
12-1	7	4.25	3.63	1.23	1.91	0.45
1-2	8	3.97	3.34	1.20	1.70	0.43
2-3	9	2.16	1.83	1.17	0.53	0.25
3-4	10	1.31	1.09	1.13	0	0
4-5	8	0.14	-	1.20	-	-
7-8	-12	0.04	-	-	-	-
8-9	-9	0.52	0.44	1.75	0	0
9-10	-3	1.12	0.95	1.55	0	0
10-11	3	4.15	3.52	1.36	1.72	0.41
11-12	4	3.52	3.04	1.33	1.36	0.39
Table 1.		ΣK=21.22			Σq=7.22	

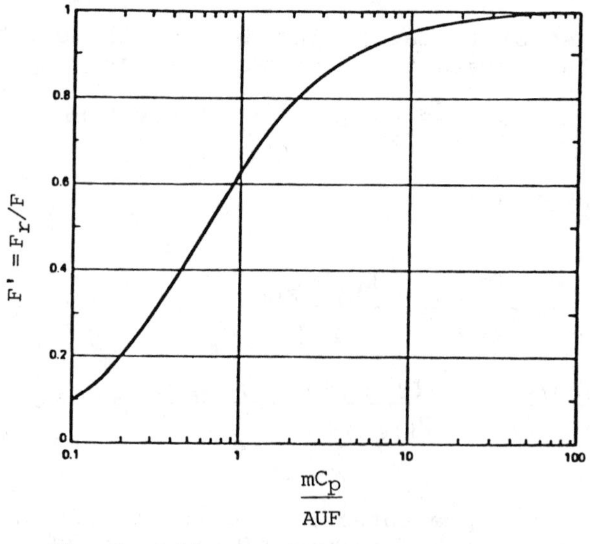

Fig. 1 Collector flow coefficient F' as a function of mC_p/AUF.

Solution: The flow coefficient for the solar receiver can be determined by:

$$F' = \frac{F_r}{F} = \frac{\dot{m}C_p}{AUF}\left[1 - e^{-(AUF/\dot{m}C_p)}\right] \quad \text{or from fig. 1} \qquad (1)$$

where $\frac{\dot{m}C_p}{AUF}$ is a dimensionless ratio and is equal to

$$\frac{(0.04\ kg/s)(4174\ J/kg°C)}{(3m^2)(9.0\ W/m^2°C)(0.852)} = 7.26$$

F_r = the energy loss coefficient = $F'F$

$$F' = 7.26\left[1 - e^{-0.138}\right] = 0.934$$

$$F_r = (0.934)(0.852) = 0.796$$

The energy loss from 12 - 1 is calculated as:

$$U(T_i - T_a) = 9.0\ W/m^2 \cdot °C(45 - 7) \times 3600\ sec/hr = 1.23\ MJ/m^2$$

and the applicable heat absorbed per unit area is

$$q = \frac{Q}{A} = F_r\left[M - U(T_i - T_a)\right]$$

$$= 0.796\ [3.63 - 1.23] = 1.91\ MJ$$

The effective coefficient from 12-1 is then calculated as:

$$\eta = \frac{Q}{AK} = \frac{q}{K} = \frac{1.91}{4.25} = 0.45$$

and the overall efficiency for the day is:

$$\eta = \frac{\Sigma q}{\Sigma K} = \frac{7.22}{21.22} = 0.34$$

The overall applicable heat absorbed by 12 solar receivers is

$$Q = NA\Sigma q = 12 \times 3 \times 7.22 \cong 260\ MJ$$

• PROBLEM 8-38

A spherical body, in space, is at a distance of 1 astronomical unit from the sun. Assume the body to be at a uniform temperature and having negligible internal heat generation. Determine the quality of the surface required such that the temperature of the body remains at 300°K.

Total Emittances and Solar Absorptances of Surfaces

	Total Normal Emittance	Extraterrestrial Solar Absorptance
Alumina, flame sprayed	0.80 (− 10°F)	0.28
Aluminum foil, as received	0.04 (70°F)	
Aluminum foil, bright dipped	0.025 (70°F)	0.10
Aluminum, vacuum deposited on duPont mylar	0.025 (70°F)	0.10
Aluminum alloy, 6061, as received	0.03 (70°F)	0.37
Aluminum alloy, 75S-T6 weathered 20,000 hr on a DC6 aircraft	0.16 (150°F)	0.54
Aluminum, hard anodized, 6061-T6, 35 amp/ft² at 45 volts in 20°F sulfuric acid solution, 1 mil thick	0.84 (− 10°F)	0.92
Aluminum, soft anodized *Reflectal* aluminum alloy, 5 amp/ft² for 2 hr in 40°F, 10% H_2SO_4 solution	0.79 (− 10°F)	0.23
Aluminum, 7075-T6, sandblasted with 60 mesh silicon carbide grit	0.30 (70°F)	0.55
Aluminized silicone resin paint Dow Corning XP-310	0.20 (200°F) 0.22 (800°F)	0.27
Beryllium	0.18 (300°F) 0.21 (700°F) 0.30 (1100°F)	0.77
Beryllium, anodized	0.90 (300°F) 0.88 (700°F) 0.82 (1100°F)	

Solution: The steady state internal heat transfer from a spherical body of radius R with uniform temperature and uniform surface finish is expressed as

$$Q = \varepsilon(T_w)\sigma T_w^4 \ (4\pi r^2) - \alpha(T_w, T_s)(I_s d_{ws})(\pi r^2)$$

where subscripts w and s refer to the spherical body and the sun, respectively. ε and α are the emissivity and absorptivity, respectively.

The irradiance on a surface normal to the solar radiation at a distance of 1 astronomical unit and outside the earth's atmosphere is known as the solar constant and is given by

$$I_s d_{ws} = 130 \text{ W/m}^2$$

The value of Q is zero as there is no internal heat generation or dissipation to the outside of the sphere. Therefore the equation simplifies to

$$\varepsilon(T_w)\ \sigma(T_w)^4(4\pi r^2) = \alpha(T_w\ T_s)(I_s d_{ws})(\pi r^2)$$

or
$$\frac{\alpha}{\varepsilon} = \frac{4\sigma T_w^4}{I_s d_{ws}}$$

$$= \frac{4 \times (5.667 \times 10^{-8}) \times (300)^4}{(1380)}$$

$$= 1.33$$

Again, the ratio of absorptivity and emissivity is also expressed as

$$\frac{\alpha}{\varepsilon} = \frac{F_1\alpha_1 + (1 - F_1)\alpha_2}{F_1\varepsilon_1 + (1 - F_1)\varepsilon_2} = 1.33$$

The view factor F_1 can be expressed as

$$F_1 = \frac{1.33\ \ \ \ \ \alpha_2}{\alpha_1 - 1.33\varepsilon_1 - \alpha_2 + 1.33\varepsilon_2}$$

The basic material for the sphere may (refer to table) be assumed as aluminium for which $\varepsilon=0.1$ and $\alpha=0.03$, and the surface may be painted with white potassium zirconium silicate for which $\alpha=0.13$ and $\varepsilon=0.86$.

Introducing these values in the equation for F_1 yields

$$F_1 = \frac{1.33(0.86) - 0.13}{0.10 - 0.04 - 0.13 + 1.33(0.86)}$$

$$= 0.943$$

• PROBLEM 8-39

A solar energy collector, aperture area $15m^2$, is installed on top of a house at an angle of 25° from the horizontal and located at latitude 35°N. The angle of acceptance is 24°. The normal radating beam on a day at a particular hour was 900 W/m^2. Calculate the solar flow to the panel.

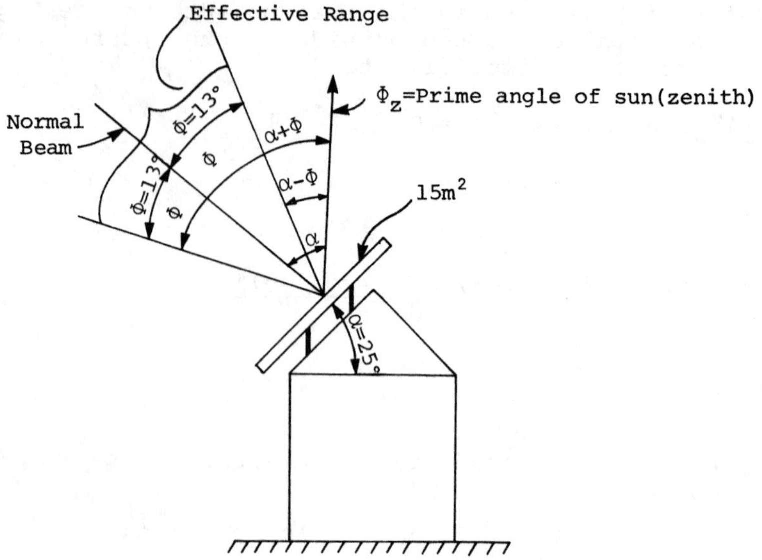

Fig. 1

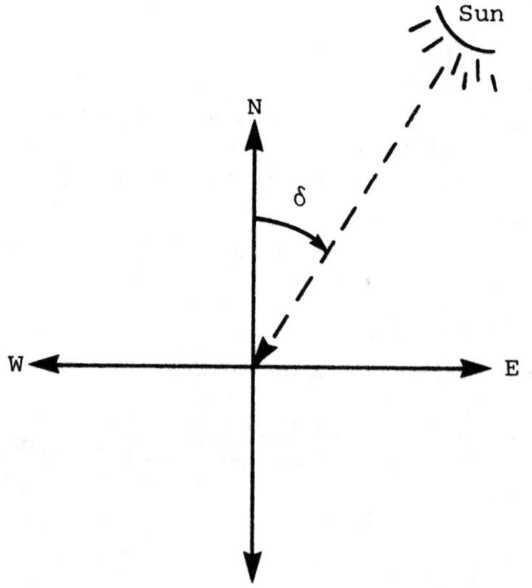

Fig. 2 Solar incident angle measured counter-clockwise from North to South (azimuth angle).

Solution: As shown in figures (1) and (2), the following condition must be satisfied for energy from the incident beam of the sun to be effectively absorbed:

$$(\alpha-\Phi) \leq \tan^{-1}(\tan\Phi_z \cos\gamma) \leq (\alpha + \Phi) \qquad (1)$$

To check for the effectiveness, a decision factor (F) is introduced such that

$$F = \begin{cases} 1 & \text{if equation (1) is valid} \\ 0 & \text{otherwise} \end{cases}$$

For the angle of zenith angle Φ_z,

$$\cos\Phi_z = \cos\delta \cos\phi \cos\omega + \sin\delta \sin\phi$$

gives $\cos\Phi_z = 0.851$ and $\Phi_z = 31.6°$

Now, solar azimuthal angle is

$$\sin\gamma = \frac{\cos\delta \sin\omega}{\sin\Phi_z} = \frac{\cos(17.9) \times \sin(-30)}{\sin(31.6)}$$

$$\sin\gamma = -0.908 \therefore \gamma = -65.2°$$

For the solar panel, $\alpha - \Phi = 25 - 12 = 13°$

and $\alpha + \Phi = 25 + 12 = 37°$

$$\therefore \tan^{-1}(\tan\Phi_z \cos\gamma) = \tan^{-1}[\tan 31.6 \cos(-65.2)]$$

$$= 14.5°$$

Since 14.5° is between 13° and 37°, the decision factor is 1.

Angle of incidence for the beam radiation on the aperture is

$$\cos\theta = \cos(L - \alpha) \cos\delta \cos\omega + \sin(L - \alpha) \sin\delta$$

$$= \cos(35\text{-}25) \cos 17.9 \cos 30 + \sin(35\text{-}25) \sin 17.9$$

$$= 0.865$$

The effective energy rate of the beam radiant incident on the solar panel is

$$G_e = G_{bn} \times F \times \cos\theta$$

$$= 900 \times 1 \times 0.865 = 778.5 \text{ W/m}$$

$\therefore$ for aperture area of $15m^2$, effective beam radiation

$$= 778.5 \times 15 = 11677.5 \text{ W}$$

$$= 11.68 \text{ kW.}$$

A flat glass plate with a thickness t=3.5 mm is positioned at an angle of 55° over a solar collector. The glass is of poor quality and has an absorption factor K=35/m. Determine the transmissivity, reflectivity, and the absorptivity of the glass.

Solution: When the incident beam on the glass is at an angle of 55°, the refraction angle of the glass is:

$$\theta_2 = \sin^{-1}\left[\frac{\sin 55°}{1.526}\right] = 32.5°$$

The transmittance is determined by:

$$\tau_a = \frac{I\tau}{I_0} = e^{-Kt/\cos\theta_2}$$

where $Kt/\cos\theta_2$ = the length of the beam.

$$= 35(0.0035)/\cos 32.5° = 0.145$$

Then $\tau_a = e^{-0.145} = 0.865$

For parallel and perpendicular elements of polarization, the transmittance is calculated using

$$\tau = \frac{\tau_a (1-r)^2}{1-(r\tau_a)^2} = \tau_a \frac{1-r}{1+r} \quad \frac{1-r^2}{1-(r\tau_a)^2}$$

where r = polarization coefficient of glass

In this case $r_{p\ell} = 0.185$ for parallel polarization and and $r_{pr} = 0.001$ for perpendicular polarization.

(1) Calculating the mean transmissivity, we obtained

$$\tau = \frac{0.865}{2}\left[\frac{1-0.185}{1+0.185} \quad \frac{1-0.185^2}{1-(0.865\times 0.185)^2} + \frac{1-0.001}{1+0.001} \quad \frac{1-0.001^2}{1-(0.915\times 0.001)^2}\right]$$

$$= \frac{0.590 + 0.864}{2} = 0.727$$

The reflectivity is obtained by averaging the sum of parallel and perpendicular elements of polarization, or

$$\rho = r + \frac{(1-r)^2\tau_a^2 r}{1-(r\tau_a)^2} = r(1+\tau_a\tau)$$

$$\rho = \frac{r_{p\ell}(1 + \tau_a \tau_{P\ell}) + r_{Pr}(1 + \tau_a \tau_{Pr})}{2}$$

$$= \frac{0.185(1 + 0.865 \times 0.59) + 0.001(1 + 0.865 \times 0.864)}{2}$$

$$= 0.141$$

The absorptivity is calculated as:

$$\alpha = \frac{1}{2} \quad (1 - \tau_a) \quad \frac{1 - r_{p\ell}}{1 - r_{p\ell}\tau_a} + (1 - \tau_a) \quad \frac{1 - r_{Pr}}{1 - r_{Pr}\tau_a}$$

$$= \frac{1}{2} \quad (1 - 0.865) \quad \frac{1 - 0.185}{1 - 0.185 \times 0.865} + (1 - 0.865) \quad \frac{1 - 0.001}{1 - 0.001 \times 0.865}$$

$$\cong 0.133$$

Another Method:

$$\tau \cong \tau_a \tau_r \qquad \text{and } \tau_r = \frac{1}{2} \quad \frac{1 - r_{P\ell}}{1 + r_{P\ell}} + \frac{1 - r_{Pr}}{1 - r_{Pr}}$$

We can get the approximate values for transmissivity, reflectivity, and absorptivity for the glass:

$$\tau = \frac{0.865}{2} \quad \frac{1 - 0.185}{1 + 0.185} + \frac{1 - 0.001}{1 + 0.001} \quad = 0.729$$

$$\alpha = 1 - \tau_a = 1 - 0.865 = 0.135$$

$$\rho = 1 - \tau - \alpha = 1 - 0.719 - 0.135 = 0.136$$

• PROBLEM 8-41

200 Kg of water at T_i=35°C is kept in an auxiliary reservoir to be heated by solar energy. The solar collector used has a surface area of A=5 m^2. The heat removal factor and the overall heat transfer coefficient are F_R=0.80 and U_L=8.0 $W/m^2.°C$, respectively. The ambient temperature T_a=30°C. The product of the heat transfer coefficient and reservoir area is (UA)=1.80 W/°C. During an emergency, the water will be used at a rate $\dot{m}$=15 kg/hr and is refilled at a temperature T_{rf}=20°C. Determine the efficiency of this operation from 7 am to 5 pm. Refer to table 1 for hourly solar data.

Table 1

Time	I_T MJ/m²	S MJ/m²	T_a °C	$U_L(T_i - T_a)$ MJ/m²	$q = F_r S - U_L(T_i - T_a)$ MJ/m²	$\eta = \frac{q}{I_T}$
7-8	0.03	-	-12	1.35	-	-
8-9	0.49	0.40	-9	1.27	0	0
9-10	1.12	0.92	-3	1.09	0	0
10-11	4.02	3.26	3	1.01	1.80	0.45
11-12	3.45	3.08	4	0.89	1.75	0.51
12-1	4.15	3.39	7	0.81	2.20	0.53
1-2	3.96	3.20	8	0.78	1.94	0.49
2-3	1.98	1.58	9	0.75	0.67	0.34
3-4	1.27	1.01	10	0.72	0.23	0.18
4-5	0.04	-	8	0.78	-	-
	ΣI_T 20.51				Σq=8.59	

I_T - energy gained per hour by the solar panel.

T_a - ambient temperature each hour.

S - Incident heat flux on the solar panel.

The overall efficiency is

$$\eta_{day} = \frac{\Sigma q}{\Sigma I_T} = \frac{8.59}{20.51} = 0.42 \text{ or } 42\%$$

Table 2

Time	I_T MJ/m²	S MJ/m²	T_a °C	T_s^+ °C	Q MJ	Loss MJ	Load MJ
Begin				35.0			
7-8	0.03	-	-12	33.2	0	0.021	0.83
8-9	0.49	0.40	-9	31.5	0	0.01	0.72
9-10	1.12	0.92	-3	30.0	0	0	0.63
10-11	4.02	3.26	3	39.8	8.75	0.064	1.24
11-12	3.45	3.08	4	45.4	6.12	0.10	1.60
12-1	4.15	3.39	7	52.8	7.85	0.15	2.06
1-2	3.96	3.20	8	57.9	5.42	0.18	2.38
2-3	1.98	1.58	9	54.3	0	0.16	2.15
3-4	1.27	1.01	10	51.6	0	0.14	1.98
4-5	0.04	-	8	48.6	0	0.12	1.80
	ΣI_T =20.51				ΣQ =28.14	Losses =0.945	ΣLoads =15.39

<u>Solution:</u> The temperature of the reservoir each hour can be calculated using

$$\frac{dT_s}{d\tau} = \frac{(T_s^+ - T_s)}{\Delta\tau}$$

The temperature each hour is then written in the form

$$T_s^+ = T_s + \frac{\Delta\tau}{(mC_p)_s} \quad AF_R \left[S - U_L(T_i - T_a)\right]^+ - (UA)(T_i - T_a') \tag{1}$$

$$- (\dot{m}C_p)_{rf}(T_s - T_{rf})$$

Assume $\Delta\tau$ = 1 hour. Introducing the values into eq. (1) yields:

$$T_s^+ = T_s + \frac{1}{200 \times 4190} \quad 5 \times 0.8 \left[S - 8.0 \times 3600(T_s - T_a)\right]^+$$

$$- 1.8 \times 3600 \ (T_s - 30) - 15 \times 4190(T_s - 20)$$

Simplifying, we obtain:

$$T_s^+ = T_s + 1.20 \quad 4.0 \left[S - 0.0288(T_s - T_a)\right]^+$$

$$- 0.0065(T_s - 30) - 0.0628(T_s - 20)$$

The terms in the equation are the applicable energy released by the collector, the energy loss from the reservoir, and the heat transferred to the mass water flowing per hour, respectively.

The computed values are compiled in table 2.

The energy balance of the system may be expressed as:

$$\Sigma E = Q - \Sigma \text{Losses} - \Sigma \text{Loads}$$

$$\dot{m}C_p(T_s^+ - T_s) \approx \Sigma Q - \Sigma \text{Losses} - \Sigma \text{Loads}$$

$$(200 \text{ kg})(4190 \text{ J/kg.°C})(48.6 - 35°\text{C}) \approx 28.14 - 0.945 - 15.39$$

This concludes that the energy absorbed by the water is approximately the heat flux to the solar panel.

The performance of the system is:

$$\eta = \frac{\Sigma Q}{\Sigma I_T A} = \frac{28.14}{20.51(5)} = 0.275$$

$$= 27.5\%$$

CALCULATION OF HEAT RADIATION BY NETWORK METHOD

• PROBLEM 8-42

Compute the rate of radiant heat transfer to a spherical meat ball, 8 in. dia., at a temperature of 70°F, placed in a cubical oven, of side 2 ft., at 400°F. The emissivity of the walls of the oven is 0.8. The meat ball is wrapped in an aluminium foil having an emissivity of 0.1.

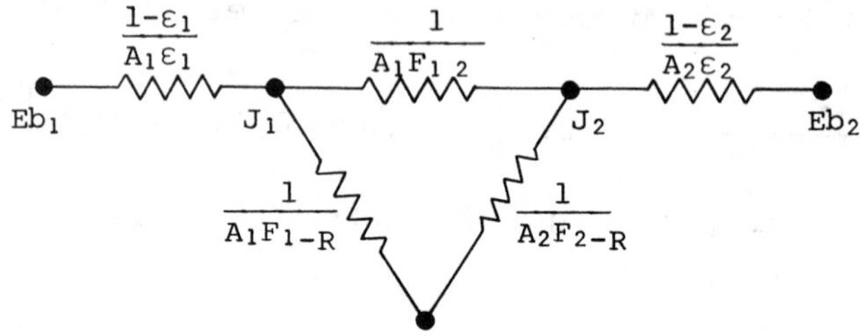

Fig. Electrical analogy for a two-gray surface system enclosed by reradiating walls.

Solution: Using the electrical analogy, the problem can be represented in the form of an electrical circuit as shown above. The notations used are as explained below.

E_b is the radiosity of the heat source (oven) and the meat ball

A is the surface area involved in the radiation heat transfer

F_{i-j} is the shape factor

ε is the emissivity of the surfaces

The subscripts 1, 2 and R correspond to the heater surface, meat ball and the other 5 walls of the oven, respectively.

The net radiant heat transfer rate can be computed by solving the electrical circuit using the equation

$$q = \frac{A_1\,\sigma(T_1^4 - T_2^4)}{\dfrac{1-\varepsilon_1}{\varepsilon_1} + \left[\dfrac{1}{F_{1-2} + \dfrac{1}{\dfrac{1}{F_{1-R}} + \dfrac{A_1}{A_2F_{2-R}}}}\right] + \dfrac{1-\varepsilon_2}{\varepsilon_2}\,\dfrac{A_1}{A_2}}$$

$T_1 = 400 + 460 = 860°R$

$T_2 = 70 + 460 = 530°R, \quad \sigma = 0.171 \times 10^{-8}$

$\varepsilon_1 = 0.8, \quad \varepsilon_2 = 0.1$

The shape factor F_{2-1} can be explained as follows:

The radiations of the meat ball fall on all the walls of the oven, therefore,

$$F_{2-1} = \frac{1}{6} \quad \text{and}$$

applying the reciprocal theorem

$$A_2F_{2-1} = A_1F_{1-2}$$

or

$$F_{1-2} = \frac{A_2}{A_1} F_{2-1}$$

$$= \frac{\pi \left(\frac{8}{12}\right)^2}{2 \times 2} \times \frac{1}{6} = 0.058$$

Again, the view factor $F_{1-R} = 1.0 - F_{1-2}$

$$= 1 - 0.058 = 0.942$$

and $F_{2-R} = \frac{5}{6}(1.0) = \frac{5}{6} = 0.833$

$$\text{Now,} \left[\frac{1}{F_{1-2} + \dfrac{1}{\dfrac{1}{F_{1-R}} + \dfrac{A_1}{A_2 F_{2-R}}}}\right] = \left[\frac{1}{0.058 + \dfrac{1}{\dfrac{1}{0.942} + \dfrac{4}{\pi\left(\frac{8}{12}\right)^2} \times \dfrac{1}{0.833}}}\right]$$

$$= 3.569$$

$$\frac{1 - \varepsilon_1}{\varepsilon_1} = \frac{1 - 0.8}{0.8} = 0.25$$

$A_1 = 2 \times 2 = 4\text{sq.ft.}$

$$\frac{1 - \varepsilon_2}{\varepsilon_2} = \frac{1 - 0.1}{0.1} = 9 \qquad \frac{A_1}{A_2} = \frac{4}{\pi\left(\frac{8}{12}\right)^2} = 2.865$$

$$T_1^4 - T_2^4 = 860^4 - 530^4 = 4.68103 \times 10^{11}$$

Substituting the values in the equation,

$$q = \frac{(4) \times (0.171 \times 10^{-8})(4.68103 \times 10^{11})}{(0.25) + (3.569) + (9 \times 2.865)}$$

$$q \cong 108 \text{ Btu/hr.}$$

• PROBLEM 8-43

The temperature of two 1 m^2 square boards, placed parallel and 0.5 m apart, are $T_1 = 1000°K$ and $T_2 = 400°K$, with their respective emissivities $\varepsilon_1 = 0.8$ and $\varepsilon_2 = 0.5$. Formulate the node relations, by network analysis, when:

(a) the boards are enclosed in a housing at 300°K, and

(b) a re-radiating wall, perfectly insulated at its outer surface, connects the two surfaces.

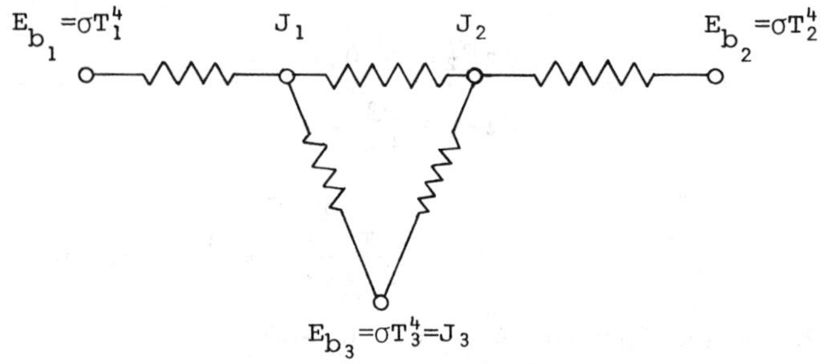

Fig. 1 Radiation network of the three body system.

Radiation shape factor for radiation between parallel rectangles.

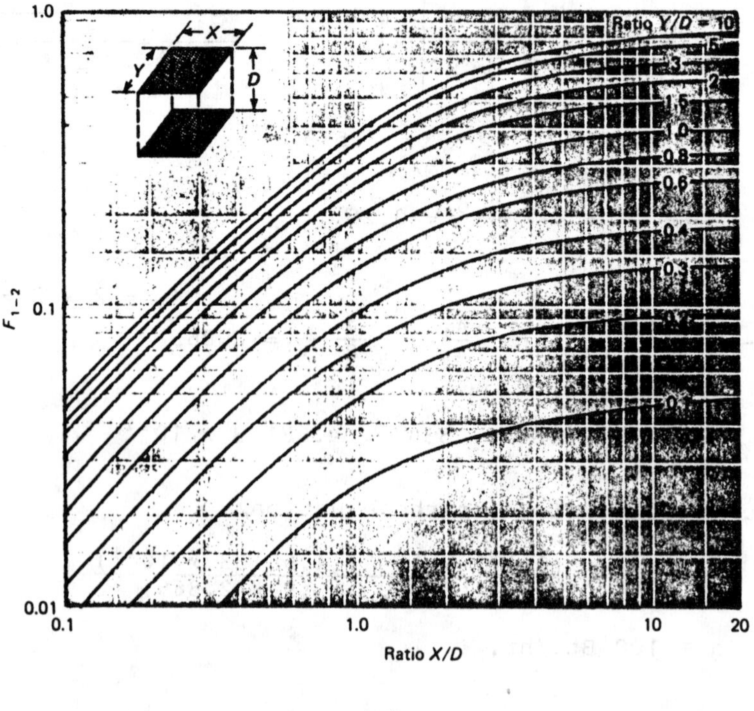

Solution: This is a three-body problem, i.e., the two boards, surfaces 1 and 2, and the receptacle, surface 3. The equivalent electrical network of the system is shown in fig.1.

Referring to fig. 2, the ratios needed to determine the shape factors are:

$$Y/D = 1.0/0.5 = 2.0, \qquad X/D = 1.0/0.5 = 2.0$$

and the shape factors are:

$$F_{1-2} = 0.46 = F_{2-1}$$

$$F_{1-1} = 0 = F_{2-2}$$

$$F_{1-3} = 1 - F_{1-2} = 1 - 0.46 = 0.54$$

$$F_{2-3} = 1 - F_{2-1} = 1 - 0.46 = 0.54$$

(a) The energies radiated per unit time and per unit area by the surfaces are:

$$E_{b1} = \sigma T_1^4 = (5.669 \times 10^{-8})(1000°K)^4 = 56690 \text{ W/m}^2$$

$$E_{b2} = \sigma T_2^4 = (5.669 \times 10^{-8})(400°K)^4 = 1451 \text{ W/m}^2$$

$$E_{b3} = \sigma T_3^4 = (5.669 \times 10^{-8})(300°K)^4 = 459 \text{ W/m}^2$$

Since the area of the receptacle is very large, the radiation shape factor from the receptacle surface to the boards will reach zero.

Using the concept of reciprocity yields:

$$A_1F_{1-3} = A_3F_{3-1}, \quad F_{3-1} = \frac{A_1F_{1-3}}{\infty} = 0$$

$$A_2F_{2-3} = A_3F_{3-2}, \quad F_{3-2} = \frac{A_2F_{2-3}}{\infty} = 0$$

For surfaces which may see themselves, the general relation is:

$$\sum_{j=1}^{n} F_{ij} = 1.0$$

Applying this relation yields:

$$F_{3-1} + F_{3-2} + F_{3-3} = 1.0, \quad F_{3-3} = 1.0$$

The relations for nodes 1, 2, and 3 are expressed by:

$$J_i - 1 - \varepsilon_i \sum_j F_{ij} J_j = \varepsilon E_{bi}$$

Node 1: $J_1 - (1 - \varepsilon_1)(F_{1-1}J_1 + F_{1-2}J_2 + F_{1-3}J_3) = \varepsilon_1 E_{b1}$

Node 2: $J_2-(1-\varepsilon_2)(F_{2-1}J_1+F_{2-2}J_2+F_{2-3}J_3) = \varepsilon_2 E_{b2}$ (1)

Node 3: $J_3-(1-\varepsilon_3)(F_{3-1}J_1+F_{3-2}J_2+F_{3-3}J_3) = \varepsilon_3 E_{b3}$

Substituting in the corresponding values for equation set (1) yields:

$$J_1-(1-0.8)\left[(0)J_1+(0.46)J_2+(0.54)J_3\right] = (0.8)(56690)$$

$$J_2-(1-0.5)\left[(0.46)J_1+(0)J_2+(0.54)J_3\right] = (0.5)(1451)$$

$$J_3-(1-\varepsilon_3)\left\lfloor(0)J_1+(0)J_2+(1.0)J_3\right\rfloor = \varepsilon_3(459)$$

Since the receptacle is very large, the radiation emitted toward it will be completely absorbed. But, the receptacle surface will still radiate heat per unit time and per unit area. Therefore, node three gives $J = 459\ W/m^2$

Simplifying the node relations yields:

$$J_1-0.092J_2-0.108J_3 = 45{,}352$$

$$-0.23J_1+J_2-0.27J_3 = 725.5$$

$$J_3 = 459$$

Solving for J_1 and J_2 yields:

$$J_1 = 46462.87\ W/m^2 = 46463\ W/m^2$$

$$J_2 = 11535.885\ W/m^2 = 11536\ W/m^2$$

The energies radiated for boards 1 and 2 are calculated as:

$$q_1 = \frac{A_1\varepsilon_1}{(1-\varepsilon_1)}(E_{b1}-J_1) = \frac{(1.0)(0.8)}{(1-0.8)}(56690-46463) = 40908\ W$$

$$q_2 = \frac{A_2\varepsilon_2}{(1-\varepsilon_2)}(E_{b2}-J_2) = \frac{(1.0)(0.5)}{(1-0.5)}(1451-11536) = -10085\ W$$

The total energy assimilated by the receptacle is:

$$q_3 = q_1+q_2 = 40908-10085 = 30823\ W$$

(b) The area of the plate joined to the board exteriors is $4.0\ m^2$.

Applying the concept of reciprocity yields:

$$A_1F_{1-3} = A_3F_{3-1},\quad F_{3-1} = \frac{A_1F_{1-3}}{A_3} = \frac{(1.0)(0.54)}{4.0} = 0.135$$

$$A_2F_{2-3} = A_3F_{3-2},\quad F_{3-2} = \frac{A_2F_{2-3}}{A_3} = \frac{(1.0)(0.54)}{4.0} = 0.135$$

and $F_{3-1}+F_{3-2}+F_{3-3} = 1.0$, $F_{3-3} = 1.0-0.135-0.135 = 0.73$

The relations for nodes 1, 2, and 3 are the same as in part (a).

Substituting in the corresponding values yields:

$$J_1 - (1 - 0.8)[(0)J_1 + (0.46)J_2 + (0.54)J_3] = (0.8)(56690)$$

$$J_2 - (1 - 0.5)[(0.46)J_1 + (0)J_2 + (0.54)J_3] = (0.5)(1451)$$

$$J_3 - (1 - \varepsilon_3)[(0.135)J_1 + (0.135)J_2 + (0.73)J_3 = \varepsilon_3 J_3$$

Simplifying the three equations yields:

$$J_1 - 0.092J_2 - 0.018J_3 = 45,352$$

$$-0.23J_1 + J_2 - 0.27J_3 = 725.5$$

$$-0.135J_1 - 0.135J_2 + 0.27J_3 = 0$$

Solving the three equations simultaneously for J_1, J_2, and J_3 yields:

$$J_1 \cong 51,424 \text{ W/m}^2$$

$$J_2 \cong 22,535 \text{ W/m}^2$$

$$J_3 \cong 36,980 \text{ W/m}^2$$

The energies radiated from the boards are:

$$q_1 = \frac{A_1\varepsilon_1}{1 - \varepsilon_1}(E_{b1} - J_1) = \frac{(1.0)(0.8)}{1 - 0.8}(56690 - 51424) = 21064 \text{ W}$$

$$q_2 = \frac{A_1\varepsilon_1}{1 - \varepsilon_1}(E_{b2} - J_2) = \frac{(1.0)(0.5)}{1 - 0.5}(1451 - 22535) \cong -21064 \text{ W}$$

It is obvious that the energies radiated from the boards compliment each other since the plate connected to them is protected from heat loss. The plate temperature may be calculated from

$$J_3 = E_{b3} = \sigma T_3^4 = 36,980 \text{ W/m}^2$$

$$T_3 = \left[\frac{36,980 \text{ W/m}^2}{5.669 \times 10^{-8}}\right]^{\frac{1}{4}} = 899 \text{ K}$$

• PROBLEM 8-44

Two rectanglular plates (0.5 m x 1 m) placed parallel to each other, 0.5 m apart, are at temperatures 1273°K and 773°K, respectively. They are placed in a huge room, which is at 300°K. Given the emissivities of the two plates as 0.2 and 0.5, respectively, calculate the net radiant heat lost by the two plates and also check for the overall energy balance in the room between the plates and the room.

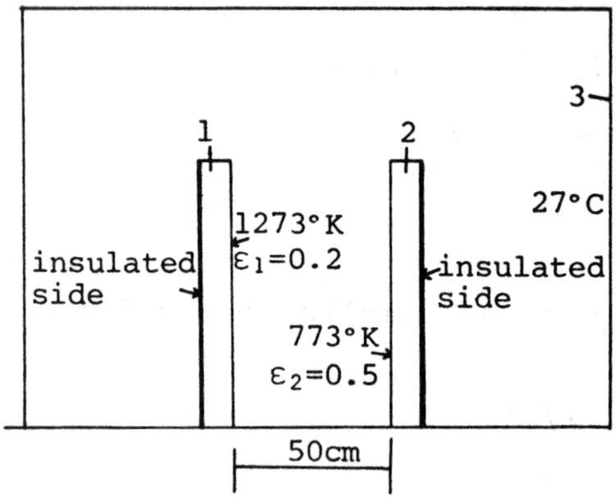

Fig. 1

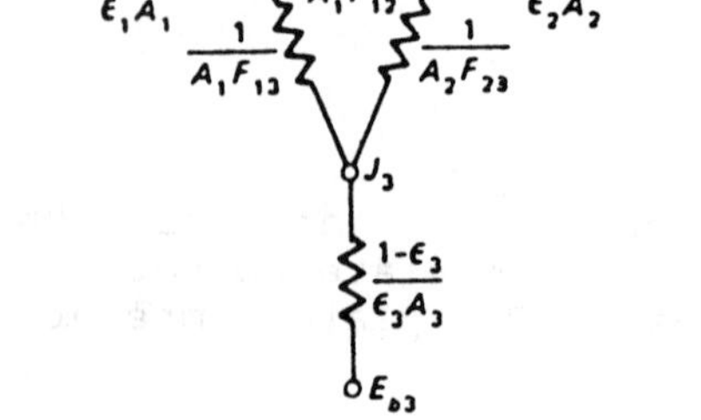

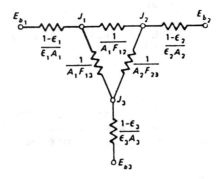

Fig. 2 Radiation network for three surfaces.

Solution: The network representation of the problem is as shown in the figure 2. In this case three bodies are involved in the complete heat transfer, i.e., the two surfaces of the plate and the six surfaces of the room. As the size of the room is very large compared to the surface area of each plate, the resistance term, expressed as

$\frac{1-\varepsilon_3}{\varepsilon_3}\frac{1}{A_3}$ where the subscript 3 corresponds to the room,

is very small and can be neglected. Then the radiosity $E_{b3} = J_3$.

Analyzing the network node by node, and considering the equilibrium at the nodes,

$$\text{for node } J_1: \frac{E_{b1}-J_1}{\frac{1-\varepsilon_1}{\varepsilon_1}\frac{1}{A_1}} + \frac{J_2-J_1}{\frac{1}{A_1F_{12}}} + \frac{E_3-J_1}{\frac{1}{A_1F_{13}}} = 0 \qquad (1)$$

$$\text{for node } J_2: \frac{E_{b3}-J_2}{\frac{1}{A_2F_{23}}} + \frac{J_1-J_2}{\frac{1}{A_1F_{12}}} + \frac{E_{b2}-J_2}{\frac{1-\varepsilon_2}{\varepsilon_2}\frac{1}{A_2}} = 0 = 0 \qquad (2)$$

The shape factor for two parallel surfaces corresponding to ratios $\frac{b}{c} = \frac{0.5}{0.5} = 1.0$ and $\frac{a}{c} = \frac{1.0}{0.5} = 2.0$ is $F_{12}=0.285$, from figure 3. Then, $F_{13}=1-F_{12}= 1-0.285 - 0.715$. Also $F_{23}= 1-F_{21}= 1-0.285=0.715$.

$E_b = \sigma T^4$, where σ is Stefan-Boltzmann constant $= 5.670 \times 10^{-8}$

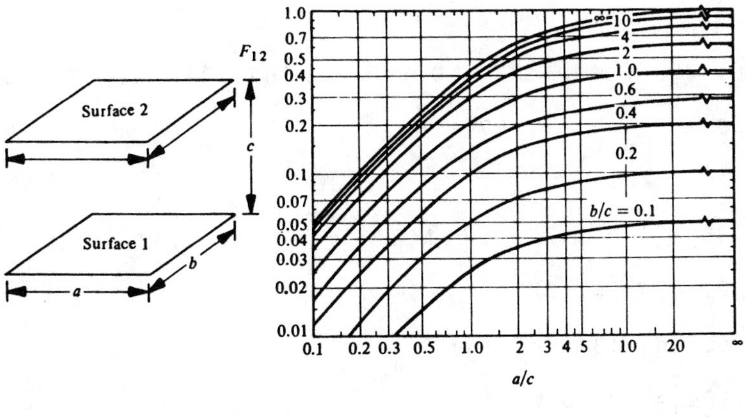

Fig. 3 The shape factor between two rectangles of equal size.

and T is the absolute temperature in Kelvin scale

$A_1 = A_2 = 0.5 \times 1 = 0.5\ m^2$

Now, $\frac{1-\varepsilon_1}{\varepsilon_1 A_1} = \frac{(1-0.2)}{(0.2)(0.5)} = 8$

$$\frac{1}{A_1 F_{12}} = \frac{1}{0.2 \times 0.285} = 7.018$$

$$\frac{1}{A_1 F_{13}} = \frac{1}{(0.5) \times (0.715)} = 2.797$$

$$\frac{1}{A_2 F_{23}} = \frac{1}{(0.5) \times (0.715)} = 2.797$$

$$\frac{1-\varepsilon_2}{\varepsilon_2 A_2} = \frac{(1-0.5)}{(0.5)(0.50)} = 2.0$$

$$E_{b1} = \sigma T_1^4 = 5.67 \times 10^{-8} \times (1273)^4$$

$$= 148900\ W/m^2$$

$$E_{b2} = \sigma T_2^4 = 5.67 \times 10^{-8} \times (773)^4$$

$$= 20244\ W/m^2$$

$$E_{b3} = \sigma T_3^4 = 5.67 \times 10^{-8} \times (300)^3$$

$$= 459\ W/m^2$$

Substituting the values in equations (1) and (2),

$$\frac{148900 - J_1}{8} + \frac{J_2 - J_1}{7.018} + \frac{459 - J_1}{2.797} = 0$$

$$\frac{459 - J_2}{2.797} + \frac{J_1 - J_2}{7.018} + \frac{20244 - J_2}{2} = 0$$

Simplifying the above equations,

$$18612.5 - 0.125J_1 + 0.142J_2 - 0.142J_1 + 164.104 - 0.358J_1 = 0$$

gives $0.625J_1 - 0.142J_2 = 18776.604$ (3)

and

$$164.104 - 0.358J_2 + 0.142J_1 - 0.142J_2 + 10122 - 0.5J_2 = 0$$

gives $J_2 - 0.142J_1 = 10286.104$ (4)

Solving equations (3) and (4),

$J_1 = 33470.104 \simeq 33470\ W/m^2$

and $J_2 = 15038.858 \simeq 15039\ W/m^2$

Total heat lost by plate 1

$$q_1 = \frac{E_{b1} - J_1}{\frac{1 - \varepsilon_1}{\varepsilon_1 A_1}}$$

$$= \frac{148900 - 33470}{8} = 14428.75 \text{ W}$$

and similarly, total heat lost by plate 2

$$q_2 = \frac{E_{b2} - J_2}{\frac{1 - \varepsilon_2}{\varepsilon_2 A_2}} = \frac{20244 - 15039}{2}$$

$$= 2602.5 \text{ W}$$

Total heat received by the room is

$$q_3 = \frac{J_1 - J_3}{\frac{1}{A_1 F_{13}}} + \frac{J_2 - J_3}{\frac{1}{A_2 F_{23}}} \quad (J_3 = E_{b3})$$

$$= \frac{33470 - 459}{2.797} + \frac{15039 - 459}{2.797}$$

$$q_3 \cong 17015 \text{ W}$$

From an overall energy balance standpoint

$$q_3 \cong q_1 + q_2,$$

for the net energy lost by both the plates must be absorbed by the room.

• PROBLEM 8-45

A box (fig. 1) with the top, bottom, and right surfaces as black bodies are kept at constant temperatures of $T_1 = 680°R$, $T_2 = 770°R$, and $T_3 = 880°R$, respectively. The left panel is a re-emitting wall. The front and rear panels are considered as one re=emitting area. Calculate (a) the radiant energy of each wall and (b) the radiant energy of each wall if the surfaces are non-black bodies with emissivities of $\varepsilon_1 = 0.75$, $\varepsilon_2 = 0.65$, and $\varepsilon_3 = 0.55$.

Solution: (a) The radiation shape factors for the system may be obtained by referring to figs. 2 and 3.

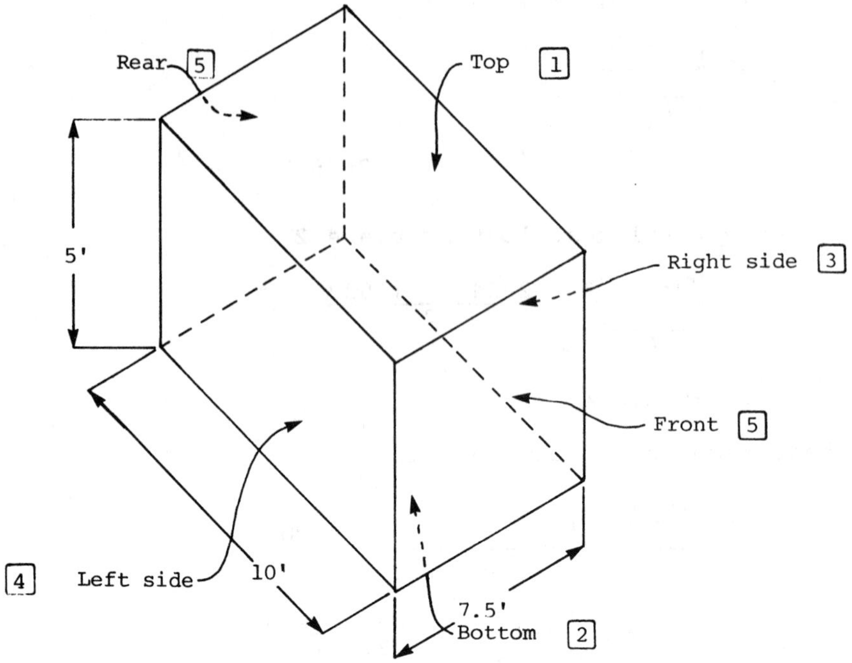

Fig 1

Figure 2. The radiation shape factor for perpendicular rectangles with a common edge.

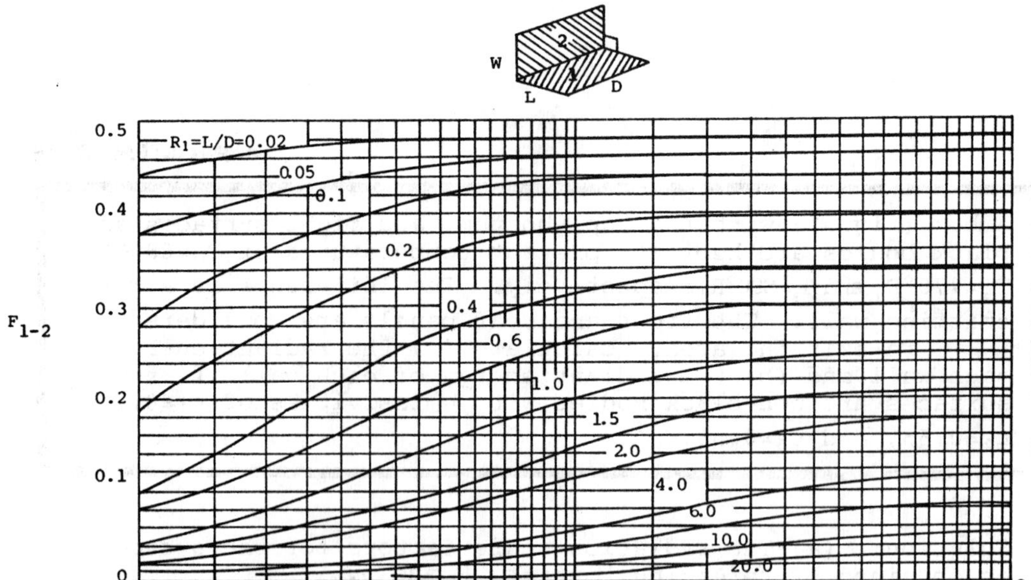

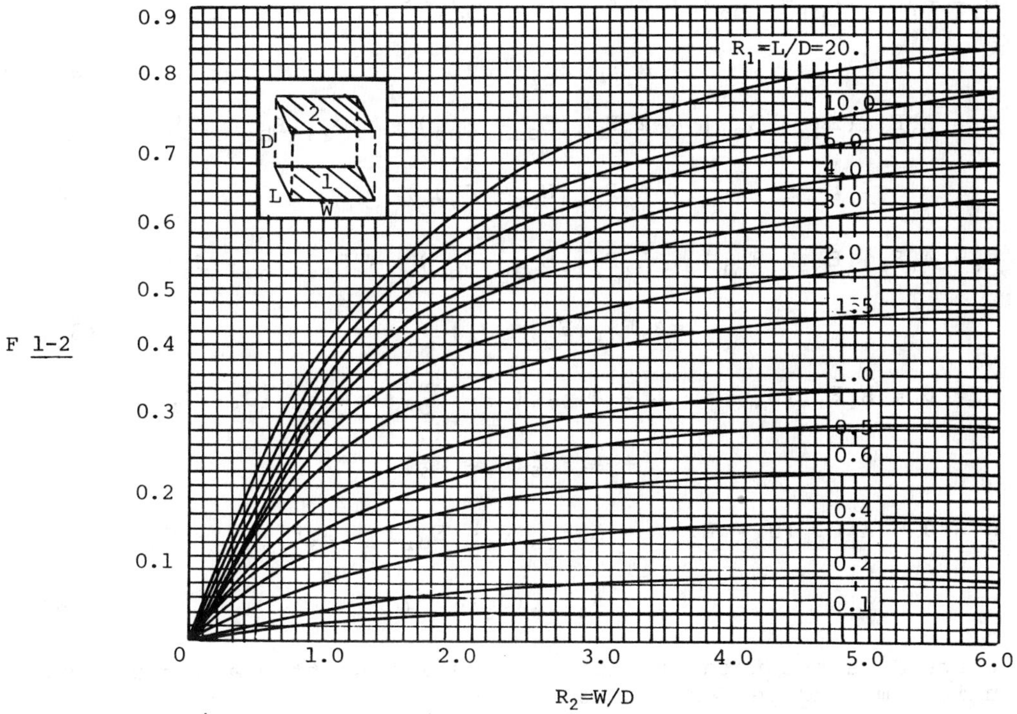

Figure 3. The radiation shape factor for parallel, directly opposed, rectangles.

For the parallel surfaces, we have:

$$F_{12} = 0.36 = F_{21}, \quad F_{34} = 0.17 = F_{43}$$

For the adjacent surfaces, we have:

$$F_{13} = F_{23} = F_{14} = F_{24} = 0.185$$

$$F_{15} = F_{25} = 0.27$$

$$F_{35} = 0.27$$

$$F_{54} = 0.18 = F_{45}$$

The emissive powers of each surface are:

$$E_{b1} = \sigma T_1^4 = (0.1714 \times 10^{-8})(680°R)^4 = 366 \text{ Btu/hr.ft}^2$$

$$E_{b2} = \sigma T_2^4 = (0.1714 \times 10^{-8})(770°R)^4 = 603 \text{ Btu/hr.ft}^2$$

$$E_{b3} = \sigma T_3^4 = (0.1714 \times 10^{-8})(880°R)^4 = 1028 \text{ Btu/hr.ft}^2$$

The energy emitting from a surface is equal to the sum of radiant transfers between itself and to the other surfaces.

$$q_1 = A_1F_{12}(E_{b1} - E_{b2}) + A_1F_{13}(E_{b1} - E_{b3}) + A_1F_{15}(E_{b1} - J_5) + A_1F_{14}(E_{b1} - J_4) \quad (1a)$$

$$q_2 = A_1F_{12}(E_{b2} - E_{b1}) + A_2F_{23}(E_{b2} - E_{b3}) + A_2F_{25}(E_{b2} - J_5) + A_2F_{24}(E_{b2} - J_4) \quad (2a)$$

$$q_3 = A_1F_{13}(E_{b3} - E_{b1}) + A_2F_{23}(E_{b3} - E_{b2}) + A_3F_{35}(E_{b3} - J_5) + A_3F_{34}(E_{b3} - J_4) \quad (3a)$$

$$q_4 = 0 = A_1F_{14}(J_4 - E_{b1}) + A_2F_{24}(J_4 - E_{b2}) + A_3F_{34}(J_4 - E_{b3}) + A_5F_{45}(J_4 - J_5) \quad (4a)$$

$$q_5 = 0 = A_1F_{15}(J_5 - E_{b1}) + A_2F_{25}(J_5 - E_{b2}) + A_3F_{35}(J_5 - E_{b3}) + A_5F_{45}(J_5 - J_4) \quad (5a)$$

where J is the total radiation reradiating from a surface per unit time and per unit area

where $A_1 = (7.5)(10) = 75\ ft^2$

$A_2 = (7.5)(10) = 75\ ft^2$

$A_3 = (5)(10) = 50\ ft^2$

$A_4 = (5)(10) = 50\ ft^2$

$A_5 = 2(5)(7.5) = 75\ ft^2$

Substituting the values,

$$q_1 = (75)(0.36)[366 - 603] + (75)(0.185)(366 - 1028) + (75)(0.27)[366 - J_5] + (75)(0.185)[366 - J_4] \quad (1b)$$

$$q_2 = (75)(0.36)[603 - 366] + (75)(0.185)[603 - 1028] + (75)(0.27)[603 - J_5] + (75)(0.185)[603 - J_4] \quad (2b]$$

$$q_3 = (75)(0.185)[1028 - 366] + (75)(0.185)[1028 - 603] + (50)(0.27)[1028 - J_5] + (50)(0.17)[1028 - J_4] \quad (3b)$$

$$0 = (75)(0.185)[J_4 - 336] + (75)(0.185)[J_4 - 603] + (50)(0.17)[J_4 - 1028] + (75)(0.18)[J_4 - J_5] \quad (4b)$$

$$0 = (75)(0.27)[J_5 - 366] + (75)(0.27)[J_5 - 603] + (50)(0.27)[J_5 - 1028] + (75)(0.18)[J_5 - J_4] \quad (5b)$$

On simplification,

$$q_1 = -3094.5 - 20.25J_5 - 13.875J_4 \quad (1c)$$

$$q_2 = 21079.5 - 20.25J_5 - 13.875J_4 \quad (2c)$$

$$q_3 = 37698.125 - 13.5J_5 - 8.5J_4 \quad (3c)$$

$$0 = -22182.875 + 49.75J_4 - 13.5J_5 \quad (4c)$$

$$0 = -33500.25 + 67.5J_5 - 13.5J_4 \quad (5c)$$

Solving for J_4 and J_5 in equations (4c) and (5c),

$$J_4 = 613.877 \text{ Btu/hr.ft}^2 \simeq 614 \text{ Btu/hr.ft}^2$$

and $$J_5 = 619.075 \text{ Btu/hr.ft}^2 \simeq 619 \text{ Btu/hr.ft}^2$$

Substituting the values of J_4 and J_5 in equations (1c), (2c) and (3c),

$$q_1 = -24148.5 \text{ Btu/hr}$$

$$q_2 = +25.5 \text{ Btu/hr}$$

and $$q_3 = +24113.6 \text{ Btu/hr}$$

(b) The radiant energy from a surface is represented as:

$$q = A(J - G)$$

$$= A\frac{\varepsilon}{1-\varepsilon}(E_b - J) \quad (1)$$

where G is the irradiation or the total radiation incident upon a surface per unit time and per unit area.

Or, in a more general relation:

$$q_i = \sum_{j=i}^{n} A_i F_{ij} (J_i - J_j) + A_i C_i (J_i - E_{bi}) \quad (2)$$

Applying eq. (2) and using the given emissivities, we have:

$$q_1 = 27(J_1 - J_2) + 13.875(J_1 - J_3) + 13.875(J_1 - J_4) + 20.25(J_1 - J_5) + (75)(3.0)(J_1 - 366) = 0$$

$$q_2 = 27(J_2 - J_1) + 13.875(J_2 - J_3) + 13.875(J_2 - J_4) + 20.25(J_2 - J_5) + (75)(1.86)(J_2 - 603.0) = 0$$

$$q_3 = 9.25(J_3 - J_1) + 9.25(J_3 - J_2) + 8.5(J_3 - J_4) + 13.5(J_3 - J_5) + 50(1.22)(J_3 - 1028) = 0$$

$$q_4 = 9.25(J_4 - J_1) + 9.25(J_4 - J_2) + 8.5(J_4 - J_3) + 9.0(J_4 - J_5) = 0$$

$$q_5 = 20.25(J_5 - J_1) + 20.25(J_5 - J_2) + 20.25(J_5 - J_3) + 13.5(J_5 - J_4) = 0$$

Simplifying, we have:

$$300J_1 - 27J_2 - 13.875J_3 - 13.875J_4 - 20.25J_5 = 82350$$

$$-27J_1 + 214.5J_2 - 13.875J_3 - 13.875J_4 - 20.25J_5 = 84118.5$$

$$-9.25J_1 - 9.25J_2 + 101.5J_3 - 8.5J_4 - 13.5J_5 = 62708$$

$$-9.25J_1 - 9.25J_2 - 8.5J_3 + 36.0J_4 - 9.0J_5 = 0$$

$$-20.25J_1 - 20.25J_2 - 20.25J_3 - 13.5J_4 + 74.25J_5 = 0$$

Using the reduced echelon technique to solve the equations, we obtained

$J_1 = 385$ Btu/hr.ft^2

$J_2 = 564$ Btu/hr.ft^2

$J_3 = 784$ Btu/hr.ft^2

$J_4 = 551$ Btu/hr.ft^2

$J_5 = 546$ Btu/hr.ft^2

Substituting the values into eq. (1), we obtain:

$$q_1 = A_1C_1(E_{b1} - J_1) = 75(3.0)(366 - 385) = -4275 \text{ Btu/hr}$$

$$q_2 = A_2C_2(E_{b2} - J_2) = 75(1.86)(603.0 - 564) = 5440 \text{ Btu/hr}$$

$$q_3 = A_3C_3(E_{b3} - J_3) = 50(1.22)(1028 - 784) = 14884 \text{ Btu/hr}$$

(1) Where $$C_i = \frac{\varepsilon_i}{1 - \varepsilon_i}$$

$$C_1 = \frac{\varepsilon_1}{1 - \varepsilon_1} = \frac{0.75}{1 - 0.75} = 3$$

$$C_2 = \frac{\varepsilon_2}{1 - \varepsilon_2} = \frac{0.65}{1 - 0.65} = 1.86$$

and $$C_3 = \frac{\varepsilon_3}{1 - \varepsilon_3} = \frac{0.55}{1 - 0.55} = 1.22$$

CALCULATION OF HEAT RADIATION BY GEBHART METHOD

• PROBLEM 8-46

A very long heat duct (see fig. 1) has its surfaces 1, 2, and 3 emitting heat at temperatures of $T_1 = 1100°R$, $T_2 = 1300°R$, and $T_3 = 1500°R$, respectively. Surface 4 is the reradiating gray surface with an emissivity of $\varepsilon_4 = 0.25$. Surfaces 1, 2, and 3 have emissivities of $\varepsilon_1 = 0.65$, $\varepsilon_2 = 0.75$, and $\varepsilon_3 = 0.82$, respectively. Calculate the energy radiated by each of the surfaces. The radiation shape factors of the system are given as:

$F_{1-1} = F_{2-2} = F_{3-3} = F_{4-4} = 0$

$F_{1-2} = F_{2-1} = F_{1-3} = F_{3-1} = F_{2-4} = F_{4-2} = F_{3-4} = F_{4-3} = 0.290$

$F_{1-4} = F_{4-1} = F_{2-3} = F_{3-2} = 0.41$

and surface area of each face = 2.25 ft²

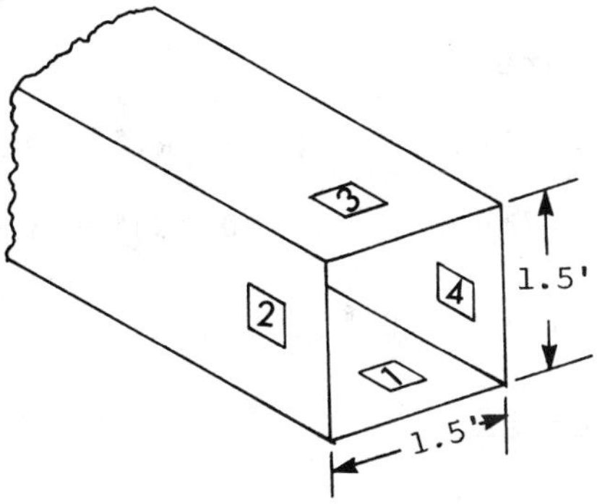

Fig. 1

Solution: The problem is approached by Gebhart and Hottel's technique. First, the total radiation incident upon a surface per unit time and per unit area can be expressed by:

$$G_k A_K = (E_1 + J_1)F_{11}A_1 + J_2 F_{21} A_2 + \ldots J_n F_{n1} A_n \tag{1}$$

When there are n surfaces involved, the following relations are introduced:

$$\rho_1 \left[(E_1 + J_1)F_{11}A_1 + J_2F_{21}A_2 + \ldots + J_n F_{n1} A_n\right] = A_1 J_1$$

$$\rho_2 \left[(E_1 + J_1)F_{12} A_1 + J_2F_{22}A_2 + \ldots + J_n F_{n2} A_n\right] = A_2 J_2 \tag{2}$$

$$\vdots \qquad\qquad \vdots$$

$$\rho_n \left[(E_1 + J_1)F_{1n}A_1 + J_2 F_{2n}A_2 + \ldots + J_n F_{nn} A_{nn}\right] + A_n J_n$$

Equation set (2) may be expressed as an augmented martrix by allowing $F_{jk}A_j = G_{jk}$:

$G_{11} = F_{11}A_1 = 0(2.25) = 0$, $G_{12} = F_{12}A_1 = 0.29(2.25) = 0.6525$

$G_{13} = F_{13}A_1 = 0.29(2.25) = 0.6525$, $G_{14} = F_{14}A_1 = 0.41(2.25) = 0.9225$

$G_{21} = F_{21}A_2 = 0.29(2.25) = 0.6525$, $G_{22} = F_{22}A_2 = 0(2.25) = 0$

$G_{23} = F_{23}A_2 = 0.41(2.25) = 0.9225$, $G_{24} = F_{24}A_2 = 0.29(2.25)$

$= 0.6525$

$G_{31} = F_{31}A_3 = 0.29(2.25) = 0.6525$, $G_{32} = F_{32}A_3 = 0.41(2.25)$

$= 0.9225$

$G_{33} = F_{33}A_3 = 0.(2.25) = 0$, $G_{34} = F_{34}A_3 = 0.29(2.25) = 0.6525$

$G_{41} = F_{41}A_4 = 0.41(2.25) = 0.9225$, $G_{42} = F_{42}A_4 = 0.29(2.25)$

$= 0.6525$

$G_{43} = F_{43}A_4 = 0.29(2.25) = 0.6525$, $G_{44} = F_{44}A_4 = 0(2.25) = 0$

These values of G_{jh} will be used in the matrix.

The emissive powers are:

$E_{b1} = \varepsilon_1 \sigma T_1^4 = 0.65(0.1714 \times 10^{-8})(1100°R)^4 = 1.63 \times 10^3$ Btu/hr.ft^2

$E_{b2} = \varepsilon_2 \sigma T_2^4 = 0.75(0.1714 \times 10^{-8})(1300°R)^4 = 3.67 \times 10^3$ Btu/hr.ft^2

$E_{b3} = \varepsilon_3 \sigma T_3^4 = 0.82(0.1714 \times 10^{-8})(1500°R) = 7.11 \times 10^3$ Btu/hr.ft^2

The matrix is written as:

J_1	J_2	J_3	J_n	Constant
$G_{11}-(A_1/\rho_1)$	G_{21}	G_{31}	G_{n1}	$-G_{11}\varepsilon_1 E_{b1}$
G_{12}	$G_{22}-(A_2/\rho_2)$	G_{32}	G_{n2}	$-G_{12}\varepsilon_1 E_{b1}$
G_{13}	G_{23}	$G_{33}-(A_3/\rho_3)$	.	.
.	.	.	.	.
.	.	.	.	.
G_{1n}	G_{2n}	G_{3n}	$G_{nn}-(A_n/\rho_n)$	$-G_{1n}\varepsilon_1 E_{b1}$

The reflectivities of the system are:

$\rho_1 = 1 - \varepsilon_1 = 1 - 0.65 = 0.35, \quad \rho_2 = 1 - \varepsilon_2 = 1 - 0.75 = 0.25$

$\rho_3 = 1 - \varepsilon_3 = 1 - 0.82 = 0.18, \quad \rho_4 = 1 - \varepsilon_4 = 1 - 0.25 = 0.75$

In order to solve for the n unknowns, the determinant D and ${}_1D_2'$ will be found first. The determinant ${}_1D_2'$ is when the 2nd column of the matrix is replaced by the constant.

Substituting the values into the matrix and solving for the determinants yield:

$$D = \begin{vmatrix} -6.43 & 0.6525 & 0.6525 & 0.9225 \\ 0.6525 & -9.0 & 0.9225 & 0.6525 \\ 0.6525 & 0.9225 & -12.5 & 0.6525 \\ 0.9225 & 0.6525 & 0.6525 & -3.0 \end{vmatrix} = -1488$$

$${}_1D_2' = \begin{vmatrix} -6.43 & 0 & 0.6525 & 0.9225 \\ 0.6525 & -69.1 & 0.9225 & 0.6525 \\ 0.6525 & -69.1 & -12.5 & 0.6525 \\ 0.9225 & -977.4 & 0.6525 & -3.0 \end{vmatrix} = -189$$

The transfer factor f is calculated as:

$$f_{1-2} = \frac{\alpha_2 \varepsilon_1 A_2}{\rho_2 A_1} \frac{{}_1D_2}{D} = \frac{(0.75)(0.65)(2.25)}{(0.25)(2.25)} \times \frac{(-189)}{(-1488)} = 0.2476$$

Next, the determinant of ${}_1D_3'$ and ${}_1D_4'$ will be calculated in order to find f_{1-3} and f_{1-4}. Since the procedure is the same as for $f_{1\ 2}$, it will not be shown.

$f_{1-1} = 0.2895$ and $f_{1-4} = 0.875$

Also, f_{1-1} can be evaluated from the following as:

$f_{1-1} = \varepsilon_1 - f_{1-2} - f_{1-3} - f_{1-4} = 0.0254$

Using the concept of reciprocity, ${}_{1-2}A_1 = {}_{2-1}A_2$;

$$\text{therefore,} \quad f_{2-1} = \frac{f_{1-2}A_1}{A_2} = \frac{(0.2476)(2.25)}{(2.25} = 0.2476$$

and $f_{3-1} = 0.3777$, $f_{4-1} = 0.1445$

Following this same procedure, the rest of the transfer factors are found to be:

$f_{1-1} = 0.0254$	$f_{2-1} = 0.2476$	$f_{3-1} = 0.2895$	$f_{4-1} = 0.0875$
$f_{1-2} = 0.2476$	$f_{2-2} = 0.1421$	$f_{3-2} = 0.3112$	$f_{4-2} = 0.0491$
$f_{1-3} = 0.2895$	$f_{2-3} = 0.3112$	$f_{3-3} = 0.1975$	$f_{4-3} = 0.0218$
$f_{1\ 4} = 0.0875$	$f_{2-4} = 0.0491$	$f_{3-4} = 0.0218$	$f_{4-4} = 0.0916$
$\varepsilon_1 = 0.65$	$\varepsilon_2 = 0.75$	$\varepsilon_3 = 0.82$	$\varepsilon_4 = 0.25$

The sum of the transfer factors must equal the corresponding emissivity.

The net energy radiated by each emitting source is:

$$q_{net1} = q_{2\rightleftharpoons 1} + q_{3\rightleftharpoons 1} + q_{4\rightleftharpoons 1}$$

The temperature of surface 4 is found by:

$$q_{net4} = 0\ q_{1\rightleftharpoons 4} + q_{2\rightleftharpoons 4} + q_{3\rightleftharpoons 4} + q_{4\rightleftharpoons 4}$$

$$= f_{1-4}A_4(E_{b4} - E_{b1}) + f_{2-4}A_4(E_{b4} - E_{b2})$$

$$+ f_{3-4}A_4(E_{b4} - E_{b3}) + f_{4-4}A_4(E_{b4} - E_{b4})$$

$$= \varepsilon_4 E_{b4} - f_{1-4}E_{b1} - f_{2-4}E_{b2} - f_{3-4}E_{b3} - f_{4-4}E_{b4}$$

$$E_{b4} = \frac{f_{1-4}E_{b1} + f_{2-4}E_{b2} + f_{3-4}E_{b3}}{\varepsilon_4 - f_{4-4}}$$

$$= \frac{0.0875(1.63 \times 10^3) + 0.0491(3.67 \times 10^3) + 0.0218(7.11 \times 10^3)}{0.25 - 0.0916}$$

$$= 3016.54 \text{ Btu/hr. ft}^2$$

$$\text{Hence, } T_4^4 = \frac{E_4}{\varepsilon_4 \sigma} = \frac{3016.54}{0.25(0.1714 \times 10^{-8})} = 7.04 \times 10^{12}$$

$$T_4 = 1628.9°R$$

Finally, for

$$q_{net1} = f_{1-2}A_1[E_{b2} - E_{b1}] + f_{1-3}A_1(E_{b3} - E_{b1})$$

$$+ f_{1-4}A_1\left[E_{b4} - E_{b1}\right]$$

$$= (0.2476)(2.25)[3.67 \times 10^3 - 1.63 \times 10^3]$$

$$+ (0.2895)(2.25)[7.11 \times 10^3 - 1.63 \times 10^3]$$

$$+ (0.0875)(2.25)[3.016 \times 10^3 - 1.63 \times 10^3]$$

$$q_{net1} \simeq 4954 \text{ Btu/hr.}$$

$$q_{net2} = q_{1\rightleftharpoons 2} + q_{2\rightleftharpoons 2} + q_{3\rightleftharpoons 2} + q_{4\rightleftharpoons 2}$$

$$q_{net2} = f_{1-2}A_2(E_{b2} - E_{b1}) + f_{3-2}A_2(E_{b2} - E_{b3})$$

$$+ f_{4-2}A_2(E_{b2} - E_{b4})$$

$$=(0.2476)(2.25)[3.67 \times 10^3 - 1.63 \times 10^3)$$

$$+ (0.3112)(2.25)[3.67 \times 10^3 - 7.11 \times 10^3]$$

$$+ (0.0491)(2.25)[3.67 \times 10^3 - 3.016 \times 10^3]$$

$$q_{net2} \simeq -1200 \text{ Btu/hr}$$

and $q_{net3} = q_{1 \rightleftharpoons 3} + q_{2 \rightleftharpoons 3} + q_{3 \rightleftharpoons 3} + q_{4 \rightleftharpoons 3}$

$$q_{net3} = f_{1-3}A_3(E_{b3} - E_{b1}) + f_{2-3}A_3(E_{b3} - E_{b2}) + 0$$

$$f_{4-3}A_3 (E_{b3} - E_{b4})$$

$$= (0.2895)(2.25)(7.11 \times 10^3 - 1.63 \times 10^3)$$

$$+ (0.3112)(2.25)(7.11 \times 10^3 - 3.67 \times 10^3)$$

$$+ (0.0218)(2.25)(7.11 \times 10^3 - 3.016 \times 10^3)$$

$$\therefore q_{net3} \simeq 6154 \text{ Btu/hr}$$

For a more simplified solution, the emissivity of the reradiating panel can be taken as zero. But the temperature of the reradiating panel cannot be calculated.

• PROBLEM 8-47

Determine the net energy transferred for a square channel (fig. 1) at each surface of different material. The properties of the surfaces are given in table 1.

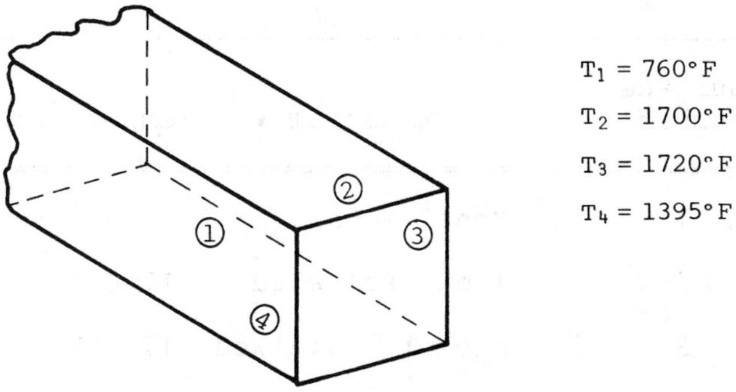

Fig. 1 Channel for semigray problem.

Solution: The problem to be solved is a Four-Node System. Each surface will have the node energy depicted as:

(1,0,0,0), (0,1,0,0,), (0,0,1,0), and (0,0,0,1).

The radiation shape factors of each surface will be equal to its corresponding emissivity:

$$F_{11} + F_{12} + F_{13} + F_{14} = \varepsilon_1$$

or, 0.0598 + 0.0909 + 0.2336 + 0.0557 = 0.440

$$F_{21} + F_{22} + F_{23} + F_{24} = \varepsilon_2$$

0.0683 + 0.0491 + 0.1765 + 0.0461 = 0.34

$$F_{31} + F_{32} + F_{33} + F_{34} = \varepsilon_3$$

0.1552 + 0.1791 + 0.2880 + 0.0877 = 0.7100

$$F_{41} + F_{42} + F_{43} + F_{44} = \varepsilon_4$$

0.074 + 0.0632 + 0.1135 + 0.0189 = 0.27

Next, the emissive powers of the walls are

$$E_{b1} = \varepsilon_1 \sigma T_1^4 = 0.44(.1714 \times 10^{-8})(760°F + 460°R)^4$$

$$= 1.67 \times 10^3 \text{ Btu/hr. ft}^2$$

$$E_{b2} = \varepsilon_2 \sigma T_2^4 = 0.34(.1714 \times 10^{-8})(1700°F + 460°R)^4$$

$$= 1.268 \times 10^4 \text{ Btu/hr.ft}^2$$

$$E_{b3} = \varepsilon_3 \sigma T_3^4 = 0.71(.1714 \times 10^{-8})(1720°F + 460°R)^4$$

$$= 2.748 \times 10^4 \text{ Btu/hr.ft}^2$$

$$E_{b4} = \varepsilon_4 \sigma T_4^4 = 0.27(.1714 \times 10^{-8})(1395°F + 460°R)^4$$

$$= 5.48 \times 10^3 \text{ Btu/hr. ft}^2$$

TABLE 1

Surface Number	Material	Temperature T	Emissivity (ε)
1	Monel Metal	760°F	0.44
2	Iron, Polished	1700°F	0.34
3	Nickel, Oxidized	1720°F	0.71
4	Mild steel	1395°F	0.27

The energy transferred by each surface may be represented by:

$$q_j = \sum_{k=1}^{4} F_{kj}E_{bk} - \varepsilon_j E_{bj} \tag{1}$$

Expanding eq. (1) yields:

$$q_1 = F_{11}E_{b1} + F_{21}E_{b2} + F_{31}E_{b3} + F_{41}E_{b4} - \varepsilon_1 E_{b1}$$

$= 0.0598(1.67 \times 10^3) + 0.0683(1.268 \times 10^4) + 0.1552(2.748 \times 10^4)$

$+ 0.0744(5.48 \times 10^3) - 0.44(1.67 \times 10^3) = 4.9 \times 10^3$ Btu/hr.ft^2

$$q_2 = F_{12}E_{b1} + F_{22}E_{b2} + F_{32}E_{b3} + F_{42}E_{b4} - \varepsilon_2 E_{b2}$$

$= 0.0909(1.67 \times 10^3) + 0.0491(1.268 \times 10^4) + 0.1791(2.748 \times 10^4)$

$+ 0.0632(5.48 \times 10^3) - 0.34(1.268 \times 10^4) = 1.73 \times 10^3$ Btu/hr.ft^2

$$q_3 = F_{13}E_{b1} + F_{23}E_{b2} + F_{33}E_{b3} + F_{43}E_{b4} - \varepsilon_3 E_{b3}$$

$= 0.2336(1.67 \times 10^3) + 0.1765(1.268 \times 10^4) + 0.2880(2.748 \times 10^4)$

$+ 0.1135(5.48 \times 10^3) - 0.71(2.748 \times 10^4) = -8.35 \times 10^3$ Btu/hr.ft^2

$$q_4 = F_{14}E_{b1} + F_{24}E_{b2} + F_{34}E_{b3} + F_{44}E_{b4} - \varepsilon_4 E_{b4}$$

$= 0.0557(1.67 \times 10^3) + 0.0461(1.26 \times 10^4) + 0.0877(2.748 \times 10^4)$

$+ 0.0189(5.48 \times 10^3) - 0.27(5.48 \times 10^3) = 1.71 \times 10^3$ Btu/hr.ft^2

• PROBLEM 8-48

A gas burner encloser (see fig. 1) is protected from heat loss on the top surface and the side walls, except the bottom panel and the door. The temperatures of the bottom panel and the door are T_1=1400°F and T_2=750°F, respectively. The emissivities of the bottom and door surfaces are $\varepsilon_1 = \varepsilon_2 = 0.68$. Consider the top panel and side walls as one surface and reradiating, with emissivity as $\varepsilon_3 = 0$. Determine the net energy radiating from the bottom panel.

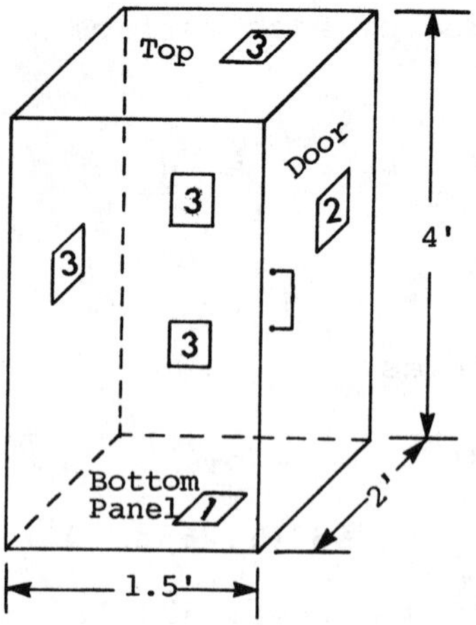

Fig. 1

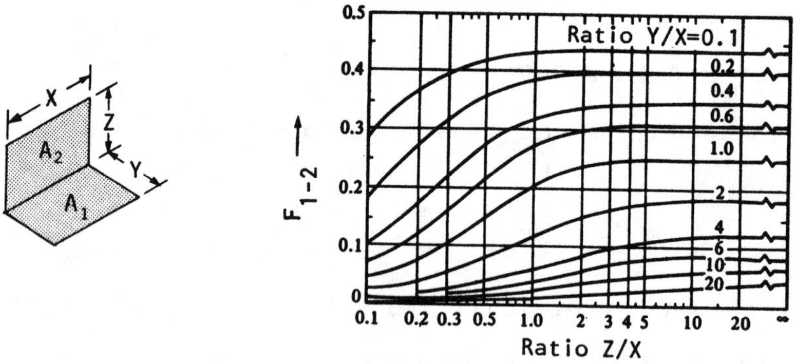

Fig. 2 Radiation shape factor for radiation between perpendicular rectangles with a common edge.

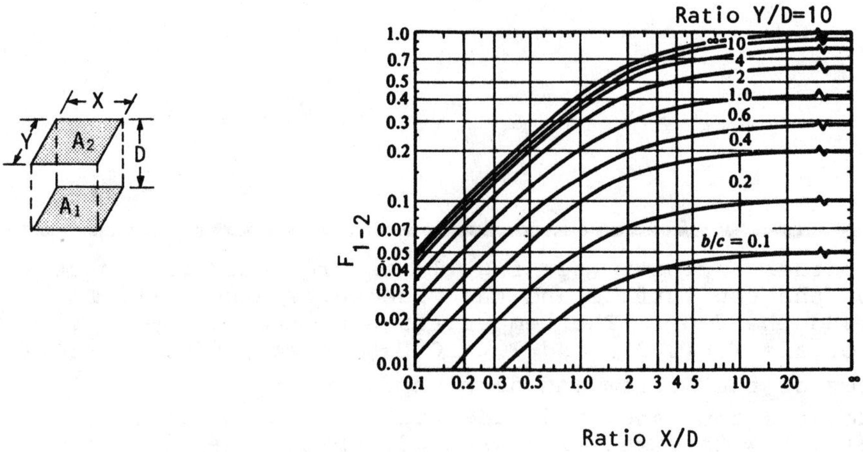

Fig. 3 Radiation shape factor for radiation between parallel rectangles.

Solution: The problem can be solved by Gelhart's technique. First the radiation shape factors for the system will be obtained from graphs in figures (3) and (4).

For the adjacent surfaces, referring to figure 2,

for $\frac{Z}{X} = \frac{4}{2} = 2$ and $\frac{Y}{X} = \frac{1.5}{2} = 0.75$

$$F_{12} = F_{21} = 0.26$$

$F_{11} = F_{22} = 0$ as the surfaces cannot see themselves.

For enclosed surfaces the following relations apply:

$$F_{11} + F_{12} + F_{13} = 1.0 \quad \therefore \ F_{13} = 1 - 0 - 0.26 = 0.74$$

$$F_{21} + F_{22} + F_{33} = 1.0 \quad \therefore \ F_{23} = 1 - 0 - 26 - 0 = 0.74$$

For parallel surfaces, referring to figure 3,

$$\frac{X}{D} = \frac{1.5}{4} = 0.375 \text{ and } \frac{Y}{D} = \frac{2}{4} = 0.5$$

$$\therefore \ F_{31} = F_{32} = 0.58$$

Also $F_{31} + F_{32} + F_{33} = 1 \quad \therefore F_{33} = 1 - 0.058 - 0.058 = 0.884.$

The absorption factor B_{nj} is represented by the following equation set:

$$\begin{aligned} B_{1j} &= \varepsilon_j F_{ij} + \rho_1 F_{11} B_{1j} + \rho_2 F_{12} B_{2j} + \ldots \rho_n F_{1n} B_{nj} \\ B_{2j} &= \varepsilon_j F_{2j} + \rho_1 F_{21} B_{1j} + \rho_2 F_{22} B_{2j} + \ldots \rho_n F_{2n} B_{nj} \\ B_{nj} &= \varepsilon_j F_{nj} + \rho_1 F_{n1} B_{1j} + \rho_2 F_{n2} B_{2j} + \ldots \rho_n F_{nn} B_{nj} \end{aligned} \qquad (1)$$

where n = number of surfaces = 3

j = surface under consideration, in this case it is 2

For n = 3 and j = 2, equation set (1) may be written as:

$$(\rho_1 F_{11} - 1)B_{12} + \rho_2 F_{12} B_{22} + \rho_3 F_{13} B_{32} = -\varepsilon_2 F_{12}$$

$$\rho_1 F_{21} B_{12} + (\rho_2 F_{22} - 1)B_{22} + \rho_3 F_{23} B_{32} = -\varepsilon_2 F_{22}$$

$$\rho_1 F_{31} B_{12} + \rho_2 F_{32} B_{22} + (\rho_3 F_{33} - 1)B_{32} = -\varepsilon_2 F_{32}$$

The reflectivities are calculated as:

$\rho_2 = \rho_1 = 1 - \varepsilon_1 = 1 - 0.68 = 0.32$, $\rho_3 = 1 - \varepsilon_3 = 1 - 0 = 1.0$

Introducing the values into equations in (2),

$$(0.32 \times 0 - 1)B_{12} + (0.32 \times 0.26)B_{22} + (1 \times 0.74)B_{32} = -0.68 \times 0.26$$

$$(0.32 \times 0.26)B_{12} + (0.32 \times 0 - 1)B_{22} + (1 \times 0.74)B_{32} = -0.68 \times 0$$

$$(0.32 \times 0.058)B_{12} + (0.32 \times 0.058)B_{22} + (1 \times 0.884 - 1)B_{32}$$

$$= -0.68 \times 0.058$$

Simplifying,

$$-B_{12} + 0.0832B_{22} + 0.74B_{32} = -0\ 1768$$

$$0.0832\ B_{12} - B_{22} + 0.74B_{32} = 0$$

$$0.01856B_{12} + 0.01856B_{22} - 0.116B_{32} = -0.03944$$

Solving for B_{12}, B_{22} and B_{32},

$$B_{12} = 0.5816097 \simeq 0.582$$

$$B_{22} = 0.4183896 \simeq 0.418$$

$$B_{32} = 0.4999995 \quad 0.0500$$

The emissive powers are

$$E_1 = \varepsilon_1 \sigma T_1^4 = 0.68 \times 0.1714 \times 10^{-8} \times (1400 + 460)^4 \quad 13950 \text{ Btu/hr.ft}^2$$

$$E_2 = \varepsilon_2 \sigma T_2^4 = 0.68 \times 0.1714 \times 10^{-8} \times (750 + 460)^4 \quad 2498 \text{ Btu/hr.ft}^2$$

$$E_3 = \varepsilon_3 \sigma T_3^4 = 0$$

∴ Net heat radiated by the bottom panel is given by

$$q_{net\ 1} = E_1 A_1 - B_{12} E_1 A_1 - B_{22} E_2 A_2 - B_{32} E_3 A_3$$

Substituting the values,

$$q_{net1} = 13950 \times (2 \times 1.5) - 0.582 \times 13950 \times (2 \times 1.5)$$

$$- 0.418 \times 2498 \times (2 \times 4) - 0.500 \times 0 \times (A_3)$$

$$9140 \text{ Btu/hr}$$

• PROBLEM 8-49

A refrigerator (see fig. 1) uses Freon-12 to keep the bottom surface constant at -7°F. The top of the regrigerator is at 75°F. The top and bottom surfaces have an emissivity of 0.8 and 0.95, respectively. The remainder of the sides are in radiant balance $\varepsilon_2=0$. Assuming all the interior surfaces to be non-black bodies, calculate the radiation heat transfer rate between the Freon-12 and the bottom panel.

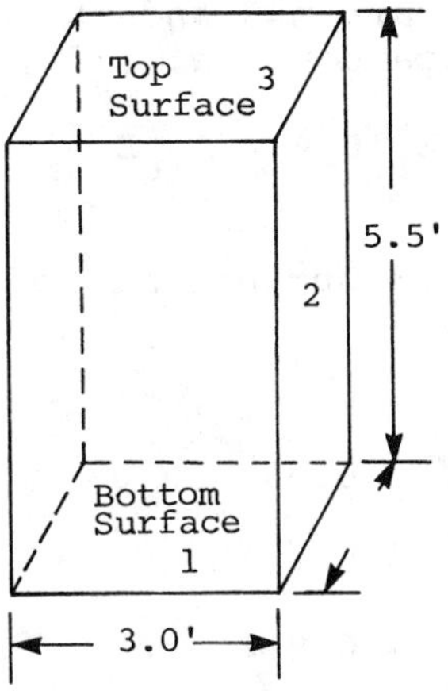

Fig. 1 Refrigerator, surface 2 is all the surfaces remaining after surfaces 1 and 3.

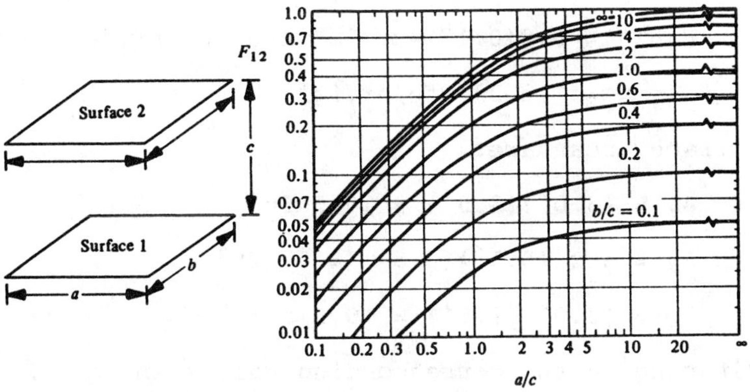

Fig. 2 Radiation factor for parallel surfaces.

Solution: The heat lost by surface 1 is equal to the energy gained by the Freon-12. The problem can be solved by Grebhart's technique. The energy transferred by the jth surface is:

$$q_j = E_j A_j - \sum_{i=1}^{n} B_{ij} E_i A_i \qquad (1)$$

where E_j = emissive power = $\varepsilon_j \sigma T_j^4$

B = absorption factor

A = area of surface

The number of surfaces are n=3 and the bottom panel is in consideration j=1. Expanding eq. (1) yields:

$$q_1 = E_1A_1 - B_{11}E_1A_1 - B_{21}E_2A_2 - B_{31}E_3A_3 \quad (2)$$

The reflectivity of each surface may be found by:

$$\rho_i = 1 - \varepsilon_i$$

$$\rho_1 = 1 - \varepsilon_1 = 1 - 0.95 = 0.05$$

$$\rho_2 = 1 - \varepsilon_2 = 1 - 0 = 1.0$$

$$\rho_3 = 1 - \varepsilon_3 = 1 - 0.8 = 0.20$$

The emissive powers are:

$$E_1 = \varepsilon_1\sigma T_1^4 = 0.95(0.1714 \times 10^{-8}\ \text{Btu/hr.ft}^2.\text{°R}^4)(453\text{°R})^4$$

$$= 68.57\ \text{Btu/hr.ft}^2$$

$$E_2 = \varepsilon_2\sigma T_2^4 = 0$$

$$E_3 = \varepsilon_3\sigma T_3^4 = 0.8(0.1714 \times 10^{-8}\ \text{Btu/hr.ft}^2.\text{°R}\ )(535\text{°R})^4$$

$$= 112.33\ \text{Btu/hr.ft}^2$$

The surface areas are:

$$A_1 = (3.0')(3.5') = 10.5\ \text{ft}^2$$

$$A_2 = 4 \times (5.5')(3.5') = 77\ \text{ft}^2$$

$$A_3 = (3.0')(3.5') = 10.5\ \text{ft}^2$$

Substituting in the corresponding values in eq. (2) yields:

$$q_1 = (68.57)(10.5) - B_{11}(68.57)(10.5) - B_{21}(0)(77) - B_{31}(112.33)(10.5)$$

$$= 720(1 - B_{11}) - 1180B_{31} \quad (2(a))$$

The equation set need for the n unknown abosrption factors is:

$$(K_{11} - 1)B_{1j} + K_{12}B_{2j} + K_{13}B_{3j} + \dots + K_{1n}B_{nj} = -\varepsilon_jF_{1j}$$

$$K_{21}B_{1j} + (K_{22} - 1)B_{2j} + K_{23}B_{3j} + \dots + K_{2n}B_{nj} = -\varepsilon_jF_{2j}$$

$$K_{31}B_{1j} + K_{32}B_{2j} + (K_{33} - 1)B_{3j} + \dots + K_{3n}B_{nj} = -\varepsilon_jF_{3j}$$

$$\vdots$$

$$K_{n1}B_{1j} + K_{n2}B_{2j} + K_{n3}B_{3j} + \dots (K_{nn} - 1)B_{nj} = \varepsilon_jF_{nj}$$

Since n=3 and j=1, we have:

$$\begin{aligned} (K_{11}-1)B_{11} + K_{12}B_{21} + K_{13}B_{31} &= -\varepsilon_1 F_{11} \\ K_{21}B_{11} + (K_{22}-1)B_{21} + K_{23}B_{31} &= -\varepsilon_1 F_{21} \qquad (3) \\ K_{31}B_{11} + K_{32}B_{21} + (K_{33}-1)B_{31} &= -\varepsilon_1 F_{31} \end{aligned}$$

The radiation shape factors are obtained form fig. 2.

$$\frac{a}{c} = 0.55, \quad \frac{b}{c} = 0.64$$

$$F_{1-3} = F_{3-1} = 0.95$$

Using the radiation factor relations in an enclosure yields:

$$F_{1-2} = F_{3-2} = 1 - F_{1-3} = 1 - F_{3-1} = 1 - 0.95 = 0.05$$

Since the surfaces cannot see themselves, we have:

$$F_{11} = F_{33} = 0$$

Applying the concept of reciprocity:

$$A_1F_{1-2} = A_2F_{2-1}, \; F_{2\;1} = \frac{A_1F_{1-2}}{A_2} = \frac{10.5(0.05)}{77} = 0.007$$

$$A_3F_{3-2} = A_2F_{2-3}, \; F_{2-3} = \frac{A_3F_{3-2}}{A_2} = \frac{10.5(0.05)}{77} = 0.007$$

$$F_{2-1} - F_{2-2} + F_{2-3} = 1.0, \; F_{2-2} = 1.0 - 0.007 - 0.007 = 0.986$$

The factor K_{ij} in equation set (3) is equal to $F_{ij}\rho_j$:

$$K_{1-1} = F_{1-1}\rho_1 = (0)(0.05) = 0, \; K_{1-2} = F_{1-2}\rho_2 = (0.05)(1.0) = 0.05$$

$$K_{1-3} = F_{1-3}\rho_3 = (0.95)(0.20) = 0.19$$

$$K_{2-1} = F_{2-1}\rho_1 = (0.007)(0.05) = 0.00035$$

$$K_{2-2} = F_{2-2}\rho_2 = (0.986)(1.0) = 0.986$$

$$K_{2-3} = F_{2-3}\rho_3 = (0.007)(0.20) = 0.0014$$

$$K_{3-1} = F_{3-1}\rho_1 = (0.95)(0.05) = 0.0475$$

$$K_{3-2} = F_{3-2}\rho_2 = (0.05)(1.0) = 0.05$$

$$K_{3-3} = F_{3-3}\rho_3 = (0)(0.20) = 0$$

Introducing these values in equation set (3) yields:

$$\begin{aligned} -B_{1-1} + 0.05B_{2-1} + 0.19B_{3-1} &= 0 \\ 0.00035B_{1-1} - 0.014B_{2-1} + 0.0014B_{3-1} &= -0.00665 \\ 0.0475B_{1-1} + 0.05B_{2-1} - B_{3-1} &= -0.9025 \end{aligned}$$

Solving the three equations simultaneously yields:

$B_{1-1} = 0.1318879 \simeq 0.132$

$B_{3-1} = 0.6423087 \simeq 0.642$

Substituting in equation (2a),

$$q_1 = (68.57)(10.5) - (0.132)(68.57)(10.5) - (112.33)(10.5)(0.642)$$

$$= -132.269 \text{ Btu/hr.}$$

Heat transfer rate = 132.269 Btu/hr.

CHAPTER 9

COMBINED MODES OF HEAT TRANSFER

Basic Attacks and Strategies for Solving Problems in this Chapter. **See pages 573 to 626 for step-by-step solutions to problems.**

Although conduction, convection and radiation have been considered separately, any two of these modes of heat transfer (or all three) may be important in a given physical situation.

Conduction and convection are typically both important when heat transfer takes place between two fluids through a solid which separates the two fluids. An example is the wall of a refrigerator. Heat is transferred from the warmer air outside the refrigerator, through the wall of the refrigerator, to the colder air inside the refrigerator. Calculations for this type of problem are handled by treating the conductive resistance of the wall and the convective resistances at the two surfaces as heat transfer resistances in series and using an electrical analogy. This type of problem was also considered in chapter 2. A major part of the problem may be determination of the convective heat transfer coefficients on each side of the wall.

Convection and radiation are typically both important for heat loss from a hot surface to a cooler fluid. Heat would be given off by both convection and radiation in this case. If the heat transfer rate is to be determined, the convective rate and radiative rate can each be calculated independently and the two simply added together. In some cases it is helpful to use a combined heat transfer coefficient, which is the sum of the convective coefficient and a radiative coefficient. The definition of the radiative coefficient and its calculation are shown in problem 9-7. The use of a combined heat transfer coefficient is especially useful when an unknown temperature is to be determined with both convection and radiation being important.

A situation in which conduction, convection, and radiation are all included is heat loss from a hot fluid inside a thick-walled container such as a pipe to a cooler fluid outside the pipe. The heat transfer will include convection from the hot fluid to the inner surface of the pipe, conduction through the pipe wall, and both convection and radiation from the outer pipe surface to the outside fluid. In many cases the radiation effect is neglected, but radiation heat transfer rate becomes rapidly larger at higher temperatures, so if the temperature of the outer pipe surface is quite high, the radiation heat transfer contribution may be important. This type of situation would be handled in the same way as combined convection and radiation transfer discussed above.

Step-by-Step Solutions to Problems in this Chapter, "Combined Modes of Heat Transfer"

CONDUCTION AND CONVECTION

• PROBLEM 9-1

Derive an analytical solution for slug flow considering the element shown in figure 1.

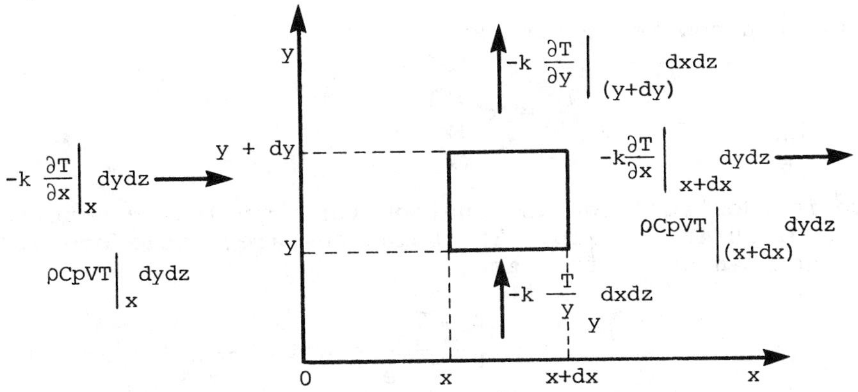

Fig. 1 Energy balance across the element.

Solution: Heat tranfer occurs across the boundaries of the element by conduction and convection. For large values of V and x in the x-direction, conductive heat transfer is very small compared to convective heat transfer, and thus may be neglected.

Considering an energy balance across the element,

$$Q_{in} = Q_{out}$$

or

$$\rho c_p VT\Big|_x dydz - k\frac{\partial T}{\partial y}\Big|_y dxdz = \rho c_p VT\Big|_{x+dx} dydz - k\frac{\partial T}{\partial y}\Big|_{y+dy} dxdz$$

Using a Taylor series, expansion of the terms on the L.H.S. of the differential equation yields

$$\rho c_p VT\Big|_{x+dx} dydz - k\frac{\partial T}{\partial y}\Big|_{y+dy} dxdz = \rho c_p VT\Big|_x dydx +$$

$$\rho c_p V\frac{\partial T}{\partial x}\Big|_x dxdydz + \ldots - k\frac{\partial T}{\partial y}\Big|_y dxdz - \frac{\partial}{\partial y}(k\frac{\partial T}{\partial y})dydxdz + \ldots$$

Assuming k to be constant,

$$\rho c_p V \frac{\partial T}{\partial x} = k \frac{\partial^2 T}{\partial y^2} \tag{1}$$

As $t = \frac{x}{V}$ or $\frac{dt}{dV} = \frac{1}{V}$,

$$\rho c_p V\frac{\partial T}{\partial x} = \rho c_p V\frac{\partial T}{\partial t}\frac{\partial t}{\partial x} = \rho c_p V\frac{\partial T}{\partial t}\frac{1}{V} = \rho c_p \frac{\partial T}{\partial t} \tag{2}$$

Eq. (1) can now be written as

$$\rho c_p \frac{\partial T}{\partial t} = k \frac{\partial^2 T}{\partial y^2}$$

which is identical to the equation for transient conduction in a semi-infinite slab. The solutions are, therefore, identical and can be written as

$$T^* = \frac{T - T_e}{T_s - T_e} = \text{erfc}\ (\frac{y}{\sqrt{4\alpha t}})$$

or

$$T^* = \text{erfc}\left(\frac{y}{\sqrt{\frac{4\alpha x}{V}}}\right) = \text{erfc}\ (\eta) \tag{3}$$

Since $\text{erfc}\ (\eta) = 1 - \frac{2}{\sqrt{\pi}} \int_0^\eta e^{-u^2} du,$

eq. (3) can be graphically represented as shown in fig. (2).

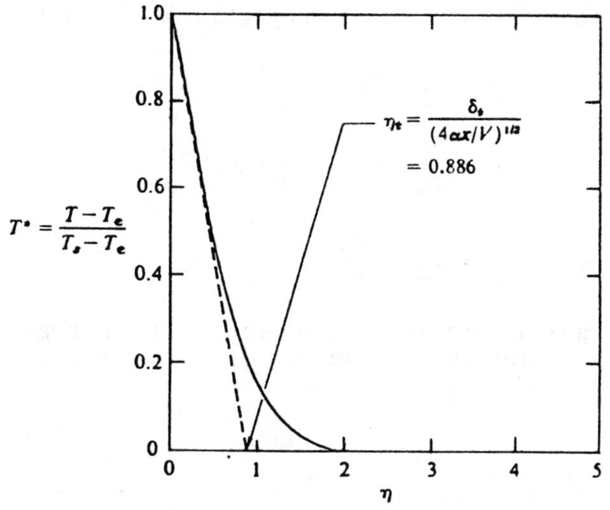

Fig. 2 Thermal boundary layer for slug flow over a heated wall.

Here, the value of the non-dimensional temperature decreases asymptotically from a value of 1 at $\eta = 0$ to zero at $\eta \simeq 2$.

The tangent to the curve crosses the η axis at some point. The intercept between this point and the point represented by $(\eta, T^*) = (0, 0)$ may be used to characterize the thickness of the heated layer. This can be done as follows:

$$\eta = \frac{y}{\sqrt{\frac{4\alpha x}{V}}}$$

$$\text{or} \quad y = \eta\sqrt{\frac{4\alpha x}{V}}$$

$$= \delta_t = 1.772\sqrt{\frac{\alpha x}{V}}$$

Heat flux, as given by the Fourier law, is

$$q = -k\left.\frac{\partial T}{\partial y}\right|_{y=0} = k(T_s - T_e)\frac{2}{\sqrt{\pi}}\frac{1}{\sqrt{\frac{4\alpha x}{V}}}$$

$$= \frac{k}{\sqrt{\pi}}(T_s - T_e)\sqrt{\frac{V}{\alpha x}}$$

This is a linear function of $(T_s - T_e)$.

Hence, by Newton's law of cooling, at low velocities

$$q = h_c (T_s - T_e)$$

and $$q = \frac{k}{\sqrt{\pi}} (T_s - T_e) \sqrt{\frac{V}{\alpha x}}$$

Therefore, $$h_c = \frac{k}{\sqrt{\pi}} \sqrt{\frac{V}{\alpha x}} .$$

This equation gives the local convective heat transfer coefficient at x. The average value of h_c over the length L of the wall or plate is

$$h_{avg} = \frac{1}{L} \int_0^L h_c dx$$

$$= \frac{1}{L} \int_0^L \frac{k}{\sqrt{\pi}} \sqrt{\frac{V}{\alpha x}} dx.$$

Integration of h_{avg}, assuming V and k to be constant over the length L, yields

$$h_{avg} = \frac{2k}{\sqrt{\pi}} \left(\frac{V}{\alpha L}\right)^{1/2}$$

$$= 1.128 \, k \left(\frac{V}{\alpha L}\right)^{1/2}$$

• PROBLEM 9-2

A horizontal steel tube is insulated with 1.5 in. of glass wool and is placed in a room at 70°F. Water vapor at 400 psia and 600°F flows through the tube at a rate of 1,000 ft/min. Calculate

(a) the overall heat transfer coefficient U.

(b) the temperature of the outer surface of the insulation.

(c) the heat loss per unit tube length.

The following data is given:

tube $k_{tube} = 25$ BTu/hr.ft.$^{\circ}$F

$k_{ins} = 0.022$ BTu/hr.ft.$^{\circ}$F

Referring to fig. 1, the numbers are assigned to identify the relevant locations.

The diameters are:

$$D_2 = 0.505 \text{ ft}$$

$$D_3 = 0.552 \text{ ft}$$

$$D_4 = 0.802 \text{ ft}$$

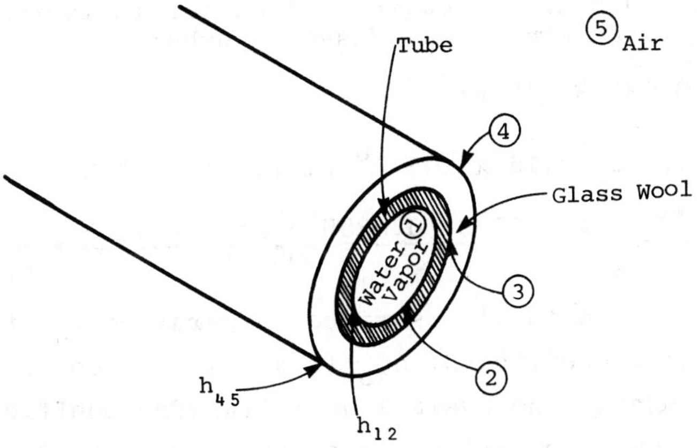

Fig. 1 Tube and insulation.

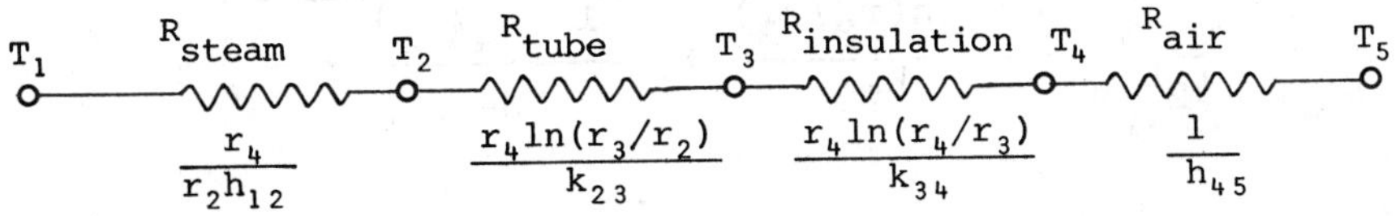

Fig. 2 Electrical analogy of resistances.

Solution:

a) From the information given, it is known that $T_1 = 600^\circ F$ and $T_5 = 70^\circ F$. The thermal properties of water vapor at 400 psia and 600°F are:

$$\rho = 0.677 \text{ lb}_m/\text{ft}^3$$

$$k_{vap} = 0.0255 \text{ BTu/hr.ft.}^\circ F$$

$$\mu = 0.0494 \ lb_m/ft.hr.$$

$$Pr = 1.12$$

The convective heat transfer coefficient of the inside surface, h_{12}, is evaluated as follows:

the Reynolds number of the flow is calculated as

$$Re = \frac{VD\rho}{\mu} = \frac{(1000 \ ft/min)(60 \ min/hr)(0.505 \ ft^2)(0.677 \ lb_m/ft^3)}{0.0444 \ lb_m/ft.hr.}$$

$$= 4.15 \times 10^5$$

The flow is obviously turbulent. Then the following relation can be used to determine the Nusselt number:

$$Nu = \frac{hD}{k} = 0.023 \ Re^{0.8} \ Pr^{0.3}$$

$$= 0.023 \ (4.15 \times 10)^{0.8} \ (1.12)^{0.3} = 743$$

Thus, $h_{12} = \frac{Nu \ k_{vap}}{D_2} = \frac{(743)(0.0255 \ Btu/hr.ft.^{\circ}F)}{0.505 \ ft} = 37.52 \ Btu/hr.ft^2 \ ^{\circ}F$

To approximate the outside surface temperature T_4, the outside convection coefficient h_{45} is assumed to be 1.0 Btu/ $hr \cdot ft^2 \cdot {}^{\circ}F$. Hence, the overall heat transfer coefficient U_4 based on the outside surface is calculated as (referring to fig. 2):

$$U_4 = \frac{1}{\Sigma R} = \frac{1}{\frac{r_4}{r_2 h_{12}} + \frac{r_4 \ln(r_3/r_2)}{k_{23}} + \frac{r_4 \ln(r_4/r_3)}{k_{34}} + \frac{1}{h_{45}}}$$

$$= \frac{1}{\frac{0.802}{(0.505)(37.52)} + \frac{0.802}{(2)(25)} \ln \frac{0.552}{0.505} + \frac{0.802}{(2)(0.04)} \ln \frac{0.802}{0.552} + \frac{1}{1}}$$

$$= \frac{1}{0.0423 + 0.00142 + 3.745 + 1.0}$$

$$= 0.209 \ Btu/hr \cdot ft^2 \cdot {}^{\circ}F.$$

An initial approximation of the outer surface temperature may be found from equating the heat transferred from the outside surface to the air to the heat transferred from the inner surface to the outside surface.

Hence

$$h_{45}(T_4 - T_5) = U_4(T_1 - T_5)$$

and

$$T_4 = \frac{U_4(T_1 - T_5)}{h_{45}} + T_5$$

$$= \frac{(0.209\ \text{Btu/hr}\cdot\text{ft}^2\cdot{}^\circ\text{F})(600^\circ - 70^\circ\text{F})}{1.0\ \text{Btu/hr}\cdot\text{ft}^2\cdot{}^\circ\text{F}} + 70^\circ\text{F}$$

$$= 180.77^\circ\text{F}$$

b) Using this value of T_4, the outside convection coeffcicient h_{45} may be recalculated as follows:

$$\Delta T_{45} = (T_4 - T_5) = (180.77 - 70^\circ\text{F}) = 110.77^\circ\text{F}$$

$$T_{m_{45}} = \frac{180.77^\circ\text{F} + 70^\circ\text{F}}{2} = 125.4^\circ\text{F}$$

The volume coefficient of expansion β , for ideal gases may be determined from the expression $\beta = \frac{1}{T}$, where T is the absolute temperature of the gas in degrees Rankine. Then

$$\beta = \frac{1}{T^\circ\text{F} + 460} = \frac{1}{70^\circ\text{F} + 460} = \frac{1}{530^\circ\text{R}} = 0.001887 \text{ per }^\circ\text{R}.$$

The thermal properties of air at $T_m = 125.4\ ^\circ\text{F}$ are:

$$\rho = 0.0678\ \text{lb}_m/\text{ft}^3$$

$$k = 0.0163\ \text{Btu/hr-ft}\ -{}^\circ\text{F}$$

$$\mu = 0.0476\ \text{lb}_m/\text{ft-hr}$$

$$Pr = 0.702$$

The Grashof number is defined as a dimensionless group representing the ratio of the buoyancy forces to the viscous forces in a free-convection flow system.

$$Gr = \frac{D^3\rho^2\beta\Delta T}{\mu^2}$$

$$= \frac{(0.802\text{ft})^3\ (0.0678\ \text{lb}_m/\text{ft}^3)^2(32.2\text{ft/sec}^2)(3600\text{sec/hr})^2}{(0.0476\ \text{lb}_m/\text{ft-hr})^2} \times$$

$$\underline{(0.001887)(110.77^\circ\text{F})}$$

$$= 9.12 \times 10^7$$

The Nusselt number for the flow inside horizontal pipes may be determined from

$$Nu = \frac{hD}{k} = 0.525\ (Gr\ Pr)^{\frac{1}{4}}$$

$$= 0.525\left[(9.96 \times 10^7)(0.702)\right]^{\frac{1}{4}} = 47$$

The convection coefficient is

$$h_{45} = \frac{Nu\ k_{air}}{D_4} = \frac{47(0.0163\ BTu/hr\cdot ft\cdot {}^0F)}{(0.802\ ft)} = 0.955\ Btu/hr\text{-}ft^2\text{-}{}^0F$$

This new value of h_{45} may now be used to calculate a new value of U_4. This is

$$U_4 = \frac{1}{0.0423 + 0.00142 + 3.745 + 1.046} = 0.2077\ Btu/hr\text{-}ft^2\text{-}\ {}^0F$$

The revised outer film temperature is, then,

$$T_4 = \frac{(0.2077\ Btu/hr\cdot ft^2\cdot {}^0F)(530^0F)}{0.976\ Btu/hr\cdot ft^2\cdot {}^0F} + 70^0F = 182.8^0F$$

This is significantly close to the initial value of T_4 = 180.77^0F, so the iteration process may be terminated.

c) The heat transfer rate is

$$\frac{q}{l} = \frac{A_4}{L} U_4(T_1 - T_5) = \frac{\pi D_4 L}{L} U_4(T_1 - T_5)$$

$$= \pi(0.802\ ft)(0.2077\ BTu/hr\cdot ft^2\cdot {}^0F)(530^0F)$$

$$= 277.2\ Btu/hr\text{-}ft$$

• PROBLEM 9-3

Air flows through a horizontal tube of length 0.4 m and diameter 0.025 m. The flow is measured at 0.3 m/s, and is at 27°C and 1 atm. If the wall surface temperature is 140°C, what is the heat transfer coefficient?

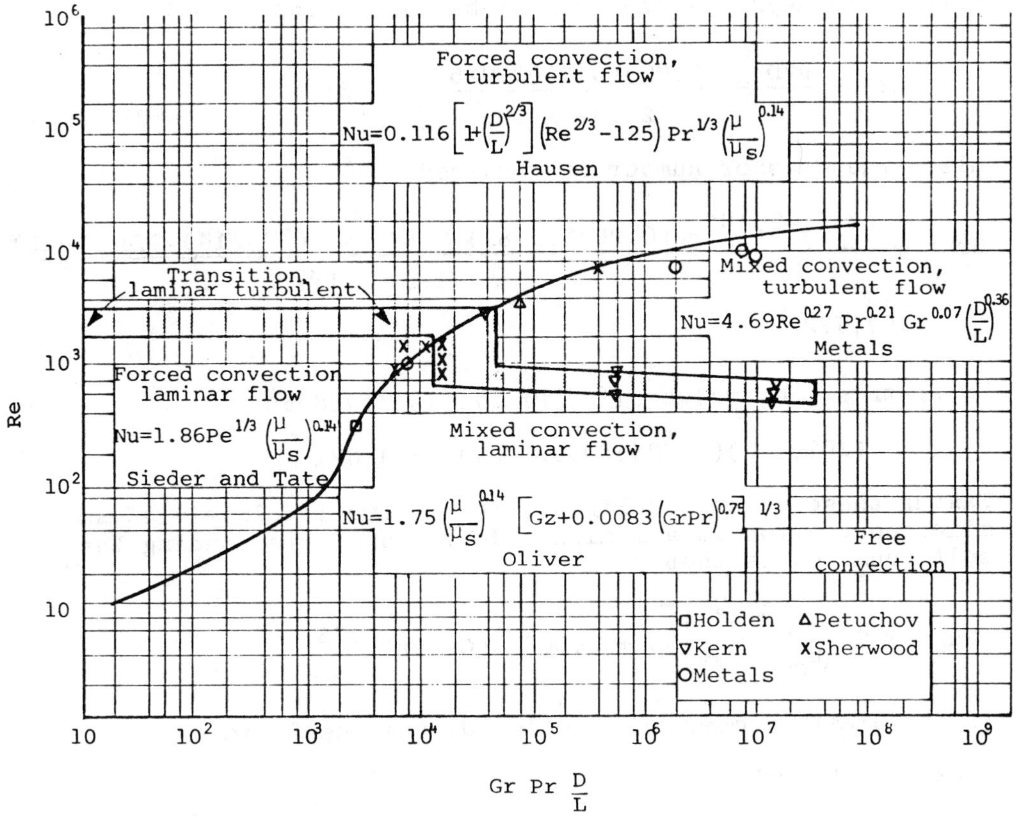

Fig. 1. Regimes of free, forced, and mixed convection for flow through horizontal tubes.

Solution: The mean temperature is $\frac{140 + 27}{2} = 83.5^{\circ}C = 356.5^{\circ}K$. From the air tables at this temperature,

$\mu = 2.102 \times 10^{-5}$ kg/m-sec $\quad k = 0.0305\ W/m\text{-}^{\circ}C \quad Pr = 0.695$

while at the surface temperature

$$\mu_s = 2.337 \times 10^{-5} \text{ kg/m-sec}$$

and at the fluid temperature

$$\mu_f = 1.983 \times 10^{-5} \text{ kg/m-sec}$$

Also, at the mean temperature, values for

$$\beta = \frac{1}{T} = \frac{1}{356.5} = 2.805 \times 10^{-3}/^{\circ}K$$

and

$$\rho = \frac{P}{RT} = \frac{1.0132 \times 10^5}{(287)(356.5)} = 0.99 \text{ kg/m}^3$$

are found.

First, the Reynolds number must be calculated.

$$Re = \frac{\rho VD}{\mu} = \frac{(0.99)(0.3)(0.025)}{2.102 \times 10^{-5}} = 353$$

Also, the Grashof number is required.

$$Gr = \frac{\rho^2 g\beta(T_s - T_f)D^3}{\mu^2} = \frac{(0.99)^2(9.81)(2.805 \times 10^{-3})(140-27)(0.025)^3}{(2.102 \times 10^{-5})^2}$$

$$= 107{,}772$$

Using this, the value of Gr Pr D/L is found as

$$(107{,}772)(0.695)(0.025/0.4) = 4{,}681$$

Taking these values to Fig. 1, the regime is classified as mixed convection in a laminar flow. The corresponding Nusselt equation is then

$$Nu = 1.75\left(\frac{\mu}{\mu_s}\right)^{0.14}\left[Gz + 0.0083\,(Gr\ Pr)^{0.75}\right]^{1/3}$$

If the Graetz number is calculated, the result is

$$Gz = Re\ Pr\ D/L$$

$$= (353)(0.695)(0.025/0.4) = 15.33$$

Substituting in the Nusselt equation gives a value of

$$Nu = 1.75\left(\frac{2.102}{2.337}\right)^{0.14}\left\{15.33 + 0.0083\left[(107{,}772)(0.695)\right]^{0.75}\right\}^{1/3}$$

$$= 6.47$$

Then the heat transfer coefficient may be found as

$$h = \frac{Nu\ k}{D} = \frac{(6.47)(0.0305)}{0.025} = 7.9\ W/m^2\text{-}^{\circ}C$$

Now, if entirely forced convection in a laminar regime had been assumed, the Nusselt number would have been found with

$$Nu = 1.86(Re\ Pr)^{1/3}\ \frac{\mu}{\mu_s}^{0.14}\ \frac{D}{L}^{1/3}$$

Here, the heat transfer coefficient would be

$$h = 1.86\left[(353)(0.695)\right]^{1/3}\left(\frac{2.102}{2.337}\right)^{0.14}\left(\frac{0.025}{0.4}\right)^{1/3}\left(\frac{0.0305}{0.025}\right)$$

$$= 5.55\ W/m^2\text{-}^{\circ}C$$

This involves an error of about 30%.

An aluminum alloy fin 0.25 in. thick and 1.5 in. wide protrudes from a wall. The base temperature is 260 $^{\circ}$F, and the ambient temperature is 80 $^{\circ}$F. Compute the fin efficiency and the rate of heat transfer per unit length, neglecting the temperature variation along the width of the fin.

Use:

$$h = 16 \text{ Btu/hr. ft.}^2 \ ^{\circ}\text{F}$$

$$k = 28 \text{ Btu/hr. ft. } ^{\circ}\text{F}$$

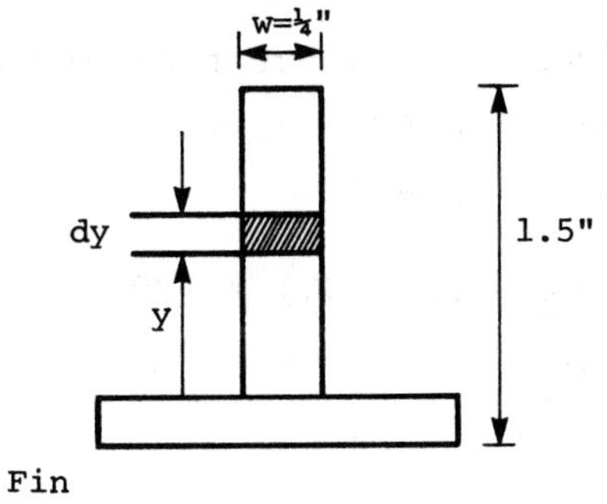

Fig. 1

Solution: From the law of conservation of energy, the heat energy flowing into the infinitesimal element of thickness dy (at a distance y from the base) and width w, per unit length, is equal to the sum of the heat energies conducted and convected from the surface of the element under consideration.

Thus, the rate of heat conducted in is

$$-kw \frac{dt}{dy}$$

and the rate of heat conducted out is

$$-kw \left[\frac{d^2t}{dy^2} dy + \frac{dt}{dy}\right]$$

The rate of heat convected from the surface is

$$h\ (2dy)(t-80)$$

Summing these,

$$-kw \frac{dt}{dy} = h\ (2dy)(t-80) - kw \left[\frac{d^2t}{dy^2} dy + \frac{dt}{dy}\right]$$

which simplifies to

$$\frac{d^2t}{dy^2} - \frac{2h}{kw}(t-80) = 0.$$

This differential equation may be assumed to have a solution of the following form:

$$(t-80) = C_1 e^{\alpha y} + C_2 e^{-\alpha y} \tag{1}$$

where

$$\alpha = \sqrt{\frac{2h}{kw}}$$

C_1 and C_2 are constants of integration and can be evaluated from the boundary conditions.

At $y = 0$, $t = 260\,^\circ F$.

Then $(t-80) = C_1 + C_2 = (260-80) = 180$

or $C_1 = 180 - C_2$.

Again, at the end of the fin, the heat conducted is equal to the heat convected to air.

Therefore, at y = 1.5 in.,

$$-kw\left(\frac{dt}{dy}\right) = hw(t-80)$$

$$\text{or} \quad \frac{dt}{(t-80)} = -\frac{h}{k}\,dy. \tag{2}$$

from eqs. (1) and (2),

$$dt = -\frac{h}{k}\left(C_1 e^{\alpha y} + C_2 e^{-\alpha y}\right) dy.$$

Now, for y = 1.5 in.,

$$\alpha y = \sqrt{\frac{2 \times 16}{28 \times \frac{1}{4 \times 12}}} \times \frac{1.5}{12}$$

$$= 0.926$$

and

$$\frac{dt}{dy} = -\frac{h}{k}\left(C_1 e^{0.926} + C_2 e^{-0.926}\right) \tag{3}$$

Differentiating eq. (1) and substituting the value of $\frac{dt}{dy}$ into eq. (3) yields

$$-k\left(\alpha C_1\ e^{\alpha y} - \alpha C_2\ e^{-\alpha y}\right) = h\left(C_1\ e^{0.926} + C_2\ e^{0.926}\right)$$

or

$$-C_1\ e^{0.926} + C_2\ e^{-0.926} = \frac{h}{\alpha k}\left(C_1\ e^{0.926} + C_2\ e^{-0.926}\right)$$

$$\frac{h}{\alpha k} = \frac{16}{\sqrt{28 \times \frac{1}{4x12} \times 28}}$$

$$= 0.077.$$

As $C_1 = 180 - C_2$,

$$\frac{(180-C_2)}{C_2} = \frac{e^{-0.926}}{e^{0.926}}\left[\frac{1 - 0.077}{1 + 0.077}\right]$$

which gives

$$C_1 = 21.33 \text{ and } C_2 = 158.66.$$

The total heat conducted into the fin is given by

$$q = -kA\left(\frac{dt}{dy}\right)_{y=0}$$

$$= -16 \times \frac{1}{4x12} \times \alpha\,(21.33-158.66)$$

$$= -\frac{16}{4x12} \times \sqrt{\frac{2 \times 16}{28 \times \frac{1}{4x12}}}\ (21.33-158.66)$$

$$= 339.04 \text{ Btu/hr. ft.}$$

The total heat loss is

$$q = hA\ (260-80)$$

$$= 16 \times \frac{(1.5x2 + \frac{1}{4})}{12}\ (180)$$

$$= 780 \text{ Btu/hr. ft.}$$

The fin efficiency is defined as the ratio of the total heat conducted into the fin to the total heat lost from the fin

$$\eta = \frac{339.04}{780} = 0.435$$

or 43.5%.

A semi-infinite cylinder of 6 cm. diameter is at a temperature of 190°C. It is suddenly exposed to an ambient temperature of 75°C. Determine the temperature at a radius of 1.3 cm. after 1.2 minutes have elapsed. Also, compute the total heat transfer per unit length of the cylinder. The following data is given:

$k = 216\ W/m^{\circ}C$ $\rho = 2700\ kg/m^3$

$h = 520\ W/m^{2\,\circ}C$ $C_p = 0.92\ kJ/kg^{\circ}C$

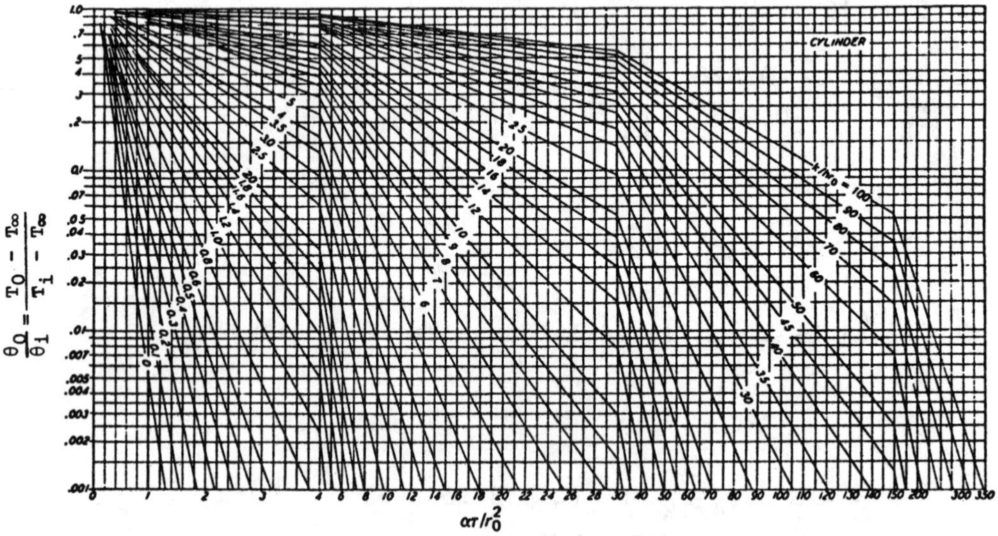

Fig. 1 Axis temperature for an infinite cylinder of radius r_0.

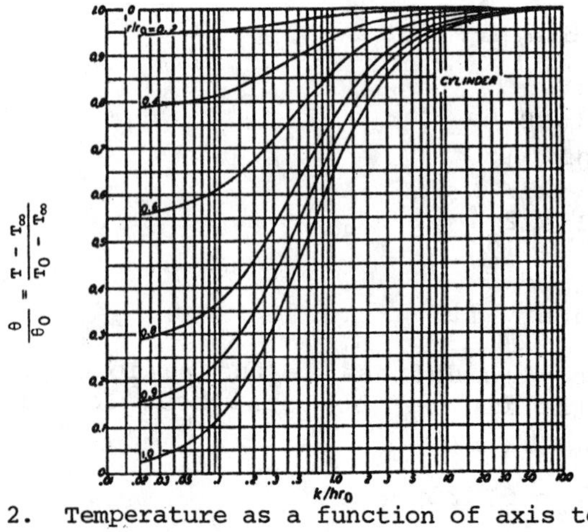

Fig 2. Temperature as a function of axis temperature in an infinite cylinder of radius r_0.

Solution: This problem is solved with the use of figures 1 and 2.

$$\frac{k}{hr_o} = \frac{216}{520 \times (0.03)} = 13.85$$

where r_o (the radius of the cylinder) = 0.03 m.

$$\frac{\alpha\tau}{r_o^2} = \frac{k}{\rho C_p} \times \frac{\tau}{r_o^2}$$

$$= \frac{216}{(2700) \times (0.92) \times 10^3} \times \frac{1.2 \times 60}{(0.03)^2} = 6.96$$

From fig. 1,

$$\frac{\theta_o}{\theta_i} = 0.35 \qquad (1)$$

Now,

$$\frac{r}{r_o} = \frac{1.3}{3} = 0.433$$

From figure 2, $\frac{\theta}{\theta_o} = 0.98$ (2)

From eqs. (1) and (2)

$$\frac{\theta_o}{\theta_i} \times \frac{\theta}{\theta_o} = \frac{\theta}{\theta_i} = 0.35 \times 0.98 = 0.343$$

or

$$\theta = 0.343\ \theta_i = 0.343\ (190-75)$$

$$= 39.45$$

But $\theta = T - T_\infty = 39.45$

Therefore, $T = 39.45 + 75 = 114.45^\circ C$

The heat transfer per unit length of cylinder is expressed as

$$Q_o = \rho C_p V\ \theta_i = \rho C_p \pi r_o^2 \theta_i$$

$$= 2700 \times (0.92)\pi(3)^2(190-75)$$

or $Q_o = 8.07 \times 10^6$ J/m

The heat transfer through the cylindrical surface at any radius during any instant of time can be obtained by using figure 3.

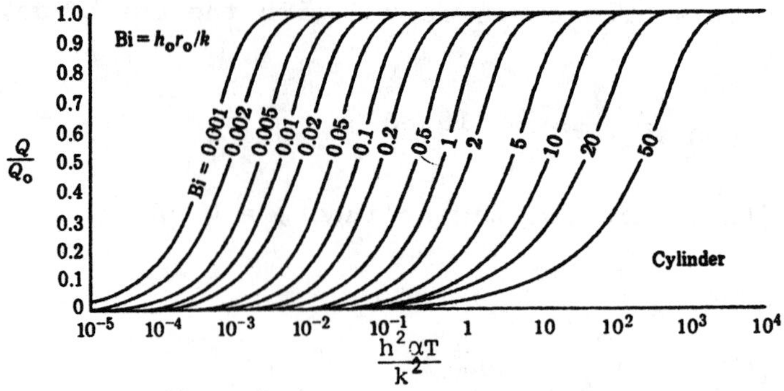

Fig. 3 Dimensionless heat loss Q/Q_0 of an infinite cylinder of radius r_0 with time.

$$\frac{k}{hr_o} = 13.85 \text{ or } \frac{hr_o}{k} = 0.072$$

$$\frac{h^2 \alpha \tau}{k^2} = (520)^2 \frac{216}{(2700)(0.92)\ 10^3} \times \frac{1.2 \times 60}{(216)^2}$$

$$= 0.036$$

From figure 3,

$$\frac{Q}{Q_o} = 0.5$$

or $$Q = 0.5\ Q_o$$

Therefore, $Q = 0.5 \times 8.07 \times 10^6$

$$= 4.035 \times 10^6 \text{ J/m.}$$

CONVECTION AND RADIATION

• PROBLEM 9-6

Compute the rate of heat loss per unit area from a circular hot air duct (a) by convection and (b) by simultaneous convection and radiation. Assume the duct has a diameter of 3 ft. and is made of a polished metal (emissivity $\varepsilon = 0.1$). The surface temperature of the metal is 150°F, and the ambient temperature is 65°F. (c) Repeat part (b), assuming the metal is painted white on the outside ($\varepsilon = 0.95$).

Solution:

(a) For a large surface, the rate of heat loss by convection is given by

$$q_C = 0.3\theta^{1.25} \text{ Btu/ft}^2\text{-hr}$$

with
$$\theta = T_2 - T_1$$

where T_2 is the temperature of the surface in ^0R

$$= 150 + 460 = 610^0R$$

and T_1 is the temperature of the surroundings in ^0R

$$= 65 + 460 = 525^0R$$

The heat lost by convection is,

then,
$$q_C = 0.30(610 - 525)^{1.25}$$

$$= 77.4 \text{ Btu/ft}^2\text{-hr}$$

(b) The rate of heat loss by radiation is given by

$$q_R = \varepsilon\sigma(T_2{}^4 - T_1{}^4)$$

where ε - surface emissivity

σ - Stefan-Boltzmann constant

$= 0.1714 \times 10^{-8}$ Btu/hr-ft^2-$^oR^4$

Thus,

$$q_R = 0.1 \times 0.1714 \times 10^{-8} \times \left[(610)^4 - (525)^4\right] \text{ Btu/ft}^2\text{-hr}$$
$$= 10.7 \text{ Btu/ft}^2\text{-hr.}$$

The combined heat loss due to convection and radiation is

$$q = q_C + q_R$$

$$= 77.4 + 10.7 = 88.1 \text{ Btu/ft}^2\text{-hr.}$$

(c) When the metal is painted white on the outside, the radiant heat loss changes as the emissivity changes. However, the convective heat loss remains the same.

$$q_R = 0.95 \times 0.1714 \times 10^{-8} \times \left[(610)^4 - (525)^4\right]$$

$$= 101.8 \text{ Btu/ft}^2\text{-hr.}$$

Therefore, the total heat loss is

$$q = q_C + q_R = 77.4 + 101.8$$
$$= 179.2 \text{ Btu/ft}^2\text{-hr.}$$

• PROBLEM 9-7

A cylinder of 6.625 in. diameter with a surface temperature of 200°F is placed horizontally in air at 80°F. Determine

(a) the convective heat transfer coefficient h_c

(b) the radiant heat transfer coefficient h_r

(c) the combined heat transfer coefficient h_{cr}

The surface emissivity may be taken as $\varepsilon = 0.94$.

Table 1: Physical Properties of Air at 101.325 kPa (1 atm Abs)

T (°F)	ρ $\left(\frac{lb_m}{ft^3}\right)$	c_p $\left(\frac{Btu}{lb_m \cdot °F}\right)$	μ (centipoise)	k $\left(\frac{Btu}{h \cdot ft \cdot °F}\right)$	N_{Pr}	$\beta \times 10^3$ (1/R)	$g\beta\rho^2/\mu^2$ (1/R·ft³)
0	0.0861	0.240	0.0162	0.0130	0.720	2.18	4.39×10^6
32	0.0807	0.240	0.0172	0.0140	0.715	2.03	3.21×10^6
50	0.0778	0.240	0.0178	0.0144	0.713	1.96	2.70×10^6
100	0.0710	0.240	0.0190	0.0156	0.705	1.79	1.76×10^6
150	0.0651	0.241	0.0203	0.0169	0.702	1.64	1.22×10^6
200	0.0602	0.241	0.0215	0.0180	0.694	1.52	0.840×10^6
250	0.0559	0.242	0.0227	0.0192	0.692	1.41	0.607×10^6
300	0.0523	0.243	0.0237	0.0204	0.689	1.32	0.454×10^6
350	0.0490	0.244	0.0250	0.0215	0.687	1.23	0.336×10^6
400	0.0462	0.245	0.0260	0.0225	0.686	1.16	0.264×10^6
450	0.0437	0.246	0.0271	0.0236	0.674	1.10	0.204×10^6
500	0.0413	0.247	0.0280	0.0246	0.680	1.04	0.163×10^6

Solution:

(a) The average film temperature is

$$T_f = \frac{T_s + T_a}{2} = \frac{200 + 80}{2} = 140°F.$$

The convective heat transfer coefficient is given by

$$h_c = Nu \frac{k}{d} \quad (1)$$

where

$$Nu = a(Gr\ Pr)^m \quad (2)$$

a and m are constants whose values are given in table 2, and Gr is the Grashof number

$$Gr = \frac{L^3 \rho^2 g\beta \Delta T}{\mu^2}$$

where β is the coefficient of expansion, equal to

$\frac{1}{T_f}$ (T_f is the film temperature).

Table 2: Constants for Nusselt Number

Physical Geometry	$(N_{Gr}N_{Pr})$	*a*	*m*
Vertical planes and cylinders [vertical height $L < 1$ m (3 ft)]			
	$<10^4$	1.36	$\frac{1}{5}$
	10^4–10^9	0.59	$\frac{1}{4}$
	$>10^9$	0.13	$\frac{1}{3}$
Horizontal cylinders [diameter D used for L and $D < 0.20$ m (0.66 ft)]			
	$<10^{-5}$	0.49	0
	10^{-5}–10^{-3}	0.71	$\frac{1}{25}$
	10^{-3}–1	1.09	$\frac{1}{10}$
	1–10^4	1.09	$\frac{1}{5}$
	10^4–10^9	0.53	$\frac{1}{4}$
	$>10^9$	0.13	$\frac{1}{3}$
Horizontal plates			
Upper surface of heated plates or lower surface of cooled plates	10^5–2×10^7	0.54	$\frac{1}{4}$
	2×10^7–3×10^{10}	0.14	$\frac{1}{3}$
Lower surface of heated plates or upper surface of cooled plates	10^5–10^{11}	0.58	$\frac{1}{5}$

At $T_f = 140^\circ F$, the following values are obtained by interpolation:

$$k = 0.0166 \text{ Btu/hr.ft.}^\circ F$$

$$Pr = 0.7026, \quad \frac{g\beta\rho^2}{\mu^2} = 1.328 \times 10^6\ {}^\circ R^{-1} ft^{-3}$$

Then

$$Gr = (L^3 \Delta T)\ \frac{g\beta\rho^2}{\mu^2}$$

$$= (6.625)^3 \left(\frac{1}{12}\right)^3 \times (200-80) \times 1.328 \times 10^6 \text{ per}^\circ R\ ft^3$$

$$= 2.68 \times 10^7 \text{ per } {}^\circ R\ ft^3$$

$$Gr\ Pr = 2.68 \times 10^7 \times 0.7026$$

$= 1.88 \times 10^7$

From table 2, for Gr Pr = 1.88×10^7, the values of a and m are

$a = 0.53$ and $m = \frac{1}{4}$.

Therefore, using eq. (2),

$$Nu = 0.53\,(1.88 \times 10^7)^{\frac{1}{4}}$$

$$= 34.9$$

The convective heat transfer coefficient is then given by eq. (1) as

$$h_c = \frac{34.9 \times 0.0166}{6.625 \times 1/12} = 1.05 \text{ Btu/hr.ft}^2.^\circ\text{F}$$

(b) The radiant heat transfer coefficient is given by

$$h_r = \sigma\varepsilon(T_s+T_a)(T_s^2+T_a^2)$$

where

σ is the Stefan-Boltzman constant

$= 0.1714 \times 10^{-8}$ Btu/hr.ft.^{2} $^\circ R^4$

ε is the emissivity of the surface = 0.94

T_s is the temperature of the surface

$= 200 + 460 = 660^\circ R$

T_a is the temperature of the surrounding fluid

$= 80 + 460 = 540^\circ R$

Therefore,

$$h_r = 0.1714 \times 10^{-8} \times 0.94 \times (660+540)(660^2+540^2)$$

$$= 1.406 \text{ Btu/hr.ft.}^2\ ^\circ\text{F}.$$

(c) The combined heat transfer coefficient is given by

$$h_{cr} = h_c + h_r$$

$$= 1.05 + 1.406$$

$$= 2.456 \text{ Btu/hr.ft.}^2\ ^\circ\text{F}$$

• PROBLEM 9-8

Steam at 60 kg/cm^2 and $1000^{\circ}K$ flows through a pipe 6 cm in diameter and 100 cm in length. The velocity of the flow is measured as 600 m/min, and the pipe surface is painted black. Considering only 6.3 μm (k = 1) and 2.7 μm (k = 2) bands, calculate the average coefficient of heat transfer due to both radiation and convection. Assume a trubulent flow. The properties of the pipe material are:

$k = 0.09\ W/m^{\circ}K;\ \nu = 2.6 \times 10^{-6}\ m^2/s;\ \ Pr = 0.9.$

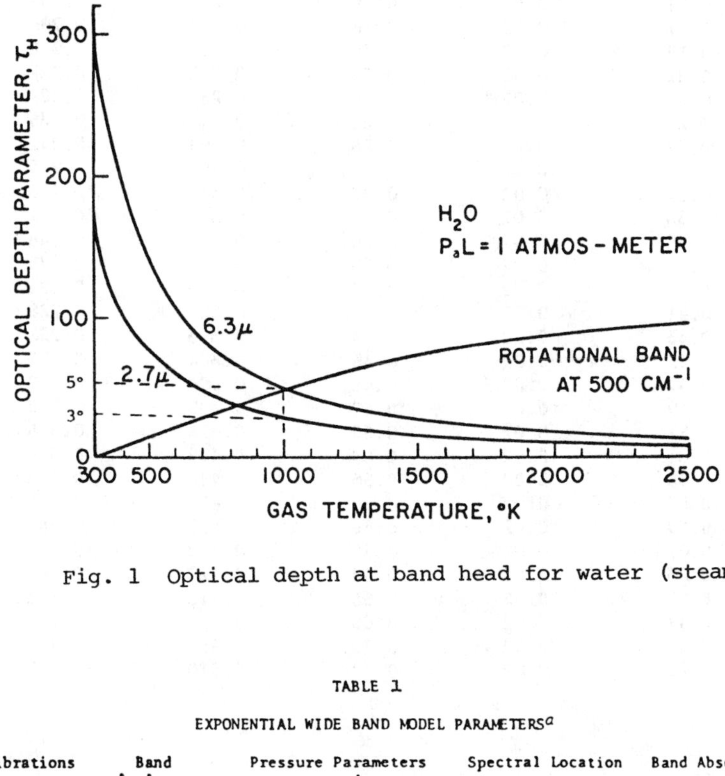

Fig. 1 Optical depth at band head for water (steam).

TABLE 1

EXPONENTIAL WIDE BAND MODEL PARAMETERS[a]

Gas	Vibrations ν_h (cm^{-1})	Band $\delta_1, \delta_2, \ldots$	Pressure Parameters n	b (T_o = 100 K)	Spectral Location ν_i (cm^{-1})	ν_c (cm^{-1})	ν_n (cm^{-1})	Band Absorption Parameters α_o ($\frac{cm^{-1}}{gm\ m^{-2}}$)	β_o	ω_o (cm^{-1})
(1) H_2O	m = 3	(1) Rotational								
	ν_1 = 3652		1	$8.6(T_o/T)^{½}+0.5$	0^b			5200.0^b	0.14311^b	28.4^b
	ν_2 = 1595	0,0,0								
	ν_3 = 3756	(2) 6.3 μm		$8.6(T_o/T)^{½}+0.5$		1600		41.2	0.09427	56.4
	g_1 = 1	0,1,0	1							
	g_2 = 1	(3) 2.7 μm								
	g_3 = 1	0,2,0						0.19		
		1,0,0	1	$8.6(T_o/T)^{½}+0.5$		3760		2.30	0.13219	60.0
		0,0,1						22.40		
		(4) 1.87 μm	1	$8.6(T_o/T)^{½}+0.5$		5350		3.0	0.08169	43.1
		0,1,1								
		(5) 1.38 μm	1	$8.6(T_o/T)^{½}+0.5$		7250		2.5	0.11628	32.0
		1,0,1								

TABLE 2

VALUES OF λT AT SPECIFIED VALUES OF THE EXTERNAL AND INTERNAL FRACTIONS OF BLACKBODY RADIATION

f	Δf	ΣΔf	$(\lambda T)_e$ cm K	$(\lambda T)_i$ cm K
0.0025	0.005	0.005	0.1230	0.1073
0.0075	0.005	0.010	0.1395	0.1209
0.0125	0.005	0.015	0.1495	0.1291
0.0175	0.005	0.02	0.1573	0.1352
0.025	0.01	0.03	0.1662	0.1423
0.035	0.01	0.04	0.1762	0.1501
0.045	0.01	0.05	0.1848	0.1565
0.055	0.01	0.06	0.1922	0.1625
0.07	0.02	0.08	0.202	0.1701
0.09	0.02	0.10	0.214	0.1791
0.11	0.02	0.12	0.225	0.1874
0.13	0.02	0.14	0.235	0.1949
0.15	0.02	0.16	0.245	0.202
0.17	0.02	0.18	0.254	0.209
0.19	0.02	0.20	0.263	0.216
0.21	0.02	0.22	0.272	0.222
0.23	0.02	0.24	0.281	0.229
0.25	0.02	0.26	0.290	0.235
0.27	0.02	0.28	0.299	0.242
0.29	0.02	0.30	0.308	0.248
0.31	0.02	0.32	0.317	0.255
0.33	0.02	0.34	0.326	0.261
0.35	0.02	0.36	0.335	0.268
0.37	0.02	0.38	0.344	0.275
0.39	0.02	0.40	0.353	0.281
0.41	0.02	0.42	0.363	0.288
0.43	0.02	0.44	0.373	0.295
0.45	0.02	0.46	0.384	0.303
0.47	0.02	0.48	0.394	0.310
0.49	0.02	0.50	0.405	0.318
0.51	0.02	0.52	0.416	0.326
0.53	0.02	0.54	0.428	0.334
0.55	0.02	0.56	0.440	0.342
0.57	0.02	0.58	0.453	0.351
0.59	0.02	0.60	0.467	0.361
0.61	0.02	0.62	0.482	0.371
0.63	0.02	0.64	0.497	0.382
0.65	0.02	0.66	0.513	0.393
0.67	0.02	0.68	0.531	0.405
0.69	0.02	0.70	0.550	0.417
0.71	0.02	0.72	0.570	0.430

Solution: The average heat transfer coefficient due to radiation and convection together can be computed from the Nusselt number expression

$$\bar{h} = \frac{k}{D} \text{Nu}$$

where $\text{Nu} = \text{Nu}_R + \text{Nu}_C$, with Nu_R and Nu_C as the Nusselt numbers for radiation and convection processes, respectively.

For a radiation mode,

$$\overline{Nu}_R = \frac{D}{K_m} \sum_{k=1}^{n} \omega_k B_k' A_s^* \overline{\tau}_{WL,k} \qquad (1)$$

where ω_k is the band width parameter.

τ is the transmissivity or optical depth parameter.

A is the band absorption.

B is the absorption factor.

The band width parameter ω_o corresponding to the reference temperature $T_o = 100^oK$, is obtained from table 1. Then the value for ω_k can be obtained from the equation

$$\omega_k = 1.2\ \omega_o(T/T_o)^{1/2}$$

The value of ω_k is increased by 20% to account for spectral width.

Therefore, for k = 1, $\omega_1 = (1.2)(56.4)(\frac{1000}{100})^{1/2} = 214$ per cm.

For k = 2, $\omega_2 = (1.2)(60)(\frac{1000}{100})^{1/2} = 228$ per cm.

Again, the figure is used to obtain the optical depth parameter for a pressure of 60 kg/cm² for k = 1 and 2.

$$t_{D,1} = 60 \times \frac{60}{1000} \times 50 = 180 \qquad \text{for } k = 1$$

$$t_{D,2} = 60 \times \frac{60}{1000} \times 30 = 100 \qquad \text{for } k = 2.$$

For $\eta_k >> 1$, the band absorption $A_s^*(t_k)$ is expressed as

$$A_s^*(t_k) = \ln t_k + \gamma + 1/2$$

where γ the Euler constant = 0.577.

Hence, for k = 1

$$A_{s,1}^* = \ln \frac{180}{2} + 0.577 + 0.5$$

$$= 5.58$$

and for k = 2,

$$A_{s,2}^* = \ln \frac{100}{2} + 0.577 + 0.5$$

$$= 4.99$$

The wall layer transmissivity $\overline{\tau}_{H,k}$ is expressed as

$$\overline{\tau}_{H,k} = \frac{\ln(1 + t_{D,k}) - \ln(1 + t_{D,k}\ \gamma_{WL})}{\ln(1 + t_{D,k})}$$

and the ratio of the thickness of the wall layer to the hydraulic diameter is expressed as

$$\bar{\gamma}_{WL} = \frac{1.37}{Re^{0.37}} \left[1 - \exp(-(x/10D)^{0.6}) \right]$$

$$\text{where } Re = \frac{VD}{\nu} = \frac{10 \times 60 \times 10^{-3}}{2.6 \times 10^{-6}}$$

$$= 2.3 \times 10^{5}.$$

Therefore,

$$\bar{\gamma}_{WL} = \frac{1.37}{(2.3 \times 10^{5})^{0.37}} \left[1 - e^{-(1000/600)^{0.6}} \right] = 0.011$$

and

$$\bar{\tau}_{WL,1} = \frac{\ln(1 + 180) - \ln(1 + 1.9)}{\ln(1 + 180)} = 0.80$$

Also

$$\bar{\tau}_{WL,2} = \frac{\ln(1 + 100) - \ln(1 + 1.1)}{\ln(1 + 100)} = 0.84$$

The product of ω, A and B can be expressed as

$$\omega_k A_s^* B_k' = 4\sigma T^3 \Delta f_{i,k}$$

where $\Delta f_{i,k}$ is given in table 2 as a function of $\lambda T = T/\nu$. The upper and lower limits of ν are to be fixed.

Now, from table 1, for the 6.3 μm band, the spectral value is centered at 1600 cm^{-1}, and for the 2.7 μm band, at 3760 cm^{-1}.

Thus,

$$\nu_{1,u} = \nu_{1,c} + \frac{1}{2}\,\omega_1 A^*_{s,1} = 1600 + \frac{1}{2}(214)(5.58) = 2197 \text{ cm}^{-1}$$

$$\nu_{1,\ell} = \nu_{1,c} - \frac{1}{2}\,\omega_1 A^*_{s,1} = 1600 - \frac{1}{2}(214)(5.58) = 1003 \text{ cm}^{-1}$$

$$\nu_{2,u} = 3760 + \frac{1}{2}(228)(4.99) = 4329 \text{ cm}^{-1}$$

$$\nu_{2,\ell} = 3760 - \frac{1}{2}(228)(4.99) = 3191 \text{ cm}^{-1}$$

The limits of λT and f_i for bands 1 and 2 can be written as

$$\left.\begin{aligned} \lambda T &= 0.455 \text{ to } 0.997 \\ f_i &= 0.741 \text{ to } 0.965 \\ \Delta f_i &= 0.224 \end{aligned}\right\} \text{band 1}$$

$$\left.\begin{aligned} \lambda T &= 0.231 \text{ to } 0.313 \\ f_i &= 0.237 \text{ to } 0.478 \\ \Delta f_i &= 0.241 \end{aligned}\right\} \text{band 2}$$

Introducing the known values in equation (1) yields

$$\overline{Nu}_R = \frac{60 \times 10^{-3}}{0.09} (4)(5.67 \times 10^{-8})(1000^3)\left[0.224(0.80) + 0.241(0.84)\right]$$

$$= 57.7$$

The Nusselt number for the convection process is expressed as

$$\overline{Nu}_C = 0.02\ Re^{0.8}\ Pr^{0.33}\ \overline{F}$$

where

$$\overline{F} = 1 + \frac{0.88}{1 + 0.4\frac{x}{D}}$$

For x = 1 meter,

$$\overline{F} = 1 + \frac{0.88}{7.67} = 1.11$$

Therefore, $\overline{Nu}_C = 0.02(2.3 \times 10^5)^{0.8}(0.9)^{0.33} \times 1.11$

$$= 419.3$$

As $Nu = \overline{Nu}_R + \overline{Nu}_C$

$$= 57.7 + 419.3$$

$$= 477,$$

the average heat transfer coefficient is

$$\overline{h} = \frac{k}{D} Nu = \frac{0.09}{60 \times 10^{-3}} (477)$$

$$= 715\ W/m^{2\,\circ}K.$$

A 10 in. thick vertical iron slab is placed in a room at 70°F. If the initial temperature of the slab is 1100°F, estimate the center plane and surface temperatures after a time elapse of 4 hours. The following data is given:

$k = 27$ Btu/hr.ft.°F $\quad C_p = 0.14$ Btu/lb, $\rho = 490$ lb/ft³,

$\varepsilon = 0.7$.

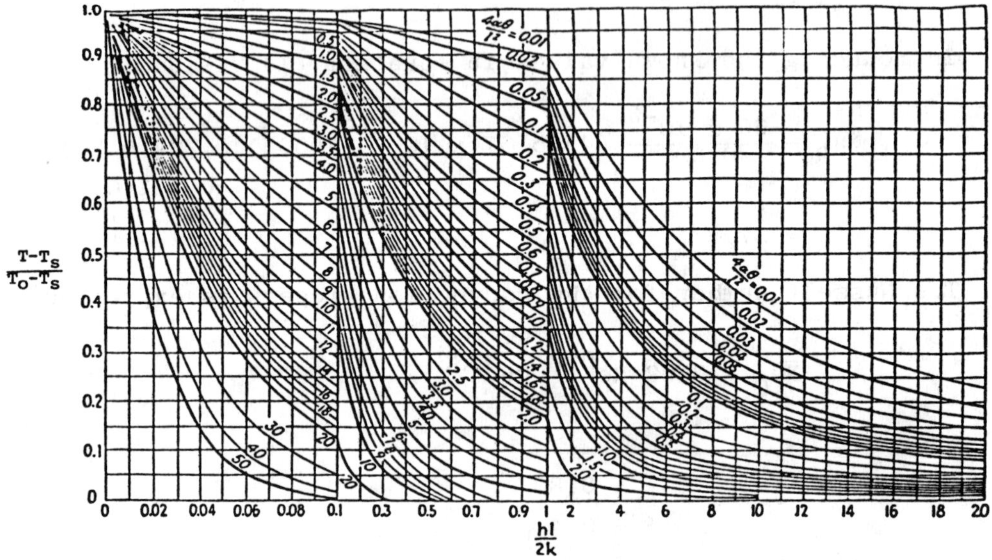

Fig. 1. Chart for determining the temperature history of points on the surfaces of rectangular shapes.

Solution: The slab is cooled in air by the combined action of convection and radiation. Thus, the combined heat transfer coefficient must be evaluated,

$$h = h_c + h_r.$$

The subscripts c and r refer to convection and radiation, respectively.

The radiation heat transfer coefficient is defined as:

$$h_r = \varepsilon\sigma(T_i+T_a)(T_i^2+T_a^2)$$

where subscripts i and a refer to the initial temperature of the slab and the air temperature in the room, respectively.

T is the absolute temperature, in ^{O}R and σ is the Stefan-Boltzman constant.

Substituting,

$$h_r = 0.7 \times 0.1714 \times 10^{-8}(1560+530)(1560^2+530^2)$$

$$= 6.9 \text{ Btu/hr.ft}^{2\,o}\text{F}$$

The coefficient for natural convection in a vertical plane is expressed as

$$h_c = 0.3\ (T_i - T_a)^{\frac{1}{4}}$$

$$= 0.3\ (1100-70)^{\frac{1}{4}}$$

$$= 1.7 \text{ Btu/hr.ft}^{2\,o}\text{F}$$

The total heat transfer coefficient is

$$h = h_c + h_r$$

$$= (6.9+1.7) \text{ Btu/hr.ft}^{2\,o}\text{F}$$

$$= 8.6 \text{ Btu/hr.ft}^{2\,o}\text{F}$$

After 4 hours, the surface temperature is assumed to be $500^{O}F$. Now, the coefficients of heat transfer are

$$h_r = 0.7 \times 0.714 \times 10^{-8}(960+530)(960^2+530^2)$$

$$\text{or} \quad h_r = 2.2$$

$$\text{and} \quad h_c = 0.3\ (500-70)^{\frac{1}{4}}$$

$$= 1.35$$

$$h = h_r + h_c$$

$$= (2.2+1.35) = 3.55 \text{ Btu/hr.ft}^{2\,o}\text{F}$$

The average coefficient of heat transfer is obtained by taking the mean of the initial and final values, or

$$h = \frac{(8.6+3.55)}{2} = 6.1 \text{ Btu/hr.ft}^{2\,o}\text{F}$$

The surface temperature of the slab can be determined using fig. 1 with the average value of the heat transfer coefficient.

$$\frac{hl}{2k} = \frac{6.1 \times (10/12)}{2 \times 27}$$

$$= 0.094$$

$$\frac{4\alpha\theta}{l^2} = 4\,\frac{k}{C\rho} \times \frac{4}{(10/12)^2}$$

$$= 4 \times \frac{27}{0.14 \times 490} \times \frac{4}{(10/12)}$$

$$= 9.1$$

From fig. 1,

$$y = \frac{T - T_s}{T_o - T_s} = 0.42$$

or $T = T_s + (T_o - T_s)\, 0.42$

$$= 70 + (1100-70)\, 0.42$$

$$= 502^\circ F$$

which is very close to the assumed value of the surface temperature. Hence no further iteration is necessary, and the calculated value of surface temperature may be taken as the true temperature.

The center plane temperature of the slab can be computed by using fig. 2.

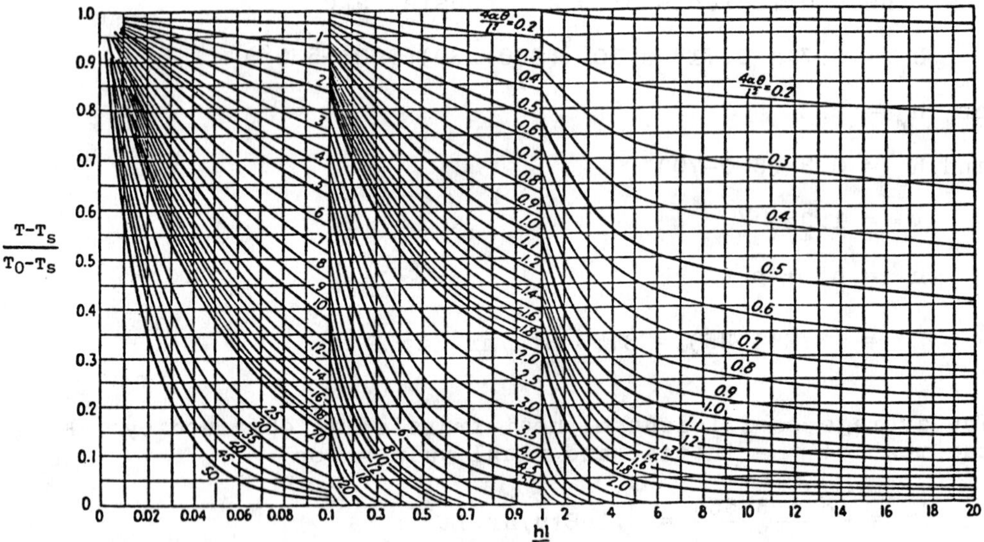

Fig. 2. Chart for determining the temperature history of points at the center of rectangular shapes.

From fig. 2, corresponding to

$$\frac{hl}{2k} = 0.094 \text{ and } \frac{4\alpha\theta}{l^2} = 9.1.$$

is

$$y = \frac{T - T_s}{T_o - T_s} = 0.43$$

Therefore, $T = T_s + (T_o - T_s)\ 0.43$

$= 70 + (1100-70)\ 0.43$

$= 513^{\circ}F$

• PROBLEM 9-10

Air at 500 °F flows through the annular space between two concentric pipes. Find the ratio of the heat radiated to the smaller pipe to that transferred to it by convection. The temperature of the outer surface of the inner pipe is 120 °F, and the air-to-metal heat transfer coefficient is 5 Btu/hr·ft^2·°F Assume that the outer surface of the outer pipe is perfectly insulated.

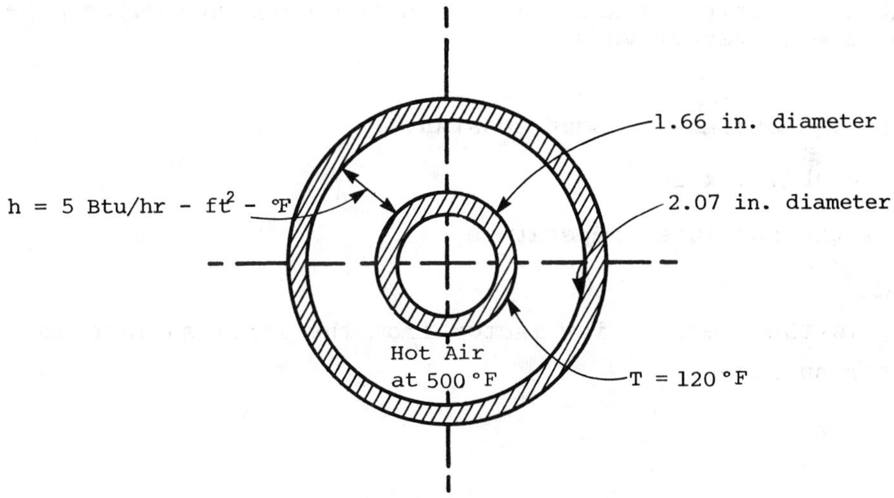

Fig. 1

Solution: Assuming a perfect insulation, the heat received by the inner surface of the pipe is radiated back to the outer surface of the inner pipe.

The heat transfer rate to the inner surface of outer pipe is expressed as

$$q_o = hA_o(T_a - T_o)$$

where A_o is the surface area of the outer pipe (inner surface),

T_a is the absolute air temperature,

and

T_o is the absolute temperature of the inner surface of the outer pipe.

$$\text{Area ratio } \frac{A_o}{A_i} = \frac{2.07}{1.66} = 1.246$$

$$\text{or } A_o = 1.246\, A_i$$

Therefore,

$$q_o = 5 \times (1.246\, A_i)(960 - T_o) \text{ Btu/hr.}$$

Now, the heat radiated from the inner surface of outer pipe to the outer surface of inner pipe is expressed as

$$q_r = \sigma\, A_i\, f_{io} \left[T_o^4 - T_i^4\right]$$

where subscripts i and o refer to the inner and outer pipe surfaces, respectively.

σ is the Stefan-Boltzman constant

$= 0.1714 \times 10^{-8}$,

T is the absolute temperature,

and

f_{io} is the overall view factor from the inner surface to the outer surface.

As

$$\frac{i}{f_{io}} = \frac{1}{F_{io}} + \frac{1}{\varepsilon_i} - 1 + \frac{A_i}{A_o}\left(\frac{1}{\varepsilon_o} - 1\right)$$

where ε is the emissivity of the surface (assumed 0.9 for both surfaces),

Then

$$q_r = \frac{0.1714 \times 10^{-8} \times A_i \left[T_o^4 - (460+120)^4\right]}{\frac{1}{1} + \frac{1}{0.9} - 1 \left(\frac{1.66}{2.077}\right)\left(\frac{1}{0.9} - 1\right)} .$$

As $q_r = q_o$,

then,

$$5 \times 1.246 \times A_i\ (960-T_o) = \frac{0.1714 \times 10^{-8} \times A_1 \left[T_o{}^4 - 580\right]}{1 + \frac{1}{0.9} - 1 + \left(\frac{1}{1.246}\right)\left(\frac{1}{0.9} - 1\right)}$$

which gives $T_o \approx 862^\circ R$

$= 402^\circ F$.

The rate of heat transfer to the inner pipe surface by convection is expressed as

$$q_i = hA_i\ (T_a - T_i)$$

$$= 5 \times A_i \times (960-580)$$

$$= 1900\ A_i$$

and by radiation,

$$q_r = 5 \times 1.246 \times A_i\ (960-862)$$

$$= 610\ A_i$$

Therefore,

$$\frac{q_r}{q_i} = \frac{610\ A_i}{1900} = 0.321$$

• PROBLEM 9-11

Estimate the rate of heat transfer from the 5% CO_2 gas at a temperature of 2000 $^\circ$F, contained in a 3 ft. diameter stack and at atmospheric pressure. The emissivity and surface temperature of the refractory is 1.0 and 1900 $^\circ$F, respectively. Assume the convective heat transfer coefficient between the gas and the refractory surface as 1.5 Btu/hr.ft^2 $^\circ$F.

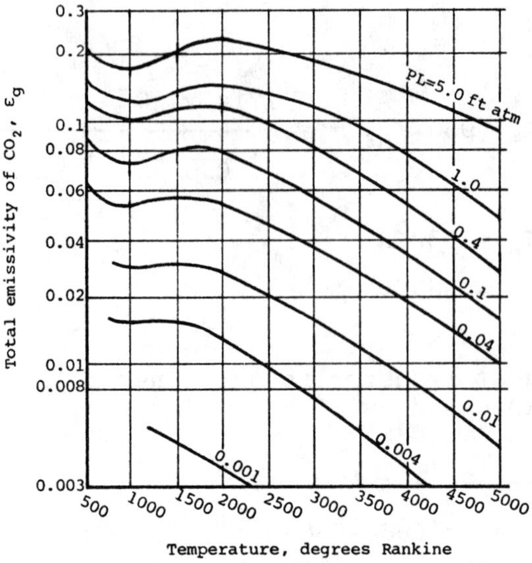

Fig. 1 Total emissivity of carbon dioxide.

Table EQUIVALENT BEAM LENGTHS FOR GAS RADIATION†

Shape	Equivalent beam length L
Sphere	0.60 × diameter
Cylinder of infinite length	0.90 × diameter
Space between infinite parallel planes	2.0 × distance between planes
Space outside an infinite bank of tubes with centers on equilateral triangles; tube diameter equals clearance	2.8 × clearance

Solution: The convective heat transfer is expressed as

$$q_c = h_c \; \Delta T \; A \quad \text{Btu/hr.}$$

or $$\frac{q_c}{A} = h_c \; \Delta T \; \text{Btu/hr.ft}^2.$$

$$\frac{q_c}{A} = 1.5 \times (2000-1900)$$

$$= 150 \; \text{Btu/hr.ft}^2.$$

The radiative heat transfer is expressed as

$$q_r = 0.171 \times 10^{-8} \left[\varepsilon T_1^4 - \alpha \; T_2^4 \right] A$$

where 0.171×10^{-8} is the Stefan-Boltzmann constant

and

T_1 and T_2 are the absolute temperature of the gas and the refractory surface

$(T_1 = 2000+460 = 2460^{\circ}R,$

$T_2 = 1900+460 = 2360^{\circ}R)$

The emissivity ε and absorptivity α depend on the mean beam length of the stack. This may be obtained from the table as

$$L = 0.9 \times \text{diameter}$$

$$= 0.9 \times 3 = 2.7$$

The parameter PL is

$$PL = 0.05 \times 2.7$$

$$= 0.135$$

The gas emissivity is obtained from the figure corresponding to a gas temperature of $2000 + 460 = 2460^{\circ}R$ and the parameter PL.
The value is $\varepsilon = 0.07$.

The total absorptivity corresponding to the parameter $\frac{PL\ T_2}{T_1}$ = $0.135 \left(\frac{2360}{2460}\right) = 0.130$.
and temperature $T_2 = 2360^{\circ}R$, from the figure, is 0.072.

Then

$$\alpha = 0.072 \left(\frac{T_1}{T_2}\right)^{0.65}$$

$$= 0.072 \left[\frac{2460}{2360}\right]^{0.65} = 0.074$$

Thus, the radiant heat transfer is

$$\frac{q_r}{A} = 0.171 \times 10^{-8}\left[0.07(2460)^4 - 0.074(2360)^4\right]$$

$$= 458\ \text{Btu/hr.ft}^2$$

The total heat transferred is

$$\frac{q}{A} = \frac{q_c}{A} + \frac{q_r}{A} = 150 + 458 = 608\ \text{Btu/hr.ft}^2$$

• PROBLEM 9-12

Air at a velocity of 10 ft/sec flows through a wide duct with an inner wall temperature of 1,200 ^{0}F. A thermocouple of diameter 1/25 in and emissivity 0.5, is placed in the duct, normal to the airstream. The thermocouple indicates a temperature of 1,400^{0}F Find the true value of the air temperature. For calculating the convection between the thermocouple and the air, take the Nusselt number as

$$Nu = 0.8\ Re^{0.38}\ ,$$

and the diameter of the thermocouple as the characteristic linear dimension. Also, the values of k and μ are to be taken at the mean film temperature. If the air velocity is doubled, what would be the approximate reading of the thermocouple?

Solution: The thermocouple attains equilibrium when the heat it gains from the air by convection equals the heat it loses to the walls by radiation. Evidently, the true gas temperature t is above 1,400 ^{0}F.

Assuming the true gas temperature to be 1,500^{0}F, the mean film temperature is

$$T = \frac{1,500 + 1,400}{2} = 1,450^0F.$$

Then

$$\rho = 0.020\ lb/ft^3$$

and the Reynolds number is

$$Re = \frac{10 \times 60 \times 60 \times 0.020}{25 \times 12 \times 0.10}$$

$$= 24.$$

This is a laminar flow. Hence,

$$Nu = \frac{Hd}{k\theta} = 0.8\ Re^{0.38}$$

$$= 0.8 \times 24^{0.38}$$

$$= 2.68$$

and $$H/\theta = 2.68 \times 0.039 \times 25 \times 12$$

$$= 31.5\ Btu/ft^2 \cdot hr \cdot {}^0F$$

Since the surface area of the wire is small compared to its surroundings, the emissivity of the surroundings can be taken as unity for purposes of calculating the radiation from the thermocouple to the duct walls.

Performing an energy balance gives

$$31.5(t-1,400) = 1.73 \times 10^{-9} \times 0.5(1,860^4 - 1,660^4)$$

$$= 0.5(20,800 - 13,100)$$

from which

$$t = 1,400 + \frac{0.5 \times 7,700}{31.5}$$

$$= 1,522^{\circ}F.$$

The value of the gas temperature is close to the estimated value of 1,500°F (within 1.5%).

If the velocity were doubled, the coefficient of convection would be

$$= 31.5 \times 2^{0.38}$$

$$= 41 \text{ Btu/ft}^2 \text{ hr.}^{\circ}F$$

and the temperature (t') indicated by the thermocouple would be given by

$$41(1,522-t') = 1.73 \times 10^{-9} \times 0.5\left[(t'+460)^4-1,660^4\right],$$

which is solved by iteration. The result temperature is t' = 1,419°F.

• PROBLEM 9-13

A solar collector with an efficiency of 48% has dimensions of 1.22 m width and 2.44 length. The flow of water averages 65.55°C, and the outdoor ambient air is at 10°C. Calculate the following, if the irradiation is 788 W/m²:

a) the temperature rise, if the water flow is $25.2\text{m}^3/\text{sec}$

b) the Reynolds number, if the water flow is distributed in parallel to eight 6.35 mm ID tubes with spacing of 152.4 mm.

c) the Reynolds number, if the water flow is through the same tubes in series.

d) the Reynolds number, if the inside diameter of the tubes is 25.4 mm.

e) the temperature rise, if the cooling fluid is air flowing at the rate of 0.03776 m^3/sec.

f) the Reynolds number, if the air flows up the collector in a 12.7 mm x 1.22 m channel behind the absorber plate.

Solution:

(a) The heat absorbed by the collector is

$$q = 788\ W/m^2 \times 0.48 = 378\ W/m^2.$$

The flow rate of the water is

$$\overset{o}{m} = \frac{25.2\mu m^3}{sec} \times \frac{10^{-6}}{\mu} \times \frac{980\ kg}{m^3}$$

$$= 0.0247\ kg/sec.$$

The specified heat is $C_p = 4.19\ kJ/kg^oC$.

The surface area of the collector is

$$A = 1.22\ m \times 2.44\ m$$

$$= 2.98\ m^2.$$

The temperature difference (ΔT) is given by

$$\Delta T = \frac{Q}{\overset{o}{m}C_p}$$

$$= \frac{378\ W}{m^2} \times 2.98\ m^2 \times \frac{sec}{0.0247} \times \frac{kg^oC}{4.19\ kJ} \times \frac{k}{10^3}$$

$$= 10.88^oC.$$

(b) The flow rate through each of the eight tubes is

$$\overset{o}{m} = \frac{0.0247\ kg}{sec} \times \frac{1}{8}$$

$$= 0.00309\ kg/sec$$

The Reynolds number expression is

$$Re = \frac{4\ \overset{o}{m}}{\mu \pi D}$$

$$= \frac{4}{\pi} \times \frac{0.00309\ kg}{sec} \times \frac{m.sec}{0.432\ g} \times \frac{1}{6.35\ mm} \times \frac{m}{10^{-3}} \times \frac{10^3}{k}$$

$= 1,434$

(c) Here, the Reynolds number is given by

$$Re = \frac{4 \times 0.0247 \times 10^6}{\pi \times 6.35 \times 0.432}$$

$$= 11,464$$

(d) If the parallel tubes have a diameter of 25.4 mm (1.0 inch), the Reynolds number is

$$Re = \frac{4\,\overset{o}{m}}{\mu\pi D}$$

$$= \frac{4}{\pi} \times \frac{0.00309 \text{ kg}}{\text{sec}} \times \frac{\text{m.sec}}{0.432 \text{ g}} \times \frac{1}{25.4 \text{ mm}} \times \frac{\text{m}}{10^{-3}} \times \frac{10^3}{\text{k}}$$

$$= 359$$

(e) If the cooling fluid is air, then the flow rate is

$$\overset{o}{m} = \frac{0.03776 \text{ m}^3}{\text{sec}} \times \frac{1.033 \text{ kg}}{\text{m}^3} = 0.03902 \text{ kg/sec}$$

$$\therefore \quad \Delta T = \frac{Q}{\overset{o}{m}C_p}$$

$$= \frac{378 \text{ W}}{\text{m}^2} \times 2.98 \text{ m}^2 \times \frac{\text{sec}}{0.03902 \text{ kg}} \times \frac{\text{kg}^\circ\text{C}}{1.02 \text{ kJ}} \times \frac{\text{k}}{10^3}$$

$$= 28.27^\circ\text{C}$$

(f) The length of the channel is

$$P = 2[1.22 + (12.7/11,000)] = 2.465 \text{ m.}$$

The Reybolds number is

$$Re = \frac{4\,\overset{o}{m}}{\mu P}$$

$$= 4 \times \frac{0.03902 \text{ kg}}{\text{sec}} \times \frac{1}{2.465 \text{ m}} \times \frac{\text{m. sec}}{0.021 \text{ g}} \times \frac{10^3}{\text{k}}$$

$$= 3,015$$

• PROBLEM 9-14

Saturated steam at 90 psia is flowing through a 2.37 in diameter pipe enclosed in a dark room at 60°F. Estimate the amount of steam condensed per hour per unit length of pipe. The convective heat transfer coefficient may be taken as 2.5 Btu/hr.ft.^{2}°F. Assume the pipe surface acts as a gray body.

TABLE 1: Saturated Steam: Pressure Table

		Specific volume			Enthalpy			Entropy		
Abs. Press. lb/in.^{2} p	Temp. °F t	Sat. Liquid v_f	Evap. v_{fg}	Sat. Vapor v_g	Sat. Liquid h_f	Evap. h_{fg}	Sat. Vapor h_g	Sat. Liquid s_f	Evap. s_{fg}	Sat. Vapor s_g
0.0886	32.00	0.01602	3305.7	3305.7	0	1075.1	1075.1	0	2.1865	2.1865
0.125	40.69	0.01602	2383.7	2383.7	8.74	1070.2	1078.9	0.0176	2.1388	2.1564
0.250	59.31	0.01603	1235.8	1235.8	27.38	1059.5	1086.9	0.0542	2.0414	2.0956
0.500	79.58	0.01607	641.71	641.73	47.60	1048.0	1095.6	0.0924	1.9434	2.0358
1	101.76	0.01614	333.77	333.79	69.72	1035.5	1105.2	0.1326	1.8443	1.9769
5	162.25	0.01641	73.584	73.600	130.13	1000.7	1130.8	0.2347	1.6090	1.8437
10	193.21	0.01659	38.445	38.462	161.17	982.1	1143.3	0.2834	1.5042	1.7876
14.696	212.00	0.01672	26.811	26.828	180.07	970.3	1150.4	0.3120	1.4446	1.7566
15	213.03	0.01672	20.303	20.320	181.11	969.6	1150.7	0.3135	1.4413	1.7548
20	227.96	0.01683	20.093	20.110	196.16	959.9	1156.1	0.3356	1.3959	1.7315
30	250.34	0.01700	13.746	13.763	218.83	945.2	1164.0	0.3680	1.3312	1.6992
40	267.24	0.01715	10.489	10.506	236.02	933.7	1169.7	0.3919	1.2844	1.6763
50	281.01	0.01727	8.505	8.522	250.09	923.9	1174.0	0.4110	1.2473	1.6583
60	292.71	0.01738	7.162	7.179	262.10	915.4	1177.5	0.4271	1.2166	1.6437
70	302.92	0.01748	6.193	6.210	272.61	907.9	1180.5	0.4409	1.1905	1.6314
80	312.03	0.01757	5.458	5.476	282.02	901.1	1183.1	0.4532	1.1677	1.6209
90	320.27	0.01766	4.880	4.898	290.57	894.8	1185.4	0.4641	1.1472	1.6113
100	327.83	0.01774	4.415	4.433	298.43	888.9	1187.3	0.4741	1.1287	1.6028
110	334.79	0.01782	4.032	4.050	305.69	883.3	1189.0	0.4832	1.1118	1.5950
120	341.26	0.01789	3.710	3.728	312.46	878.1	1190.6	0.4916	1.0963	1.5879
130	347.31	0.01796	3.437	3.455	318.81	873.2	1192.0	0.4995	1.0820	1.5815
140	353.03	0.01803	3.202	3.220	324.83	868.5	1193.3	0.5069	1.0686	1.5755
150	358.43	0.01809	2.998	3.016	330.53	863.9	1194.4	0.5138	1.0560	1.5698
160	363.55	0.01815	2.816	2.834	335.95	859.6	1195.5	0.5204	1.0442	1.5646
170	368.42	0.01821	2.656	2.674	341.11	855.2	1196.3	0.5266	1.0327	1.5593
180	373.08	0.01827	2.514	2.532	346.07	851.1	1197.2	0.5325	1.0220	1.5545
190	377.55	0.01833	2.386	2.404	350.83	847.2	1198.0	0.5382	1.0119	1.5501

Solution: The pipe surface temperature will be assumed to be that of the condensing steam. At 90 psia, the steam tables give a saturation temperature of 320.27°F = 320°F.

The net heat transfer from the pipe surface is the sum of the heats transferred due to radiation and convection.

The radiation heat transfer is given by

$$q_r = \sigma A \varepsilon \left[T_p^4 - T_a^4 \right]$$

where subscripts p and a refer to the pipe surface and air, respectively,

σ is the Stefan-Boltzmann constant,

ε is the emissivity of the pipe surface (assumed 0.95), and

A is the surface area of pipe per unit length.

Then

$$q_r = 0.1714 \times 10^{-8} \times \frac{\pi \times 2.37}{12} \times 0.95 \left[(320 + 460)^4 - (60+460)^4\right]$$

$$= 300 \text{ Btu/hr.ft. length of pipe.}$$

Next, the convective heat transfer per unit length of pipe surface is calculated

$$q_c = h_c A(T_p - T_a)$$

$$= 2.5 \times \frac{\pi \times 2.37}{12} (320 - 60)$$

$$= 403 \text{ Btu/hr.ft. length of pipe.}$$

Thus, the total heat transfer is

$$q = q_r + q_c$$

$$= 300 + 403$$

$$= 703 \text{ Btu/hr.ft. length of pipe}$$

The rate of condensation of steam is given by

$$\overset{\circ}{m} = \frac{q}{h_{fg}}$$

where h_{fg} is the enthalpy of vaporization at 90 psia.

From steam table, that value is 894.8 Btu/lb_m.

Then

$$\overset{\circ}{m} = \frac{q}{h_{fg}}$$

$$= \frac{703}{894.8}$$

$$= 0.78 \text{ } lb_m\text{/hr.ft. length of pipe.}$$

• PROBLEM 9-15

The surface temperature of a 3 ft square panel is maintained at 200°F. Calculate the heat lost by the panel

(a) when attached to the ceiling, facing downwards in a room at 80°F, and

(b) when attached to the floor, facing upwards in the same room.

The condition through the back of the panel may be neglected. Take $\varepsilon = 0.9$ for the panel surface.

Solution: Heat transfer occurs both by radiation and convection.

(a) Heat transfer due to convection can be found by using the relation

$$q_c = \frac{0.12(\Delta T)^{1.25}}{(\ell)^{0.25}}$$

$$= \frac{0.12(200 - 80)^{1.25}}{3^{0.25}}$$

$$= 36.2 \text{ Btu/hr.ft}^2.$$

and the heat transfer by radiation as

$$q_r = \sigma\varepsilon(T_1{}^4 - T_2{}^4)$$

where σ is the Stefan-Boltzmann constant,

ε is the emissivity of the surface,

T_1 is the absolute temperature of the panel, and

T_2 is the absolute temperature of the room.

Then

$$q_r = 0.1714 \times 10^{-8} \times 0.9(660^4 - 540^4)$$

$$= 161.5 \text{ Btu/hr.ft}^2.$$

Therefore, the total heat transfer is

$$q = q_c + q_r$$

$$= 36.2 + 161.5$$

$$= 197.7 \text{ Btu/hr.ft}^2.$$

(b) When the panel is placed facing upwards, the product of Grashof anf Prandtl number is $Gr\ Pr = 8.6 \times 10^8$, giving the heat transfer by convection as

$$q_c = 0.35(\Delta T)^{1.25}$$

$$= 0.35(200 - 80)^{1.25}$$

$$= 139 \text{ Btu/hr.ft}^2.$$

It is obvious that the radiation heat transfer would remain the same as before. Thus, the total heat transfer is

$$q = q_c + q_r$$

$$= 139 + 161.5$$

$$= 300.5 \text{ Btu/hr.ft}^2.$$

CONDUCTION AND RADIATION

• PROBLEM 9-16

An insulated horizontal pipe at 400°F (dimensions are as shown in figure 1) is exposed to air at 70°F. Estimate the heat transfer per foot length of pipe surface, assuming a thermal conductivity for the insulation of k = 0.035 Btu/hr-ft-°F.

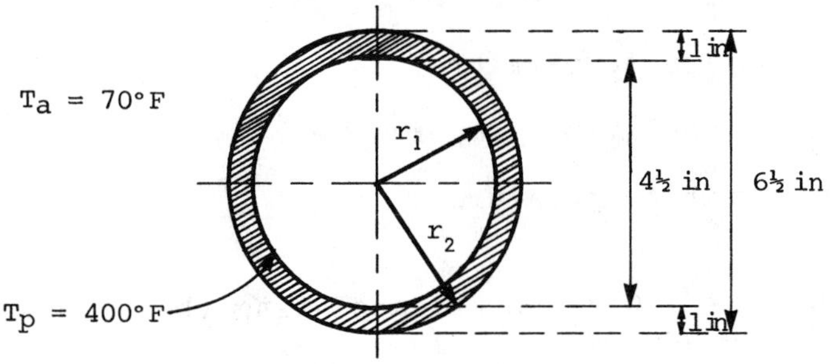

Fig. 1 Insulated Pipe

TABLE 1

HEAT TRANSFER FROM HORIZONTAL TUBES TO STILL AIR AT ORDINARY ROOM TEMPERATURE

Combined coefficients $(h_c + h_r)$ in B hr^{-1} ft^{-2} F^{-1}

Outer Diameter D, Inches	Temperature Difference Δt[F]													
	50	100	150	200	250	300	350	400	450	500	550	600	650	700
1.3	2.26	2.50	2.73	3.00	3.29	3.60	3.95	4.34	4.73	5.16	5.60	6.05	6.51	6.98
3.5	2.05	2.25	2.47	2.73	3.00	3.31	3.69	4.03	4.43	4.85	5.26	5.71	6.19	6.66
5.6	1.95	2.15	2.36	2.61	2.90	3.20	3.54	3.90	4.30	...	...	...	...	...
10.75	1.87	2.07	2.29	2.54	2.82	3.12	3.47	3.84	...	...	...	...	...	...

Solution: The heat transfer from the pipe surface to the insulation occurs by a conduction process, and from the insulation to the surroundings by convection and radiation.

The insulation temperature will be assumed, and the conductive heat transfer equated to the combined heat transfer from the insulation surface.

Thus,

(Heat Conduction) = (Combined heat transfer from insulation surface)

or

$$\frac{k\ 2\pi(T_1 - T_2)}{\ln \frac{r_2}{r_1}} = (h_c + h_r)\ 2\pi r_2(T_2 - T_3)$$

where k - thermal conductivity of insulation

(h_c+h_r) - combined heat transfer coefficient obtained from the table corresponding to the surface temperature of the insulation

T_1 - pipe temperature

T_2 - insulation temperature

T_3 - air temperature

Assuming $T_2 = 110^0F$ and extrapolating the values in the table, the equation becomes

$$h_c + h_r = \frac{0.352\ (400 - 110)}{\frac{6.5}{2x12}\ \ln\left(\frac{6.5}{4.5}\right)(110 - 70)} = 2.55$$

From the table, for $T_2 = 110^{O}F$ and D = 6.5 in., $(h_c + h_r)$ is 1.9 Btu/hr-ft^2-^{O}F. Then T_2 will be assumed to be 120^OF. The new value of $\left(h_c + h_r\right)$ is 1.97 Btu/hr-ft^2-^{O}F. Corresponding to $T_2 = 120^O$F and D = 6.5 in., the table value of $(h_c + h_r)$ is 1.9, which is quite close to the calculated value. Taking $(h_c + h_r) = 1.9$ Btu/hr-ft^2-^{O}F and $T_2 = 120^O$F, the heat transfer from the pipe can be computed using the equation

$$q = (h_c + h_r)\ A\ (T_2 - T_3)$$

$$= 1.9 \times \pi \frac{6.5}{12}\ (120 - 70)$$

$$= 162 \text{ Btu/hr.ft. length of pipe}$$

What is the thickness b_B required for a 90% effective fin, given $k = 200$ W/m-^oK, $\ell = 5$ cm., $T_B = 450^oK$, $T_r = 225^oK$., and $\varepsilon = 0.903$? Also, how should b/b_B vary with x/ℓ?

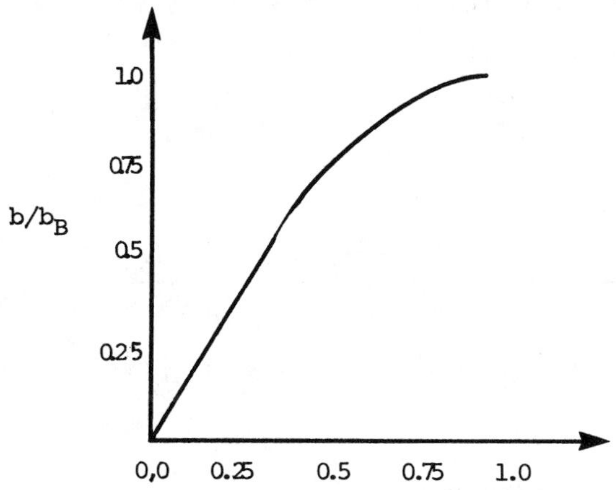

Fig. 1

Solution: The fin effectiveness is expressed as

$$\eta = \frac{(1 - T_o^{*5}) - 5\beta^4(1 - T_o^{*5})}{5(1 - \beta^4)(1 - T_o^*)}$$

where

$$\beta = \frac{T_r}{T_B} = \frac{225}{450}$$

$$= 0.5,$$

and

$$\eta = 0.9.$$

Therefore,

$$0.9 = \frac{(1 - T_o^{*5}) - 5(0.5)^4(1 - T_o^*)}{5(1 - 0.5^4)(1 - T_o^*)}$$

which gives $T_o^* = 0.95076$.

Now, the parameter N_B in terms of T_o^* is expressed as

$$N_B = \frac{5(1 - T_o^*)^2}{(1-T_o^{*5})-5\beta^4(1-T_o^*)}$$

Therefore,

$$N_B = \frac{5(1 - 0.95076)^2}{(1 - 0.95076^5) - 5(0.5)^4(1 - 0.95076)}$$

$$= 0.05836$$

Another expression for this parameter is

$$N_B = \frac{\varepsilon\sigma T_B^3 \ell^2}{kb_B}$$

Substituting,

$$0.05836 = \frac{0.903(5.67 \times 10^{-8})(450)^3 (0.05)^2}{200\ b_B}$$

or $b_B = 0.996$ mm.

The fin profile $A^* = b/b_B$ is expressed in terms of T^* and T_o^* as

$$A^* = \frac{(T^{*5} - T_o^{*5}) - 5(T^* - T_o^*)}{(1 - T_o^{*5}) - 5(1 - T_o^*)}$$

where

$$T^* = T_o^* + (1 - T_o^*)x^*$$

Substituting values of x^* from 1 to 0 yields

x^*	b/b_B
1.00	1.00
0.80	0.958
0.60	0.835
0.40	0.633
0.20	0.354
0.0	0.0

The accompanying figure shows this variation graphically.

• PROBLEM 9-18

0.228 cm thick circumferential fins with a radial depth of 2.54 cm are welded to a 15.24 cm dia. steel barrel. The fins are spaced at a distance of 0.456 cm apart. The barrel wall temperature is 393°K, and the ambient air temperature is 293°K. Determine the rate at which heat is rejected from the finned barrel. Also, find the rate of heat rejection, assuming that the barrel is not finned. The convective heat transfer coefficient is 284 W/m^2°K, and the thermal conductivity of steel may be taken as 48 W/m-°K.

Table 1: Selected values of the modified Bessel functions of the first and second kinds, orders zero and one.

x	$I_0(x)$	$I_1(x)$	$(2/\pi)K_0(x)$	$(2/\pi)K_1(x)$
0.0	1.000	0.0000	$+\infty$	$+\infty$
0.2	1.0100	0.1005	1.1158	3.0405
0.4	1.0404	0.2040	0.70953	1.3906
0.6	1.0920	0.3137	0.49498	0.82941
0.8	1.1665	0.4329	0.35991	0.54862
1.0	1.2661	0.5652	0.26803	0.38318
1.2	1.3937	0.7147	0.20276	0.27667
1.4	1.5534	0.8861	0.15512	0.20425
1.6	1.7500	1.0848	0.11966	0.15319
1.8	1.9896	1.3172	0.92903×10^{-1}	0.11626
2.0	2.2796	1.5906	0.72507	0.89041×10^{-1}
2.2	2.6291	1.9141	0.56830	0.68689
2.4	3.0493	2.2981	0.44702	0.53301
2.6	3.5533	2.7554	0.35268	0.41561
2.8	4.1573	3.3011	0.27896	0.32539
3.0	4.8808	3.9534	0.22116	0.25564
3.2	5.7472	4.7343	0.17568	0.20144
3.4	6.7848	5.6701	0.13979	0.15915
3.6	8.0277	6.7028	0.11141	0.12602
3.8	9.5169	8.1404	0.8891×10^{-2}	0.9999×10^{-2}
4.0	11.3019	9.7595	0.7105	0.7947
4.2	13.4425	11.7056	0.5684	0.6327
4.4	16.0104	14.0462	0.4551	0.5044
4.6	19.0926	16.8626	0.3648	0.4027
4.8	22.7937	20.2528	0.2927	0.3218
5.0	27.2399	24.3356	0.2350	0.2575
5.2	32.5836	29.2543	0.1888	0.2062
5.4	39.0088	35.1821	0.1518	0.1653
5.6	46.7376	42.3283	0.1221	0.1326
5.8	56.0381	50.9462	0.9832×10^{-3}	0.1064
6.0	67.2344	61.3419	0.7920	0.8556×10^{-3}
6.2	80.7179	73.8859	0.6382	0.6879
6.4	96.9616	89.0261	0.5146	0.5534
6.6	116.537	107.305	0.4151	0.4455
6.8	140.136	129.378	0.3350	0.3588
7.0	168.593	156.039	0.2704	0.2891
7.2	202.921	188.250	0.2184	0.2331
7.4	244.341	227.175	0.1764	0.1880

Solution: The heat transfer from a fin is given by the equation

$$Q_f = 2\pi l_1 bk \sqrt{\beta}\Delta t \frac{I_1(l_2\sqrt{\beta})K_1(l_1\sqrt{\beta}) - K_1(l_2\sqrt{\beta})I_1(l_1\sqrt{\beta})}{I_1(l_2\sqrt{\beta})K_0(l_1\sqrt{\beta}) + K_1(l_2\sqrt{\beta})I_0(l_1\sqrt{\beta})} \quad (1)$$

where

$\beta = 2h/kb$.

Here, b is the fin thickness

l_1 is the outer radius of the barrel

= radius of the base of the fin

= 7.62 cm,

l_2 = radius of the end of the fin

= 10.16 cm,

and I_0, K_0, I_1 and K_1 are the Bessel functions of the 1st and 2nd kind, respectively. These may be obtained from the table.

Also, Δt is the temperature difference

$= 393 - 293 = 100^\circ K$.

Substituting,

$$\beta = \frac{2h}{kb} = \frac{2 \times (284)\ 100}{48(0.228)} = 5{,}200$$

or $\sqrt{\beta} = 72$ per meter.

From the table,

for $x = 7.3$, $I_1 = 207$ and $K_1 \simeq 0$;

for $x = 5.5$, $I_1 = 38.5$ and $K_1 = 0.0023$;

for $x = 5.5$, $I_0 = 43.0$ and $K_0 = 0.0021$.

Substituting the known values in eq. (1) yields

$$Q_f = 2\pi \times \frac{7.62}{100} \times \frac{0.228}{100}\ 48 \times 72 \times 100 \times \frac{207 \times 0.0023}{207 \times 0.0021}$$

$= 414$ W.

When the barrel is not finned, the rate of heat rejection is given by

$$Q = 2\pi l_1 ah\Delta t$$

$$= 2\pi \frac{7.62}{100} \times \frac{0.684}{100} \times 284 \times 100$$

$$= 92.8 \text{ W}$$

• PROBLEM 9-19

Compute the heat transfer per hour per 100 ft length for a horizontal pipe carrying steam at 370°F. The pipe is covered with 2 in. of insulation, and the ambient temperature is 80°F. Assume a thermal conductivity of 0.04 Btu/hr-ft²- (°F/ft) for the insulation. Also, compute the rate of heat transfer in the absence of insulation. The nominal diameter of pipe is 2 in., and outside diameter is 2.38 in.

Table 1: VALUES OF $h_c + h_r$

(For horizontal bare or insulated standard steel pipe of various sizes in a room at 80°F)

Nominal pipe diam, in.	(Δt)$_s$, temperature difference, deg F, from surface to room														
	50	100	150	200	250	300	400	500	600	700	800	900	1000	1100	1200
½	2.12	2.48	2.76	3.10	3.41	3.75	4.47	5.30	6.21	7.25	8.40	9.73	11.20	12.81	14.65
1	2.03	2.38	2.65	2.98	3.29	3.62	4.33	5.16	6.07	7.11	8.25	9.57	11.04	12.65	14.48
2	1.93	2.27	2.52	2.85	3.14	3.47	4.18	4.99	5.89	6.92	8.07	9.38	10.85	12.46	14.28
4	1.84	2.16	2.41	2.72	3.01	3.33	4.02	4.83	5.72	6.75	7.89	9.21	10.66	12.27	14.09
8	1.76	2.06	2.29	2.60	2.89	3.20	3.88	4.68	5.57	6.60	7.73	9.05	10.50	12.10	13.93
12	1.71	2.01	2.24	2.54	2.82	3.13	3.83	4.61	5.50	6.52	7.65	8.96	10.42	12.03	13.84
24	1.64	1.93	2.15	2.45	2.72	3.03	3.70	4.48	5.37	6.39	7.52	8.83	10.28	11.90	13.70

Solution: Heat transfer in the absence of insulation will be analysed first. The only resistance offered is that due to the material of the pipe wall and the convective and radiative properties of the outer surface of the pipe. The conductive resistance of the pipe wall material is much smaller than the convective and radiative resistances. Therefore,

$$q = (h_c + h_r)A\Delta T$$

where h_c and h_r are the convective and radiative heat transfer coefficients, respectively. They may be obtained from the accompanying table by interpolation for a 2 in. nominal diameter and the specified temperature difference.

For $\Delta T = 370 - 80 = 290^\circ F$, the table gives

$$(h_c + h_r) = 3.4 \text{ Btu/hr-ft}^2\text{-}^\circ F.$$

Then

$$q = 3.4 \times A \times (370 - 80)$$

$$= 3.4 \times \pi \times \frac{2.38}{12} \times 100 \times 290$$

$$= 61{,}300 \text{ Btu/hr.}$$

Now, when insulation is present, the total resistance to heat transfer is the sum of the separate resistances due to the insulation, the pipe wall material and the convective and radiative properties. The pipe wall material resistance is negligibly small as compared to the other resistances present. As the relative contribution of these two types of resistances is not known, the temperature difference between insulation surface and the air will first be approximated by a trial and error process. The heat transfer is expressed as

$$q = \frac{\Delta T}{\frac{t}{k\,A_m} + \frac{1}{(h_c + h_r)A}}$$

where $\Delta T = 370 - 80 = 290^{\circ}F$,

t - thickness of the insulation,

$= 2/12$ ft

k - thermal conductivity of the insulation

$= 0.04$ Btu/hr-ft^2- ($^{\circ}$F/ft),

and

A_m = logarithmic-mean area

$$= \frac{A_2 - A_1}{\ln(A_2/A_1)} \quad ,$$

where

$$A_2 = \pi D_o l = \pi \times \frac{6.38}{12} \times 100 \quad ft^2,$$

and

$$A_1 = \pi D_i l = \pi \times \frac{2.38}{12} \times 100 \quad ft^2.$$

Then

$$A_m = \frac{100\pi}{12} \frac{6.38 - 2.38}{2.3 \log_{10}\left(\frac{6.38}{2.38}\right)}$$

$$= 105.5 \text{ ft}^2.$$

Now, assuming $h_c + h_r = 1.8$ for a nominal diameter of 6.38 in,

$$q = \frac{290}{\frac{2/12}{0.04 \times 105.5} + \frac{1}{1.8 \times \pi \times 6.38 \times 100}}$$

$= 6{,}780$ Btu/hr.

The surface resistance with this assumption is 0.00333, which is just 7.8% of the total resistance of

$$\frac{\frac{2}{12}}{0.04 + 105.5} + \frac{1}{1.8 \times \pi \times 6.38 \times 100} = 0.0428.$$

The corresponding temperature difference is

$$\Delta T = 0.078\ (370 - 80)$$

$$= 23^{\circ}F.$$

Now, from the table, the value of $\left(h_c + h_r\right)$ for a diameter of 6.38 in. and $\Delta T = 23^{\circ}F$ is approximately 1.8 (by interpolation). Thus, the assumed value is acceptable, and the calculated heat transfer rate, i.e., 6,780 Btu/hr, is assumed correct,

Comparing the two rates of heat transfer (i.e., with and without insulation), it is seen that rate of heat transfer is reduced by $\left(\frac{61{,}300 - 6{,}780}{61{,}300}\right) \times 100 = 89\%$ when the insulation is used.

CONDUCTION, CONVECTION AND RADIATION

• PROBLEM 9-20

The heat loss due to convection and radiation from the surface of a 3 cm long, 2 mm diameter electric cable is 40% of the total I^2R power loss. The balance (60%) is conducted through the connectors, which act as fins. Compute the surface temperature of the cable when it is

(a) insulated

(b) uninsulated

if its power rating is one watt.

The physical and thermal properties of the core and insulation are:

$\Omega = 2M\Omega$; $k = 0.105\ W/m^{\circ}K$; $h = 17.3\ W/m^{2\circ}K$.

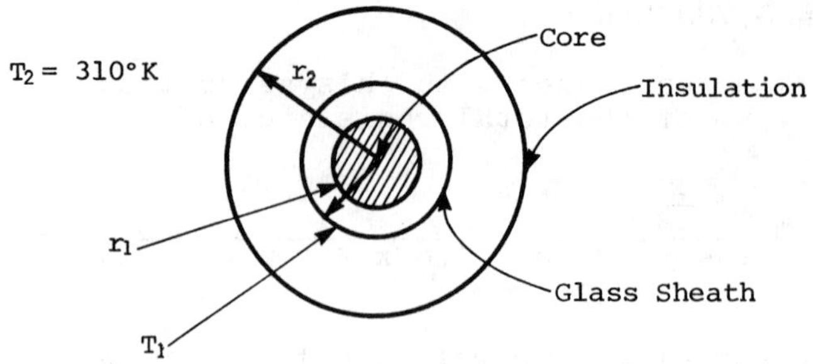

Fig. 1

Solution: For a coaxial cylindrical system, the heat transfer is expressed as

$$\dot{Q} = \frac{T_1 - T_2}{\frac{1}{2\pi Lh}\left[\frac{1}{r_2} + \frac{h}{k}\ln\left(\frac{r_2}{r_1}\right)\right]} \tag{1}$$

$$= \frac{N}{D(r_2)}$$

The critical radius of insulation for the maximum amount of heat transfer is obtained by differentiating this expression with respect to r_2 as

$$\frac{d\dot{Q}}{dr_2} = -\frac{N}{D^2}\frac{dD}{dr_2} = -\frac{N}{D^2}\left[\frac{h}{kr_2} - \frac{1}{r_2^2}\right]$$

$$= -\frac{N}{r_2 D^2}\left[\frac{h}{k} - \frac{1}{r_2}\right]$$

Now, the LHS will be zero when

$$\frac{h}{k} - \frac{1}{r_2} = 0 \quad \text{or} \quad r_2 = \frac{k}{h}.$$

This value of r_2 $(= \frac{k}{h})$ is the critical radius r_c of the insulation. It will be a maximum if the second derivative of $\dot{Q}$ is negative. Checking this,

$$\frac{d^2Q}{dr_2^2} = -\frac{N}{D^2}\frac{d^2D}{dr_2{}^2}\Bigg|_{r_2 = r_{cr}} = -\frac{N}{D^2} - \frac{h}{kr_c^2} + \frac{2}{r_c^3}$$

$$= -\frac{N}{r_c{}^2 D^2}\ \frac{1}{r_c}$$

It is a maximum. Therefore,

$$r_2 = r_c = \frac{k}{h} = \frac{0.105}{17.3}$$

$$= 6.07 \times 10^{-3} m.$$

In an electrical analogy, the surface convective heat transfer and the insulation resistances are in series, and the heat transfer rate may be found using eq.(1).

With

$r_1 = 1$ mm or 10^{-3} m,

$\dot{Q} = 40\% \times 1W$, $L = 3$ cm or 0.03 m,

$k = 0.105$ w/m ^oK, and $T_2 = 310^oK$,

eq. (1) may be rearranged and substituted into as

$$T_1 = T_2 + \dot{Q}\left[\frac{\ln (r_2/r_1)}{2\pi kL} + \frac{1}{2\pi hLr_2}\right]$$

$$= 310 + 0.4\left[\frac{\ln \frac{6.07 \times 10^{-3}}{10^{-3}}}{2\pi \times 0.105 \times 0.03} + \frac{1}{2\pi \times 17.3 \times 0.03 \times 6.07 \times 10^{-3}}\right]$$

$$= 310 + 56.6$$

$$= 366.6^oK.$$

For the case of an uninsulated cable, eq. (1) reduce to

$$T_1 = T_2 + \dot{Q}\left[\frac{1}{2\pi hr_1L}\right] {}^oK$$

Substituting,

$$T_1 = 310 + 0.4 \times \frac{1}{2\pi \times 17.3 \times 10^{-3} \times 0.03}$$

$$= 310 + 122.7$$

$$= 432.7\ {}^oK$$

A flat roof consists of 3 mm. of steel (k_s=52 W/m°K), 38 mm. of insulation (k_i=0.035 W/m°K), and 9 mm. of asphalt (k_a=0.17 W/m°K), as shown in fig. 1. The roof measures 12.5 m. x 22 m. Inside, the air is at a temperature of 21°C, while outside, it is at -1°C. The inside and outside convective heat transfer coefficients are 11 W/m²°K and 34 W/m²°K, respectively. The roof acts as a perfect black body, collecting solar radiation at the rate of 785 W/m². What is the rate of heat transfer through the roof? Assume a space temperature of -20°C, and that conditions are at a steady state.

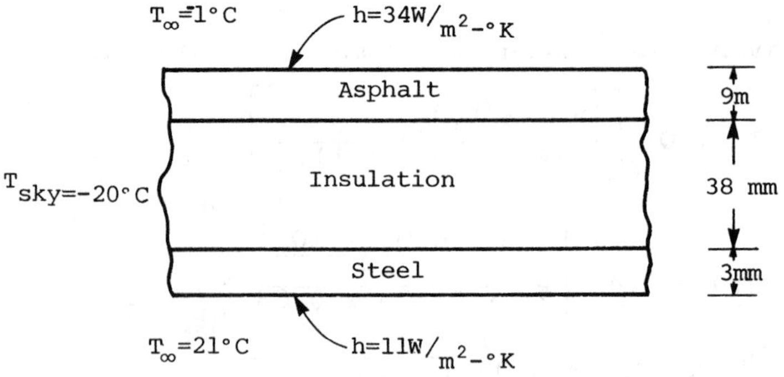

Fig. 1

Solution: The total roof area is

$$A_T = 12.5 \times 22 = 275 \text{ m}^2.$$

The incident solar energy is $Q_{in} = q_{rad}\, A_T = 785 \times 275$

$$= 2.16 \times 10^5 \text{ watt.}$$

The convective heat given off is

$$Q_{conv} = h_o\, A_T\, (T_s - T_\infty)$$

$$= 34 \times 275\, (T_s + 1)$$

$$= (9350\, T_s + 9350) \text{ W.}$$

The thermal resistance of the roof is

$$\Sigma R = \frac{L_s}{k_s A_T} + \frac{L_i}{k_i A_T} + \frac{L_A}{k_A A_T} + \frac{1}{h_i A_T},$$

where L is the thickness of each layer, $1/h_iA_T$ is the inside convection resistance, and subscripts s, i, and A refer to the steel insulation and asphalt, respectively.

Then,

$$\Sigma R = \left[\frac{0.003}{52} + \frac{0.038}{0.035} + \frac{0.009}{0.17} + \frac{1}{11}\right]\frac{1}{275}$$

$$= 4.47 \times 10^{-3}\,{}^{\circ}C/W$$

Conduction through the roof is

$$Q_{cond} = \frac{T_s - T_{\infty i}}{\Sigma R}$$

where $T_{\infty i}$ is the inside air temperature. Then

$$Q_{cond} = \left(\frac{T_s - 21}{4.47 \times 10^{-3}}\right) \text{ W}$$

The radiant heat transfer is

$$Q_{rad} = A_T\,\varepsilon\,(e_{bs} - e_{bsky})$$

where ε - emissivity of the roof surface = 1 (for a black body)

e - radiation heat transfer per unit area.

$= \sigma\, T^4$, where $\sigma = 5.668 \times 10^{-8}$ $W/m^2\,{}^{\circ}K^4$

and T is the absolute temperature.

$$e_{bs} = (5.668 \times 10^{-8})(T_s + 273)^4$$

$$e_{bsky} = (5.668 \times 10^{-8})(-20 + 273)^4$$

$$= (5.668 \times 10^{-8})(253)^4$$

Substituting,

$$Q_{rad} = 275 \times 1 \times 5.668 \times 10^{-8}\left[(T_s + 273)^4 - (253)^4\right]$$

$$= \{1.558 \times 10^{-5}\left(T_s + 273\right)^4 - 6.38 \times 10^4\}\text{W}$$

Now, performing an energy balance on the top surface of the roof gives

$$Q_{in} = Q_{conv} + Q_{cond} + Q_{rad}$$

Substituting,

$$2.16 \times 10^5 = 9350\,T_s + 9350 + \frac{T_s - 21}{4.47 \times 10^{-3}}$$

$$+ (1.558 \times 10^{-5})(T_s + 273)^4 - 638 \times 10^4$$

or

$$965 = 41.8\ T_s + 41.8 + T_s - 21 + 6.96 \times 10^{-8} \times (T_s + 273)^4 - 285$$

This is a fourth order equation, and must be solved by iteration.

Assuming $T_s = 17^\circ C$,

$$1229 = 42.8(17) + (6.96 \times 10^{-8})(17 + 273)^4$$

$$= 1220$$

Next, a value of

$T_s = 18^\circ C$ is chosen.

$$1229 = 42.8(18) + (6.96 \times 10^{-8})(18 + 273)^4$$

$$= 1269$$

Thus, the value lies between $17^\circ c$ and $18^\circ c$.

With $17.2^\circ C$,

$$1229 = 42.8(17.2) + (6.96 \times 10^{-8})(17.2 + 273)$$

$$1229 = 1229.78$$

This degree of accurace is acceptable if this value of $T_s = 17.2^\circ C$ is substituted into the conduction equation, the result is

$$Q_{cond} = \frac{17.2 - 21}{4.47 \times 10^{-3}} = -850 \text{ W.}$$

The negative sign indicates that the heat flows outwards.

CHAPTER 10

HEAT TRANSFER WITH PHASE CHANGE

Basic Attacks and Strategies for Solving Problems in this Chapter. See pages 627 to 675 for step-by-step solutions to problems.

The phase changes to be considered in this chapter are condensation, boiling, evaporation, and solidification.

For condensation the heat transfer coefficient may be determined in a manner similar to that used for free convection. There are empirical equations and graphs available for different physical configurations for the condensation surface, such as a vertical pipe, vertical wall, horizontal pipe, horizontal wall, or bank of horizontal tubes. For each physical configuration, there may be several equations or graphs to handle laminar & turbulent flow and perhaps ranges of values of other parameters such as Prandtl number. As with free and forced convection, there are equations for local heat transfer coefficient and for average heat transfer coefficient over a surface. The average coefficient is needed to calculate heat transfer rate for the entire surface. The rate of heat transfer for condensation is calculated from an equation similar to Newton's law of cooling for convection. It is equal to the product of the heat transfer coefficient, the heat transfer surface area, and the difference between the saturation temperature of the condensing fluid and the heat transfer surface temperature:

$$q = h_{av} A\,(T_{sat} - T_{surf})$$

In addition to the heat transfer rate, other parameters such as the mass rate of condensate formed, the thickness of the condensate film, and the velocity of flow in the condensate film may be calculated. The mass rate of condensation may be calculated by dividing the heat transfer rate by the heat of vaporization of the fluid. The film thickness and film velocity may be determined from empirical correlations.

Some knowledge of the different classifications of boiling heat transfer is needed to interpret problem statements. Boiling on a heat transfer surface may be classified as pool boiling (when a pool of unmoving liquid is in contact with a heated surface) or forced convection boiling (when the liquid is flowing past a heated surface). Several different types of pool boiling may occur depending upon the difference between the temperature of the heated surface and the saturation temperature of the liquid, $T_s - T_{sat}$. At low temperature differences, free convections boiling occurs, in which vapor bubbles form slowly at the heated surface and condense in the liquid *before reaching the liquid surface. At higher temperature differences nucleate boiling occurs, in which bubbles form more rapidly at the heated surface and rise to the liquid surface without condensing. At still higher temperature differences a significant portion of the heated surface is covered with bubbles and the rate of heat transfer decreases because of the need to transfer the heat through the vapor bubbles. Film boiling is the term used for the situation when the temperature difference is high enough to maintain a layer of vapor over the entire heated surface.*

There will be a maximum heat flux (heat transfer rate per unit area) which occurs at the upper end of the nucleate boiling range. At higher temperature differences the heat flux decreases due to the effect of the bubbles on the surface. The maximum heat flux, also called the burnout heat flux, can be calculated from fluid properties as illustrated in problems 10-14 and 10-16.

Correlations for the boiling heat flux in terms of temperature difference and fluid properties are available as illustrated in the boiling heat transfer examples in this chapter.

Evaporation is essentially the same as boiling heat transfer. A hot surface is used to evaporate liquid in contact with it or flowing past it. Heat transfer coefficients for convection and for boiling may be added together to obtain the evaporation heat transfer coefficient.

Calculations involving solidification are less common than condensation or evaporation. An example of solidification calculations is given in problem 10-21.

Step-by-Step Solutions to Problems in this Chapter, "Heat Transfer with Phase Change"

CONDENSATION

• PROBLEM 10-1

> Laminar film condensation of super-saturated steam occurs on a vertical surface. The steam is at a pressure of 1.43 psi and a saturation temperature of 114°F. If the wall temperature is 105°F, what is the coefficient of heat transfer at a point 3 in. from the top?

Solution: For film condensation in a laminar regime, the Nusselt number may be found as

$$Nu_x \equiv \left[\frac{h_x x}{k_l} = \frac{\rho_l g(\rho_l - \rho_v)h_{fg}x^3}{4\mu_l k_l(T_{sat}-T_s)}\right]^{1/4} \quad (1)$$

Since the density of the liquid is so much greater than that of the vapor, the approximation

$$\rho_l(\rho_l - \rho_v) \simeq \rho_l^2 \quad (2)$$

may be used. At the saturation conditions, the steam tables give the latent heat of condensation as

$$h_{fg} = 1{,}029 \text{ Btu/lbm}$$

For a mean temperature of

$$T_m = \frac{105 + 114}{2} = 109.5^\circ F,$$

the water tables list

$\mu = 1.486$ lbm/ft-hr $\qquad$ $k = 0.3656$ Btu/hr-ft-°F

$\rho_1 = 62\ \text{lbm/ft}^3$

Then substituting eq. (2) into eq. (1) and the values into eq. (1),

$$Nu_x = \left[\frac{(62)^2(32.2)(1{,}029)(3/12)^3}{4(1.486/3{,}600)(0.3656/3{,}600)(114 - 105)}\right]^{1/4}$$

$$= 1{,}071.6$$

∴ The local heat transfer coefficient is

$$h_x = \frac{Nu_x k}{x} = \frac{(1{,}071.6)(0.3656)}{(3/12)}$$

$$= 1{,}567\ \text{Btu/hr-ft}^2\text{-}^0\text{F}$$

This is the local coefficient of heat transfer value at a point 3 in. from the top.

• PROBLEM 10-2

Saturated steam condenses, at 1 atm pressure, on a vertical wall 1 m high maintained at a uniform surface temperature of 80°C. At the mid-point of the wall, find

(a) the film thickness,

(b) the average velocity,

(c) the local coefficient of heat transfer,

(d) the average coefficient of heat transfer over the total length,

(e) the rate of heat flow per unit width,

(f) the rate of condensation per hour, and

(g) the average coefficient of heat transfer, when the wall is tilted at an angle of 30° with the horizontal.

Solution: For a vertical surface, if

$$(4.52)\left\{\frac{\rho_1 g(\rho_1 - \rho_v)k_1^3(T_{sat} - T_s)^3}{\mu_1^5 h_{fg}^3}L^3 \sin\phi\right\}^{1/4} < 1800 \qquad (1)$$

the flow is laminar.

At a saturation temperature of $100^{\circ}C$, the steam tables give

$$h_{fg} = 2,257 \text{ kJ/kg},$$

while at the mean temperature of

$$T_m = \frac{80 + 100}{2} = 90^{\circ}C,$$

the water tables list

$\mu = 0.315 \times 10^{-3}$ kg/m-sec k = 0.6727 W/m-$^{\circ}$C

$\rho = 959.2$ kg/m^3

Substituting into eq. (1) and taking $\rho_l(\rho_l - \rho_v) \simeq \rho^2$,

$$4.52 \left[\frac{(959.2)^2(9.81)(0.6727)^3(100-80)^3(1)^3 \sin 90^{\circ}}{(0.315 \times 10^{-3})^5(2,257)^3(1,000)^3}\right]^{1/4}$$

$$= 712.2$$

As $712.2 < 1800$, the flow is laminar.

(a) For laminar film condensation, the thickness is given by

$$\delta = \left[\frac{4\mu k(T_{sat} - T_s)x}{g\rho^2 h_{fg}}\right]^{1/4} \tag{2}$$

Substituting into eq. (2) yields

$$\delta = \left[\frac{4(0.315 \times 10^{-3})(0.6727)(100-80)(0.5)}{(9.81)(959.2)^2(2,257)(1,000)}\right]^{1/4}$$

$$= 1.43 \times 10^{-4} \text{ m.}$$

(b) Since the Reynolds number is defined as

$$Re = \frac{4\mu_m \rho\delta}{\mu},$$

the mean velocity may be solved for as

$$U_m = \frac{Re\mu}{4\rho\delta} \tag{3}$$

Here, the Reynolds number referred to is that evaluated at a distance of 0.5 m from the top. Substituting this length into eq. (1), an identical calculation yields

$$Re = 423.5$$

If this value is used in eq. (3), the result is

$$U_m = \frac{(423.5)(0.315 \times 10^{-3})}{4(959.2)(1.43 \times 10^{-4})}$$

$$= 0.24 \text{ m/sec.}$$

(c) The local coefficient of heat transfer is found from the simple relation

$$h_x = \frac{k}{\delta} \qquad (4)$$

$$= \frac{0.6727}{1.43 \times 10^{-4}}$$

$$= 4{,}704.2 \text{ W/m}^2\text{-}^\circ\text{C.}$$

(d) The average coefficient of heat transfer is

$$h_{av} = \left[0.943 \frac{\rho^2 g h_{fg} k^3}{\mu L(T_{sat} - T_s)}\right]^{1/4} \qquad (5)$$

$$= 0.943 \left[\frac{(959.2)^2(9.81)(2{,}257)(1{,}000)(0.6727)^3}{(0.315 \times 10^{-3})(1)(100 - 80)}\right]^{1/4}$$

$$= 5{,}282 \text{ W/m}^2\text{-}^\circ\text{C.}$$

(e) The rate of heat flow per unit width is

$$q = h_{av} A(T_{sat} - T_s) \qquad (6)$$

$$= (5{,}282)(1)(1)(100 - 80)$$

$$= 105{,}640 \text{ W}$$

This assumes a width of 1 m.

(f) From thermodynamics, the rate of condensation is the total heat flux divided by the latent heat of vaporization, or

$$\dot{m} = \frac{q}{h_{fg}} \qquad (7)$$

$$= \frac{(105{,}640)(3{,}600)}{(2{,}257)(1{,}000)}$$

$$= 168.5 \text{ kg/hr.}$$

(g) Finally, when a plate is tilted at an angle ϕ with the horizontal, the average heat transfer coefficient as given in eq. (5) is modified by the term

$$(\sin \phi)^{1/4}.$$

As $\phi = 30^{\circ}$, the average value is

$$h_{av} = 5,282 \, (\sin 30)^{1/4}$$

$$= 4,441.6 \ W/m^2\text{-}^{\circ}C$$

• PROBLEM 10-3

> Steam condenses on a vertical surface. The steam is saturated and is at a temperature of 262°F. The plate is 12 ft. high and 1 ft. wide, with a surface temperature of 162°F. Assuming the flow to be turbulent and neglecting the vapor shear stress, calculate the rate of steam condensation.

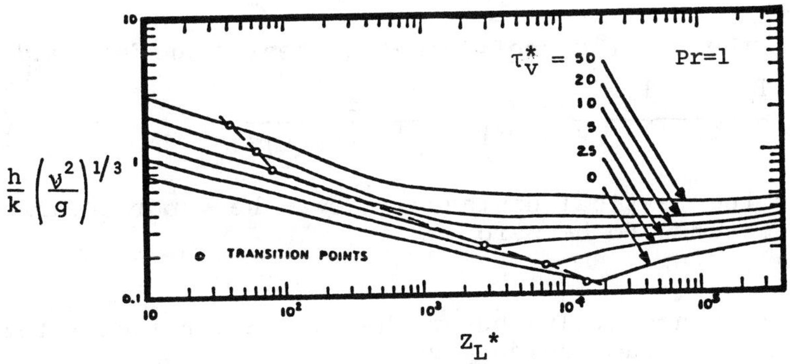

Fig. 1

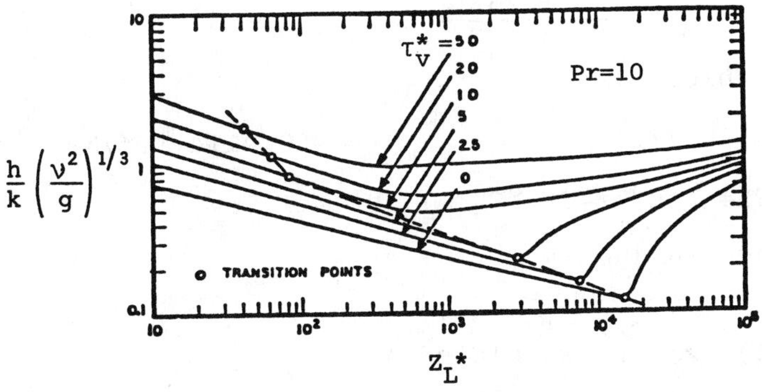

Fig. 2 Effect of turbulence and vapor shear stress on condensation.

Solution: The mass flow rate of a condensate is given by

$$\dot{m} = \frac{q}{h_{fg}} \tag{1}$$

where q --heat flow

h_{fg} --latent heat of vaporization

and $$q = hA(T_{sat} - T_s) \tag{2}$$

To get the heat flow, it will be necessary to first find the coefficient of heat transfer. Referring to the graphs supplied (figures 1 and 2), h^* is plotted versus z_L^* for varying amounts of vapor shear stress τ_v^*. This is done at Prandtl numbers of 1 and 10 in figure 1 and 2, respectively. The terms defined in the graph are

$$h^* = \frac{h}{k}\left(\frac{\nu^2}{g}\right)^{1/3} \tag{3}$$

where h is the desired coefficient of heat transfer, and

$$z_L^* = \frac{4z(T_{sat} - T_s)}{Pr}\frac{C_p}{h_{fg}}\left(\frac{g}{\nu^2}\right)^{1/3}\frac{1}{(1 - \rho_v/\rho_l)} \tag{4}$$

where z is the vertical distance. Now, the vapor shear stress is assumed to be zero.

The properties are evaluated at the saturation temperature of 262°F; The steam tables give

$h_{fg} = 937.25$ Btu/lbm and $\rho_v = 0.087$ lbm/ft³.

At the mean film temperature of

$$T_m = \frac{162 + 262}{2} = 212^\circ F,$$

the water tables list

$\rho_l = 59.8$ lbm/ft³ $\qquad \nu = 0.0114$ ft²/hr

$k = 0.3914$ Btu/hr-ft-°F $\qquad Pr = 1.76$

$C_p = 1.007$ Btu/lbm-°F

From eq. (4), z_L^* is calculated as

$$z_L^* = \frac{4(12)(262 - 162)}{1.76}\left(\frac{1.007}{937.25}\right)\left[\frac{(32.2)}{(0.0114/3{,}600)^2}\right]^{1/3}\frac{1}{\left[1-\left(\frac{0.087}{59.8}\right)\right]}$$

$$= 43{,}293$$

For this z_L^* value, fig. 1 gives

$$h^* = 0.15$$

at $\tau_v^* = 0$, for Pr = 1. Similarly, fig. 2 gives

$$h^* = 0.45$$

at $\tau_v^* = 0$, for Pr = 10.

Interpolation between these values gives

$$h^* = 0.175$$

for Pr = 1.76.

Substituting into eq. (3) and rearranging yields

$$h = h^*k\left(\frac{g}{\nu^2}\right)^{1/3} = (0.175)(0.3914)\left[\frac{(32.2)}{(0.0114/3,600)^2}\right]^{1/3}$$

$$= 1,010.52 \text{ Btu/hr-ft}^2\text{-}^{\circ}\text{F}.$$

Finally, combining eqs. (1) and (2) and substituting gives the mass flow rate of condensate,

$$\dot{m} = \frac{hA(T_{sat} - T_s)}{h_{fg}}$$

$$= \frac{(1,010.52)(12)(1)(262 - 162)}{937.25}$$

$$= 1,293.8 \text{ lbm/hr}.$$

• PROBLEM 10-4

Steam condenses on the outside of a pipe, while cool water flows inside it. The bulk mean water temperature is 65 °F, and the water side coefficient of heat transfer is 300 Btu/hr-ft^2-°F. The steam is at a pressure of 3 in. Hg abs., and the steam side coefficient of heat transfer is 1,500 Btu/hr-ft^2-°F. The pipe is a 3/4 in. 18-gage standard size, and it is made of brass. If the thermal conductivity of brass is 66 Btu/hr-ft-°F, calculate the mass flow rate of steam condensed per unit length of pipe.

<u>Solution</u>: The mass flow rate of a condensate is

$$\dot{m} = \frac{q}{h_{fg}} \tag{1}$$

At a pressure of 3 in. Hg abs. (or 1.4736 psia), the saturated steam pressure table gives the temperature of the saturated steam as 115.5°F. Also, the latent heat of vaporization is 1,028.5 Btu/lbm. Now, the heat flow per unit length of pipe is to be calculated. Considering the entire system, the heat flow conducted from the steam through to the water inside may be expressed as

$$q/L = \frac{2\pi(T_s - T_w)}{\dfrac{1}{r_2 h_{12}} + \dfrac{\ln(r_3/r_2)}{k_{23}} + \dfrac{1}{r_3 h_{34}}} \tag{2}$$

This accounts for the overall temperature drop. In the denominator, the first term, $\frac{1}{r_2 h_{12}}$, refers to the convection from the water, the second term, $\frac{\ln(r_3/r_2)}{k_{23}}$, to the conduction through the pipe walls, and the third, $\frac{1}{r_3 h_{34}}$, to the convection from the steam.

Referring to the table of standard pipe sizes for a 3/4 in. 18-gage pipe, the outside diameter is 0.75 in. and the inside diameter is 0.657 in.

$\therefore \quad r_2 = 0.652/(2)(12) = 0.02717$ ft. and

$r_3 = 0.75/(2)(12) = 0.03125$ ft. Substituting the values in eq. (2),

$$q/L = \frac{2\pi(115.5 - 65)}{\dfrac{1}{(0.02717)(300)} + \dfrac{\ln\left[\dfrac{0.03125}{0.02717}\right]}{66} + \dfrac{1}{(0.03125)(1,500)}}$$

$$= 2,171.25 \text{ Btu/hr-ft.}$$

Finally, from eq. (1),

$$\dot{m} = \frac{2,171.25}{1,028.5}$$

$$= 2.11 \text{ lbm/hr-ft.}$$

• **PROBLEM 10-5**

Film condensation occurs on the outside of a vertical pipe. The pipe has an internal diameter of 2 in., and the all surface temperature is 162°F. The saturated steam temperature is 262°F. If the rate of condensation is 1,450 lbm/hr, calculate the required length of the pipe.

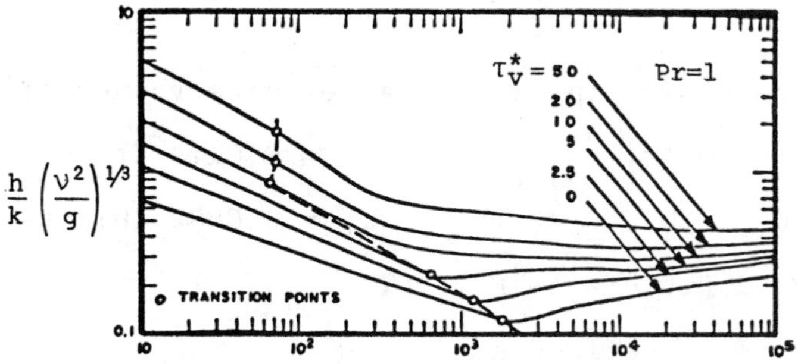

Fig. 1 (a)

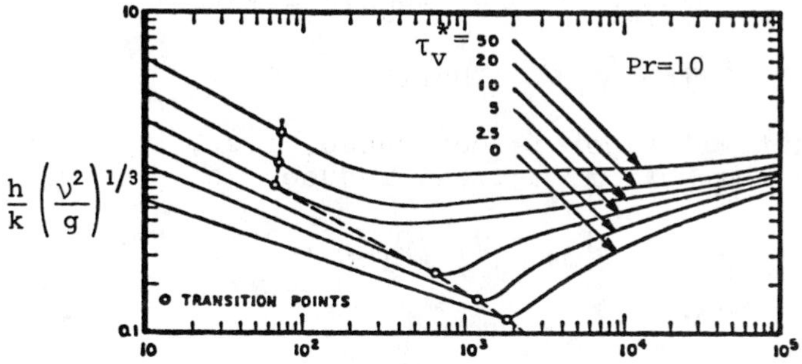

Fig. 1 (b)

Solution: Heat flow is defined as

$$q = h\pi dL(T_{sat} - T_s) \tag{1}$$

and mass flow rate of a condensate as

$$\dot{m} = \frac{q}{h_{fg}} \tag{2}$$

Properties are evaluated at $T_{sat} = 262^{\circ}F$, from the steam tables,

$h_{fg} = 937.25$ Btu/lbm $\qquad \rho_v = 0.087$ lbm/ft³

$\mu_v = 0.032$ lbm/ft-hr

The mean film temperature is

$$T_m = \frac{162 + 262}{2} = 212^{\circ}F.$$

At this temperature, the water tables give values of

$\rho_1 = 59.8$ lbm/ft³ $\qquad \nu = 0.0114$ ft²/hr

$Pr = 1.76$ $\qquad \mu = 0.6832$ lbm/ft-hr

$k = 0.3914$ Btu/hr-ft-°F

Mass flow rate is defined as

$$\dot{m} = \rho VA = GA \tag{3}$$

where G is the mass velocity. In the case of a vertical tube, the average mass velocity for the flow rate of condensate at the top and at the bottom is given by

$$G_{vm} = 0.4(G_{Top} + 1.56\ G_{Bottom}) \tag{4}$$

Assuming that all vapor is condensed, i.e., there is no flow of vapor at the bottom, eq. (4) reduces to

$$G_{vm} = 0.4\ G_{Top}\ ,$$

and from eq. (6),

$$G_{vm} = \frac{0.4\ \dot{m}}{A}$$

$$= \frac{0.4(1,450)}{\frac{\pi}{4}\left(\frac{2}{12}\right)^2}$$

$$= 26,585 \text{ lbm/hr-ft}^2$$

Using this, the Reynolds number of the vapor is

$$Re_v = \frac{G_{vm}d}{\mu}$$

$$= \frac{(26,585)(2/12)}{0.032}$$

$$= 138,464$$

Now, $\frac{\sigma_w}{\sigma}\frac{\Gamma}{\rho}$ must be found. The value of σ_w/σ may be assumed to be 1. Then

$$\Gamma = \frac{\dot{m}}{2\pi d}$$

$$= \frac{1,450}{2\pi\left(\frac{2}{12}\right)}$$

$$= 1,384.65 \text{ lbm/hr-ft}$$

Using this,

$$\frac{\sigma_w}{\sigma}\frac{\Gamma}{\rho} = \frac{(1.0)(1,384.65)}{59.8}$$

$$= 23.2$$

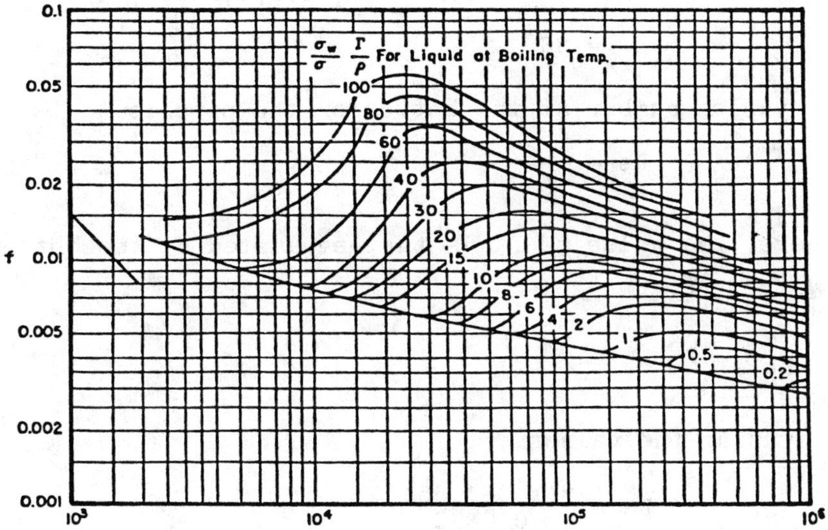

$$Re_v = \frac{G_{vm} d}{\mu_v}$$

Fig. 2 Friction factor for gas (air) flowing in a tube with a liquid layer on the wall.

With these values, fig. 2 gives a value of

$$f = 0.015$$

∴ The shear stress τ_v is given by

$$\tau_v = \frac{f\,G_{vm}^2}{2g\rho_v} \quad . \tag{5}$$

Substituting the values,

$$\tau_v = \frac{(0.015)(26,585)^2}{2(32.2)(3,600)^2(0.087)}$$

$$= 0.146 \text{ lbf/ft}^2$$

The mean shear stress τ_v^* is given by

$$\tau_v^* = \frac{\tau_v}{(\rho - \rho_v)(\nu^2/g)^{1/3}} \tag{6}$$

Substituting the values,

$$\tau_v^* = \frac{0.146}{(59.8 - 0.087)\left[\dfrac{(0.0114)^2}{32.2(3,600)^2}\right]^{1/3}}$$

$$= 36.1$$

The Reynolds number at the bottom of the pipe is

$$\text{Re} = \frac{4\Gamma_e}{\mu} \tag{7}$$

But here, the value of Γ_e is that evaluated at the bottom, i.e., $\Gamma_e = 2\Gamma$.

$$\therefore \quad \text{Re} = \frac{8\Gamma}{\mu}$$

Substituting the values,

$$\text{Re} = \frac{8(1,384.65)}{0.6832}$$

$$= 16,214.$$

From fig. 1(a), for Re = 16,214 and Pr = 1, h* = 0.4. From fig. 1(b), for Pr = 10, h* = 1. Interpolating between these two values gives h* = 0.45 at a Prandtl number of 1.76. The mean coefficient of heat transfer is given by

$$h^* = \frac{h_m}{k}\left(\frac{\nu^2}{g}\right)^{1/3} \tag{8}$$

On rearranging,

$$h_m = kh^*\left(\frac{g}{\nu^2}\right)^{1/3}$$

Substituting the values,

$$h_m = (0.3914)(0.45)\left[\frac{(32.2)(3,600)^2}{(0.0114)^2}\right]^{1/3}$$

$$= 2,599 \text{ Btu/hr-ft}^2\text{-}^{\circ}\text{F}.$$

Writing eq. (2) in terms of heat flow "q" and combining eqs. (2) and (1) gives

$$q = \dot{m}h_{fg} = h\pi dL(T_{sat} - T_s)$$

Rearranging in terms of the length (L) gives

$$L = \frac{\dot{m}h_{fg}}{h\pi d(T_{sat} - T_s)}$$

Substituting the values,

$$L = \frac{(1,450)(937.25)}{(2,599)\pi\left(\frac{2}{12}\right)(262 - 162)}$$

$$= 9.99 \text{ ft} \simeq 10 \text{ ft}.$$

• PROBLEM 10-6

Saturated steam, at 100°F, condenses on the outer surface of a vertical pipe, 1 ft long and 6 in OD, maintained at 80°F. Calculate

a) the average coefficient of heat transfer,

b) the rate of heat transfer, and

c) the mass flow rate of the condensate, assuming laminar flow.

Solution: The properties of steam at 100°F, from tables are

$h_{fg} = 1,037.1$ Btu/lbm and $\rho_v = 0.00285$ lbm/ft³.

Water tables give

$C_p = 0.998$ Btu/lbm-°F $\qquad \rho_l = 62.12$ lbm/ft³

$\mu = 1.842$ lbm/ft-hr $\qquad k = 0.3572$ Btu/hr-ft-°F

$\nu = 0.02965$ ft²/hr and $\qquad Pr = 5.15$

a) The local Nusselt number for laminar flow on a vertical plate is given by

$$Nu_x \equiv \frac{h_x x}{k_l} = \left[\frac{\rho_l g(\rho_l - \rho_v)h_{fg}x^3}{4\mu_l k_l(T_{sat}-T_s)}\right]^{1/4}$$

By integrating the local value of conductance over the entire height L of the plate, we get the average heat-transfer coefficient,

$$\bar{h} = \frac{4}{3} h_L = (0.943)\left[\frac{\rho_l g(\rho_l - \rho_v)h_{fg}k_l^3}{\mu_l L(T_{sat} - T_s)}\right]^{1/4}$$

Substituting the values,

$$\bar{h} = (0.943)\left[\frac{(62.12) \times (32.2) \times (62.12 - 0.00285) \times (1,037.1)}{(1.842 \times 3,600)(1)(100 - 80)} \cdot \frac{(0.3572 \times 3600)^3}{}\right]^{1/4}$$

$\simeq$ 1,131 Btu/hr-ft^2-$^\circ$F

Experimental results have shown that this equation is conservative, yielding results approximately 20 percent lower than measured values. Therefore, the recommended relation for inclined (including vertical) plates is

$$\bar{h} = (1.13)\left[\frac{\rho_l g(\rho_l - \rho_v)h_{fg}k_l^3}{\mu_l L(T_{sat} - T_s)} \sin\phi\right]^{1/4}$$

where for vertical plate $\phi = 90^\circ$

$$\therefore \sin\phi = 1$$

$\therefore$ The actual average heat transfer coefficient is

$$\bar{h} = 1.2 \times (0.943)\left[\frac{\rho_l g(\rho_l - \rho_v)h_{fg}k_l^3}{\mu_l L(T_{sat} - T_s)}\right]^{1/4}$$

$$= 1.2 \times 1,131 = 1,357.2 \text{ Btu/hr-ft}^2\text{-}^\circ\text{F}$$

b) The total rate of heat transfer is given by

$$q = \bar{h} \times \pi dL(T_{sat} - T_s)$$

$$= (1,357.2) \times \pi \times \left(\frac{6}{12}\right) \times (1) \times (100 - 80)$$

$$\simeq 42,638 \text{ Btu/hr.}$$

c) The mass rate flow of condensate is given by

$$\dot{m} = \frac{q}{h_{fg}} = \frac{42,638}{1,037.1}$$

$$\simeq 41 \text{ lbm/hr.}$$

• PROBLEM 10-7

Steam at 15.3 KN/m^2 pressure condenses at a rate of 25 kg/hr on a vertical pipe, 56 mm OD. The temperature drop across the film is 5°C. Calculate

a) the coefficient of heat transfer, and

b) the required length of the pipe

Solution: Referring to steam tables at a pressure of 15.3 KN/m^2,

Saturation temperature T_{sat} = 54.4°C

Latent heat of vaporization = h_{fg} = 2,372.3 KJ/kg.

The physical properties at the mean film temperature $(54.3 - \frac{5}{2})$ = 51.8°C are: ρ_1 = 993.6 kg/m^3; μ = 0.532 x 10^{-3}kg/m-sec.; and k = 0.6424 W/m-°C.

a) The Reynolds number is given by

$$Re = \frac{\rho U d}{\mu}$$

and

$$\dot{m} = \rho UA = \rho U \frac{\pi}{4} d^2.$$

Substituting the value of $\dot{m}$ in the equation for the Reynolds number gives

$$Re = \frac{4\dot{m}}{\pi d \mu} \quad (1)$$

Substituting the values,

$$Re = \frac{4 \times (25)}{\pi(0.056) \times (0.532 \times 10^{-3} \times 3600)} = 297.$$

This value is less than the critical value (1,800), hence the flow is laminar.

$\therefore$ the coefficient of heat transfer is given by

$$h = 1.47\left(\frac{\rho_l(\rho_l - \rho_v)gk^3}{Re\ \mu^2}\right)^{1/3} \qquad (2)$$

As $\rho_l >> \rho_v$, taking $\rho_l(\rho_l - \rho_v) \simeq \rho_l^2$, eq. (2) becomes

$$h = (1.47)\left(\frac{\rho_l^2 gk^3}{Re\mu^2}\right)^{1/3}$$

Substituting the values,

$$h = (1.47)\left[\frac{(993.6)^2 \times (9.81) \times (0.6424)^3}{(297) \times (0.532 \times 10^{-3})^2}\right]^{1/3}$$

$$= 4{,}595\ W/m^2\text{-}^\circ C.$$

Allowing 20% for the discrepancy between the theoretical value and the observed higher experimental value,

$$h = 1.2 \times 4{,}595 = 5{,}514\ W/m^2\text{-}^\circ C.$$

(b) The equation for heat balance taken over the pipe section is given by

$$q = h\pi dL\Delta T = \dot{m}\ h_{fg} \qquad (3)$$

Rearranging for L and substituting yields

$$L = \frac{\dot{m}\ h_{fg}}{h\pi d\Delta T}$$

$$= \frac{(25/3{,}600) \times (2{,}372 - 3 \times 10^3)}{(5{,}514) \times \pi \times (0.056) \times (5)}$$

$$\simeq 3.4\ m.$$

• PROBLEM 10-8

Steam condenses on the outside of a vertical cylinder of length 2.54 cm. and diameter 30.5 cm. The saturated steam is at a pressure of 68.9 kPa. The surface temperature of the cylinder is mainatined at 86°C. Assuming laminar flow, calculate the average coefficient of heat transfer.

Solution: From steam tables, the properties of steam at 68.9 kPa are:

$T_{sat} = 89.5^{\circ}C \qquad h_{fg} = 2,284.5 \text{ kJ/kg}$

$\rho_v = 0.415 \text{ kg/m}^3$

The mean film temperature is

$$T_m = \frac{86 + 89.5}{2} = 87.75^{\circ}C$$

At this temperature, the water tables give

$C_p = 4.202 \text{ kJ/kg-}^{\circ}C \qquad \rho_1 = 966.6 \text{ kg/m}^3$

$\mu_1 = 0.324 \times 10^{-3} \text{ kg/m-sec} \qquad k_1 = 0.6714 \text{ W/m-}^{\circ}C$

$Pr = 2.02$

Since

$$\rho_1 >> \rho_v ,$$

the approximation

$$\rho_1(\rho_1 - \rho_v) \simeq \rho_1^2 \tag{1}$$

can be used.

The Jakob number is given by

$$Ja = \frac{C_p \Delta T}{h_{fg}} \tag{2}$$

Substituting the values,

$$Ja = \frac{4.202\ (89.5 - 86)}{2,284.5}$$

$$= 0.00644$$

As Pr > 0.5 and Ja < 1.0, McAdam's relation for a laminar film condensate on a vertical cylinder applies.

$$Nu = 1.13 \left[\frac{g\rho_1(\rho_1 - \rho_v)h'_{fg}L^3}{\mu_1\ k_1\ \Delta T}\right]^{1/4} \tag{3}$$

where $h'_{fg} = h_{fg}(1 + 0.68\ Ja)$ (4)

$$\therefore \quad h'_{fg} = 2,284.5\left[1 + 0.68(0.00644)\right]$$

$$= 2,294.5 \text{ kJ/kg}$$

Using the relation (1) and substituting the values,

$$Nu_L = 1.13\left[\frac{(9.81)(966.6)^2(2,284.5 \times 1,000)(30.5/100)^3}{(0.324 \times 10^{-3})(0.6714)(89.5 - 86)}\right]^{1/4}$$

$$\simeq 5,972.$$

Since

$$Nu_L = \frac{hL}{k},$$

the coefficient of heat transfer is

$$h = \frac{Nu_L k}{L}$$

$$= \frac{(5,972.0)(0.6714)}{(30.5/100)}$$

$$\simeq 13,146 \text{ W/m}^2\text{-}^\circ\text{C}$$

• PROBLEM 10-9

> A vertical tube, 2 in OD, 10 ft high, with the outer surface temperature 200°F, is used for heating an oil flowing through it. Heat is supplied to the tube at a rate of 92,000 Btu/hr by the condensation of saturated steam on the outside. Evaluate the temperature of the saturated steam.

Solution: An iterative procedure is used to evaluate the saturated steam temperature. Let the assumed value be $T_{sat} = 200^\circ F$. Steam tables at $T_{sat} = 200^\circ F$ give $h_{fg} = 977.9$ Btu/lbm. Water tables at $T_s = 200^\circ F$ give

$$\mu = 0.734 \text{ lbm/ft-hr} \qquad \rho = 60.11 \text{ lbm/ft}^3$$

and $k = 0.3897$ Btu/hr-ft-°F.

An energy balance equation for the pipe is given by

$$q = \dot{m}\, h_{fg} = \pi d \Gamma h_{fg} \tag{1}$$

$$\therefore \quad \Gamma = \frac{q}{\pi d h_{fg}}$$

and the Reynolds number is given by

$$Re = \frac{\Gamma}{\mu} = \frac{(q/\pi d h_{fg})}{\mu} \tag{2}$$

Substituting the values,

$$Re = \frac{92,000}{\pi \times \left(\frac{2}{12}\right) \times (977.9) \times (0.734)} \simeq 245$$

This is a laminar flow.

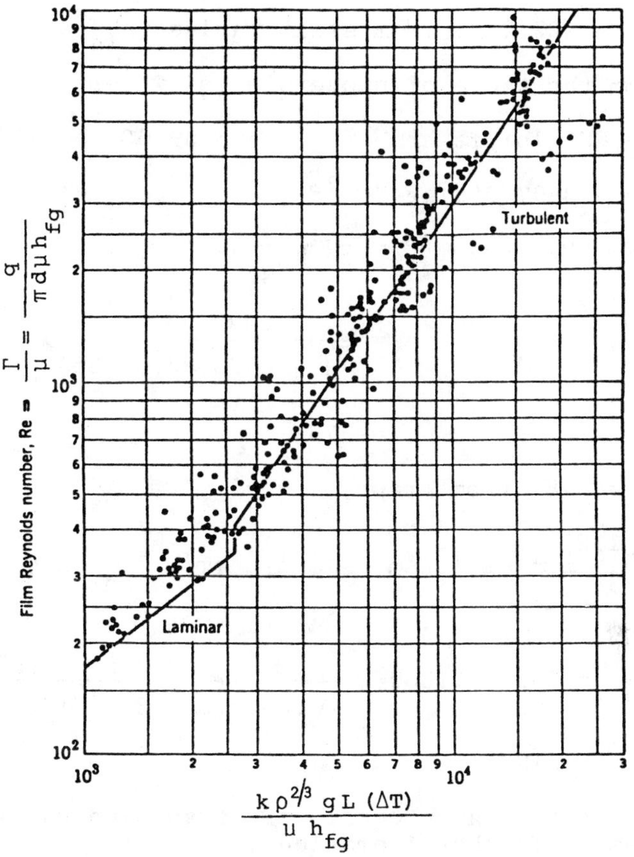

Fig. 1 Correlation of heat-transfer data for film condensation of pure vapors on vertical surfaces.

From figure 1 for Re = 245, the value for

$$\frac{k\rho^{2/3} g^{1/3} \Delta T \times L}{\mu^{5/3} h_{fg}} = 1,700 \quad \text{where } \Delta T = T_{sat} - T_s.$$

$$\therefore \quad \Delta T = (1,700) \times \frac{\mu^{5/3} \times h_{fg}}{k \times \rho^{2/3} \times g^{1/3} \times L} = T_{sat} - T_s$$

Substituting the values,

$$T_{sat} = 200 + (1,700) \times \frac{(0.734)^{5/3}(977.9)}{(0.3897)(60.11)^{2/3}[(32.2)(3600)^2]^{1/3}(10)}$$

$$= 222.2^{\circ}F$$

The calculated value is different from the assumed. Hence $T_{sat} = 222.2^{\circ}F$ is assumed for the second iteration.

At $T_{sat} = 222.2^{\circ}F$, from steam tables, $h_{fg} = 963.9$ Btu/lbm.

and from the table of properties of water,

$\mu = 0.655$ lbm/ft-hr $\qquad \rho = 59.63$ lbm/ft^3

and $k = 0.3897$ Btu/hr-ft-$^{\circ}$F

$\therefore$ From eq. (2), $Re = \frac{q}{\pi d\, h_{fg}\mu}$

$$= \frac{92,000}{\pi\left(\frac{2}{12}\right)(963.9)(0.655)}$$

$$= 278$$

For Re = 278, from graph,

$$\left(\frac{k\rho^{2/3}g^{1/3}L\Delta T}{\mu^{5/3}\, h_{fg}}\right) = 1,800 \qquad \text{where } \Delta T = T_{sat} - T_s.$$

$$\therefore \quad T_{sat} = 200 + 1,800 \times \frac{(0.655)^{5/3}(963.9)}{(0.394)(59.63)^{2/3}[32.2x(3600)^2]^{1/3}(10)}$$

$$\simeq 219.1$$

This value is nearly equal to the assumed value, and hence requires no further iterations.

$\therefore$ Saturated steam temperature is $\simeq 221^{\circ}F$

• PROBLEM 10-10

Steam condenses on a vertical tube bank. The bank consists of four horizontal tubes, each with a length of 2 ft. and a diameter of 1 in. The saturated steam is at 216°F. If the wall surface temperature is at 208°F, calculate (a) the mass flow rate of condensate and (b) the rate at which heat is being transferred.

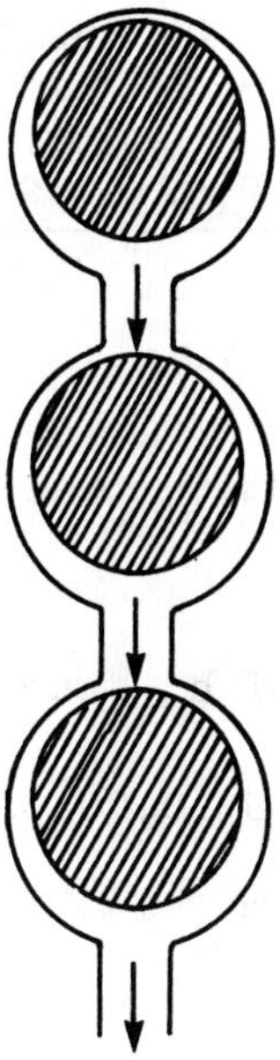

Vertical Banking

Fig. 1

Solution: For saturated steam at 216°F (from steam tables),

h_{fg} = 967.8 Btu/lbm $\qquad\qquad$ ρ_v = 0.0402 ft³/lbm

The mean film temperature is

$$T_m = \frac{208 + 216}{2} = 212^\circ F$$

At this temperature, the water tables give

C_p = 1.007 Btu/lbm-°F $\qquad\qquad$ ρ = 59.81 lbm/ft³

μ = 0.6832 lbm/ft-hr $\qquad\qquad$ k = 0.3914 Btu/hr-ft-°F

Pr = 1.76

∴ The Jakob number is given by $Ja = \dfrac{C_p(\Delta T)}{h_{fg}}$ (1)

$$\therefore \quad Ja = \frac{(1.007)(216 - 208)}{967.8}$$

$$= 0.00832$$

Now, as $Pr > 0.5$ and $Ja < 1.0$, the Nusselt number relation for the horizontal tube is

$$Nu = \frac{hD}{k} = 0.728\left(\frac{g\rho_1(\rho_1 - \rho_v)h'_{fg}D^3}{\mu_1 k_1 \Delta t}\right)^{1/4} \qquad (2)$$

where $h'_{fg} = h_{fg}(1 + 0.68Ja)$ (3)

and $\Delta T = T_{sat} - T_s$.

Substituting the values of h_{fg} and Ja in eq. (3),

$$h'_{fg} = 967.8[1 + 0.68(0.00832)]$$

$$= 973.3 \text{ Btu/lbm}$$

As

$$\rho_1 >> \rho_v ,$$

the approximation

$$\rho_1(\rho_1 - \rho_v) \simeq \rho_1^{\,2} \qquad \text{can be used.}$$

Substituting into eq. (2),

$$Nu_d = 0.728\left(\frac{32.2 \times (3{,}600)^2\ (59.81)^2\ (973.3)(1/12)^3}{(0.6832)(0.3914)(216 - 208)}\right)^{1/4}$$

$$= 576.4$$

The condensate from one tube drips onto that of the next, and so on. Thus, the film thickness is greater for the lower tubes, increasing the resistance to heat transfer. For a vertical stack of N tubes, Nusselt showed that the average coefficient is reduced according to

$$\frac{(Nu)_N}{(Nu)_1} = \left(\frac{1}{N}\right)^{1/4} \qquad (3)$$

where the subscripts N and 1 refer to N tubes and a single tube.

$$\therefore \quad (Nu)_{(N=4)} = (576.4)\left(\frac{1}{4}\right)^{1/4}$$

$$= 407.6$$

Since

$$Nu = \frac{hd}{k},$$

the heat transfer coefficient is

$$h = \frac{Nuk}{d}$$

$$= \frac{(407.6)(0.3914)}{(1/12)}$$

$$= 1,914.4 \text{ Btu/hr-ft}^2\text{-}^0\text{F}$$

(a) A heat balance equation over the bank of tubes yields

$$q = \dot{m}\, h_{fg} = 4\pi dLh(T_{sat} - T_s) \qquad (4)$$

Solving for the mass flow rate and substituting,

$$\dot{m} = \frac{4\pi dLh(T_{sat} - T_s)}{h_{fg}}$$

$$= \frac{4\pi(1/12)(2)(1,914.4)(216 - 208)}{967.8}$$

$$= 33.14 \text{ lbm/hr.}$$

(b) For the rate of heat flow, either side of eq. (4) may be used. Choosing the left side,

$$q = \dot{m}\, h_{fg}$$

$$= (33.14)(967.8)$$

$$= 32,076 \text{ Btu/hr}$$

• PROBLEM 10-11

Steam condenses on the outer surface of a vertical bank of 16 tubes. The tubes, 1 in. dia, are horizontally placed. The saturated steam is 212^0F. If the average wall surface temperature is 200^0F, determine the mean coefficient of heat transfer. Use the chart provided in figure 1.

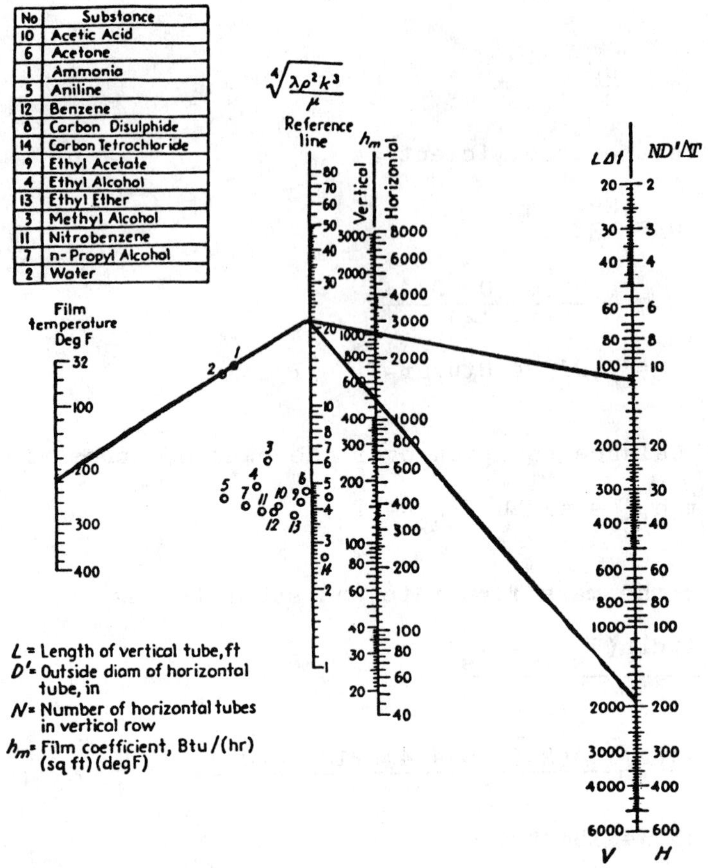

Fig. 1 Alignment chart for film-type condensation of single vapors, with streamline flow of condensate, based on Nusselt's theoretical relations.

Solution: The chart shown provides a quick alternative to the sometimes lengthy calculations involved in finding the coefficient of heat transfer. This chart may be used for any single-vapor substance listed. In this case, water is the component. It is denoted by the number 2.

To use the chart, first the mean film temperature is calculated. From that point on the left-side temperature scale, a line is drawn through the component number point, 2 for this problem, to the reference line. Now, the right-side line has two scales: for vertical tubes and for horizontal tubes (denoted by V and H, respectively). Since this problem deals with horizontal tubes, the right-side scale is utilized, and the quantity $ND'\Delta T$ is calculated. Connecting the previously-found point on the reference line to the new point on the right, a point of intersection on the film coefficient column is found. This gives the value of the mean coefficient of heat transfer.

To begin with, the mean film temperature is

$$T_m = \frac{200 + 212}{2} = 206^{\circ}F.$$

The point on the temperature line is found. Connecting this point and the free point labeled 2 (for water) and extending the line gives a reference number of 21.5. Now, a single tube has the value

$$ND'\Delta T = (1)(1)(212 - 200) = 12$$

Connecting this point with the reference number 21.5 intersects the h_m line at approximately 2,500 Btu/hr-ft²-°F. But for the entire tube bank, the value of the right-hand column is

$$ND'\Delta T = (16)(1)(212 - 200) = 192$$

This point corresponds to a value of

$$h_m \simeq 1{,}250 \text{ Btu/hr-ft}^2\text{-}^{\circ}F,$$

which is half the value for a single tube. This value can also be gotten by calculation from the value of the single tube, i.e.,

$$h_m = 2{,}500/(16)^{0.25} = 1{,}250 \text{ Btu/hr-ft}^2\text{-}^{\circ}F.$$

• PROBLEM 10-12

Steam condenses on the outside of a tube bank. Each tube can be considered thin enough such that the inside diameter is approximately equal to the outside diameter, which is 1 in. The saturated steam is at 212°F, and has a heat transfer coefficient of 1,000 Btu/hr-ft²-°F. Water is flowing through the tubes with a velocity of 6 ft/sec, entering at 50°F and leaving at 100°F. The heat transfer coefficient on the water side is 800 Btu/hr-ft²-°F. If the rate of condensation is 10,000 lbm/hr, determine the required number of tubes in the bank, and also the length of each tube.

Solution: Considering an energy balance over the bank, the heat given up to the water must equal the heat required to condense the steam, or

$$q = N\dot{m}C_p(\Delta T_b) = \dot{m}\,h_{fg} \qquad (1)$$

where

$$\dot{m} = \rho U \frac{\pi}{4} d^2$$

and

N = number of tubes.

The value of N may be found by rearranging eq. (1) as

$$N = \frac{4\dot{m}\, h_{fg}}{\rho U \pi d^2 C_p (\Delta T_b)} \tag{2}$$

The mean bulk temperature is

$$T_b = \frac{T_{bin} + T_{bout}}{2} = \frac{50 + 100}{2}$$

$$= 75^{\circ}F$$

At this temperature, the water tables list

$$C_p = 0.999\ \text{Btu/lbm-}^{\circ}\text{F} \qquad \rho = 62.77\ \text{lbm/ft}^3$$

At a saturation temperature of 212°F, the steam tables give a latent heat of vaporization of

$$h_{fg} = 970.3\ \text{Btu/lbm}$$

Substituting these values into eq. (2) gives

$$N = \frac{4(10{,}000)(970.3)}{(62.77)(6)(3{,}600)\pi(1/12)^2(0.999)(100 - 50)}$$

$$= 26.3$$

A number of 27 pipes is taken.

The overall heat transfer coefficient is given by

$$h = \frac{1}{\dfrac{1}{h_{(water)}} + \dfrac{1}{h_{(steam)}}} \tag{3}$$

Substituting the values,

$$h = \frac{1}{\dfrac{1}{800} + \dfrac{1}{1{,}000}}$$

$$= 444.44\ \text{Btu/hr-ft}^2\text{-}^{\circ}\text{F}$$

The energy balance equation is given by

$$h(\Delta T_m)A = \dot{m}\, h_{fg} \tag{4}$$

Solving for the area in eq. (4) and substituting yields

$$A = \frac{\dot{m}\, h_{fg}}{h(\Delta T_m)}$$

$$= \frac{(10,000)(970.3)}{(444.44)\left[212 - \left(\frac{50 + 100}{2}\right)\right]}$$

$$= 159.4 \text{ ft}^2$$

Dividing this by the circumference πd gives the total length as

$$L_T = \frac{A}{\pi d}$$

$$= \frac{159.4}{\pi(1/12)}$$

$$= 608.7 \text{ ft.}$$

The length per tube is

$$L = \frac{L_T}{N}$$

$$= \frac{608.7}{27}$$

$$= 22.54 \text{ ft.}$$

• PROBLEM 10-13

Steam, at 212°F, is condensing at 198°F on the outer surface of a bank of pipes with 20 pipes horizontally placed in a row. The diameter of each 3 ft long pipe is 1 in. Calculate the rate of condensate.

Solution: The properties at the saturation temperature of 212°F are

$$h_{fg} = 970.3 \text{ Btu/lbm} \qquad \rho_v = 0.0373 \text{ lbm/ft}^3$$

The mean film temperature is

$$T_m = \frac{198 + 212}{2} = 205°F$$

At this temperature, the water tables list

C_p = 1.006 Btu/lbm-°F $\qquad$ ρ = 59.99 lbm/ft³

μ = 0.7125 lbm/ft-hr $\qquad$ k = 0.3905 Btu/hr-ft-°F

Pr = 1.84

The Jakob number is given by

$$Ja = \frac{C_p \Delta T}{h_{fg}}$$

Substituting the values,

$$Ja = \frac{1.006(212 - 198)}{970.3}$$

$$= 0.01452$$

Since Pr > 0.5 and Ja < 1.0, the relevant Nusselt number eq. is

$$Nu = \frac{hd}{k} = 0.728\left[\frac{g\rho_1(\rho_1 - \rho_v)h'_{fg}(Nd)^3}{\mu_1 k_1(\Delta T)}\right]^{1/4}$$

where N = no. of tubes

$$h'_{fg} = h_{fg}(1 + 0.68Ja)$$

and $\Delta T = T_{sat} - T_s$.

Substituting the values,

$$h'_{fg} = 970.3\left[1 + 0.68(0.01452)\right]$$

$$= 979.9 \text{ Btu/lbm}$$

Since

$$\rho_1 >> \rho_v ,$$

the approximation

$$\rho_1(\rho_1 - \rho_v) \simeq \rho_1^2$$

will be used. Solving for the coefficient of heat transfer by substituting the values yields

$$h = 0.728\left[\frac{g\rho_l^{\,2}h'_{fg}(Nd)^3}{\mu_l k_l(T_{sat}-T_s)}\right]^{1/4}\left(\frac{k}{d}\right)$$

$$= 0.728\left[\frac{32.2 \times (3{,}600)^2\ (59.99)^2(979.9)(20 \times 1/12)^3}{(0.7125)(0.3905)(212 - 198)}\right]^{1/4}$$

$$\times \frac{0.3905}{1/12}$$

$$= 22{,}061.6\ \text{Btu/hr-ft}^2\text{-}^\circ\text{F}$$

Finally, from the energy balance equation, $h\pi dL(T_{sat} - T_s) = \dot{m}\, h_{fg}$, the mass flow rate is

$$\dot{m} = \frac{h\pi dL(T_{sat} - T_s)}{h_{fg}}$$

$$= \frac{(22{,}061.6)\pi(1/12)(3)(212 - 198)}{970.3}$$

$$= 250\ \text{lbm/hr}$$

BOILING

• PROBLEM 10-14

A copper kettle with a flat bottom, 9 in dia, is boiling water at a rate of 40 lbm/hr. Calculate

a) the bottom surface temperature, and

b) the burn-out heat flux for nucleate boiling.

Solution:

a) For a pool of boiling regime, Rohsenow's heat flux correlation is

$$\frac{q}{A} = \mu_l h_{fg}\left[\frac{g(\rho_l - \rho_v)}{g_c\sigma}\right]^{1/2}\left[\frac{C_{pl}\,\Delta T}{C_{sf}h_{fg}Pr_l^s}\right]^3 \qquad (1)$$

For water, s = 1, and for polished copper, $C_{sf} = 0.013$. These are merely constants pertaining to the material components.

The properties are to be evaluated at the saturation temperature which is assumed to be 212°F. At this value, the steam tables list

$h_{fg} = 970.3$ Btu/lbm $\qquad \rho_v = 0.0373$ lb/ft³

and the water tables give

$C_{pl} = 1.0074$ Btu/lbm-°F $\qquad \rho_l = 59.81$ lbm/ft³

$\mu_l = 0.6832$ lbm/ft-hr $\qquad Pr = 1.76$

The heat flux may also be expressed as

$$q = \dot{m}\, h_{fg} \tag{2}$$

Substituting into eq. (2) yields

$$q = 40(970.3)$$

$$= 38{,}812 \text{ Btu/hr}$$

If eq. (1) is rearranged to solve for the temperature difference, the result is

$$\Delta T = \left\{\frac{q}{A\mu_l h_{fg}}\left[\frac{g(\rho_l - \rho_v)}{g_c\sigma}\right]^{-1/2}\right\}^{1/3}\left(\frac{Pr_l^s h_{fg} C_{sf}}{c_{pl}}\right) \tag{3}$$

$$A = \frac{\pi}{4}d^2 = \frac{\pi}{4}\left(\frac{9}{12}\right)^2$$

$$= 0.442 \text{ ft}^2$$

The surface tension σ for water may be taken as 40.3×10^{-4} lbf/ft. Substituting the known values into eq. (3) gives

$$\Delta T = \left\{\frac{(38{,}812)}{(0.442)(0.6832)(970.3)}\left[\frac{32.2(59.81 - 0.0373)}{32.2(40.3 \times 10^{-4})}\right]^{-1/2}\right\}^{1/3}$$

$$\times \left[\frac{1.76(970.3)(0.013)}{1.0074}\right]$$

$$= 22.66^{\circ}\text{F}$$

And since

$$\Delta T = T_s - T_{sat},$$

$$T_s = T_{sat} + \Delta T$$

$$= 212 + 22.66$$

$$= 234.66^{\circ}\text{F}$$

b) For the burn-out heat flux for nucleate boiling,

$$\frac{q}{A} = \frac{\pi}{24} h_{fg}\rho_v \left(\frac{\sigma g\, g_c(\rho_1 - \rho_v)}{\rho_v^2}\right)^{1/4} \left(1 + \frac{\rho_v}{\rho_1}\right)^{1/2} \quad (4)$$

This describes the peak heat flux. Substituting the values into eq. (4),

$$\frac{q}{A} = \frac{\pi}{24}(970.3)(0.0373)\left(\frac{(40.3 \times 10^{-4})\left[(32.2)^2 \times (3{,}600)^4\right]}{(0.0373)^2}\right.$$

$$\left. \times \frac{(59.81 - 0.0373)}{}\right)^{1/4} \times \left(1 + \left(\frac{0.0373}{59.81}\right)\right)^{1/2}$$

$$= 351{,}169.6 \text{ Btu/hr-ft}^2$$

To get the flux in Btu/hr, multiply by the area,

$$q = 351{,}169.6(0.442)$$

$$= 155{,}217 \text{ Btu/hr}$$

• PROBLEM 10-15

Water film boiling occurs on top of a horizontal plate. The water is at a pressure of 1 atm. and the surface temperature of the plate is 554°C. Calculate the heat flux given off by water assuming a characteristic length (L_c) of 1 cm.

Solution: Properties at the saturation temperature of 100°C, from steam tables are

$$h_{fg} = 2{,}257 \text{ kJ/kg}$$

and at the mean film temperature of

$$T_m = \frac{100 + 554}{2}$$

$$= 327^\circ\text{C},$$

C_p = 2.025 kJ/kg-°C $\qquad \rho_v$ = 0.362 kg/m³

μ_v = 21.4 x 10^{-6} kg/m-sec $\qquad k_v$ = 46.45 x 10^{-3} W/m-°C

Pr = 0.935

In the water tables at $100^\circ C$,

$$\rho_1 = 958.1 \text{ kg/m}^3$$

The Jakob number for the vapor film is given by

$$Ja_{(v)} = \frac{C_{p(v)}(T_s - T_{sat}}{h_{fg}}$$

Substituting the values,

$$Ja_{(v)} = \frac{(2.025)(554 - 100)}{2,257}$$

$$= 0.4073$$

The kinematic viscosity is $\nu = \frac{\mu}{\rho}$

$$\therefore \quad \nu = \frac{21.4 \times 10^{-6}}{0.362}$$

$$= 5.91 \times 10^{-5} \text{ m}^2/\text{sec}$$

The average thickness of the vapor film is expressed as

$$\delta_v = \left[\left(\frac{\nu_v{}^2 L_c}{g}\right)\left(\frac{\rho_v}{\rho_1 - \rho_v}\right)\left(\frac{1}{Pr_{(v)}}\right)\left(\frac{Ja_{(v)}}{1 + 0.4\ Ja_{(v)}}\right)\right]^{1/4}$$

Substituting the values yields

$$\delta_v = \left\{\left[\frac{(5.91 \times 10^{-5})^2(1/100)}{9.81}\right]\left(\frac{0.362}{958.1 - 0.362}\right)\left(\frac{1}{0.935}\right)\right.$$

$$\left. \times \left(\frac{0.4073}{1 + 0.4(0.4073)}\right)\right\}^{1/4}$$

$$= 1.498 \times 10^{-4} \text{m}$$

Berenson's Nusselt number correlation for the coefficient of heat transfer is $Nu = \frac{h\delta_{(v)}}{k_{(v)}} = 0.425$ and when substituted into the heat flux equation $\frac{q}{A} = h(T_s - T_{sat})$, the heat flux expression becomes

$$\frac{q}{A} = 0.425 \frac{k_v}{\delta_v}(T_s - T_{sat}) \qquad (5)$$

Substituting the known values into eq. (5) gives

$$\frac{q}{A} = 0.425 \frac{(46.45 \times 10^{-3})}{(1.498 \times 10^{-4})} (554 - 100)$$

$$= 59{,}830 \text{ W/m}^2$$

or $\quad 5.983 \text{ W/cm}^2$.

• PROBLEM 10-16

> Nucleate boiling occurs on a surface at temperature 242°F. If the substance is water, find (a) the maximum heat flux per unit area, and (b) the coefficient of heat transfer.

Solution:

(a) The maximum heat flux is given by the correlation of Zuber,

$$\left(\frac{q}{A}\right)_{max} = \frac{\pi}{24} h_{fg}\rho_v \left[\frac{\sigma g g_c(\rho_1 - \rho_v)}{\rho_v^2}\right]^{1/4} \left(1 + \frac{\rho_v}{\rho_1}\right)^{1/2} \qquad (1)$$

The properties must be evaluated at the saturation temperature of 212°F. The steam tables list

$h_{fg} = 970.3$ Btu/lbm $\qquad \rho_v = 0.0373$ lbm/ft³

$\rho_1 = 59.81$ lbm/ft³

The value of σ(the surface tension) may be found from

$$\sigma = 0.00528 (1 - 0.0013T) \qquad (2)$$

$$= 0.00528 \left[1 - (0.0013)(212)\right]$$

$$= 3.83 \times 10^{-3} \text{ lbf/ft}$$

Substituting into eq. (1) yields

$$\left(\frac{q}{A}\right)_{max} = \frac{\pi}{24} (970.3)(0.0373) \times$$

$$\left[\frac{(3.83 \times 10^{-3})(32.2)^2 \times (3{,}600)^4 (59.81 - 0.0373)}{(0.0373)^2}\right]^{1/4}$$

$$\times \; 1 + \left[\left(\frac{0.0373}{59.81}\right)\right]^{1/2}$$

$$= 346{,}729 \text{ Btu/hr-ft}^2$$

(b) Since the heat flux is defined as

$$\frac{q}{A} = h(T_s - T_{sat}) \quad . \qquad (3)$$

the corresponding coefficient of heat transfer is

$$h = \frac{q/A}{(T_s - T_{sat})}$$

$$= \frac{346{,}729}{(242 - 212)}$$

$$= 11{,}558 \text{ Btu/hr-ft}^2\text{-}^\circ\text{F}$$

• PROBLEM 10-17

Consider the stainless steel jacketed kettle shown in fig. 1. Water boils at a pressure of 1 atm. absolute, while steam condenses at a temperature of 120 °C inside the jacket. The kettle is assumed to be flat-bottomed. The bottom is 3.2 mm. thick, with a thermal conductivity of 16.27 W/m-°K. If the coefficient of heat transfer of the steam is 10,200 W/m²-°K, calculate the coefficient for the boiling water at the bottom.

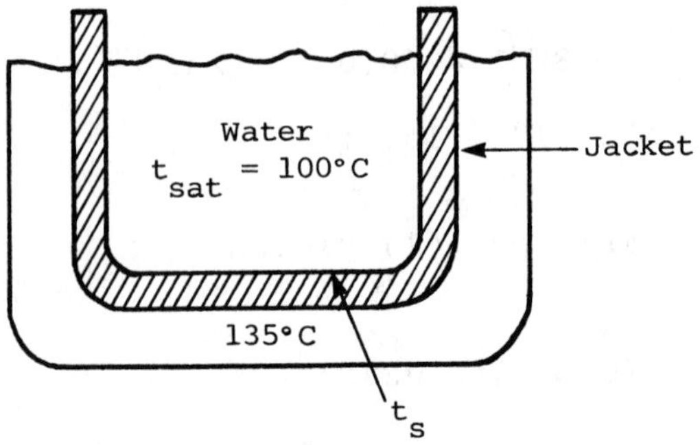

Fig. 1

Solution: For water boiling on a submerged horizontal surface at atmospheric pressure, for calculating the coefficient of heat transfer, simplified empirical correlations are given as follows:

$$h = 1{,}043(\Delta T)^{1/3} \text{ W/m}^2\text{-}^\circ\text{K} \quad \text{for } \frac{q}{A} < 16 \text{ kW/m}^2 \tag{1}$$

$$h = 5.56(\Delta T)^3 \text{ W/m}^2\text{-}^\circ\text{K} \quad \text{for } 16 \frac{\text{kW}}{\text{m}^2} < \frac{q}{A} < 240 \frac{\text{kW}}{\text{m}^2} \tag{2}$$

Now, if the temperature difference $(T_s - T_{sat})$ is known, using one of the above equations, the coefficient of heat transfer may be calculated. The conditions to use the relations may be checked as

$$\frac{q}{A} = h\Delta T \tag{3}$$

But the surface temperature of the kettle is not known. A value is assumed and the resulting heat transfer coefficient is used to calculate the actual temperature as follows:

$$T_s = T_{sat} + \frac{R_{water}}{R_{water} + R_{wall} + R_{steam}}(T_{steam} - T_{sat}) \tag{4}$$

where the resistances are defined as

$$R_{water} = \frac{1}{h_{water} A_{water}} \tag{5}$$

$$R_{wall} = \frac{\Delta x}{k_{wall} A_{wall}} \tag{6}$$

and

$$R_{steam} = \frac{1}{h_{steam} A_{steam}} \tag{7}$$

To begin the iteration, a value of $T_s = 108^\circ C$ is assumed. Eq. (2) is selected at random. Since the saturated steam is at $100^\circ C$,

$$h = 5.56(108 - 100)^3$$

$$= 2{,}846.7 \text{ W/m}^2\text{-}^\circ\text{K}$$

Then from eq. (3),

$$\frac{q}{A} = 2{,}846.7\ (108 - 100)$$

$$= 22.77 \text{ kW/m}^2$$

Since this is between the limits of eq. (2), its use is justified.

Equal areas of 1 m^2 will be assumed for each resistance. Substituting into eq. (5) yields

$$R_{water} = \frac{1}{(2,846.7)(1)}$$

$$= 3.513 \times 10^{-4} \, {}^{o}K/W$$

Similarly, substituting into eqs. (6) and (7) gives

$$R_{wall} = \frac{3.2/1,000}{16.27(1)}$$

$$= 1.967 \times 10^{-4} \, {}^{o}K/W$$

and

$$R_{steam} = \frac{1}{10,200\ (1)}$$

$$= 9.804 \times 10^{-5} \, {}^{o}K/W$$

Bringing these values to eq. (4),

$$T_s = 100 + \left(\frac{3.513 \times 10^{-4}}{(3.513 \times 10^{-4} + 1.967 \times 10^{-4} + 9.804 \times 10^{-5})}\right)$$

$$\times (120 - 100)$$

$$= 110.9 {}^{o}C$$

This is slightly higher than the assumed value of 108°C. If the average of the two values is used (109.5^{o}C) and the iteration process is repeated, the results are:

$$h = 5.56(109.5 - 100)^3$$

$$= 4,767 \ W/m^2\text{-}{}^{o}K$$

$$\frac{q}{A} = 4,767(109.5 - 100)$$

$$= 45.3 \ KW/m^2$$

$$R_{water} = \frac{1}{(4,767)(1)}$$

$$= 2.098 \times 10^{-4} \, {}^{o}K/W$$

R_{wall} and R_{steam} remain the same. Then

$$T_s = 100 + \left[\frac{2.098 \times 10^{-4}}{(2,098 \times 10^{-4} + 1.967 \times 10^{-4} + 9.804 \times 10^{-5})}\right]$$

$$\times (120 - 100)$$

$$= 108.3^{\circ}C$$

This is close enough to the original assumed value, and hence no further iterations are required. The average value of T_s is taken as $108.9^{\circ}C$

SUBCOOLING

• PROBLEM 10-18

The core of a nuclear reactor is to be designed for minimum capital cost. The specifications and limitations are:

(a) the core must produce 100 MW of power,

(b) the fuel element must contain an 18-rod bank, each rod of 0.6 in. diameter, enclosed in a 4 in. diameter flow channel,

(c) the coolant is to be boiling water at 1000 psia, entering with $8^{\circ}F$ of subcooling and a mass velocity of 2.2×10^6 lbm/hr.ft.^{2},

(d) the maximum allowable core length is 10 ft. and

(e) the cost per fuel channel is \$1,000 per ft., with the cost of fittings for each channel at \$10,000. Considering the limitation of the critical heat flux, what is the optimum core length?

Solution: Macbeth's rod cluster relation gives the critical channel heat flux as a function of length.

$$\left(\frac{q}{A}\right)_{crit} \times 10^{-6} = \frac{X + Y\,(\Delta h_{SUB})}{Z + 1} \qquad (1)$$

where

$$X = 67.6\, D_h^{0.83}\,(G \times 10^{-6})^{0.57}, \qquad (2)$$

$$Y = 0.25\ D_h\ (G \times 10^{-6}) \tag{3}$$

and

$$Z = 47.3\ D_h^{0.57}\ (G \times 10^{-6})^{0.27} \tag{4}$$

D_h is the hydraulic diameter, defined as

$$D_h = \frac{4A}{P} \tag{5}$$

Since

$$A = \frac{\pi}{4}\left[d^2_{channel} - (\text{number of rods})(d_{rod})^2\right]$$

$$= \frac{\pi}{4}\left[4^2 - (18)(0.6)^2\right]$$

$$= 7.477\ \text{in.}^2$$

and

$$P = (\text{number of rods}) \times \pi d_{rod}$$

$$= (18)\pi(0.6)$$

$$= 33.93\ \text{in.},$$

then

$$D_h = \frac{4(7.477)}{33.93}$$

$$= 0.8815\ \text{in.}$$

The mass velocity G is given as 2.2×10^6 lbm/hr-ft.^{2}. Substituting into eq. (2) yields

$$X = 67.6\ (0.8815)^{0.83}\ (2.2)^{0.57}$$

$$= 95.425$$

Similarly, eqs. (3) and (4) give

$$Y = 0.25(0.8815)(2.2)$$

$$= 0.485$$

and

$$Z = 47.3\ (0.8815)^{0.57}\ (2.2)^{0.27}$$

$$= 54.46$$

Now, the saturation temperature for water at 1000 psia is 544.6°F. With 8°F of subcooling, the steam tables give a change in enthalpy of

$$\Delta h_{SUB} = 12.33 \text{ Btu/lbm.}$$

These values may be substituted into eq. (1). Since is the channel length (in.),

$$\frac{q}{A}_{crit} \times 10^{-6} = \frac{95.425 + 0.485\ (12.33)}{54.46 + 1}$$

$$= \frac{101.4}{54.46 + 1} \qquad (6)$$

The critical channel power is

$$P = \pi \times (\text{number of rods}) \times d_{rod} \times 1(\text{ft.}) \times \left(\frac{q}{A}\right)_{crit} \times$$

$$\left(\frac{1054.5\text{W}}{1 \text{ Btu/sec}}\right) \times \left(\frac{1 \text{ KW}}{1000 \text{ W}}\right) \times \left(\frac{1 \text{ hr.}}{3600 \text{ sec}}\right) = \text{kW channel power.} \quad (7)$$

Substituting the values into eq. (7),

$$P = \pi(18)\left(\frac{0.6}{12}\right) \times 1 \times \left(\frac{q}{A}\right)_{crit} \times (1054.4) \times (1/1000) \times$$

$$(1/3600)$$

$$= 0.0008281\left(\frac{q}{A}\right)_{crit}$$

Now, a channel length l is assumed. Eq. (6) is used to find the critical heat flux, and then eq. (7) is used to get the corresponding channel power. This process is repeated over a range of values, and then the optimum length is found.

To begin with, for l = 0 ft.,

$$\left(\frac{q}{A}\right)_{crit} = \frac{101.4}{54.56}$$

$$= 1.86 \times 10^{6} \text{ Btu/hr-ft}^{2}$$

but

$$P = 0$$

as there is no channel length.

For l = 2 ft.,

$$\left(\frac{q}{A}\right)_{crit} = \frac{101.4}{54.46 + 2(12)}$$

$$= 1.29 \times 10^6 \text{ Btu/hr-ft}^2$$

and

$$P = 0.000828(2)(1.29 \times 10^6)$$

$$= 2136 \text{ kW}$$

The number of channels will be found as

$$N = \text{number of channels} = \frac{100 \text{ MW}}{P(\text{MW})} \tag{9}$$

(rounded to the next highest integer), and the total cost will be

$$\text{Cost}_T = N \times \left[\$10{,}000 + l \times (\$1{,}000)\right] \tag{10}$$

So, for l = 2 ft.,

$$N = \frac{100(1000)}{2136}$$

$$= 47$$

and

$$\text{Cost}_T = 47 \times \left[10{,}000 + 2(1000)\right]$$

$$= \$564{,}000$$

Calculations are made for various values of length (l) from 2 ft. to 10 ft. in steps of 2 ft. for finding the critical heat flux, critical channel power, number of channels for 100 MW and the total capital cost, as shown in the table.

Channel Length (ft)	Critical Heat Flux $\left(\frac{q}{a}\right) \times 10^{-6}$ Btu/h.ft^2	Critical Channel Power KW	Number of Channels for 100 MW	Total Capital Cost $ (in thousands)
0	1.860	0	-	-
2	1.290	2136.0	47	564
4	0.990	3277.6	31	434
6	0.802	3984.0	26	416
8	0.674	4464.6	23	414
10	0.580	4802.4	21	420

From the table, it can be observed that for a channel length of 8 ft., the cost is the lowest and hence that can be selected.

EVAPORATION

• PROBLEM 10-19

A closed tank contains a saturated liquid, 60 mole percent water and 40 mole percent alcohol. The system is heated at a pressure of 1 atm, vaporizing liquid until the remaining liquid is at 180°F. Referring to fig. 1, calculate the amount of liquid vaporized and the amount of heat added to the system.

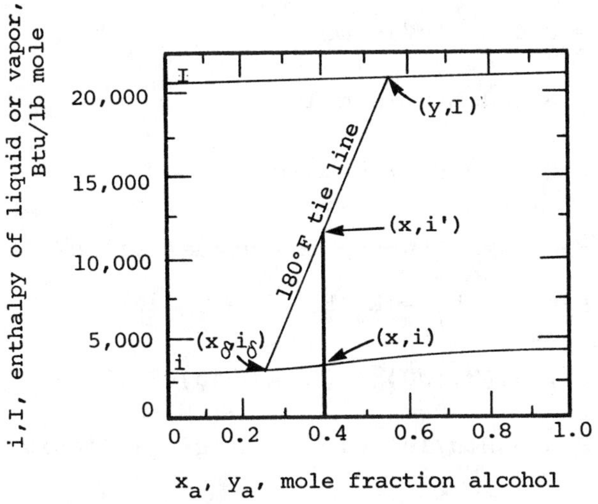

Fig. 1 Saturated liquid, vapor enthalpy curve at 180°F.

Solution: It is assumed that equilibrium conditions prevail. Figure 1 gives a plot of enthalpy versus mole fraction of alcohol. The slanted line follows the change in state from liquid to vapor at 180°F. The lower line is the saturated liquid curve, and the upper one is the saturated vapor curve.

The 180°F tie line crosses the saturated liquid line at the point (x_δ, i_δ). The corresponding mole fraction of alcohol is 26.5 mole percent. The remaining liquid in the system must have this concentration. Similarly, the mole fraction of alcohol in the vapor is 56.5 mole percent.

Now, the amount of the original liquid that was vaporized may be found by using the lever-arm rule. This is

$$\frac{L_1 - L_v}{L_1} = \frac{x - x_\delta}{y - y_\delta} = \frac{0.4 - 0.265}{0.565 - 0.265}$$

$$= 0.45$$

Thus, 45% of the original liquid was vaporized.

To find the amount of heat added to the system, an enthalpy balance must be written. This is

$$L_1 h + Q = L_v h_\delta + (L_1 - L_v)H$$

where the points are as labeled on the graph and Q is the heat added.

Reading off values,

$$h = 3{,}400 \text{ Btu/lb mole}$$

$$h_\delta = 3{,}100 \text{ Btu/lb mole}$$

$$H = 21{,}000 \text{ Btu/lb mole}$$

Solving for Q and substituting the values yields

$$Q = L_v h_\delta + (L_1 - L_v)H - L_1 h$$

$$= (0.45)(3{,}100) + (1 - 0.45)(21{,}000) - (1)(3{,}400)$$

$$= 9{,}545 \text{ Btu/lb mole of original liquid.}$$

• PROBLEM 10-20

Water flows through a tube at the rate of 0.0375 kg/sec. It is evaporated at a pressure of 1.186 bar. If the tube diameter is 12.5 mm, calculate the heat flux for a superheat temperature difference of (a) 2.78°C, (b) 11.1°C, and (c) 22.2°C. The quality of the steam is 0.05. Use Chen's correlation and the accompanying graphs.

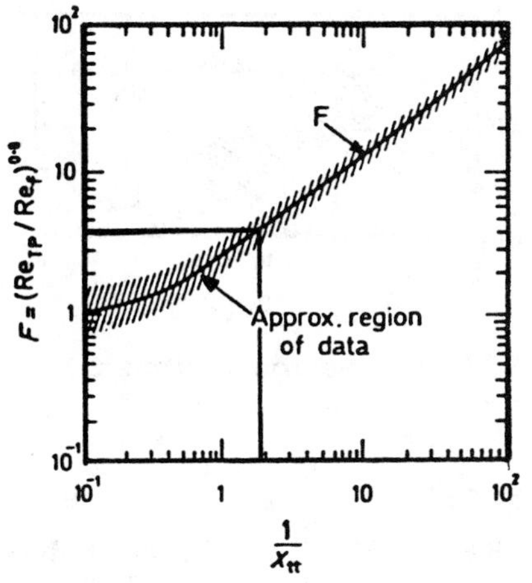

Fig. 1 Reynolds number factor, F.

Solution: Heat flux is defined as

$$\frac{q}{A} = h(\Delta T) \qquad (1)$$

where h is the combined heat transfer coefficient for boiling and convection, and ΔT is the amount of superheat (above the saturation temperature).

The heat transfer coefficients are to be found and added together. For the convection mode, the coefficient may be found as

$$h_c = 0.023\ Re_l^{0.8} Pr_l^{0.4} \left(\frac{k_l}{d}\right) F \qquad (2)$$

where F is the Reynolds number factor. It is plotted vs. $1/X_{tt}$ in fig. 1. F may be found for a corresponding value of $1/X_{tt}$, which is

$$1/X_{tt} = \left(\frac{x}{1 - x}\right)^{0.9} \left(\frac{\nu_v}{\nu_l}\right)^{0.5} \left(\frac{\mu_v}{\mu_l}\right)^{0.1} \qquad (3)$$

where x is the quality. This expression may be evaluated when properties are evaluated at the given conditions.

Now, the heat transfer coefficient for the boiling mode must be found. It is given by the expression

$$h_b = \frac{0.00122\ k_l^{0.79}\ C_{pl}^{\ 0.45}\ \rho_l^{\ 0.49}\ S(\Delta T)^{0.24}(\Delta p)^{0.75}}{\sigma^{0.5}\ \mu_l^{\ 0.29}\ i_{lv}^{\ 0.24}\ \rho_v^{\ 0.24}} \qquad (4)$$

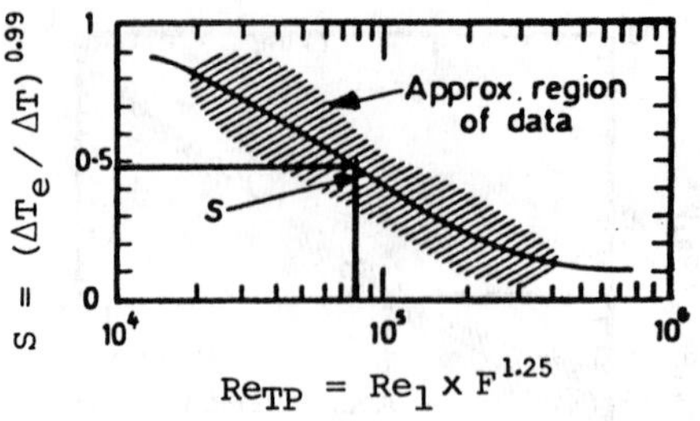

Fig. 2 Supression factor, S.

S is the suppression factor. It is plotted in figure 2 vs. the quantity $Re_{TP} = Re_1 \times F^{1.25}$. Since F has been found previously, S may be read off the graph and h_b may be evaluated. Then the total heat transfer coefficient may be calculated as

$$h = h_c + h_b \tag{5}$$

This may be brought to eq. (1) and the heat flux solved for.

(a) At a pressure of 1.186 bar, water is saturated at a temperature of 104.5°C. The water tables give values of

$k_1 = 0.679$ W/m-°C $\qquad C_{pl} = 4.22$ kJ/kg-°C

$\rho_1 = 954.7$ kg/m³ $\qquad \mu_1 = 2.71 \times 10^{-4}$ kg/m-sec

$Pr = 1.69$ $\qquad \nu_1 = 1.021 \times 10^{-3}$ m²/sec

The steam tables give

$i_{lv} = 2{,}245$ kJ/kg $\qquad \rho_v = 0.687$ kg/m³

$\nu_v = 1.45$ m²/sec $\qquad \mu_v = 1.23 \times 10^{-5}$ kg/m-sec

The Reynolds number is given by

$$Re_1 = \frac{dG(1 - x)}{\mu_1}$$

where

$$G = \frac{\dot{m}}{A} = \frac{4\dot{m}}{\pi d^2}$$

$$\therefore \quad Re_1 = \frac{4\dot{m}(1 - x)}{\pi d \mu_1}$$

$$= \frac{4(0.0375)(1 - 0.05)}{\pi(12.5/1{,}000)(2.71 \times 10^{-4})}$$

$$= 13{,}390$$

Substituting into eq. (3),

$$1/X_{tt} = \left(\frac{0.05}{1 - 0.05}\right)^{0.9} \left(\frac{1.45}{1.021 \times 10^{-3}}\right)^{0.5} \left(\frac{1.23 \times 10^{-5}}{2.71 \times 10^{-4}}\right)^{0.1}$$

$$= 1.954$$

From fig. 1, $F = 4.3$

Now the convective heat transfer coefficient may be calculated using eq. (2).

$$h_c = 0.023(13{,}390)^{0.8}(1.69)^{0.4}\left(\frac{0.679}{12.5/1{,}000}\right)(4.3)$$

$$= 13{,}266 \text{ W/m}^2\text{-}^\circ\text{C}$$

For the boiling coefficient of heat transfer, the surface tension is needed. The standard value is

$$\sigma = 0.058 \text{ N/m}$$

If the amount of superheat is 2.78°C, the temperature is 104.5 + 2.78 = 107.28°C. At this value, the saturation pressure is 1.304 bar. All that is needed now is the suppression factor. Since

$$Re_{TP} = Re_1 \times F^{1.25}$$

$$= (13{,}390)(4.3)^{1.25}$$

$$= 82{,}912$$

At this value, fig. 2 gives $S = 0.44$.

From eq. (4),

$$h_b = \frac{0.00122(0.679)^{0.79}(4{,}220)^{0.45}(954.7)^{0.49}(0.44)(2.78)^{0.24}}{(0.058)^{0.5}(2.71 \times 10^{-4})^{0.29}(2{,}245 \times 10^3)^{0.24}(0.687)^{0.24}}$$

$$\times \left[(1.304 - 1.186) \times 10^5\right]^{0.75}$$

$$= 1{,}039 \text{ W/m}^2\text{-}^\circ\text{C}$$

From eq. (5),

$$h = 13{,}266 + 1{,}039$$

$$= 14{,}305 \text{ W/m}^2\text{-}^{\circ}\text{C}$$

Finally, using eq. (1) gives

$$\frac{q}{A} = 14{,}305(2.78)$$

$$= 39{,}768 \text{ W/m}^2$$

(b) Now the superheat is 11.1°C, which brings the water to a temperature of 104.5 + 11.1 = 115.6°C. This corresponds to a saturation pressure of 1.721 bar. All calculations will remain the same except for that of h_b. The new value is

$$h_b = \frac{0.00122(0.679)^{0.79}(4{,}220)^{0.45}(954.7)^{0.49}(0.44)(11.1)^{0.24}}{(0.058)^{0.5}(2.71 \times 10^{-4})^{0.29}(2{,}245 \times 10^3)^{0.24}(0.687)^{0.24}}$$

$$\times \left(\frac{(1.721 - 1.186) \times 10^5}{}\right)^{0.75}$$

$$= 4{,}500 \text{ W/m}^2\text{-}^{\circ}\text{C}$$

Then from eq. (5),

$$h = 13{,}266 + 4{,}500$$

$$= 17{,}766 \text{ W/m}^2\text{-}^{\circ}\text{C}$$

and from eq. (1),

$$\frac{q}{A} = 17{,}766(11.1)$$

$$= 197{,}203 \text{ W/m}^2$$

(c) A similar procedure is followed for a superheat of 22.2°C. As the saturation pressure at T = 104.5 + 22.2 = 126.7°C is 2.44 bar,

$$h_b = \frac{0.00122(0.679)^{0.79}(4{,}220)^{0.45}(954.7)^{0.49}(0.44)(22.2)^{0.24}}{(0.058)^{0.5}(2.71 \times 10^{-4})^{0.29}(2{,}245 \times 10^3)^{0.24}(0.687)^{0.24}}$$

$$\times \frac{(2.44 - 1.186) \times 10^5}{}^{\;0.75}$$

$$= 10{,}068 \text{ W/m}^2\text{-}^{\circ}\text{C}$$

and

$$h = 13{,}266 + 10{,}068$$

$$= 23{,}334 \text{ W/m}^2\text{-}^\circ\text{C}$$

Then

$$\frac{q}{A} = 23{,}334(22.2)$$

$$= 518{,}015 \text{ W/m}^2$$

SOLIDIFICATION

• PROBLEM 10-21

A droplet of molten lead of average 1/8 in dia, at 750°F, falls from a height of 60 ft, and solidifies by the cooling effect of the surrounding air. The surrounding temperature is 70°F. If the lead solidifies at 621°F, calculate the coefficient of heat transfer. For lead,

$$C_p = 0.031 \text{ Btu/lbm-}^\circ\text{F}$$

$$\rho = 710 \text{ lbm/ft}^3$$

$$h_{fusion} = 10.6 \text{ Btu/lbm}$$

Solution: The process involved is in two stages: first, the lead solidifes, with no change in temperature; second, it cools to the final temperature.

The time for both cooling and solidifying may be found from the simple mechanics relation

$$y_o = \tfrac{1}{2}gt^2 \tag{1}$$

$$t = \sqrt{\frac{2y_o}{g}}$$

$$= \sqrt{\frac{2(60)}{32.2}}$$

$$= 1.93 \text{ sec} = 0.000536 \text{ hr.}$$

Now, the solidifcation process involves the transfer of heat to the surroundings. This heat is

$$Q_s = h(T_f - T_{amb})At_s \tag{2}$$

where t_s is the time of solidification. Rearranging in terms of t_s,

$$t_s = \frac{Q_s}{h(T_f - T_{amb})A} \qquad (3)$$

where

$$Q_s = h_{fusion}\, m = h_{fusion}(Vol)\rho$$

$$= (10.6)\left[\frac{4}{3}\pi\left(\frac{1/8}{2 \times 12}\right)^3\right](710)$$

$$= 0.004454 \text{ Btu}$$

Substituting this into eq. (3) gives

$$t_s = \frac{0.004454}{h(621 - 70)4\pi\left(\frac{1/8}{2 \cdot 12}\right)^2}$$

$$= \frac{0.0237}{h}$$

The cooling part of the process is described by

$$-(Vol)\gamma C_p \frac{dT}{dt} = h(T - T_{amb})A$$

or

$$-\frac{4}{3}\pi r^3 \gamma C_p \frac{dT}{dt} = h(T - T_{amb})4\pi r^2 \qquad (4)$$

which states that the heat given off is equal to the heat lost. Simplifying and separating the variables gives

$$\frac{dT}{T - T_{amb}} = \frac{-3h}{r\gamma C_p}\, dt$$

Integrating from $T = T_i$ to $T = T_f$,

$$\ln\left(\frac{T_f - T_{amb}}{T_i - T_{amb}}\right) = \frac{-3ht}{r\gamma C_p} \qquad (5)$$

The time involved is that of cooling. Solving for the time and substituting yields

$$t_c = \ln\left(\frac{T_f - T_{amb}}{T_i - T_{amb}}\right)\left(\frac{-r\gamma C_p}{3h}\right)$$

$$= \ln\left(\frac{621 - 70}{750 - 70}\right)\left(-\frac{\frac{1/8}{2 \quad 12}(710)(0.031)}{3h}\right)$$

$$= \frac{0.00804}{h}$$

Since the total time is known, the cooling and solidification times may be added and set equal to it.

$$t_s + t_c = 0.000536$$

$$\frac{0.0237}{h} + \frac{0.00804}{h} = 0.000536$$

Solving for the coefficient of heat transfer,

$$h = \frac{0.0237 + 0.00804}{0.000536}$$

$$= 59.2 \text{ Btu/hr-ft}^2\text{-}^\circ\text{F}$$

If the cooling and solidification times are desired, they may be solved for as

$$t_c = \frac{0.00804}{59.2}$$

$$= 0.0001358 \text{ hr} = 0.489 \text{ sec}$$

and

$$t_s = \frac{0.0237}{59.2}$$

$$= 0.0004003 \text{ hr} = 1.44 \text{ sec.}$$

CHAPTER 11

HEAT EXCHANGERS

Basic Attacks and Strategies for Solving Problems in this Chapter. **See pages 676 to 759 for step-by-step solutions to problems.**

The purpose of a heat exchanger is to transfer heat from a hot fluid to a cold fluid. The primary purpose may be to heat up the cold fluid or to cool down the hot fluid, but the temperatures of both will, of course, be changed.

One of the simplest types of heat exchanger is the double pipe configuration, one pipe inside another. In this case one fluid flows through the inner pipe (the tube side) and the other flows through the annular space between the two pipes (the shell side). Both fluids may flow in the same direction, called concurrent flow, or they may flow in opposite directions, called countercurrent flow. For a double pipe heat exchanger, the general equation for heat transfer rate is:

$$q = UA\,(LMTD)\,, \tag{1}$$

where A is the heat transfer surface area, U is the overall heat transfer coefficient, and LMTD is the log mean temperature difference. The LMTD is calculated from the temperature differences between hot and cold fluid at each end of the heat exchanger. This equation applies to either concurrent or countercurrent flow. The overall heat transfer coefficient, U, may be calculated from the convection coefficients on the inside and outside surfaces and the thermal conductivity and thickness of the heat transfer wall. Calculation of overall heat transfer coefficient is illustrated in chapter 1 and in several problems in this chapter. In order to determine the overall heat transfer coefficient it may be necessary to determine the tube side and/or shell side heat transfer coefficients using methods covered in Chapter 7.

For any type of heat exchanger the rate of heat transfer is also related to the mass flow rate, heat capacity and temperature change of either fluid as follows:

$$q = m_h\,Cp_h\,\Delta T_h \tag{2}$$

$$q = m_c\,Cp_c\,\Delta T_c\,, \tag{3}$$

where the subscript h refers to the hot fluid and the subscript c refers to the cold fluid.

A variety of problems can be solved with the above three equations. Three unknowns can be determined since there are three equations. For example, the required heat transfer rate to cool a given mass flow rate of oil from its inlet temperature to its desired outlet temperature can be calculated from equation (2). The required flow rate of cooling water can be determined from equation (3) if the inlet temperature and desired outlet temperature are known. The required heat exchanger surface area can then be determined from equation (1) if U can be determined because q was calculated and all of the temperatures needed to calculate LMTD were specified. There are many other combinations of three unknowns which could be posed in a problem and solved with the above three equations.

Shell and tube heat exchangers, which consist of a bundle of small tubes inside a larger pipe or "shell", are often used rather than a simple double pipe heat exchanger in order to increase the heat transfer area. Shell and tube heat exchangers are characterized by the number of tube passes and the number of shell passes. For example a 1, 2 heat exchanger has 1 shell pass and 2 tube passes; a 2,4 heat exchanger has 2 shell passes and 4 tube passes, etc. Diagrams in several problems such as 11-6, 11-9, and 11-12 illustrate these configurations. Equation (1) above is modified for shell and tube exchangers by including a correction factor F as follows:

$$q = FUA\,(LMTD)\,.$$

The value of the correction factor, F, is typically obtained from a generalized graph for the particular number of shell passes and tube passes. Examples of such graphs are given in several problems in this chapter, including 11-6, 11-8, and 11-9.

Another approach for heat exchanger configurations other than the double pipe is the use of the number of transfer units, NTU, and the effectiveness. Generalized graphs for each configuration are used for this method also. This approach is illustrated in problems 11-8 and 11-11.

Step-by-Step Solutions to Problems in this Chapter, "Heat Exchangers"

VARIOUS TYPES OF HEAT EXCHANGES AND COMPARISONS BETWEEN THEM

• PROBLEM 11-1

Steam (with a heat transfer coefficient of 6050 W/m^2-°K) is used to heat air (which has a much smaller coefficient of 61 W/m^2-°K) within a heat exchanger. Determine the rate of heat transfer per unit length of tube for the following arrangements: (a) Air flows through the shell side while steam flows through the tube side. Take the outside surface area to be 0.06 m^2/m and the inside surface area to be 0.052 m^2/m. (b) Fins are added to the set-up of (a), giving a total outside surface area of 0.165 m^2/m. (c) Steam flows through the shell side, and air flows through the finned tubes. For all three arrangements, assume that the overall temperature difference is maintained at 50°K and that the tube walls offer a negligible resistance.

Solution: The heat flow can be calculated from the following equation:

$$q = \frac{\Delta T}{\Sigma R} \qquad (1)$$

where ΣR is the sum of all resistances.

(a) First, the resistances must be determined.

$$R_s = \frac{1}{h_s A_i}$$

$$= \frac{1}{(6050)(0.052)} = 0.00318°K/W$$

$$R_a = \frac{1}{h_a A_0} = \frac{1}{(61)(0.06)} = 0.273°K/W$$

$$\Sigma R = R_s + R_a = 0.00318 + 0.273 = 0.276°K/W$$

Substituting into eq. (1), the heat flow is

$$q = \frac{50}{0.276} = 181.16 \text{ W}$$

(b) For the finned tubing arrangement, the resistance of the steam (on the tube side) remains the same. However, the resistance of the air changes, due to an increased surface area. It becomes

$$R_a = \frac{1}{h_a A_T} = \frac{1}{(61)(0.165)} = 0.099°K/W$$

$$\Sigma R = 0.00318 + 0.099 = 0.10218 \ {}^0K/W$$

From eq. (1), the heat flow is

$$q = \frac{50}{0.10218} = 489.3 \text{ W}$$

(c) For the finned tubing with steam in the shell side and air inside the tubes, the following resistances are present:

$$R_{st} = \frac{1}{h_s A_t} = \frac{1}{(6050)(0.165)} = 0.00102 \ {}^0K/W$$

$$R_a = \frac{1}{h_a A_i} = \frac{1}{(61)(0.052)} = 0.315 \ {}^0K/W$$

These sum to

$$\Sigma R = 0.00102 + 0.315 = 0.31602 \ {}^0K/W$$

Finally, eg. (1) gives a heat flow of

$$q = \frac{50}{0.31602} = 158.22 \text{ W}$$

• PROBLEM 11-2

Air (C_p = 0.24 Btu/lbm-°F) with a flow rate of 2400 lbm/hr is to be cooled from 150 to 100°F. Water is to be the coolant, entering at 70°F and leaving at 90°F. The overall heat transfer coefficient U = 30 Btu/hr-ft²-°F. Calculate the size of the intercooler for (a) concurrent (parallel) flow and (b) countercurrent flow.

Solution: a) The total heat transferred from the air to the water is

$$q = m_a C_{Pa} \Delta T_a \tag{1}$$

$$= (2400)(0.24)(150 - 100) = 28{,}800 \text{ Btu/hr}$$

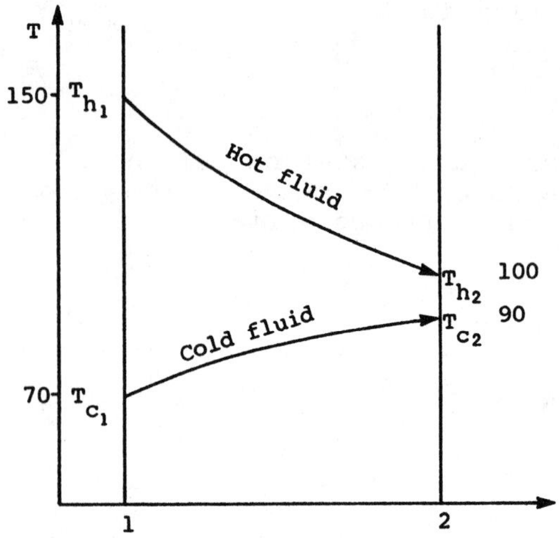

Fig. 1 Temperature plots for concurrent flow

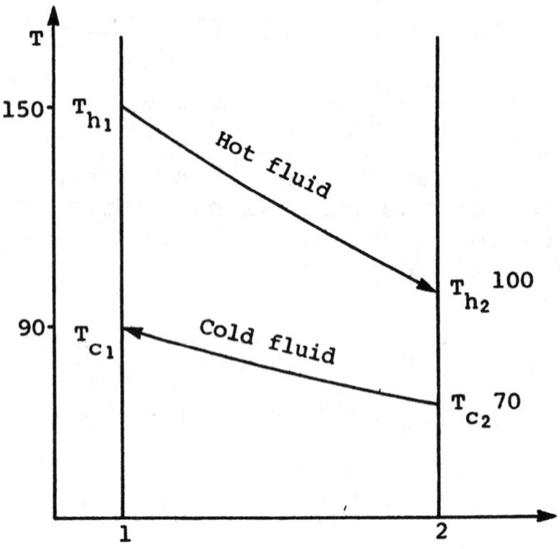

Fig. 2 Temperature plots for countercurrent flow

The log mean temperature difference (LMTD) approach to heat-exchanger analysis is useful when the inlet and outlet temperatures are known or are easily determined. The LMTD is expressed as

$$\text{LMTD} = \frac{(T_{h_2} - T_{c_2}) - (T_{h_1} - T_{c_1})}{\ln\ (T_{h_2} - T_{c_2})/(T_{h_1} - T_{c_1})} \tag{2}$$

Referring to fig. 1, the appropriate values are substituted into eq. (2) to give

$$\text{LMTD} = \frac{(100 - 90) - (150 - 70)}{\ln[(100 - 90)/(150 - 70)} = 33.66°\text{F}$$

Since the heat flow is expressed as

$$q = A\,U\,(\text{LMTD}) \tag{3}$$

the area of the parallel-flow heat exchanger can be determined as

$$A = \frac{q}{U\,(\text{LMTD})} = \frac{28{,}800}{(30)(33.66)} = 28.52\ \text{ft}^2$$

b) Similarly, for the counter-flow heat exchanger the log mean temperature difference is

$$\text{LMTD} = \frac{(100 - 70) - (150 - 90)}{\ln[(100 - 70)/(150 - 90)]} = 43.28°\text{F}$$

Then the area of the counter-flow heat exchanger is

$$A = \frac{q}{U\,(\text{LMTD})} = \frac{28{,}800}{(30)(43.28)} = 22.18\ \text{ft}^2$$

• PROBLEM 11-3

Kerosene (Cp = 0.6 Btu/lbm-°F) at an initial temperature of 450°F is used to heat engine oil (Cp = 0.455 Btu/lbm-°F), which flows at the rate of 2300 lbm/hr. The oil temperature is raised 130°F from its original value of 80°F. If the final temperature of the kerosene is 230°F, calculate the surface area over which the required heat transfer may be accomplished, and the necessary flow rate for a) parallel flow and b) counterflow. The overall coefficient of heat transfer is 80 Btu/hr-ft^2-°F.

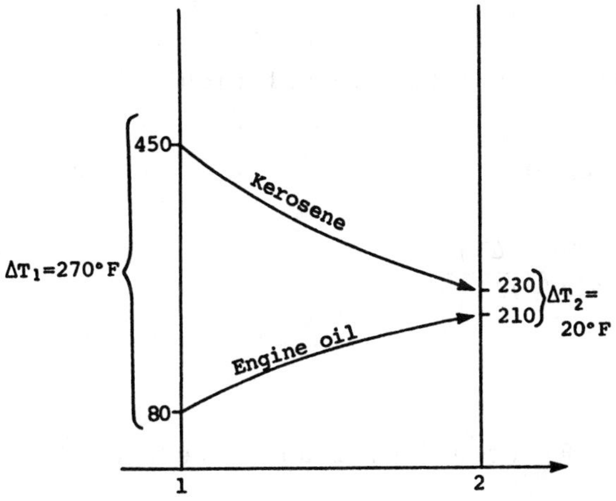

Fig. 1(a).

Parallel flow

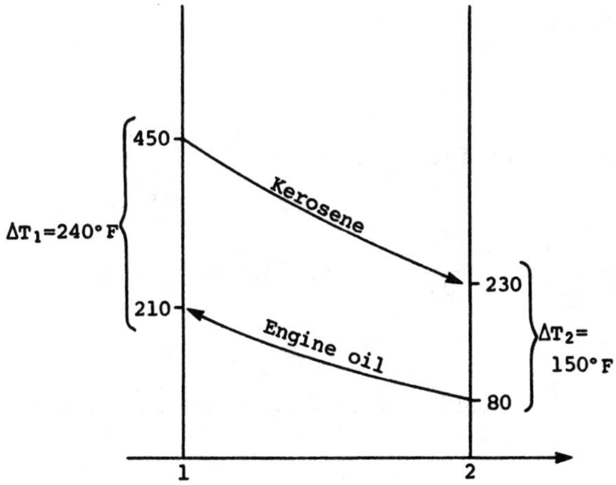

Fig. 1(b).

Counter flow

<u>Solution</u>: (a) The total heat flow is calculated as:

$$q_T = \dot{m}_{oil} C_{P\ oil} \Delta T_{oil}$$

$$= (2300)(0.455)(130) = 136{,}045 \text{ Btu/hr}$$

The kerosene flow rate may then be determined as

$$\dot{m}_{kerosene} = \frac{q_T}{C_{P\ oil} \Delta T_{kerosene}} = \frac{136{,}045}{(0.60)(450 - 230)} = 1031 \text{ lbm/hr}$$

The pipe area is found from the relation

$$q_T = UA\ (LMTD)$$

$$LMTD = \frac{\Delta T_2 - \Delta T_1}{\ln(\Delta T_2 / \Delta T_1)}$$

where

Knowing ΔT_1 and ΔT_2 (see fig. 1(a)), the area is calculated as

$$A = \frac{q_T}{U\left[\frac{\Delta T_1 - \Delta T_2}{\ln(\Delta T_1 / \Delta T_2)}\right]} = \frac{136{,}045}{(80)\left[\frac{370 - 20°F}{\ln(370/20)}\right]}$$

$$= 14.18 \text{ ft}^2$$

(b) Now, the kerosene flow rate is

$$m = \frac{136{,}045}{(0.6)(450 - 230)} = 1031 \text{ lbm/hr}$$

From fig. 1(b), $\Delta T_1 = 240°F$ and $\Delta T_2 = 30°F$. Then the pipe area is

$$A = \frac{136{,}045}{(80)\left[\frac{240 - 150}{\ln(240/150)}\right]} = 8.88 \text{ ft}^2$$

• PROBLEM 11-4

Water (C_{pw} = 4181 J/Kg-°K) at 20°C flows at the rate of 4,000 Kg/hr. It is used to cool engine oil (C_{po}= 1900J/Kg-°K) at 205°C, which flows at 11,000 Kg/hr. The overall heat transfer coefficient is 300 W/m²-°K. Calculate the oil outlet temperature if the intercooler functions in (a) a concurrent flow (see fig. 1), and (b) in a countercurrent flow (see fig. 2).

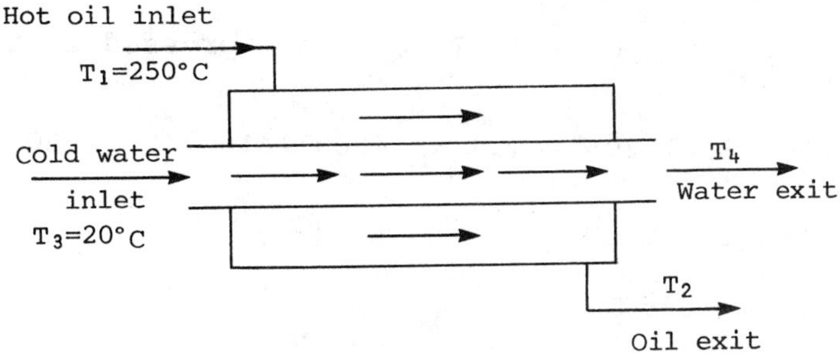

Fig. 1 concurrent flow

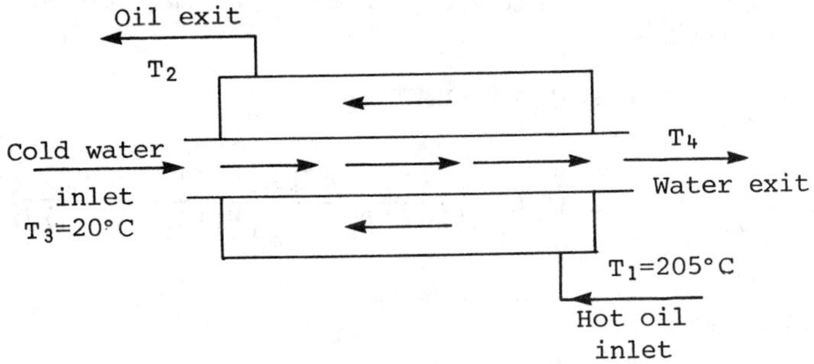

Fig. 2. Contercurrent flow.

Solution: The heat absorbed by the water is equal to the heat dissipated by the oil, which also equals the total heat transferred over the surface area of the heat exchanger. Therefore, the heat flow equation is

$$q = \dot{m}_h C_{Ph} \, (T_1 - T_2) = \dot{m}_c C_{Pc} (T_4 - T_3) \tag{1}$$

or

$$q = U_0 A_0 \, (LMTD) \tag{2}$$

The log mean temperature difference for the parallel flow heat exchanger is:

$$(LMTD)_P = \frac{(T_1 - T_3) - (T_2 - T_4)}{\ln[(T_1 - T_3)/(T_2 - T_4)]} \tag{3}$$

while for a countercurrent flow heat exchanger it is given as:

$$(LMTD)_C = \frac{(T_1 - T_4) - (T_2 - T_3)}{\ln[(T_1 - T_4)/(T_2 - T_3)]} \tag{4}$$

(a) Sutstituting into eq. (1), T_4 can be expressed in terms of T_2.

$$(11{,}000)(1900)(205 - T_2) = (4000)(4181)(T_4 - 20)$$

Rearranging gives

$$T_4 = 276 - 1.25\,T_2 \tag{5}$$

This is valid for both the concurrent flow and counter-current flow case.

Now, for the concurrent flow case, eq. (1) may be set equal to eq. (2) to give

$$\dot{m}_h C_P (T_1 - T_2) = U_0 A_0 \; (LMTD)_P$$

$$= U_0 A_0 \left\{\frac{(T_1 - T_3) - (T_2 - T_4)}{\ln[(T_1 - T_3)/(T_2 - T_4)]}\right\}$$

Inserting the values,

$$(11{,}000)(1900)(205 - T_2)$$

$$= (300)(17.5) \; \frac{(205 - 20) - (T_2 - T_4)}{\ln[(205 - 20)/(T_2 - T_4)} \; (3600 \text{ sec})$$

Substituing for T_4, and simplifying yields

$$\ln \frac{185}{2.25T_2 - 276} = \frac{8.7 \times 10^9 - 4.25 \times 10^7 T_2}{4.28 \times 10^9 - 2.09 \times 10^7 T_2}$$

$$= \frac{4.25(2.04 - T_2)}{2.09(2.04 - T_2)} = 2.03$$

Taking the exponential of both sides,

$$\frac{185}{2.25T_2 - 276} = e^{2.03} = 7.61$$

Solving for T_2,

$$T_2 = \frac{(185/7.61) + 276}{2.25} = 133.5°C$$

Substituting this into eq. (5) gives

$$T_4 = 276 - 1.25(133.5) = 109.1°C$$

(b) The same method is used for the counterflow function, and the corresponding LMTD is utilized. Therefore,

$$\dot{m}_h C_P (T_1 - T_2) = U_0 A_0 \left\{\frac{(T_1 - T_4) - (T_2 - T_3)}{\ln[(T_1 - T_4)/(T_2 - T_3)]}\right\}$$

Substituting,

$$(11,000)(1900)(205 - T_2)$$

$$= (300)(17.5) \left\{\frac{(205 - T_4) - (T_2 - 20)}{\ln[(205 - T_4)/(T_2 - 20)]}\right\} (3600 \text{ sec})$$

Similar algebraic manipulations give a value of

$$T_2 = 115°C$$

which in turn yields

$$T_4 = 132.25°C.$$

It can be seen that the countercurrent flow heat exchanger is more effective in cooling the oil ($T_2 > T_4$). In the concurrent heat exchanger, T_2 will be greater than or equal to T_4 only when the surface area becomes infinite. This occurs with a continually decreasing forcing function ($T_h - T_c$) that is usable for heat transfer as the fluid follows its course.

A mixed crossflow intercooler with an effective area of 16.5 ft^2 uses water to cool a supply of engine oil (C_p = 0.5 Btu/lbm-°F) flowing at 4000 lbm/hr. The initial temperature of the oil is 200°F, and the water (C_p = 1.0 Btu/lbm-°F) is pumped in at 2200 lbm/hr and 50°F on the shell side. If the overall heat transfer coefficient is 45 Btu/hr-ft^2-°F, determine the oil and water outlet temperatures.

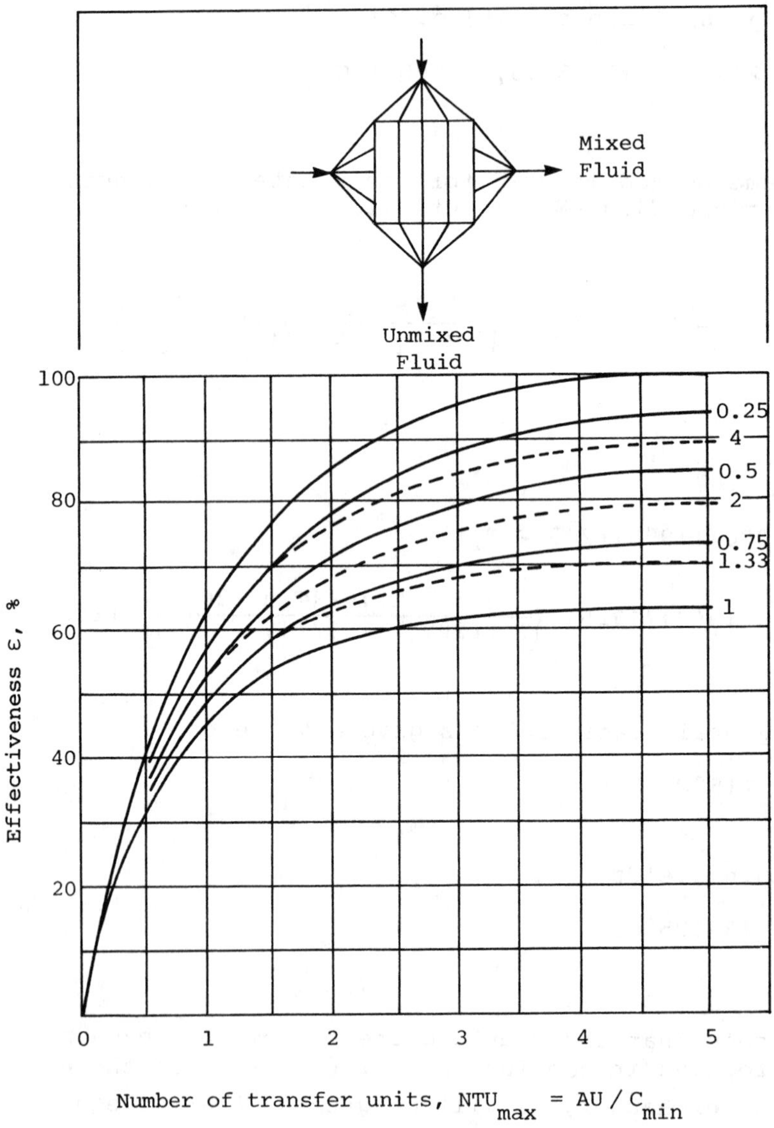

Fig. 1.

Crossflow exchanger effectiveness with one fluid mixed.

Solution: Since the oil and water outlet temperatures are unknown, it is not feasible to use the LMTD approach. That method will require a trial and error technique. Therefore, it is better to use the effectiveness--NTU method.

First, it is necessary to determine the minimum fluid by calculating the capacity coefficients of the oil and water.

$$C_{oil} = \dot{m}_{oil} \, C_{P\,oil} = (400)(0.5) = 200 \text{ Btu/hr-°F}$$

and

$$C_w = \dot{m}_w \, C_{Pw}$$

$$= (2200)(1.0) = 2200 \text{ Btu/hr-°F}$$

Since $C_{oil} < C_w$, C_{oil} is the minimum fluid.

As $C_{mixed}/C_{unmixed} = 2200/2000 = 1.1$ and

$$NTU = \frac{UA}{C_{min}} = \frac{(45)(16.5)}{2000} = 0.371,$$

Fig. 1 gives an effectiveness $\varepsilon \cong 0.3 = 30\%$

The total heat transfer rate is then calculated as:

$$q = \varepsilon C_{min} (T_{h\ in} - T_{c\ in})$$

$$= 0.3(2000)(200 - 50) = 90{,}000 \text{ Btu/hr}$$

The outlet temperatures can then be calculated using the equations

$$q = C_c (T_{c,o} - T_{c,i}) \tag{1}$$

and

$$q = C_h (T_{h,i} - T_{h,o}) \tag{2}$$

Using eq. (1), the water outlet temperature is

$$T_{c,o} = T_{c,i} + \frac{q}{C_c} = 50 + \frac{90{,}000}{2200} = 90.9°F$$

and using eq. (2), the oil outlet temperatue is

$$T_{h,o} = T_{h,i} - \frac{q}{C_h} = 200 - \frac{90{,}000}{2000} = 155°F$$

• PROBLEM 11-6

Steam at 700°F enters the shell side of a heat exchanger and it exits at 500°F. This heats the tube side fluid from 300 to 400°F. Assuming that the exchanger is insulated, calculate the mean overall temperature difference for (a) a concentric pipe counterflow heat exchanger (see fig. 1), and (b) a one-shell-pass, two-tube-pass counterflow heat exchanger (see fig. 2).

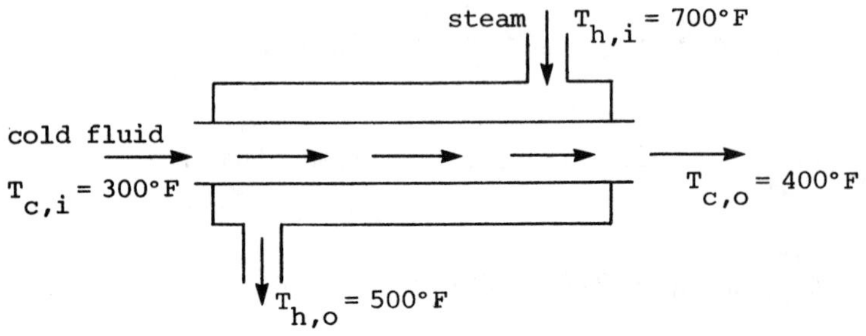

Fig. 1.

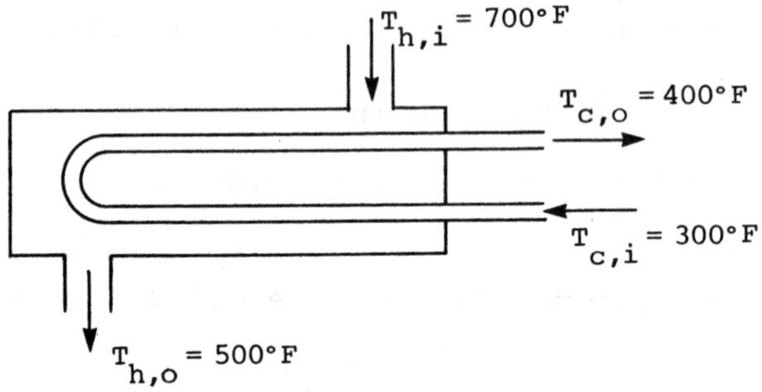

Fig. 2.

Solution: (a) The log mean temperature difference for a concentric pipe counterflow heat exchanger is given as:

$$LMTD = \frac{(T_{h,i} - T_{c,o}) - (T_{h,o} - T_{c,i})}{\ln[(T_{h,i} - T_{c,o})/(T_{h,o} - T_{c,i})]} \tag{1}$$

Inserting the corresponding temperatures into eq. (1) gives

$$LMTD = \frac{(700 - 400) - (500 - 300)}{\ln[(700 - 400)/(500 - 300)]} = 246.6°F$$

(b) If a heat exchanger other than the concentric pipe type is used, a dimensionless correction factor F must be used to determine the relationship between the LMTD and the true mean temperature difference ΔT. It is defined as

$$F = \frac{\Delta T_m}{LMTD} \tag{2}$$

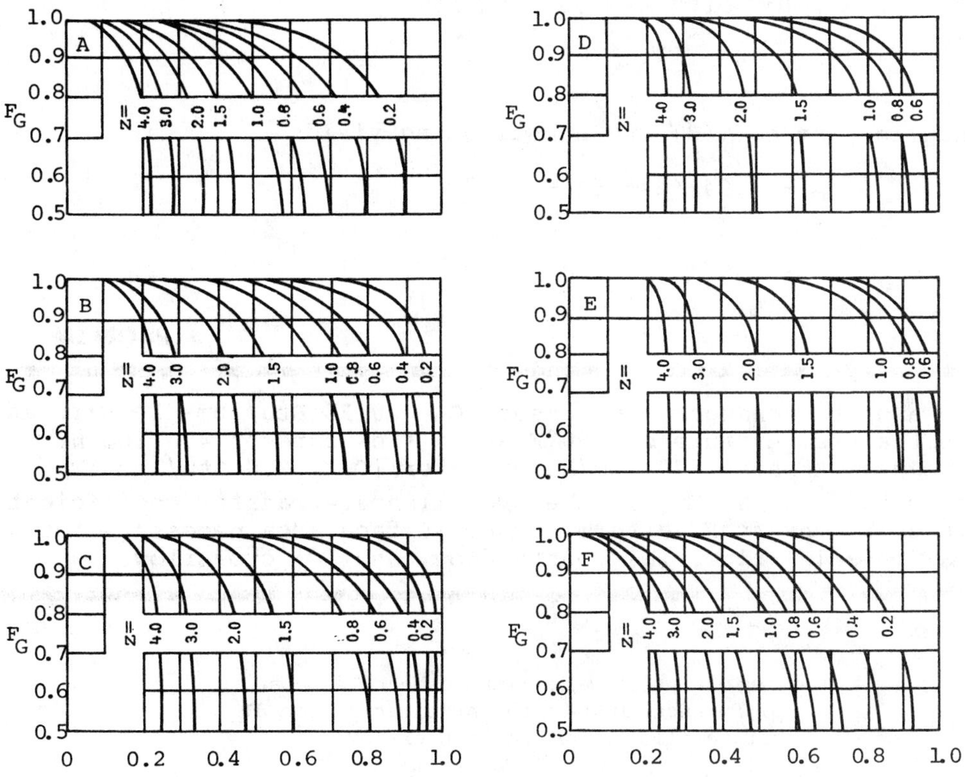

$$Y = \frac{T_{c2} - T_{c2}}{T_{h1} - T_{c1}} \qquad Z = \frac{T_{h1} - T_{h2}}{T_{c2} - T_{c1}}$$

Fig. 3.

Mean Temperature Difference in Reversed-Current Exchangers (shell tube fluid well mixed at a given cross section)

A) One shell pass and 2,4,6, etc., tube passes.
B) Two shell passes, and 4,8,12,etc., tube passes.
C) Three shell passes, and 6,12,18,etc., tube passes.
D) Four shell passes, and 8,16,24, etc., tube passes.
E) Six shell passes, and 12,24,36,etc., tube passes.
F) One shell pass, and 3,6,9,etc., tube passes.

It is immaterial wheter the warmer fluid flows through the shell or the tubes.

Referring to fig. 3, the ratio Y and Z must first be calculated, as

$$Y = \frac{T_{c_2} - T_{c_1}}{T_{h_1} - T_{c_1}} = \frac{400 - 300}{700 - 300} = 0.25$$

$$Z = \frac{T_{h_1} - T_{h_2}}{T_{c_2} - T_{c_1}} = \frac{700 - 500}{400 - 300} = 2.0$$

Then the value of F may be read as

$$F = 0.945$$

Rearranging eq. (2) and substituting yields

$$\Delta T_m = 0.945(246.6) = 233°C$$

• PROBLEM 11-7

In a heat exchanger flue gases (C_p = 0.25 Btu/lbm-°F) with an initial temperature of 350°F and a flow rate of 450 lbm/hr are used to heat 120 lbm/hr of water (C_p = 1.0 Btu/lbm-°F) from 50°F to 180°F. If the overall heat-transfer coefficient is 20 Btu/hr-ft^{2}F, determine the surface area necessary for (a) parallel flow, (b) counterflow, and (c) crossflow.

Ratio a of the Logarithmic Mean Temperature Difference LMTD to the Arithmetic Mean ΔT_m for Parallel Flow and Counterflow

$$LMTD = a\Delta T_m$$

$\frac{\Delta T_i}{\Delta T_o}$	a	$\frac{\Delta T_i}{\Delta T_o}$	a	$\frac{\Delta T_i}{\Delta T_o}$	a	$\frac{\Delta T_i}{\Delta T_o}$	a
1.0	1.000	2.0	0.962	3.0	0.910	6.0	0.798
1.2	0.998	2.2	0.952	3.5	0.889	7.0	0.770
1.4	0.991	2.4	0.942	4.0	0.867	8.0	0.748
1.6	0.981	2.6	0.928	4.5	0.846	9.0	0.729
1.8	0.971	2.8	0.918	5.0	0.829	10.0	0.710

Table 1.

Solution: Consider first a heat balance for the system. The heat absorbed by the water is

$$q_w = \dot{m}_w C_{Pw} \Delta T_w \tag{1}$$

Ratio a of the Logarithmic Mean Temperature Difference LMTD to the Arithmetic Mean ΔT_m for Crossflow

$LMTD = a\Delta T_m$

$\frac{m_1c_1}{m_2c_2}$ \ $\frac{\Delta T_o}{\Delta T_i}$	0.9	0.8	0.7	0.6	0.5	0.4	0.3	0.2	0.1
0	0.998	0.993	0.986	0.978	0.962	0.939	0.902	0.836	0.707
0.2	0.998	0.993	0.987	0.981	0.971	0.958	0.936	0.902	0.848
0.5	0.998	0.993	0.988	0.982	0.977	0.968	0.945	0.935	0.913
1	0.998	0.993	0.989	0.983	0.978	0.974	0.961	0.948	0.933

$\frac{m_1c_1}{m_2c_2}$ \ $\frac{\Delta T_o}{\Delta T_i}$	0	-0.1	-0.2	-0.3	-0.4	-0.5	-0.6	-0.7	-0.8
0	0								
0.2	0.762	0.630							
0.5	0.874	0.819	0.750	0.632	0.519				
1	0.912	0.876	0.835	0.766	0.710	0.618	0.500	0.380	0.220

Table 2.

$$= (120)(1.0)(180 - 50) = 15,600 \text{ Btu/hr}$$

The heat transferred by the gases is equal to this quantity if the heat exchanger is insulated. Assuming that this is true, the final temperature of the gases may be determined, as

$$q_w = q_g \qquad (2)$$

$$q_w = \dot{m}_g C_p \Delta T_g$$

$$15,600 = (450)(0.25)(350 - T_0)$$

This gives $T_0 = 211.33°F$

Now to find the required surface area, the following equation will be used:

$$q = AU\ (LMTD) \qquad (3)$$

The value of the log mean temperature difference for each case will be found from

$$LMTD = a\Delta T_m \qquad (4)$$

where a is a constant. Since the value of

$$\Delta T_m = \frac{\Delta T_i + \Delta T_0}{2} \tag{5}$$

may be calculated, it only remains to find the value of a to calculate the LMTD. For a given temperature ratio, a may be read from either Table 1 or 2, depending on the type of flow being analyzed. Finally, when the LMTD is found, the surface area may be calculated form eq. (3).

(a) For the case of parallel flow,

$$\Delta T_i = 350 - 50 = 300°F$$

and

$$\Delta T_0 = 211.33 - 180 = 31.33\ °F$$

From eq. (5),

$$\Delta T_m = \frac{300 + 31.33}{2} = 165.66\ °F$$

The ratio

$$\Delta T_i/\Delta T_0 = 300/31.33 = 9.575$$

gives

$$a = 0.718$$

through interpolation form Table 1.

The LMTD may now be calculated using eq. (4):

$$\text{LMTD} = a\Delta T_m = 0.718(165.66) = 118.9°F$$

Rearranging eq. (3) and substituting yields the effective heat transfer area,

$$A = \frac{q_w}{U\ (\text{LMTD})} = \frac{15,600}{(20)(118.9)} = 6.56\ \text{ft}^2$$

(b) Using the same method for the counterflow case,

$$\Delta T_i = 350 - 180 = 170°F$$

$$\Delta T_0 = 211.33 - 50 = 161.33°F$$

From eq. (5)

$$\Delta T_m = \frac{170 + 161.33}{2} = 165.66°F$$

$$\Delta T_i/\Delta T_0 = 170/161.33 = 1.054$$

From Table 1,

$$a = 0.9995$$

Then from eq. (4),

$$\text{LMTD} = 0.9995\,(165.66) = 165.58°\text{F}$$

and from eq. (3),

$$A = \frac{15{,}600}{(20)(165.58)} = 4.7\ \text{ft}^2$$

(c) Using the same method for the cross flow case gives

$$\Delta T_i = 350 - 50 = 300°\text{F}$$

$$\Delta T_0 = 211.33 - 180 = 31.33°\text{F}$$

From eq. (5),

$$\Delta T_m = \frac{300 + 31.33}{2} = 165.66°\text{F}$$

Table 2 must be used now. The appropriate ratio is

$$\Delta T_0/\Delta T_i = 31.33/300 = 0.104$$

In addition, the value of

$$\frac{\dot{m}_w C_{Pw}}{\dot{m}_g C_{Pg}} = \frac{(120)(1.0)}{(450)(.25)} = 1.067$$

must be used. Then

$$a = 0.933$$

From eq. (4),

$$\text{LMTD} = 0.933\,(165.66) = 154.56°\text{F}$$

and from eq. (3), the effective heat transfer area is

$$A = \frac{15600}{20(154.56)} = 5.05\ \text{ft}^2$$

• **PROBLEM 11-8**

Water (C_p = 1.0 Btu/lbm°F) at 95°F and a flow rate of 9,000 lbm/hr is used to cool hot water from 205°F to 155°F, which flows at the same rate. If the overall heat transfer coefficient is 400 Btu/hr-ft^2-°F, calculate the effective area for a (a) parallel flow, (b) counterflow, and (c) crossflow heat exchanger. Solve the problem first using the LMTD approach, and then use the NTU-ε charts.

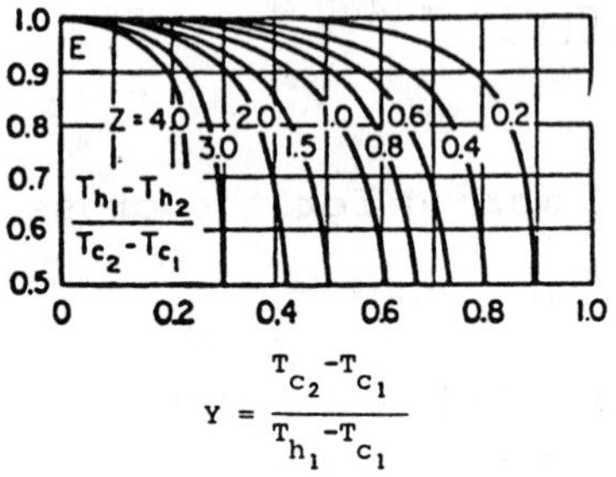

Fig. 1.

Solution: (i) The LMTD approach to heat exchanger analysis is useful when the inlet and outlet temperatures are known.

The heat flow equation for the process is written as:

$$q = \dot{m}_C C_p (T_{C2} - T_{C1}) = \dot{m}_h C_p (T_{h1} - T_{h2}) \tag{1}$$

The heat transferred to the cold water is

$$q = \dot{m}_h C_p (T_{h1} - T_{h2}) = (9,000)(1.0)(205 - 155)$$

$$= 450,000 \text{ Btu/hr}$$

Then the cold water outlet temperature T_{C2} can then be calculated by substituting the value of q into eq. (1).

$$450,000 = (9,000)(1.0)(T_{C2} - 95)$$

Solving for T_{C_2} gives

$$T_{C_2} = 145°F$$

Now that the inlet and outlet temperatures are known, the area is easily determine from the relation

$$A = \frac{q}{U\ (LMTD)} \tag{2}$$

(a) The LMTD for parallel flow is defined as:

$$LMTD = \frac{(T_{h_1} - T_{c_1}) - (T_{h_2} - T_{c_2})}{\ln[(T_{h_1} - T_{c_1})/(T_{h_2} - T_{c_2})]} \tag{3}$$

Inserting the corresponding temperatures into eq. (3),

$$LMTD = \frac{(205 - 95) - (155 - 145)}{\ln[(205 - 95)/(155 - 145)]} = 41.7°F$$

and from eq. (2), the area is found to be

$$A = \frac{450,000}{(400)(41.7)} = 26.98 \text{ ft}^2$$

(b) Since the two fluids have the same flow rates, the temperature difference ΔT over the counterflow heat exchanger is consistent at $(T_{h_2} - T_{c_1})$, which is $(155 - 95) = 60°F$. If this temperature difference is used in eq. (2), the resulting area is

$$A = \frac{450,000}{(400)(60)} = 18.75 \text{ ft}^2$$

(c) For this case, a cross-flow exchanger must be a one-shell-pass, one-tube-pass arrangement, with hot water mixed in the shell side. The temperature difference for the cross-flow exchanger is $(T_{h_2} - T_{c_1}) = 60°F$, since the two fluids have the same flow rates.

Now the correction factor F must be found to obtain the true mean temperature difference ΔT_m as:

$$\Delta T_m = F \Delta T \tag{4}$$

Referring to fig. 1,the following dimensionless ratios are calculated:

$$Y = \frac{T_{c_2} - T_{c_1}}{T_{h_1} - T_{c_1}} \tag{5}$$

$$= \frac{145 - 95}{205 - 95} = 0.455$$

$$Z = \frac{T_{h_1} - T_{h_2}}{T_{c_2} - T_{c_1}} \tag{6}$$

$$= \frac{205 - 155}{145 - 95} = 1.0$$

This gives a correction factor F = 0.88

Therefore, $\Delta T_m = (0.88)(60) = 52.8°F$

and the area is calculated from eq. (2) as

$$A = \frac{(450,000)}{(400)(52.8)} = 21.3 \text{ ft}^2$$

(ii) When the inlet or outlet temperatures are to be determined for a given heat exchanger, the solution frequently requires an iterative procedure because of the logarithmic function in the LMTD. In these cases the analysis is done more easily by applying a method based on the effectiveness of the heat exchanger in transferring a given amount of heat. The effectiveness- NTU method is also useful for the analysis of problems in which a comparison between the different types of exchangers is to be made, in order to select the most efficient one. In order to use the NTU charts, the following quantities are to be calculated first:

$$C_{min} = C_{max} = \dot{m}C_p \tag{7}$$

$$= (9,000)(1.0) = 9,000 \text{ Btu/hr-°F}$$

And

$$C_{min}/C_{max} = 1.0$$

The effectiveness is calculated as

$$\varepsilon = \frac{T_{C2} - T_{C1}}{T_{h1} - T_{C1}} \tag{8}$$

$$= \frac{145 - 95}{205 - 95} = 0.455 \times 100\% = 45.5\%$$

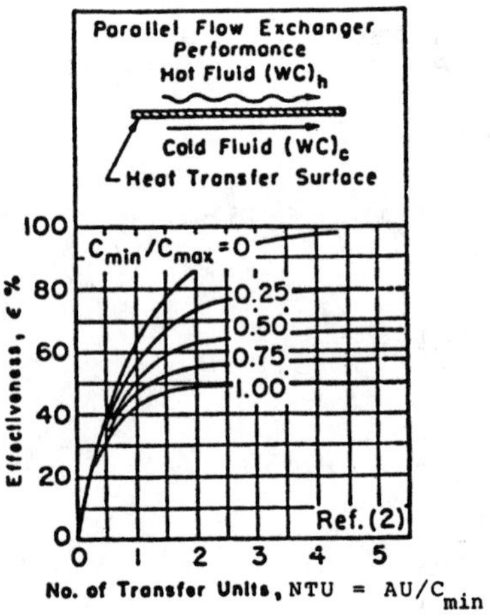

Fig. 2.

(a) Referring to fig. 2, NTU = 1.3

the area is calculated from the equation

$$A = \frac{(NTU)C_{min}}{U} \tag{9}$$

$$= \frac{1.3\ (9{,}000)}{(400)} = 29.25\ ft^2$$

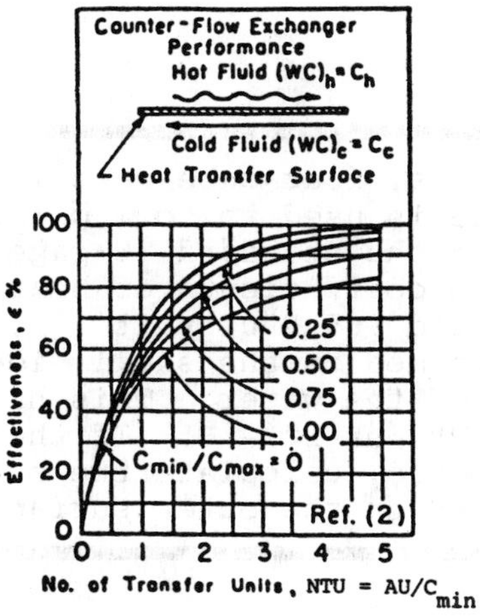

Fig. 3.

(b) Referring to fig. 3, NTU = 0.8. Then the area is

$$A = \frac{0.8\ (9{,}000)}{(400)} = 18.0\ ft^2$$

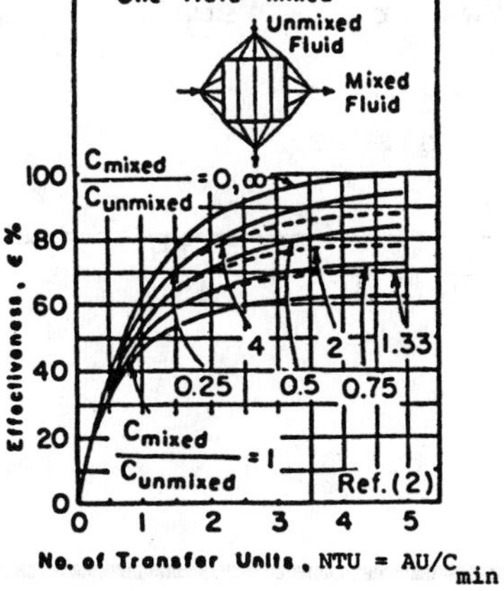

Fig. 4.

(c) Referring to fig. 4, NTU = 1.0,

and the area is

$$A = \frac{1.0\ (9,000)}{(400)} = 22.5\ ft^2.$$

• PROBLEM 11-9

A two-shell-pass, four-tube-pass intercooler with water on the shell side is used to cool oil in the tube side. The tubes used are the standard 16-gage, with a 1 in. outside diameter. The coefficients of heat transfer for the oil and water are 45 and 165 Btu/hr-ft^2-°F, respectively. A scale deposit has formed on the shell side due to the high impurity content of the water. This has a heat transfer coefficient of 450 Btu/hr-ft^2-°F. If the resistance of the metal wall is neglected, calculate the overall heat transfer coefficient based on the outer surface.

Solution: For a 1 in. 16-gage tube, the inside diameter is 0.87 in. The ratio of the outside area to the inside area is:

$$\frac{A_o}{A_i} = \frac{D_o}{D_i} = \frac{1\ in}{.8\ in} = 1.15$$

Assuming that the inside of the tube does not corrode, the overall heat transfer coefficient will be

$$U_0 = \cfrac{1}{\cfrac{1}{h_{water}} + \cfrac{1}{h_{scale}} + \cfrac{A_o}{A_i\ h_{oil}}}$$

Substituting, this is

$$U_0 = \cfrac{1}{\cfrac{1}{165} + \cfrac{1}{450} + \cfrac{1.15}{45}} = 29.55\ Btu/hr\text{-}ft^2\text{-}°F$$

• PROBLEM 11-10

In a heat exchanger, an oil stream A in fig. 1 is to be cooled from 160°F to 100°F. It flows at the rate of 130,000 lbm/hr, and its specific heat is 0.75 Btu/lbm-°F. The cooling action is to be achieved using water at 60°F, and it is desired to use a one shell-pass, two-tube-pass heat exchanger. The exit temperature of the water must be kept below 80°F to avoid environmental disturbances, since it empties into a river. The overall heat transfer coefficient is 136 Btu/ft^2-hr-°F, and the available tube size is a standard 12-gage 1.25 in. outside diameter. There are 166 tubes. What is the effective area for the heat exchanger? Also, what must the length of each tube be?

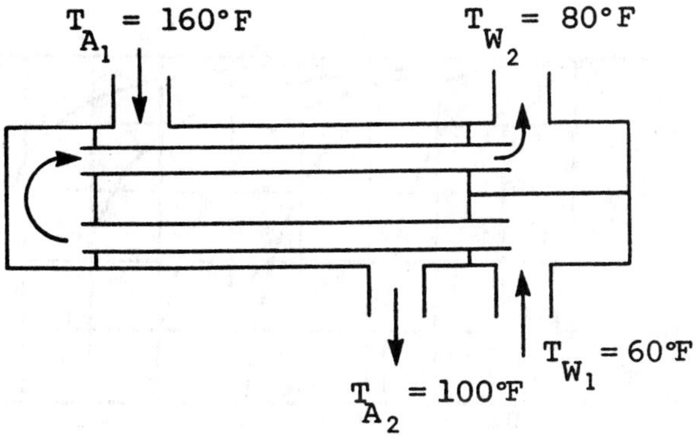

Fig. 1.

Solution: The heat flow of the oil is

$$q = \dot{m}_A C_{PA} \Delta T_A$$

$$= (130,000)(0.75)(160 - 100)$$

$$= 5.85 \times 10^6 \text{ Btu/hr}$$

Using this value, the flow rate of water may be found as

$$\dot{m}_W = \frac{q}{C_{Pw} \Delta T_w} = \frac{5.85 \times 10^6}{(1.0 \text{ Btu/lbm°F})(80 - 60)}$$

$$= 2.925 \times 10^5 \text{ lbm/hr}$$

The log mean temperature difference is

$$\text{LMTD} = \frac{(T_{A1} - T_{W2}) - (T_{A2} - T_{W1})}{\ln[(T_{A1} - T_{W2})/(T_{A2} - T_{W1})]} = \frac{(160 - 80) - (100 - 60)}{\ln[(160-80)/(100 - 60)}$$

$$= 57.7°\text{F}$$

Fig. 2 must be used to find the appropriate correction factor Y. The necessary ratios are:

$$Z = \frac{\tilde{T}_1 - \tilde{T}_2}{T_2 - T_1} = \frac{T_{A1} - T_{A2}}{T_{W2} - T_{W1}} = \frac{160 - 100}{80 - 60} = 3$$

$$X = \frac{T_2 - T_1}{\tilde{T}_1 - T_1} = \frac{T_{W2} - T_{W1}}{T_{A1} - T_{W1}} = \frac{80 - 60}{160 - 60} = 0.2$$

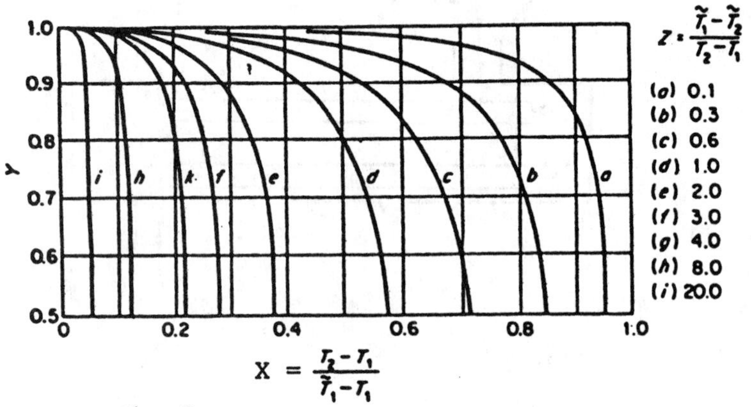

Fig. 2.

Correction factor for shell and tube exchanger with one shell pass and an even number of tube passes.

This gives a value of $Y = 0.93$.

Since the heat-transfer equation is

$$q = UAY\,(LMTD)$$

the area can be calculated as

$$A = \frac{q}{U\,Y\,(LMTD)} = \frac{5.85 \times 10^6}{(136)(0.93)(57.7)} = 801.6 \text{ ft}^2$$

From this, the individual tube length must be

$$L = \frac{A}{A_S\, N_{tubes}}$$

For a 12-gage tube with a 12.5 in. outside diameter, the outside surface area per unit length is 0.3272 ft²/ft. Then

$$L = \frac{801.6}{(0.3272)(166)}$$

$$= 14.76 \text{ ft.}$$

• PROBLEM 11-11

In a condenser, steam at a temperature of 127°F is used to heat salt water (C_p = 0.95 Btu/lb F) flowing at a rate of 20,000 lbm/hr from 50°F to 120°F. It is a one-shell-pass, six-tube-pass exchanger with 20 brass tubes (k = 60 Btu/hr ft F) of 0.875 in. inside diameter and 1.00 in. outside diameter. The average convection coefficients $\bar{h}_0$ and $\bar{h}_i$ at the steam and water sides are approximately 600 and 300 Btu/hr ft^{2}F, respectively. Determine the necessary tube length.

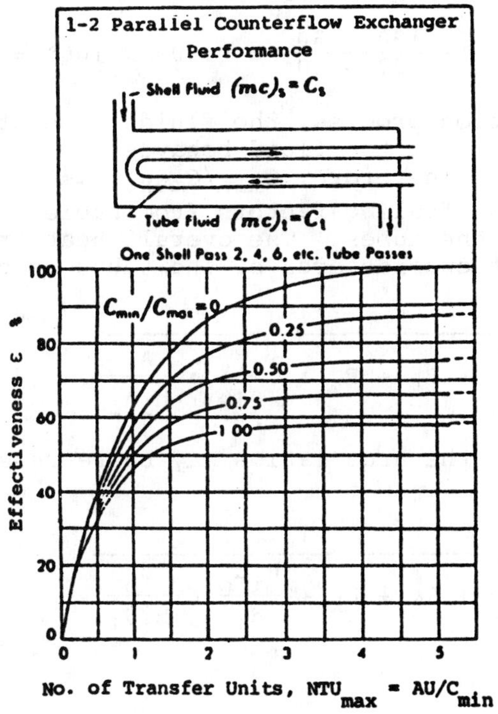

Fig. 1

Solution: If the effectiveness - NTU relations of fig. 1 are to be used, the effectiveness (ε) must be calculated as

Table of Normal Fouling Factors

Type of Fluid	Fouling Resistance (hr F sq. ft./Btu)
Sea water below 125°F	0.0005
Sea water above 125°F	0.001
Treated boiler feed water above 125°F	0.001
East River water below 125°F	0.002-0.003
Fuel oil	0.005
Quenching oil	0.004
Alcohol vapors	0.0005
Steam, non-oil bearing	0.0005
Industrial air	0.002
Refrigerating liquid	0.001

Fig. 2.

$$\varepsilon = \frac{T_{h2} - T_{c2}}{T_{h1} - T_{c1}}$$

$$= \frac{120 - 50}{127 - 50} = 0.909 \times 100\% = 90.9\%$$

In a condensation process, the fluid temperature stay essentially constant, or the fluid behaves as if it had infinite specific heat. Therefore, $C_{min}/C_{max} \rightarrow 0$. From fig. (1), NTU = 2.5. The fouling factors from table (1) are 0.0005 for both sides of the tubes. The overall heat-transfer coefficient per unit outside area of the tubes is given by:

$$U_0 = \frac{1}{\frac{1}{\bar{h}_0} + R_0 + R_k + \frac{R_i A_0}{A_i} + \frac{A_0}{\bar{h}_i A_i}}$$

Since A = πdL, the area ratios may be represented by diameter ratios, giving

$$U_0 = \frac{1}{\frac{1}{\bar{h}_0} + R_0 + R_k + \frac{R_i d_o}{d_i} + \frac{d_o}{\bar{h}_i d_i}}$$

Substituting,

$$U_0 = \left[\frac{1}{600} + 0.0005 + \frac{1.00}{(2)(12)(60)} \ln \frac{1.00}{0.875} + \frac{(0.0005)(1.00)}{0.875} + \frac{1.00}{(300)(0.875)}\right]^{-1}$$

$$= 150.59 \text{ Btu/hr-ft}^2\text{-F}$$

The total outside area is

$$A_0 = n\pi D_0 L,$$

and from fig. 1,

$$U_0 A_0 / C_{min} = NTU = 2.5$$

since $\quad C_{min} = \dot{m}C_p,$

solving for the length and substituting yields

$$L = \frac{(NTU)\dot{m}C_p}{n\pi d_0 U}$$

$$= \frac{(2.5)(20,000)(0.95)(12)}{(20)(3.14)(1.00)(150.59)} = 60.24 \text{ ft}$$

• PROBLEM 11-12

Hot water at a inlet temperature of 116°C and an outlet temperature of 49°C is used to heat water flowing at 2.5 Kg/s from 21°C to 55°C in a one-shell-pass, two-tube-pass regenerator (see fig. 1(a)). The outside surface area of the tubes is found to be $A_0 = 9.5 \text{ m}^2$.

a) Determine the mean temperature difference ΔT_m, and the overall heat-transfer coefficient U_o

b) Determine ΔT_m for a two-shell-pass, four-tubes-pass regenerator (see fig. 1b) under same conditions.

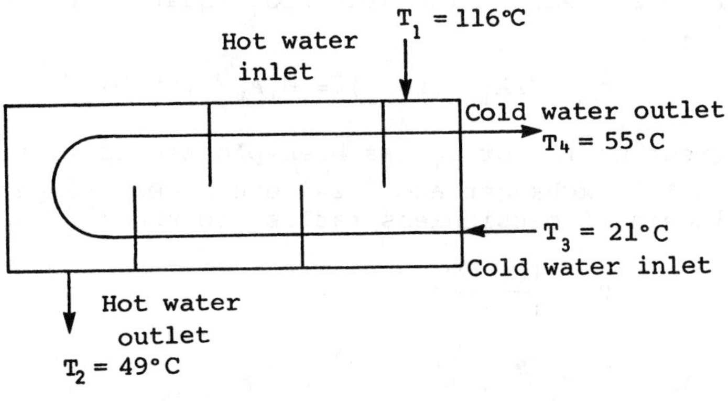

Fig. 1. (a) 1-2 exchanger

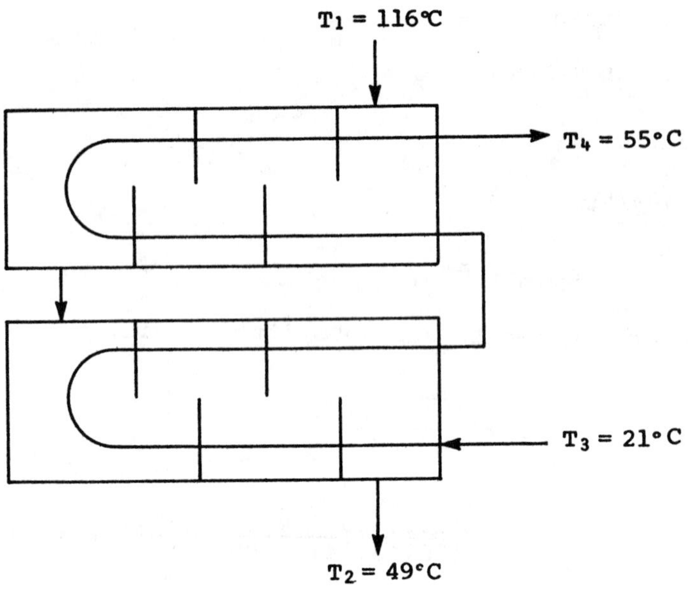

Fig. 1(b).   2-4 exchanger

Solution: The LMTD for a double-pipe heat exchanger of the one-shell-pass, one-tube-pass variety in a countercurrent flow or concurrent flow is

$$\text{LMTD} = \frac{(T_1 - T_4) - (T_2 - T_3)}{\ln\left[(T_1 - T_4)/(T_2 - T_3)\right]}$$

This relation does not hold true for a multi-pass arrangement, which has a combination of both counter and parallel flow. In that case, a correction factor F_T will be used to modify the LMTD in order to obtain the correct average temperature difference. The heat flow equation for the exchanger is

$$q = U_i A_i F_T(\text{LMTD}) = U_0 A_0 F_T(\text{LMTD})$$

The correction factor F_T has been plotted in figs. 2a and 2b for a 1-2 exchanger and a 2-4 exchanger, respectively. The following dimensionless ratios are used:

$$Z = \frac{T_1 - T_2}{T_4 - T_3}$$

$$Y = \frac{T_4 - T_3}{T_1 - T_3}$$

Assuming water has a specific heat C_p = 4187 J/Kg-°K, the heat absorbed by the cold water is:

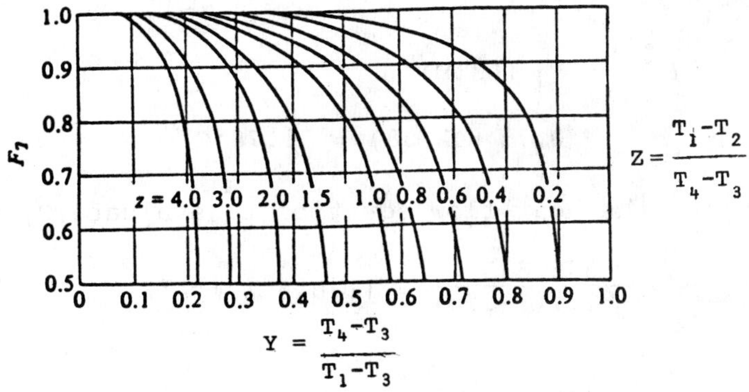

Fig. 2(a).

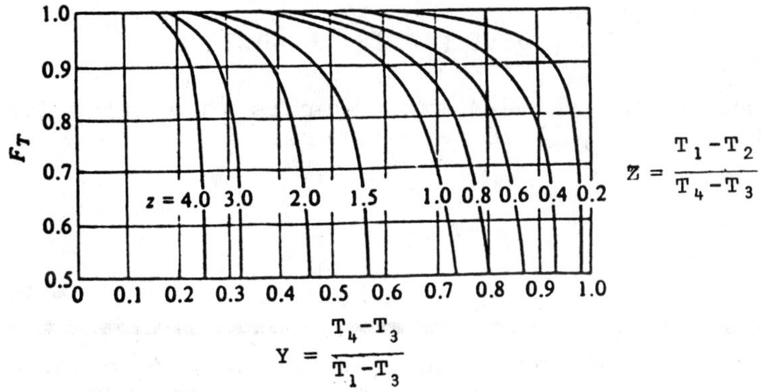

Fig. 2(b).

Correction factor F_T to log mean temperature difference:

(a) 1-2 exchangers

(b) 2-4 exchangers

$$q = mC_p(T_4 - T_3) = (2.5)(4187)(55 - 21) = 355{,}895\text{ W}$$

a) The LMTD is calculated as:

$$\text{LMTD} = \frac{(116 - 55) - (49 - 21)}{\ln\left[(116 - 55)/(49 - 21)\right]} = 42.38°\text{C}$$

The dimensionless ratios are calculated as

$$Z = \frac{116 - 49}{55 - 21} = 1.97$$

$$Y = \frac{55 - 21}{116 - 21} = 0.358$$

Referring to fig. 2(a), the correction factor $F_T = 0.75$.

Then

$$\Delta T_m = F_T \text{ (LMTD)}$$

$$= 0.75 \text{ (42.38)} = 31.78°C$$

Rearranging the heat flow equation to evaluate U_0 gives

$$U_0 = \frac{q}{A_0 \Delta T_m} = \frac{355,895}{(9.5)(31.78)} = 1178.6 \text{ W/m}^2\text{-°K}$$

b) Referring to fig. 2(b), a 2-4 exchanger gives a correction factor of $F_T = 0.95$. Then

$$\Delta T_m = 0.95 \text{ (42.38)} = 40.26°C$$

In conclusion, the 2-4 exchanger is more effective in utilizing the heating fluid temperature.

• PROBLEM 11-13

Two counterflow heat exchangers connected in series are used to heat 283.2 L^3/sec of air from 22°C to 30°C, using water at 35°C. Determine the effectiveness and the quantity UA for each exchanger.

Solution: The final effectiveness may be determined from the relation

$$\varepsilon_2 = \frac{T_{c2} - T_{c1}}{T_{h1} - T_{c1}}$$

since the cold fluid is minimum. Substituting into eq. (1) yields

$$\varepsilon_2 = \frac{30 - 22}{35 - 22} = 0.615$$

Now, the quantity UA may be calculated using

$$(UA)_T = C_{min} \text{ NTU} \qquad (2)$$

where the heat capacity ratio is

$$C_{min} = (\dot{m}C_P)_{air} \qquad (3)$$

and

$$\text{NTU} = \ell n \frac{1}{1 - t_2} \qquad (4)$$

The volume flow rate is given. The mass flow rate is found as

$$\dot{m} = Q\rho$$

At the mean air temperature of

$$T_m = \frac{22 + 30}{2} = 26°C$$

the air tables give values of

$\rho = 1.187\ Kg/m^3$ $\qquad\qquad$ $C_p = 1.01\ KJ/Kg\text{-}°C$

Substituting eq. (5) into eq. (3) and inserting the values gives

$$C_{min} = (Q\rho\, C_p)_{air}$$

$$= (283.2)(1.187)(1.01) = 339.5\ W/°C$$

From eq. (4),

$$NTU = \ln \frac{1}{1 - 0.615} = 0.955$$

Using eq. (2), the total UA is

$$(UA)_T = (339.5)(0.955) = 324\ W/°C$$

There are two heat exchangers. Then for each one, the value of UA is

$$UA = \frac{324}{2} = 162\ W/°C$$

If the value of the effectiveness for the first heat exchanger is desired, it may be found as

$$\varepsilon_1 = \frac{\varepsilon_2}{2 - \varepsilon_2}$$

$$= \frac{0.615}{2 - 0.615} = 0.444$$

• PROBLEM 11-14

A fluid flows at the rate of 20,000 lbm/hr and is comprised of 5% by weight sodium hydroxide in water. It is to be condensed to a 20% solution by a two-shell-pass parallel flow heat excahnger (see fig. 1). The solution enters the tubes at 80°F while saturated steam at 220°F enters the first shell, with a pressure of 1.275 psia preserved in the second shell. The overall heat transfer coefficients in the first and second shells are U_1 = 265 Btu/hr-ft^2-°F and U_2 = 200 Btu/hr-ft^2-°F, respectively. Assume equal heat transfer areas for each shell. Calculate the heat transfer area in each shell, the amount of steam required, and the amount of vapor leaving the second shell.

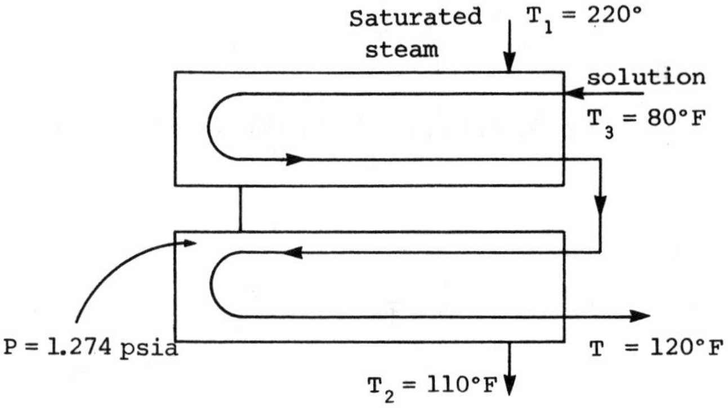

Fig. 1

Two-shell-pass parallel feed evaporator

Solution: All values used in this problem will be found either in the steam tables or in fig. 2, since one component is a sodium hydroxide solution.

At 1.275 psia, T_{sat} = 110°F. From fig. 2, for a 20% solution, the exit temperature must be 120°F. This corresponds to a pressure equal to that of the water. Now, the overall temperature difference is

$$\Delta T = 220 - 110 = 110°F$$

The concentration in the first shell is one-quarter of that in the second. Since

$$\Delta T_{boil\ 2} = 120 - 110 = 10°F,$$

then

$$\Delta T_{boil\ 1} = \frac{10}{4} \simeq 3°F$$

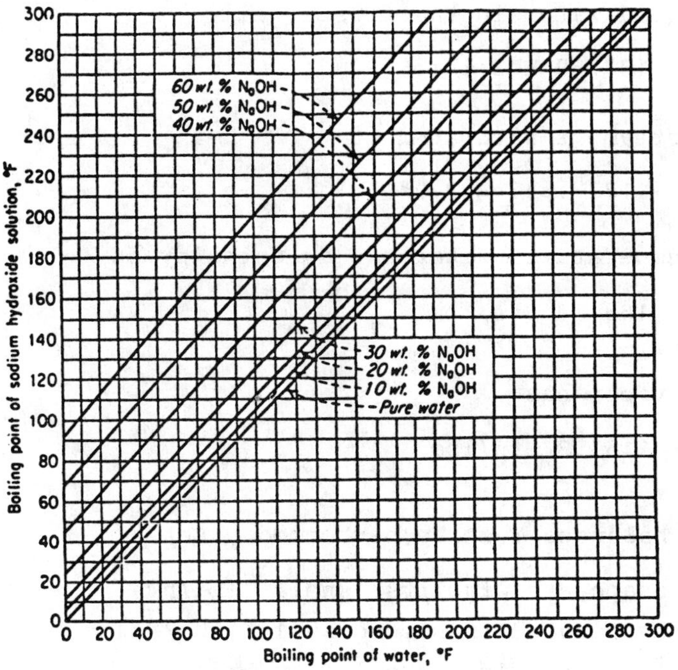

Fig. 2.

Dühring lines for aqueous solutions of sodium hydroxide.

The sum of the temperature differences is

$$\Delta T_1 + \Delta T_2 + 3 + 10 = 110°F$$

It will be assumed that

$$\Delta T_1 = 47°F \qquad \text{and } \Delta T_2 = 50°F$$

Now, all enthalpies will be found from the steam tables, corresponding to the given temperatures. For vapor in the second shell (considering a 45% quality with $\Delta T_{boil\ 2} = 10°F$),

$$h_{g_2} = 1108.8 + 0.45(10) = 1113.3 \text{ Btu/lbm}$$

In the first shell, the vapor is saturated at

$$T_{sat} = 120 + 50 = 170°F$$

and since $\Delta T_{boil\ 1} = 3°F$, the liquid and vapor in the first shell is at 173°F. For this value,

$$h_{g_1} = 1134 + 0.45(3) = 1135.35 \text{ Btu/lbm.}$$

For saturated water in the first shell,

$$h_f = 137.89 \text{ Btu/lbm.}$$

Here the steam is saturated at

$$T_{g1} = 173 + 47 = 220°F$$

At this temperature, water has an enthalpy of

$$h_f = 188.14 \text{ Btu/lbm}$$

while the vapor has

$$h_g = 1153.3 \text{ Btu/lbm}$$

Now, heat balances must be performed. Let

x = water evaporated in first shell (lbm/hr)

y = water evaporated in second shell (lbm/hr)

w = steam given to first shell (lbm/hr)

Energy into the first shell is

$$q_{st} + q_{sol} = wh_g + \dot{m}_{sol} \Delta T_{sol} [\text{conc}] \tag{1}$$

$$= w(1153.3) + (20{,}000)(80 - 32)(0.95)$$

$$= 1153.3w + 912{,}000$$

Since at 32°F, water has a zero energy value. The energy leaving the first shell is

$$q_{vap} + q_{sol} = xh_{g1} + \dot{m}_{sol} \Delta T_{sol}[\text{conc}] \tag{2}$$

$$= x(1135.35) + (20{,}000 - x)(173 - 32)(0.92)$$

$$= 1005.63x + 2{,}594{,}400$$

(if the mixture is concentrated to an 8% by weight solution).

Finally, the condensed steam contains energy as

$$q_{cond} = wh_f = w(188.14) \tag{3}$$

For the first shell, the energy balance will consist of eq. (1) set equal to the sum of eqs. (2) and (3), or

$$1153.3w + 912{,}000 = 1005.63x + 2{,}594{,}400 + 188.14w \tag{4}$$

Continuing on to the second shell, the energy flowing in

must equal that flowing out of the first shell. The energy out will be

$$q_{vap} + q_{sol} = yh_{g_2} + \dot{m}_{sol}\ \Delta T_{sol}[conc] \quad (5)$$

$$= y(1113.3) + (20{,}000 - x - y)(120 - 32)(0.8)$$

$$= 1042.9\,y - 70.4\,x + 1{,}408{,}000$$

and

$$q_{cond} = xh_f = x(137.89) \quad (6)$$

Performing the heat balance gives

$$1005.63x + 2{,}594{,}400 = 1042.9y + 67.49\,x + 1{,}408{,}000 \quad (7)$$

By stoichiometry, the total amount of water evaporated must be

$$x + y = 20{,}000 - \frac{(20{,}000)(5\%)}{20\%} \quad (8)$$

$$= 15{,}000$$

Equations (4), (7), and (8) may now be solved simultaneously for the 3 unknowns. The results are

$x = 7298$ lbm/hr,

$y = 7702$ lbm/hr,

and $w = 9347$ lbm/hr.

Now, the heat given off from the steam initially supplied to the first shell is

$$q_1 = w(h_g - h_f)_{220°F} = (9347)(1153.3 - 188.14)$$

$$= 9{,}021{,}351 \text{ Btu/hr}$$

The heat given off from the evaporated water in the second shell is

$$q_2 = x(h_g - h_f)_{173°F} = 7298(1135.35 - 137.89)$$

$$= 7{,}279{,}463 \text{ Btu/hr}$$

Since

$$q = UA\Delta T,$$

The corresponding heat transfer areas may be found as

$$A_1 = \frac{q_1}{U_1 \Delta T_1} = \frac{9,021,351}{(265)(47)}$$

$$= 724.3 \text{ ft}^2$$

and

$$A_2 = \frac{q_2}{U_2 \Delta T_2} = \frac{7,279,463}{(200)(50)}$$

$$= 727.95 \text{ ft}^2$$

To check the assumption of an 8% concentration, the sodium hydroxide in water leaving the first shell is

$$[NaOH] = \frac{(20,000)(5\%)}{20,000 - 7298}$$

$$= 7.87\%$$

This is an acceptable value. Since the heat transfer areas are almost equal, the temperature difference assumptions are also adequate.

DESIGN / PERFORMANCE

• PROBLEM 11-15

Steam at 230°F in the shell side of a multi-pass regenerator is used to heat oil flowing at 550 lbm/hr in 0.625 in OD steel pipes, 16-gage and 8 ft long. The oil is heated from 80 to 130°F. Assume that the thermal conductivity K and the specific heat C_p of the oil remains stable at 0.080 Btu/hr-ft-°F and 0.50 Btu/lbm-°F, respectively. The viscosity μ of the oil will differ with an increase in temperature as shown in the table. Determine the required number of passes to effectively heat the oil.

T,°F	68	80	90	100	110	120	130	140	220
μ,lbm/ft.hr	55.66	48.4	43.56	39.2	36.3	32.67	29.04	26.62	8.71

Solution: The Reynold number will be determined in relation to the highest bulk temperature, 130°F. According to the table, the viscosity of oil at 130°F is μ = 29.04 lbm/ft-hr. The wall thickness of a 16-gage pipe is 0.065 in. Therefore, the inside diameter of the 0.625 OD tube is

$$D_i = 0.625 - 2(0.065) = 0.495 \text{ in} = 0.4125 \text{ ft}$$

The Reynold number is calculated as

$$Re = \frac{DG}{\mu} = \frac{4\dot{m}}{\pi D_i \mu} = \frac{4(550)}{\pi(0.04125)(29.04)} = 585$$

Since Re < 2100, the following equation can be applied to all passes:

$$\frac{hD}{K_b}\left[\frac{\mu_s}{\mu}\right]^{0.14} = 1.86\left[\frac{4\dot{m}C_p}{\pi KL}\right]^{1/3} = 1.86\left[\left(\frac{DG}{\mu}\right)\left(\frac{C_p\mu}{K}\right)\left(\frac{D}{L}\right)\right]^{1/3} \qquad (1)$$

The dimensionless term $4\dot{m}C_p/\pi KL$ in eq. (1) is evaluated as

$$\frac{4\dot{m}C_p}{\pi KL} = \frac{4(550)(0.5)}{\pi(0.080)(8)} = 547$$

Eq. (1) will be used to calculate a convection heat transfer coefficient h for each successive pass.

A trial and error method is required for the solution.

To begin, assume that the oil exit temperature for the first pass T_1 = 95°F, and the shell side temperature has decreased to T_s = 220°F. Then the average temperature for the first pass is

$$T_{av} = \frac{T_i + T_1}{2} = \frac{80 + 95}{2} = 87.5\ {}^0F$$

The viscosity μ at T_{av} = 87.5°F is 44.77 lbm/ft-hr. Since the viscosity at 220°F is 8.71 lbm/ft-hr,

Then $\qquad (\mu/\mu_s)^{0.14} = 1.26$

Using eq. (1), h is calculated as

$$h = \frac{1.86K}{D_i}\left[\frac{\mu}{\mu_s}\right]^{0.14}\left[\frac{4\dot{m}C_p}{\pi KL}\right]^{1/3} = \frac{(1.86)(0.08)(1.26)(547)^{1/3}}{0.04125}$$

$$= 37.17 \text{ Btu/hr-ft}^2\text{-°F}$$

The difference between the shell side temperature and the average tube temperature is calculated as

$$\Delta T_{av} = T_s - T_{av} = 220 - 87.5°F = 132.5°F$$

The assumed temperature difference for the first pass

$$\Delta T_1 = T_1 - T_i = 95 - 80 = 15°F$$

may be checked using the following equation:

$$\Delta T_1 = \frac{hA_i\ T_{av}}{\dot{m}C_p} \tag{2}$$

ΔT_1 is calculated as

$$\Delta T_1 = \frac{(37.17)(\pi \times 0.0412 \times 8)(132.5)}{(550)(0.5)} = 18.5°F$$

which is adequately close to the assumed temperature difference of 15°F.

The assumed temperature difference on the shell side

$$\Delta T_{s1} = 230 - 220 = 10°F$$

may be checked using the following equation:

$$\Delta T_{s1} = q\Sigma R \tag{3}$$

where

$$q = \dot{m}C_p\Delta T_1 = (550)(0.5)(18.5) = 5087.5 \text{ Btu/hr}$$

Assuming values of

$h = 1000$ Btu/hr-ft²-°F and $k = 26$ Btu/hr-ft-°F
for the sealing on the pipe wall and the pipe wall itself.

The resistances are calculated as

$$R_{\text{steam side}} = \frac{1}{h_s A_0} = \frac{1}{(1000)(\pi \times 0.625/12 \times 8)} =$$

$$= 0.000764 \text{ hr-°F/Btu}$$

$$R_{\text{scale}} = \frac{\Delta x}{KA_i} = \frac{(0.065/12)}{(26)(\pi \times 0.04125 \times 8)} = 0.000201 \text{ hr-°F/Btu}$$

Then

$$\Sigma R = R_{\text{steam side}} + R_{\text{scale}} = 0.000764 + 0.000201$$

$$= 0.000965 \text{ hr-°F/Btu}$$

Inserting the values into eq. (3),

ΔT_{s1} = (5087.5)(0.000965) = 6°F, which is also close to the the assumed value of 10°F.

From the results of the first calculation, the oil will enter the second pass at 98.5°F and exit at a new temperature T_2, which will be used in the next calculation. The method will be continued until the exit temperature of the oil from the last pass equals or surpass the desired temperature of 130°F. The number of passes needed will then be known. Following the procedure outlined above, the number of passes required is found to be 5.

• PROBLEM 11-16

> A feedwater heater supplying hot water at 15,000 lbm/hr and 200°F on the shell side is used to heat water coming in at 30,000 lbm/hr from 100°F to 130°F. The diameter of each pipe is 0.75 in. and the mean velocity is 1.2 ft/sec. The pipe length must not exceed 8 ft. due to the limitations of space. If the overall heat transfer coefficient is 250 Btu/hr-ft²-°F and the exchanger is of the one-shell-pass type, determine the required number of passes, the number of pipes per pass, and the individual pipe length.

Solution To begin the analysis, first assume that it is a one-pipe-pass exchanger, and see if it meets the given requirements. The heat flow equation is

$$q = \dot{m}_c C_{pc} \Delta T_c = \dot{m}_h C_{ph} \Delta T_h \qquad (1)$$

Assuming that $C_{pc} = C_{ph}$ = 1.0 Btu/lbm-°F, the temperature difference of the hot water is

$$\Delta T_h = \frac{\dot{m}_c C_{pc} \Delta T_c}{\dot{m}_h C_{ph}}$$

$$= \frac{(30,000)(1.0)(130 - 100)}{(15,000)(1.0)} = 60°F$$

Therefore, the hot water outlet temperature is

$$T_{h,out} = 200 - 60 = 140°F$$

From eq. (1), the total heat absorbed by the cold water is

$$q = (30,000)(1.0)(130 - 100) = 900,000 \text{ Btu/hr}$$

For a counterflow heat exchanger, the LMTD is found as

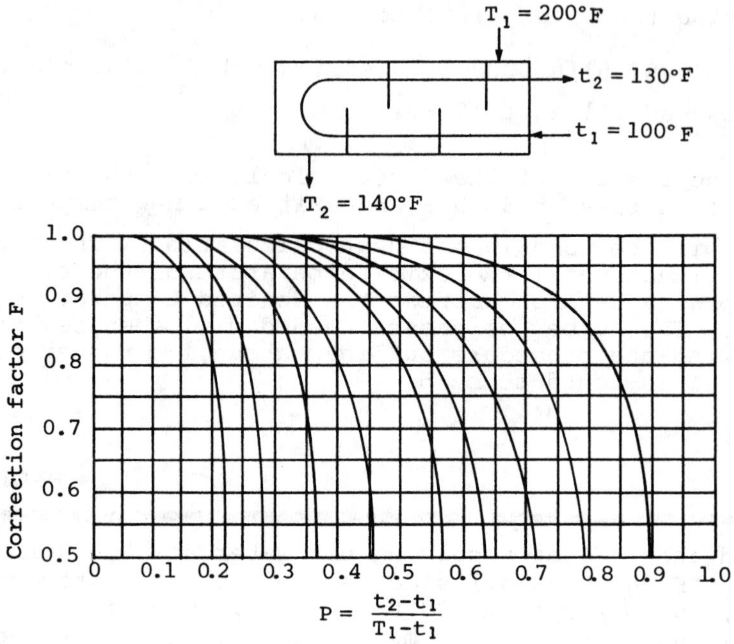

Correction-factor plot for exchanger with one shell pass and two, four, or any multiple of tube passes.

$$\text{LMTD} = \frac{(T_{h,in} - T_{c,out}) - (T_{h,out} - T_{c,in})}{\ln\left[(T_{h,in} - T_{c,out})/(T_{h,out} - T_{c,in})\right]} \tag{2}$$

$$= \frac{(200 - 130) - (140 - 100)}{\ln[(200 - 130)/(140 - 100)]} = 53.6°\text{F}.$$

Using this, the area of a one-pipe-pass feedwater heater is then calculated as

$$A_c = \frac{q}{U\,(\text{LMTD})}$$

$$= \frac{900{,}000}{250(53.6)} = 67.16\ \text{ft}^2$$

The total flow area of the pipes may be calculated from the expression

$$\dot{m}_c = \rho v A_t$$

If $\rho = 62.4\ \text{lbm/ft}^3$,

then $A_t = \dfrac{\dot{m}}{\rho v}$

$$= \frac{30,000}{(62.4)(1.2)(3600)} = 0.111 \text{ ft}^2$$

The number of pipes needed would simply be the total flow area divided by the flow area per pipe, or

$$n = \frac{A_t}{\frac{\pi}{4} d^2}$$

$$= \frac{0.111}{\frac{\pi}{4}\left(\frac{0.75}{12}\right)^2} = 36.2$$

so the number of pipes required is 36.

The length of this one-pipe-pass heater may be calculated as the total heat transfer area divided by the product of the number of pipes and the outside area per pipe per ft, or

$$L = \frac{A_c}{n \pi d}$$

$$= \frac{67.16}{(36)(\pi)\left(\frac{0.75}{12}\right)} = 9.5 \text{ ft}$$

Since this is longer than the specified length (8 ft.), a multi-pass heater must be used. This will require the use of a correction factor f (see fig. 1) to decrease the LMTD, thus increasing the total heat transfer area. Using fig. 1 and assuming a two-pipe-pass heater, the dimensionless ratios are:

$$R = \frac{T_1 - T_2}{t_2 - t_1}$$

$$= \frac{200 - 140}{130 - 100} = 2.0$$

and

$$P = \frac{t_2 - t_1}{T_1 - t_1}$$

$$= \frac{130 - 100}{200 - 100} = 0.3$$

Then the correction factor is F = 0.88. The increased heat transfer area is then calculated as

$$A_c = \frac{q}{UF\,(LMTD)}$$

$$= \frac{900,000}{(250)(0.88)(53.6)} = 76.32 \text{ ft}^2$$

The number of pipes per pass will remain at the original

value of 36, as the velocity remains the same. The individual pipe length is now

$$L_c = \frac{A_c}{2n\pi d}$$

$$= \frac{76.32}{2(36)(\pi)\left[\frac{0.75}{12}\right]} = 5.4 \text{ ft.}$$

As this is within the tolerance range, the acceptable design is a one-shell-pass, two-pipe-pass heater with 36 tubes of 5.4 ft. length per pass.

• PROBLEM 11-17

A shell-and-tube heat exchanger is to be constructed for the purpose of heating air in the tubes from 60°F to 160°F. This will be accomplished by the condensation of steam at 220°F (see Fig. 1). The air flows at the rate of 50,000 lbm/hr, with a mass velocity of 7000 lbm/hr-ft². If the tubes are steel, with a 2 in. inside diameter, calculate (a) the required number of tubes, (b) the effective heat transfer area, and (c) the length of each pipe. Also, find (d) the pressure drop in the tube, assuming that the air enters at atmospheric pressure.

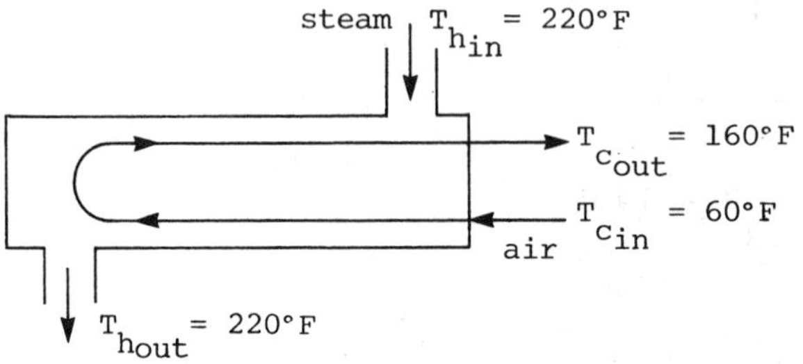

Fig. 1.

Solution: The airflow must be analyzed, so first the properties must be evaluated. The mean bulk temperature is

$$T_m = \frac{60 + 160}{2} = 110°\text{F}.$$

From the air tables,

$C_p = 0.24$ Btu/lbm-°F $\qquad$ $k = 0.016$ Btu/hr-ft-°F

$\mu = 0.047$ lbm/ft-hr $\qquad$ $Pr = 0.71$

(a) The number of tubes necessary can be determined as

$$nA_i = \frac{\dot{m}_a}{G_a} \tag{1}$$

where A_i = cross-sectional flow area of the tube = $\frac{\pi d_i^2}{4}$

$$= \frac{\pi(2/12)^2}{4} = 0.0218 \text{ ft}^2$$

Solving for n gives

$$n = \frac{\dot{m}_a}{A_i G_a} = \frac{(50,000)}{(0.0218)(7000)} = 327.6 \text{ or } 328 \text{ tubes.}$$

(b) The Reynolds number must be calculated.

$$Re = \frac{G_a d_i}{\mu} = \frac{(7000)(2/12)}{(0.047)} = 24,823$$

This is a turbulent flow.

Therefore, the following relation is recommended:

$$Nu = \frac{hd}{k} = 0.023\ Re^{0.8}\ Pr^n \tag{2}$$

The thermal properties in eq. (2) are determined at the fluid bulk temperature, and the exponent n may be

$$n = \begin{cases} 0.4 \text{ for heating} \\ 0.3 \text{ for cooling} \end{cases}$$

In this instance, n will take on the value n = 0.4.

Using eq. (2), the convective heat transfer coefficient of the air flow is calculated as

$$h_{air} = \frac{0.023\ k(Re)^{0.8}(Pr)^{0.4}}{d_i} = \frac{0.023(0.016)(24,823)^{0.8}(0.71)^{0.4}}{(2/12)}$$

$$= 6.32 \text{ Btu/hr-ft}^2\text{-}^0\text{F}$$

Since the resistances of the pipe wall and the steam are small compared to the air resistance, they will be neglected. The overall heat transfer coefficient U_i is then approximately equal to the convection heat transfer coefficient of the air, h_{air}.

Assuming that the inlet and outlet temperatures of the steam are equal ($T_{h\,in} = T_{h\,out}$), the log mean temperature difference (LMTD) for counterflow conditions may be calculated as

$$\text{LMTD} = \frac{(T_{h\ in} - T_{c\ out}) - (T_{h\ out} - T_{c\ in})}{\ln\left[(T_{h\ in} - T_{c\ out})/(T_{h\ out} - T_{c\ in})\right]}$$

$$= \frac{(220 - 160) - (220 - 60)}{\ln[(220 - 160)/(220 - 60)} = 102°\text{F}.$$

From the heat flow equation

$$q = A_i U_i\ (\text{LMTD}),$$

The effective heat transfer area based on the inside surface is

$$A_i = \frac{q}{U_i\ (\text{LMTD})} \tag{3}$$

Now, q is the heat absorbed by the air, and is calculated as

$$q = \dot{m}_a C_P\ (T_{c\ out} - T_{c\ in}) = (50{,}000)(0.24)(160 - 60)$$

$$= 1.2 \times 10^6\ \text{Btu/hr}$$

Using eq. (3),

$$A_i = \frac{1.2 \times 10^6}{(6.32)(102)} = 1861.5\ \text{ft}^2$$

(c) The length of each pipe is found from the relation

$$L = \frac{A_i}{n\pi d_i} = \frac{1861.5}{(328)(\pi)(2/12)} = 10.84\ \text{ft}$$

(d) The densities of air at 60°F and 160°F are 0.0763 lbm/ft^3 and 0.064 lbm/ft^3, respectively. The density at the mean bulk temperature 110°F is 0.0696 lbm/ft^3.

For the range $5{,}000 < Re < 30{,}000$, the friction factor is given as

$$f = \frac{0.079}{Re^{0.25}} = \frac{0.079}{(24{,}823)^{0.25}} = 0.0063$$

The equation for the pressure drop in a horizontal tube is

$$P_1 - P_2 = \frac{G^2}{\alpha g}\left[\left(\frac{1}{\rho_2} - \frac{1}{\rho_1}\right) + \frac{4fL}{2d\rho f}\right] \tag{4}$$

where the void fraction α is equal to 1 for two flows having almost the same velocity. Assuming that this is true and substituting the values into eq. (4),

$$P_1 - P_2 = \left(\frac{7000}{3600}\right)^2 \left(\frac{1}{32.2}\right) \left[\left(\frac{1}{0.064}\right) - \left(\frac{1}{0.0763}\right) + \frac{4(0.0063)(10.84)}{2(2/12)(0.0696)}\right]$$

$$= 1.678 \text{ psfa}$$

The pressure drop can be modified by assuming the cross-sectional area of the entrance and exit regions to be twice the cross-sectional area of the pipes, or

$$\frac{A_1}{A_0} = \frac{A_2}{A_3} = 0.5$$

The velocities of the airflow at sections 0, 1, 2, and 3 are calculated as follows:

$$V_1 = \frac{G_a}{\rho_1} = \frac{7000/3600}{0.0763} = 25.48 \text{ ft/sec}$$

$$V_2 = \frac{G_a}{\rho_2} = \frac{7000/3600}{0.064} = 30.38 \text{ ft/sec}$$

$$V_0 = 0.5V_1 = 0.5(25.48) = 12.74 \text{ ft/sec}$$

$$V_3 = 0.5V_2 = 0.5(30.38) = 15.19 \text{ ft/sec}$$

Referring to fig. 2, $K_e = 0.21$, and $K_c = 0.31$.

The pressure drop equations for non-uniform cross-sectional tubes are

$$\frac{P_0 - P_1}{\rho_{01}} + \frac{V_0^2 - V_1^2}{2g} = K_c \frac{V_1^2}{2g} \tag{5}$$

and

$$\frac{P_2 - P_3}{\rho_{23}} + \frac{V_2^2 - V_3^2}{2g} = K_e \frac{V_2^2}{2g} \tag{6}$$

The contraction and enlargement is shown in fig. 3.

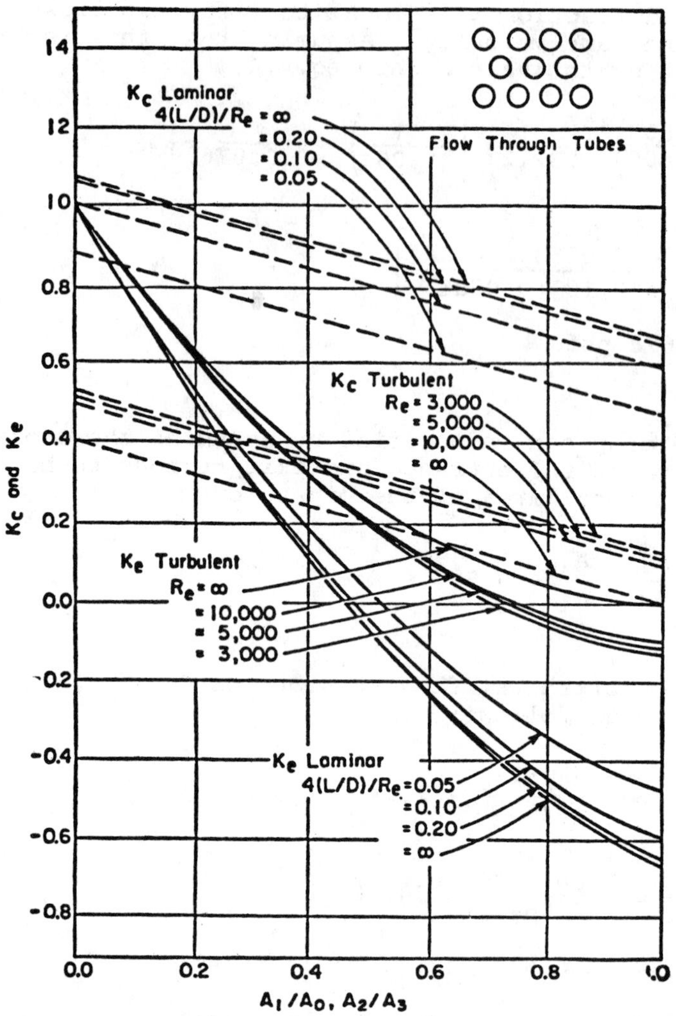

Fig. 2.

Values of K_c and K_e for a tube bundle (Kays (14b)).

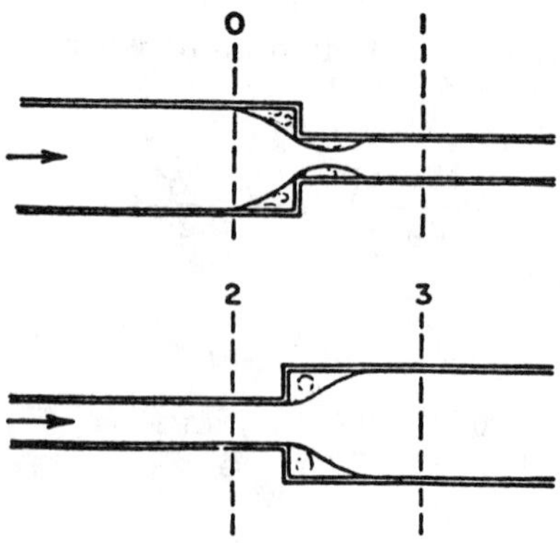

Fig. 3.

Sudden contraction and enlargement

Rearranging eqs. (5) and (6) and inserting the corresponding values gives

$$P_0 - P_1 = \frac{\rho_{01}}{2g}\left[K_C V_1^2 + V_1^2 - V_0^2\right]$$

$$= \frac{0.0763}{2(32.2)}\left[0.31(25.48)^2 + (25.48)^2 - (12.74)^2\right]$$

$$= 0.815 \text{ psfa}$$

and

$$P_2 - P_3 = \frac{\rho_{23}}{2g}\left[K_e V_2^2 - V_2^2 + V_3^2\right]$$

$$= \frac{0.064}{2(32.2)}\left[0.21(30.38)^2 - (30.38)^2 + (15.12)^2\right]$$

$$= -0.5 \text{ psfa}$$

Hence,

$$P_0 - P_3 = (P_1 - P_2) + (P_0 - P_1) + (P_2 - P_3)$$

$$= 1.678 + 0.815 - 0.5 = 1.993 \text{ psfa.}$$

• PROBLEM 11-18

A corrosive fluid flowing at 3000 gal/hr inside the tubes of a one-shell-pass, two-tube-pass intercooler is to be cooled from 250°F to 180°F by water flowing at 5000 gal/hr. The heat exchanger is constructed of a 20 in. inside diameter shell with 158 stainless steel tubes of 1 in. outside diameter, 0.834 in. inside diameter, and 16 ft. length, which are arranged on a 1.25 in. square pitch. The baffle spacing is 6 in. Determine if the intercooler specifications are adequate to meet the cooling demand.

Solution: The problem can be approached by calculating the required heat exchanger area based on the given data, and then comparing the value with the actual surface area available.

The mass flow rate of the corrosive fluid is

$$\dot{m}_{fluid} = (3000)(8.34 \text{ lbm/gal}) = 25020 \text{ lbm/hr}$$

If the corrosive fluid has the same specific heat as water (C_p = 1.0 Btu/lbm.°F), the total heat transferred from the fluid to the water is

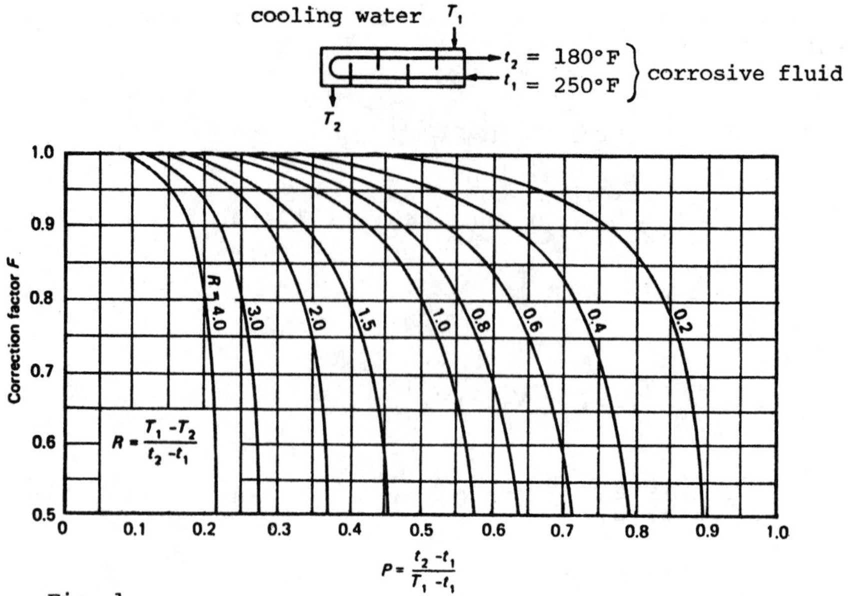

Fig. 1.

Correction-factor plot for exchanger with one shell pass and two, four, or any multiple of tube passes.

$$q = \dot{m}_{fluid} C_P (t_2 - t_1) = (25020)(1.0)(250 - 180)$$

$$= 1{,}751{,}400 \text{ Btu/hr}$$

Accordingly, the rise in temperature of the water is

$$(T_2 - T_1) = \frac{q}{\dot{m}_{water}} = \frac{1{,}751{,}400}{(5000)(8.34)} = 42°F$$

Assuming that the water inlet temperature $T_1 = 80°F$, the outlet temperature is $T_2 = \Delta T + T_1 = 42 + 80 = 122°F$. Since the inlet and outlet temperatures are known, the log mean temperature difference (LMTD) of the intercooler may be calculated directly.

$$(\text{LMTD})_{\substack{\text{counterflow}\\ \text{exchanger}}} = \frac{(t_1 - T_2) - (t_2 - T_1)}{\ln\left[(t_1 - T_2)/(t_2 - T_1)\right]}$$

$$= \frac{(250 - 122) - (180 - 80)}{\ln\left[(250 - 122)/(180 - 80)\right]}$$

$$= 113.4°F$$

Referring to fig. 1, the dimensionless ratios are found to be

$$P = \frac{t_2 - t_1}{T_1 - t_1} = \frac{180 - 250}{80 - 250} = 0.412$$

and,

$$R = \frac{T_1 - T_2}{t_2 - t_1} = \frac{80 - 122}{180 - 250} = 0.6$$

which gives a correction factor of 0.97.

The LMTD then becomes

$$(\text{LMTD})_{\text{corr}} = F(\text{LMTD}) = (0.97)(113.4) = 110°\text{F}$$

The convective heat transfer coefficient for the pipe flow will be calculated using the following procedure:

Cross-sectional area per pipe $A_i = \frac{\pi d_i^2}{4}$

$$= \frac{\pi(0.834/12)^2}{4} = 0.00379 \text{ ft}^2$$

total flow area of pipe $A_T = \frac{(\text{no. tubes})A_i}{\text{no. passes}}$

$$= \frac{(158)(0.00379)}{2} = 0.3 \text{ ft}^2$$

mass velocity of fluid $G = \frac{\dot{m}}{A_T} = \frac{25020}{0.3} = 83400 \text{ lbm/hr-ft}^2$

The thermal properties of the corrosive fluid at

$t_{av} = \frac{250 + 180}{2} = 215°\text{F}$ are:

$\mu = 0.605$ lbm/hr.ft $\qquad C_p = 1.0$ Btu/lbm-°F

$k = 0.402$ Btu/hr.ft.°F

The Reynolds and Prandtl number are calculated as

$$Re = \frac{d_i G_{tube}}{\mu} = \frac{(0.834/12)(83{,}400)}{0.605} = 9581$$

$$Pr = \frac{\mu C_p}{k} = \frac{(1.0)(0.605)}{0.402} = 1.5, \text{ respectively.}$$

From the Reynolds number, the corrosive fluid flow is seen to be turbulent.

The length-to-diameter ratio of the tube is

$$L/D = \frac{16}{(0.834/12)} = 230.$$

Referring to fig. 2, the heat transfer factor is $j = 0.0038$.

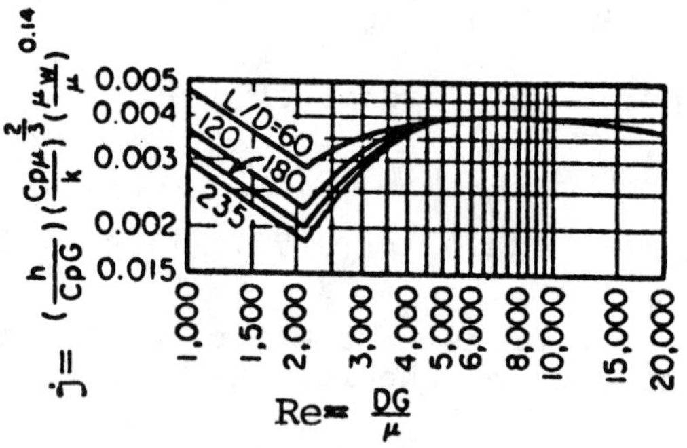

Fig. 2. Heat transfer factors for fluids flowing inside tubes.

The convective heat transfer coefficient of the tube side may be determined as

$$j = \left[\frac{h}{C_pG}\right] (Pr)^{2/3} \left[\frac{\mu_{wall}}{\mu}\right]^{0.14} \tag{1}$$

Assuming that $(\mu_{wall}/\mu)^{0.14} = 1.0$ for the first calculation and solving eq. (1) for h_i gives

$$h_i = \frac{jC_pG_{tube}}{(Pr)^{2/3}\left[\frac{\mu_{wall}}{\mu}\right]^{0.14}}$$

$$h_i = \frac{(0.0038)(1.0)(83,400)}{(1.5)^{2/3}(1.0)} = 242 \text{ Btu/hr-ft}^2\text{-°F}$$

Now, the convective heat transfer coefficient for the shell side will be determined as follows:

mass flow rate of water $\dot{m}_{water} = (5000)(8.34) = 41,700$ lbm/hr

For 1.0 in. O.D. tubes on a 1.25 in. square pitch, a clearance of 0.25 in. is given for the water flow. The number of pipes possible in the middle of the heat exchanger is calculated by dividing the shell diameter by the pipe pitch and subtracting one from the resulting value, or

$$\text{no. middle tubes} = \frac{20}{1.25} - 1 = 15$$

If the length of the shell side flow area is considered to be equal to the baffle spacing, the minimum flow area for the shell side fluid is calculated as

$$A_{shell} = (\text{no. middle tubes})(\text{clearance})(\text{baffle spacing})$$

$$= (15)(0.25/12)(6.0/12) = 0.15625 \text{ ft}^2$$

Then mass velocity

$$G_{water} = \frac{41,700}{0.15625} = 266,880 \text{ lbm/hr.ft}^2$$

The thermal properties of water at $T_{av} = \frac{80 + 122}{2} = 101°F$ are:

$\mu = 1.645$ lbm/hr.ft $\qquad C_p = 1.0$ Btu/lbm-°F

$k = 0.35$ Btu/hr.ft.°F

The Reynold and Prandtl number for the water flow are calculated as

$$Re = \frac{d_0 G_{water}}{\mu} = \frac{(1.0/12)(266,880)}{1.645} = 13520$$

$$Pr = \frac{(1.0)(1.645)}{0.35} = 4.7, \text{ respectively.}$$

From the Reynolds number, the shell side flow is also found to be turbulent.

Assuming again that $(\mu_{wall}/\mu)^{0.14} = 1.0$ and referring to fig. 3, the heat transfer factor is $j = 0.0085$.

Solving eq. (1) for h_0 gives

$$h_0 = \frac{(0.0085)(1.0)(266,880)}{(4.7)^{2/3}(1.0)} = 808 \text{ Btu/hr.ft}^2.°F$$

Assuming that the cooling is only 60% effective,

$$h_0 = (0.60)(808) = 484.8 \text{ Btu/hr-ft}^2\text{-°F}$$

Neglecting the fouling factor, the resistances of the exchanger are calculated as

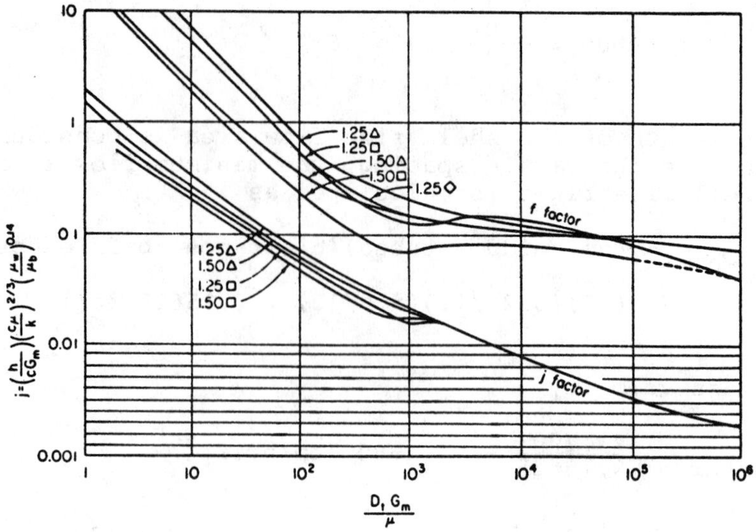

Fig. 3.

Heat transfer and friction factors for cross flow in tube bundles.

$$R_{\substack{\text{corrosive}\\\text{fluid}}} = \frac{1}{h_i} = \frac{1}{242} = 0.004132 \text{ hr-°F/Btu}$$

$$R_{\substack{\text{tube}\\\text{wall}}} = \frac{\Delta X A_{ci}}{k_{tube}\bar{A}}$$

where ΔX = pipe thickness = 0.00691 ft

A_{ci} = inside surface area of tube = $\pi d_i L = \pi(0.834/12)(16)$

$= 3.5 \text{ ft}^2$

A_{co} = outside surface area of tube $= \pi d_0 L = \pi(1.0/12)(16)$

$= 4.189 \text{ ft}^2$

$\bar{A}$ = the average of the inside and outside surface area of the tube

$$= \frac{A_{ci} + A_{co}}{2} = \frac{3.5 + \pi(1.0/12)(16)}{2} = 3.84 \text{ ft}^2,$$

and k_{wall} = conductivity of pipe = 9.4 Btu/hr-ft-°F.

Therefore,

$$R_{\text{tube wall}} = \frac{(0.00691)(3.5)}{(9.4)(3.84)} = 0.00067 \text{ hr-°F/Btu}.$$

Also,

$$R_{water} = \frac{A_{ci}}{h_0 A_{co}} = \frac{3.5}{(484.8\ \text{Btu/hr-ft}^2\text{-°F})(4.189\ \text{ft}^2)}$$

$$= 0.001723\ \text{hr-°F/Btu}$$

The sum of the resistances is $\Sigma R = R_{\substack{corrosive\\fluid}} + R_{\substack{tube\\wall}} + R_{water}$

$$= \frac{1}{U_i} = 0.006525\ \text{hr-°F/Btu}$$

Then U_i (the overall heat transfer coefficient based on the inside surface area of the tubes) is 153.26 Btu/hr-°F-ft^2.

With the overall heat transfer coefficient, the surface convection coefficients may be corrected for $(\mu_{wall}/\mu)^{0.14}$ not equal to one. The outside pipe surface temperature is determined by establishing the ratios of the resistances and the driving forces based on the coefficients evaluated in the first calculation, or

$$\frac{R_{water}}{\Sigma R} = \frac{\Delta T_{tube\ outside}}{t_{av} - T_{av}}$$

Rearranging,

$$\Delta T_{tube\ outside} = \frac{R_{water}(t_{av} - T_{av})}{\Sigma R}$$

$$= \frac{(0.001723)(215 - 101)}{0.006525} = 30.1°F$$

Hence, the mean temperature of the pipe outside surface is

$$T_{tube\ outside} = T_{av} + \Delta T_{tube\ outside}$$

$$= 101 + 30.1°F = 131.1°F.$$

At 131.1°F, μ_{wall} = 1.21 Btu/hr.ft.

Accordingly,

$$\frac{\mu_{wall}}{\mu}^{0.14} = \frac{1.21}{1.645}^{0.14} = 0.958$$

Correcting the shell side coefficient,

$$(h_0)_{corr} = \frac{h_0}{0.958} = \frac{484.8}{0.958} = 506.1\ \text{Btu/hr-ft}^2\text{-°F}.$$

The same procedures are used to correct the pipe-flow coefficient.

$$\Delta T_{\text{tube inside}} = \frac{R_{\text{corrosive}}(t_{av} - T_{av})}{\Sigma R} = \frac{(0.004132)(114)}{0.006525}$$

$$= 72.2°F$$

and $T_{\text{tube inside}} = 215 - 72.2°F = 142.8°F$, corresponding to

$$\mu_{\text{wall}} = 1.09 \text{ Btu/hr-ft}$$

Accordingly,

$$\left[\frac{\mu_{\text{wall}}}{\mu}\right]^{0.14} = \left[\frac{1.09}{0.605}\right]^{0.14} = 1.086$$

Correcting the tube side coefficient,

$$(h_i)_{\text{corr}} = \frac{h_i}{1.086} = \frac{242}{1.086} = 222.8 \text{ Btu/hr-ft}^2\text{-°F}.$$

The resistances are recalculated as

$$R_{\substack{\text{corrosive}\\\text{fluid}}} = \frac{1}{222.8} = 0.004488 \text{ hr-°F/Btu}$$

$$R_{\text{tube wall}} = 0.00067 \text{ hr-°F/Btu}$$

and $$R_{\text{water}} = \frac{3.5}{(506.1)(4.189)} = 0.001651 \text{ hr-°F/Btu}$$

Then $$\Sigma R_{\text{corr}} = 0.00681 \text{ hr-°F/Btu}$$

which gives

$$(U_i)_{\text{corr}} = 146.9 \text{ Btu/hr-ft}^2\text{°F}$$

Using the corrected U_i and neglecting the fouling factor, the necessary heat exchanger area is:

$$A_{\text{exchanger}} = \frac{q}{U_i F\,(LMTD)} = \frac{1{,}751{,}400}{(146.9)(110)} = 108.4 \text{ ft}^2.$$

The heat transfer area readily available in the exchanger is:

$$A_{\text{actual}} = (\text{no. tubes})L\pi d_i = (158)(16)(\pi)(0.834/12) = 552 \text{ ft}^2.$$

Therefore, the necessary area is significantly less than the actual area, so that even if the fouling factor is included, the heat exchanger can still execute the specified cooling.

The maximum tolerable scale resistance can be determined using the available surface area as

$$\frac{1}{U_{i\ scale}} = \frac{A_{actual}F(LMTD)}{q} = \frac{(552)(110)}{1,751,400}$$

$$= 0.03467 \text{ hr-ft}^2\text{-}^0\text{F/Btu}$$

Hence,

$$R_{scale} = \frac{1}{U_{i\ scale}} - \frac{1}{U_{i\ clean}} = 0.03467 - 0.00681$$

$$= 0.02786 \text{ hr-}^\circ\text{F/Btu}$$

• PROBLEM 11-19

A heat exchanger that is used for cooling oil is constructed with a bank of 200 tubes of outside diameter $d_0 = 0.5$ in, inside diameter $d_i = 0.428$ in, and length L = 4 ft. The tubes are enclosed inside a shell of inside diameter $D_i = 10$ in. Water at 65°F is flowing at a rate of 10,000 gal/hr inside the tubes, and oil at 190°F is flowing at 1,500 gal/hr in the shell side. Determine the exit temperature of the oil.

Solution: In the tube flow, the thermal properties of the fluid are obtained at the mean bulk temperature. If this is assumed to be 65°F, the thermal properties will be:

$$k = 0.346 \text{ Btu/hr-ft-}^\circ\text{F} \qquad C_p = 0.998 \text{ Btu/lbm-}^\circ\text{F}$$

$$\mu = 2.54 \text{ lbm/ft-hr}$$

A small correction may be necessary later if the water temperature increases significantly.

The total cross-sectional area of the water flow is

$$A_i = (\text{no. tubes})\ \frac{\pi d_i^2}{4} = \frac{(200)(\pi)(0.428/12)^2}{4} = 0.2 \ \text{ft}^2$$

Using the conversion 1 gallon of water = 10 lbm, the Reynold number is calculated as

$$Re = \frac{\dot{m} d_i}{A_i \mu}$$

$$= \frac{(10{,}000)(10)(0.428/12)}{(0.2)(2.54)} = 7021$$

The Prandtl number is

$$Pr = \frac{\mu C_p}{k}$$

$$= \frac{(2.54)(0.998)}{(0.346)} = 7.33$$

The water flow is found to be turbulent, and the convection heat transfer coefficient h_{water} can be determined by the relation

$$Nu = \frac{hd}{k} = 0.023\ Re^{0.8}\ Pr^{0.4} \qquad (1)$$

Rearranging and solving for h gives

$$h_{water} = \frac{0.023 Re^{0.8} Pr^{0.4} k}{d_i} = \frac{0.023(7021)^{0.8}\ (7.33)^{0.4}\ (0.346)}{(0.428/12)}$$

$$= 591\ Btu/hr\text{-}ft^2\text{-}°F$$

or it can be based on the outer surface of the tube, giving

$$h_{water} = h(d_i/d_0)$$

$$= (591)(0.428/0.5) = 506\ Btu/hr\text{-}ft^2\text{-}°F$$

Assuming an oil exit temperature of 170°F, the mean bulk temperature of the oil is $T_b = \frac{170 + 190}{2} = 180°F$. At this temperature, the thermal properties are

$k = 0.1$ Btu/hr-ft-°F $\qquad C_p = 0.509$ Btu/lbm-°F

$\mu = 65$ lbm/ft-hr $\qquad$ sp. gr. = 0.93

The total cross-sectional area of the oil flow is

$$A_{oil} = \frac{\pi}{4}\left[D_i^2 - (\text{no. tubes}) d_0^2\right] = \frac{\pi}{4}\left[(10/12)^2 - (200)(0.5/12)^2\right]$$

$$= 0.273\ ft^2$$

Since the shell is not really of circular cross section, it is recommended that the heat-transfer correlations be based on the hydraulic diameter D_H, defined as

$$D_H = \frac{4 \text{ x cross-sectional area}}{\text{wetted perimeter}} = \frac{4A_{oil}}{P}$$

where $P = \pi D_i + (\text{no. tubes})\pi d_0 = (\pi)(10/12) + (200)(\pi)(0.5/12)$

$= 28.8$ ft

Then $D_H = \frac{4(0.273)}{(28.8)} = 0.0379$ ft

The Reynolds number for the oil flow is calculated as

$$Re_{oil} = \frac{(1500)(10)(0.93)(0.0379)}{(0.273)(65)} = 29.8$$

and the Prandtl number is

$$Pr = \frac{(65)(0.509)}{0.1} = 330.85$$

Since the oil flow is seen to be laminar, the convection heat transfer coefficient h_{oil} can be calculated from the following relation:

$$Nu = \frac{hd}{k} = 1.86(Re\ Pr)^{1/3} \left[\frac{d}{L}\right]^{1/3} \left[\frac{\mu}{\mu_w}\right]^{0.14} \tag{2}$$

The thermal properties in eq (2) are evaluated at the average bulk temperature of the fluid, except for μ_w, which is obtained at the wall temperature.

The thermal resistances of the water and the tube wall can be neglected because they are small compared with the resistance of the oil.

For an assumed mean wall temperature of $T_w = 67°F$,

$$\mu_w = 1{,}800 \text{ lb/ft-hr}$$

Rearranging and solving eq. (2) for h gives

$$h_{oil} = 1.86 \frac{k}{d} (Re\ Pr)^{1/3} \left[\frac{d}{L}\right]^{1/3} \left[\frac{\mu}{\mu_w}\right]^{0.14}$$

$$= 1.86 \frac{0.1}{0.0379} [(29.8)(330.85)]^{1/3} \left[\frac{0.0378}{4}\right]^{1/3} \left[\frac{65}{1800}\right]^{0.14}$$

$$= 13.98 \text{ Btu/hr-ft}^2\text{-°F}$$

The overall heat transfer coefficient U is calculated as

$$U = \frac{1}{\frac{1}{h_{water}} + \frac{1}{h_{oil}}} = \frac{1}{\frac{1}{506} + \frac{1}{13.98}} = 13.6 \text{ Btu/hr-ft}^2\text{-°F}$$

The average temperature difference of the oil to water is:

$$\Delta T_{av} = (180 - 65) = 115°\text{F}$$

The total heat transfer area based on the tube outside diameter is:

$$A_{Total} = (\text{no. tubes})\pi d_0 L = (200)(\pi)(0.5/12)(4) = 104.7 \text{ ft}^2$$

Therefore, the total heat transferred is

$$q = UA\Delta T_{av} = (13.6)(104.7)(115) = 163{,}751 \text{ Btu/hr}$$

The temperature drop of the oil is calculated as

$$\Delta T_{oil} = \frac{q}{\dot{m}_{oil} C_P} = \frac{163{,}751}{(1500)(10)(0.93)(0.509)} = 23°\text{F}$$

instead of the assumed value of 20°F.

The exit temperature of the oil is, then,

$$T_{h\ out} = T_{h\ in} - \Delta T_{oil} = 190 - 23 = 167°\text{F}$$

The assumed average temperature of the water and tube wall can be verified as follows:

$$\Delta T_{water} = \frac{q}{\dot{m}_{water} C_P} = \frac{163{,}751}{(10{,}000)(10)(0.998)}$$

$$= 1.64°\text{F}$$

Therefore, the actual mean water temperature is

$$T_{water,\ mean} = \frac{65 + (65 + 1.64)}{2} = 65.8\ °\text{F}$$

Since this is close to the assumed value of 65°F, no correction is required.

The average oil temperature is

$$T_{oil,\ mean} = \frac{190 + 167}{2} = 178.5°\text{F}$$

Equating the heat flow of water and oil,

$$h_{water}\left(T_{wall} - 65.8\right) = h_{oil}\left(178.5 - T_{wall}\right) \qquad (3)$$

If the areas of heat transfer are equal, the wall temperature is obtained from eq. (3) as

$$T_{wall} = \frac{178.5\ h_{oil} + 65.8\ h_{water}}{h_{oil} + h_{water}}$$

$$= \frac{(178.5)(13.98) + (65.8)(506)}{13.98 + 506}$$

$$= 68.8\ ^\circ F$$

This is close to the assumed value of 67°F. Once again, no correction is required.

• PROBLEM 11-20

A shell-and-tube regenerator with a total outside tube area of 48 ft² is shown in fig. 1. Temperatures are measured as shown. It is constructed of 98 brass tubes, each with a 0.375 in. outside diameter and a 0.05 in wall thickness. The shell has a 6 in inside diameter, with 11 half-moon baffles spaced 4.3 in apart. This allows a minimum area of 6.7 in² between the outside water channel and the tubes. The flow rates of the water through the tubes and shell are 18,000 lbm/hr and 22,000 lbm/hr, respectively. Determine the overall heat transfer coefficient a) from the given data, and b) using empirical relations.

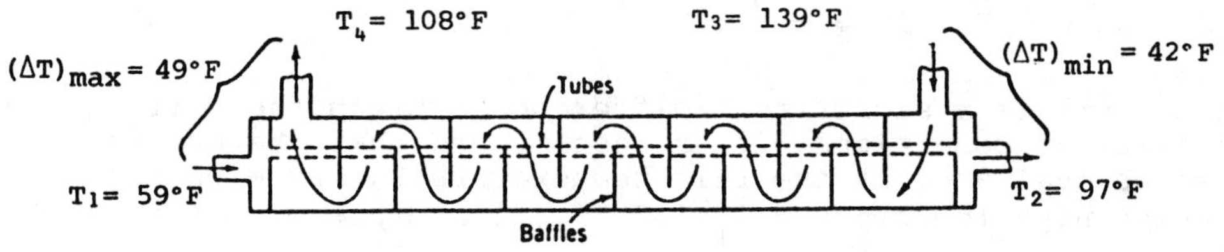

Fig. 1. Arrangement of a shell and tube heat exchanger with baffles.

Solution: a) The derivation for the log mean temperature difference (LMTD), if the pipe sections between any two adjacent baffles are differential components, brings the same result as for countercurrent flow. Assuming this to occur in the immediate case, the LMTD may be calculated as

$$\text{LMTD} = \frac{(\Delta T)_{max} - (\Delta T)_{min}}{\ln\left[\frac{(\Delta T)_{max}}{(\Delta T)_{min}}\right]} = \frac{(108-59) - (139-97)}{\ln\left[\frac{(108 - 59)}{(139 - 97)}\right]} = 45.4°\text{F}$$

The rate at which heat is transferred is

$$q = UA\,(\text{LMTD})$$

$$= U\,(48)(45.4)$$

This can also be defined as

$$q = \dot{m}_c C_c\,(T_2 - T_1)$$

where T_2 and T_1 are the outlet and inlet temperatures on the tube side, respectively.

Therefore,

$$q = (18,000)(1.0)(97 - 59) = 684,000 \text{ Btu/hr}$$

Then the overall heat-transfer coefficient is

$$U = \frac{q}{A(\text{LMTD})} = \frac{684,000}{(48)(45.4)} = 313.9 \text{ Btu/ft}^2\text{-hr-°F}$$

b) Considering the flow in the shell as a flow normal to the tube channels, the heat transfer coefficient h_{34} can be calculated using the expression

$$h = \frac{Nuk}{D}$$

where $Nu = 0.33(Re)^{0.6}\,(Pr)^{0.33}$

and $Re = \frac{vD\rho}{\mu}$, $Pr = \frac{\mu C_P}{k}$

When there is a significant difference between the wall and free-stream conditions, it is recommended that the properties be evaluated at the film temperature T_f assuming the average pipe temperature $\tilde{T}_t$=115°F, the average film temperature is:

$$T_f = \tilde{T}_s + \frac{\tilde{T}_t - \tilde{T}_s}{2}$$

where $\tilde{T}_s = \frac{T_3 + T_4}{2} = \frac{139 + 108}{2} = 123.5°\text{F}$

Then $$T_f = 123.5 + \frac{115 - 123.5}{2} = 119.25°F$$

At this temperature, the thermal properties of water are

$\rho = 1.917$ slug/ft^3 $\quad$ $C_P = 32.2$ Btu/slug-°F

$\mu = 0.043$ slug/ft-hr $\quad$ $k = 0.366$ Btu/hr-ft-°F

Also, we have

$$D_3 = \frac{0.375}{12} = 0.03125 \text{ ft}$$

and $$v = \frac{\dot{m}}{\rho A}$$

$$= \frac{22{,}000}{(61.7)\left(\frac{6.7}{144}\right)} = 7663.5 \text{ ft/hr}$$

Then the Reynolds number and the Prandtl number are calculated as

$$Re = \frac{vD_3\rho}{\mu} = \frac{(7663.5)(0.03125)(1.917)}{0.043} = 10{,}677$$

$$Pr = \frac{\mu C_P}{k} = \frac{(0.043)(32.2)}{0.366} = 3.78$$

Solving for the Nusselt number gives

$$Nu = 0.33(10{,}677)^{0.6}(3.78)^{0.33} = 133.7$$

Therefore, the convection coefficient for the shell side is

$$h_{34} = \frac{(133.7)(0.366)}{0.03125} = 1566 \text{ Btu/hr-ft}^2\text{-°F}$$

For fully developed turbulent flow in smooth tubes, the heat transfer coefficient for the tube flow may be found from

$$Nu = \frac{hD}{k} = 0.023(Re)^{0.8}(Pr)^n$$

where $n = \begin{cases} 0.4 \text{ for heating} \\ 0.3 \text{ for cooling} \end{cases}$

The average temperature of the water in the tube side is

$$T_{av} = \frac{T_1 + T_2}{2} = \frac{59 + 97}{2} = 78°F$$

The properties of water at this temperature are

$\rho = 1.93$ slug/ft^3 $\qquad C_p = 32.2$ Btu/slug-°F

$\mu = 0.0662$ slug/ft-hr $\qquad k = 0.348$ Btu/hr-ft-°F

Also,

$$D_2 = \frac{0.277}{12} = 0.0231 \text{ ft}$$

and

$$v = \frac{\dot{m}}{(\text{no. tubes})\rho A}$$

$$= \frac{18,000}{98(62.2)\ \frac{\pi}{4}\ (0.0231)^2} = 7046 \text{ ft/hr}$$

Then the Reynolds number and the Prandtl number are

$$Re = \frac{(7046)(0.0231)(1.93)}{0.0662} = 4745$$

and

$$Pr = \frac{(0.0662)(32.2)}{0.348} = 6.125$$

Solving for the Nusselt number again gives

$$Nu = 0.023(4745)^{0.8}(6.125)^{0.4} = 41.45$$

Therefore, the convective heat transfer coefficient for the tubes is

$$h_{12} = \frac{(41.45)(0.348)}{0.0231} = 624.4 \text{ Btu/hr-ft}^2\text{-°F}$$

Assuming the thermal conductivity of brass is 58 Btu/hr-ft-°F, the overall heat transfer coefficient can be determined from

$$U = \frac{1}{\dfrac{r_3}{h_{12} r_2} + \dfrac{r_3 \ln(r_3/r_2)}{k_{23}} + \dfrac{1}{h_{34}}}$$

$$= \frac{1}{\left(\frac{0.03125}{0.0231}\right)\left(\frac{1}{624.4}\right) + \left(\frac{0.03125}{2}\right)\left(\frac{1}{58}\right)\ln\left(\frac{0.03125}{0.0231}\right) + \left(\frac{1}{1566}\right)}$$

$$= 346.5 \text{ Btu/hr-ft}^2\text{-°F}$$

The empirical value is 10.4% larger than the true value. Considering the assumptions made, the deviation is acceptable. The brass tube wall represents only about 3% of the total resistance, and since the mean temperature difference between the two fluids is approximately 45°F, only 1.3°F is due to the thermal resistance of the tube wall. The following is a check to see if the assumed mean tube temperature of 115°F is valid:

If the mean temperature of the tube outer surface is $\tilde{T}_t$, then $\tilde{T}_t$ - 1.3°F would be for the inner surface of the tubes. Next, balancing the heat flow between the hot water and the wall, and that between the wall and the cold water will yield

$$h_{34} A_3 \ (\tilde{T}_t - 123.5) = h_{12} A_2 [(T_2 + T_1) - (\tilde{T}_t - 1.3)]$$

or, substituting in the values,

$$156(0.03125)\pi L(\tilde{T}_t - 123.5) = 624.4(0.0231)\pi L \ [78 - (\tilde{T}_t - 1.3)]$$

This gives

$$\tilde{T}_t = 113.4°F,$$

which is close to the assumed value of 115°F. If a substantial difference is present, the calculation will have to be repeated with another approximation of $\tilde{T}_t$.

• PROBLEM 11-21

A two-tube-pass heat exchanger (see fig. 1) with water at 78°F and an average velocity of 4.6 ft/sec flowing in the tubes is used to cool steam at 2 psia. There are 126 tubes per pass, and each tube is made of stainless steel with d_0 = 0.625 in, d_i = 0.527 in, and L = 6 ft. The convective heat transfer coefficient of the outside surface of the tubes is 2050 Btu/hr-ft²-°F. Calculate the amount of condensed steam in pounds.

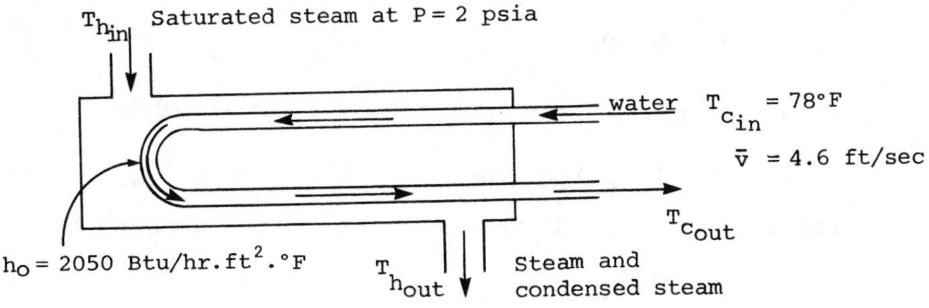

Fig. 1

Solution: To begin the analysis, a listing of all the equations necessary will be given.

The heat flow equation from which the mass of the steam condensed can be calculated from

$$Q = U_0 A_0 \text{ (LMTD)} \tag{1}$$

The overall heat transfer coefficient U_0 based on the outside area A_0 of the tubes can be calculated from

$$\frac{1}{U_0} = \frac{1}{h_0} + \frac{r_0}{k} \ln \frac{d_0}{d_i} + \frac{d_0}{d_i h_i} \tag{2}$$

The convective heat transfer coefficient h_i inside the tubes can be calculated by the expression developed by Petukhov for fully developed turbulent flow in smooth tubes,

$$\text{Nu} = \text{Nusselt number} = \frac{h_i d_i}{k} = \frac{(f/8)\text{RePr}}{1.07 + 12.7(f/8)^{\frac{1}{2}}\,(\text{Pr}^{\frac{2}{3}} - 1)} \left[\frac{\mu_b}{\mu_w}\right]^n \tag{3}$$

where $n = 0.11$ for $T_{wall} > T_{bulk}$,

$n = 0.25$ for $T_{wall} < T_{bulk}$,

and $n = 0$ for constant heat flux or for gases.

In this case, $n = 0.11$ is used. The thermal properties will be determined at the film temperature $T_f = (T_{wall} + T_{bulk})/2$ except the dynamic viscosities μ_b and μ_w, which will be determined at their respective temperatures.

Eq. (3) can be used under the following conditions:

$0.5 < \text{Pr} < 200$ for 6% accuracy

$200 < \text{Pr} < 2000$ for 10% accuracy

$10^4 < \text{Re} < 5 \times 10^6$

$0 < \mu_b/\mu_w < 40$

The friction factor f in eq. (3) can be calculated by

$$f = (1.82 \log_{10} \text{Re} - 1.64)^{-2} \tag{4}$$

The outside tube area A_0 in eq. (1) can be determined as

$$A_0 = \pi d_i L N_p N_t \tag{5}$$

The log mean temperature difference (LMTD) in eq. (1) is given as:

$$\text{LMTD} = \frac{(T_{h\ in} - T_{c\ out}) - (T_{h\ out} - T_{c\ in})}{\ln[(T_{h\ in} - T_{c\ out})/(T_{h\ out} - T_{c\ in})]} \tag{6}$$

The solution to this problem is obtained by a trial and error method. Assuming the cold fluid exit temperature $T_{c\,out} = T_{wall} = 100°F$, then the film temperature at which the thermal properties will be obtained is $T_f = (100 + 78)/2 = 89°F$. The Reynolds and Prandtl numbers in eq. (3) are calculated by

$$Re = \frac{\rho \bar{v} d_i}{\mu b} \tag{7}$$

and
$$Pr = \frac{\mu_b C_c}{k} \tag{8}$$

The mass flow rate of the water is found from

$$\dot{m}_c = \rho A_i \bar{v} \tag{9}$$

where A_i is the cross-sectional area in the tube, and is defined as

$$A_i = \frac{\pi d_i^2}{4} \text{ (no. tubes)} \tag{10}$$

Now, all the equations required are present. The procedure will be to calculate the heat flow Q from eq. (1), and then we use the value to determine $T_{c\,out}$ from

$$Q = \dot{m}_c C_c \left[T_{c\,out} - T_{c\,in}\right] \tag{11}$$

The calculated value of $T_{c\,out}$ can then be compared to the value assumed. This procedure will be repeated until the value of $T_{c\,out}$ converges with the assumed value. When this is satisfied, the mass flow rate of steam can then be calculated from the relation

$$\dot{m}_{steam} h_{fg} = \dot{m}_c C_c \left[T_{c\,out} - T_{c\,in}\right] \tag{12}$$

where h_{fg} = latent heat of vaporization of water

The temperature and the latent heat of saturated steam at 2 psia can be found from the steam tables:

$$T_{h\,in} = T_{h\,out} = 126°F, \quad h_{fg} = 1022.1 \text{ Btu/lbm}$$

The thermal properties of water at the film temperature $T_f = 89°F$ are:

$C_c = 0.997$ Btu/lbm °F $\qquad \mu_b = 1.875$ lmb/ft-hr

$\rho = 62.11$ lbm/ft³ $\qquad k = 0.359$ Btu/hr-ft-°F

The dynamic viscosity of water at the wall temperature

$$T_{wall} = T_{h\,in} = T_{h\,out} = 126°F \text{ is}$$

$\mu_w = 1.297$ lbm/ft-hr

Using eq. (5), the total outside tube area is calculated as

$$A_0 = \pi \left[\frac{0.625}{12}\right] (6)(126)(2) = 247 \text{ ft}^2;$$

and using eq. (10), the total cross-sectional flow area inside the tube is found to be

$$A_i = \frac{\pi}{4} \left[\frac{0.527}{12}\right]^2 (126) = 0.191 \text{ ft}^2$$

Since we assumed that $T_{h\,out} = T_{h\,in} = 126°F$ and $T_{c\,out} = 100°F$, all the inlet and outlet temperatures are known. The LMTD can be calculated, using eq. (6):

$$\text{LMTD} = \frac{(126 - 100°F) - (126 - 78°F)}{\ln[(126 - 100°F)/(126 - 78°F)} = 35.88°F$$

Using eq. (9), the mass flow rate of the water is

$$\dot{m}_c = (62.11)(0.19)(4.6)(3600) = 196{,}452 \text{ lbm/hr}$$

Using eq. (7) and (8), respectively, the Reynolds and the Prandtl numbers are calculated as

$$Re = \frac{(62.11)(4.5)(3600)\left[\frac{0.527}{12}\right]}{1.875} = 23{,}567$$

$$Pr = \frac{(1.875)(0.997)}{0.359} = 5.2$$

Inserting the value of Re into eq. (4) to calculate the friction factor f yields

$$f = [1.82 \log_{10}(23{,}567) - 1.64]^{-2} = 0.025$$

Next, the Nusselt number can be calculated using eq. (3).

$$Nu = \frac{(0.025/8)(23{,}567)(5.2)}{1.07 + 12.7\,(0.025/8)^{1/2}(5.2^{2/3}-1)} \left[\frac{1.875}{1.297}\right]^{0.11} = 160$$

The the inside convection heat transfer coefficient h_i is calculated as

$$h_i = \frac{Nuk}{d_i} = \frac{(160)(0.359)}{0.527/12} = 1308 \text{ Btu/hr-ft}^2\text{-°F}$$

Using eq. (2), the overall heat transfer coefficient U_0 based on the outside surface area of the tubes is calculated as

$$\frac{1}{U_0} = \frac{1}{2050} + \frac{\left[\frac{0.625}{2 \times 12}\right]}{(0.359)} \ln\left[\frac{0.625}{0.527}\right] + \left[\frac{0.625}{0.527}\right]\left[\frac{1}{1308}\right]$$

$$U_0 = [0.0004878 + 0.01237 + 0.000907]^{-1}$$

$$= 72.65 \text{ Btu/hr-ft}^2\text{-°F}$$

The mean temperature of the outside of the tubes is calculated using

$$h_0 \left[T_{steam} - T_{0,mean}\right] = U_0 \left[T_{steam} - T_{c,mean}\right]$$

Substituting,

$$\left[2050\ 126 - T_{0,mean}\right] = 72.65 \left[126 - \frac{78 + 100}{2}\right]$$

$$T_{0,mean} = 124.7°F$$

Equating eqs. (1) and (11),

$$\dot{m}_c C_c \left(T_{c\,out} - T_{c\,in}\right) = Q = U_0 A_0 (LMTD)$$

From this equation, $T_{c\,out}$ can be calculated as

$$(196{,}452)(0.997)(T_{c\,out} - 78) = (72.65)(247)(35.88)$$

$$T_{c\,out} = 81.29°F$$

The calculations should be repeated until the calculated value of $T_{c\,out}$ is equal to the assumed value of $T_{c\,out}$.

To obtain an approximation of how much steam is condensed, the actual exit temperature of the water will be taken as the average of the assumed and calculated values.

$$T_{c,out} = \frac{T_{assumed} + T_{calculated}}{2} = \frac{100 + 81.29}{2} = 90.65°F$$

Using eq. (12), the steam will condense at a rate of

$$\dot{m}_{steam} = \frac{m_c C_c \left[T_{c\,out} - T_{c\,in}\right]}{h_{fg}} = \frac{(196{,}452)(0.997)(90.65 - 78)}{1022.1}$$

$$= 2424.1 \text{ lbm/hr}$$

The performance of a heat exchanger is specified according to the accompanying figure. Using properties at 38°C for the large stream (denoted by ℓ) and at 66°C for the small stream (denoted by s), calculate the overall heat transfer coefficient (UA).

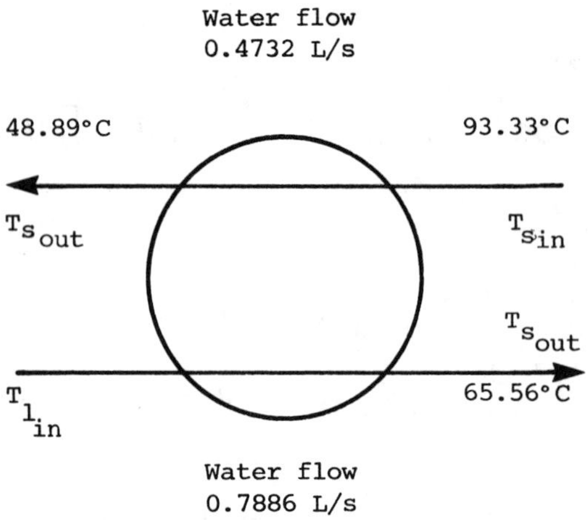

Fig. 1 Diagram of a countercurrent heat exchanger showing performance data.

Solution: The properties must be evaluated at each temperature. At 66°C,

$$\rho = 0.979 \text{ Kg/L} \qquad C_p = 4.188 \text{ kJ/Kg-°C}$$

and at 38°C,

$$\rho = 0.993 \text{ Kg/L} \qquad C_p = 4.179 \text{ kJ/Kg-°C}$$

Heat capacity rate is defined as

$$HCR = \dot{m}C_p$$

Since $\dot{m} = \rho VA$, for the small stream,

$$(HCR)_s = (0.4732)(0.979)(4.188)(1000)$$

$$= 1940.15 \text{ W/°C}$$

and for the large stream,

$$(HCR)_\ell = (0.7886)(0.993)(4.179)(1000)$$

$$= 3272.5 \text{ W/°C}$$

The inlet temperature for the large stream may be calculated using the following relation:

$$\frac{(HCR)_s}{(HCR)_\ell} = \frac{T_{\ell,out} - T_{\ell,in}}{T_{s,in} - T_{s,out}}$$

Rearranging to solve for $T_{\ell,in}$ and substituting yields

$$T_{\ell,in} = T_{\ell,out} - (T_{s,in} - T_{s,out}) \frac{(HCR)_s}{(HCR)_\ell}$$

$$= 65.56 - (93.33 - 48.89) \left[\frac{1940.15}{3272.5}\right]$$

$$= 39.2°C$$

Heat exchanger effectiveness is defined as

$$\varepsilon = \frac{T_{s,in} - T_{s,out}}{T_{s,in} - T_{\ell,in}}$$

This is

$$\varepsilon = \frac{93.33 - 48.89}{93.33 - 39.2} = 0.821$$

The number of transfer units may then be calculated.

$$NTU = \frac{\ell n\left[\left(1 - \varepsilon \frac{(HCR)_s}{(HCR)_\ell}\right) \div (1-\varepsilon)\right]}{1 - \frac{(HCR)_s}{(HCR)_\ell}}$$

$$= \frac{\left[\left(\ell n\ 1 - 0.821 \left(\frac{1940.15}{3272.5}\right)\right) \div (1 - 0.821)\right]}{1 - \left(\frac{1940.15}{3272.5}\right)}$$

$$= 2.587$$

This is also equal to

$$NTU = \frac{UA}{(HCR)_s}$$

Solving for UA and substituting yields

$$UA = NTU(HCR)_s$$

$$= 2.587(1940.15) = 5019.2 \text{ W/°C}.$$

• PROBLEM 11-23

> A cross-flow heat exchanger is shown in fig. 1. Air at a flow rate of 5000 lbm/hr enters at 60°F. It is used to cool liquid nitrogen j entering at 3500 lbm/hr, with an average temperature of -9.4°F. If no mixing of the two streams occurs, what is the exit temperature of the air?

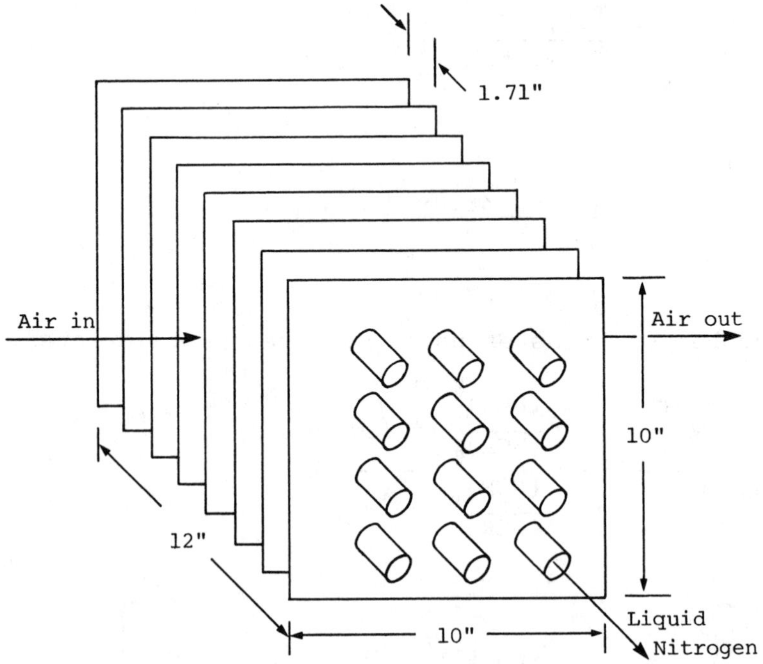

Fig. 1.

Solution: All dimensions are given in fig. 1. With those, the wetted perimeter, cross-sectional area and surface area for each fluid stream may be calculated, and also the heat transfer area. Proceeding,

Wetted perimeter:

on air side $P_a = 2(1.71) + 2(10) = 23.42$ in or 1.95 ft

on nigrogen side $P_n = 4\pi(0.875) = 10.99$ in or 0.92 ft

Cross-sectional area per passage:

for air $A_a = 1.71(10) = 17.1\ \text{in}^2$ or $0.12\ \text{ft}^2$

for nitrogen $A_n = \frac{\pi}{4}(0.875)^2 = 0.6\ \text{in}^2$ or $0.0042\ \text{ft}^2$

Surface area:

for air $S_a = \pi(1)(12)(12) = 452.39\ \text{in}^2$ or $3.14\ \text{ft}^2$

for nitrogen $S_n = \pi(0.875)(12)(12) = 395.84\ \text{in}^2$ or $2.75\ \text{ft}^2$

Heat transfer area:

$2[7(10)(10) - 12(\pi)(1)] = 1324.6\ \text{in}^2$ or 9.2 ft

Now, if the channel through which the fluid flows is not of circular cross-section (as in the air passages shown in fig. 1), all calculations should be based on the hydraulic diameter, which is

$$D_H = \frac{4A}{P}$$

For the air, this is

$$D_H = \frac{4A_a}{P_a} = \frac{4(0.12)}{1.95} = 0.246 \text{ ft}$$

The passage through which the nitrogen flows is circular, and the hydraulic diameter is merely the inside diameter (0.875 in or 0.073 ft).

Since both fluids will vary in temperature along the duct, an average temperature must be approximated first, and then the outlet temperature will be refined. At the average liquid nitrogen temperature, tables give

$$\mu_n = 10.36 \times 10^{-6} \text{ lbm/ft-sec}$$

Assuming an average air temperature of 55°F, air tables list

$$\mu_a = 12.0 \times 10^{-6} \text{ lbm/ft-sec}$$

The mass velocities are

$$\left[\frac{\dot{m}}{A}\right]_a = \frac{5000}{(3.14)(0.12)} = 13{,}270 \text{ lbm/hr-ft}^2$$

and $$\left[\frac{\dot{m}}{A}\right]_n = \frac{3500}{(2.75)(0.0042)} = 303{,}030 \text{ lbm/hr-ft}^2$$

With these values the Reynolds numbers of the flows may be evaluated.

$$Re = \frac{(\dot{m}/A)D_H}{\mu}$$

For air,

$$Re_a = \frac{(13{,}270)(0.246)}{(12.0 \times 10^{-6})(3600)} = 75{,}565$$

For nitrogen,

$$Re_n = \frac{(303{,}030)(0.073)}{(10.36 \times 10^{-6})(3600)} = 593{,}125$$

These are both in the turbulent flow regime. Then the heat transfer coefficients may be found using the relation

$$h = \left[0.023 \left(\frac{k}{D_H}\right) Re^{0.8} \; Pr^{0.33}\right] \left[1 + \left(\frac{D_H}{L}\right)^{0.7}\right]$$

Referring again to the air tables,

$$k_a = 0.0145 \text{ Btu/hr-ft-°F} \qquad Pr_a = 0.715$$

Also, from the nitrogen tables,

$$k_n = 0.0129 \text{ Btu/hr-ft-°F} \qquad Pr_n = 0.73$$

Then the coefficient for air is:

$$h_a = \left[0.023 \frac{0.0145}{0.246} (75{,}565)^{0.8} (0.715)^{0.33}\right]\left[1 + \frac{0.246}{12/12}^{0.7}\right]$$

$$= 13.33 \text{ Btu/hr-ft}^2\text{-°F}$$

while the coefficient for nitrogen is

$$h_n = \left[0.023 \frac{0.0129}{0.073} (593{,}125)^{0.8} (0.73)^{0.33}\right]\left[1 + \left(\frac{0.073}{10/12}\right)^{0.7}\right]$$

$$= 179.88 \text{ Btu/hr-ft}^2\text{-°F}$$

These values are to be used to calculate the overall heat transfer coefficient. If the wall resistance is deemed negligible, this is

$$UA = \frac{1}{\frac{1}{h_a A} + \frac{1}{h_n A}} = \frac{1}{\frac{1}{(13.33)(9.2)} + \frac{1}{(179.88)(9.2)}}$$

$$= 114.2 \text{ Btu/hr-°F}$$

Since the air has a smaller flow rate, the number of transfer units (NTU) will be

$$NTU = \frac{UA}{C_{min}} = \frac{UA}{(\dot{m}C_P)_a}$$

Taking the average value $C_{Pa} = 0.24$ Btu/lbm-°F,

$$NTU = \frac{114.2}{5000(0.24)} = 0.0952$$

Now, if C_{min}/C_{max} is calculated, fig. 2 may be used to find the heat exchanger effectiveness. So,

$$\frac{C_{min}}{C_{max}} = \frac{(\dot{m}C_P)_n}{(\dot{m}C_P)_a} = \frac{(3500)(0.25)}{(5000)(0.24)}$$

$$= 0.73$$

If the value of C_{Pn} is 0.25 Btu/lbm-°F.

Referring to fig. 2,

effectiveness $\varepsilon = 12\%$.

Finally, the exit temperature of the air may be calculated using

$$T_{a,out} = T_{a,in} + \frac{C_n}{C_a} \varepsilon \Delta T_{max}$$

Substituting,

$$T_{a,out} = 60 + (0.73)(0.12)(-9.4 - 60)$$

$$= 53.92°F.$$

• PROBLEM 11-24

A regenerator (see fig. 1 for dimensions) used for increasing the efficiency of a steam turbine facility has the following functioning conditions: a supply of water at $\dot{m}_w$=400,000 lbm/hr and T_{w1}=60°F enters the heat exchanger as shown in fig. 1. A crossflow of hot air enters the exchanger at $\dot{m}_a$=200,000 lbm/hr, T_{a1}=260°F, P_{a1}=39.7 psia, and a moisture content of ω_a=0.015 lbs H_2O/lbm. The type of heat exchange surface used is 11.32 - 0.737-SR. Using the given operating data and the surface features of the heat exchanger, determine (a) the adequacy of the regenerator, and (b) the water air flow pressure changes.

Solution: The surface characteristics are obtained from fig. 5 for the air side as follows:

hydraulic radius of flow channel r_a= 0.00288 ft

air side heat transfer area/total volume α_a= 270 ft²/ft³

area of fin/total area α_f= 0.845

flow area/frontal area σ_a= 0.78

thickness of fin δ = 0.00033 ft

conductivity of aluminum fin k = 100 Btu/hr-ft-°F

length of fin equals ½ spacing between tubes $L_{fin} = \frac{1}{2}\left[\frac{0.45}{12}\right]$

$= 0.01875$ ft

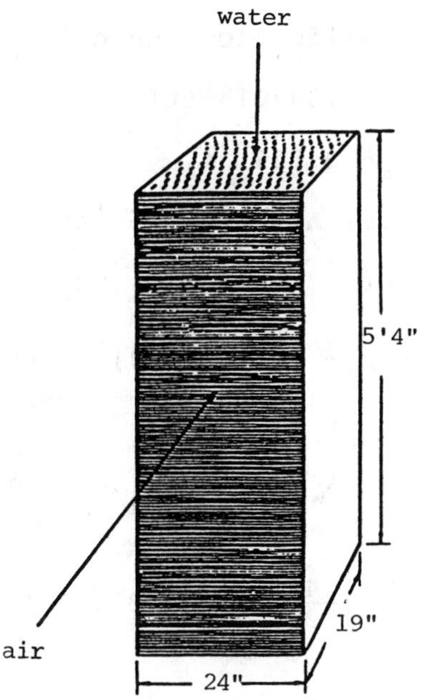

Fig. 1(a)

Regenerator dimensions

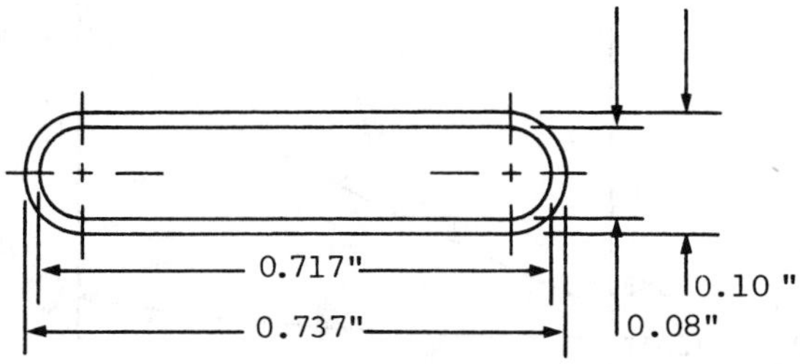

Fig. 1(b). Tube Dimensions

flow area = 0.000389 ft^2.

heat transfer area/total volume,

$\alpha = 42.1\ ft^2/ft^3$

flow passage hydraulic diameter,

$4r_h = 0.01224$ ft.

frontal area = 0.003014 ft^2

flow area/frontal area, $\sigma = 0.129$

inside surface area = 0.0106 ft^2

Referring to fig. 1b for the water side tube dimensions,

flow area of tube $A_{cw} = 0.000389\ ft^2$

heat transfer area of water side/total volume $\alpha_w = 421 ft^2/ft^3$

hydraulic radius of water-side flow channel $r_w = 3.06 \times 10^{-3} ft$

Frontal area of tube $A_{fr} = 0.003014\ ft^2$

flow area/frontal area $\sigma_w = 0.129$

inside surface area of tube $A_{si} = 0.0106\ ft^2$

Also fig. 1a gives the core dimensions:

$$\text{frontal area of air side } A_a = \left[\frac{24}{12}\right](5.33) = 10.66\ ft^2$$

$$\text{frontal area of water side } A_w = \left[\frac{24}{12}\right]\left[\frac{19}{12}\right] = 3.167\ ft^2$$

$$\text{heat exchanger volume } V = \left[\frac{24}{12}\right]\left[\frac{19}{12}\right](5.33) = 16.89\ ft^3$$

To begin we will estimate an air exit temperature of 75°F and a water exit temperature of 80°F for the first assumption. These estimations will be confirmed later.

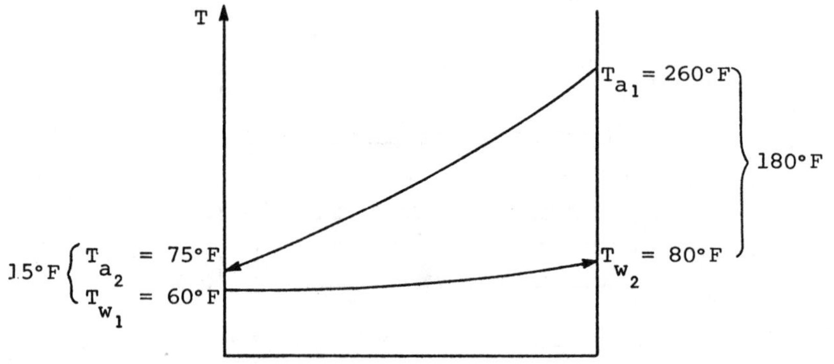

Fig. 2.

Temperature profile for countercurrent regenerator

In fig. 2, the temperature profile is plotted for a counter-flow exchanger (as is the case, even though the exchanger has a crossflow configuration). The thermal properties are obtained at the average bulk temperatures. These, are calculated as

$$T_a = \frac{80 + 60}{2} + \frac{180 - 15}{\ln(180/15)} = 136.4°F$$

$$T_w = \frac{80 + 60}{2} = 70°F$$

From the air tables,

$\mu_a = 0.0482$ lbm/hr-ft $\qquad$ $C_{Pa} = 0.24$ Btu/lbm-°F

$P_{ra} = 0.70$

Fig. 3. **Humidity correction factors for density and specific heat.**
$\rho = X_{d,\omega}\rho_a \qquad c_p = X_{c,\omega}c_{pa}$

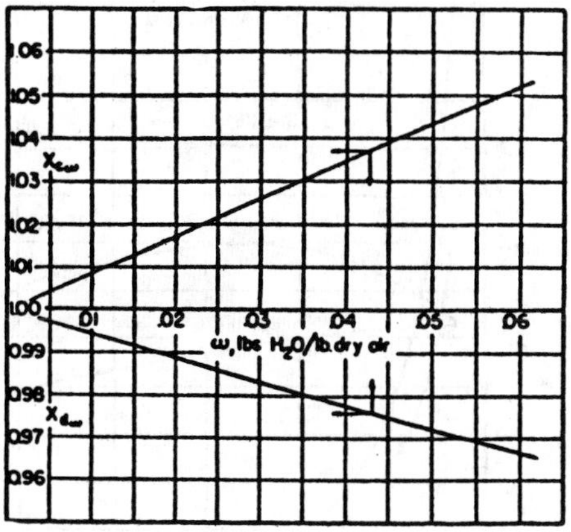

The use of fig. 3 C_p for a humidity of 0.015 lbs H_2O/lbm gives $X_{c,w} = 1.013$, and the corrected specific heat is $C_{Pa} = 1.013(0.24) = 0.243$ Btu/lbm.°F. If a 1% pressure drop is assumed, the air exit pressure $P_{a2} = 0.99(39.7) = 39.3$psia (this will be confirmed later).

Applying the perfect gas equation of state and the density correction factor for the humidity, the inlet and outlet air specific volumes can be determined as

$$v_1 = \frac{1}{\rho_1} = \frac{1}{X_{d,w}} \frac{RT_{a1}}{MP_{a1}}$$

where $X_{d,w} = 0.992$ from fig. 3
R = gas constant = 53.34 ft-lb/lbm-°R

$$= \frac{(53.34)(260 + 460)}{(0.992)(39.7)(144)} = 6.77 \text{ ft}^3/\text{lbm}$$

$$v_2 = \frac{1}{\rho_2} = \frac{(53.34)(75 + 460)}{(0.992)(39.3)(144)} = 5.083 \text{ ft}^3/\text{lbm}$$

The ratio of the mean air specific volume to the inlet air specific volume is

$$\frac{v_m}{v_1} = \frac{P_{a1}}{P_{average}} \frac{T_a}{T_{a1}} = \frac{(39.7)(136.4 + 460)}{\left[\frac{39.7 + 39.3}{2}\right](720)} = 0.8325$$

Fig. 4.

Transport properties of saturated liquid water.

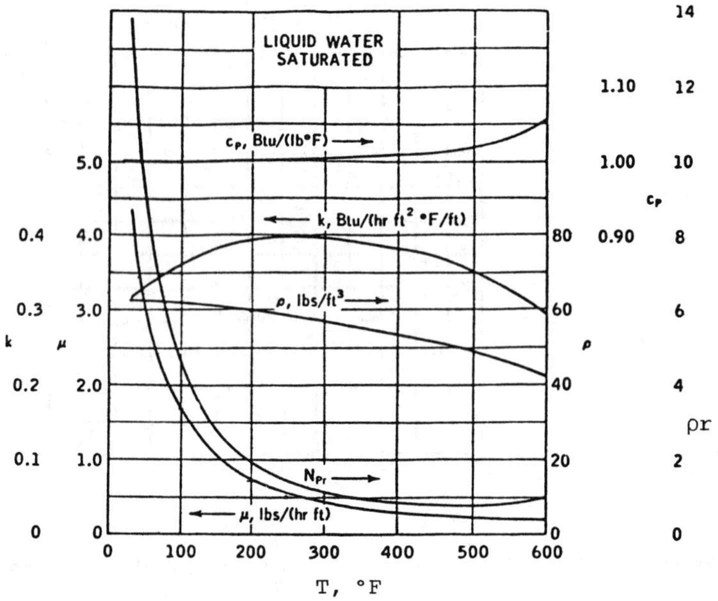

From fig. 4, the water properties at T_W=70°F are:

k = 0.347 Btu/hr-ft-°F $\quad$ C_{Pw}= 1.0 Btu/lmb-°F

μ_W=2.36 lbm/hr-ft $\quad$ v = 0.0163 ft³/lbm

P_{rw} = 6.8

The Nusselt number for the water flow inside the tubes is determined from fig. 6:

Fig. 5.

Finned flat tubes, surface 11.32-0.737-SR.

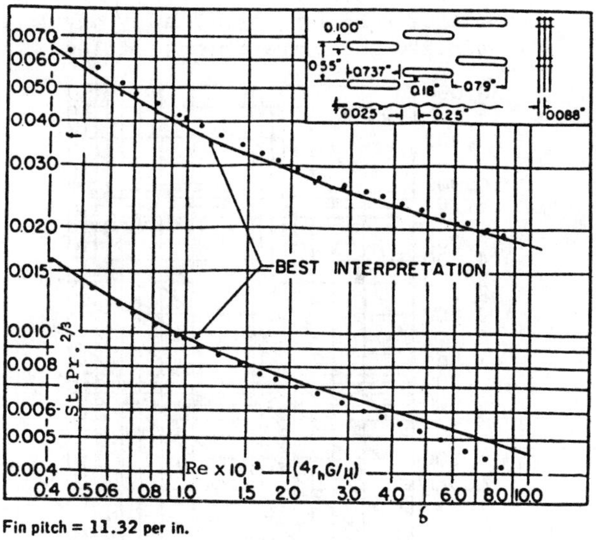

Fin pitch = 11.32 per in.

Flow passage hydraulic diameter, $4r_h$ = 0.01152 ft

Fin metal thickness = 0.004 in., copper

Free-flow area/frontal area, σ = 0.780

Total heat transfer area/total volume, $\alpha = 270\ ft^2/ft^3$

Fin area/total area = 0.845

For R_{ew}= 5073 and P_{rw}= 6.8, Nu= 55

If the viscosity at the wall temperature is approximately equal to the viscosity at the mean temperature of the water side, no adjustment is required for a viscosity change with temperature.

From fig. 7, the friction factor for the tube flow is f_w= 0.0094.

The convection film coefficient for the air is

$$h_a = (St)(G_a)(C_p) = (0.00685)(24030)(0.24)$$

$$= 39.5\ Btu/hr\text{-}ft^2\ °F$$

and for the water side,

$$h_w = \frac{Nuk}{4r_w} = \frac{(55)(0.347)}{4(3.06 \times 10^{-3})} = 1560\ Btu/hr\text{-}ft^2\text{-}°F$$

Fig. 6.

Nussell numbers for turbulent flow in a circular tube with constant heat rate per unit of tube length and fully developed velocity and temperature profiles.

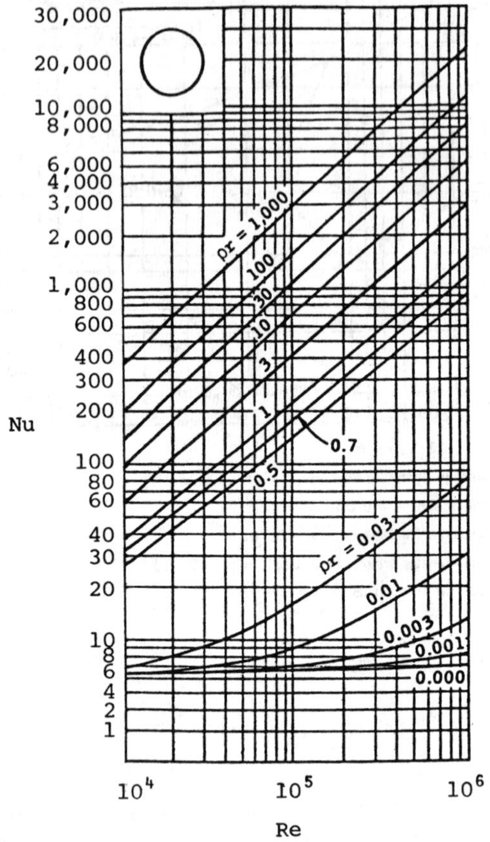

Fig. 7.

Friction factors for fully developed turbulent flow in a smooth-walled circular tube. Turbulent flow friction factors for rectangular and annular tubes do not differ significantly.

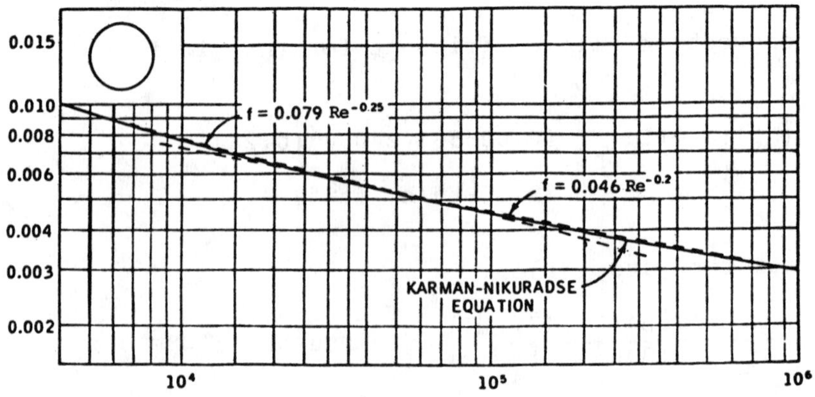

The Reynolds numbers are calculated as:

$$\text{mass velocity of air} = G_a = \frac{\dot{m}_a}{A_{ca}} = \frac{\dot{m}_a}{A_a \sigma_a} = \frac{200{,}000}{(10.66)(0.78)}$$

$$= 24{,}030 \frac{\text{lbm}}{\text{hr-ft}^2}$$

A_{ca} = free flow area of the air side

$$R_{ea} = \frac{4r_a G_a}{\mu_a} = \frac{(4)(0.00288)(240{,}030)}{0.0482} = 5743$$

$$\text{mass velocity of water} = G_w = \frac{\dot{m}_w}{A_w \sigma_w} = \frac{400{,}000}{(3.167)(0.129)}$$

$$= 978{,}162 \text{ lbm/hr.ft}^2$$

$$R_{ew} = \frac{4r_w G_w}{\mu_w} = \frac{(4)(3.06 \times 10^{-3})(978{,}162)}{2.36} = 5073$$

The Stanton number and the friction factor are determined from fig. 5 for the 11.32 -0.737-SR (air side) surface as

$$(St)P_r^{2/3} = 0.0054$$

$$f_a = 0.021$$

and $$St = \frac{0.0054}{\left[P_{ra}\right]^{2/3}} = \frac{0.0054}{\left[0.70\right]^{2/3}} = 0.00685$$

the fin efficiency of the air side will be obtained using fig. 8.

$$m = \left[\frac{2h}{k\delta}\right]^{\frac{1}{2}}$$

where m = fin effectiveness parameter

h = convection heat transfer coefficient of air side
= 40.2 Btu/hr-ft²-°F

$$\text{Hence, } m = \left[\frac{2(40.2)}{(100)(0.00033 \text{ ft})}\right]^{\frac{1}{2}} = 49.36 \text{ ft}^{-1}$$

and $m(L_{fin}) = 49.36(0.225/26) = 0.9255$. The fin efficiency is then $\eta_f = 0.79$

The overall surface efficiency of the air side is

$$\eta_0 = 1 - \alpha_f(1 - \eta_f)$$

$$= 1 - (0.845)(1 - 0.79) = 0.823$$

Neglecting the small wall resistance, the overall heat transfer coefficient based on the air side area is

$$\frac{1}{U_a} = \frac{1}{\eta_0 h_a} + \frac{1}{(\alpha_w/\alpha_a)h_w}$$

$$= \frac{1}{(0.823)(39.5)} + \frac{1}{[(42.1)/(270)]\ (1560)}$$

$$= 0.03487$$

or $U_a = 28.67$ Btu/hr-ft^2-°F

Fig. 8.

Heat transfer effectiveness of straight and circular fins.

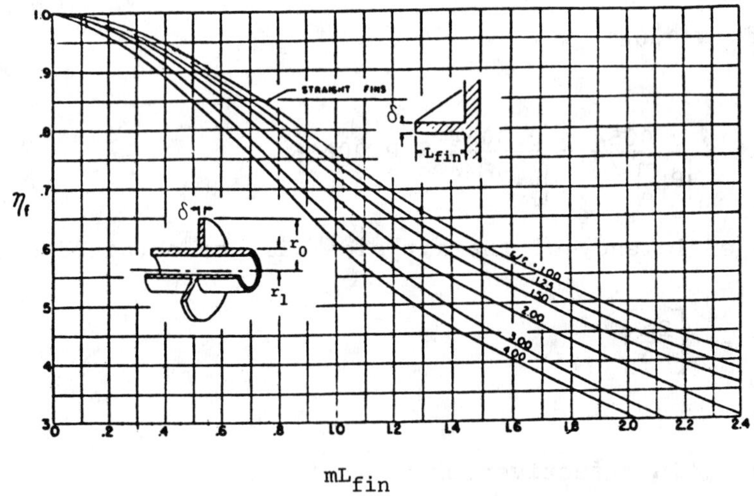

In order to obtain the effectiveness ε from fig. 9, the number of transfer units (NTU) and the minimum capacity rate C_{min} must first be calculated.

Fig. 9

Heat transfer effectiveness as a function of number of transfer units and capacity rate ratio; crossflow exchanger with fluids unmixed.

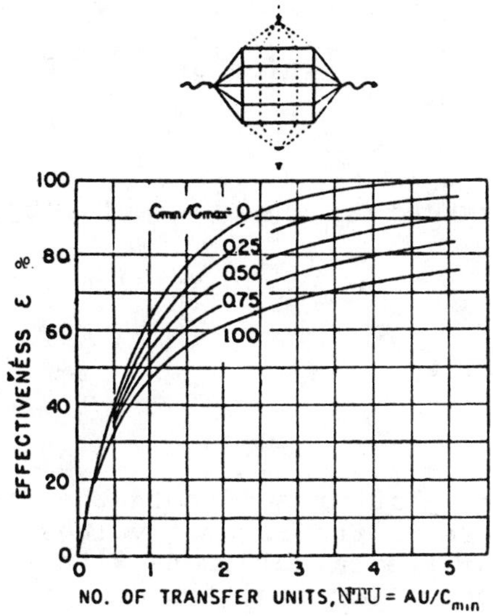

$C_a = \dot{m}_a C_{Pa} = (200{,}000)(0.243) = 48{,}600$ Btu/hr-°F

$C_w = (400{,}000)(1.0) = 400{,}000$ Btu/hr-°F

Therefore, the air side capacity rate C_a is C_{min}.

The NTU is calculated as

$NTU = \dfrac{A_{at}U_a}{C_{min}}$ where A_{at} = total heat transfer area of the air side

$= \alpha_a V = (270)(16.89) = 4563$ ft^2

and $NTU = \dfrac{(4563)(28.67)}{48{,}600} = 2.69$

Also, $\dfrac{C_a}{C_w} = \dfrac{C_{min}}{C_{max}} = \dfrac{48{,}600}{400{,}000}$ 0.1215

Now, referring to fig. 9 gives the efficiency of the intercooler as ε = 90%.

The air exit temperature T_a can be calculated from the following relation:

$$\varepsilon = \frac{C_a\left[T_{a1} - T_{a2}\right]}{C_a\left[T_{a1} - T_{w1}\right]}$$

$$0.90 = \frac{260 - T_{a2}}{260 - 60} \text{, which gives } T_{a2} = 80°F.$$

The water exit temperature T_{wo} is calculated by:

$$T_{w2} - T_{w1} = \frac{C_a}{C_w}\left(T_{a1} - T_{a2}\right) = 0.1215(260-80)$$

$$T_{w2} = 81.87°F$$

The calculated exit temperatures of both the air and water are significantly close to the assumed values. Hence, it is not required to repeat the calculations with the new values. The thermal properties will not differ much from those applied in the first trial. If the head loss for the entrance and exit are neglected, the pressure drop can be determined from the following equation:

$$\frac{\Delta P}{P_1} = \frac{G^2}{2g}\frac{v_1}{P_1}\left[(1 + \sigma^2)\left[\frac{v_2}{v_1} - 1\right] + f\frac{A_t}{A_c}\left[\frac{v_m}{v}\right]\right]$$

Substituting the values for the air side,

$$\frac{\Delta P}{P_1} = \left[\frac{24030}{3600}\right]^2 \frac{1}{2(32.2)}\frac{6.77}{(39.7)(144)}\left[(1 + 0.78^2)\right.$$

$$\left. \times \left[\frac{5.083}{6.77} - 1.0\right] + 0.021\left[\frac{4563}{8.323}\right]\left(0.8325\right)\right]$$

$= 0.0075$ or 0.75% of the initial air pressure, which is approximately the assumed value of 1.0%.

Now, for the water side, since the specific volume of the water side is constant, the pressure drop is defined as

$$\Delta P = \left[\frac{G_w^2}{2g}\right] v\, f\left[\frac{A_{wt}}{A_{wc}}\right]$$

where A_{wt} = total heat transfer area of the water side

$$= \alpha_w V = (42.1)(16.89) = 711.5 \text{ ft}^2$$

$$A_{wc} = A_w \sigma_w = (3.167)(0.129) = 0.409 \text{ ft}^2$$

so $$\Delta P = \frac{978,162}{3600}^2 \frac{1}{2(32.2)} (0.01603)(0.0094) \left[\frac{711.5}{0.409}\right]$$

$$= 301.4 \text{ psfa or } 2.09 \text{ psia.}$$

APPENDIX

Heat Transfer Properties of Liquid Water

(English Units)

T (°F)	ρ $\left(\frac{lb_m}{ft^3}\right)$	c_p $\left(\frac{btu}{lb_m \cdot °F}\right)$	$\mu \times 10^3$ $\left(\frac{lb_m}{ft \cdot s}\right)$	k $\left(\frac{btu}{h \cdot ft \cdot °F}\right)$	N_{Pr}	$\beta \times 10^4$ $(1/°R)$	$(g\beta\rho^2/\mu^2) \times 10^{-6}$ $(1/°R \cdot ft^3)$
32	62.4	1.01	1.20	0.329	13.3	−0.350	
60	62.3	1.00	0.760	0.340	8.07	0.800	17.2
80	62.2	0.999	0.578	0.353	5.89	1.30	48.3
100	62.1	0.999	0.458	0.363	4.51	1.80	107
150	61.3	1.00	0.290	0.383	2.72	2.80	403
200	60.1	1.01	0.206	0.393	1.91	3.70	1,010
250	58.9	1.02	0.160	0.395	1.49	4.70	2,045
300	57.3	1.03	0.130	0.395	1.22	5.60	3,510
400	53.6	1.08	0.0930	0.382	0.950	7.80	8,350
500	49.0	1.19	0.0700	0.349	0.859	11.0	17,350
600	42.4	1.51	0.0579	0.293	1.07	17.5	30,300

(SI Units)

T (°C)	T (K)	ρ (kg/m³)	c_p (kJ/kg·K)	$\mu \times 10^3$ (Pa·s, or kg/m·s)	k (W/m·K)	N_{Pr}	$\beta \times 10^4$ (1/K)	$(g\beta\rho^2/\mu^2) \times 10^{-8}$ (/K·m³)
0	273.2	999.6	4.229	1.786	0.5694	13.3	−0.630	
15.6	288.8	998.0	4.187	1.131	0.5884	8.07	1.44	10.93
26.7	299.9	996.4	4.183	0.860	0.6109	5.89	2.34	30.70
37.8	311.0	994.7	4.183	0.682	0.6283	4.51	3.24	68.0
65.6	338.8	981.9	4.187	0.432	0.6629	2.72	5.04	256.2
93.3	366.5	962.7	4.229	0.3066	0.6802	1.91	6.66	642
121.1	394.3	943.5	4.271	0.2381	0.6836	1.49	8.46	1,300
148.9	422.1	917.9	4.312	0.1935	0.6836	1.22	10.08	2,231
204.4	477.6	858.6	4.522	0.1384	0.6611	0.950	14.04	5,308
260.0	533.2	784.9	4.982	0.1042	0.6040	0.859	19.8	11,030
315.6	588.8	679.2	6.322	0.0862	0.5071	1.07	31.5	19,260

Properties of Saturated Water

Temperature t °F	Density ρ lb_m/ft^3	Dynamic Viscosity μ lb_m/hr ft	Kinematic Viscosity ν ft^2/hr	Specific Heat c_p B/lb_m F	Thermal Conductivity k B/hr ft F	Thermal Diffusivity α ft^2/hr	Prandtl Number N_{Pr}
32	62.4	4.33	0.0694	1.005	0.320	0.00510	13.61
40	4	3.73	598	1.005	26	520	11.50
50	4	3.16	506	1.00	32	532	9.51
60	62.4	2.71	0.0434	1.00	0.337	0.00540	8.04
70	3	36	379	1.00	43	551	6.88
80	2	08	334	1.00	49	561	5.95
90	1	1.85	298	1.00	54	570	5.23
100	0	66	268	1.00	59	579	4.63
110	61.9	1.49	0.0241	1.00	0.363	0.00586	4.11
120	7	36	220	1.00	67	595	3.70
130	6	24	201	1.00	72	604	3.33
140	4	14	186	1.00	76	612	3.04
150	2	05	172	1.00	80	621	2.77
160	61.0	0.970	0.0159	1.00	0.383	0.00628	2.53
170	60.8	0.900	148	1.005	86	632	2.34
180	6	0.840	139	1.005	88	637	2.18
190	4	0.786	130	1.005	90	642	2.02
200	1	0.738	123	1.01	92	646	1.90
210	59.9	0.693	0.0116	1.015	0.393	0.00646	1.80
220	7	54	110	1.015	4	650	1.69
230	4	18	104	1.01	5	658	1.58
240	1	0.585	0990	1.01	6	663	1.49
250	58.8	55	0944	1.015	6	663	1.42
260	58.6	0.528	0.00901	1.02	0.396	0.00663	1.36
270	3	04	864	1.02	6	666	1.30
280	0	0.482	831	1.025	6	666	1.25
290	57.7	63	802	1.025	6	670	1.20
300	4	48	780	1.03	6	670	1.16

Heat Transfer Properties of Water Vapor

1 Atm Abs (English Units)

T (°F)	ρ $\left(\frac{lb_m}{ft^3}\right)$	c_p $\left(\frac{btu}{lb_m \cdot °F}\right)$	$\mu \times 10^3$ $\left(\frac{lb_m}{ft \cdot s}\right)$	k $\left(\frac{btu}{h \cdot ft \cdot °F}\right)$	N_{Pr}	$\beta \times 10^3$ $(1/°R)$	$g\beta\rho^2/\mu^2$ $(1/°R \cdot ft^3)$
212	0.0372	0.451	0.870	0.0145	0.96	1.49	0.877×10^6
300	0.0328	0.456	1.000	0.0171	0.95	1.32	0.459×10^6
400	0.0288	0.462	1.130	0.0200	0.94	1.16	0.243×10^6
500	0.0258	0.470	1.265	0.0228	0.94	1.04	0.139×10^6
600	0.0233	0.477	1.420	0.0257	0.94	0.943	82×10^3
700	0.0213	0.485	1.555	0.0288	0.93	0.862	52.1×10^3
800	0.0196	0.494	1.700	0.0321	0.92	0.794	34.0×10^3

(SI Units)

T (°C)	T (K)	ρ (kg/m^3)	c_p $(kJ/kg \cdot K)$	$\mu \times 10^5$ $(Pa \cdot s$, or $kg/m \cdot s)$	k $(W/m \cdot K)$	N_{Pr}	$\beta \times 10^3$ $(1/K)$	$g\beta\rho^2/\mu^2$ $(1/K \cdot m^3)$
100.0	373.2	0.596	1.888	1.295	0.02510	0.96	2.68	0.557×10^8
148.9	422.1	0.525	1.909	1.488	0.02960	0.95	2.38	0.292×10^8
204.4	477.6	0.461	1.934	1.682	0.03462	0.94	2.09	0.154×10^8
260.0	533.2	0.413	1.968	1.883	0.03946	0.94	1.87	0.0883×10^8
315.6	588.8	0.372	1.997	2.113	0.04448	0.94	1.70	52.1×10^5
371.1	644.3	0.341	2.030	2.314	0.04985	0.93	1.55	33.1×10^5
426.7	699.9	0.314	2.068	2.529	0.05556	0.92	1.43	21.6×10^5

Specific Heat of Water

Temp. °C	Thermal Capacity		Enthalpy	
	Cal./g/°C	Joules/g/°C	Cal./g	Joules/g
0	1.00738	4.2177	0.0245	0.1026
1	1.00652	4.2141	1.0314	4.3184
2	1.00571	4.2107	2.0376	8.5308
3	1.00499	4.2077	3.0429	12.7400
4	1.00430	4.2048	4.0475	16.9462
5	1.00368	4.2022	5.0515	21.1498
6	1.00313	4.1999	6.0549	25.3508
7	1.00260	4.1977	7.0578	29.5496
8	1.00213	4.1957	8.0602	33.7463
9	1.00170	4.1939	9.0621	37.9410
10	1.00129	4.1922	10.0636	42.1341
11	1.00093	4.1907	11.0647	46.3255
12	1.00060	4.1893	12.0654	50.5155
13	1.00029	4.1880	13.0659	54.7041
14	1.00002	4.1869	14.0660	58.8916
15	.99976	4.1858	15.0659	63 0779
16	.99955	4.1849	16.0655	67.2632
17	.99933	4.1840	17.0650	71.4476
18	.99914	4.1832	18.0642	75.6312
19	.99897	4.1825	19.0633	79.8141
20	.99883	4.1819	20.0622	83.9963
21	.99869	4.1813	21.0609	88.1778
22	.99857	4.1808	22.0596	92.3589
23	.99847	4.1804	23.0581	96.5395
24	.99838	4.1800	24.0565	100.7196
25	.99828	4.1796	25.0548	104.8994
26	.99821	4.1793	26.0530	109.0788
27	.99814	4.1790	27.0512	113.2580
28	.99809	4.1788	28.0493	117.4369
29	.99804	4.1786	29.0474	121.6157
30	.99802	4 1785	30.0455	125 7943
31	.99799	4.1784	31.0435	129.9727
32	.99797	4.1783	32.0414	134.1510
33	.99797	4.1783	33.0394	138.3293
34	.99795	4.1782	34.0374	142.5076
35	.99795	4.1782	35.0353	146.6858
36	.99797	4.1783	36.0333	150.8641
37	.99797	4.1783	37.0312	155.0423
38	.99799	4.1784	38.0292	159.2207
39	.99802	4.1785	39.0272	163.3991
40	.99804	4.1786	40.0253	167.5777
41	.99807	4.1787	41.0233	171.7563
42	.99811	4.1789	42.0214	175.9351
43	.99816	4.1791	43.0195	180.1141
44	.99819	4.1792	44.0177	184.2933
45	.99826	4.1795	45.0159	188.4726
46	.99830	4.1797	46.0142	192.6522
47	.99835	4.1799	47.0125	196.8320
48	.99842	4.1802	48.0109	201.0120
49	.99847	4.1804	49.0094	205.1923
50	.99854	4.1807	50.0079	209.3729
51	.99862	4.1810	51.0065	213.5538
52	.99871	4.1814	52.0051	217.7350
53	.99878	4.1817	53 0039	221.9166
54	.99885	4.1820	54.0027	226.0984
55	.99895	4.1824	55.0016	230.2806
56	.99905	4.1828	56.0006	234.4632
57	.99914	4.1832	56.9997	238.6462
58	.99924	4.1836	57.9989	242.8296
59	.99933	4.1840	58.9982	247.0134
60	.99943	4.1844	59.9975	251.1976
61	.99955	4.1849	60.9970	255.3822
62	.99964	4.1853	61.9966	259.5673
63	.99976	4.1858	62.9963	263.7529
64	.99988	4.1863	63.9962	267.9390
65	1.00000	4.1868	64.9961	272.1256
66	1.00014	4.1874	65.9962	276.3127
67	1.00026	4.1879	66.9964	280.5003
68	1.00041	4.1885	67.9967	284.6885
69	1.00053	4.1890	68.9972	288.8772
70	1.00067	4.1896	69.9977	293.0665
71	1.00081	4.1902	70.9985	297.2564
72	1.00096	4.1908	71.9994	301.4469
73	1.00112	4.1915	73.0004	305.6381
74	1.00127	4.1921	74.0016	309.8299
75	1.00143	4.1928	75.0030	314.0224
76	1.00160	4.1935	76.0045	318.2155
77	1.00177	4.1942	77.0062	322.4094
78	1.00194	4.1949	78.0080	326.6039
79	1.00213	4.1957	79.0101	330.7992
80	1.00229	4.1964	80.0123	334.9952
81	1.00248	4.1972	81.0147	339.1920
82	1.00268	4.1980	82.0172	343.3897
83	1.00287	4.1988	83.0200	347.5881
84	1.00308	4.1997	84.0230	351.7873
85	1.00327	4.2005	85.0262	355.9874
86	1.00349	4.2014	86.0295	360.1883
87	1.00370	4.2023	87.0331	364.3902
88	1.00392	4.2032	88.0369	368.5929
89	1.00416	4.2042	89.0410	372.7966
90	1.00437	4.2051	90.0452	377.0012
91	1.00461	4.2061	91.0497	381.2068
92	1.00485	4.2071	92.0545	385.4135
93	1.00509	4.2081	93.0594	389.6211
94	1.00535	4.2092	94.0647	393.8297
95	1.00561	4.2103	95.0701	398.0395
96	1.00588	4.2114	96.0759	402.2503
97	1.00614	4.2125	97.0819	406.4622
98	1.00640	4.2136	98.0882	410.6753
99	1.00669	4.2148	99.0947	414.8895
100	1.00697	4.2160	100.1015	419.1049

Properties of Dry Air at Atmospheric Pressure (14.7 lb/sq in.)

Temperature t °F	Density ρ lb_m/ft^3	Dynamic Viscosity μ lb_m/hr ft	Kinematic Viscosity ν ft^2/hr	Specific Heat c_p B/lb_m F	Thermal Conductivity k B/hr ft F	Thermal Diffusivity α ft^2/hr	Prandtl Number N_{Pr}
−100	0.1103	0.0324	0.294	0.2395	0.0107	0.405	0.726
−80	045	338	323	95	112	447	23
−60	0.0993	352	355	96	117	492	21
−40	946	366	387	96	122	538	19
−20	903	380	421	97	127	587	17
0	0.0864	0.0394	0.456	0.2397	0.0132	0.637	0.716
20	828	408	493	98	137	690	15
40	795	421	530	99	141	739	14
60	764	434	568	0.2400	146	796	13
80	736	447	607	02	151	854	12
100	0.0710	0.0460	0.649	0.2403	0.0155	0.909	0.711
120	685	473	691	05	160	971	10
140	662	485	733	06	165	1.036	09
160	641	497	775	08	169	095	08
180	621	508	818	11	173	156	07
200	0.0602	0.0520	0.864	0.2413	0.0178	1.225	0.706
220	584	532	911	16	182	290	05
240	568	543	956	19	186	354	04
260	552	554	1.004	22	191	429	03
280	537	565	052	25	195	498	02
300	0.0523	0.0576	1.101	0.2429	0.0199	1.567	0.703
320	509	588	155	32	203	640	04
340	497	599	205	36	207	709	05
360	485	610	258	41	211	782	05
380	474	620	308	46	215	855	05
400	0.0463	0.0631	1.363	0.2450	0.0219	1.931	0.705
420	452	641	418	55	223	2.009	06
440	442	651	473	60	227	088	06
460	432	661	530	66	231	169	06

Note. **In order to convert pound-mass units into slug units divide by 32.16. Values of N_{Pr} are smoothed.**

Thermal Conductivity of Douglas Fir at a Mean Temperature of 75 °F

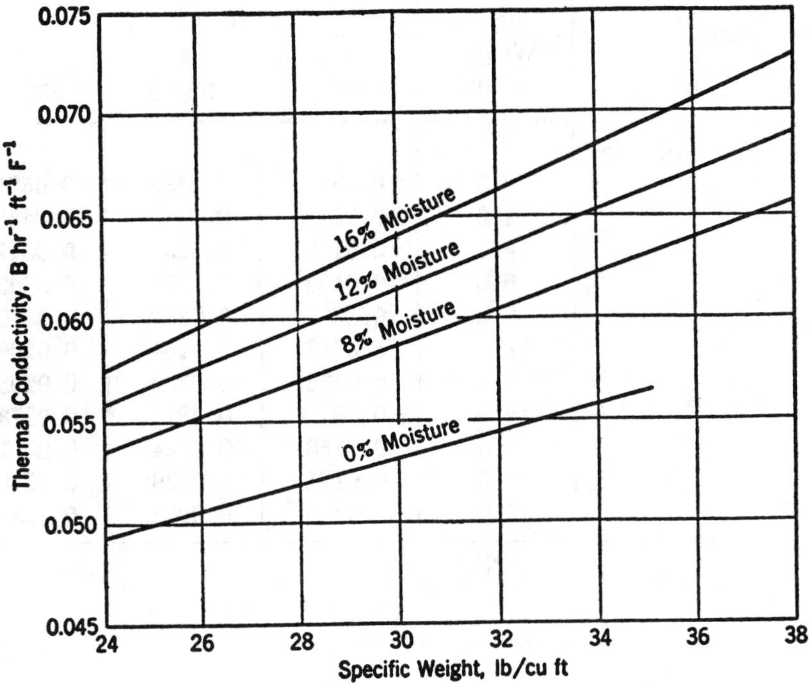

Thermal Conductivity for Selected Building Materials at About 70 °F

Material	Apparent Specific Weight lb/cu ft	Thermal Conductivity B hr^{-1} ft^{-1} F^{-1}
Asphalt	130	0.40
Brick, dry	105	0.30
1% moisture by volume	105	0.40
2% moisture by volume	105	0.60
Concrete, dry	120	0.45
10% moisture by volume	140	0.70
reinforced	140	0.75
Glass	...	0.5 to 0.6
Gypsum, dry	80	0.25
Limestone, fine grain, dry	105	0.40
fine grain, 15% moisture by volume	105	0.55
coarse grain, dry	125	0.55
Linoleum	75	0.11
Plaster, with 2% moisture by volume	115	0.5 to 0.6
Rubber, vulcanized, soft, with 40% pure rubber	...	0.17
with 90% pure rubber	...	0.10
Sand, dry	95	0.20
10% moisture by volume	100	0.60
Soil, dry	125	0.30
fresh clay, 30% moisture by volume	125	1.40

Thermal Conductivity of Selected Insulating Materials

Insulation	Apparent Specific Weight lb/cu ft	Thermal Conductivity $[B\ hr^{-1}\ ft^{-1}\ F^{-1}]$ at 20° F	100° F	200° F
Corrugated asbestos (4 plies per inch)	13.0	0.0354	0.0430	0.0525
Glass wool	1.5	0.0217	0.0313	0.0435
	4.0	0.0179	0.0239	0.0317
	6.0	0.0163	0.0218	0.0288
Hair felt	8.2	0.0237	0.0269	0.0310
	11.4	0.0212	0.0254	0.0299
	12.8	0.0233	0.0262	0.0295
"85% magnesia"	16.9	0.0321	0.0344	0.0373
Rock wool	4.0	0.0150	0.0224	0.0317
	8.0	0.0171	0.0228	0.0299
	.12.0	0.0183	0.0226	0.0281

Thermal Conductivity of Powdered Diatomaceous Earth

Apparent Specific Weight lb/cu ft	Thermal Conductivity $[B\ hr^{-1}\ ft^{-1}\ F^{-1}]$ at 100° F	200° F	300° F	400° F	500° F	600° F
10	0.024	0.029	0.034	0.038	0.044	0.048
14	0.030	0.033	0.036	0.039	0.043	0.046
18	0.038	0.040	0.043	0.045	0.048	0.049

Thermal Conductivity of Selected Materials

Metal	Impurities %	Thermal Conductivity $[B\ hr^{-1}\ ft^{-1}\ F^{-1}]$ at −58° F	32° F	392° F	752° F	1112° F
Aluminum	0.3	134	132	131	131	...
	1.0	116	115	114	...	...
Copper	0.1	...	220	215	210	206
Iron	0.1	...	43	36	28	22
Lead	0.0	22	21	19	...	...
Nickel	0.1	...	...	42	34	...
	1.0	34	35	34	34	34
Platinum	0.05	41	40	42	45	47
Silver	0.0	243	242	...	...	...
	0.1	236	233	217	204	225
Zinc	0.2	67	66	62	54	...

Thermal Conductivity of Selected Non-Ferrous Alloys

Material	Thermal Conductivity $[B\ hr^{-1}\ ft^{-1}\ F^{-1}]$ at −58° F	32° F	392° F	752° F
Brass: 70% Cu, 30% Zn	55	60	75	80
"Chromel A": 80% Ni, 20% Cr	...	...	9	11
"Chromel P": 90% Ni, 10% Cr	...	...	12	14
Gunbronze: 86% Cu, 9% Sn, 4% Zn	...	...	38	44
Monel: 67% Ni, 29% Cu	...	...	18	20

Thermal Conductivities, Densities and Heat Capacities of Selected Metals

Material	t (°C)	ρ $\left(\frac{kg}{m^3}\right)$	c_p $\left(\frac{kJ}{kg \cdot K}\right)$	$k(W/m \cdot K)$		
Aluminum	20	2707	0.896	202 (0°C)	206 (100°C)	215 (200°C)
				230 (300°C)		
Brass (70–30)	20	8522	0.385	97 (0°C)	104 (100°C)	109 (200°C)
Cast iron	20	7593	0.465	55 (0°C)	52 (100°C)	48 (200°C)
Copper	20	8954	0.383	388 (0°C)	377 (100°C)	372 (200°C)
Lead	20	1137	0.130	35 (0°C)	33 (100°C)	31 (200°C)
Steel 1% C	20	7801	0.473	45.3 (18°C)	45 (100°C)	45 (200°C)
				43 (300°C)		
308 stainless	20	7849	0.461	15.2 (100°C)	21.6 (500°C)	
304 stainless	0	7817	0.461	13.8 (0°C)	16.3 (100°C)	18.9 (300°C)
Tin	20	7304	0.227	62 (0°C)	59 (100°C)	57 (200°C)

Normal Total Emissivities of Surfaces

Surface	K	ε	Surface	K	ε
Aluminum			Lead, unoxidized	400	0.057
highly oxidized	366	0.20	Nickel, polished	373	0.072
highly polished	500	0.039	Nickel oxide	922	0.59
	850	0.057	Oak, planed	294	0.90
Aluminum oxide	550	0.63	Paint		
Asbestos board	296	0.96	aluminum	373	0.52
Brass, highly	520	0.028	oil (16 different, all colors)	373	0.92–0.96
polished	630	0.031	Paper	292	0.924
Chromium, polished	373	0.075	Roofing paper	294	0.91
Copper			Rubber (hard, glossy)	296	0.94
oxidized	298	0.78	Steel		
polished	390	0.023	oxidized at 867 K	472	0.79
Glass, smooth	295	0.94	polished stainless	373	0.074
Iron			304 stainless	489	0.44
oxidized	373	0.74	Water	273	0.95
tin-plated	373	0.07		373	0.963
Iron oxide	772	0.85			

Thermal Conductivity of Selected Metals

T,K	Aluminum 99.996+% ρ_0 = 0.00315 μohm cm	Copper 99.999+% ρ_0 = 0.000851 μohm cm	Gold 99.999+% ρ_0 = 0.0055 μohm cm	Iron 99.998+% ρ_0 = 0.0327 μohm cm	Manganin	Platinum 99.999% ρ_0 = 0.0106 μohm cm	Silver 99.999+% ρ_0 = 0.00062 μohm cm	Tungsten 99.99+% ρ_0 = 0.0017 μohm cm
0	0	0	0	0	0	0	0	0
1	7.8	28.7	4.4	0.75	0.0007	2.31	39.4	14.4
2	15.5	57.3	8.9	1.49	0.0018	4.60	78.3	28.7
3	23.2	85.5	13.1	2.24	0.0031	6.79	115	42.6
4	30.8	113	17.1	2.97	0.0046	8.8	147	55.6
5	38.1	138	20.7	3.71	0.0062	10.5	172	67.1
6	45.1	159	23.7	4.42	0.0078	11.8	187	76.2
7	51.5	177	26.0	5.13	0.0095	12.6	193	82.4
8	57.3	189	27.5	5.80	0.0111	12.9	190	85.3
9	62.2	195	28.2	6.45	0.0128	12.8	181	85.1
10	66.1	196	28.2	7.05	0.0145	12.3	168	82.4
11	69.0	193	27.7	7.62	0.0162	11.7	154	77.9
12	70.8	185	26.7	8.13	0.0180	10.9	139	72.4
13	71.5	176	25.5	8.58	0.0197	10.1	124	66.4
14	71.3	166	24.1	8.97	0.0215	9.3	109	60.4
15	70.2	156	22.6	9.30	0.0232	8.4	96	54.8
16	68.4	145	20.9	9.56	0.0250	7.6	85	49.3
18	63.5	124	17.7	9.88	0.0285	6.1	66	40.0
20	56.5	105	15.0	9.97	0.0322	4.9	51	32.6
25	40.0	68	10.2	9.36	0.0410	3.15	29.5	20.4
30	28.5	43	7.6	8.14	0.0497	2.28	19.3	13.1
35	21.0	29	6.1	6.81	0.0583	1.80	13.7	8.9
40	16.0	20.5	5.2	5.55	0.067	1.51	10.5	6.5
45	12.5	15.3	4.6	4.50	0.075	1.32	8.4	5.07
50	10.0	12.2	4.2	3.72	0.082	1.18	7.0	4.17
60	6.7	8.5	3.8	2.65	0.097	1.01	5.5	3.18
70	5.0	6.7	3.58	2.04	0.110	0.90	4.97	2.76
80	4.0	5.7	3.52	1.68	0.120	0.84	4.71	2.56
90	3.4	5.14	3.48	1.46	0.127	0.81	4.60	2.44
100	3.0	4.83	3.45	1.32	0.133	0.79	4.50	2.35
150	2.47	4.28	3.35	1.04	0.156	0.762	4.32	2.10
200	2.37	4.13	3.27	0.94	0.172	0.748	4.30	1.97
250	2.35	4.04	3.20	0.865	0.193	0.737	4.28	1.86
273	2.36	4.01	3.18	0.835	0.206	0.734	4.28	1.82
300	2.37	3.98	3.15	0.803	0.222	0.730	4.27	1.78
350	2.40	3.94	3.13	0.744	0.250	0.726	4.24	1.70
400	2.40	3.92	3.12	0.694	(0.279)	0.722	4.20	1.62
500	2.37	3.88	3.09	0.613	(0.338)	0.719	4.13	1.49
600	2.32	3.83	3.04	0.547	(0.397)	0.720	4.05	1.39
700	2.26	3.77	2.98	0.487		0.723	3.97	1.33
800	2.20	3.71	2.92	0.433		0.729	3.89	1.28
900	2.13	3.64	2.85	0.380		0.737	3.82	1.24
1000	[0.93]**	3.57	(2.78)	0.326		0.748	(3.74)	1.21
1100	[0.96]	3.50	(2.71)	0.297		0.760	(3.66)	1.18
1200	[0.99]	3.42	(2.62)	0.282		0.775	(3.58)	1.15
1300	[1.02]	(3.34)†	(2.51)	0.299		0.791		1.13
1400				0.309		0.807		1.11
1500				0.318		0.824		1.09
1600				(0.327)		0.842		1.07
						0.860		1.05
						0.877		1.03
						(0.895)		1.02
						(0.913)		1.00
								0.98
								0.96
								0.94
								0.925
								0.915
								0.905
								0.900
								(0.895)

* In the table the third significant figure is given only for the purpose of comparison and for smoothness and is not indicative of the degree of accuracy.
** Values in square brackets are for liquid state.
† Values in parentheses are extrapolated.
‡ Estimated.

Thermal Conductivity k, Specific Heat c, Density ρ, and Thermal Diffusivity α of Selected Metals and Alloys

Material	*k*(Btu/hr ft °F)				*c*(Btu/lbm °F)	*ρ*(lbm/cu ft)	*α*(ft²/hr)
	32°F	212°F	572°F	932°F	32°F	32°F	32°F
Metals							
Aluminum	117	119	133	155	0.208	169	3.33
Bismuth	4.9	3.9	....	....	0.029	612	0.28
Copper, pure	224	218	212	207	0.091	558	4.42
Gold	169	170	....	....	0.030	1203	4.68
Iron, pure	35.8	36.6	....	....	0.104	491	0.70
Lead	20.1	19	18	....	0.030	705	0.95
Magnesium	91	92	....	....	0.232	109	3.60
Mercury	4.8	...	....	....	0.033	849	0.17
Nickel	34.5	34	32	....	0.103	555	0.60
Silver	242	238	....	....	0.056	655	6.6
Tin	36	34	....	....	0.054	456	1.46
Zinc	65	64	59	....	0.091	446	1.60
Alloys							
Admiralty metal ..	65	64					
Brass, 70% Cu, 30% Zn	56	60	66	....	0.092	532	1.14
Bronze, 75% Cu, 25% Sn	15	....	....	....	0.082	540	0.34
Cast iron							
Plain	33	31.8	27.7	24.8	0.11	474	0.63
Alloy	30	28.3	27	....	0.10	455	0.66
Constantan, 60% Cu, 40% Ni	12.4	12.8	....	....	0.10	557	0.22
18-8 Stainless steel,							
Type 304.......	8.0	9.4	10.9	12.4	0.11	488	0.15
Type 347.......	8.0	9.3	11.0	12.8	0.11	488	0.15
Steel, mild, 1% C ..	26.5	26	25	22	0.11	490	0.49

Physical Properties of Selected Non-Metallic Materials

Material	Average temperature (°F)	k (Btu/hr ft °F)	c (Btu/lbm°F)	ρ (lbm/cu ft)	α (ft²/hr)
Insulating materials					
Asbestos	32	0.087	0.25	36	~0.01
	392	0.12	....	36	~0.01
Cork	86	0.025	0.04	10	~0.006
Cotton, fabric	200	0.046			
Diatomaceous earth, powdered	100	0.030	0.21	14	~0.01
	300	0.036	....		
	600	0.046	....		
Molded pipe covering	400	0.051	....	26	
	1600	0.088	....		
Glass wool					
Fine	20	0.022	....		
	100	0.031	....	1.5	
	200	0.043	....		
Packed	20	0.016	....		
	100	0.022	....	6.0	
	200	0.029	....		
Hair felt	100	0.027	....	8.2	
Kaolin insulating					
Brick	932	0.15	....	27	
	2102	0.26	....		
Firebrick	392	0.05	....	19	
	1400	0.11	....		
85% magnesia	32	0.032	....	17	
	200	0.037	....	17	
Rock wool	20	0.017	....	8	
	200	0.030	....		
Rubber	32	0.087	0.48	75	0.0024
Building Materials					
Brick					
Fire clay	392	0.58	0.20	144	0.02
	1832	0.95			
Masonry	70	0.38	0.20	106	0.018
Zirconia	392	0.84	....	304	
	1832	1.13	....		
Chrome brick	392	0.82	....	246	
	1832	0.96			
Concrete					
Stone	~70	0.54	0.20	144	0.019
10% moisture	~70	0.70	....	140	~0.025
Glass, window	~70	~0.45	0.2	170	0.013
Limestone, dry	70	0.40	0.22	105	0.017
Sand					
Dry	68	0.20	....	95	
10% H_2O	68	0.60	....	100	
Soil					
Dry	70	~0.20	0.44	....	~0.01
Wet	70	~1.5	....	....	~0.03
Wood					
Oak ⊥ to grain	70	0.12	0.57	51	0.0041
∥ to grain	70	0.20	0.57	51	0.0069
Pine ⊥ to grain	70	0.06	0.67	31	0.0029
∥ to grain	70	0.14	0.67	31	0.0067
Ice	32	1.28	0.46	57	0.048

Thermal Conductivity of Some Gases and Vapors

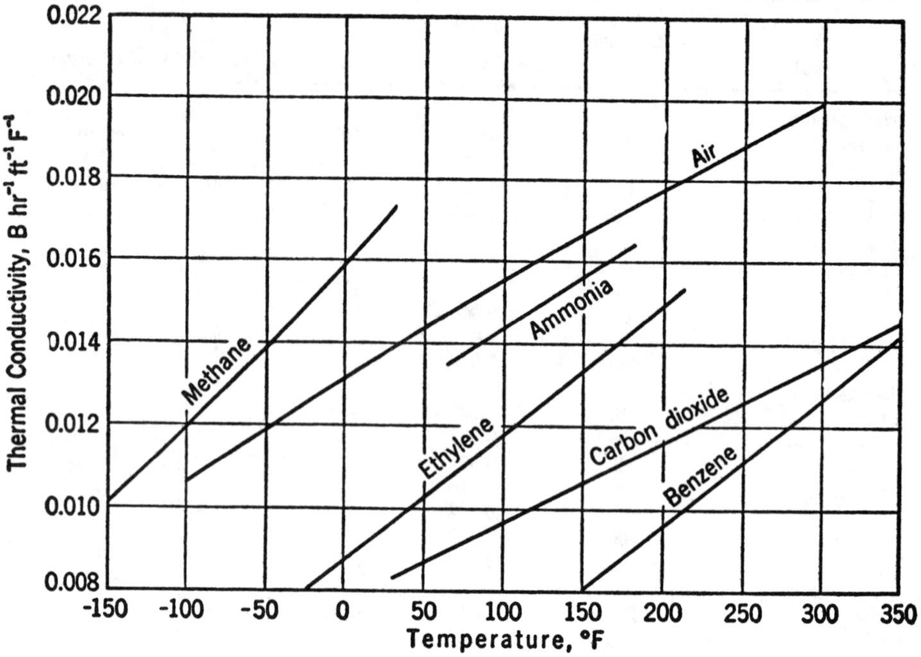

Thermal Conductivity of Water

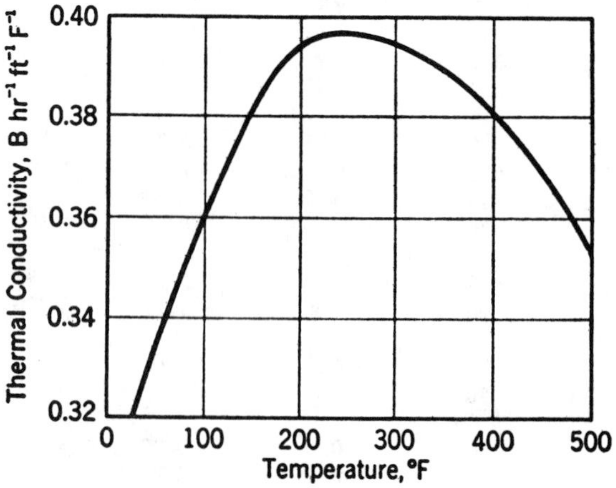

Thermal Conductivity of Gases, Liquids, and Solids

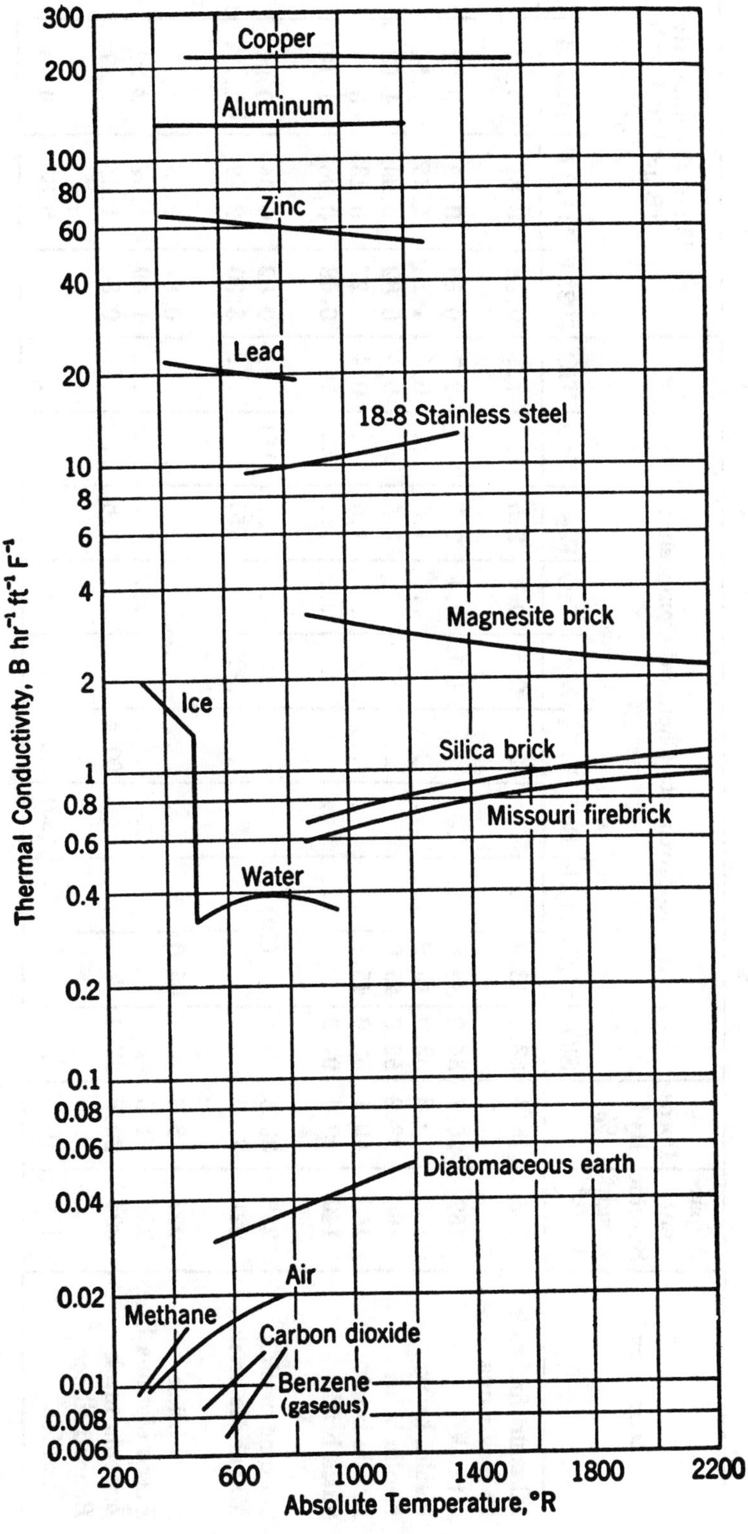

Thermal Conductivity of Refractory Materials

Refractory	Apparent Specific Weight lb/ft^3	Porosity† %	Approximate Chemical Composition %										Thermal Conductivity [B hr^{-1} ft^{-1} F^{-1}] at			
			SiO_2	Al_2O_3	Cr_2O_3	SiC	ZrO_2	MgO	CaO	Fe_2O_3	FeO	TiO	392° F	1112° F	1832° F	2552° F
Missouri firebrick	165	18.4	53.1	43.3	...	...	...	0.5	0.6	2.5	...	...	0.58	0.84	0.94	1.00
Pennsylvania firebrick	162	26.7	54.2	38.8	...	...	...	1.1	0.1	2.7	...	2.7	0.58	0.72	0.82	0.89
Kaolin brick	166	10.8	52.0	45.9	...	...	...	...	...	1.5	...	0.4	1.13	1.35	1.54	1.69
Kaolin brick	168	23.2	52.0	45.9	...	...	...	...	...	1.5	...	0.4	0.82	0.99	1.08	1.16
Kaolin brick	156	49.1	52.0	45.9	...	...	...	...	...	1.5	...	0.4	0.27	0.43	0.53	0.58
Silica brick	140	30.4	97.0	...	...	...	...	...	...	...	...	...	0.68	0.96	1.16	1.30
Chrome brick	246	30.5	(5)‡	(60)		...	...	(15)	...	...	(15)	...	0.82	0.94	0.96	0.99
Magnesite brick	221	31.6	...	...	...	...	...	(90)	...	(5)	...	...	3.30	2.48	2.17	2.04
Spinel brick	227	36.3	...	65.0	...	...	...	(26)		...	...	...	0.87	1.05	1.13	1.23
Fused alumina brick	229	21.3	...	...	...	...	...	...	...	...	...	...	1.49	1.97	2.29	2.53
Zirconia brick	304	29.5	27.3	7.8	...	...	60.4	...	...	1.6	...	...	0.84	1.05	1.11	1.18
Bonded silicon carbide	158	16	...	...	...	(89)	...	...	...	...	...	...	...	8.83	9.00	9.24

† Porosity is defined as the ratio of the volume of the void space to the total volume.

Forced Convection: Turbulent Flow

Configuration	L/D	$Re_D = V_\infty D/\nu$	C_D
Circular cylinder, axis perpendicular to the flow 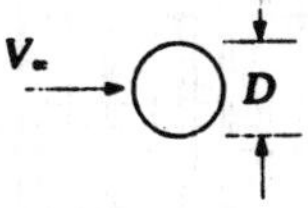	1 5 20 ∞ 5 ∞	10^5 $>5 \times 10^5$	0.63 0.74 0.90 1.20 0.35 0.33
Circular cylinder, axis parallel to the flow 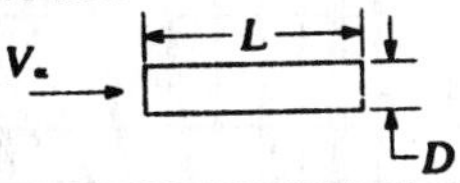	0 1 2 4 7	$>10^3$	1.12 0.91 0.85 0.87 0.99
Elliptical cylinder 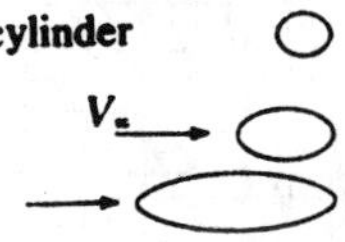 (2 : 1)* (4 : 1)* (8 : 1)*		4×10^4 10^5 2.5×10^4 to 10^5 2.5×10^4 2×10^5	0.6 0.46 0.32 0.29 0.20
Airfoil (1 : 3)†	∞	4×10^4	0.07
Rectangular plate, normal to the flow L = length D = width	1 5 20 ∞	$>10^3$	1.16 1.20 1.50 1.90
Square cylinder 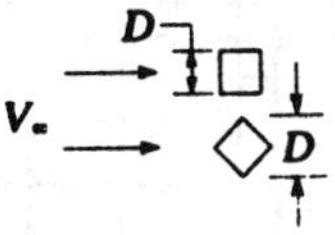		3.5×10^4 10^4 to 10^5	2.0 1.6
Triangular cylinder 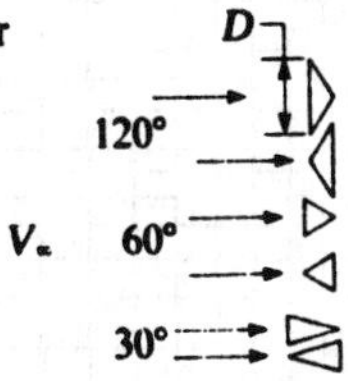		$>10^4$ $>10^5$	2.0 1.72 2.20 1.39 1.80 1.0
Hemispherical shell 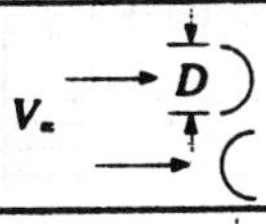		$>10^3$ 10^3 to 10^5	1.33 0.4
Circular disk, normal to the flow 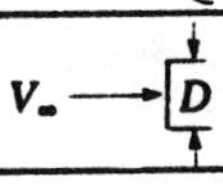		$>10^3$	1.12
Tandem disks 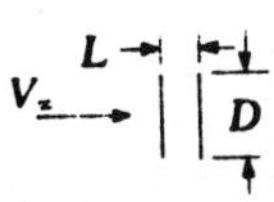	0 1 2 3	$>10^3$	1.12 0.93 1.04 1.54

*Ratio of major axis to minor axis
†Ratio of chord to span at zero angle of attack

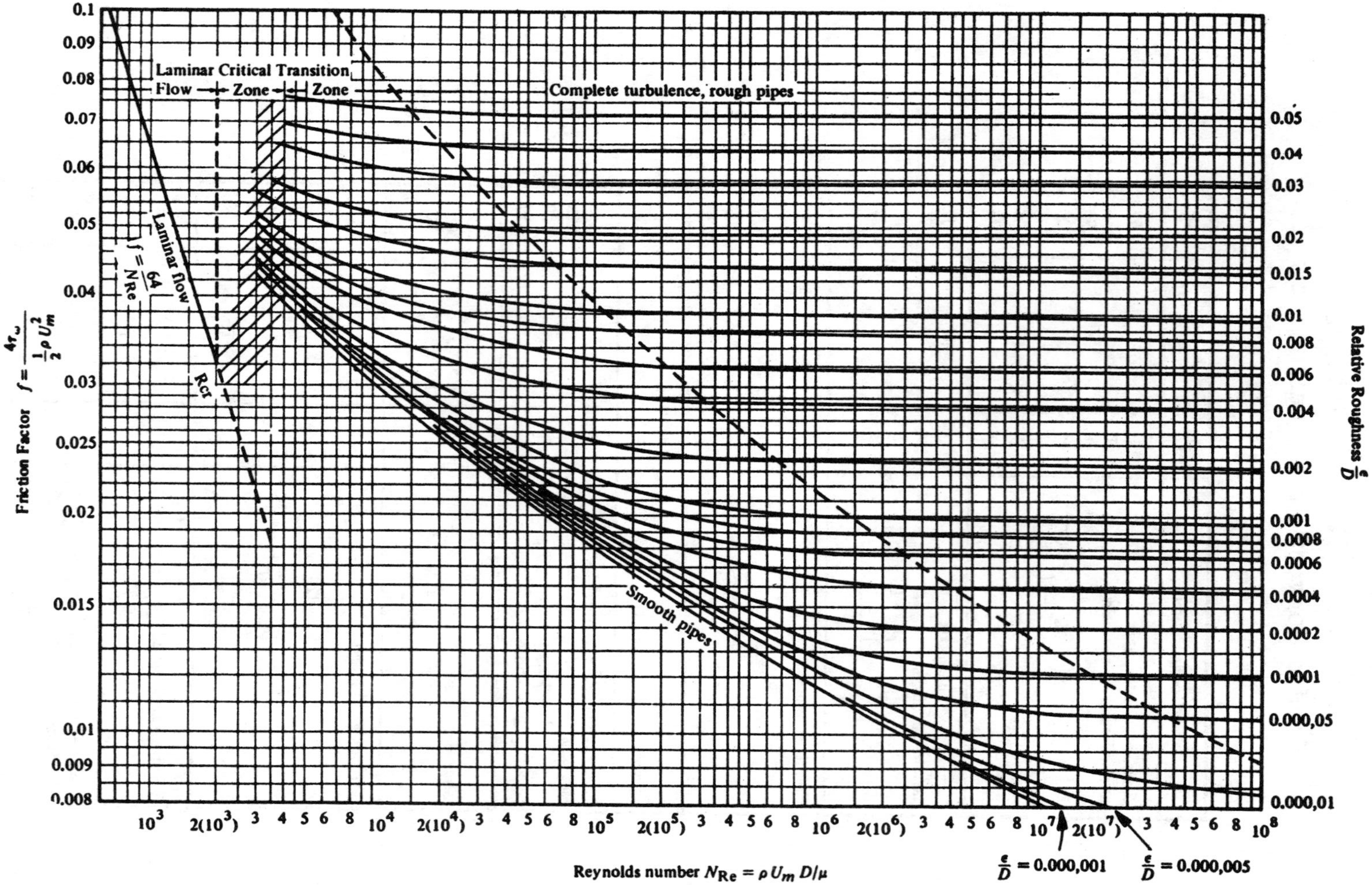

Moody Diagram, Friction Factor and Relative Roughness for Fully-Developed Flow in Pipes.

Conduction in Geometrical Configurations

Physical Description	Sketch	Conductive Shape Factor
Conduction through a material of uniform k from a horizontal isothermal cylinder to an isothermal surface	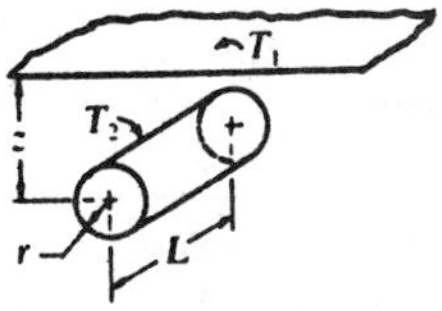	(*a*) Finite length $S = \dfrac{2\pi L}{\cosh^{-1}(z/r)}$ $\quad \dfrac{z}{L} \ll 1$ (*b*) Infinite length (per unit length) $\dfrac{S}{L} = \dfrac{2\pi}{\cosh^{-1}(z/r)}$
Conduction in a medium of uniform k from a cylinder of length L to two parallel planes of infinite width and length L	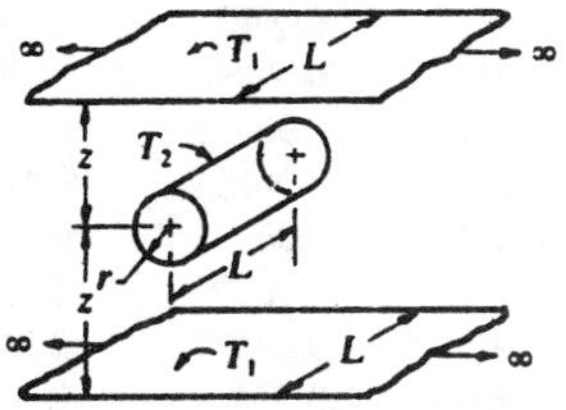	$S = \dfrac{2\pi L}{\ln(4z/r)}$
Conduction from an isothermal sphere through a material of uniform k to an isothermal surface	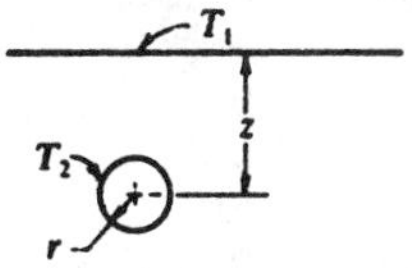	$S = \dfrac{4\pi r}{1 - (r/2z)}$
Conduction between two long isothermal parallel cylinders in an infinite medium of constant k	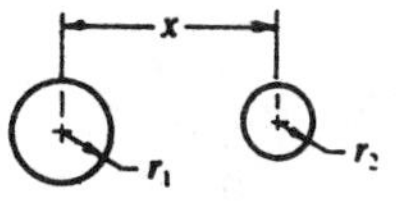	$\dfrac{S}{L} = \dfrac{2\pi}{\cosh^{-1}[(x^2 - r_1^2 - r_2^2)/2r_1r_2]}$ $L \gg r$ $L \gg x$
Conduction between a vertical isothermal cylinder in a medium of uniform k and a horizontal isothermal surface	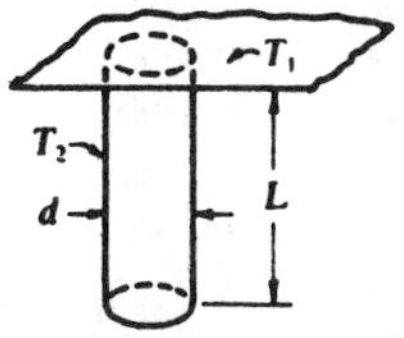	$S = \dfrac{2\pi L}{\ln(4L/d)}$ $L \gg d$
Conduction through an edge formed by intersection of two plane walls, with inner wall temperature T_1 and outer wall temperature T_2 as shown*	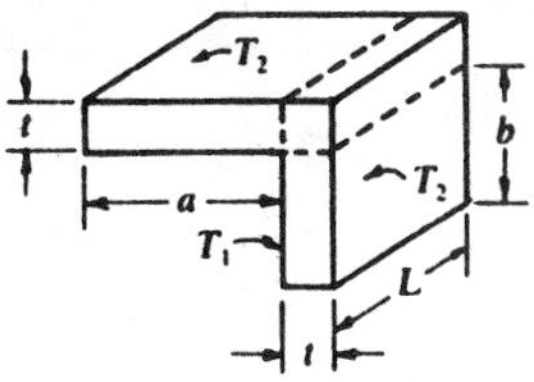	$S = 0.54\,L$ $a > t/5$ $b > t/5$
Conduction through a corner at intersection of three plane walls, each of thickness t, with uniform inner temperature T_1 and outer temperature T_2	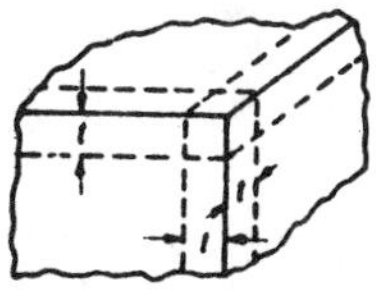	$S = 0.15\,t$ inside dimensions $> t/5$

***S for the plane wall is simply A/t, where A for the top wall shown is A = aL; for side wall, A = bL.**

From J. E. Sunderland and K. R. Johnson, Trans. ASHRAE, 10: 238-239, 1964

Dimensions of Standard Steel Pipe

Nominal Pipe Size (in.)	Outside Diameter (in.)	Schedule No.	Wall Thickness (in.)	Inside Diameter (in.)	Inside Cross-Sectional Area (ft^2)	Circumference (ft) or Surface (ft^2/ft) of Length		Capacity at 1 ft/s Velocity	
						Outside	Inside	U.S. gal/min	Water (lb_m/h)
1/8	0.405	40	0.068	0.269	0.00040	0.106	0.0705	0.179	89.5
		80	0.095	0.215	0.00025	0.106	0.0563	0.113	56.5
1/4	0.540	40	0.088	0.364	0.00072	0.141	0.095	0.323	161.5
		80	0.119	0.302	0.00050	0.141	0.079	0.224	112.0
3/8	0.675	40	0.091	0.493	0.00133	0.177	0.129	0.596	298.0
		80	0.126	0.423	0.00098	0.177	0.111	0.440	220.0
1/2	0.840	40	0.109	0.622	0.00211	0.220	0.163	0.945	472.0
		80	0.147	0.546	0.00163	0.220	0.143	0.730	365.0
3/4	1.050	40	0.113	0.824	0.00371	0.275	0.216	1.665	832.5
		80	0.154	0.742	0.00300	0.275	0.194	1.345	672.5
1	1.315	40	0.133	1.049	0.00600	0.344	0.275	2.690	1,345
		80	0.179	0.957	0.00499	0.344	0.250	2.240	1,120
1 1/4	1.660	40	0.140	1.380	0.01040	0.435	0.361	4.57	2,285
		80	0.191	1.278	0.00891	0.435	0.335	3.99	1,995
1 1/2	1.900	40	0.145	1.610	0.01414	0.497	0.421	6.34	3,170
		80	0.200	1.500	0.01225	0.497	0.393	5.49	2,745
2	2.375	40	0.154	2.067	0.02330	0.622	0.541	10.45	5,225
		80	0.218	1.939	0.02050	0.622	0.508	9.20	4,600
2 1/2	2.875	40	0.203	2.469	0.03322	0.753	0.647	14.92	7,460
		80	0.276	2.323	0.02942	0.753	0.608	13.20	6,600
3	3.500	40	0.216	3.068	0.05130	0.916	0.803	23.00	11,500
		80	0.300	2.900	0.04587	0.916	0.759	20.55	10.275
3 1/2	4.000	40	0.226	3.548	0.06870	1.047	0.929	30.80	15,400
		80	0.318	3.364	0.06170	1.047	0.881	27.70	13,850
4	4.500	40	0.237	4.026	0.08840	1.178	1.054	39.6	19,800
		80	0.337	3.826	0.07986	1.178	1.002	35.8	17,900
5	5.563	40	0.258	5.047	0.1390	1.456	1.321	62.3	31,150
		80	0.375	4.813	0.1263	1.456	1.260	57.7	28,850
6	6.625	40	0.280	6.065	0.2006	1.734	1.588	90.0	45,000
		80	0.432	5.761	0.1810	1.734	1.508	81.1	40,550
8	8.625	40	0.322	7.981	0.3474	2.258	2.089	155.7	77,850
		80	0.500	7.625	0.3171	2.258	1.996	142.3	71,150

Dimensions of Heat Exchanger Tubes

Outside Diameter (in.)	Wall Thickness BWG No.	Wall Thickness In.	Inside Cross-Diameter (in.)	Inside Sectional Area (ft^2)	Circumference (ft) or Surface (ft^2/ft) of Length: Outside	Circumference (ft) or Surface (ft^2/ft) of Length: Inside	Capacity at 1 ft/s Velocity: U.S. gal/min	Capacity at 1 ft/s Velocity: Water (lb_m/h)
$\frac{5}{8}$	12	0.109	0.407	0.000903	0.1636	0.1066	0.4053	202.7
	14	0.083	0.459	0.00115	0.1636	0.1202	0.5161	258.1
	16	0.065	0.495	0.00134	0.1636	0.1296	0.6014	300.7
	18	0.049	0.527	0.00151	0.1636	0.1380	0.6777	338.9
$\frac{3}{4}$	12	0.109	0.532	0.00154	0.1963	0.1393	0.6912	345.6
	14	0.083	0.584	0.00186	0.1963	0.1529	0.8348	417.4
	16	0.065	0.620	0.00210	0.1963	0.1623	0.9425	471.3
	18	0.049	0.652	0.00232	0.1963	0.1707	1.041	520.5
$\frac{7}{8}$	12	0.109	0.657	0.00235	0.2291	0.1720	1.055	527.5
	14	0.083	0.709	0.00274	0.2291	0.1856	1.230	615.0
	16	0.065	0.745	0.00303	0.2291	0.1950	1.360	680.0
	18	0.049	0.777	0.00329	0.2291	0.2034	1.477	738.5
1	10	0.134	0.732	0.00292	0.2618	0.1916	1.310	655.0
	12	0.109	0.782	0.00334	0.2618	0.2047	1.499	750.0
	14	0.083	0.834	0.00379	0.2618	0.2183	1.701	850.5
	16	0.065	0.870	0.00413	0.2618	0.2278	1.854	927.0
$1\frac{1}{4}$	10	0.134	0.982	0.00526	0.3272	0.2571	2.361	1,181
	12	0.109	1.032	0.00581	0.3272	0.2702	2.608	1,304
	14	0.083	1.084	0.00641	0.3272	0.2838	2.877	1,439
	16	0.065	1.120	0.00684	0.3272	0.2932	3.070	1,535
$1\frac{1}{2}$	10	0.134	1.232	0.00828	0.3927	0.3225	3.716	1,858
	12	0.109	1.282	0.00896	0.3927	0.3356	4.021	2,011
	14	0.083	1.334	0.00971	0.3927	0.3492	4.358	2,176
2	10	0.134	1.732	0.0164	0.5236	0.4534	7.360	3,680
	12	0.109	1.782	0.0173	0.5236	0.4665	7.764	3,882

Diameters of Standard-Weight Wrought-Iron Pipes

Nominal Size, Inches	External Diameter, Inches	Approximate Internal Diameter, Inches
½	0.840	0.622
1	1.315	1.049
1½	1.900	1.610
2	2.375	2.067
3	3.500	3.068
4	4.500	4.026
5	5.563	5.047
6	6.625	6.065
7	7.625	7.023
8	8.625	8.071
9	9.625	8.941
10	10.750	10.192
12	12.750	12.090

Monochromatic Intensity of Radiation for a Black Body at Various Absolute Temperatures (Planck's Law)

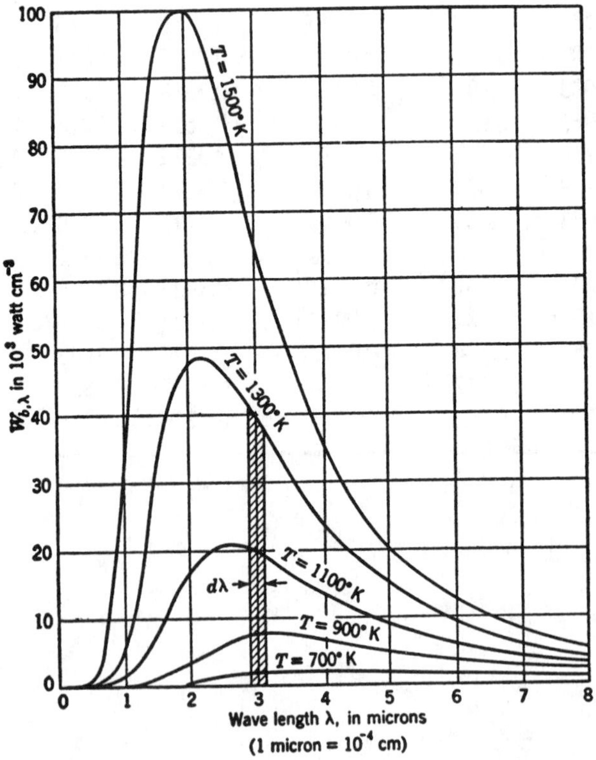

INDEX

Numbers on this page refer to PROBLEM NUMBERS, not page numbers.

Numbers on this page refer to PROBLEM NUMBERS, not page numbers.

Numbers on this page refer to PROBLEM NUMBERS, not page numbers.

Numbers on this page refer to PROBLEM NUMBERS, not page numbers.

Numbers on this page refer to PROBLEM NUMBERS, not page numbers.

Numbers on this page refer to PROBLEM NUMBERS, not page numbers.

Numbers on this page refer to PROBLEM NUMBERS, not page numbers.

The Best Test Preparation for the
GRE
GRADUATE
RECORD
EXAMINATION
GENERAL TEST
6 full-length exams
Completely Up-to-Date Based on the Current Format of the GRE
➤ All of the exam sections prepared by test experts in the particular subject fields
➤ The ONLY test preparation book with detailed explanations to every exam question to help you achieve a TOP SCORE
➤ Includes a special section of expert test tips and strategies
PLUS... a MATH REVIEW with drills to build math skills for quantitative sections, and a VOCABULARY LIST with word drills of the most frequently tested words on the GRE
REA
RESEARCH & EDUCATION ASSOCIATION